T0328616

Introduction to Fiber-Optic Communications

Introduction to Fiber-Optic Communications

Rongqing Hui

ACADEMIC PRESS

An imprint of Elsevier

ELSEVIER

Academic Press is an imprint of Elsevier
125 London Wall, London EC2Y 5AS, United Kingdom
525 B Street, Suite 1650, San Diego, CA 92101, United States
50 Hampshire Street, 5th Floor, Cambridge, MA 02139, United States
The Boulevard, Langford Lane, Kidlington, Oxford OX5 1GB, United Kingdom

Notices

Knowledge and best practice in this field are constantly changing. As new research and experience broaden our understanding, changes in research methods, professional practices, or medical treatment may become necessary.

Practitioners and researchers must always rely on their own experience and knowledge in evaluating and using any information, methods, compounds, or experiments described herein. In using such information or methods they should be mindful of their own safety and the safety of others, including parties for whom they have a professional responsibility.

To the fullest extent of the law, neither the Publisher nor the authors, contributors, or editors, assume any liability for any injury and/or damage to persons or property as a matter of products liability, negligence or otherwise, or from any use or operation of any methods, products, instructions, or ideas contained in the material herein.

Library of Congress Cataloging-in-Publication Data
A catalog record for this book is available from the Library of Congress

British Library Cataloguing-in-Publication Data
A catalogue record for this book is available from the British Library

ISBN 978-0-12-805345-4

For information on all Academic Press publications
visit our website at https://www.elsevier.com/books-and-journals

Publisher: Katey Birtcher
Acquisition Editor: Steve Merken
Editorial Project Manager: Ana Claudia Garcia
Production Project Manager: Sruthi Satheesh
Cover Designer: Victoria Pearson

Typeset by SPi Global, India

Dedication

To Erica and Andy with love

Contents

Author Biography xv
Preface xvii
Acknowledgments xxi
List of constants xxiii

SECTION 1

CHAPTER 1 Introduction 3
  1.1 Why optical fiber.. 3
  1.2 Evolution of fiber-optic communication technologies 6
  1.3 Optical systems enabled by DSP 11
  1.4 Basic units often used for fiber-optic system engineering 14

SECTION 2

CHAPTER 2 Optical fibers.. 19
         Introduction 20
  2.1 General properties of optical waves 21
  2.2 Reflection and refraction.. 24
         2.2.1 Fresnel reflection coefficients... 25
         2.2.2 Special cases of reflection angles 26
         2.2.3 Optical field phase shift between the incident and the reflected beams........ 29
         2.2.4 Brewster angle 29
  2.3 Propagation modes in optical fibers 30
         2.3.1 Geometric optics analysis 31
         2.3.2 Mode analysis using electromagnetic field theory 34
         2.3.3 Mode classification.. 39
         2.3.4 Numerical aperture 39
         2.3.5 Field distribution profile of SMF.. 43
  2.4 Optical fiber attenuation... 44
  2.5 Group velocity and dispersion 48
         2.5.1 Phase velocity and group velocity 48
         2.5.2 Group velocity dispersion 50
         2.5.3 Sources of chromatic dispersion 51
         2.5.4 Modal dispersion 53
         2.5.5 Polarization mode dispersion 53
         2.5.6 Mode division multiplexing 55

2.6 Nonlinear effects in an optical fiber 56
      2.6.1 Stimulated Brillouin scattering 56
      2.6.2 Stimulated Raman scattering 57
      2.6.3 Kerr effect nonlinearity and nonlinear Schrödinger equation 58
2.7 Different types of optical fibers 63
      2.7.1 Standard optical fibers for transmission 63
      2.7.2 Specialty optical fibers 66
2.8 Summary 69
      Problems 71
      References 75
      Further reading 76

CHAPTER 3   Light sources for optical communications **77**
      Introduction 77
3.1 Properties of semiconductor materials for light sources 79
      3.1.1 PN junction and energy diagram 79
      3.1.2 Direct and indirect semiconductors 81
      3.1.3 Spontaneous emission and stimulated emission 82
      3.1.4 Carrier confinement 83
3.2 Light-emitting diodes 84
      3.2.1 $P \sim I$ curve 85
      3.2.2 Modulation dynamics 86
3.3 Laser diodes 88
      3.3.1 Amplitude and phase conditions for self-sustained oscillation ... 88
      3.3.2 Rate equations 91
      3.3.3 Steady state solutions of rate equations 92
      3.3.4 Side-mode suppression ratio 94
      3.3.5 Modulation response 95
      3.3.6 Laser noises 99
3.4 Single-frequency semiconductor lasers 104
      3.4.1 DFB and DBR laser structures 105
      3.4.2 External cavity LDs 108
3.5 VCSELs and arrays 111
3.6 LD biasing and packaging 115
3.7 Summary 119
      Problems 121
      References 124

CHAPTER 4   Photodetectors **125**
      Introduction 125
4.1 PN and PIN photodiodes 126
4.2 Responsivity and electric bandwidth 131

4.2.1 Quantum efficiency and responsivity 131
4.2.2 Speed of photodetection response 133
4.2.3 Electrical characteristics of a photodiode 135
4.3 Photodetector noise and SNR 137
4.3.1 Sources of photodetection noise 137
4.3.2 SNR of an optical receiver 139
4.3.3 Noise-equivalent power 140
4.4 Avalanche photodiodes 141
4.4.1 APD used as a linear detector 142
4.4.2 APD used as a single-photon detector 144
4.5 Other types of photodetectors 145
4.5.1 Photovoltaic 145
4.5.2 Charge-coupled devices 148
4.6 Summary 150
Problems 151
References 154
Further reading 154

CHAPTER 5   Optical amplifiers **155**
Introduction 155
5.1 Optical gain, gain bandwidth, and saturation 157
5.2 Optical noise and noise figure 160
5.2.1 Optical noise power spectral density 160
5.2.2 Impact of ASE noise in the electrical domain 161
5.2.3 Noise figure 163
5.3 Semiconductor optical amplifiers 165
5.3.1 Steady-state analysis 165
5.3.2 Gain dynamics of SOA 169
5.3.3 All-optical signal processing based on the cross-gain and cross-phase
modulation 172
5.3.4 Wavelength conversion based on four-wave mixing in an SOA 176
5.4 Erbium-doped fiber amplifiers 179
5.4.1 Absorption and emission cross sections 181
5.4.2 Rate equations 183
5.4.3 Numerical solutions of Rate equations 186
5.4.4 Additional considerations in EDFA design 190
5.5 Raman amplifiers 195
5.6 Summary 201
Problems 203
References 206
Further reading 207

CHAPTER 6 Passive optical components **209**
    Introduction 210
    6.1 Fiber-optic directional couplers 211
      6.1.1 Basic parameters of a fiber-optic directional coupler 213
      6.1.2 Transfer matrix of a 2×2 optical coupler 217
    6.2 Optical interferometer based on two-beam interference 219
    6.3 Interferometers based on multi-pass interference 223
      6.3.1 Fabry-Perot Interferometers 223
      6.3.2 Optical ring resonators 230
    6.4 Fiber Bragg gratings 234
    6.5 WDM multiplexers and demultiplexers 239
      6.5.1 Thin-film-based interference filters 239
      6.5.2 Arrayed waveguide gratings 243
      6.5.3 Acousto-optic filters 245
    6.6 Optical isolators and circulators 249
      6.6.1 Optical isolators 250
      6.6.2 Optical circulators 252
    6.7 PLCs and silicon photonics 255
      6.7.1 Slab optical waveguides 256
      6.7.2 Rectangle optical waveguides 263
      6.7.3 Directional couplers 267
      6.7.4 Silicon photonics 270
    6.8 Optical switches 274
      6.8.1 Types of optical switches 274
    6.9 Summary 290
    Problems 291
    References 296
    Further reading 297

CHAPTER 7 External electro-optic modulators **299**
    Introduction 299
    7.1 Basic operation principle of electro-optic modulators 301
      7.1.1 EO coefficient and phase modulator 301
      7.1.2 Electro-optic intensity modulator 307
      7.1.3 Optical intensity modulation vs. field modulation 310
      7.1.4 Frequency doubling and high-order harmonic generation 311
    7.2 Optical single-sideband modulation 312
    7.3 Optical I/Q modulator 316
    7.4 Electro-optic modulator based on ring-resonators 317
    7.5 Optical modulators using electro-absorption effect 324
    7.6 Summary 329

Problems 330
References 334
Further reading 335

CHAPTER 8 Optical transmission system design **337**
Introduction 337
8.1 BER vs. Q-value for binary modulated systems 339
8.1.1 Overview of IMDD optical systems 339
8.1.2 Receiver *BER* and Q 342
8.2 Impacts of noise and waveform distortion on system Q-value 347
8.2.1 Q-calculation for optical signals without waveform distortion 347
8.2.2 Q-estimation based on eye diagram parameterization 348
8.3 Receiver sensitivity and required OSNR 352
8.3.1 Receiver sensitivity 352
8.3.2 Required OSNR 356
8.4 Concept of wavelength division multiplexing 358
8.5 Sources of optical system performance degradation 361
8.5.1 Performance degradation due to linear sources 361
8.5.2 Performance degradation due to fiber nonlinearities 372
8.5.3 Semi-analytical approaches to evaluate nonlinear crosstalks in fiber-optic
systems 376
8.6 Conclusion 408
Problems 409
References 414
Further reading 416

CHAPTER 9 Coherent optical communication systems **417**
Introduction 417
9.1 Basic principles of coherent detection 418
9.2 Receiver signal-to-noise ratio calculation of coherent detection 421
9.2.1 Heterodyne and homodyne detection 421
9.2.2 Signal-to-noise-ratio in coherent detection receivers 422
9.3 Balanced coherent detection and polarization diversity 427
9.3.1 Balanced coherent detection 428
9.3.2 Polarization diversity 429
9.4 Phase diversity and I/Q detection 430
9.5 Conclusion 435
Problems 436
References 438
Further reading 438

CHAPTER 10 Modulation formats for optical communications **439**

Introduction 439

10.1 Binary NRZ vs. RZ modulation formats 440

10.2 Generation of PRBS patterns and clock recovery 445

10.3 Polybinary, duobinary, and carrier-suppressed RZ modulation 449

10.3.1 M-ary and Polybinary coding 450

10.3.2 Duo-binary optical modulation 453

10.3.3 Carrier-suppressed return-to-zero (CSRZ) 456

10.4 BPSK and DPSK optical systems 460

10.5 High-level PSK and QAM modulation 466

10.6 Analog optical systems and radio over fiber 472

10.6.1 Analog subcarrier multiplexing and optical single-sideband modulation .. 473

10.6.2 Carrier-to-signal ratio, inter-modulation distortion and clipping 478

10.6.3 Impact of relative intensity noise 481

10.6.4 Radio-over-fiber technology 483

10.7 Optical system link budgeting 486

10.7.1 Power budgeting 486

10.7.2 OSNR budgeting 487

10.8 Summary 489

10.9 Problems 490

References 494

Further reading 495

CHAPTER 11 Application of high-speed DSP in optical communications **497**

Introduction 497

11.1 Enabling technologies for DSP-based optical transmission 498

11.2 Electronic-domain compensation of transmission impairments 499

11.2.1 Dispersion compensation 500

11.2.2 PMD compensation and polarization de-multiplexing 505

11.2.3 Carrier phase recovery in a coherent detection receiver 507

11.3 Digital subcarrier multiplexing: OFDM and Nyquist frequency-division

multiplexing 511

11.3.1 Orthogonal frequency-division multiplexing 512

11.3.2 Nyquist pulse modulation and frequency-division multiplexing 523

11.3.3 Optical system considerations for DSCM 527

11.4 Summary 549

Problems 549

References 553

CHAPTER 12 Optical networking **555**
        Introduction 555
  12.1 Layers of communication networks and optical network topologies 556
        12.1.1 Layers of an optical network 557
        12.1.2 Optical network topologies 559
        12.1.3 SONET, SDH, and IP... 567
  12.2 Optical network architectures and survivability.. 573
        12.2.1 Categories of optical networks.. 573
        12.2.2 Optical network protection and survivability 574
  12.3 Passive optic networks and fiber to the home.. 579
        12.3.1 TDM-PON 580
        12.3.2 WDM-PON 584
  12.4 Optical interconnects and datacenter optical networks 586
        12.4.1 Optical interconnection 586
        12.4.2 Datacenter optical networks... 588
  12.5 Optical switching and cross connection 596
        12.5.1 Reconfigurable optical add/drop multiplexing 596
        12.5.2 Optical cross-connect circuit switching..................................... 599
        12.5.3 Elastic optical networks 602
        12.5.4 Optical packet switching.. 604
  12.6 Electronic circuit switching based on digital subcarrier multiplexing.................. 606
  12.7 Summary... 612
        Problems 613
        References 618
        Further reading 619

Index 621

Author Biography

Rongqing Hui is a Professor of Electrical Engineering and Computer Science at the University of Kansas. He received the BS and MS degrees from Beijing University of Posts and Telecommunications in China, and the PhD degree from Politecnico di Torino in Italy, all in Electrical Engineering. Before joining the faculty of the University of Kansas in 1997, Dr. Hui was a Member of Scientific Staff at Bell-Northern Research and then Nortel, where he worked in the research and development of high-speed fiber-optic transport networks. He served as a program director in the National Science Foundation for 2 years from 2006 to 2008, where he was in charge of research programs in electronic and photonic devices. Dr. Hui has published more than 100 papers in referred engineering journals in addition to numerous papers in referred international conferences in the area of fiber-optic systems and devices and holds 15 US patents. He served as a topic editor for *IEEE Transactions on Communications* from 2001 to 2007 and served as an associate editor for *IEEE Quantum Electronics* from 2006 to 2013.

Preface

While light has been used for delivering information over free space for many centuries, modern fiber-optic communications did not begin until the early 1960s when Charles Kao theoretically predicted that high-speed information could be carried by optical waves and transmited for long distances over a narrow glass waveguide, which is now commonly referred to as the optical fiber. In 1970, a team of researchers at Corning successfully fabricated optical fibers using fused-silica with less than 20 dB/km of loss at 633 nm wavelength. The Corning breakthrough was the most significant step toward practical applications of fiber-optic communication. Over the next several years, fiber losses dropped dramatically, aided by both the improved material quality and better fabrication methods, as well as the shift to longer signal wavelengths where fused-silica has inherently lower attenuation.

Meanwhile, the prospect of fiber-optics for communication intensified R&D efforts in semiconductor lasers and other related optical devices. Near-infrared semiconductor lasers operating in 810, 1320, and 1550 nm wavelengths were developed to fit into the low loss windows of silica optical fibers. The bandwidth in 1550 nm wavelength window alone, with the elimination of the water absorption peak, can be extended to more than 160 nm, which is approximately 20 THz. In order to make full and efficient uses of this vast bandwidth, many innovative technologies have been developed, such as tunable laser diodes, dispersion-managed optical fibers, optical amplifiers, and wavelength division multiplexing, as well as advanced modulation formats and detection techniques. In addition to wide bandwidth and long-distance fiber-optic links, metropolitan and local area optical networks have also been developed, and significant research and development efforts have been especially devoted to fiber-to-the-home in order to remove the last mile bottleneck of high-speed access.

Fiber-optic communication is a subject that requires good understanding of fundamental physics, rules of engineering design, and communication system applications. This book is intended to provide a comprehensive description of fiber-optic communication technology, with a special focus on explaining engineering design rules through underlining physics principles of optical components and basic theories of telecommunications. As a rapidly growing field, fiber-optic communication has evolved significantly over the past decades with various enabling technologies, devices, and system design concepts. For example, receiver sensitivity, defined as the minimum signal optical power required to achieve the desired bit error rate, is no longer valid when multiple optical amplifiers are used in the system which generates significant optical noise. In such systems, the required optical signal to noise ratio is the most relevant performance indicator. In recent years, complex optical field modulation and coherent detection have reemerged as the dominant format for high capacity fiber-optic systems, in conjunction with the availability of adaptive digital signal processing (DSP). Traditionally, an important fiber-optic system design rule has been to perform signal processing in the optical domain whenever possible, and optical domain chromatic dispersion compensation and polarization mode dispersion (PMD) compensation have been developed and widely implemented in commercial fiber-optic systems. The availability of high-speed DSP has completely changed this design rule, and electronic domain compensation of transmission impairments has become more efficient and flexible, especially in high-speed coherent systems. This book strives to capture these new technological developments and system design concepts.

After a brief overview of fiber-optic communications in Chapter 1, Chapter 2 discusses fundamental principles of wave propagation and basic properties of optical fiber as the main transmission medium of fiber-optic communications. Chapter 3 introduces semiconductor lasers and light-emitting diodes (LED), which are commonly used as light sources for optical communications. While low-cost LED and vertical cavity surface emitting laser (VCSEL) arrays are important in short distance optical interconnection, single-frequency diode lasers are indispensable for long-distance and high-speed fiber-optic systems to maximize the spectral efficiency and to minimize the impact of chromatic dispersion. Chapter 4 discusses the basic properties of photodiodes, which perform optical-to-electric conversions in optical receivers. In addition to quantum efficiency and responsivity of photodetection, noise is also created in the photodetection process through a number of different mechanisms. Avalanche photodiodes can help magnifying photocurrents, but cannot improve the signal-to-noise ratio. Also based on revise biased semiconductor pn-junctions, similar devices, such as photovoltaic and charge-coupled devices (CCD) are also discussed in this chapter. Although not directly used in fiber-optic communications, they are very popular for renewable energy generation and digital imaging. Optical amplification discussed in Chapter 5 is an enabling technology which makes long-distance optical communication possible. Both semiconductor optical amplifiers (SOA) and erbium-doped fiber amplifiers (EDFA) are discussed and their properties are compared. While an SOA with fast carrier dynamics is good for all-optical signal processing, EDFAs are most popular as inline optical amplifiers in wavelength multiplexed (WDM) optical systems because of the slow carrier dynamics and less crosstalk between WDM channels. Fiber Raman amplifier is also discussed, which provides an additional mechanism for optical amplification. Distributed fiber Raman amplification is a promising technique to further extend the reach of a fiber-optic system beyond the capability of only using EDFAs, and the effective noise figure of a distributed fiber Raman amplifier can even be negative. We explain the reason why the negative value of the effective noise figure does not violate fundamental physics principles. Chapter 6 introduces basic structures and properties of passive optical devices often used in fiber-optic systems, which include directional couplers, interferometers, optical filters, fiber Bragg gratings, wavelength division multiplexers and de-multiplexers, optical isolators and circulators, as well as optical switches. Planar lightwave circuit (PLC) is a technique that allows the integration of a large number of functional photonic components onto a chip. Especially when silicon is used as the substrate, known as silicon photonics, large-scale integration can be possible thanks to the tight field confinements, and photonic circuits can also be monolithically integrated with CMOS electronic circuits. Chapter 7 is dedicated to the discussion of electro-optic modulators as they are uniquely important in fiber-optic systems for translating electronic signals into the optical domain. External modulation avoids frequency chirp associated with direct current modulation of laser diodes and allows both intensity modulation and complex optical field modulation of the optical signal. Three basic modulator configurations are discussed, including Mach-Zehnder, micro-resonator ring, and electro-absorption. Optical single-sideband modulation, in-phase/quadrature modulation, and the importance of modulator biasing control are also explained.

After the discussion of various optical components as building blocks, Chapter 8 introduces basic concepts of digital optical communication systems. Starting from the bit error rate (BER) and Q-value, the definition and implication of receiver sensitivity and the required optical signal to noise ratio are explained. Various signal degradation mechanisms are discussed, which originate from both random noise and waveform distortion. While random noise can be generated by the amplified spontaneous emission of optical amplifiers and by the receiver, waveform distortion is more deterministic, which can be introduced by chromatic dispersion, PMD, as well as nonlinear effects and crosstalk from other channels. Chapter 9 explores the principle and applications of coherent optical communication systems, including various modulation and detection techniques. Due to the ability of linearly translating the complex optical field into the electronic domain, a coherent system can be more flexible with much higher spectral efficiency. As a strong optical local oscillator is used in the receiver, design rules and performance evaluation of a coherent system can be quite different from systems with direct detection. Modulation format of the optical signal is another important topic in optical communications which is discussed in Chapter 10. In addition to binary intensity modulation, advanced modulation formats can be very beneficial to allow the improved spectral efficiency and the increased tolerance against performance degradation. Especially after the introduction of complex optical field modulation and coherent detection, high-level modulation in both amplitude and optical phase can be applied. Analog optical systems and radio-over-fiber are also introduced in this chapter, which are traditionally used for cable TV but can also be used to support wireless communication networks by delivering high carrier frequency radio signals between the central office and antenna sites. Optical link budgeting discussed at the end of this chapter is a comprehensive task in optical system design which determines how far the system can go while still meets the performance requirement. Link budgeting has to consider all factors impacting the system, including signal optical power level of the transmitter, attenuation by the fiber and other optical components, as well as signal quality degradation due to noise and waveform distortion. Chapter 11 discusses the application of DSP in fiber-optic systems. The introduction of high-speed DSP has brought fundamental changes in the design of optical communication systems, which allows digital filters, arbitrary waveform generation, and fast Fourier transform (FFT) to be used in the signal processing. Dispersion compensation, PMD compensation, and carrier-phase recovery in coherent receivers can be made in the electronic domain and adaptive to variations of system parameters and phase noise. Digital subcarrier multiplexing (DSCM), including orthogonal frequency division multiplexing (OFDM) and Nyquist pulse modulation, can be implemented to make fiber-optic systems more flexible and spectrally efficient. The last chapter of this book, Chapter 12, introduces basic architectures, functionalities, and applications of optical networks. Layered structures, network topologies, routing protocols, and network management algorithms are presented, and multiplexing standards of SONET, SDH, and IP are used as examples of data frame structures. Passive optical networks (PON) and data center networks are specifically addressed as emerging applications in access and high capacity data interconnection. The concept and implementation of reconfigurable optical add/drop multiplexing, optical circuit switching, and packet switching are discussed and compared. The last section of Chapter 12 introduces digital subcarrier cross-connect (DSXC) in the electronic domain. As a circuit-switching technique, DSXC can provide much finer data rate granularity and faster switching speed compared to optical circuit switching, and much lower latency and complexity compared to electronic packet switching.

Teaching ancillaries for this book are available to qualified instructors by registering at www.textbooks.elsevier.com. Once registered, visit https://textbooks.elsevier.com/web/product_details.aspx?isbn=9780128053454 for more details.

Acknowledgments

I would like to thank my friends and colleagues Dr. Maurice O'Sullivan of Ciena, Dr. Andrea Fumagalli of UT Dallas, Dr. Feng Tian of Gofoton Inc., and Dr. Chongjin Xie of Alibaba for many helpful discussions. My graduate students, soon to be Doctors, Govind Vedala and Mustafa Al-Qadi have read through several chapters and provided valuable comments.

Finally, I would like to thank my wife Jian. Without her support and tolerance, it would not have been possible for me to accomplish this task.

List of constants

Table of constants:

Constant	Name	Value	Unit
κ	Boltzmann's constant	1.28×10^{-23}	J/K
h	Planck's constant	6.62×10^{-34}	J s
q	Electron charge	1.6023×10^{-19}	C
c	Speed of light in free space	2.99792×10^{8}	m/s
T	Absolute temperature	$273\,\mathrm{K} = 0^{\circ}\mathrm{C}$	K
ε_0	Permittivity in free space	8.85×10^{-12}	F/m
μ_0	Permeability in free space	12.566×10^{-7}	$\mathrm{N/A}^2$

Conversion between units:

Unit 1	Unit 2	Conversion
Electron volt [eV]	Energy [J]	$1\,\mathrm{eV} = 1.6023 \times 10^{-19}\,\mathrm{J}$
Free space wavelength (λ [m])	Frequency (f [Hz])	$f = c/\lambda$
Attenuation (α [Neper/m])	α_{dB} [dB/km]	$\alpha_{\mathrm{dB}} = 4.343 \times 10^{3}\alpha$
Chromatic dispersion β_2 [s^2/m]	D [ps/nm/km]	$D = -(2\pi c/\lambda^2)\beta_2 \times 10^{6}$

INTRODUCTION

CHAPTER OUTLINE

1.1 Why optical fiber 3
1.2 Evolution of fiber-optic communication technologies 6
1.3 Optical systems enabled by DSP 11
1.4 Basic units often used for fiber-optic system engineering 14

Telecommunication has enabled large-volume information exchange over long distances. Nowadays, the internet has formed a giant World Wide Web that permeates into almost all areas of human society, and has fundamentally changed our lifestyles. Global internet traffic has passed one zettabyte (10^{21} bytes) in 2016, and is still growing at a fast rate. Because of its low transmission loss and ultrawide spectral bandwidth, optical fiber is an obvious choice for long distance and high-speed communication. Optical fiber, as the preferred transmission medium, provides the backbone of the broadband internet worldwide. Fiber-optic communication as a technology has evolved significantly over the last few decades along with the emergence of many enabling components, innovative communication system concepts, and network architectures. The research and development (R&D) field of fiber-optic communication is interdisciplinary, which requires the understanding of fundamental device physics, communication theory, modulation and coding, as well as telecommunication networking protocols and architectures. The major purpose of this book is to provide an up to date and comprehensive coverage of optical fiber communication technologies and their applications. As fiber-optic system/network designs and applications are based on the properties of key devices as building blocks, this book emphasizes fundamental concepts and engineering issues of optical and optoelectronic devices. With a good understanding of device physics, capabilities and limitations, reasonable tradeoffs can be made in system and network design, performance estimation, implementation, trouble shooting, and optimization.

1.1 WHY OPTICAL FIBER

The earliest telecommunication can be traced back to the first telegraph sent by Samuel F. B. Morse in 1844 based on electromagnetic waves, which may be categorized as the wireless communication we know now. The general idea of optical communication started even much earlier for visual information delivery through smoke or flag semaphore, which can be seen as the earliest version free-space optical communication. While wireless and free-space communications can be flexible and mobile, communication via wire lines can provide better reliability and security.

Introduction to Fiber-Optic Communications. https://doi.org/10.1016/B978-0-12-805345-4.00001-9

A pair of twisted wires shown in Fig. 1.1.1A is flexible and can support megahertz transmission bandwidth over relatively short distances of typically no more than 10 m. The high-frequency impedance of twisted wires is not well defined, and radiation loss and electromagnetic interference can be major problems when carrying high-frequency signals. Coaxial cables are designed to carry high-frequency signals with an electromagnetic shield to prevent the radiation as shown in Fig. 1.1.1B. The center of a coaxial cable is a solid copper wire, which is surrounded by a dielectric insulation layer acting as a buffer to keep the center conductor straight. Covering the dielectric layer is a layer of metal shield, which can be braided copper wire or aluminum foil. The impedance of a coaxial cable at high frequency is determined by the dielectric constant and the thickness of the dielectric layer. There is also an outer jacket to protect and strengthen the cable. The bandwidth of a high-speed coaxial cable with proper connectors can be up to a few tens of gigahertz, but the cable length is typically not more than a few meters because of the high transmission loss. The lack of low loss and wide bandwidth transmission medium had hindered wire line communications for many years before the introduction of optical fiber. In fact, before 1980s, most of the long-distance telecommunication was carried on microwave, and twisted wires and coaxial cables were used for short-distance interconnections such as telephone and TV.

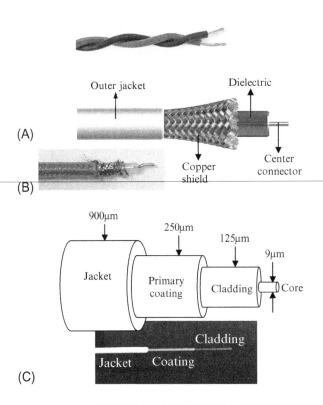

FIG. 1.1.1

Illustrations of (A) a pair of twisted wires, (B) a coaxial cable, and (C) a single-mode optical fiber.

The introduction of optical fiber in the early 1980s revolutionized the telecommunication's industry by providing an ideal transmission medium with unprecedented ultra-wide bandwidth and very low propagation loss. As shown in Fig. 1.1.1C, an optical fiber has a center core with a diameter of typically 9 μm for single-mode fiber and 50 μm for multimode fiber. Outside the core, there is a cladding layer with 125 μm diameter for both single-mode and multimode fibers. Both the core and the cladding are made of SiO_2 material, but the core has a slightly higher refractive index than the cladding through doping, so that the signal optical field is mostly confined within the core. An elastic acrylate coating layer is applied outside the cladding to provide primary protection to the fiber. A secondary layer of jacket further strengthens the fiber. Overall, the fiber diameter, including coating and jacket, is less than 1 mm, and no metal material is used at all, so that the weight per unit length of fiber is much less than for a coaxial cable.

Fig. 1.1.2 shows the attenuation of a standard single-mode fiber (Corning SMF-28) as the function of the signal wavelength. The attenuation is lower than 0.5 dB/km across a wide wavelength window from 1200 to 1600 nm, corresponding to an approximately 62.5 THz bandwidth in frequency. In the 1550-nm wavelength window commonly used for long-distance optical communication, the attenuation is as low as 0.19 dB/km. This means more than 1% of the signal optical power can still remain after propagating through 100 km of fiber.

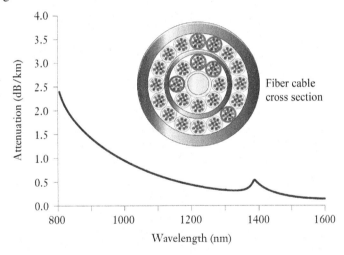

FIG. 1.1.2

Attenuation of standard single-mode fiber (Corning SMF-28). The inset shows the cross section of a fiber cable with 288 fibers.

In addition to the low attenuation and wide available bandwidth, advantages of optical fiber also include low security risk, high reliability, low weight, and low cost. As the optical signal is tightly confined within the core of optical fiber based on total internal reflection, it is extremely difficult to tap the information out of a fiber without disturbing its low loss transmission. Even if the fiber is tapped along the transmission line, it can be quite easy to find out because the fiber loss would be abnormally increased, which can cause the entire system to fail.

Because an optical fiber is made of silica which is an electric insulator, and there is no metal material involved, it is highly resistant to many environmental factors such as erosion and moisture that would affect copper wires and coaxial cables. As the signal is in the optical domain, the communication quality is immune to electromagnetic radiation and radio-frequency (RF) interference.

The lightweight of fiber is another advantage, which makes it easy to handle and to install. Thanks to the lightweight and small diameter of each fiber, a fiber cable can hold a large number of fibers arranged in many bundles. For example, a standard fiber cable with cross section shown in the inset of Fig. 1.1.2 has 24 fiber bundles with 12 fibers in each bundle, and thus a total of 288 fibers are carried by this cable. Because of the massive production and deployment, the cost of optical fiber can be much lower than wideband coaxial cable per unit length. Together with the low maintenance cost, the fiber-optic system has become the best choice for wire line communication systems and networks.

1.2 EVOLUTION OF FIBER-OPTIC COMMUNICATION TECHNOLOGIES

Fig. 1.2.1 shows the block diagram of the simplest fiber-optic communication system, which includes an optical transmitter, an optical receiver, and a transmission optical fiber. Because an optical fiber can only carry an optical signal, the electric signal from an information source has to be translated into an optical signal by the optical transmitter that performs electric-to-optical (E/O) conversion. After the optical signal is delivered to the destination by the optical fiber, it has to be converted back to the electric domain, which is accomplished by the optical receiver performing optical-to-electric (O/E) conversion.

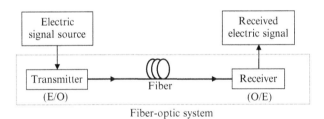

Fiber-optic system

FIG. 1.2.1

Block diagram of a basic optical fiber communication system.

In the early days of optical fiber communication, most of the key components were not mature enough, including laser diodes required for the optical transmitter, photodiodes for the optical receiver, as well as the quality of the optical fiber.

The first-generation optical fiber communication systems were developed in the late 1970s after Corning successfully reduced fiber loss to the level below 10 dB/km, and with room temperature operating semiconductor lasers made available. The first-generation fiber-optic communication systems operated in the 850-nm wavelength window mainly due to the availability of GaAs semiconductor lasers operating at that wavelength, and the first-generation multimode fibers were optimized for that wavelength. Because the loss of the optical fiber was more than 3 dB/km in the 850-nm wavelength window at that time, the signal optical power would drop by 30 dB after 10 km of fiber transmission, and thus the signal has to be regenerated for every 10 km. Direct injection-current modulation on laser diodes in the transmitter was used in the first-generation fiber-optic systems, and the modulation speed was limited to a few tens of MHz. Silicon photodiodes were used at the receiver as 850 nm is well within the detectable wavelength region of silicon. Although these first-generation fiber-optic systems do not seem very impressive now, it was already much better than coaxial cable systems that need to be regenerated every 1 km or less for this data rate.

It was clear that the major limitation on the transmission distance of the first-generation fiber-optic systems was the high fiber attenuation at 850 nm. The second generation of fiber-optic systems moved the operation wavelength from 850 to 1310 nm, mainly for two reasons: first, fiber attenuation at 1310 nm could be reduced to less than 1 dB/km (vs. 3 dB/km at 850 nm), and second, the chromatic dispersion of silica fiber in the 1310-nm wavelength window is almost zero. The realization of the second-generation fiber-optic systems was enabled by two key technological developments. The first of them was bandgap engineering of semiconductor lasers, which enabled the creation of 1310-nm laser emission based on InGaAsP semiconductor materials. The second key development was the single-mode optical fiber that eliminated the modal dispersion of multimode fibers which, to a large extent, limited the data rate increase of the first-generation fiber-optic systems. InP-based photo-detectors were also indispensible for the detection of 1310 nm optical signals, which fall outside the detectible wavelength region of silicon photodiodes. The second-generation fiber-optic systems developed in the mid-1980s were able to extend the regenerating distance to approximately 40 km with a signal data rate of up to 2 Gb/s. The improvement of the second-generation fiber-optic systems was significant with respect to the first-generation in terms of both regeneration distance and the data rate. Although fiber attenuation was reduced over time with the improved material quality and fabrication techniques, the intrinsic loss of fiber at 1310 nm is still much higher than that in the 1550-nm wavelength window. This resulted in the development of the third-generation fiber-optic systems operating in the 1550 nm wavelength window in which the fiber loss was reduced to the 0.25 dB/km level in late 1980s.

However, in the 1550-nm wavelength window single-mode optical fiber made of silica has significant chromatic dispersion on the order of 17 ps/nm/km. This means for two wavelength components of an optical signal separated by 1 nm, their differential group delay (DGD) can reach 1.7 ns after 100 km transmission. Thus, a laser source used for fiber-optic system has to have a narrow enough spectral width to prevent signal pulse spreading caused by chromatic dispersion. For a Fabry-Perot type InGaAsP semiconductor laser with a cavity length of 300 μm, the longitudinal mode spacing is on the order of 1 nm. Thus, multiple longitudinal modes from a laser diode would severely limit the maximum data rate that the system can support. This triggered intensive R&D efforts in single-longitudinal-mode diode lasers based on external cavity, or based on distributed feedback (DFB) structures. While an external cavity semiconductor laser was able to provide very narrow spectral linewidth, it was not a fully integrated device that can be mass produced, and was not stable enough to meet the telecommunication standard. DFB lasers, on the other hand, with Bragg gratings written inside laser chips are fully integrated devices that can be mass produced. Thus, DFB lasers emerged as the main light sources for fiber-optic communication systems. With rapid advances in nano-fabrication techniques over the years, the quality and yield of DFB lasers have been improved significantly. Other types of integrated single-frequency semiconductor laser structures were also created such as distributed Bragg reflector (DBR) lasers, and multi-section DBR lasers with the capability of wavelength tuning. The introduction of single-frequency semiconductor lasers and the reduced fiber attenuation in the 1550-nm wavelength window enabled the third-generation fiber-optic system to carry up to 10 Gb/s data rate over 100 km of fiber.

Another way to reduce the impact of chromatic dispersion is to change the fiber design so that the dispersion parameter can be reduced in the 1550-nm wavelength window. Dispersion-shifted fibers, with zero dispersion wavelength shifted to, or near, the 1550-nm wavelength window, were developed in late 1980s and deployed in some areas in early 1990s. With dispersion-shifted fiber, the transmission distance could be extended beyond 100 km for a 10-Gb/s binary modulated data channel.

Coherent optical detection technique was also demonstrated in the 1990s mainly for the purpose of improving the receiver sensitivity and increasing the transmission distance. Coherent detection with a strong optical local oscillator (LO) at the receiver effectively amplifies the optical signal so that a much lower signal optical power can be enough to reach an acceptable level of bit-error rate (BER). However, the coherent system was not commercially successful at that time due to the lack of low cost and narrow spectral linewidth semiconductor lasers, and the major advantage of improved receiver sensitivity soon lost its competitiveness after the introduction of optical amplifiers.

The fourth generation of fiber-optic systems was represented by wavelength-division multiplexing (WDM) and the introduction of optical amplifiers, which enabled orders of magnitude increase of both the capacity and the transmission distance of optical transmission systems. It is evident that the multi-terahertz optical bandwidth cannot be fully filled with a single wavelength channel, which is limited by the speed of electronic circuits and electro-optic modulation. By combining multiple wavelength channels into the same fiber, the wide optical bandwidth of the fiber can be more efficiently utilized. Meanwhile, erbium-doped fiber amplifiers (EDFAs) can amplify optical signals directly in the optical domain without converting to electronics. This eliminated the need for frequent regeneration of optical signals of every 100 km, which requiring O-E-O (optical-electronic-optical) conversion. Receiver sensitivity can also be significantly improved with an EDFA pre-amplifier in front of an optical receiver that amplifies the received optical signal.

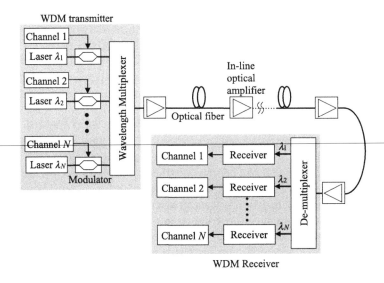

FIG. 1.2.2

Block diagram of a WDM optical system using external modulation.

Fig. 1.2.2 shows the block diagram of a WDM optical system with N wavelength channels. In this system, each data channel is modulated onto an optical carrier with a specific wavelength through an external electro-optic modulator. All the optical carriers are combined into a single output through a wavelength division multiplexer. The power of the combined optical signal is boosted by an optical fiber amplifier and sent to the transmission optical fiber. Along the fiber transmission line, the optical signal is periodically amplified by in-line optical amplifiers to overcome the transmission loss of the optical fiber. At the destination, a wavelength division de-multiplexer is used to separate optical carriers of different wavelengths, and each wavelength channel is detected separately. In this system, the WDM configuration allows the efficient use of the wide bandwidth offered by the optical fiber, and optical amplification technology significantly extended the overall reach of the optical system without electronic regeneration.

Because of the dramatically increased transmission distance, chromatic dispersion of the optical fiber became a major limitation. In addition, as multiple wavelength channels propagate in the same fiber and optical power levels in the fiber also became high due to optical amplification, other sources of performance impairments start to become significant, which include fiber nonlinearity, polarization-mode dispersion, and amplified spontaneous emission (ASE) noise generated by optical amplifiers.

Dispersion-shifted fiber can be used to reduce the impact of chromatic dispersion in a long distance transmission system. However, dispersion-shifted fibers were later found to be susceptive to nonlinear effects, especially when WDM was used. Another way to reduce the accumulated chromatic dispersion along the fiber is to insert dispersion compensating modules (DCMs) into the system, which have opposite sign of dispersion compared to the transmission fiber. A DCMs can be made of specially designed fiber known as dispersion compensating fiber (DCF), or optical all-pass filter based on fiber Bragg grating (FBG).

As multiple channels with different wavelengths propagate in the same optical fiber, the impact of fiber nonlinearity can be significant; especially for fiber systems with multiple spans involving in-line optical amplifiers. Self-phase modulation (SPM), cross-phase modulation (XPM), and four-wave mixing (FWM) are among the most severe degradation effects due to Kerr-effect fiber nonlinearity. A dispersion-shifted fiber usually has smaller core size compared to a standard single-mode fiber, and thus the power density is higher for the same total signal power. In addition, due to the reduced chromatic dispersion, differential walk off between different wavelength channels is reduced so that the strength of inter-channel crosstalk due to nonlinear effects such as FWM in dispersion-shifted fiber can be much stronger than that in a standard single-mode fiber. Standard single-mode fiber with relatively high local dispersion but compensated periodically along the system was shown to be less susceptible to fiber nonlinearity. Various types of single-mode fibers have been developed to improve optical transmission system performance, such as non-zero dispersion-shifted fiber (NZ-DSF), and large effective area fiber (LEAF).

Polarization-mode dispersion (PMD) arises from residual birefringence of single-mode fiber. PMD creates DGD between the two orthogonally polarized fundamental modes along the fiber. As the energy of the optical signal is partitioned into these two degenerate modes, their DGD can introduce pulse spreading of the optical signal. To make things more complicated, the residual birefringence of an optical fiber is quite sensitive to the environmental condition and mechanical stress, which can change randomly over time. Based on the Maxwellian distribution statistics of PMD, even with a relatively small average DGD, the instantaneous DGD can temporarily reach high values. If this instantaneous DGD reaches a level comparable to the length of data symbol carried by the optical signal, it can cause surges in bit-error rate, and system outage. The impact of PMD was not a big concern for the 1st and the 2nd generations of fiber-optic systems because of the relative low symbol rates. However, it became one of the major concerns with the increase of data symbol rates beyond 10 Gbaud/s. A number of adaptive PMD compensation techniques have been demonstrated for high-speed fiber-optic systems, mostly through feedback control of state of polarization (SOP) of the optical signal. The requirement of PMD compensation significantly increased the complexity of the system.

The successful commercialization of the 4th-generation fiber-optic systems was enabled by several key components and technologies. Single-frequency semiconductor lasers with wavelengths either locked at the ITU (international Telecommunication Union) grid for WDM, or were tunable for flexible wavelength settings. External electro-optic modulator is another important device for data encoding, which avoided the large modulation chirp inherent to direct modulation of laser diodes. Electro-optic modulators built with Mach-Zehnder configuration based on $LiNbO_3$ electro-optic material exhibits superior performance, while electro-absorption modulators based III-V semiconductor materials can be monolithically integrated with laser diodes, so that the cost could be lower.

Optical amplifiers with wide gain bandwidth and low noise figure were developed for the 4th-generation fiber-optic systems. Semiconductor optical amplifiers (SOAs) are electrically pumped with miniature size and easy for integration. However, the short carrier lifetime and fast gain dynamics on the order of sub-nanoseconds made an SOA susceptible to inter-channel crosstalk in WDM system applications. An EDFA, on the other hand, has a carrier lifetime on the order of 10 ms, which is several orders of magnitude longer than symbol lengths of modulated optical signals, and thus inter-channel crosstalk due to cross-gain saturation is not a concern for WDM system applications. EDFAs are now implemented in most high-speed commercial WDM fiber-optic systems. Meanwhile, the quality and the functionality of passive optical components have also been improved considerably over years such as WDM multiplexers (MUXs) and de-multiplexers (DE-MUXs), tunable optical filters, isolators, and circulators. For example, although thin film-based WDM components traditionally used for MUX and DE-MUX have excellent performance, their insertion loss can be increased dramatically with the increase of the number of channels, and the fabrication yield can be reduced in the assembling process based on free-space optics. Thanks to the advance in photonic integration technology, arrayed-waveguide-gratings (AWGs) based on planar lightwave circuits (PLCs) allowed the fabrication through photolithography and etching. PLC-based MUXs and DE-MUXs with high spectral selectivity, low loss, and large port count are commercially available.

From the systems point of view, significant efforts have been paid in the modeling and understanding of various degradation effects affecting fiber-optic transmission system performance. Numerical techniques such as the split-step Fourier method allows the inclusion of most linear and nonlinear effects of optical fiber for system performance simulation. Commercial simulation software packages were also made available, which provided powerful tools for optical system design and performance evaluation. Meanwhile, various analytic and semi-analytic techniques were developed for system performance evaluation, allowing better understanding of specific physical mechanisms behind system performance degradation. Analytic and semi-analytic techniques also allowed fast calculation of system performance, which is required for making real-time routing decisions for dynamic optical networks. Through the excise of optical system modeling and analysis, and the understanding of system performance degradation mechanisms, various modulation formats have been demonstrated to improve the performance of optical systems and to maximize the system capacity. In addition to non-return-to-zero (NRZ), other modulation formats such as return-to-zero (NR), carrier-suppressed RZ (CSRZ), duo-binary, quadrature phase shift keying (QPSK), and differential QPSK (DQPSK) have shown to have advantages in different aspects, such as better tolerance to chromatic dispersion, higher spectral efficiency, or better receiver sensitivity. With the introduction of inline optical amplifiers, ASE noise accumulated along the system can become the dominant noise source at the receiver. Thus, tolerance to optical noise became the most important issue in an optical receiver. Instead of using receiver sensitivity to evaluate optical receiver performance, required optical signal to noise ratio (R-OSNR) became the most relevant measure of receiver performance in an optical system with multiple optical amplifiers. By early 2000s, transmission capacity was able to reach more than 3 Tb/s with a single fiber through the combination of WDM and TDM (time division multiplexing). The per-channel data rate of commercial WDM systems went up from 2.5 to 40 Gb/s in only about 10 years.

1.3 **OPTICAL SYSTEMS ENABLED BY DSP**

Through the history of fiber-optic communication systems R&D, signal processing in the optical domain has been considered a better choice than that in the electric domain. All-optical communication networking was once considered an ultimate goal of the optical communications industry. A number of optical domain techniques have been developed, such as chromatic dispersion compensation, PMD compensation, and all-optical wavelength conversion.

With the increase of optical signal baud rate beyond 40 Gbaud/s, and the symbol length shorter than 25 ps, the transmission quality can be extremely vulnerable to system impairments. In fact, system tolerance to uncompensated dispersion and PMD is generally proportional to the square of the baud rate. The stringent requirement on the precision of dispersion compensation and PMD compensation made design and maintenance of high-speed optical systems quite complicated. Further increasing the per-channel data rate beyond 40 Gb/s appeared quite challenging.

On the other hand, signal processing based on digital electronics can be very flexible and is able to provide high spectral efficiency for radio frequency (RF) and wireless communication. Highly efficient digital signal processing (DSP) algorithms such has fast Fourier transform (FFT) and digital filtering have been developed to compensate for various degradation effects and to optimize the performance of wireless systems. However, for many years the speed of digital electronic circuits was not fast enough to handle fiber-optic communication systems, which usually have much broader bandwidth and higher data rates.

The opportunity opened recently thanks to the dramatic increase of the processing speed of CMOS electronics. As Moore's law predicted, the number of transistors in CMOS integrated circuits has doubled approximately every 2 years, and the processing speed of CMOS electronic circuits follows the same general trend as well. Cost and power consumption per digital operation in CMOS electronic circuits have also been reduced considerably. This provided an opportunity for the fiber-optic communications industry to re-examine its strategy by taking advantage of the high-speed digital electronics technologies. Thanks to the flexibility of electronic processing, shifting some of the optical domain functionalities back to the electronic domain can provide much improved capability, flexibility, and precise compensation of transmission impairments for fiber-optic systems.

Another important development in recent years is the re-introduction of complex optical field modulation and coherent detection. Although coherent detection was initially introduced and actively pursued by the research community in early 1980s, it did not result in commercial success mainly because the lack of maturity of some key components. For example, narrow spectral linewidth external-cavity semiconductor lasers developed for coherent optical systems in 1980s were too bulky and too expensive. While they could be used as laboratory equipment, it was not feasible to integrate these external-cavity lasers into optical transmitters for fiber-optic communication systems. In recent years, miniaturized external-cavity laser diodes with ∼100 kHz spectral linewidth have been developed based on the MEMS (micro-electro-mechanical systems) technology, which meets telecommunication standards and yet with relatively low cost. Integrated multi-section DFB laser diodes were also shown to be able to produce single-frequency optical emission with less than 500-kHz spectral linewidth. Based on these new devices, coherent fiber-optic systems were re-introduced in 2010s together with the more powerful DSP capabilities in the electronic domain.

With complex optical field modulation and coherent detection, information can be encoded into both the intensity and the phase of an optical carrier. High-level complex modulation formats such as QPSK (quadrature phase shift keying) and QAM (quadrature amplitude modulation) can be used for data encoding to achieve high spectral efficiency. In a coherent receiver that linearly translates the complex optical field into the electronic domain, phase information carried by the optical signal is preserved. The impact of chromatic dispersion, shown as a frequency-dependent DGD, can be electronically compensated in the receiver. PMD can also be compensated in the electronic domain through adaptive DSP, which de-correlates optical signals carried by the two polarization modes in a single-mode fiber. This also allows polarization multiplexing in the transmission fiber with each of the two orthogonally polarized modes carrying an independent data channel, so that the transmission capacity can be doubled.

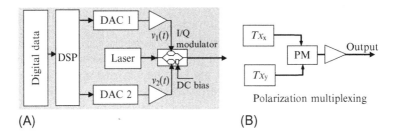

(A)   (B)

FIG. 1.3.1

(A) Block diagram of an optical transmitter with complex optical field modulation, and (B) polarization multiplexing (PM) of two orthogonally polarized optical channels into the same transmission fiber.

Fig. 1.3.1A shows an example of coherent transmitter for complex optical field modulation. Electronic signal is first processed by digital electronic circuits, and converted to analog waveforms through two DACs. These two real-valued analog waveforms are then translated into the real and the imaginary parts of an optical signal through an in-phase (I)/quadrature (Q) electro-optic modulator. As shown in Fig. 1.3.1B, signals from two optical transmitters with the same emitting wavelength can be combined through a polarization beam combiner, commonly referred to as polarization multiplexer (PM), and sent to an optical fiber for transmission. The combination of PM and WDM doubles the optical spectral efficiency compared to the system employing only WDM. As both the amplitude and the phase of the optical signal are used as information carriers, the phase noise of the laser diode has to be small enough.

In a coherent receiver, advanced DSP algorithms can help relax the requirement on the spectral linewidth of the laser diodes used in the system.

Fig. 1.3.2 shows the block diagram of a coherent optical receiver, where an optical LO has to be used as an optical frequency and phase reference. The LO operates in constant wave (CW), and its output mixes with the received optical signal in an I/Q detector consisting of a 90-degree optical hybrid and eight photodiodes to perform balanced detection of the I and the Q components of the x- and the y-polarized optical signals, which will be discussed with more details in Chapters 9 and 11. The four analog voltage waveforms v_{xI}, v_{xQ}, v_{yI}, and v_{yQ} are digitized and converted by four ADCs and processed in a DSP unit. The impact of chromatic dispersion can be compensated by two digital FIR (finite impulse response) all-pass filters, which provide differential delays on the digitized electric signals, which are linearly proportional to the real and imaginary parts of the signal optical field. The coefficients of FIR filters for dispersion compensation can be optimized based on the actual dispersion value of the fiber link, which is usually deterministic. As the DGD introduced by fiber PMD is also a linear process, it can be compensated by digital FIR filters as well in the receiver. Due to the time-dependent nature of PMD, its compensation has to be adaptive with the FIR filter coefficients updated frequently through feedback optimization based on DSP algorithms, which will be discussed in Chapter 11.

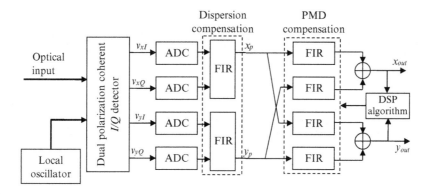

FIG. 1.3.2

Block diagram of a coherent receiver using DSP for chromatic dispersion compensation and PMD compensation.

In a coherent receiver, as a rule of thumb, the combined spectral linewidth between the laser source in the transmitter and the LO in the receiver has to be 4–5 orders of magnitude lower than the symbol rate carried by the optical signal. Within this range, carrier optical phase can be recovered through DSP in the receiver, and information carried by signal optical phase can be detected correctly.

In addition to coherent systems, direct-detection systems can also be benefited from high speed DSP through adaptive optimization of system performance. The capability of efficient spectral shaping as well as precise digital compensation of transmission impairments and cross talk has made DSP an indispensible tool for all fiber-optic systems. While WDM and optical amplification are signature technologies for the fourth generation of fiber-optic systems, the application of high-speed DSP and the reintroduction of coherent detection can be seen as the unique feature of the fifth generation of fiber-optic systems.

1.4 BASIC UNITS OFTEN USED FOR FIBER-OPTIC SYSTEM ENGINEERING

Frequency and wavelength: In the study of fiber-optic systems, wavelength λ and frequency f are often used interchangeably based on the relation:

$$f\lambda = c \tag{1.4.1}$$

where c is the speed of light in free space. For example, lasers emitting at 1550 and 1310 nm wavelengths are equivalent to optical carrier frequencies of 193.55 and 229 THz, respectively, as frequency is inversely proportional to the wavelength. In this book, values of wavelength are referenced to free space.

However, the conversion between the widths of frequency window and wavelength window is a bit more complicated. For two wavelength channels at λ_1 and λ_2, their frequency difference is

$$f_2 - f_1 = c\left(\frac{1}{\lambda_2} - \frac{1}{\lambda_1}\right) = c\left(\frac{\lambda_1 - \lambda_2}{\lambda_2\lambda_1}\right) \tag{1.4.2}$$

As we often deal with small wavelength difference with respective to the average wavelength, that is, $\lambda_1 - \lambda_2 \ll (\lambda_1 + \lambda_2)/2$, an approximation can be made with, $df/d\lambda = -c/\lambda^2$ so that, $f_2 - f_1 \approx -\frac{c}{\lambda^2}(\lambda_2 - \lambda_1)$ or,

$$\delta f \approx -\frac{c}{\lambda^2}\delta\lambda \tag{1.4.3}$$

where, $\delta f = f_2 - f_1$, $\delta\lambda = \lambda_2 - \lambda_1$, and $\lambda = (\lambda_2 + \lambda_1)/2$.

For example, for two wavelength channels separated by 1 nm in the 1550-nm wavelength region, their frequency difference is approximately 125 GHz. Table 1.4.1 shows most often used wavelength regions for optical communication, and the corresponding frequency bandwidth in THz.

Table 1.4.1 Wavelength windows used for optical communications

	O-band	S-band	C-band	L-band	U-band
Wavelength (nm)	1260–1360	1460–1530	1530–1565	1565–1625	1625–1675
Bandwidth (THz)	17.5	9.4	4.4	7	5.5

The C-band with about 4.4 THz frequency bandwidth is the most utilized wavelength window for long-distance fiber-optic systems where the fiber loss is low and EDFA has the highest gain and lowest noise figure. S-band and L-band optical communication products have been developed in recent years to expand the WDM transmission capacity. Although fiber attenuation in the L-band can be even lower than that in the C-band, optical amplifiers in the L-band are not as mature and low-cost as those in the C-band. The S-band has slightly higher fiber attenuation even with the minimum water absorption, and thus the S-band is mostly used for short distance access networks and interconnection such as passive optical networks (PONs) and fiber to the home. The major advantage for the O-band in the 1310-nm wavelength region is the zero chromatic dispersion of the standard single-mode fiber. O-band is most often used for relatively short-reach optical systems using direct modulation of laser diodes.

Photon energy and optical power: The frequency, f, of an optical signal generated from a laser diode is determined by the energy bandgap, E_g, of the semiconductor material,

$$f = E_g/h \tag{1.4.4}$$

where $h = 6.63 \times 10^{-34} \text{J} \cdot \text{s}$ is Planck's constant. For an optical signal at 1550 nm wavelength, the energy of each signal photon is $E = hc/\lambda \approx 1.28 \times 10^{-19} \text{J}$. 1 mW of signal optical power at this wavelength is equivalent to a photon flow rate of approximately of 781 trillion photons per second. If an optical system has a symbol rate of B, and requiring a minimum signal optical power of P_{sen} to achieve a targeted bit-error rate, the required number of photons per data symbol can be calculated as $N = P_{sen}\lambda/(hcB)$.

Conversion between decibel and linear units: Optical power can be measured in linear units such as W, mW, μW, nW, and so on, but electrical engineers sometimes prefer to use Decibel units. A most commonly used Decibel unit for optical power is dBm, with

$$P_{dBm} = 10 \cdot \log_{10}(P_{mW}) \tag{1.4.5}$$

where P_{mW} is the power measured in *mW*, and P_{dBm} is the corresponding power in *dBm*. For example, 1 W, 1 mW, 1 μW, and 1 nW of optical power levels are equivalent to 30, 0, −30, and −60 dBm, respectively.

It could be quite confusing for power addition or subtracting using the dBm unit. For example, when two power sources with power levels of $P_1 = -5$ dBm and $P_2 = -3$ dBm, respectively, are combined, the total power level cannot be obtained by simply adding their dBm values. Instead, the total power has to be calculated by

$$P_{total} = 10 \cdot \log_{10}\left(10^{P_1/10} + 10^{P_2/10}\right) = -0.8756 \text{dBm} \tag{1.4.6}$$

While it seems not very convenient for addition or subtracting power levels using the dBm unit, the Decibel unit is most convenient for counting loss or gain in the system. When passing through an optical device, which can be an optical fiber or an optical amplifier, the loss or gain can be measured as $\eta = P_{out}/P_{in}$, with P_{in} and P_{out} the input and out optical powers in linear unit such as W, or mW. This unitless power ratio η can be converted into dB by

$$\eta_{dB} = 10 \cdot \log_{10}(\eta) \tag{1.4.7}$$

For example, if 1 mW optical power is attenuated to 0.01 mW after propagating through an optical fiber, then $\eta = 0.01$, which corresponds to $\eta_{dB} = -20$ dB. Thus, we say this fiber has 20 dB loss. Another example is that if the power of an optical signal propagating along a fiber is reduced to half after every 10 km, the attenuation parameter of this fiber should be 3 dB per 10 km, or 0.3[dB/km]. Assume an optical signal with 10 dBm power is launched to the input of this fiber; then, the power level will be reduced to $P_{20km} = 10$ dBm $- 20 \times 0.3$ dB $= 4$ dBm after 20 km, and $P_{100km} = 10$ dBm $- 100 \times 0.3$ dB $= -20$ dBm after 100 km. More detailed discussion about fiber loss and unit conversion between dB/km and Neper/m can be found in Section 2.4.

In an optical system, an optical signal can pass through many components, and each of them can have gain or loss. If the input optical power is given in dBm, and the gain and loss of all these optical components are provided in dB, the output optical power can be simply obtained through adding or subtracting (instead of multiplying or dividing used with linear units).

An important point to remember is that while dBm is a power measure, which is converted from mW, dB is a measure of power ratio, which is unitless.

OPTICAL FIBERS

CHAPTER OUTLINE

Introduction 20
2.1 General properties of optical wave 21
2.2 Reflection and refraction 24
    2.2.1 Fresnel reflection coefficients 25
    2.2.2 Special cases of reflection angles 26
    2.2.3 Optical field phase shift between the incident and the reflected beams 29
    2.2.4 Brewster angle 29
2.3 Propagation modes in optical fibers 30
    2.3.1 Geometric optics analysis 31
    2.3.2 Mode analysis using electromagnetic field theory 34
    2.3.3 Mode classification 39
    2.3.4 Numerical aperture 39
    2.3.5 Field distribution profile of SMF 43
2.4 Optical fiber attenuation 44
2.5 Group velocity and dispersion 48
    2.5.1 Phase velocity and group velocity 48
    2.5.2 Group velocity dispersion 50
    2.5.3 Sources of chromatic dispersion 51
    2.5.4 Modal dispersion 53
    2.5.5 Polarization mode dispersion 53
    2.5.6 Mode division multiplexing 55
2.6 Nonlinear effects in an optical fiber 56
    2.6.1 Stimulated Brillouin scattering 56
    2.6.2 Stimulated Raman scattering 57
    2.6.3 Kerr effect nonlinearity and nonlinear Schrödinger equation 58
2.7 Different types of optical fibers 63
    2.7.1 Standard optical fibers for transmission 63
    2.7.2 Specialty optical fibers 66
2.8 Summary 69
Problems 71
References 75
Further reading 76

Introduction to Fiber-Optic Communications. https://doi.org/10.1016/B978-0-12-805345-4.00002-0

INTRODUCTION

Optical wave is a special category of electromagnetic waves which can propagate in free space as well as been guided with dielectric waveguides. Optical fiber is enabled by the optical field confinement mechanism of the waveguide. Low absorption of the materials that construct the optical waveguide is another important factor to enable long-distance transmission of lightwave along the waveguide. The total internal reflection at the core/cladding interface of an optical waveguide can be achieved either by a discrete change of refractive index at the interface known as the step index profile, or by a gradual index change from the core to the cladding known as the graded index profile.

For long-distance optical transmission, optical fiber is often used, and the basic structure of an optical fiber is shown in Fig. 2.1A, which has a central core, a cladding surrounding the core, and an external coating to protect and strengthen the fiber. Silicon dioxide (SiO_2) is the basic material to make optical fibers because of its ultralow absorption in the optical communications wavelength windows. Historically, the fabrication process developed by Corning in the 1970s that reduced the fiber loss to $<10\,dB/km$ made long-distance fiber-optic communication feasible, and was commonly regarded as the starting point for industrial applications of fiber-optic communication systems. In addition to optical transmission over long distance, guided-wave mechanism also enables the design of photonic devices for the processing of optical signals. In photonic devices, optical waveguides are often created on planar dielectric substrates, known as the planar lightwave circuits (PLC). One example of a PLC is illustrated in Fig. 2.1B, in which the higher index core of the waveguide is buried inside the lower index cladding, and they are both deposited on a substrate which provides the mechanical support. Waveguides based on PLC are usually created through photolithography and etching (Okamoto, 2005), and can be made into various structures to form photonic devices with the desired functionalities. As the sizes of photonic devices are often in the orders of millimeters or centimeters, the material loss requirement for PLC is not as stringent as that for optical fibers. Thus, a variety of materials, such as polymer, silicon (Si), indium phosphate (InP), and gallium nitride (GaN), have been used to make PLC to enhance their functionality, reduce the fabrication cost, and to allow photonic integration.

Although both optical fiber and PLC are based on total internal reflection between the core and the cladding, we will focus on the basic guiding mechanisms and applications of optical fibers in this chapter, while integrated optical circuits based on PLC will be discussed later in Chapter 6. This chapter starts with the introduction of wave reflection, refraction, and the condition of total internal reflection at

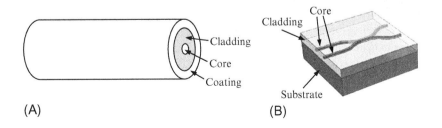

(A)   (B)

FIG. 2.1

Basic structures of an optical fiber (A) and a planar optical waveguide coupler (B).

an optical interface which is the basic requirement for wave guiding. We will then discuss propagation modes and mode classification in an optical fiber. Although single-mode optical fiber is the major fiber type for long-distance optical communications, multimode fibers (MMFs) with relatively bigger core diameters are also widely used, especially for local area and access optical networks, as well as fiber to the home applications, because of their much relaxed tolerance to misalignment and the simplicity of handling. In addition, MMFs can potentially be used for mode multiplexing in which each mode carries an independent information channel in a communication system. Basic properties of optical fibers, including attenuation, dispersion, and nonlinearity, will also be discussed. While attenuation describes the reduction of optical power level along the fiber, dispersion is a differential delay which introduces waveform distortion in a transmission system. In a MMF, in principle different spatial modes in the same fiber would travel in different speeds, which is responsible for the so-called modal dispersion. At the same time, within each spatial mode, different frequency components of an optical signal may also have different propagation velocities and this is the origin of "chromatic dispersion," which is one of the major sources of waveform distortion in a single-mode fiber (SMF). Another important parameter in an optical fiber is the Kerr-effect nonlinearity, which is caused by the power-dependent refractive index. As the amplitude of information-carrying optical signal in an optical fiber is usually not constant, the variation of the instantaneous signal optical power can modulate the refractive index of the optical fiber which causes the modulation in the propagation delay and introduces waveform distortion at the optical receiver. In addition to the Kerr-effect nonlinearity, nonlinear scatterings also exist in an optical fiber mainly through stimulated Brillouin scattering (SBS) and stimulated Raman scattering (SRS). SBS, originates from the mechanical property of the silica material, is a narrow band phenomenon, in which the optical signal creates an acoustic vibration of the material and been scattered by this acoustic wave to produce a new frequency component approximately 11 GHz away from the original signal optical frequency. On the other hand, SRS is generated by molecular vibrations of the silica material caused by the optical signal, which is broadband. The frequency shift of the scattered optical signal caused by the SRS process is in the order of 13 THz. The last section of this chapter introduces a few special fiber types which were designed to provide various different properties compared to a standard SMF in terms of their fabrication processes, chromatic dispersion parameters, polarization selectivity, and nonlinear property.

As optical fiber is the basic transmission medium of fiber-optic communication systems and networks, a good understanding of its basic properties and wave propagation mechanism is fundamental.

2.1 GENERAL PROPERTIES OF OPTICAL WAVES

Lightwave is a special format of electromagnetic wave which occupies the spectral regions from infrared (IR) to ultraviolet (UV). Popular wavelength windows of lightwave used for optical communication include 850, 1310, and 1550 nm, which are located in the near IR region. The general properties of the lightwave signal include amplitude, frequency, phase, and the state of polarization (SOP). A linearly polarized optical signal propagating in a certain direction can be expressed by a complex representation as

$$\vec{E}\left(\vec{x},t\right) = \vec{e}\,E_0\exp\left\{-j\left(\omega t - \vec{k}\cdot\vec{x}\right)\right\} \tag{2.1.1}$$

where $\vec{E}$ is the complex optical field, which is a vector, with the polarization orientation represented by a unit vector $\vec{e}$ and a position vector represented by $\vec{x}$. In Eq. (2.1.1), E_0 is the field magnitude, $\omega = 2\pi f$ is the angular frequency (f is the circular frequency), $\vec{k}$ is known as the vectoral propagation constant with its absolute value defined by $\left|\vec{k}\right| = 2\pi n/\lambda$. n is the refractive index of the medium, $\lambda = c/f$ is the vacuum wavelength of the lightwave, and c is the speed of light in vacuum. Both wavelength and frequency are often used to describe an optical signal depending on the convenience. For example, optical communication wavelengths of $\lambda = 850$, 1310, and 1550 nm correspond to the frequencies of 353, 229, and 193.5 THz, respectively. The following are a few common properties of lightwave.

Phase front (also known as wave front) refers to a surface of equal phase. That is represented by $\omega t - \vec{k} \cdot \vec{x} =$ constant. With this definition, for a fixed frequency ω and at each moment of time, $\vec{k} \cdot \vec{x}$ is a constant. The spatial distribution of this equal phase depends only on the nature of $\vec{k}$. The wave front of a point source can be spherical, while a collimated light beam often has a plane wave front perpendicular to the propagating direction.

Speed of propagation is represented by the speed of equal-phase front in the direction determined by $\vec{k}$. Based on Eq. (2.1.1), for an incremental time Δt the lightwave travels a distance Δx, the equal-phase relation is $\omega \Delta t - \vec{k} \cdot \Delta \vec{x} = 0$ and thus, $\Delta \vec{x} / \Delta t = \omega / \vec{k}$. Then, if $\Delta \vec{x}$ is in the same direction of $\vec{k}$, $|\Delta x / \Delta t| = |\omega/k| = f\lambda/n = c/n$ which is indeed the speed of light in the medium with a refractive index n. If we consider both the positive and the negative frequencies, a lightwave can also be conveniently represented as a real field, $\vec{E}\left(\vec{x}, t\right) = \vec{e} E_0 \cos\left(\omega t - \vec{k} \cdot \vec{x}\right)$.

State of polarization (SOP): although wave front can have various spatial shapes depending on the focusing or diverging nature of the light beam, the simplest wave format is the plane wave, also known as *transverse wave*, which represents a collimated light beam with a plane wave front. In this case, the electrical field $\vec{E}$ and the magnetic field $\vec{H}$ are both perpendicular to the direction of wave propagation. For example, in a Cartesian coordination for a plane wave traveling in the z-direction: $\vec{E} = \vec{e}_x E_x + \vec{e}_y E_y$ and $\vec{k} = \vec{e}_z k_z$, so that, $\vec{k} \cdot \vec{x} = k_z z$, where $\vec{e}_x$, $\vec{e}_y$, and $\vec{e}_z$ are unit vectors in the x-, y-, and z-directions, respectively.

In fact, a polarized plane wave electrical field has the general expression

$$\vec{E}(z, t) = \vec{e}_x E_x(z, t) + \vec{e}_y E_y(z, t) \tag{2.1.2}$$

where

$$E_x(z, t) = E_{0x} \cos(\omega t - kz) \tag{2.1.3a}$$

$$E_y(z, t) = E_{0y} \cos(\omega t - kz + \delta) \tag{2.1.3b}$$

with δ the relative phase difference between x and y field components. The SOP of the lightwave is determined by field magnitudes E_{x0}, E_{y0}, and differential phase δ. The SOP of a polarized light can be categorized as linear, circular, or elliptical.

Linear polarization is obtained with no phase difference between the E_x and the E_y components ($\delta = 0$ or π) as illustrated in Fig. 2.1.1A. In this case, the E_x and E_y components increase or

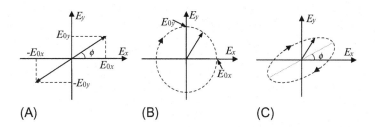

FIG. 2.1.1

Illustration of state of polarization of a plane wave. (A) Linear polarization, (B) circular polarization, and (C) elliptical polarization.

decrease at the same time, so that the tip of the vector $\vec{E}$ moves along a straight line with an angle $\varphi = \tan^{-1}(E_{0y}/E_{0x})$ with respect to the x-axis.

For a circular polarization, $E_{0x} = E_{0y}$ and the phase difference has to be $\delta = 2m\pi \pm \pi/2$, where m is an integer. In this case, when the magnitude of E_x is the maximum, E_y is zero, and vice versa. Thus, the tip of the vector $\vec{E}$ moves along a circle as shown in Fig. 2.1.1B. For $\delta = 2m\pi + \pi/2$, E_y advances E_x by $\pi/2$, so that the circle is clockwise. Otherwise for $\delta = 2m\pi - \pi/2$, the circle is counterclockwise.

The most general SOP is elliptical as shown in Fig. 2.1.1C, which has no special requirement for E_{0x}, E_{0y}, and δ. The trajectory of vector $\vec{E}$ can be expressed by (Keiser, 2011).

$$\left(\frac{E_x}{E_{0x}}\right)^2 + \left(\frac{E_y}{E_{0y}}\right)^2 - 2\frac{E_x E_y}{E_{0x} E_{0y}} \cos\delta = \sin^2\delta \qquad (2.1.4)$$

The orientation angle of the ellipse with respect to the x-axis is

$$\varphi = \frac{1}{2}\tan^{-1}\left(\frac{2E_{0x} E_{0y} \cos\delta}{E_{0x}^2 - E_{0y}^2}\right) \qquad (2.1.5)$$

Note that, the SOP is only valid for the polarized light for which there is a deterministic phase difference δ between the E_x and E_y components. If the phase difference δ is random, the SOP will also be random so that the definition of SOP becomes meaningless. Polarized light can be produced by coherent light sources such as lasers (here "coherent" means that E_x and E_y components are mutually coherent so that δ is deterministic), but many light sources in the nature, such as sun light, are not coherent, and a practical light source is often partially coherent.

Degree of polarization (DOP) is defined as the percentage of the polarized light power ($P_{polarized}$) in the total power of the light ($P_{polarized} + P_{unpolarized}$),

$$DOP = \frac{P_{polarized}}{P_{polarized} + P_{unpolarized}} \qquad (2.1.6)$$

Thus, $0 \leq DOP \leq 1$, with 0 and 1 representing completely unpolarized and completely polarized lights, respectively. In practice, light generated from a semiconductor laser for optical communication has the DOP value of generally higher than 95%.

2.2 REFLECTION AND REFRACTION

An optical interface is generally defined as a plane across which optical property discontinues. For example, water surface is an optical interface because the refractive indices suddenly change from $n=1$ in the air to $n=1.3$ in the water. To simplify our discussion, the following assumptions have been made:

1. Plane wave propagation
2. Linear medium
3. Isotropic medium
4. Smooth planar optical interface

As illustrated in Fig. 2.2.1, an optical interface is formed between two optical materials with refractive indices of n_1 and n_2, respectively. A plane optical wave is projected onto the optical interface at an incidence angle θ_1 (with respect to the surface normal). The optical wave is linearly polarized, and its field amplitude vector can be decomposed into two orthogonal components, $E_{//}^i$ and $E_{\perp}^i$ parallel and perpendicular to the incidence plane (the plane that contains the incident light ray and the interface normal). At the optical interface, part of the energy is reflected back to the same side of the interface and the other part is refracted across the interface.

$$n_2 \sin\theta_2 = n_1 \sin\theta_1 \tag{2.2.1}$$

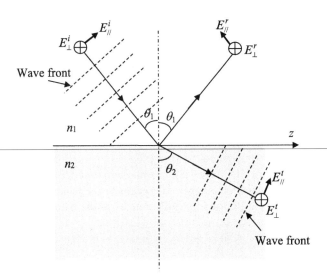

FIG. 2.2.1

Plane wave reflection and refraction at an optical interface.

is known as the Snell's law, which tells the propagation direction of the refracted lightwave with respect to that of the incident wave. The wave propagation direction change is proportional to the refractive index difference across the interface.

Snell's law is derived based on the fact that phase velocity along z-direction should be continuous across the interface. Since the phase velocities in the z-direction are $v_{p1} = \frac{c}{n_1} \frac{1}{\sin\theta_1}$ and $v_{p2} = \frac{c}{n_2} \frac{1}{\sin\theta_2}$ at the two sides of the interface, Eq. (2.2.1) can be obtained with $v_{p1} = v_{p2}$.

Because Snell's law was obtained without any assumption of lightwave polarization state and wavelength, it is independent of these parameters. An important implication of Snell's law is that $\theta_2 > \theta_1$ with $n_2 < n_1$.

2.2.1 FRESNEL REFLECTION COEFFICIENTS

To find out the strength and the phase of the optical field that is reflected back to the same side of the interface, we have to treat optical field components $E_{//}^i$ and $E_\perp^i$ separately. An important fact is that optical field components parallel to the interface must be continuous at both sides of the interface.

Let us first consider the field components $E_{//}^i$, $E_{//}^t$, and $E_{//}^r$ (they are all parallel to the incident plane but not to the interface). They can be decomposed into components parallel with and perpendicular to the interface; the parallel components are $E_{//}^i \cos\theta_1$, $E_{//}^t \cos\theta_2$, and $-E_{//}^r \cos\theta_1$, respectively, which can be derived from Fig. 2.2.2. Because of the field continuity across the interface, we have

$$\left(E_{//}^i - E_{//}^r \right) \cos\theta_1 = E_{//}^t \cos\theta_2 \tag{2.2.2}$$

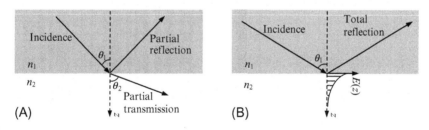

(A)   (B)

FIG. 2.2.2

Illustration of evanescent wave for (A) partial reflection with $\theta_1 < \theta_c$ and (B) total reflection with $\theta_1 > \theta_c$.

At the same time, the magnetic field components associated with $E_{//}^i$, $E_{//}^t$, and $E_{//}^r$ have to be perpendicular to the incident plane, and they are $H_\perp^i = \sqrt{\varepsilon_1/\mu_1} E_{//}^i$, $H_\perp^t = \sqrt{\varepsilon_2/\mu_2} E_{//}^t$, and $H_\perp^r = \sqrt{\varepsilon_1/\mu_1} E_{//}^r$, respectively, where ε_1 and ε_2 are electrical permittivities and μ_1 and μ_2 are magnetic permeabilities of the optical materials at two sides of the interface. Since $H_\perp^i$, $H_\perp^t$, and $H_\perp^r$ are all parallel to the interface (although perpendicular to the incident plane), magnetic field continuity requires $H_\perp^i + H_\perp^r = H_\perp^t$. Assume that $\mu_1 = \mu_2$, $\sqrt{\varepsilon_1} = n_1$, and $\sqrt{\varepsilon_2} = n_2$, and we have

$$n_1 E_{//}^i + n_1 E_{//}^r = n_2 E_{//}^t \tag{2.2.3}$$

Combine Eqs. (2.2.2) and (2.2.3) and we can find the field reflectivity:

$$\rho_{//} = \frac{E_{//}^r}{E_{//}^i} = \frac{n_1 \cos\theta_2 - n_2 \cos\theta_1}{n_1 \cos\theta_2 + n_2 \cos\theta_1} \tag{2.2.4}$$

Using Snell's law, Eq. (2.2.4) can also be written as

$$\rho_{//} = \frac{-n_2^2 \cos\theta_1 + n_1 \sqrt{(n_2^2 - n_1^2 \sin^2\theta_1)}}{n_2^2 \cos\theta_1 + n_1 \sqrt{(n_2^2 - n_1^2 \sin^2\theta_1)}} \tag{2.2.5}$$

where variable θ_2 is eliminated.

Similar analysis can also find the reflectivity for optical field components perpendicular to the incident plane as

$$\rho_\perp = \frac{E_\perp^r}{E_\perp^i} = \frac{n_1 \cos\theta_1 - n_2 \cos\theta_2}{n_1 \cos\theta_1 + n_2 \cos\theta_2} \tag{2.2.6}$$

Or, equivalently,

$$\rho_\perp = \frac{n_1 \cos\theta_1 - \sqrt{(n_2^2 - n_1^2 \sin^2\theta_1)}}{n_1 \cos\theta_1 + \sqrt{(n_2^2 - n_1^2 \sin^2\theta_1)}} \tag{2.2.7}$$

Power reflectivities for parallel and perpendicular field components are, therefore,

$$R_{//} = \left|\rho_{//}\right|^2 = \left|E_{//}^r / E_{//}^i\right|^2 \tag{2.2.8}$$

and

$$R_\perp = \left|\rho_\perp\right|^2 = \left|E_\perp^r / E_\perp^i\right|^2 \tag{2.2.9}$$

Then, according to the energy conservation, the power transmission coefficients can be found as

$$T_{//} = \left|E_{//}^t / E_{//}^i\right|^2 = 1 - \left|\rho_{//}\right|^2 \tag{2.2.10}$$

and

$$T_\perp = \left|E_\perp^t / E_\perp^i\right|^2 = 1 - \left|\rho_\perp\right|^2 \tag{2.2.11}$$

In practice, for an arbitrary incidence polarization state, the input field can always be decomposed into $E_{//}$ and $E_\perp$ components. Each can be treated independently.

2.2.2 SPECIAL CASES OF REFLECTION ANGLES

(A) Normal incidence

This is when the light is launched perpendicular to the material interface. In this case, $\theta_1 = \theta_2 = 0$ and $\cos\theta_1 = \cos\theta_2 = 1$, the field reflectivity can be simplified as

$$\rho_{//} = \rho_\perp = \frac{n_1 - n_2}{n_1 + n_2} \tag{2.2.12}$$

Note that there is no phase shift between incident and reflected field if $n_1 > n_2$ (the phase of both $\rho_{//}$ and $\rho_\perp$ is zero). On the other hand, if $n_1 < n_2$, there is a π phase shift for both $\rho_{//}$ and $\rho_\perp$ because they both become negative.

With normal incidence, the power reflectivity is

$$R_{//} = R_{\perp} = \left| \frac{n_1 - n_2}{n_1 + n_2} \right|^2 \tag{2.2.13}$$

This is a very often used equation to evaluate optical reflection. For example, reflection at an open fiber end is approximately 4%. This is because the refractive index in the fiber core is $n_1 \approx 1.5$ (silica fiber) and refractive of air is $n_1 = 1$. Therefore,

$$R = \left| \frac{n_1 - n_2}{n_1 + n_2} \right|^2 = \left| \frac{1.5 - 1}{1.5 + 1} \right|^2 = 0.2^2 = 0.04 \approx -14 dB$$

In practical optical measurement setups using optical fibers, if optical connectors are not properly terminated, the reflections from fiber-end surface can potentially cause significant measurement errors.

(B) Critical angle

Critical angle is defined as an incident angle θ_1 at which total reflection happens at the interface. According to the Fresnel Eqs. (2.2.5) and (2.2.7), the only possibility that $|\rho_{//}|^2 = |\rho_{\perp}|^2 = 1$ is to have $n_2^2 - n_1^2 \sin^2 \theta_1 = 0$ or

$$\theta_1 = \theta_c = \sin^{-1}(n_2/n_1) \tag{2.2.14}$$

where θ_c is defined as the critical angle. Obviously, the necessary condition to have a critical angle depends on the interface condition.

First, if $n_1 < n_2$, there is no real solution for $\sin^2 \theta_1 = (n_2/n_1)^2$. This means that when a light beam goes from a low index material to a high index material, total reflection is not possible.

Second, if $n_1 > n_2$, a real solution can be found for $\theta_c = \sin^{-1}(n_2/n_1)$ and therefore, total reflection can only happen when a light beam launches from a high index material to a low index material.

It is important to note that at a larger incidence angle $\theta_1 > \theta_c$, $n_2^2 - n_1^2 \sin^2 \theta_1 < 0$, and $\sqrt{n_2^2 - n_1^2 \sin^2 \theta_1}$ becomes imaginary. Eqs. (2.2.5) and (2.2.7) clearly show that if $\sqrt{n_2^2 - n_1^2 \sin^2 \theta_1}$ is imaginary, both $|\rho_{//}|^2$ and $|\rho_{\perp}|^2$ are equal to 1.

The important conclusion is that for all incidence angles satisfying $\theta_1 > \theta_c$, total internal reflection will happen with $R = 1$.

Evanescent field: Note that if the total reflection condition is satisfied, there is no optical power flow across the interface. However, because of the continuity constrain, the optical field on the other side of the interface does not suddenly reduce to zero, which is known as the evanescent field.

As illustrated in Fig. 2.2.2A, when the incidence angle is smaller than the critical angle, $\theta_1 < \theta_c$, the reflection is partial, and the optical field that propagates in the z-direction in the n_2 medium can be described by $E(z) = E_0 e^{j\beta_z z}$, where $\beta_z = (2\pi/\lambda) n_2 \cos \theta_2$ which is the propagation constant projected in the z-direction. Based on the Snell's law, this projected propagation constant can be expressed as $\beta_z = (2\pi/\lambda) \sqrt{n_2^2 - n_1^2 \sin^2 \theta_1}$. The value of β_z reduces with the increase of the incidence angle and β_z becomes zero when the incidence angle is equal to the critical angle ($\theta_1 = \theta_c$). Further increasing the incidence angle for $\theta_1 > \theta_c$ results in an imaginary β_z value, $\beta_z = j(2\pi/\lambda) \sqrt{n_1^2 \sin^2 \theta_1 - n_2^2} = j\alpha$, where

$$\alpha = \frac{2\pi}{\lambda} \sqrt{n_1^2 \sin^2 \theta_1 - n_2^2}$$

is defined as the attenuation parameter of the evanescent field on the n_2 side of the medium, and the optical field is then,

$$E(z) = E_0 e^{j\beta_z z} = E_0 e^{-\alpha z}$$

As illustrated in Fig. 2.2.2B, the penetration depth, z_e, of the evanescent field across the interface is usually defined by the distance at which the field is reduced by $1/e$, and thus $z_e = 1/\alpha$. This evanescent field penetration depth not only depends on the values of n_1 and n_2, but also depends on the wavelength of the optical signal and the wave incidence angle. As an example, Fig. 2.2.3 shows the penetration depth as the function the incidence angle for two different materials: glass ($n_1 = 1.5$) and silicon ($n_1 = 3.5$) in the air ($n_2 = 1$), for the signal wavelength of 1.5 μm. This figure indicates that the evanescent field penetration depth z_e is much shorter than the wavelength of the optical signal, especially when the index difference ($n_1 - n_2$) is large, and the incidence angle approaches $\pi/2$. This unique property of tight field concentration has been utilized to make evanescent field photonic biosensors, in which only the molecules immobilized on the waveguide surface are illuminated by the optical field (Taitt et al., 2005).

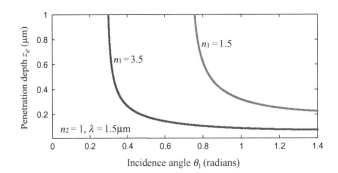

FIG. 2.2.3

Penetration depth of evanescent field as the function of incidence angle, for air/glass interface and air/silicon interface.

EXAMPLE 2.1

Assume the refractive indices of water and glass are 1.3 and 1.5, respectively. Please find critical angles on the air/water interface and air/glass interface. Is it possible to have 100% power reflection on the air/water surface when the light beam is launched from the air?

Solution

Based on Eq. (2.2.14), the critical angles on the air/water and air/glass surfaces are $\theta_{c,\ water} = \sin^{-1}(1/1.3) = 50.28$ degree and $\theta_{c,\ glass} = \sin^{-1}(1/1.5) = 41.8$ degree. If the light beam is launched from air to the water surface, $n_2 = 1.3$ and $n_1 = 1$ in Eq. (2.2.14), and thus $\theta_c = \sin^{-1}(1.3)$ which is imaginary, implying that critical angle does not exist in this case and the power reflectivity cannot be 100% for $90° > \theta > 0$. In fact, this is the reason that optical wave can only be confined with high index material, such as an optical waveguide or a fiber core which will be discussed in this chapter.

2.2.3 OPTICAL FIELD PHASE SHIFT BETWEEN THE INCIDENT AND THE REFLECTED BEAMS

(a) When $\theta_1 < \theta_c$ (partial reflection and partial transmission), both $\rho_{//}$ and $\rho_\perp$ are real and therefore, there is no phase shift for the reflected wave at the interface.

(b) When total internal reflection happens, $\theta_1 > \theta_c$, $\sqrt{n_2^2 - n_1^2 \sin^2 \theta_i}$ is imaginary. Fresnel Eqs. (2.2.5) and (2.2.7) can be written as

$$\rho_{//} = \frac{-n_2^2 \cos\theta_1 + jn_1 \sqrt{\left(n_1^2 \sin^2\theta_1 - n_2^2\right)}}{n_2^2 \cos\theta_1 + jn_1 \sqrt{\left(n_1^2 \sin^2\theta_1 - n_2^2\right)}} \tag{2.2.15}$$

$$\rho_\perp = \frac{n_1 \cos\theta_1 - j\sqrt{\left(n_1^2 \sin^2\theta_1 - n_2^2\right)}}{n_1 \cos\theta_1 + j\sqrt{\left(n_1^2 \sin^2\theta_1 - n_2^2\right)}} \tag{2.2.16}$$

Therefore, phase shift for the parallel and the perpendicular electrical field components are, respectively,

$$\Delta\Phi_{//} = \arg\left(\frac{E_{//}^r}{E_{//}^i}\right) = -2\tan^{-1}\left(\frac{n_1 \sqrt{n_1^2 \sin^2\theta_1 - n_2^2}}{n_2^2 \cos\theta_1}\right) \tag{2.2.17}$$

$$\Delta\Phi_\perp = \arg\left(\frac{E_\perp^r}{E_\perp^i}\right) = -2\tan^{-1}\left(\frac{n_2 \sqrt{\left(n_1^2/n_2^2\right)\sin^2\theta_1 - 1}}{n_1 \cos\theta_1}\right) \tag{2.2.18}$$

This optical phase shift happens at the optical interface, which has to be considered in optical waveguide design, as will be discussed later.

2.2.4 BREWSTER ANGLE

Consider a light beam launched onto an optical interface. If the input electrical field is parallel to the incidence plane, there exists a specific incidence angle θ_B at which the reflection is equal to zero. Therefore, the energy is totally transmitted across the interface. This angle θ_B is defined as the *Brewster angle*.

Consider the Fresnel Eq. (2.2.5) for parallel field components. If we solve this equation for $\rho_{//} = 0$ and use θ_1 as a free parameter, the only solution is $\tan\theta_1 = n_2/n_1$ and therefore, the Brewster angle is defined as

$$\theta_B = \tan^{-1}(n_2/n_1) \tag{2.2.19}$$

Two important points we need to note: (1) the Brewster angle is only valid for the polarization component which has the electrical field vector parallel to the incidence plane. For the perpendicular polarized component, no matter how you choose θ_1, total transmission will never happen. (2) $\rho_{//} = 0$ happens only at one angle $\theta_1 = \theta_B$. This is very different from the critical angle where total reflection happens for all incidence angles within the range of $\theta_c < \theta_1 < \pi/2$.

The Brewster angle is often used to minimize the optical reflection and it can also be used to select the polarization. Fig. 2.2.4 shows an example of optical field reflectivities $\rho_{//}$ and $\rho_{\perp}$, and their corresponding phase shifts $\Delta\Phi_{//}$ and $\Delta\Phi_{\perp}$ at an optical interface of two materials with $n_1 = 1.5$ and $n_2 = 1.4$. In this example, the critical angle is $\theta_c \approx 69$ degree and the Brewster angle is $\theta_B \approx 43$ degree.

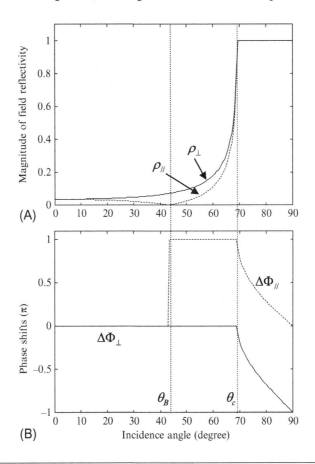

FIG. 2.2.4

Field reflectivities (A) and phase shifts (B) vs. incidence angle. Optical interface is formed with $n_1 = 1.5$ and $n_2 = 1.4$.

2.3 PROPAGATION MODES IN OPTICAL FIBERS

The cross section of a planar optic waveguide on the PLC platform is usually rectangle, and the details of PLC design will be presented in Chapter 6 where integrated photonic circuits will be discussed. This section will focus on the discussion of optical fiber which is a cylindrical glass bar with a core, a cladding, and an external coating, as shown in Fig. 2.1A. To confine and guide the lightwave signal within the fiber core, a total internal reflection is required at the core-cladding interface. According to what we have discussed in Section 2.1, this requires the refractive index of the core to be higher than that of the cladding.

Practical optical fibers can be divided into two categories: step-index fiber and graded-index fiber. The index profiles of these two types of fibers are shown in Fig. 2.3.1. In a step-index fiber, the refractive index is n_1 in the core and n_2 in the cladding; there is an index discontinuity at the core-cladding interface. A lightwave signal is bounced back and forth at this interface, which forms guided modes propagating in the longitudinal direction. On the other hand, in a graded-index fiber, the refractive index in the core gradually reduces its value along the radius. A generalized Fresnel equation indicates that in a medium with a continual index profile, a light trace would always bend toward high-refractive areas. In fact, this graded-index profile creates a self-focus effect within the fiber core to form an optical waveguide (Meumann, 1988). Although graded-index fibers form a unique category, they are usually made for multimode applications. The popular SMFs are made with step-index fibers. Because of their popularity and simplicity, we will focus our analysis on step-index fibers.

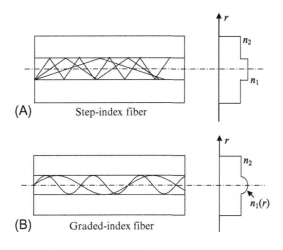

FIG. 2.3.1

Illustration of ray traces (left) and index profiles (right) of step-index (A) and graded-index (B) fibers.

Rigorous description of wave propagation in optical fibers requires solving Maxwell's equations and applying appropriate boundary conditions. In this section, we first use geometric ray trace approximation to provide a simplified analysis, which helps us understand the basic physics of wave propagation. Then, we present electromagnetic field analysis, which provides precise mode cutoff conditions.

2.3.1 GEOMETRIC OPTICS ANALYSIS

In this geometric optics analysis, different propagation modes in an optical fiber can be seen as rays traveling at different angles. There are two types of light rays that can propagate along the fiber: skew rays and meridional rays. Fig. 2.3.2A shows an example of skew rays, which are not confined to any particular plane along the fiber. Although skew rays represent a general case of fiber modes, they are difficult to analyze. A simplified special case is the meridional rays shown in Fig. 2.3.2B, which are confined to the meridian plane, which contains the symmetry axis of the fiber. The analysis of meridional rays is relatively easy and provides a general picture of ray propagation along an optical fiber.

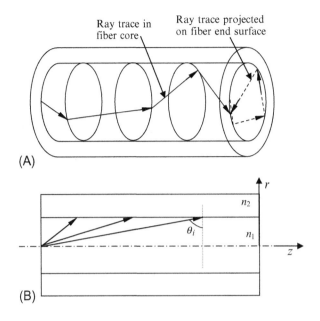

(A)

(B)

FIG. 2.3.2

Illustration of fiber propagation modes in geometric optics: (A) skew ray trace and (B) meridional ray trace.

Consider meridional rays as shown in Fig. 2.3.2B. This is a two-dimensional (2D) problem where the optical field propagates in the longitudinal direction z and its amplitude varies over the transversal direction r. We define $\beta_1 = n_1\omega/c = 2\pi n_1/\lambda$ (in rad/m) as the propagation constant in a homogeneous core medium with a refraction index of n_1. Each fiber mode can be explained as a light ray that travels at a certain angle, as shown in Fig. 2.3.3. Therefore, for ith mode propagating in the $+z$ direction, the propagation constant can be decomposed into a longitudinal component β_{zi} and a transversal component k_{i1} such that

$$\beta_1^2 = \beta_{zi}^2 + k_{i1}^2 \qquad (2.3.1)$$

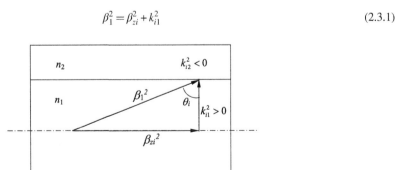

FIG. 2.3.3

Decompose propagating vector β_1 into longitudinal and transversal components.

Then, the optical field vector of the ith mode can be expressed as

$$\vec{E}_i(r,z) = \vec{E}_{i0}(r,z)\exp\{-j(\omega t - \beta_{zi}z)\} \qquad (2.3.2)$$

where $\vec{E}_{i0}(r,z)$ is the field amplitude of the mode.

Since the mode is propagating in the fiber core, both k_{i1} and β_{zi} must be real. First, for k_i to be real in the fiber core, we must have

$$k_{i1}^2 = \beta_1^2 - \beta_{zi}^2 \geq 0 \qquad (2.3.3)$$

The physical meaning of this real propagation constant in the transversal direction is that the lightwave propagates in the transverse direction but is bounced back and forth between the core-cladding interfaces. This creates a standing wave pattern in the transverse direction, like a resonant cavity. In addition, k_{i1} can only have discrete values because the standing wave pattern in the transversal direction requires phase matching. This is the reason that propagating optical modes in a fiber have to be discrete.

Now, let us look at what happens in the cladding. Because the optical mode is guided in the fiber core, there should be no energy propagating in the transversal direction in the cladding (otherwise optical signal power would be leaked). Therefore, k_i has to be imaginary in the cladding, that is,

$$k_{i2}^2 = \beta_2^2 - \beta_{zi}^2 < 0 \qquad (2.3.4)$$

where subscript 2 represents parameters in the cladding and $\beta_2 = n_2\omega/c = 2\pi n_2/\lambda$ is the propagation constant in the homogeneous cladding medium.

Note that since the optical field has to propagate with the same phase velocity in the z-direction both in the core and in cladding, β_{zi} has the same value in both Eqs. (2.3.3) and (2.3.4).

Eqs. (2.3.3) and (2.3.4) can be simplified as

$$\beta_{zi}/\beta_1 \leq 1 \qquad (2.3.5)$$

and

$$\beta_{zi}/\beta_2 > 1 \qquad (2.3.6)$$

Bringing Eqs. (2.3.5) and (2.3.6) together with $\beta_2 = \beta_1 n_2/n_1$, we can find the necessary condition for a propagation mode,

$$1 \geq \frac{\beta_{zi}}{\beta_1} > \frac{n_2}{n_1} \qquad (2.3.7)$$

It is interesting to note that in Fig. 2.3.3, θ_i is, in fact, the incidence angle of the ith mode at the core-cladding interface. The triangle in Fig. 2.3.3 clearly shows that $\beta_{zi}/\beta_1 = \sin\theta_i$. This turns Eq. (2.3.7) into $1 \geq \sin\theta_i > n_2/n_1$, which is the same as the definition of the critical angle as given by Eq. (2.2.14).

The concept of discrete propagation modes comes from the fact that the transversal propagation constant k_{i1} in the fiber core can only take discrete values to satisfy standing wave conditions in the transverse direction. Since β_1 is a constant, the propagation constant in the z-direction $\beta_{zi}^2 = \beta_1^2 - k_i^2$ can only take discrete values as well. Or equivalently the ray angle θ_i can only take discrete values within the range defined by $1 \geq \sin\theta_i > n_2/n_1$.

The geometric optics description given here is simple and it qualitatively explains the general concept of fiber modes. However, it is not adequate to obtain quantitative mode field profiles and cutoff conditions. Therefore, electromagnetic field theory has to be applied by solving Maxwell's equations and using appropriate boundary conditions, which we discuss next.

2.3.2 MODE ANALYSIS USING ELECTROMAGNETIC FIELD THEORY

Mode analysis in optical fibers can be accomplished more rigorously by solving Maxwell's equations and applying appropriate boundary conditions defined by fiber geometries and parameters. We start with classical Maxwell's equations,

$$\nabla \times \vec{E} = -\mu \frac{\partial \vec{H}}{\partial t} \tag{2.3.8}$$

$$\nabla \times \vec{H} = \varepsilon \frac{\partial \vec{E}}{\partial t} \tag{2.3.9}$$

The complex electrical and the magnetic fields are represented by their amplitudes and phases,

$$\vec{E}\left(t, \vec{r}\right) = \vec{E}_0 \exp\left\{-j\left(\omega t - \vec{k} \cdot \vec{r}\right)\right\} \tag{2.3.10}$$

$$\vec{H}\left(t, \vec{r}\right) = \vec{H}_0 \exp\left\{-j\left(\omega t - \vec{k} \cdot \vec{r}\right)\right\} \tag{2.3.11}$$

Since fiber material is passive and there is no generation source within the fiber,

$$\left(\nabla \cdot \vec{E}\right) = 0 \tag{2.3.12}$$

$$\nabla \times \nabla \times \vec{E} \equiv \nabla\left(\nabla \cdot \vec{E}\right) - \nabla^2 \vec{E} = -\nabla^2 \vec{E} \tag{2.3.13}$$

Combining Eqs. (2.3.8)–(2.3.13) yields,

$$\nabla \times \nabla \times \vec{E} = j\omega\mu\left(\nabla \times \vec{H}\right) = j\omega\mu\left(-j\omega\varepsilon \vec{E}\right) \tag{2.3.14}$$

And the Helmholtz equation,

$$\nabla^2 \vec{E} + \omega^2 \mu\varepsilon \vec{E} = 0 \tag{2.3.15}$$

Similarly, a Helmholtz equation as can also be obtained for the magnetic field $\vec{H}$

$$\nabla^2 \vec{H} + \omega^2 \mu\varepsilon \vec{H} = 0 \tag{2.3.16}$$

The next task is to solve Helmholtz equations for the electrical and the magnetic fields. Because the geometric shape of an optical fiber is cylindrical, we can take advantage of this axial symmetry to simplify the analysis by using cylindrical coordinates. In cylindrical coordinates, the electrical field can be decomposed into radial, azimuthal, and longitudinal components: $\vec{E} = \vec{a}_r E_r + \vec{a}_\varphi E_\varphi + \vec{a}_z E_z$ and $\vec{H} = \vec{a}_r H_r + \vec{a}_\varphi H_\varphi + \vec{a}_z H_z$, where $\vec{a}_r$, $\vec{a}_\varphi$, and $\vec{a}_z$ are unit vectors. With this separation, the Helmholtz Eqs. (2.3.15) and (2.3.16) can be decomposed into separate equations for E_r, E_ϕ, E_z, H_r, H_ϕ, and H_z, respectively. However, these three components are not completely independent. In fact, classic electromagnetic theory indicates that in cylindrical coordinate the transversal field components E_r, E_ϕ, H_r, and H_ϕ can be expressed as a combination of longitudinal field components E_z and H_z (Iizuka, 2002). This means that E_z and H_z need to be determined first and then we can find all other field components.

In cylindrical coordinates, the Helmholtz equation for E_z is

$$\frac{\partial^2 E_z}{\partial r^2} + \frac{1}{r}\frac{\partial E_z}{\partial r} + \frac{1}{r^2}\frac{\partial^2 E_z}{\partial \varphi^2} + \frac{\partial^2 E_z}{\partial z^2} + \omega^2 \mu \varepsilon E_z = 0 \tag{2.3.17}$$

Since $E_z = E_z(r, \varphi, z)$ is a function of both r, ϕ, and z, Eq. (2.3.17) cannot be solved analytically. We assume a standing wave in the azimuthal direction and a propagating wave in the longitudinal direction, then the variables can be separated as

$$E_z(r, \varphi, z) = E_z(r) e^{jl\varphi} e^{j\beta_z z} \tag{2.3.18}$$

where $l = 0, \pm 1, \pm 2, \dots$ is an integer. Substituting Eq. (2.3.18) into Eq. (2.3.17), we can obtain a one-dimensional (1D) wave equation:

$$\frac{\partial^2 E_z(r)}{\partial r^2} + \frac{1}{r}\frac{\partial E_z(r)}{\partial r} + \left(\frac{n^2\omega^2}{c^2} - \beta_z^2 - \frac{l^2}{r^2}\right) E_z(r) = 0 \tag{2.3.19}$$

This is commonly referred to as a *Bessel equation* because its solutions can be expressed as Bessel functions.

For a step-index fiber with a core radius a, its index profile can be expressed as

$$n = \begin{cases} n_1 & (r \le a) \\ n_2 & (r > a) \end{cases} \tag{2.3.20}$$

We have assumed that the diameter of the cladding is infinite in this expression. The Bessel Eq. (2.3.19) has solutions only for discrete β_z values, which correspond to discrete modes. The general solutions of Bessel Eq. (2.3.19) can be expressed in Bessel functions as

$$E_z(r) = \begin{cases} AJ_l(U_{lm}r) + A'Y_l(U_{lm}r) & (r \le a) \\ CK_l(W_{lm}r) + C'I_l(W_{lm}r) & (r > a) \end{cases} \tag{2.3.21}$$

where $U_{lm}^2 = \beta_1^2 - \beta_{z,\,lm}^2$ and $W_{lm}^2 = \beta_{z,\,lm}^2 - \beta_2^2$ represent equivalent transversal propagation constants in the core and cladding, respectively, with $\beta_1 = n_1\omega/c$ and $\beta_2 = n_2\omega/c$ as defined before. $\beta_{z,\,lm}$ is the propagation constant in the z-direction. This is similar to the vectorial relation of propagation constants shown in Eq. (2.3.1) in geometric optics analysis. However, we have two mode indices here, l and m. The physical meanings of these two mode indices are the amplitude maximums of the standing wave patterns in the azimuthal and the radial directions, respectively.

In Eq. (2.3.21), J_l and Y_l are the first and the second kind of Bessel functions of the lth order, and K_l and I_l are the first and the second kind of *modified* Bessel functions of the lth order. Their values are shown in Fig. 2.3.4. A, A', C, and C' in Eq. (2.3.21) are constants that need to be defined using boundary appropriate conditions.

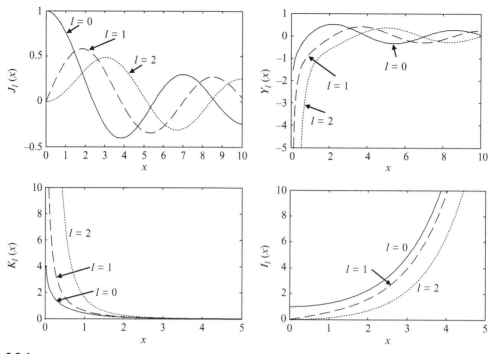

FIG. 2.3.4

Bessel function *(top)* and modified Bessel functions *(bottom), first kind (left) and second king (right).*

The first boundary condition is that the field amplitude of a guided mode should be finite at the center of the core ($r=0$). Since the special function $Y_l(0) = -\infty$, one must set $A'=0$ to ensure that $E_z(0)$ has a finite value.

The second boundary condition is that the field amplitude of a guided mode should be zero far away from the core ($r=\infty$). Since $I_l(\infty) \neq 0$, one must set $C'=0$ to ensure that $E_z(\infty)=0$. Consider $A'=C'=0$; Eq. (2.3.21) can be simplified, and for the mode index of (l, m), it becomes

$$E_{z,lm}(r, \varphi, z) = \begin{cases} AJ_l(U_{lm}r)e^{jl\varphi} \cdot e^{j\beta_{z,lm}z} & (r \le a) \\ CK_l(W_{lm}r)e^{jl\varphi} \cdot e^{j\beta_{z,lm}z} & (r > a) \end{cases} \tag{2.3.22a}$$

Similarly, we can write the magnetic field as

$$H_{z,lm}(r, \varphi, z) = \begin{cases} BJ_l(U_{lm}r)e^{jl\varphi} \cdot e^{j\beta_{z,lm}z} & (r \le a) \\ DK_l(W_{lm}r)e^{jl\varphi} \cdot e^{j\beta_{z,lm}z} & (r > a) \end{cases} \tag{2.3.22b}$$

And other field components can be found through the Maxwell's equations as (Iizuka, 2002)

$$E_{r,lm} = \begin{cases} \dfrac{j}{U_{lm}^2}\left(\beta_{z,lm}\dfrac{\partial E_{z,lm}}{\partial r} + \mu_0\dfrac{\omega}{r}\dfrac{\partial H_{z,lm}}{\partial \varphi}\right) & r \le a \\ \dfrac{-j}{W_{lm}^2}\left(\beta_{z,lm}\dfrac{\partial E_{z,lm}}{\partial r} + \mu_0\dfrac{\omega}{r}\dfrac{\partial H_{z,lm}}{\partial \varphi}\right) & r > a \end{cases} \tag{2.3.23a}$$

$$E_{\varphi,lm} = \begin{cases} \dfrac{j}{U_{lm}^2}\left(\dfrac{\beta_{z,lm}}{r}\dfrac{\partial E_{z,lm}}{\partial\varphi} + \mu_0\omega\dfrac{\partial H_{z,lm}}{\partial r}\right) & r \leq a \\ \dfrac{-j}{W_{lm}^2}\left(\dfrac{\beta_{z,lm}}{r}\dfrac{\partial E_{z,lm}}{\partial\varphi} + \mu_0\omega\dfrac{\partial H_{z,lm}}{\partial r}\right) & r > a \end{cases} \tag{2.3.23b}$$

$$H_{r,lm} = \begin{cases} \dfrac{j}{U_{lm}^2}\left(\beta_{z,lm}\dfrac{\partial H_{z,lm}}{\partial r} + \varepsilon_0 n^2\dfrac{\omega}{r}\dfrac{\partial E_{z,lm}}{\partial\varphi}\right) & r \leq a \\ \dfrac{-j}{W_{lm}^2}\left(\beta_{z,lm}\dfrac{\partial H_{z,lm}}{\partial r} + \varepsilon_0 n^2\dfrac{\omega}{r}\dfrac{\partial E_{z,lm}}{\partial\varphi}\right) & r > a \end{cases} \tag{2.3.23c}$$

and

$$H_{\varphi,lm} = \begin{cases} \dfrac{j}{U_{lm}^2}\left(\dfrac{\beta_{z,lm}}{r}\dfrac{\partial H_{z,lm}}{\partial\varphi} + \varepsilon_0 n^2\omega\dfrac{\partial E_{z,lm}}{\partial r}\right) & r \leq a \\ \dfrac{-j}{W_{lm}^2}\left(\dfrac{\beta_{z,lm}}{r}\dfrac{\partial H_{z,lm}}{\partial\varphi} + \varepsilon_0 n^2\omega\dfrac{\partial E_{z,lm}}{\partial r}\right) & r > a \end{cases} \tag{2.3.23d}$$

$E_{z,lm}$, $H_{z,lm}$, $E_{\phi,lm}$, and $H_{\phi,lm}$ have to be continuous at the core-cladding interface ($r=a$), which can be satisfied by only a set of discrete values of U_{lm} and W_{lm}. This explains the reason why the fiber modes have to be discrete.

Mathematically, the modified Bessel function fits well to an exponential characteristic $K_l(W_{lm}r) \propto \exp(-W_{lm}r)$, so that $K_l(W_{lm}r)$ represents an exponential decay of optical field over r in the fiber cladding. For a propagation mode, $W_{lm} > 0$ is required to ensure that energy does not leak through the cladding. In the fiber core, the Bessel function $J_l(U_{lm}r)$ oscillates as shown in Fig. 2.3.4, which represents a standing-wave pattern in the core over the radius direction. For a propagating mode, $U_{lm}^2 \geq 0$ is required to ensure this standing-wave pattern in the fiber core.

It is interesting to note that based on the definitions of $U_{lm}^2 = \beta_1^2 - \beta_{z,\,lm}^2$ and $W_{lm}^2 = \beta_{z,\,lm}^2 - \beta_2^2$, the requirement of $W_{lm} > 0$ and $U_{lm}^2 \geq 0$ is equivalent to $\beta_2^2 < \beta_{z,\,lm}^2 \leq \beta_1^2$ or $(n_2/n_1) < \beta_{z,\,lm}^2/\beta_1 \leq 1$. This is indeed equivalent to the mode condition (2.3.7) derived by the ray optics.

There are a few often used definitions to categorize the propagation modes in the fiber:

Transverse electric-field mode (TE mode): $E_z = 0$
Transverse magnetic-field mode (TM mode): $H_z = 0$
Hybrid mode (HE mode) $E_z \neq 0$ and $H_z \neq 0$

V-number is an important parameter of a fiber, which is defined as

$$V = a\sqrt{U_{lm}^2 + W_{lm}^2} \tag{2.3.24}$$

since

$$U_{lm}^2 = \beta_1^2 - \beta_{z,lm}^2 = \left(\frac{2\pi n_1}{\lambda}\right)^2 - \beta_{z,lm}^2$$

and

$$W_{lm}^2 = \beta_{z,lm}^2 - \beta_2^2 = \beta_{z,lm}^2 - \left(\frac{2\pi n_2}{\lambda}\right)^2$$

V-number can be expressed as

$$V = a\sqrt{U_{lm}^2 + W_{lm}^2} = \frac{2\pi a}{\lambda}\sqrt{n_1^2 - n_2^2} \tag{2.3.25}$$

In an optical fiber with large core size and large core-cladding index difference, it will support a large number of propagating modes. Approximately, the total number of guided modes in a fiber is related to the V-number as (Keiser, 2011)

$$M \approx V^2/2 \tag{2.3.26}$$

In a MMF, the number of guided modes can be on the order of several hundred. Imagine that a short optical pulse is injected into a fiber and the optical energy is carried by many different modes. Because different modes have different propagation constants $\beta_{z,\,lm}$ in the longitudinal direction and they will arrive at the output of the fiber in different times, the short optical pulse at the input will become a broad pulse at the output. In optical communications systems, this introduces signal waveform distortions and bandwidth limitations. This is the reason SMF is required in high-speed long-distance optical systems.

In a SMF, only the lowest-order mode is allowed to propagate; all higher-order modes are cutoff. In a fiber, the lowest-order propagation mode is HE_{11}, whereas the next lowest modes are TE_{01} and TM_{01} ($l=0$ and $m=1$). In fact, TE_{01} and TM_{01} have the same cutoff conditions: (1) $W_{01}=0$ so that these two modes radiate in the cladding and (2) $J_0(U_{01}a)=0$ so that the field amplitude at core/cladding interface ($r=a$) is zero. Under the first condition ($W_{01}=0$), we can find the cutoff V-number $V=a\sqrt{U_{01}^2+W_{01}^2}=aU_{01}$, whereas under the second condition ($J_0(U_{01}a)=0$), we can find $J_0(aU_{01})=J_0(V)=0$, which implies that $V=2.405$ as the first root of $J_0(V)=0$.

Therefore, the single-mode condition is

$$V = \frac{2\pi a}{\lambda}\sqrt{n_1^2 - n_2^2} < 2.405 \tag{2.3.27}$$

Note that the approximation given in Eq. (2.45) is valid only for a fiber with large number of modes so that $V \gg 2.4.5$ is valid. For a few-modes fiber, the number of modes can be counted from Fig. 2.3.5 as calculated through the actual solutions of Maxwell's equations.

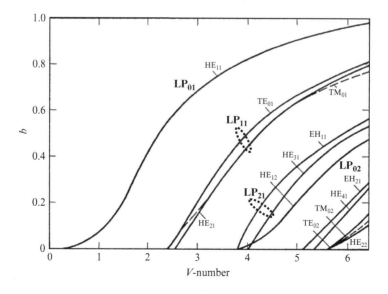

FIG. 2.3.5

Normalized propagation constant b as the function of the V-number (Gloge, 1971).

2.3.3 MODE CLASSIFICATION

In general, each mode in the fiber has its unique identification number determined by the indices l and m. Multiple solutions ($m = 1, 2, 3, \ldots$) can exist for each given l value. Meridional modes refer to the modes with $l = 0$, and the mode field is independent of azimuthal angle ϕ. Meridional modes include TE_{0m} ($E_z = 0$) and TM_{0m} ($H_z = 0$). Skew modes are commonly referred to as hybrid modes for $l \neq 0$, and they are often denoted as EH_{lm} (E_z dominant) and HE_{lm} (H_z dominant).

For a guided mode, the projection of the propagation constant in the z-direction, β_z, has to satisfy $\beta_2^2 < \beta_z^2 \leq \beta_1^2$ (see Fig. 2.3.3), where $\beta_1 = kn_1$, $\beta_2 = kn_2$, and $k = 2\pi/\lambda$. This is equivalent to

$$0 < \frac{(\beta_z/k)^2 - n_2^2}{n_1^2 - n_2^2} < 1 \tag{2.3.28}$$

In Fig. 2.3.5, the normalized propagation constant is defined as

$$b = \frac{(\beta_z/k)^2 - n_2^2}{n_1^2 - n_2^2} \tag{2.3.29}$$

Thus, $0 < b < 1$ is the necessary condition for the existence of a propagation mode, and the relation between b and the V-number is shown in Fig. 2.3.5. Mode cut-off condition is for b to approach zero. Fig. 2.3.5 indicates that the lowest-order mode in a fiber is HE_{11}, which is a hybrid mode corresponding to $l = 1$ and $m = 1$, and the electrical field is only in the transversal direction ($E_z = 0$). High-order modes (TE_{01} and TM_{01}) start to exist when the V-number reaches to 2.405.

For most practical optical fibers, the index contrast between the core and the cladding is small so that $\Delta = (n_1 - n_2)/n_1 << 1$. Under such a weak guidance condition, certain groups of propagating modes have almost identical propagation constants, so that the linear combination of these modes within a group can form a so called LP mode ("LP" stands for linearly polarized). For example, the combination of TE_{01}, TM_{01}, and HE_{21} forms the LP_{11} mode; and EH_{11} and HE_{31} forms LP_{21}.

Fig. 2.3.6 shows the transversal E and H field distributions of HE_{11}, TM_{01}, TE_{01}, and HE_{21}. While HE_{11} constitutes the LP_{01} mode by itself as shown in Fig. 2.3.6A, LP_{11} mode is formed by the superposition of HE_{21} and TM_{01}, or HE_{21} and TE_{01} as shown in Fig. 2.3.6B and C.

Because of the central symmetry of the fiber cross section, the mode structure will be identical when the fiber is rotated by 90 degree (exchange x- and y-axis). Thus, each mode shown in Fig. 2.3.6 has a degenerate mode with the orthogonal polarization orientation. As the result, there are 2 LP_{01} modes, 4 LP_{11} modes, 4 LP_{21} modes, and 2 LP_{02} modes. A detailed mode list can be found at Buck (2004).

2.3.4 NUMERICAL APERTURE

Numerical aperture is a parameter that is often used to specify the acceptance angle of a fiber. Fig. 2.3.7 shows an azimuthal cross section of a step-index fiber and a light ray that is coupled into the fiber from the left-side end surface.

For the light to be into the guided mode in the fiber, total internal reflection has to occur inside the core and $\theta_i > \theta_c$ is required, as shown in Fig. 2.3.7, where $\theta_c = \sin^{-1}(n_2/n_1)$ is the critical angle of the core-cladding interface. With this requirement on θ_i, there is a corresponding requirement on incident

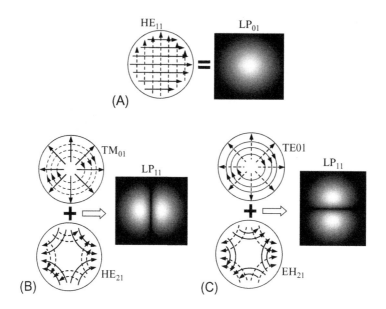

FIG. 2.3.6

Illustration of transversal electrical *(solid lines)* and magnetic *(dashed lines)* fields on the fiber cross section for different fiber modes. The images represent mode field intensity. (A) Fundamental LP01 mode, (B) LP11 mode with two light spots in the horizontal direction, and (C) LP11 mode with two light spots in the vertical direction.

angle θ_a at the fiber-end surface. It is easy to see from the drawing that $\sin\theta_1 = \sqrt{1 - \sin^2\theta_i}$, and by Snell's law,

$$n_0 \sin\theta_a = n_1 \sqrt{1 - \sin^2\theta_i} \tag{2.3.30}$$

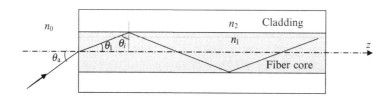

FIG. 2.3.7

Illustration of light coupling into a step-index fiber.

If total reflection happens at the core-cladding interface, which requires $\theta_i \geq \theta_c$, then $\sin\theta_i \geq \sin\theta_c = n_2/n_1$. This requires the incidence angle θ_a to satisfy the following condition:

$$n_0 \sin\theta_a \leq \sqrt{n_1^2 - n_2^2} \tag{2.3.31}$$

The definition of *numerical aperture* is

$$NA = \sqrt{n_1^2 - n_2^2} \tag{2.3.32}$$

For weak optical waveguide like a SMF, the difference between n_1 and n_2 is very small (not more than 1%). Use $\Delta = (n_1 - n_2)/n_1$ to define a normalized index difference between core and cladding, then Δ must also be very small ($\Delta \ll 1$). In this case, the expression of numerical aperture can be simplified as

$$NA \approx n_1\sqrt{2\Delta} \tag{2.3.33}$$

In most cases, fibers are placed in air and $n_0 = 1$. $\sin\theta_a \approx \theta_a$ is valid when $\sin\theta_a < < 1$ (weak waveguide); therefore, Eq. (2.3.31) reduces to

$$\theta_a \leq n_1\sqrt{2\Delta} = NA \tag{2.3.34}$$

From this discussion, the physical meaning of numerical aperture is very clear. Light entering a fiber within a cone of acceptance angle, as shown in Fig. 2.3.8, will be converted into guided modes and will be able to propagate along the fiber. Outside this cone, light coupled into fiber will radiate into the cladding. Similarly, light exits a fiber will have a divergence angle also defined by the numerical aperture. This is often used to design focusing optics if a collimated beam is needed at the fiber output.

Typically parameters of a SMF are $NA \approx 0.1 \sim 0.2$ and $\Delta \approx 0.2\% \sim 1\%$. Therefore, $\theta_a \approx \sin^{-1}(-$

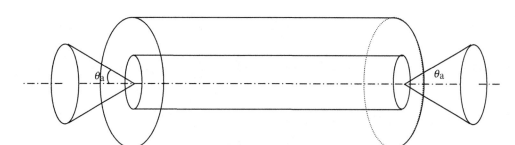

FIG. 2.3.8

Light can be coupled to an optical fiber only when the incidence angle is smaller than the numerical aperture.

$NA) \approx 5.7° \sim 11.5°$. This is a very small angle and it makes difficult to couple light into a SMF. Not only that, the source spot size has to be small ($\sim 80\,\mu m^2$) but also the angle has to be within ± 10 degree.

With the definition of the numerical aperture in Eq. (2.3.32), the V-number of a fiber can be expressed as a function of NA

$$V = \frac{2\pi a}{\lambda}NA \tag{2.3.35}$$

Another important fiber parameter is the cutoff wavelength λ_c. It is defined such that the second-lowest mode ceases to exist when the signal wavelength is longer than λ_c, and therefore, when $\lambda < \lambda_c$ a SMF will become multimode. According to Eq. (2.3.27), cutoff wavelength can be expressed as

$$\lambda_c = \frac{\pi d}{2.405}NA \tag{2.3.36}$$

where d is the core diameter of the step-index fiber. As an example, for a typical standard single-mode fiber (SSMF) with, $n_1 = 1.47$, $n_2 = 1.467$, and $d = 9\,\mu m$, the numerical aperture is

$$NA = \sqrt{n_1^2 - n_2^2} = 0.0939$$

The maximum incident angle at the fiber input is

$$\theta_a = \sin^{-1}(0.0939) = 5.38°$$

and the cutoff wavelength is

$$\lambda_c = \pi d \cdot NA/2.405 = 1.1 \; \mu m$$

EXAMPLE 2.2

To reduce the Fresnel reflection, the end surface of a fiber connector can be made non-perpendicular to the fiber axis. This is usually referred to as angled physical contact (APC) contactor. If the fiber has the core index $n_1 = 1.47$ and cladding index $n_2 = 1.467$, what is the minimum angle ϕ such that the Fresnel reflection by the fiber end facet will not become the guided fiber mode?

Solution

To solve this problem, we use ray trace method and consider three extreme light beam angles in the fiber. The surface has to be designed such that after reflection at the fiber end surface, all these three light beams will not be coupled into fiber-guided mode in the backward propagation direction.

As illustrated in Fig. 2.3.9A, first, for the light beam propagating in the fiber axial direction (z-direction), the direction of the reflected beam from the end surface has an angle θ with respect to the surface normal of the fiber sidewall: $\theta = \pi/2 - 2\varphi < \theta_c$. In order for this reflected light beam not to

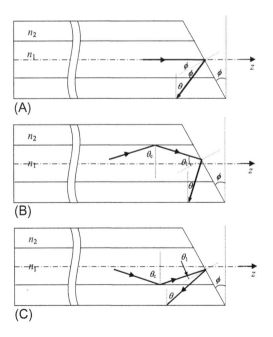

(A)

(B)

(C)

FIG. 2.3.9

Illustration of an angle-polished fiber surface. (A) Light beam in axial direction, (B) extreme light beam in the downward direction, and (C) extreme light beam in the upward direction.

become the guided mode of the fiber, $\theta < \theta_c$ is required, where θ_c is the critical angle defined by Eq. (2.2.14). Therefore, the first requirement for ϕ is

$$\varphi > \pi/4 - \theta_c/2 \qquad (2.3.37)$$

Second, for the light beam propagating at the critical angle of the fiber as shown in Fig. 2.3.9B, the beam has an angle θ_1 with respect to the surface normal of the fiber end surface, which is related to ϕ by $\theta_1 = (\pi/2 - \theta_c) + \varphi$. Then, the θ angle of the reflected beam from the end surface with respect to the fiber sidewall can be found as $\theta = \pi - \theta_c - 2\theta_1 = \theta_c - 2\varphi$. This angle also has to be smaller than the critical angle, that is, $\theta_c - 2\varphi < \theta_c$, or

$$\varphi > 0 \qquad (2.3.38)$$

In the third case as shown in Fig. 2.3.9C, the light beam propagates at the critical angle of the fiber but at the opposite side as compared to the ray trace shown in Fig. 2.3.9B. This produces the smallest θ_1 angle, which is, $\theta_1 = \varphi - (\pi/2 - \theta_c)$. In this geometry, $\pi - \theta_c + \theta + 2\theta_1 = \pi$, that is, $\theta = \theta_c - 2\theta_1 = \theta_c - 2\phi + \pi - 2\theta_c$. This corresponds to the biggest θ angle, $\theta = \pi - \theta_c - 2\phi$. Again, this θ angle has to be smaller than the critical angle, $\theta < \theta_c$, that is,

$$\varphi > \pi/2 - \theta_c \qquad (2.3.39)$$

Since in this example

$$\theta_c = \sin^{-1}\left(\frac{n_2}{n_1}\right) = \sin^{-1}\left(\frac{1.467}{1.47}\right) = 86.34° \qquad (2.3.40)$$

The three constraints given by Eqs. (2.3.37), (2.3.38), and (2.3.39) become $\varphi > 1.83°$, $\varphi > 0$, and $\varphi > 3.66°$, respectively. Obviously, in order to satisfy all these three specific conditions, the required surface angle is $\varphi > 3.66°°$. In fact, as an industry standard, commercial APC connectors usually have the surface tilt angle of $\varphi = 8°$.

2.3.5 FIELD DISTRIBUTION PROFILE OF SMF

When $V \leq 2.405$ is satisfied, the fiber only supports a single mode, which is the LP01 mode illustrated in Fig. 2.3.6. A stand SMF, such as Corning SMF-28, has the core diameter of $d = 8\,\mu m$ and the normalized core/cladding index difference $\Delta n = (n_1 - n_2)/n_2$ of approximately 0.35%. Because of the circular geometry of an optical fiber, the field distribution of the fundamental mode in a single-mode optical fiber is circularly symmetrical, which can be specified by a single parameter known as the *mode-field diameter* (MFD) and the electrical field distribution can often be assumed as Gaussian:

$$E(r) = E_0 e^0 \mathrm{xp}\left(-\frac{r^2}{W_0^2}\right) \qquad (2.3.41)$$

where r is the radius, E_0 is the optical field at $r = 0$, and W_0 is the width of the field distribution. Specifically, the MFD is defined as $2W_0$, which is (Artiglia et al., 1989)

$$2W_0 = 2\left(\frac{2\int_0^\infty r^3 E^2(r)dr}{\int_0^\infty r E^2(r)dr}\right)^{1/2} \qquad (2.3.42)$$

Fig. 2.3.10 illustrates the mode-field distribution of a SMF in which Gaussian approximation is used. The physical meaning of the MFD definition given by Eq. (2.3.42) can be explained as follows: the denominator in Eq. (2.3.42) is proportional to the integration of the power density across the entire fiber cross section, which is the total power of the fundamental mode, whereas the numerator is the integration of the square of the radial distance (r^2) weighted by the power density over the fiber cross section. Therefore, MFD defined by Eq. (2.3.42) represents a root mean square (*rms*) value of the mode distribution of the optical field on the fiber cross section. Mode field radius W_0 is proportional to the geometric core radius, a, of the fiber, but they are not equal. In fact, within $1.2 < V < 2.4$, mode field radius can be approximated as (Marcus, 1978),

$$W_0 \approx a\left(0.65 + 1.619V^{-3/2} + 2.879V^{-6}\right) \qquad (2.3.43)$$

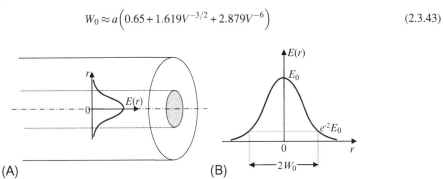

(A)   (B)

FIG. 2.3.10

(A) Illustration of mode distribution in a single-mode fiber and (B) definition of mode-field diameter under Gaussian approximation.

It is important to point out that mode-field distribution $E(r)$ in Eq. (2.3.41) represents the field distribution inside the fiber; thus, it is equivalent to the optical field distribution exactly on the output surface of the fiber. It is commonly referred to as the *near-field* (NF) distribution. NF distribution is a very important parameter of the fiber that determines the effective cross-section area of the fiber as

$$A_{eff} = \frac{2\pi\left[\int_0^\infty |E(r)|^2 rdr\right]^2}{\int_0^\infty |E(r)|^4 rdr} = \pi W_{eff}^2 \qquad (2.3.44)$$

where W_{eff} is the effective mode field radius. If the total optical power P carried by the fiber is known, the power density in the fiber core can be determined by using the effective cross-section area as $I_{density} = P/A_{eff}$. This effective cross-section area will be used later in this chapter when discussing fiber nonlinearities.

2.4 OPTICAL FIBER ATTENUATION

Optical fiber is an ideal medium that can be used to carry optical signals over long distances. Attenuation is one of the most important parameters of an optical fiber; it, to a large extent, determines how far an optical signal can be delivered at a detectable power level. There are several sources that contribute to fiber attenuation, such as absorption, scattering, and radiation.

Material absorption is mainly caused by photoinduced molecular vibration, which absorbs signal optical power and turns it into heat. Pure silica molecules absorb optical signals in UV and IR wavelength bands. At the same time, there are often impurities inside silica material such as OH^- ions, which may be introduced in the fiber perform fabrication process. These impurity molecules create additional absorption in the fiber. Typically, OH^- ions have high absorptions around 700, 900, and 1400 nm, which are commonly referred to as *water absorption peaks.*

Scattering loss arises from microscopic defects and structural inhomogeneities. In general, the optical characteristics of scattering depend on the size of the scatter in comparison to the signal wavelength. However, in optical fibers, the scatters are most likely much smaller than the wavelength of the optical signal, and in this case the scattering is often characterized as *Rayleigh scattering*. A very important spectral property of Rayleigh scattering is that the scattering loss is inversely proportional to the fourth power of the wavelength. Therefore, Rayleigh scattering loss is high in a short wavelength region.

In the last few decades, the loss of optical fiber has been decreased significantly by reducing the OH^- impurity in the material and eliminating defects in the structure. However, absorption by silica molecules in the UV and IR wavelength regions and Rayleigh scattering still constitute fundamental limits to the loss of silica-based optical fiber. Fig. 2.4.1 shows the typical absorption spectra of silica fiber. The dotted line shows the attenuation of old fibers that were made before the 1980s. In addition to strong water absorption peaks, the attenuation is generally higher than new fiber due to material impurity and waveguide scattering. Three wavelength windows have been used since the 1970s for optical communications in 850, 1310, and 1550 nm where optical attenuation has local minimums. In the early days of optical communication, the first wavelength window in 850 nm was used partly because of the availability of GaAs-based laser sources, which emit in that wavelength window. The advances in longer wavelength semiconductor lasers based on InGaAs and InGaAsP pushed optical communications toward the second and the third wavelength windows in 1310 and 1550 nm where optical losses are significantly reduced and optical systems can reach longer distances without regeneration.

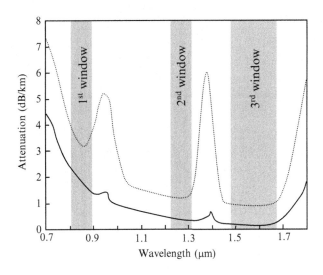

FIG. 2.4.1

Attenuation of old *(dotted line)* and new *(solid line)* silica fibers. The shaded regions indicate the three telecommunication wavelength windows.

Another category of optical loss that may occur in optical fiber cables is radiation loss. It is mainly caused by fiber bending. Micro-bending, usually caused by irregularities in the pulling process, may introduce coupling between the fundamental optical mode and high-order radiation modes and thus creating losses. On the other hand, macro-bending, often introduced by cabling processing and fiber handling, causes the spreading of optical energy from fiber core into the cladding. For example, for a SSMF, bending loss starts to be significant when the bending diameter is smaller than approximately 30 cm.

Mathematically, the complex representation of an optical wave propagating in the z-direction is

$$E(z,t) = E_0 \exp[-j(\omega t - kz)] \tag{2.4.1}$$

where E_0 is the complex amplitude, ω is the optical frequency, and k is the propagation constant. Considering attenuation in the medium, the propagation constant should be complex:

$$k = \beta + j\frac{\alpha}{2} \tag{2.4.2}$$

where $\beta = 2\pi n/\lambda$ is the real propagation constant and α is the power attenuation coefficient. By separating the real and the imaginary parts of the propagation constant, Eq. (2.4.1) can be written as

$$E(z,t) = E_0 \exp[-j(\omega t - \beta z)] \cdot \exp\left(-\frac{\alpha}{2}\right)z \tag{2.4.3}$$

The average optical power can be simply expressed as

$$P(z) = P_0 e^{-\alpha z} \tag{2.4.4}$$

where P_0 is the input optical power. Note here the unit of α is in Neper per meter.

This attenuation coefficient α of an optical fiber can be obtained by measuring the input and the output optical power:

$$\alpha = \frac{1}{L}\ln\left[\frac{P_0}{P(L)}\right] \tag{2.4.5}$$

where L is the fiber length and $P(L)$ is the optical power measured at the output of the fiber.

However, engineers use decibel (dB) to describe fiber attenuation and use dB/km as the unit of attenuation coefficient. If we define α_{dB} as the attenuation coefficient which has the unit of dB/km, then the optical power level along the fiber length can be expressed as

$$P(z) = P_0 \times 10^{-\frac{\alpha_{dB}}{10}z} \tag{2.4.6}$$

Similar to Eq. (2.4.5), for a fiber of length L, α_{dB} can be estimated using

$$\alpha_{dB} = \frac{1}{L}10\log_{10}\left[\frac{P_0}{P(L)}\right] \tag{2.4.7}$$

Comparing Eqs. (2.4.5) and (2.4.7), the relationship between α and α_{dB} can be found as

$$\frac{\alpha_{dB}}{\alpha} = \frac{10\log_{10}[P_0/P(L)]}{\ln[P_0/P(L)]} = 10\log(e) = 4.343 \tag{2.4.8}$$

or simply, $\alpha_{dB} = 4.343\alpha$.

α_{dB} is a simpler parameter to use for evaluation of fiber loss. For example, for an 80-km-long fiber with $\alpha_{dB} = 0.25$ dB/km attenuation coefficient, the total fiber loss can be easily found as $80 \times 0.25 = 20$ dB. On the other hand, if complex optical field expression is required to solve wave propagation equations, α needs to be used instead. In practice, people always use the symbol α to represent optical-loss coefficient, no matter in [Neper/m] or in [dB/km]. But one should be very clear which unit to use when it comes to finding numerical values.

The following is an interesting example that may help to understand the impact of fiber numerical aperture and attenuation.

EXAMPLE 2.3

A very long step-index, SMF has a numerical aperture $NA = 0.1$ and a core refractive index $n_1 = 1.45$. Assume that the fiber end surface is ideally antireflection coated and fiber loss is only caused by Rayleigh scattering. Find the reflectivity of the fiber $R_{ref} = P_b/P_i$, where P_i is the optical power injected into the fiber and P_b is the optical power that is reflected back to the input fiber terminal, as illustrated in Fig. 2.4.2A.

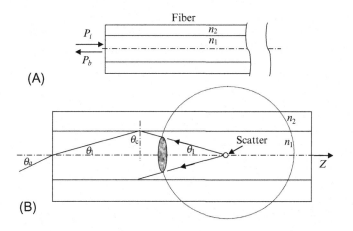

FIG. 2.4.2

Illustrations of fiber reflection (A) and scattering in fiber core (B).

Solution

In this problem, Rayleigh scattering is the only source of attenuation in the fiber. Each scattering source scatters the light into all directions uniformly, which fills the entire 4π solid angle. However, only a small part of the scattered light with the angle within the numerical aperture can be converted into guided mode within the fiber core and travels back to the input. The rest will be radiated into cladding and lost. For simplicity, we assume that scatters are only located along the center of the fiber core, as shown in Fig. 2.4.2B. Therefore, the conversion efficiency from the scattered light to that captured by the fiber is

$$\eta = \frac{2\pi(1 - \cos\theta_1)}{4\pi} \tag{2.4.9}$$

where the numeritor is the area of the spherical cap shown as the shaded area in Fig. 2.4.2B, whereas the denominator 4π is the total area of the unit sphere. θ_1 is the maximum trace angle of the guided mode with respect to the fiber axis. Since the numerical aperture is defined as the maximum acceptance angle θ_a at the fiber entrance as given in Eq. (2.3.34), θ_1 can be expressed as a function of NA as

$$\theta_1 = \frac{n_0}{n_1} NA = \frac{1}{1.45} \times 0.1 = 0.069 \tag{2.4.10}$$

where $n_0 = 1$ is used for the index of air. Substitute the value of θ_1 into Eq. (2.4.9), the ratio between the captured and the total scattered optical power can be found as $\eta = 1.19 \times 10^{-3} = -29$ dB.

Now let us consider a short fiber section Δz located at position z along the fiber. The optical power at this location is $P(z) = P_i e^{-\alpha z}$, where α is the fiber attenuation coefficient. The optical power loss within this section is

$$\Delta P(z) = \frac{dP(z)}{dz} \Delta z = -\alpha P_i e^{-\alpha z} \Delta z \tag{2.4.11}$$

Since we assumed that the fiber loss is only caused by Rayleigh scattering, this power loss $\Delta P(z)$ should be equal to the total scattered power within the short section. $\eta \Delta P(z)$ is the amount of scattered power that is turned into the guided mode that travels back to the fiber input.

However, during the traveling from location z back to the fiber input, attenuation also applies, and the total amount of power loss is, again, $e^{-\alpha z}$. Considering that the fiber is composed of many short sections and adding up the contributions from all sections, the total reflected optical power is

$$P_b = \sum \eta |\Delta P(z)| e^{-\alpha z} = \int_0^\infty \eta \alpha P_i e^{-2\alpha z} dz = \frac{\eta P_i}{2} \tag{2.4.12}$$

Therefore, the reflectivity is

$$R_{ref} = \frac{P_b}{P_i} = \frac{\eta}{2} = 5.95 \times 10^{-4} = -32 dB \tag{2.4.13}$$

This result looks surprisingly simple. The reflectivity, sometimes referred to as return loss, only depends on the fiber numerical aperture and is independent of the fiber loss. The physical explanation is that since Rayleigh scattering is assumed to be the only source of loss, increasing the fiber loss will increase both scattered signal generation and its attenuation. In practical SMFs, this approximation is quite accurate, and the experimentally measured return loss in a standard SMF is between -31 and -34 dB.

2.5 GROUP VELOCITY AND DISPERSION

When an optical signal propagates along an optical fiber, not only is the signal optical power attenuated but also different frequency components within the optical signal propagate in slightly different speeds. This frequency dependency of propagation speed is commonly known as the *chromatic dispersion*.

2.5.1 PHASE VELOCITY AND GROUP VELOCITY

Neglecting attenuation, the electric field of a single-frequency plane optical wave propagating in the z-direction is often expressed as

$$E(z, t) = E_0 \exp[-j\Phi(t, z)] \tag{2.5.1}$$

where $\Phi(t, z) = (\omega_0 t - \beta_0 z)$ is the optical phase, ω_0 is the optical angular frequency, and $\beta_0 = 2\pi n/\lambda = n\omega_0/c$ is the propagation constant.

The phase front of this lightwave is the plane where the optical phase is constant:

$$(\omega_0 t - \beta_0 z) = \text{constant} \tag{2.5.2}$$

The propagation speed of the phase front is called *phase velocity*, which can be obtained by differentiating both sides of Eq. (2.5.2)

$$v_p = \frac{dz}{dt} = \frac{\omega_0}{\beta_0} \tag{2.5.3}$$

Now consider that this single-frequency optical wave is modulated by a sinusoid signal of frequency $\Delta\omega$. Then, the electrical field is

$$E(z,t) = E_0 \exp[-j(\omega_0 t - \beta z)] \cos(\Delta\omega t) \tag{2.5.4}$$

This modulation splits the signal frequency lightwave into two frequency components. At the input ($z=0$) of the optical fiber, the optical field is

$$E(0,t) = E_0 e^{-j\omega_0 t} \cos(\Delta\omega t) = \frac{1}{2} E_0 \left(e^{-j(\omega_0 + \Delta\omega)t} + e^{-j(\omega_0 - \Delta\omega)t} \right) \tag{2.5.5}$$

Since wave propagation constant $\beta = n\omega/c$ is linearly proportional to the frequency of the optical signal, the two frequency components at $\omega_0 \pm \Delta\omega$ will have two different propagation constants $\beta_0 \pm \Delta\beta$. Therefore, the general expression of the optical field is

$$E(z,t) = \frac{1}{2} E_0 \left\{ e^{-j[(\omega_0 + \Delta\omega)t - (\beta_0 + \Delta\beta)z]} + e^{-j[(\omega_0 - \Delta\omega)t - ((\beta_0 - \Delta\beta)z]} \right\}$$
$$= E_0 e^{-j(\omega_0 t - \beta_0 z)} \cos(\Delta\omega t - \Delta\beta z) \tag{2.5.6}$$

where $E_0 e^{-j(\omega_0 t - \beta_0 z)}$ represents an optical carrier, which is identical to that given in Eq. (2.5.4), whereas $\cos(\Delta\omega t - \Delta\beta z)$ is an envelope that is carried by the optical carrier. In fact, this envelope represents the information that is modulated onto the optical carrier. The propagation speed of this information-carrying envelope is called *group velocity*. Similar to the derivation of phase velocity, one can find group velocity by differentiating both sides of $(\Delta\omega t - \Delta\beta z) = $ constant, which yields

$$v_g = dz/dt = \Delta\omega/\Delta\beta$$

With infinitesimally low modulation frequency, $\Delta\omega \to d\omega$ and $\Delta\beta \to d\beta$, so the general expression of group velocity is

$$v_g = \frac{d\omega}{d\beta} \tag{2.5.7}$$

In a nondispersive medium, the refractive index n is a constant that is independent of the frequency of the optical signal. In this case, the group velocity is equal to the phase velocity: $v_g = v_p = c/n$. However, in many practical optical materials, the refractive index $n(\omega)$ is a function of the optical frequency and therefore, $v_g \neq v_p$ in these materials.

Over a unit length of 1 m, the propagation phase delay is equal to the inverse of the phase velocity:

$$\tau_p = \frac{1}{v_p} = \frac{\beta_0}{\omega_0} \tag{2.5.8}$$

And similarly, the propagation group delay over a 1-m length is defined as the inverse of the group velocity:

$$\tau_g = \frac{1}{v_g} = \frac{d\beta}{d\omega} \tag{2.5.9}$$

2.5.2 GROUP VELOCITY DISPERSION

To understand group velocity dispersion, we consider that two sinusoids with the frequencies $\Delta\omega \pm \delta\omega/2$ are modulated onto an optical carrier of frequency ω_0. The spectrum of the modulated signal is illustrated in Fig. 2.5.1. When propagating along a fiber, each modulating frequency will have its own group velocity; then over a unit fiber length, the group delay difference between these two frequency components can be found as,

$$\delta\tau_g = \frac{d\tau_g}{d\omega}\delta\omega = \frac{d}{d\omega}\left(\frac{d\beta}{d\omega}\right)\delta\omega = \frac{d^2\beta}{d\omega^2}\delta\omega \qquad (2.5.10)$$

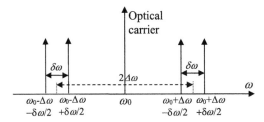

FIG. 2.5.1

Spectrum of two-tone modulation on an optical carrier, where ω_0 is the carrier frequency and $\Delta\omega \pm \delta\omega/2$ are the modulation frequencies.

Obviously, this group delay difference is introduced by the frequency dependency of the propagation constant. In general, the frequency-dependent propagation constant $\beta(\omega)$ can be expended in a Taylor series around a central frequency ω_0:

$$\beta(\omega) = \beta(\omega_0) + \frac{d\beta}{d\omega}\bigg|_{\omega=\omega_0}(\omega - \omega_0) + \frac{1}{2}\frac{d^2\beta}{d\omega^2}\bigg|_{\omega=\omega_0}(\omega - \omega_0)^2 + \cdots\cdots$$

$$= \beta(\omega_0) + \beta_1(\omega - \omega_0) + \frac{1}{2}\beta_2(\omega - \omega_0)^2 + \cdots\cdots \qquad (2.5.11)$$

where

$$\beta_1 = \frac{d\beta}{d\omega} \qquad (2.5.12)$$

represents the group delay and

$$\beta_2 = \frac{d^2\beta}{d\omega^2} \qquad (2.5.13)$$

is the group delay dispersion parameter.

If the fiber is uniform with length L, use Eq. (2.5.10), we can find the relative time delay between two frequency components separated by $\delta\omega$ as

$$\Delta\tau_g = \delta\tau_g L = \beta_2 L \delta\omega \qquad (2.5.14)$$

Sometimes it might be convenient to use wavelength separation $\delta\lambda$ instead of frequency separation $\delta\omega$ between the two frequency (or wavelength) components. In this case, the relative delay over a unit fiber length can be expressed as

$$\delta\tau_g = \frac{d\tau_g}{d\lambda}\delta\lambda \equiv D\delta\lambda \tag{2.5.15}$$

where $D = d\tau_g/d\lambda$ is another group delay dispersion parameter. The relationship between these two dispersion parameters D and β_2 can be found as

$$D = \frac{d\tau_g}{d\lambda} = \frac{d\omega}{d\lambda} \cdot \frac{d\tau_g}{d\omega} = -\frac{2\pi c}{\lambda^2}\beta_2 \tag{2.5.16}$$

For a fiber of length L, we can easily find the relative time delay between two wavelength components separated by $\delta\lambda$ as

$$\Delta\tau_g = D \cdot L \cdot \delta\lambda \tag{2.5.17}$$

In practical fiber-optic systems, the relative delay between different wavelength components is usually measured in picoseconds; wavelength separation is usually expressed in nanometers; and fiber length is usually measured in kilometers. Therefore, the most commonly used units for β_1, β_2, and D are [ps/nm], [ps^2/km], and [ps/nm-km], respectively.

2.5.3 SOURCES OF CHROMATIC DISPERSION

The physical reason of chromatic dispersion is the wavelength-dependent propagation constant $\beta(\lambda)$. Both material property and waveguide structure may contribute to this wavelength dependency of $\beta(\lambda)$, which are referred to as material dispersion and waveguide dispersion, respectively.

Material dispersion is originated by the wavelength-dependent material refractive index $n = n(\lambda)$; thus, the wavelength-dependent propagation constant is $\beta(\lambda) = 2\pi n(\lambda)/\lambda$.

For a unit fiber length, the wavelength-dependent group delay is

$$\tau_g = \frac{d\beta(\lambda)}{d\omega} = -\left(\frac{\lambda^2}{2\pi}\right)\frac{d\beta(\lambda)}{d\lambda} = \frac{1}{c}\left[n(\lambda) - \lambda\frac{dn(\lambda)}{d\lambda}\right] \tag{2.5.18}$$

The group delay dispersion between two wavelength components separated by $\delta\lambda$ is then

$$\delta\tau_g = \frac{d\tau_g}{d\lambda}\delta\lambda = \frac{-1}{c}\left[\lambda\frac{d^2 n(\lambda)}{d\lambda^2}\right]\delta\lambda \tag{2.5.19}$$

Therefore, material-induced dispersion is proportional to the second derivative of the refractive index.

Waveguide dispersion can be explained as the wavelength-dependent angle of the light ray propagating inside the fiber core, as illustrated by Fig. 2.3.3. Under weak guidance condition, the normalized propagation constant defined in Eq. (2.3.29) can be linearized as

$$b = \frac{(\beta_z/k)^2 - n_2^2}{n_1^2 - n_2^2} \approx \frac{(\beta_z/k) - n_2}{n_1 - n_2} \tag{2.5.20}$$

The actual propagation constant in the z-direction, β_z can be expressed as a function of b as

$$\beta_z(\lambda) = kn_2(b\Delta + 1) \tag{2.5.21}$$

where $\Delta = (n_1 - n_2)/n_2$ is the normalized index difference between the core and the cladding. Then, the group delay can be found as

$$\tau_g = \frac{d\beta_z(\lambda)}{d\omega} = \frac{n_2}{c}\left(1 + b\Delta + k\frac{db}{dk}\Delta\right)$$

Fig. 2.3.5 shows the variation of b as a function of V for different modes. Since V is inversely proportional to λ, as defined by Eq. (2.3.25), τ_g varies with λ as well.

In general, material dispersion is difficult to modify, because doping other materials into silica might introduce excess attenuation, but waveguide dispersion can be modified by index profile design.

The overall chromatic dispersion in an optical fiber is the combination of material dispersion and waveguide dispersion. In general, different types of fiber have different dispersion characteristics. However, for a SSMF, the dispersion parameter D can usually be described by a Sellmeier equation (Meumann, 1988):

$$D(\lambda) = \frac{S_0}{4}\left(\lambda - \frac{\lambda_0^4}{\lambda^3}\right) \tag{2.5.22}$$

where S_0 is the dispersion slope, which ranges from 0.08 to 0.1 ps/nm²-km and λ_0 is the zero-dispersion wavelength, which is around 1315 nm.

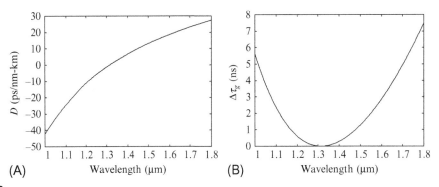

FIG. 2.5.2

(A) Chromatic dispersion D vs. wavelength and (B) relative group delay vs. wavelength. $S_0 = 0.09$ ps/nm²-km, $\lambda_0 = 1315$ nm.

Fig. 2.5.2A shows the dispersion parameter D vs. wavelength for standard single-mode fiber (SSMF), which has dispersion slope $S_0 = 0.09$ ps/nm²-km and zero-dispersion wavelength $\lambda_0 = 1315$ nm. $D(\lambda)$ is generally nonlinear; however, if we are only interested in a relatively narrow wavelength window, it is often convenient to linearize this parameter. For example, if the central frequency of an optical signal is at $\lambda = \lambda_a$, then $D(\lambda)$ can be linearized in the vicinity of λ_a as

$$D(\lambda) \approx D(\lambda_a) + S(\lambda_a) \cdot (\lambda - \lambda_a) \tag{2.5.23}$$

where

$$D(\lambda_a) = \frac{S_0}{4}\left(\lambda - \frac{\lambda_0^4}{\lambda_a^3}\right) \tag{2.5.24}$$

and the local dispersion slope at λ_a is

$$S(\lambda_a) = \frac{S_0}{4}\left(1 + \frac{3\lambda_0^4}{\lambda_a^4}\right)$$

(2.5.25)

In general, $S(\lambda_a) \neq S_0$, except when the optical signal is near the zero-dispersion wavelength $\lambda_a = \lambda_0$.

As a consequence of wavelength-dependent dispersion parameter, the group delay is also wavelength-dependent. Considering the definition of dispersion $D = d\tau_g/d\lambda$ as given by Eq. (2.5.15), the wavelength-dependent group delay $\tau_g(\lambda)$ can be found by integrating $D(\lambda)$ over wavelength. Based on Eq. (2.5.22), the group delay can be derived as

$$\tau_g(\lambda) = \int D(\lambda)d\lambda = \tau_0 + \frac{S_0}{8}\left(\lambda - \frac{\lambda_0^2}{\lambda}\right)^2$$

(2.5.26)

Fig. 2.5.2B shows the relative group delay $\Delta\tau_g(\lambda) = \tau_g(\lambda) - \tau_0$ vs. wavelength. The group delay is not sensitive to wavelength change around $\lambda = \lambda_0$ because where the dispersion is zero.

2.5.4 MODAL DISPERSION

Chromatic dispersion discussed earlier specifies wavelength-dependent group velocity within one optical mode. If a fiber has more than one mode, different modes will also have different propagation speeds; this is called modal *dispersion*. In a MMF, the effect of modal dispersion is typically much stronger than the chromatic dispersion within each mode; therefore, chromatic dispersion is usually neglected.

Modal dispersion depends on the number of propagation modes that exist in the fiber, which, in turn, is determined by the fiber core size and the index difference between the core and the cladding. By using geometric optics analysis as described in Section 2.3, we can find the delay difference between the fastest propagation mode and the slowest propagation mode. Obviously, the fastest mode is the one that travels along the fiber longitudinal axis, whereas the ray trace of the slowest mode has the largest angle with the fiber longitudinal axis or the smallest θ_i shown in Fig. 2.3.2B. This smallest angle θ_i is limited by the condition of total reflection at the core-cladding interface, $\theta_i > \sin^{-1}(n_2/n_1)$. Since the group velocity of the fastest ray trace is c/n_1 (here we assume n_1 is a constant), the group velocity of the slowest ray trace should be $(c/n_1)\sin\theta_i = (cn_2/n_1^2)$. Therefore, for a fiber of length L, the maximum group delay difference is approximately

$$\delta T_{max} = \frac{n_1 L}{c}\left(\frac{n_1 - n_2}{n_2}\right)$$

(2.5.27)

This expression does not consider the core size of the fiber, and therefore, it only provides an absolute maximum of the modal dispersion.

2.5.5 POLARIZATION MODE DISPERSION

Polarization mode dispersion (PMD) is a special type of modal dispersion that exists in SMFs. It is worth noting that there are actually two fundamental modes that coexist in a SMF. As shown in Fig. 2.5.3, these two modes are orthogonally polarized. In an optical fiber with perfect cylindrical symmetry, these two modes have the same cutoff condition and they are referred to as *degenerate modes*.

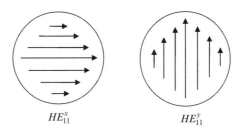

HE_{11}^x HE_{11}^y

FIG. 2.5.3

Illustration of optical field vector across the core cross section of a single-mode fiber. Two degenerate modes exist in a single-mode fiber.

However, practical optical fibers might not have perfect cylindrical symmetry due to birefringence; therefore, these two fundamental modes may propagate in different speeds. Birefringence in an optical fiber is usually caused by small perturbations of the structure geometry as well as the anisotropy of the refractive index. The sources of the perturbations can be categorized as intrinsic and extrinsic. *Intrinsic perturbation* refers to permanent structural perturbations of fiber geometry, which are often caused by errors in the manufacturing process. The effect of intrinsic perturbation include (1) noncircular fiber core, which is called *geometric birefringence* and (2) nonsymmetric stress originated from the nonideal perform, which is usually called *stress birefringence*. On the other hand, *extrinsic perturbation* usually refers to perturbations due to random external forces in the cabling and installation process. Extrinsic perturbation also causes both geometric and stress birefringence.

The effect of birefringence is that the two orthogonal polarization modes HE_{11}^x and HE_{11}^y experience slightly different propagation constants when they travel along the fiber; therefore, their group delays become different. Assuming that the effective indices in the core of a birefringence fiber are n_x and n_y for the two polarization modes, their corresponding propagation constants will be $\beta_x = \omega n_x/c$ and $\beta_y = \omega n_y/c$, respectively. Due to birefringence, β_x and β_y are not equal and their difference is

$$\Delta\beta = (\beta_x - \beta_y) = \frac{\omega}{c}\Delta n_{eff} \tag{2.5.28}$$

where $\Delta n_{eff} = n_{//} - n_{\perp}$ is the effective differential refractive index of the two modes.

For a fiber of length L, the relative group delay between the two orthogonal polarization modes is

$$\Delta\tau_g = \frac{(n_{//} - n_{\perp})}{c}L = \frac{L\Delta n_{eff}}{c} \tag{2.5.29}$$

This is commonly referred to as *differential group delay* (DGD).

As a result of fiber birefringence, the SOP of the optical signal will rotate while propagating along the fiber because of the accumulated relative phase change $\Delta\Phi$ between the two polarization modes:

$$\Delta\Phi = \frac{\omega\Delta n_{eff}}{c}L \tag{2.5.30}$$

According to Eq. (2.5.30), when an optical signal is launched into a birefringence fiber, its polarization evolution can be accomplished by the changes of either the fiber length L, the differential refractive index Δn_{eff}, or the lightwave signal angular frequency ω.

At a certain fiber length $L = L_p$, if the SOP of the optical signal completes a full $\Delta\Phi = 2\pi$ rotation, L_p is, therefore, defined as the birefringence *beat length*. On the other hand, at a fixed fiber length L, the polarization state of the optical signal can also be varied by changing the frequency. For a complete polarization rotation, the change of optical frequency should be

$$\Delta\omega_{cycle} = \frac{2\pi c}{L\Delta n_{eff}} = \frac{2\pi}{\Delta\tau_g} \qquad (2.5.31)$$

When the fiber length is short enough, energy coupling between the two orthogonally polarized modes is negligible, and the DGD shown in Eq. (2.5.29) is linearly proportional to the fiber length L [Eq. (2.5.29) is also valid for polarization-maintaining fiber in which there no coupling between the two modes.] However, when the fiber is long enough and the fiber is not polarization maintaining (PM), energy carried by the two polarization modes may exchange between each other, known as mode coupling. This mode coupling is often random due to the random nature of perturbation such as bending and stressing along the fiber. As a result, the overall DGD scales with the fiber length by $\sqrt{L}$, so that $\Delta\tau_g \approx \left(\Delta n_{eff}/c\right)\sqrt{L}$.

In modern high-speed optical communications using SMF, PMD has become one of the most notorious sources of transmission performance degradation. Due to the random nature of the perturbations that cause birefringence, PMD in an optical fiber is a stochastic process. Standard SMF for telecommunications in 1550 nm wavelength usually has the unit-length DGD value less than $0.1\ ps/\sqrt{km}$.

EXAMPLE 2.4

A 1550 nm optical signal from a multi longitudinal mode laser diode has two discrete wavelength components separated by 0.8 nm. There are two pieces of optical fiber; one of them is a SSMF with chromatic dispersion parameter $D=17\ ps/nm/km$ at 1550 nm wavelength, and the other is a step-index MMF with core index $n_1 = 1.48$, cladding index $n_2 = 1.46$, and core diameter $d = 50\ \mu m$. Both of these two fibers have the same length of 20 km. Find the allowed maximum signal data rate that can be carried by each of these two fibers.

Solution

For the SMF, chromatic dispersion is the major source of pulse broadening of the optical signal. In this example, the chromatic dispersion-induced pulse broadening is

$$\Delta t_{SMF} = D \cdot L \cdot \Delta\lambda = 17 \times 20 \times 0.8 = 272 ps$$

For the MMF, the major source of pulse broadening is modal dispersion. Using Eq. (2.5.27), this pulse broadening is

$$\Delta t_{MMF} \approx \frac{n_1 L}{c}\left(\frac{n_1 - n_2}{n_2}\right) = 1.35\mu s$$

Obviously, MMF introduces pulse broadening more than three orders of magnitude higher than the SMF. The data rate of the optical signal can be on the order of 5 Gb/s if the SMF is used, whereas it is limited to less than 1 Mb/s with the MMF.

2.5.6 MODE DIVISION MULTIPLEXING

Propagation modes in a MMF are mutually orthogonal, and ideally there is no energy exchange between different modes during propagation. Under this assumption, each mode can provide an independent communication channel for information transmission through the fiber. However, energy exchange between modes, known mode coupling, always exist in practical optical fibers due to intrinsic and extrinsic perturbations. The major difficulty arises from the randomness of the mode coupling which makes the cross talk between modes unpredictable. Despite this difficulty, polarization division multiplexing (PDM), the simplest form of mode division multiplexing, has been successfully applied in fiber-optic transmission systems which utilizing two orthogonal polarization states of the fundamental

LP_{01} mode to carry independent channels. The practical application of PM was enabled by the availability of high-speed digital electronics for signal processing which adaptively tracks and corrects the random cross talk between the two polarization modes. Mode division multiplexing is an active area of research to further extend the capacity of optical transmission systems. Transmission using few-mode fiber with up to 5 or 6 modes has been shown to be possible (Randel et al., 2011). As the number of modes increases, the difficulty of separating these modes at the receiver and removing the cross talk between them increases exponentially. The complexity of electronic digital signal processing and the power consumption will become the major concern challenging the practicality of mode-division multiplexing.

2.6 NONLINEAR EFFECTS IN AN OPTICAL FIBER

Fiber parameters we have discussed so far, such as attenuation, chromatic dispersion, and modal dispersion, are all linear effects. The values of these parameters do not change with the change in the signal optical power. On the other hand, the effects of fiber nonlinearity depend on the optical power density inside the fiber core. The typical optical power carried by an optical fiber may not seem very high, but since the fiber core cross-section area is very small, the power density can be very high to cause significant nonlinear effects. For example, for a SSMF, the cross-section area of the core is about $80\,\mu m^2$. If the fiber carries 10 mW of average optical power, the power density will be as high as $12.5\,kW/cm.^2$ SBS, SRS, and the Kerr effect are the three most important nonlinear effects in silica optical fibers.

2.6.1 STIMULATED BRILLOUIN SCATTERING

SBS is originated by the interaction between the signal photons and the traveling sound waves, also called *acoustic phonons* (Boyd, 1992). It is just like that when you blow air into an open-ended tube: A sound wave may be generated. Because of the SBS, the signal lightwave is modulated by the traveling sound wave. Stokes photons are generated in this process, and the frequency of the Stokes photons is downshifted from that of the original optical frequency. The amount of this frequency shift can be estimated approximately by

$$\Delta f = 2f_0 \frac{V}{(c/n_1)} \tag{2.6.1}$$

where n_1 is the refractive of the fiber core, c/n_1 is the group velocity of the lightwave in the fiber, V is the velocity of the sound wave, and f_0 is the original frequency of the optical signal. In a silica-based optical fiber, sound velocity along the longitudinal direction is $V = 5760$ m/s. Assume a refractive index of $n_1 = 1.47$ in the fiber core at 1550 nm signal wavelength, this SBS frequency shift is about 11 GHz.

SBS is highly directional and narrowband. The generated Stokes photons only propagate in the opposite direction of the original photons, and therefore, the scattered energy is always counter propagating with respect to the signal. In addition, since SBS relies on the resonance of a sound wave, which has very narrow spectral linewidth, the effective SBS bandwidth is as narrow as 20 MHz. Therefore, SBS efficiency is high only when the linewidth of the optical signal is narrow.

When the optical power is high enough, SBS turns signal optical photons into frequency-shifted Stokes photons that travel in the opposite direction. If another optical signal travels in the same fiber, in the same direction and at the same frequency of this Stokes wave, it can be amplified by the SBS process. Based on this, the SBS effect has been used to make optical amplifiers; however, the narrow amplification bandwidth nature of SBS limits their applications. On the other hand, in optical fiber communications, the effect of SBS introduces an extra loss for the optical signal and sets an upper limit for the amount of optical power that can be used in the fiber. In commercial fiber-optic systems, an effective way to suppress the effect of SBS is to frequency modulate the optical carrier and increase the spectral linewidth of the optical source to a level much wider than 20 MHz.

2.6.2 STIMULATED RAMAN SCATTERING

SRS is an inelastic process where a photon of the incident optical signal (pump) stimulates molecular vibration of the material and loses part of its energy. Because of the energy loss, the photon reemits in a lower frequency (Smith, 1972). The introduced vibrational energy of the molecules is referred to as an *optical phonon*. Instead of relying on the acoustic vibration as in the case of SBS, SRS in a fiber is caused by the molecular-level vibration of the silica material. Consequently, through the SRS process, pump photons are progressively absorbed by the fiber, whereas new photons, called *Stokes photons*, are created at a downshifted frequency.

Unlike SBS, where the Stokes wave only propagates in the backward direction, the Stokes waves produced by the SRS process propagate in both forward and backward directions. Therefore, SRS can be used to amplify both co- and counterpropagated optical signals if their frequencies are within the SRS bandwidth. Also, the spectral bandwidth of SRS is much wider than that of SBS. As shown in Fig. 2.6.1, in silica-based optical fibers, the maximum Raman efficiency happens at a frequency shift of about 13.2 THz, and the bandwidth can be as wide as 10 THz. Optical amplifiers based on the SRS effect have become popular in the recent years because of their unique characteristics compared to other type of optical amplifiers. Mode details of fiber Raman amplification will be discussed in Chapter 5.

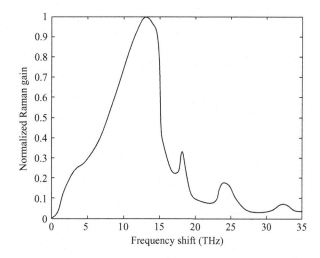

FIG. 2.6.1

A normalized Raman gain spectrum of a silica fiber.

On the other hand, SRS also may create interchannel crosstalk in wavelength-division multiplexed (WDM) optical systems. In a fiber carrying multiple wavelength channels, SRS effect may create energy transfer from short wavelength (higher-frequency) channels to long wavelength (lower-frequency) channels.

2.6.3 KERR EFFECT NONLINEARITY AND NONLINEAR SCHRÖDINGER EQUATION

Kerr effect nonlinearity is introduced by the fact that the refractive index of an optical material is often a weak function of the optical power density:

$$n = n_0 + n_2 \frac{P}{A_{eff}} \tag{2.6.2}$$

where n_0 is the linear refractive index of the material, n_2 is the nonlinear index coefficient, P is the optical power, and A_{eff} is the effective cross-section area of the optical waveguide. P/A_{eff} represents optical power density.

Considering both linear and nonlinear effects, a nonlinear differential equation is often used to describe the envelope of optical field propagating along an optical fiber (Agrawal, 2001):

$$\frac{\partial A(t,z)}{\partial z} + \frac{j\beta_2}{2}\frac{\partial^2 A(t,z)}{\partial t^2} + \frac{\alpha}{2}A(t,z) - j\gamma|A(t,z)|^2 A(t,z) = 0 \tag{2.6.3}$$

This equation is known as the *nonlinear Schrödinger (NLS) equation*. $A(t,z)$ is the complex amplitude of the optical field. The fiber parameters β_2 and α are group delay dispersion and attenuation, respectively. γ is defined as the *nonlinear parameter*:

$$\gamma = \frac{n_2 \omega_0}{c A_{eff}} \tag{2.6.4}$$

On the left side of Eq. (2.6.3), the second term represents the effect of chromatic dispersion; the third term is optical attenuation; and the last term represents a nonlinear phase modulation caused by the Kerr effect nonlinearity. To understand the physical meaning of each term in the NLS equation, we can consider dispersion and nonlinearity separately.

First, we only consider the dispersion effect and assume

$$\frac{\partial A}{\partial z} + \frac{j\beta_2}{2}\frac{\partial^2 A}{\partial t^2} = 0 \tag{2.6.5}$$

This equation can easily be solved in Fourier domain as

$$\widetilde{A}(\omega L) = \widetilde{A}(\omega 0)\exp\left(j\frac{\omega^2}{2}\beta_2 L\right) = \widetilde{A}(\omega 0)e^{j\Phi(\omega)} \tag{2.6.6}$$

where L is the fiber length, $\widetilde{A}(\omega, L)$ is the Fourier transform of $A(t,L)$, and $\widetilde{A}(\omega, 0)$ is the optical field at the fiber input. The differential phase between frequency components ω and ω_0 at the end of the fiber is

$$\delta\Phi = \frac{\omega^2 - \omega_0^2}{2}\beta_2 L \approx \omega_0(\omega - \omega_0)\beta_2 L \tag{2.6.7}$$

where we have assumed that $|\omega - \omega_0| < <\omega_0$, so that $\omega + \omega_0 \approx 2\omega_0$. If we let $\delta\Phi = \omega_0\delta t$, where δt is the arrival time difference at the fiber end between these two frequency components, Δt can be found as $\delta t \approx \delta\omega\beta_2 L$ with $\delta\omega = \omega - \omega_0$. Then, if we convert β_2 into D using Eq. (2.5.16), we find

$$\delta t \approx D \cdot L \cdot \delta\lambda \tag{2.6.8}$$

where $\delta\lambda = \delta\omega\lambda^2/(2\pi c)$ is the wavelength separation between these two components. In fact, Eq. (2.6.8) is identical to Eq. (2.5.17).

Now let us neglect dispersion and only consider fiber attenuation and Kerr effect nonlinearity. Then, the NLS equation is simplified to

$$\frac{\partial A(t, z)}{\partial z} + \frac{\alpha}{2}A(t, z) = j\gamma|A(t, z)|^2 A(t, z) \tag{2.6.9}$$

If we start by considering the optical power, $P(z, t) = |A(z, t)|^2$, Eq. (2.6.9) gives $P(z, t) = P(0, t)e^{-\alpha z}$. Then, we can use a normalized variable $E(z, t)$ such that

$$A(z, t) = \sqrt{P(0, t)} \exp\left(\frac{-\alpha z}{2}\right) E(z, t) \tag{2.6.10}$$

Eq. (2.6.9) becomes

$$\frac{\partial E(z, t)}{\partial z} = j\gamma P(0, t)e^{-\alpha z} E(z, t) \tag{2.6.11}$$

And the solution is

$$E(z, t) = E(0, t) \exp[j\Phi_{NL}(t)] \tag{2.6.12}$$

where

$$\Phi_{NL}(t) = \gamma P(0, t) \int_0^L e^{-\alpha z} dz = \gamma P(0, t) L_{eff} \tag{2.6.13}$$

with

$$L_{eff} = \frac{1 - e^{-\alpha L}}{\alpha} \approx \frac{1}{\alpha} \tag{2.6.14}$$

known as the nonlinear length of the fiber, which only depends on the fiber attenuation (where $e^{-\alpha L} < <1$ is assumed). For a SSMF operating in a 1550-nm wavelength window, the attenuation is about 0.25 dB/km (or 5.8×10^{-5} Np/m) and the nonlinear length is approximately 17.4 km.

According to Eq. (2.6.13), the nonlinear phase shift $\Phi_{NL}(t)$ follows the time-dependent change of the optical power. The corresponding optical frequency change can be found by

$$\delta f(t) = \frac{1}{2\pi}\gamma L_{eff} \frac{\partial}{\partial t}[P(0, t)] \tag{2.6.15}$$

Or the corresponding signal wavelength modulation:

$$\delta\lambda(t) = -\frac{\lambda^2}{2\pi c}\gamma L_{eff} \frac{\partial}{\partial t}[P(0, t)] \tag{2.6.16}$$

Fig. 2.6.2 illustrates the waveform of an optical pulse and the corresponding nonlinear phase shift. This phase shift is proportional to the signal waveform, an effect known as *self-phase modulation* (SPM) (Stolen and Lin, 1978).

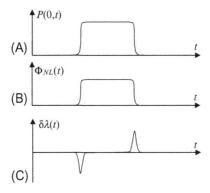

FIG. 2.6.2

(A) Optical pulse waveform, (B) nonlinear phase shift, and (C) wavelength shift introduced by self-phase modulation.

If the fiber has no chromatic dispersion, this phase modulation alone would not introduce optical signal waveform distortion if optical intensity is detected at the fiber output. However, if the fiber chromatic dispersion is considered, wavelength deviation created by SMP at the leading edge and the falling edge of the optical pulse, as shown in Fig. 2.6.2, will introduce group delay mismatch between these two edges of the pulse, therefore, creating waveform distortion. For example, if the fiber has anomalous dispersion ($D > 0$), short wavelength components will travel faster than long-wavelength components. In this case, the blue-shifted pulse falling edge travels faster than the red-shifted leading edge; therefore, the pulse will be squeezed by the SMP process. On the other hand, if the fiber dispersion is normal ($D < 0$), the blue-shifted pulse falling edge travels slower than the red-shifted leading edge and this will result in pulse spreading.

In the discussion of SPM, we have only considered one wavelength channel in the fiber, and its optical phase is affected by the intensity of the same channel. If there is more than one wavelength channel traveling in the same fiber, the situation becomes more complicated and cross talk-created channels will be created by Kerr effect nonlinearity.

Now let us consider a system with only two wavelength channels; the combined optical field is

$$A(z, t) = A_1(z, t)e^{-j\theta_1} + A_2(z, t)e^{-j\theta_2} \tag{2.6.17}$$

where A_1 and A_2 are the optical field amplitude of the two wavelength channels and $\theta_1 = n\omega_1/c$ and $\theta_2 = n\omega_2/c$ are optical phases of these two optical carriers. Substituting Eq. (2.6.17) into the NLS Eq. (2.6.3) and collecting terms having $e^{-j\theta_1}$ and $e^{-j\theta_2}$, respectively, will result in two separate equations:

$$\frac{\partial A_1}{\partial z} + \frac{j\beta_2}{2}\frac{\partial^2 A_1}{\partial t^2} + \frac{\alpha}{2}A_1 = j\gamma|A_1|^2 A_1 + j2\gamma|A_2|^2 A_1 + j\gamma A_1^2 A_2^* e^{j(\theta_1 - \theta_2)} \tag{2.6.18}$$

$$\frac{\partial A_2}{\partial z} + \frac{j\beta_2}{2}\frac{\partial^2 A_2}{\partial t^2} + \frac{\alpha}{2}A_2 = j\gamma|A_2|^2 A_2 + j2\gamma|A_1|^2 A_2 + j\gamma A_2^2 A_1^* e^{j(\theta_2 - \theta_1)} \tag{2.6.19}$$

Each of these two equations describes the propagation characteristics of an individual wavelength channel. On the right side of each equation, the first term represents the effect of SPM as we have described; the second term represents *cross-phase modulation* (XPM); and the third term is responsible for another nonlinear phenomenon called *four-wave mixing* (FWM).

XPM is originated from the nonlinear phase modulation of one wavelength channel by the optical power change of the other channel (Islam et al., 1987). Similar to SPM, it requires chromatic dispersion of the fiber to convert this nonlinear phase modulation into waveform distortion. Since signal waveforms carried by these wavelength channels are usually not synchronized with each other, the precise time-dependent characteristic of cross talk is less predictable. Statistical analysis is normally used to estimate the effect of XPM-induced cross talk.

FWM can be better understood as two optical carriers copropagating along an optical fiber; the beating between the two carriers modulates the refractive index of the fiber at the frequency difference between them. Meanwhile, a third optical carrier propagating along the same fiber is phase modulated by this index modulation and then creates two extra modulation sidebands (Hill et al., 1978; Inoue, 1992). If the frequencies of three optical carriers are ω_j, ω_k, and ω_l, the new frequency component created by this FWM process is

$$\omega_{jkl} = \omega_j + \omega_k - \omega_l = \omega_j - \Delta\omega_{kl} \tag{2.6.20}$$

where $l \neq j$, $l \neq k$, and $\Delta\omega_{kl}$ are the frequency spacing between channel k and channel l. If there are only two original optical carriers involved as in our example, the third carrier is simply one of the two original carriers ($j \neq k$), and this situation is known as *degenerate FWM*. Fig. 2.6.3 shows the wavelength relations of degenerate FWM where two original carriers at ω_j and ω_k beat in the fiber, creating an index modulation at the frequency of $\Delta\omega_{jk} = \omega_j - \omega_k$. Then, the original optical carrier at ω_j is phase modulated at the frequency $\Delta\omega_{jk}$, creating two modulation sidebands at $\omega_i = \omega_j - \Delta\omega_{jk}$ and $\omega_k = \omega_j + \Delta\omega_{jk}$. Similarly, the optical carrier at ω_k is also phase modulated at the frequency $\Delta\omega_{jk}$ and creates two sidebands at $\omega_j = \omega_k - \Delta\omega_{jk}$ and $\omega_l = \omega_k + \Delta\omega_{jk}$. In this process, the original two carriers become four, and this is probably where the name "four-wave mixing" cames from.

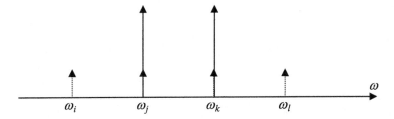

FIG. 2.6.3

Degenerate four-wave mixing, where two original carriers at ω_j and ω_k create four new frequency components at ω_i, ω_j, ω_k, and ω_l.

FWM is an important nonlinear phenomenon in optical fiber; it introduces interchannel cross talk when multiple wavelength channels are used. The buildup of new frequency components generated by FWM depends on the phase match between the original carriers and the new frequency component when they travel along the fiber. Therefore, the efficiency of FWM is very sensitive to the dispersion-induced relative walk-off between the participating wavelength components. In general, the optical field amplitude of the FWM component created in a fiber of length L is

$$A_{jkl}(L) = j\gamma\sqrt{P_j(0)P_k(0)P_l(0)} \int_0^L e^{-\alpha z}\exp\left(j\Delta\beta_{jkl}z\right)dz \tag{2.6.21}$$

where $P_j(0)$, $P_k(0)$, and $P_l(0)$ are the input power of the three participating optical carriers and

$$\Delta\beta_{jkl} = \beta_j + \beta_k - \beta_l - \beta_{jkl} \tag{2.6.22}$$

is a propagation constant mismatch, $\beta_j = \beta(\omega_j)$, $\beta_k = \beta(\omega_k)$, $\beta_l = \beta(\omega_l)$, and $\beta_{jkl} = \beta(\omega_{jkl})$. Expanding β as $\beta(\omega) = \beta_0 + \beta_1(\omega - \omega_0) + (\beta_2/2)(\omega - \omega_0)^2$ and using the frequency relation given in Eq. (2.6.20), we can find

$$\Delta\beta_{jkl} = -\beta_2\left(\omega_j - \omega_l\right)\left(\omega_k - \omega_l\right) \tag{2.6.23}$$

Here for simplicity we neglected the dispersion slope and considered that the dispersion value is constant over the entire frequency region. This propagation constant mismatch can also be expressed as the functions of the corresponding wavelengths:

$$\Delta\beta_{jkl} = \frac{2\pi cD}{\lambda^2}\left(\lambda_j - \lambda_l\right)\left(\lambda_k - \lambda_l\right) \tag{2.6.24}$$

where dispersion parameter β_2 is converted to D using Eq. (2.5.16). The integration of Eq. (2.6.21) yields

$$A_{jkl}(L) = j\gamma\sqrt{P_j(0)P_k(0)P_l(0)}\frac{e^{(j\Delta\beta_{jkl}-\alpha)L}-1}{j\Delta\beta_{jkl}-\alpha} \tag{2.6.25}$$

The power of the FWM component is then

$$P_{jkl}(L) = \eta_{FWM}\gamma^2 L_{eff}^2 P_j(0)P_k(0)P_l(0) \tag{2.6.26}$$

where $L_{eff} = (1 - e^{-\alpha L})/\alpha$ is the fiber nonlinear length and

$$\eta_{FWM} = \frac{\alpha^2}{\Delta\beta_{jkl}^2 + \alpha^2}\left[1 + \frac{4e^{-\alpha L}\sin^2\left(\Delta\beta_{jkl}L/2\right)}{(1 - e^{-\alpha L})^2}\right] \tag{2.6.27}$$

is the FWM efficiency. In most of the practical cases, when the fiber is long enough, $e^{-\alpha L} << 1$ is true. The FWM efficiency can be simplified as

$$\eta_{FWM} \approx \frac{\alpha^2}{\Delta\beta_{jkl}^2 + \alpha^2} \tag{2.6.28}$$

In this simplified expression, FWM efficiency is no longer dependent on the fiber length. The reason is that as long as $e^{-\alpha L} << 1$, for the fiber lengths far beyond the nonlinear length the optical power is significantly reduced and thus the nonlinear contribution. Consider the propagation constant mismatch given in Eq. (2.6.24), the efficiency of FWM can be reduced either by the increase of fiber dispersion or by the increase of channel separation. Fig. 2.6.4 shows the FWM efficiency for several different fiber dispersion parameters calculated with Eqs. (2.6.24) and (2.6.28), where the fiber loss is $\alpha = 0.25$ dB/km and the operating wavelength is 1550 nm. Note in these calculations, the unit of attenuation α has to be in Np/m when using Eqs. (2.6.25)–(2.6.28). As an example, if two wavelength channels with 1 nm channel spacing, the FWM efficiency increases for approximately 25 dB when the chromatic dispersion is decreased from 17 to 1 ps/nm/km. As a consequence, in WDM optical systems, if dispersion-shifted fiber (DSF) is used, interchannel cross talk introduced by FWM may become a legitimate concern, especially when the optical power is high.

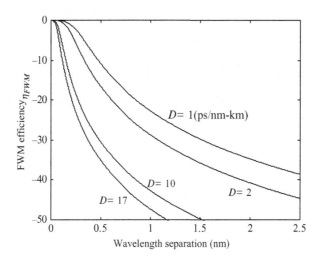

FIG. 2.6.4

Four-wave mixing efficiency η_{FWM}, calculated for $\alpha = 0.25$ dB/km, $\lambda = 1550$ nm. Unit of D is in [ps/nm-km]

2.7 DIFFERENT TYPES OF OPTICAL FIBERS

As we all know, optical fiber is a cylindrical waveguide that supports low-loss propagation of optical signals. The general properties of optical fibers have been discussed in Chapter 1. In the recent years, numerous fiber types have been developed and optimized to meet the demand of various applications. Some popular fiber types that are often used in optical communication systems have been standardized by the International Telecommunication Union (ITU-T). The list includes graded index multimode fiber (G.651), nondispersion shifted fiber (NDSF) also known as standard single-mode fiber (G.652), DSF (G.653), and nonzero dispersion shifted fiber (NZDSF) (G.655). In addition to fibers designed for optical transmission, there are also various specialty fibers for optical signal processing, such as dispersion compensating fibers (DCF), PM fibers, photonic crystal fibers (PCFs), and rare-earth-doped active fibers for optical amplification. Unlike transmission fibers, these specialty fibers are less standardized.

2.7.1 STANDARD OPTICAL FIBERS FOR TRANSMISSION

The ITU-T G.651 MMF has a 50-μm core diameter and a 125-μm cladding diameter. The attenuation parameter for this fiber is on the order of 0.8 dB/km at 1310 nm wavelength. Because of its large core size, MMF is relatively easy to handle with large misalignment tolerance for optical coupling and connection. However, due to its large modal dispersion, MMF is often used for short-reach and low data-rate optical communication systems. Although this fiber is optimized for use in the 1300 nm band, it can also operate in the 850 and 1550-nm wavelength bands [Corning Website; OFS Website]).

The ITU-T G.652 fiber, also known as standard single mode fiber (SSMF), is the most commonly deployed fiber in optical communication systems. This fiber has a simple step-index structure with a 9-μm core diameter and a 125-μm cladding diameter. It is single mode with a zero-dispersion wavelength around $\lambda_0 = 1310$ nm. The typical chromatic dispersion value at 1550 nm is about 17 ps/nm-km. The attenuation parameter for G.652 fiber is typically 0.5 dB/km at 1310 nm and 0.2 dB/km at 1550 nm. An example of this type of fiber is Corning SMF-28.

Although standard SMF has low loss in the 1550-nm wavelength window, which makes it suitable for long-distance optical communications, it shows relatively high chromatic dispersion values in this wavelength window. In high-speed optical transmission systems, chromatic dispersion introduces significant waveform distortion, which may significantly degrade system performance. The trend of shifting the transmission wavelength window from 1310 to 1550 nm in the early 1990s initiated the development of DSF. Through proper cross-section design, DSF shifts the zero-dispersion wavelength λ_0 from 1310 nm to approximately 1550 nm, making the 1550-nm wavelength window have both the lowest loss and the lowest dispersion. The core diameter of the DSF is about 7 μm, which is slightly smaller than the standard SMF. DSF is able to significantly extend the dispersion-limited transmission distance if there is only a single optical channel propagating in the fiber, but people soon realized that DSF is not suitable for multiwavelength WDM systems due to the high-level nonlinear cross talk between optical channels through FWM and XPM. For this reason, the deployment of DSF did not last very long.

To reduce chromatic dispersion while maintaining reasonably low nonlinear cross talk in WDM systems, nonzero dispersion-shifted fibers (NZDSF) were developed. NZDSF moves the zero-dispersion wavelength outside the 1550 nm window so that the chromatic dispersion for optical signals in the 1550 nm wavelength is less than that in standard SMF but higher than that of DSF. The basic idea of this approach is to keep a reasonable level of chromatic dispersion at 1550 nm, which prevents high nonlinear cross talk from happening but without the need of dispersion compensation in the system. There are several types of NZDSF depending on the selected value of zero-dispersion wavelength λ_0. In addition, since λ_0 can be either longer or shorter than 1550 nm, the dispersion at 1550 nm can be either negative (normal) or positive (anomalous). The typical chromatic dispersion for G.655 fiber at 1550 nm is 4.5 ps/nm-km. Although NZDSF usually has core sizes smaller than standard SMF, which increases the nonlinear effect, some designs such as Corning large effective area fiber (LEAF) have the same effective core area as standard SMF, which is approximately 80 μm².

Fig. 2.7.1 shows typical dispersion vs. wavelength characteristics for several major fiber types that have been offered for long-distance links. The detailed specifications of these fibers are listed in Table 2.1. In optical communication system applications, the debate on which fiber has the best performance can never settle down, because data rate, optical modulation formats, channel spacing, the number of WDM channels, and the optical powers used in each channel may affect the conclusion. In general, standard SMF has high chromatic dispersion so that the effect of nonlinear cross talk between channels is generally small; however, a large amount of dispersion compensation must be used, introducing excess loss and requiring higher optical gain in the amplifiers. This in turn will degrade optical signal-to-noise ratio (OSNR) at the receiver. On the other hand, low dispersion fibers may reduce the requirement of dispersion compensation but at the risk of increasing nonlinear cross talk.

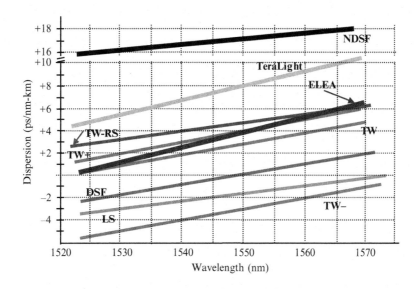

FIG. 2.7.1

Chromatic dispersions of non-dispersion-shifted fiber (NDSF) and various different nonzero-dispersion-shifted fibers (NZDSF).

Table 2.1 Important parameters for nondispersion shifted fiber (NDSF), long-span (LS) fiber, truewave (TW) fiber, truewave reduced-slope (TW-RS) fiber, large effective area fiber (LEAF), and teralight fiber

Fiber type	ITU	Dispersion @ 1550 nm (ps/nm/km)	Dispersion slope (ps/km/nm²)	Effective area (A_{eff}) (μm²)
NDSF(SMF-28)	NDSF/G.652	16.70	0.06	86.6
LS	NZDSF/G.655	−1.60	0.075	50
Truewave (TW)	NZDSF/G.655	2.90	0.07	55.42
TW–RS	NZDSF/G.655	4.40	0.042	55.42
LEAF	NZDSF/G.655	3.67	0.105	72.36
TERALIGHT	NZDSF/G.655	8.0	0.058	63

2.7.2 SPECIALTY OPTICAL FIBERS

In addition to the fibers designed for optical transmission, a number of specialty fibers have also been developed for various purposes ranging from linear and nonlinear optical signal processing to interconnection between equipment. The following are a few examples of specialty optical fibers that have been widely used.

DCF is a widely used specialty fiber that usually provides a large value of negative (normal) dispersion in a 1550-nm wavelength window. It is developed to compensate for chromatic dispersion in optical transmission systems that are based primarily on standard SMF. The dispersion coefficient of DCF is typically on the order of $D = -95$ ps/nm-km at a 1550-nm wavelength window. Therefore, approximately 14 km DCF is required to compensate for the chromatic dispersion of 80 km standard SMF in an amplified optical span. For practical system applications, DCFs can be packaged into modules, which are commonly referred to as dispersion compensating module (DCM).

Compared to other types of dispersion compensation techniques such as fiber Bragg gratings, the distinct advantage of DCF is its wide wavelength window, which is critical for WDM applications, and its high reliability and negligibly small dispersion ripple over the operating wavelength. In addition, DCF can be designed to compensate the slope of chromatic dispersion, thus making it an ideal candidate for WDM applications involving wide wavelength windows. However, due to the limited dispersion value per unit length, DSF usually has relatively higher attenuation compared to fiber gratings, especially when the required total dispersion is high. In addition, because a large value of waveguide dispersion is needed to achieve normal dispersion in 1550-nm wavelength window, the effective core area of DCF can be as small as $A_{eff} \approx 15$ µm^2, which is less than one-fifth that of a standard SMF. Therefore, the nonlinear effect in DCF is expected to be significant, which has to be taken into account in designing a measurement setup involving DCF.

PM fiber is another important category of specialty fibers. It is well known that in an ideal SMF with circular cross-section geometry, two degenerate modes coexist, with mutually orthogonal polarization states and identical propagation constants. The effect of external stress may cause the fiber to become birefringent, and the propagation constants of these two degenerate modes will become different. The partitioning of the propagating optical signal into the two polarization modes not only depends on the coupling condition from the source to the fiber but also on the energy coupling between the two modes while propagating in the fiber, which is usually random. As a consequence, the polarization state of the output optical signal is usually random, even after only a few meters of propagation length in the fiber; the mode coupling and the output polarization state are very sensitive to external perturbations such as temperature variation, mechanical stress change, and micro- and macro-bending.

It is also known that energy coupling between the two orthogonal polarization modes can be minimized if the difference between the propagation constants of these two modes is large enough. This can be accomplished by incorporating extra elements in the fiber cladding that apply asymmetric stress to the fiber core. Due to the difference in the thermal expansion coefficients of various materials, the unidirectional stress in the fiber core can be achieved from the manufacturing process when the fiber is drawn from a preform. Depending on the shape of the stress-applying parts (SAPs), PM fibers can be classified as "Panda" and "Bowtie," as illustrated in Fig. 2.7.2. The bowtie structure is depicted in which the SAPs are arranged in an arced manner around the fiber core; the Panda structure is named based on its similarity to the face of a panda bear. In the direction of the SAPs, which is parallel to the field of tension (horizontal in Fig. 2.7.2), the fiber core has a slightly higher refractive index so that this axis is also referred to as the *slow axis* because the horizontally polarized mode propagates slower than the vertically polarized mode. The principle axis perpendicular to the slow axis is thus called the *fast axis*.

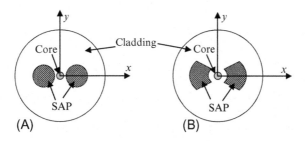

FIG. 2.7.2

Cross sections of polarization fibers: (A) Panda fiber and (B) Bowtie fiber. *SAPs:* stress-applying parts.

It is important to note that a PM fiber is simply a highly birefringent fiber, in which the coupling between the two orthogonally polarized propagation modes is minimized. However, for a PM fiber to maintain the polarization state of an optical signal, the polarization state of input optical signal has to be aligned to either the slow or the fast axis of the PM fiber. Otherwise, both of the two degenerate modes will be excited, although there is minimum energy coupling between them; their relative optical phases will still be affected by the fiber perturbation, and the output polarization state will not be maintained because of the vectorial summation of these two mode fields. To further explain, assume that the input optical field is linearly polarized with an orientation angle θ with respect to the birefringence axis of the PM fiber. The optical field E_0 will be split into the two polarization modes such that $E_x = E_0 \cos\theta$ and $E_y = E_0 \sin\theta$. At the output of the fiber, the composite optical field vector is $\vec{E} = E_0 e^{-\alpha L}\left(\vec{a}_x \cos\theta + \vec{a}_y \sin\theta \cdot e^{j\Delta\phi}\right)$, where $\vec{a}_x$ and $\vec{a}_y$ are the unit vectors, L is the fiber length, and α is the attenuation parameter. $\Delta\phi = (\beta_x - \beta_y)L$ with β_x and β_y the propagation constants of the two modes. It is important to note that the differential phase $\Delta\phi$ is very sensitive to external perturbations of the fiber. When $\Delta\phi$ varies, the polarization state of the output optical field $\vec{E}$ also varies if $\theta \neq 0$ (or 90°), as illustrated in Fig. 2.7.3. In this case, since the phase difference $\Delta\phi$ between the two mode fields E_x and E_y are considered random, the variation of the output polarization orientation can be as high as $\Delta\Phi$, as shown in Fig. 4.1.3.

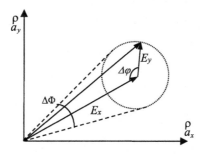

FIG. 2.7.3

Illustration of the output field vector of a PM fiber when the input field polarization state is not aligned to the principle axis of the fiber.

Therefore, in an optical system, if a PM fiber is used, one has to be very careful in the alignment of the signal polarization state at the fiber input. Otherwise the PM fiber could be even worse than a SSMF in terms of the output polarization stability. Another issue regarding the use of PM fibers is the difficulty of connecting and splicing. When connecting between two PM fibers, we must make sure that their birefringence axes are perfectly aligned. Misalignment between the axes would cause the same problem as the misalignment of input polarization state, as previously discussed. To provide the functionality of precisely controlled axes rotation and alignment, a PM fiber splicer can be five times more expensive than a conventional fiber splicer due to its complexity.

PCF, also known as photonic bandgap fiber, is an entirely new category of optical fibers because of its different wave-guiding mechanisms. As shown in Fig. 2.7.4, a PCS fiber usually has large number of air holes periodically distributed in its cross section; for that reason, it is also known as the "holey" fiber. The guiding mechanism of PCF is based on the Bragg resonance effect in the transversal direction of the fiber; therefore, the low-loss transmission window heavily depends on the bandgap structure design.

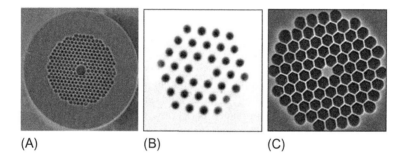

(A)   (B)   (C)

FIG. 2.7.4

Cross-sectional view of photonic crystal fibers: (A) hollow-core PCF, (B) large core area PCF, and (C) highly nonlinear PCF (Thorlabs: crystal fiber).

Fig. 2.7.4A shows the cross-section view of a hollow core PCF in which the optical signal is guided by the air core. In contrast to the conventional optical wave-guiding mechanism where high refractive index solid dielectric material is required for the core, the photonic bandgap structure in the cladding of PCF acts as a virtual mirror that confines the propagating lightwave to the hollow core. In general, the periodical photonic bandgaps structure of PCF is formed by the index contrast between silica and air by incorporating the holes into a silica matrix. In most of the hollow core PCFs, more than 95% of the optical power exists in the air; therefore, the interaction between signal optical power and the glass material is very small. Because the nonlinearity of the air is approximately three orders or magnitude lower than in the silica, hollow core PCF can have extremely low nonlinearity and can be used to deliver optical signals with very high optical power. In addition, because of the design flexibility of the photonic bandgap structures, by varying the size and the pitch of the holes, zero dispersion can be achieved at virtually any wavelength. One drawback for the hollow core PCF is its relatively narrow transmission window, which is typically on the order of 200 nm. This is the consequence of the very strong resonance effect of the periodic structure that confines the signal energy in the hollow core.

Another category of PCF as shown in Fig. 2.7.4B is referred to as *large core area PCF*. This type of fiber has a solid silica core to guide the optical signal, whereas the periodic holes in the cladding are used to facilitate optical wave guiding and help confine signal optical power in the core area. Generally, SSMF with a 10 μm core diameter has a cutoff wavelength at approximately 1100 nm, below which the fiber will have multiple propagation modes. Large core PCF, on the other hand, allows for single-mode operation in a very wide wavelength window—for example, from 750 to 1700 nm—while maintaining a very large core area. Compared to hollow core PCF, large core area PCF has much wider low-loss window. Although a large core area PCF has a lower nonlinear parameter than SSMF, its nonlinear parameter is typically much higher than a hollow core PCF.

Highly nonlinear PCF, as shown in Fig. 2.7.4C, is a very useful fiber for optical signal processing and has very small solid core cross section; therefore, the power density in the core is extremely high. For example, for a highly nonlinear PCF with zero-dispersion wavelength at $\lambda_0 = 710$ nm, the core diameter is as small as 1.8 μm and the nonlinear parameter is $\gamma > 100$ W^{-1} km^{-1}, which is 40 times higher than that of a SSMF. Highly nonlinear PCFs are usually used in nonlinear optical signal processing, such as parametric amplification and supercontinuum generation.

PCFs are considered high-end fiber types that are usually expensive due to their complex fabrication processes. They are generally sold in meters instead of kilometers because relatively short PCF fibers are enough for most of the applications for which PCFs are designed. They are also delicate and difficult to handle; for example, surface treatment, termination, connection, and fusion splicing are not straightforward because of the existence of air holes in the fiber.

Plastic optical fiber (POF) is a low-cost type of optical fiber that is also easy to handle. The core material of POF is typically made of polymethyl methacrylate (PMMA), which is a general-purpose resin, whereas the cladding is usually made of fluorinated polymer, which has a lower refractive index than the core. The cross-section design of POF is more flexible than silica fibers and various core sizes and core/cladding ratio can be easily obtained. For example, in a large-diameter POF, 95% of the cross section is the core that allows the transmission of light. The fabrication of POF does not need an expensive MOCVD process, which is usually required for the fabrication of silica-based optical fibers; this is one of the reasons for the low cost of the POFs. Although silica-based fiber is widely used for telecommunications, POF has found more and more applications due to its low cost and high flexibility. The cost associated with POF connection and installation costs is also low, which are especially suitable for fiber to the home application. On the other hand, POFs have transmission losses on the order of 0.25 dB/m, which is almost three orders of magnitude higher than silica fibers. This excluded POFs from being used in long-distance optical transmission. In addition, the majority of POFs are multimode, and therefore, they are often used in low-speed, short-distance applications such as fiber-to-the-home networks, optical interconnections, networks inside automobile, and flexible illumination and instrumentation.

2.8 SUMMARY

In this chapter, we have discussed basic concept of wave propagation in an optical waveguide. Based on the free-space Fresnel refraction and refraction on a dielectric interface, we have introduced the concepts of total internal reflection, optical phase shift upon reflection, as well as evanescent wave penetrating into the waveguide cladding. Then, based on the roundtrip phase matching conditions in the transversal direction of the waveguide and the standing wave principle, we have explained the physical meaning and the definition of propagation modes in an optical waveguide. Although geometric ray tracing in free space is simple and intuitive in explaining the basic concept, precise mode field analysis in an optical fiber has to be accomplished based on the solutions of Maxwell equations in a cylindrical coordinate with appropriate boundary conditions. In addition to mode-field distributions, several simple, but very important, parameters are resulted from this electromagnetic analysis include mode cut-off conditions, V-number, single-mode condition, cut-off wavelength, numerical aperture, and propagation constant of each mode.

Silica-based optical fiber has low loss windows around 1320 and 1550 nm. The typical fiber losses in these two wavelength windows are approximately 0.5 and 0.25 dB/km, respectively. With advances in material science and fabrication techniques, the ultimate fiber loss is eventually limited by the Rayleigh scatter. As a common practice of engineering, propagation loss in an optical fiber is defined in [dB/km], which allows straightforward estimation of signal optical power along the fiber with the knowledge of the signal power in [dBm] at the fiber input. On the other hand, when complex optical field expression, for example, in Eq. (2.4.3), is needed in the calculation, the attenuation parameter α needs to be expressed in [Neper/m]. A conversion factor between [dB/km] and [Neper/m] is 4.343×10^3.

Dispersion is an effect of differential delay. For a MMF, if a signal optical pulse is carried by multiple propagation modes with each mode propagating in a different speed, the narrow pulse at the fiber input would be converted into a broader pulse, or even multiple pulses, at the fiber output. This is the effect of modal dispersion. In addition, even within a single mode, different frequency components of the optical signal may propagate in different speeds, and thus a short optical pulse at the fiber input can also be converted into a broader pulse at the fiber output due to chromatic dispersion. Usually for a MMF, the differential delay caused modal dispersion can be several orders of magnitude bigger than that of chromatic dispersion within each mode. For a SMF, chromatic dispersion is the dominate mechanism and the differential delay is directly proportional to the spectral width of the optical signal. Note that even for a SMF, two degenerate modes exist with mutually orthogonal states of polarizations. Differential propagation delay between these two polarization modes is known as the polarization mode dispersion (PMD). PMD, originated from the birefringence and the noncircular core of the fiber, is usually random and easily perturbed by environmental conditions, and often specified by a statistical Maxwellian distribution.

We have also discussed various nonlinear effects in an optical fiber, their origins, and their impacts in optical transmission systems. Kerr-effect nonlinearity, caused by the power-dependent refractive index in the fiber, is responsible for various nonlinear effects such as SPM, XPM, and FWM, which have been seen to cause performance degradation in fiber-optic systems. Especially in multiwavelength optical networks, cross talk caused by XPM and FWM can be quite significant when the signal optical power is high. Nonlinear scattering, including SBS and SRS, can cause optical signal energy shift from short wavelength to longer wavelength components. For a transmission system with only a single wavelength, nonlinear scattering causes a nonlinear loss of optical signal power, while for a fiber system carrying multiple wavelengths, nonlinear cross talk can introduce a nonlinear cross talk. Although nonlinear effects are limiting factors for the maximum optical power level than can be used in the fiber system, they can also be utilized to perform nonlinear optic signal processing. One such example is nonlinear pulse compression through SPM which is the basic mechanism to create optical soliton. Another example is to introduce stimulated Raman amplification, in which a strong optical pump at a short wavelength transfers its energy to an optical signal at a longer wavelength through the SRS process, which will be further discussed in Chapter 5.

In the last section of this chapter, we have presented different types of optical fibers. In addition to SSMF and standard multimode fiber (SMMF), which are most commonly used in optical communication systems, there are various other fiber types which are often referred to as specialty fibers. DSF was an early effort to reduce the chromatic dispersion in a SMF by shifting the zero-dispersion wavelength to the communication window around 1550 nm. DSF was then found to be susceptible to nonlinear cross talk in multiwavelength systems due to FWM because of the slow walk-off between different wavelength channels. Then, other types of DSFs were designed by changing the zero-dispersion wavelength away from the communications wavelength window to make trade-offs between chromatic dispersion and nonlinear cross talk. Dispersion compensation fiber with large normal dispersion value in the 1550-nm wavelength window was developed to compensate the accumulated anomalous dispersion in optical systems with SSMF. LEAF is another transmission fiber developed to reduce the power density in the fiber core so that the nonlinear effect can be reduced. Other specialty fibers such as PM fiber and PCF are usually used in special circumstances with relatively short lengths.

With the advances in material science, the maturity of fiber fabrication techniques, and the emergence of new applications, new materials and fiber types will be developed in the future. Good understanding of fundamental physics, wave guiding mechanics, and key parameters will be essential in the design of optical fibers, as well as the design and performance estimation of fiber-optic systems and networks.

PROBLEMS

1. Consider a plane wave with a relative phase difference $\delta = \pi/3$ between the E_x and E_y components. If the magnitude of the x-component is twice as large as the y-component ($E_{0x} = 2E_{0y}$), and assume $E_{0y} = 1$, please find the orientation angle ϕ of this polarization ellipse, and find the lengths of the long axis and the short axis of the ellipse.

2. Consider the air/water interface shown in the following figure.
   (a) A collimated light beam launches from the air to the water at an angle $\theta = \pi/6$ with respect to the surface normal as shown in Fig. P2.2A, where $h_a = 1.5$ m and $h_b = 1$ m. The refractive index of air is $n_0 = 1$, and assume the refractive index of water is $n_1 = 1.3$.

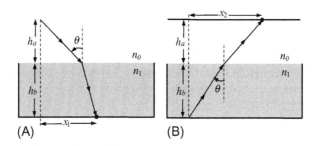

(A)   (B)

FIG. P2.2

Figure for problem 2.

Please find the distance $x_1 = ?$

In order to eliminate reflection from the water surface, what should be the angle θ, and in which direction the light should be polarized?

(b) If the light is launched from the bottom of the water tank as shown in Fig. P2.2B, and the incidence angle is still $\theta=\pi/6$, please find the distance $x_2 = ?$.

What is the angle θ to achieve total reflection on the water/air surface?

3. A light beam is launched to a thin semiconductor film with a thickness of $d=2\,\mu m$. The refractive index of air is $n_0=1$, and assume the semiconductor film has no loss and its refractive index is $n_1=3.5$. The wavelength of the light is $\lambda=633\,nm$ which is parallel polarized on the plane of the paper, and the incidence angle is $\theta=45°$.

Please find the relative amplitude $|E_1|/|E_2|$ and the phase difference $[\text{Phase}(E_1)-\text{Phase}(E_2)]$ of the two reflected beams at the reference plane (Fig. P2.3).

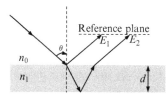

FIG. P2.3

Figure for problem 3.

4. A light beam with $\lambda=0.63\,\mu m$ travels inside a slab optical waveguide by bouncing back and forth between the two interfaces. The refractive index of the waveguide is $n_1=1.5$, and the thickness is $d=3\,\mu m$. Assume the lightwave is vertically polarized (E field goes into the paper) with respect to the incidence plane. For the incidence angle $\theta=60°$, what is the optical phase shift after each roundtrip? What is the evanescent field penetration depth near the glass/air interface? (Fig. P2.4).

FIG. P2.4

Figure for problem 4.

5. A step-index optical fiber has the core refractive index $n_1=1.50$ and the cladding refractive index $n_2=1.497$. The core diameter of the fiber is $d=8\,\mu m$.

(a) What is the numerical aperture of this fiber?

(b) What is the cut-off wavelength λ_c of this fiber? (The fiber operates in single mode when the operation wavelength satisfies $\lambda>\lambda_c$)

(c) To carry a blue light at $0.4\,\mu m$ wavelength, approximately how many modes exist in this fiber?

6. For a single-mode step-index fiber, the core index $n_1=1.49$ and the cladding index $n_2=1.487$, and the core radius is $a=4.5\,\mu m$.

   (a) Please find the wavelength at which $V=2$.

   (b) According to Fig. 2.3.5, for $V=2$, the normalized propagation constant is approximately $b\approx0.4$. Please find the longitudinal propagation constant of the fundamental mode β_z at this b-value, and compare this β_z-value with free-space β-value corresponding to the refractive index of the core.

   (c) Approximately how far away from the core/cladding interface the field amplitude is reduced to 1% compared to its value at core/cladding interface?

7. Again based on Fig. 2.3.5, for $V=3$, the b-values of LP01 and LP11 modes are 0.7 and 0.16, respectively. Assume the step-index fiber the core index $n_1=1.49$ and the cladding index $n_2=1.487$, and the core radius is $a=4.5\,\mu m$. What is the propagation constant, β_z, difference between LP01 and LP11 modes? For a fiber length of $L=10\,km$, what is the arrival time difference between these two modes?

8. For a standard SMF, the core index is $n_1=1.47$, the normalized index difference is $\Delta=0.35\%$, and the core diameter is $d=8\,\mu m$. Please use Eq. (2.3.43) to find MFD $2W_0$ at 1310 and 1550 nm wavelengths. Please explain why the MFD is smaller at 1310 nm wavelength than that at 1550 nm wavelength.

9. Optical field propagating along an optical fiber can be expressed as $E(z,t)=E_0\exp[-j(\omega t-\beta z)]\cdot\exp(-\alpha)z$, where $\alpha=1.15\times10^{-4}m^{-1}$.

   1. What is the power attenuation parameter in dB/km?

   2. If the fiber length is 30 km and the input optical power is 10 dBm, what is the optical power at the fiber output?

10. A fiber-optic system consists of three fibers with lengths $L_1=10\,km$, $L_2=20\,km$, and $L_3=30\,km$, and loss parameters $\alpha_1=0.5\,dB/km$, $\alpha_2=0.25\,dB/km$, and $\alpha_3=0.2\,dB/km$, respectively. These three fibers are spliced together to form a 60-km optical link. Neglect the splicing loss between fibers, what is the total loss of this fiber link? If the input optical power is 4 mW, what is the output optical power in mW?

11. Chromatic dispersion can be represented by a frequency-dependent propagation constant $\beta(\omega)$. Suppose the propagation constant of a SMF is $\beta(\omega)=A\omega_0+B(\omega-\omega_0)+C(\omega-\omega_0)^2$, where $\omega_0=1.216\times10^{15}$ Hz (corresponding to $\lambda=1550$ nm wavelength).

   (a) If at the optical frequency ω_0, the fiber has a group velocity $v_g=2\times10^8$ m/s and a chromatic dispersion parameter $D=15$ ps/(nm × km), please find the values of B and C in the $\beta(\omega)$ expression.

   (b) Assume an optical pulse is simultaneously carried by two wavelengths separated by 3 nm in the 1550-nm wavelength window. After 20 km transmission through this fiber, approximately what is the pulse separation in the time domain caused by the chromatic dispersion?

12. In a SMF system the light source has two discrete wavelength components with the same power but separated by 1.2 nm. The fiber dispersion parameter in the laser diode wavelength range is $D=17$ ps/(nm × km).

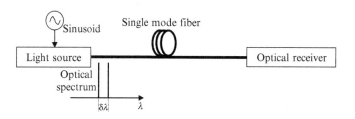

FIG. P2.12

Figure for problem 12.

The light source is amplitude modulated by a sinusoid so that the total optical power can be expressed as $P_{opt}(t)=P_0(1+\cos(2\pi ft))$ [note: power carried by each wavelength component is $P_1(t)=P_2(t)=0.5P_0(1+\cos(2\pi ft))$ at the fiber input]. After transmitting for a certain distance along the fiber, the amplitude modulation on the total optical power may disappear, this is commonly referred to as "carrier fading."

(a) Explain the physical reason why this carrier fading may occur?
(b) If the modulation frequency is $f=10\,\text{GHz}$, find the fiber length at which carrier fading happens?
(c) If $P_0=4\,\text{mW}$, fiber loss coefficient is $\alpha=0.5\,\text{dB/km}$, and the fiber length is 50 km, what is the average optical power at the fiber output? Does this power change with the modulation frequency?

13. Sellmeier Eq. (2.5.22) may be linearized as $D(\lambda)=S_0(\lambda-\lambda_0)$ for simplicity. A SMF has zero-dispersion wavelength $\lambda_0=1310\,\text{nm}$ and the dispersion slop $S_0=0.092\,\text{ps}/(\text{nm}^2\text{km})$ near the zero-dispersion wavelength.
   (a) Please find the chromatic dispersion parameters D in [ps/(nm-km)] and β_2 in [ps^2/km] at $\lambda=1550\,\text{nm}$ using Sellmeier equation and the linearized version.
   (b) Calculate group delay difference between 1310 and 1550 nm wavelengths using Sellmeier equation.

14. Reconsider problem 7 (step-index fiber with core index $n_1=1.49$, and cladding index $n_2=1.487$), but using geometric ray trace technique (Eq. 2.5.27) to calculate the maximum modal dispersion. Discuss why the result calculated from ray trace equation is higher than that from Fig. 2.3.5. Which one is more accurate?

15. (a) PMD of a standard SMF is specified by its mean unit-length differential group delay DGD_{mean}. For $DGD_{mean}=0.06\ \text{ps}/\sqrt{\text{km}}$, what is the average differential delay $\Delta\tau_g$ after 1000 km of fiber?
   (b) A polarization-maintaining (PM) fiber is highly birefringent, and its birefringence is specified by the beat length. For a PM fiber with 5 mm beat length at 1550 nm wavelength, what is the differential index $|n_x-n_y|$?

16. SMF has loss parameter $\alpha_{dB}=0.23\,\text{dB/km}$ at 1550 nm wavelength. Based on Eq. (2.6.14), please plot the nonlinear effective length L_{eff}, as the function of the actual fiber length L from 1 to 100 km. Discuss at what fiber length the simplified formula, $L_{eff}\approx 1/\alpha$ has less than 10% error.

17. A DSF has the effective cross-section area $A_{eff}=52\,\mu\text{m}^2$, a nonlinear index $n_2=2.2\times 10^{-20}\ \text{m}^2/\text{W}$, and a loss parameter $\alpha_{dB}=0.23\,\text{dB/km}$ at 1550 nm wavelength. If the fiber length is 5 km and the optical signal is continuous wave (CW), what is the optical power P required at the fiber input to produce nonlinear phase shift of $\Phi_{NL}=1$ rad? What are the powers required if the fiber length is 100 and 500 km (to produce $\Phi_{NL}=1$ rad)?

18. A trapezoid-shaped optical pulse shown in the following figure is injected into an optical fiber with the peak power $P_0=300$ mW, pulse width $T_0=1$ ns, and equal length of leading edge and trailing edge of 0.1 ns ($t_L=t_T=0.1$ ns). A SSMF is used with $A_{eff}=80\,\mu m^2$, a nonlinear index $n_2=2.2\times10^{-20}$ m^2/W, and a loss parameter $\alpha_{dB}=0.23$ dB/km at 1550 nm wavelength. The fiber length is 80 km.

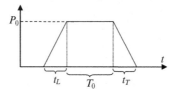

FIG. P2.18

Figure for problem 18.

(a) Draw and label the waveforms of nonlinear phase shift, $\Phi_{NL}(t)$, and optical frequency shift $\delta f(t)$ at the fiber output

(b) If the fiber chromatic dispersion is 16 ps/(nm-km), please estimate the arrival time difference between the pulse leading edge and the trailing edge. (To simplify the problem, only consider the dispersion effect on the nonlinear frequency shifted leading edge and the trailing edge. Also, consider the nonlinear effect only happens from fiber input to L_{eff}, and dispersion has the effect only from L_{eff} to the end of fiber.) Which edge of the pulse travels faster?

19. Consider three discrete frequency components ω_j, ω_k, and ω_l. The frequency differences $\delta\omega_{jk}=(\omega_j-\omega_k)$, and $\delta\omega_{kl}=(\omega_k-\omega_j)$ are not equal. How many FWM components can be created by them? Within which how many are degenerated FWM components?

20. When the fiber length is much longer than the effective length ($L\gg L_{eff}$), normalized FWM efficiency η_{FWM} is only related to fiber loss, dispersion, and channel spacing. Consider a degenerate FWM caused by two frequency components in the 1550-nm wavelength window separated by 25 GHz, and assume the fiber loss is $\alpha_{dB}=0.2$ dB/km. Plot FWM efficiency normalized FWM efficiency η_{FWM} as the function of dispersion parameter D [from 0 to 20 ps/(nm-km)]. At which dispersion value, η_{FWM} is 1%?

REFERENCES

Agrawal, G.P., 2001. Nonlinear Fiber Optics, third ed. Academic Press.

Artiglia, M., Coppa, G., Di Vita, P., Potenza, M., Sharma, A., 1989. Mode field diameter measurements in single-mode optical fibers. J. Lightwave Technol. 7 (8), 1139–1152.

Boyd, R., 1992. Nonlinear Optics. Academic Press.

Buck, J.A., 2004. Fundamentals of Optical Fibers, second ed Wiley Chapter 3.

Gloge, D., 1971. Weakly guiding fibers. Appl. Opt. 10 (10), 2252–2258.

Hill, K.O., Johnson, D.C., Kawasaki, B.S., MacDonald, R.I., 1978. cw three-wave mixing in single-mode optical fibers. *J. Appl. Phys.* 49, 5098.

Iizuka, K., 2002. Elements of Photonics, Volume II: For Fiber and Integrated Optics. Wiley Section 11.2.

Inoue, K., 1992. Polarization effect on four-wave mixing efficiency in a single-mode fiber. IEEE J. Quantum Electron. 28, 883–894.

Islam, M.N., et al., 1987. Cross-phase modulation in optical fibers. Opt. Lett. 12 (8), 625.

Keiser, G., 2011. Optical Fiber Communications, third ed. McGraw-Hill.

Marcus, D., 1978. Gaussian approximation of the fundamental modes of graded-index fibers. J. Opt. Soc. Am. 68, 103.

Meumann, E.G., 1988. Single-Mode Fibers Fundamentals. Springer Series in Optical Science57Springer-Verlag. ISBN: 3-54018745-6.

Okamoto, K., 2005. Fundamentals of Optical Waveguides, second ed. Academic Press.

Randel, S., et al., 2011. 6×56-Gb/s mode-division multiplexed transmission over 33-km few-mode fiber enabled by 6×6 MIMO equalization. Opt. Express 19 (17), 16697–16707.

Smith, R.G., 1972. Optical power handling capacity of low loss optical fibers as determined by stimulated Raman and Brillouin scattering. Appl. Opt. 11 (11), 2489.

Stolen, R.H., Lin, C., 1978. Self-phase modulation in silica optical fibers. Phys. Rev. A 17 (4), 1448.

Taitt, C.R., Anderson, G.P., Ligler, F.S., 2005. Evanescent wave fluorescence biosensors. Biosens. Bioelectron. 20 (12), 2470–2487.

FURTHER READING

Demarest, K., Richards, D., Allen, C., Hui, R., 2002. Is standard single-mode fiber the fiber to fulfill the needs of tomorrow's long-haul networks? In: National Fiber Optic Engineers Conference (NFOEC)pp. 939–946.

www.corning.com

www.crystal-fibre.com

www.ofsoptics.com

LIGHT SOURCES FOR OPTICAL COMMUNICATIONS

CHAPTER OUTLINE

Introduction 77
3.1 Properties of semiconductor materials for light sources 79
    3.1.1 PN junction and energy diagram 79
    3.1.2 Direct and indirect semiconductors 81
    3.1.3 Spontaneous emission and stimulated emission 82
    3.1.4 Carrier confinement 83
3.2 Light-emitting diodes 84
    3.2.1 $P \sim I$ curve 85
    3.2.2 Modulation dynamics 86
3.3 Laser diodes 88
    3.3.1 Amplitude and phase conditions for self-sustained oscillation 88
    3.3.2 Rate equations 91
    3.3.3 Steady state solutions of rate equations 92
    3.3.4 Side-mode suppression ratio 94
    3.3.5 Modulation response 95
    3.3.6 Laser noises 99
3.4 Single-frequency semiconductor lasers 104
    3.4.1 DFB and DBR laser structures 105
    3.4.2 External cavity LDs 108
3.5 VCSELs and arrays 111
3.6 LD biasing and packaging 115
3.7 Summary 119
Problems 121
References 124

INTRODUCTION

In an optical communication system, information is carried by photons and thus optical source is an indispensible part which has important impact in both the cost and the performance of the system. Although there are various types of optical sources, semiconductor-based devices such as

Introduction to Fiber-Optic Communications. https://doi.org/10.1016/B978-0-12-805345-4.00003-2

light-emitting diodes (LEDs) and laser diodes (LDs) are most often used in fiber-optic systems because of their miniature size and high-power efficiency and superior reliability. Semiconductor LD and LED are based on the forward-biased PN junctions, and their output optical powers are proportional to the injection electric currents. Thus, optical power emitted from a LED or a LD can be electrically modulated through carrier injection with the speed primarily determined by the carrier lifetime, which is commonly referred to as direct modulation. As semiconductor materials have energy *bands* instead of discrete energy *levels* as in the case of gases, carrier transition between these energy bands allows photon emission in a relatively wide wavelength window. Thus, semiconductor-based light-emitting sources are uniquely suited to support wavelength division multiplexing (WDM) in fiber-optic systems to make efficient use of the wide bandwidth of the fiber. On the flip side, the broadband nature of semiconductor material also makes it difficult to achieve single-frequency operation. An LED light source with a broad spectral width suffers from chromatic dispersion when used in a fiber system, and thus it is limited to applications in short-distance optical systems. In addition, since an LED is based on the spontaneous emission which propagates in all directions filling an entire 4π Euler spatial angle, the external efficiency cannot be high because total internal reflection traps a large portion of optical power inside the device. In comparison, a LD is based on the stimulated emission which produces coherent light-wave in a selected propagation direction. Thus, total internal reflection is avoided and the external efficiency of a LD can be much higher than an LED.

To ensure single longitudinal mode operation in a LD with narrow spectral linewidth, special photonic structures such as distributed feedback (DFB), distributed Bragg reflector (DBR), or external cavity have to be used. In addition, because the refractive index of the semiconductor material inside the laser cavity is a function of the carrier density, the fluctuation of carrier density caused by spontaneous emission not only introduces the change of the optical power, but also changes the signal optical frequency through the change of the cavity resonance condition. Thus, the spectral linewidth of a diode laser is typically an order of magnitude broader than that predicted by the classic Schawlow-Townes formula (Schawlow and Townes, 1958). This carrier-dependent refractive index also introduces signal optical phase modulation when the injection current is directly modulated, known as the frequency chirp, which can significantly broaden the signal spectral linewidth, and thus external optical modulation has to be used in long-distance fiber-optic systems which will be further discussed in Chapter 7.

The performance of a diode laser can be modeled by a set of rate equations describing the interaction between the carrier density and the photon density inside the laser cavity. Rate equation analysis can help understand the impact of various physical parameters on the static and dynamic properties of laser emission.

Historically, the evolution of communication wavelengths from 850 to 1310 nm and then to 1550 nm were largely dictated by the availability of laser sources. Advances in semiconductor material growth and bandgap engineering and innovative nano-fabrication techniques have tremendous impacts in the laser design, performance improvement, cost reduction, as well as their functionality enhancement for applications in optical systems and networks. One such example is the availability of miniaturized narrow linewidth diode laser modules, which brought coherent optical communication into mainstream in the industry with commercial success. Another example is the vertical-cavity surface-emitting laser (VCSEL) array which can be made low cost and high performance, enabling massive parallel optical interconnections in datacenters and short-distance fiber-optic systems.

The last section of this chapter discusses LED and LD biasing and packing. Both LED and LD have an exponential current-voltage relation, similar to a conventional junction diode, and the output optical power is linearly proportional to the injection current. Constant current sources, instead of voltage sources, have to be used in the biasing circuits to avoid devices damaging due to over-driving.

3.1 PROPERTIES OF SEMICONDUCTOR MATERIALS FOR LIGHT SOURCES

3.1.1 PN JUNCTION AND ENERGY DIAGRAM

In a doped semiconductor, the Fermi level E_F depends on the doping density as

$$E_F = E_{Fi} + kT \ln(n/n_i) \tag{3.1.1}$$

where k is the Boltzmann's constant, T is the absolute temperature, n is the electron density, E_{Fi} is the intrinsic Fermi level, and n_i is the intrinsic electron density. Eq. (3.1.1) can also be expressed as the function of the hole density p and the intrinsic hole density p_i as $E_F = E_{Fi} - kT\ln(p/p_i)$. This is because under thermal equilibrium, the product of electron and hole densities is a constant, $np = n_i^2$, and the intrinsic electron and hole densities are always equal, $n_i = p_i$. As intrinsic electron (hole) density is usually low at room temperature, carrier density of a typical doped semiconductor is determined by the doping densities, such that $n \approx N_d$, for n-type doping and $p \approx N_a$ for p-type doping, where N_d and N_a are electron and hole doping densities, respectively. As the result, the Fermi level of a doped semiconductor is largely determined by doping condition, that is, $E_F = E_{Fi} + kT\ln(N_d/n_i)$, or $E_F = E_{Fi} - kT\ln(N_a/p_i)$. As illustrated in Fig. 3.1.1A, the Fermi level is closer to the conduction band in an n-type semiconductor because $N_d \gg n_i$, and it is closer to the valence band in a p-type semiconductor because $N_a \gg p_i$.

A pn junction is formed when the n- and p-type semiconductors have intimate contact. Under thermal equilibrium, the Fermi level will be unified across the pn junction structure as shown in Fig. 3.1.1B. This happens because high-energy free electrons diffuse from n- to p-side and low-energy holes diffuse in the opposite direction; as a result, the energy level of the p-type side is increased compared to that in the n-type side. Meanwhile, because free electrons migrate from n- to p-type-side, uncovered protons left over at the edge of the n-type semiconductor create a positively charged layer on the n-type side. Similarly, a negatively charged layer is created at the edge of the p-type semiconductor due to the loss of holes. Thus, built-in electrical field and hence a potential barrier is created at the *pn* junction, which pulls the diffused free electrons back to the n-type side and holes back to the p-type-side, a process commonly referred to as *carrier drift*. Because of this built-in electrical field, neither free electrons nor holes exist at the junction region, and therefore, this region is called the *depletion region* or *space-charged region*. Without an external bias, there is no net carrier flow across the pn junction due to the exact balance between carrier diffusion and carrier drift. The built-in electrical potential barrier across the pn junction is related to the electron doping density N_d on the n-type side and hole doping density N_a on the p-type side, and can be expressed as

$$\Delta V_{pn} = \frac{kT}{q}\left[\ln\left(\frac{N_d}{n_i}\right) + \ln\left(\frac{N_a}{p_i}\right)\right] = \frac{kT}{q}\ln\left(\frac{N_d N_a}{n_i^2}\right)$$

where q is the electron charge.

When the *pn*-junction is forward biased as illustrated in Fig. 3.1.1C, excess electrons and holes are injected into the *n*- and *p*-type sections, respectively. This carrier injection reduces the built-in potential barrier so that the balance between the electron drift and diffusion is broken. The external biasing source pushes excess electrons and holes to diffuse across the junction area. In this process, excess electrons and holes recombine inside the depletion region to generate photons. The recombination can be *nonradiative* which turns the energy difference between the participating electron and the hole into heat. It can also be *radiative*, in such a case a photon is created and the photon energy is equal to the energy of the bandgap, which will be discussed in the next section.

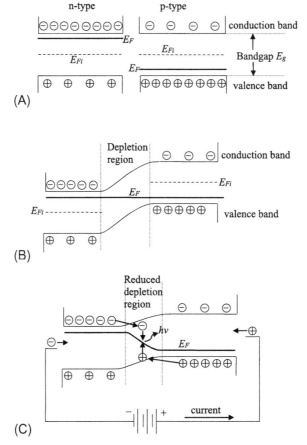

FIG. 3.1.1

(A) Band diagram of separate n- and p-type semiconductors, (B) energy diagram of a pn junction under equilibrium, and (C) band diagram of a pn junction with forward bias.

3.1.2 DIRECT AND INDIRECT SEMICONDUCTORS

One important rule of radiative recombination process is that both energy and momentum must be conserved. Depending on the shape of their band structure, semiconductor materials can be generally classified as having direct bandgap or indirect bandgap, as illustrated in Fig. 3.1.2, where E is the energy and k is the momentum.

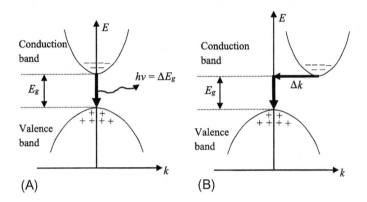

FIG. 3.1.2

Illustration of direct bandgap (A) and indirect bandgap (B) of semiconductor materials.

For direct semiconductors, holes at the top of the valence band have the same momentum as the electrons at the bottom of the conduction band. In this case, electrons directly recombine with the holes to emit photons, and the photon energy is equal to the bandgap,

$$hv = \frac{hc}{\lambda} = E_g \tag{3.1.2}$$

where h is the Planck's constant, E_g is the bandgap of the semiconductor material, $v = c/\lambda$ is the optical frequency, c is the speed of light, and λ is the optical wavelength. Examples of direct bandgap semiconductor materials include GaAs, InAs, InP, AlGaAs, and InGaAsP. The desired bandgap can be obtained through compositional design of the material, as well as structural engineering in the nanometer scales such as quantum well and quantum dots.

For indirect semiconductors, on the other hand, holes at the top of the valence band and electrons at the bottom of the conduction band have different momentum. Any recombination between electrons in the conduction band and holes in the valence band would require significant momentum change. Although a photon can have considerable energy hv, its momentum hv/c is much smaller, which cannot compensate for the momentum mismatch between the electrons and holes. Therefore, radiative recombination is considered impossible in indirect semiconductor materials unless a third particle (e.g., a phonon created by crystal lattice vibration) is involved and provides the required momentum. Silicon is a typical indirect bandgap semiconductor material, and therefore, it cannot be directly used to make light-emitting devices.

3.1.3 SPONTANEOUS EMISSION AND STIMULATED EMISSION

As discussed, radiative recombination between electrons and holes creates photons, but this is a random process. The energy is conserved in this process, which determines the frequency of the emitted photon as $v = \Delta E/h$, where ΔE is the energy gap between the conduction band electron and the valence band hole that participated in the process. h is the Planck's constant. However, the phase of the emitted light-wave is not predictable. Indeed, since semiconductors are solids, carriers are not on discrete energy levels; instead they are continuously distributed within energy bands following the Fermi-Dirac distribution, as illustrated in Fig. 3.1.3. Although the nominal value of ΔE is in the vicinity of the material bandgap E_g, its distribution depends on the width of the associated energy bands and the Fermi-Dirac distribution of carriers within these bands.

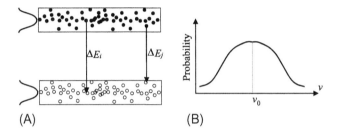

FIG. 3.1.3

Illustration of an energy band in semiconductors and the impact on the spectral width of radiative recombination. (A) Energy distributions of electrons and holes. (B) Probability distribution of the frequency of emitted photons.

Fig. 3.1.3A shows that different electron-hole pairs may be separated by different energy gaps and ΔE_i might not be equal to ΔE_j. Recombination of different electron-hole pairs will produce emission at different wavelengths. The spectral width of the emission is determined by the statistical energy distribution of the carriers, as illustrated in Fig. 3.1.3B.

Spontaneous emission is created by the spontaneous recombination of electron-hole pairs. The photon generated from each recombination event is independent, although statistically the emission frequency falls into the spectrum shown in Fig. 3.1.3B. The frequencies, phases, and direction of propagation of the emitted photons are not correlated. This is illustrated in Fig. 3.1.4A.

Stimulated emission, on the other hand, is created by stimulated recombination of electron-hole pairs. In this case, the recombination is induced by an incoming photon, as shown in Fig. 3.1.4B. Both the frequency and the phase of the emitted photon are identical to those of the incoming photon. Therefore, photons generated by the stimulated emission process are coherent, which results in narrow spectral linewidth. It is also important to know that when carriers at a certain energy level are depleted due to stimulated recombination, carriers at other energy levels will quickly come in to fill the gap through thermally induced relaxation, and thus carrier distribution as the function of energy level within the band will not change. As a consequence, carrier depletion at one particular energy level will result in the carrier density reduction of all energy levels within the same band. This is commonly referred to as *homogeneous broadening*. In contrast, *inhomogeneous broadening* may happen with some materials such as gases in which carrier depletion at an energy level cannot be filled by carriers from other energy levels. This inhomogeneous broadening may result in a phenomenon called spectral hole burning, which means that the material gain can be reduced at a certain frequency due to the strong emission at that frequency.

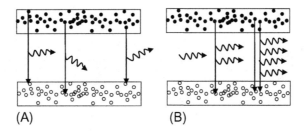

(A)   (B)

FIG. 3.1.4

Illustration of spontaneous emission (A) and stimulated emission (B).

3.1.4 CARRIER CONFINEMENT

In addition to the requirement of using direct bandgap semiconductor material, another important requirement for LEDs is the carrier confinement. In early LEDs, materials with the same bandgap were used at both sides of the pn junction, as shown in Fig. 3.1.1. This is referred to as *homojunction*. In this case, carrier recombination happens over the entire depletion region with the width of 1–10 μm depending on the diffusion lengths of the electrons and the holes. This wide depletion region makes it difficult to achieve high spatial concentration of carriers. To overcome this problem, double heterojunction was introduced in which a thin layer of semiconductor material with a slightly narrower bandgap is sandwiched in the middle of the junction region between the *p*- and *n*-type sections. This concept is illustrated in Fig. 3.1.5, where $E_{g'} < E_g$.

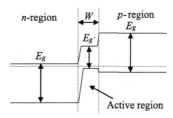

FIG. 3.1.5

Illustration of semiconductor double heterostructure.

In this structure, the thin layer with slightly smaller bandgap than other regions attracts the concentration of carriers when the junction is forward biased; therefore, this layer is referred to as the *active region* of the device. The carrier confinement is a result of bandgap discontinuity. The width, *W*, of the low bandgap layer can be precisely controlled in the layer deposition process which is typically on the order of ~0.1 μm. This is several orders of magnitude thinner than the depletion region of a homojunction; therefore, very high levels of carrier concentration can be realized at a certain injection current.

In addition to providing carrier confinement, another advantage of double heterostructure is the ability of providing efficient photon confinement which is essential for initiating stimulated emission. By using a material with slightly higher refractive index for the sandwich layer, a dielectric waveguide is formed. This dielectric optical waveguide provides a mechanism to maintain spatial confinement of photons within the active layer, and therefore, very high photon density can be achieved.

3.2 LIGHT-EMITTING DIODES

Light emission in an LED is based on the spontaneous emission of a forward-biased semiconductor *pn* junction. The basic structures are shown in Fig. 3.2.1 for surface- and edge-emitting LEDs.

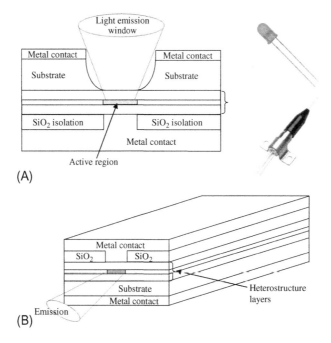

FIG. 3.2.1

Illustration of surface emission (A) and edge emission (B) LEDs.

For surface-emitting diodes, light emits in the perpendicular direction of the active layer. The brightness of the active area is usually uniform and the emission angle is isotopic, which is commonly referred to as *Lambertian*. In such a case, optical power emission patterns can be described by $P(\theta) = P_0 \cos\theta$, where θ is the angle between the emitting direction and the surface normal of the emitter, and P_0 is the optical power viewed from the direction of surface normal. By properly designing the shape of the bottom metal contact, the active emitting area can be made circular to maximize the coupling efficiency to an optical fiber.

For edge-emitting diodes, on the other hand, light emits in a direction parallel to the plane of the active layer. In this case, a waveguiding mechanism is usually required in the design, where the active layer has slightly higher refractive index than the surrounding layers, so that a higher power density in the active layer can be achieved. Compared to surface-emitting diodes, the emitting area of edge-emitting diodes is usually much smaller and asymmetric, which is determined by the width and thickness of the active layer, or the cross section of the waveguide.

3.2.1 *P~I* CURVE

For an LED, the emitting optical power P_{opt} is linearly proportional to the injected electrical current, as shown in Fig. 3.2.2. This is commonly referred to as the *P~I curve*. In the idea case, the recombination of each electron-hole generates a photon. If we define the power efficiency dP_{opt}/dI as the ratio between the emitting optical power P and the injected electrical current I, we have

$$\frac{dP_{opt}}{dI} = \frac{hv}{q} = \frac{hc}{\lambda q} \tag{3.2.1}$$

where q is the electron charge, h is the Plank's constant, c is the speed of light, v is the optical frequency, and λ is the wavelength. In a practical light-emitting device, not all recombination events are radiative, and some of them do produce photons, and thus, an internal quantum efficiency can be defined as

$$\eta_q = \frac{R_r}{R_r + R_{nr}} \tag{3.2.2}$$

where R_r and R_{nr} are the rates of radiative and nonradiative recombinations, respectively. A higher recombination rate is equivalent to a shorter carrier lifetime and thus the radiative and nonradiative carrier times are defined as $\tau_r = 1/R_r$ and $\tau_{nr} = 1/R_{nr}$, respectively, and the overall carrier lifetime is $\tau = (1/\tau_r + 1/\tau_{nr})^{-1}$. Another factor that reduces the slope of the $P~I$ curve is that not all the photons generated through radiative recombination are able to exit the active layer. Various effects contribute to this efficiency reduction, such as internal material absorption, and total interface reflection at certain regions of emission angle. This is characterized as an external efficiency defined by

$$\eta_{ext} = \frac{R_{emit}}{R_r} \tag{3.2.3}$$

where R_{emit} is the rate of the generated photons that actually exit the LED.

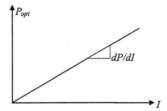

FIG. 3.2.2

LED emitting power is linearly proportional to the injection current.

Considering both internal quantum efficiency and external efficiency, the slope of the $P\sim I$ curve should be

$$dP_{opt}/dI = \eta_q \eta_{ext} \frac{hc}{\lambda q} \qquad (3.2.4)$$

Because of the linear relationship between optical power and the injection current, the emitted optical power of an LED is

$$P_{opt} = \eta_q \eta_{ext} \frac{hc}{\lambda q} I \qquad (3.2.5)$$

In general, the internal quantum efficiency of an LED can be on the order of 70%. However, since an LED is based on the spontaneous emission and the photon emission is isotropic, its external efficiency is usually less than 5%. As a rule of thumb, for $\eta_q = 75\%$, $\eta_{ext} = 2\%$, and a wavelength of $\lambda = 1550\,nm$, the output optical power efficiency is approximately $12\,\mu W/mA$.

EXAMPLE 3.1

For a surface-emitting LED shown in Fig. 3.2.3, what is the probability of a spontaneously generated photon inside the active layer to escape into the air above the LED? Assume the $n_1 = 3.5$, $n_2 = 3.45$, and $n_0 = 1$.

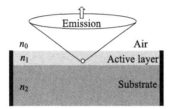

FIG. 3.2.3

Illustration of LED external efficiency reduction due to total internal reflection.

Solution

Spontaneous emission has random emission angle which fills 4π. Total internal reflection between active layer and air is $\theta = \sin^{-1}(n_0/n_1) - 0.29 rad. = 16.6°$.

The probability of escaping is $\eta = \frac{2\pi(1-\cos\theta_1)}{4\pi} = \frac{0.0418}{2} = 0.029 = 2.9\%$

Note that the critical angle on the n_1/n_2 interface is much larger than that between n_1 and n_0. Thus no reflection from n_1/n_2 interface can turn back to escape through n_1/n_0 interface.

This example indicates that the external efficiency of an LED is usually very small, mainly because of the random emission angle of spontaneous emission and a relatively narrow escape angle due to total internal reflection of the active layer.

3.2.2 MODULATION DYNAMICS

Within the active layer of an LED, the increase of carrier population is proportional to the rate of external carrier injection minus the rate of carrier recombination. Therefore, the rate equation of carrier population N_t is

$$\frac{dN_t(t)}{dt} = \frac{I(t)}{q} - \frac{N_t(t)}{\tau} \qquad (3.2.6)$$

where τ is referred to as *carrier lifetime*. In general, τ is a function of carrier density, and the rate equation does not have a closed-form solution. To simplify, if we assume τ is a constant, Eq. (3.2.6) can be easily solved in the frequency domain as

$$\widetilde{N}_t(\omega) = \frac{\widetilde{I}(\omega)\tau/q}{1+j\omega\tau} \tag{3.2.7}$$

where $\widetilde{N}_t(\omega)$ and $\widetilde{I}(\omega)$ are the Fourier transforms of $N_t(t)$ and $I(t)$, respectively. Eq. (3.2.7) demonstrates that carrier population can be modulated through injection current modulation, and the 3-dB modulation bandwidth is $B_{3dB}=1/\tau$. Because the photons are created by radiative carrier recombination, and the optical power is proportional to the carrier population, and thus the modulation bandwidth of the optical power also has the bandwidth of $1/\tau$.

$$P_{opt}(\omega) = \frac{P_{opt}(0)}{\sqrt{1+(\omega\tau)^2}} \tag{3.2.8}$$

where $P_{opt}(0)$ is the optical power at DC. Typical carrier lifetime of an LED is in the order of nanoseconds and therefore, the modulation bandwidth is in the $100\,MHz$–$1\,GHz$ level, depending on the structure of the LED.

It is worth noting that since the optical power is proportional to the injection current, the modulation bandwidth can be defined as either electrical or optical. Here comes a practical question: to fully support an LED with an optical bandwidth of B_{opt}, what electrical bandwidth B_{ele} is required for the driver circuit? Since optical power is proportional to the electrical current, $P_{opt}(\omega) \propto I(\omega)$, and the driver electrical power is proportional to the square of the injection current, $P_{ele}(\omega) \propto I^2(\omega)$, therefore, driver electrical power is proportional to the square of the LED optical power $P_{ele}(\omega) \propto P_{opt}^2(\omega)$, that is,

$$\frac{P_{ele}(\omega)}{P_{ele}(0)} = \frac{P_{opt}^2(\omega)}{P_{opt}^2(0)} \tag{3.2.9}$$

If at a frequency the optical power is reduced by $3\,dB$ compared to its DC value, the driver electrical power is supposed to be reduced by only $1.5\,dB$ at that frequency. That is, $3\,dB$ electrical bandwidth is equivalent to $6\,dB$ optical bandwidth.

EXAMPLE 3.2

Consider an LED emitting at $\lambda=1550\,nm$ wavelength window. The internal quantum efficiency is 70%, the external efficiency is 2%, the carrier lifetime is $\tau=20\,ns$, and the injection current is $20\,mA$. Find:
 (a) the output optical power of the LED, (b) the $3\,dB$ optical bandwidth, and (c) the required driver electrical bandwidth

Solution
(a) Output optical power is

$$P_{opt} = \eta_q\eta_{ext}\frac{hc}{\lambda q}I = 0.7 \times 0.02 \times \frac{6.63 \times 10^{-34} \times 3 \times 10^8 \times 20 \times 10^{-3}}{1550 \times 10^{-9} \times 1.6 \times 10^{-19}} = 0.225\,mW$$

(b) To find the $3\,dB$ optical bandwidth, we use

$$P_{opt}(\omega) = \frac{P_{opt}(0)}{\sqrt{1+\omega^2\tau^2}}$$

at $\omega=\omega_{3dB,opt}$, $\dfrac{1}{\sqrt{1+\omega_{3dB,opt}^2\tau^2}} = \dfrac{1}{2}$, that is, $\omega_{3dB,opt}^2\tau^2 = 3$.

Therefore, the angular frequency of optical bandwidth is $\omega_{3dB,opt} = \sqrt{3}/\tau = 86.6$ Mrad/s, which corresponds to a circular frequency $f_{3dB,opt} = \sqrt{3}/(2\pi\tau) \approx 13.8$ MHz.

(c) To find the 3 dB electric bandwidth, we use

$$P_{ele}(\omega) = |P_{opt}(\omega)|^2 = \frac{P_{opt}^2(0)}{|1 + \omega^2\tau^2|}$$

at $\omega = \omega_{3dB, ele}$, $\frac{1}{1 + \omega_{3dB,ele}^2\tau^2} = \frac{1}{2}$, Therefore, $\omega_{3dB, ele} = 1/\tau = 50$ Mrad/s and $f_{3dB, ele} \approx 8$ MHz.

The relation between electrical and optical bandwidth is

$$\omega_{3dB,opt}/\omega_{3dB,ele} = \sqrt{3}$$

3.3 LASER DIODES

Semiconductor LDs are based on the *stimulated* emission of forward-biased semiconductor pn junction. Compared to LEDs, LDs have higher spectral purity and higher external efficiency because of the spectral and spatial coherence of stimulated emission.

3.3.1 AMPLITUDE AND PHASE CONDITIONS FOR SELF-SUSTAINED OSCILLATION

One of the basic requirements of LDs is optical feedback. Consider an optical cavity of length L as shown in Fig. 3.3.1, where the semiconductor material in the cavity provides an optical gain g and an optical loss α per unit length, the refractive index of the material in the cavity is n, and the reflectivity of the facets is R_1 and R_2.

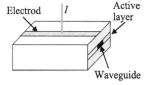
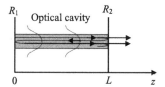

FIG. 3.3.1

A laser cavity with facet reflectivity R_1 and R_2, optical gain g, optical loss α, and refractive index n.

The lightwave travels back and forth in the longitudinal ($\pm z$) direction in the cavity. The ratio of optical field before and after each roundtrip in the cavity is

$$\frac{E_{i+1}}{E_i} = \sqrt{R_1}\sqrt{R_2}\exp(\Delta G + j\Delta\Phi) \tag{3.3.1}$$

where the total roundtrip phase shift of the optical field is

$$\Delta\Phi = \frac{2\pi}{\lambda}2nL \tag{3.3.2}$$

and the net roundtrip optical gain coefficient of the optical field is

$$\Delta G = (\Gamma g - \alpha)2L \tag{3.3.3}$$

where g is the optical gain coefficient in $[cm^{-1}]$ and α is the material loss coefficient, also in $[cm^{-1}]$. $0 < \Gamma < 1$ is a confinement factor. Since not all the optical field is confined within the active region of the waveguide, Γ is defined as the ratio between the optical field in the active region and the total optical field.

To support a self-sustained oscillation, the optical field has to repeat itself after each roundtrip, and thus,

$$\sqrt{R_1}\sqrt{R_2}\exp(\Delta G + j\Delta\Phi) = 1 \tag{3.3.4}$$

This is a necessary condition of oscillation, which is also commonly referred to as the *threshold condition*. Eq. (3.3.4) can be further decomposed into a phase condition and a threshold gain condition.

The phase condition is that after each roundtrip the optical phase change must be multiples of 2π, that is, $\Delta\Phi = 2m\pi$, where m is an integer. One important implication of this phase condition is that it can be satisfied by multiple wavelengths,

$$\lambda_m = \frac{2nL}{m} \tag{3.3.5}$$

This explains the reason that a laser may emit at multiple wavelengths, which are generally referred to as *multiple longitudinal modes*.

EXAMPLE 3.3

For an InGaAsP semiconductor laser operating in 1550 nm wavelength window, if the effective refractive index of the waveguide is $n \approx 3.5$ and the laser cavity length is $L = 300$ μm, find the wavelength spacing between adjacent longitudinal modes.

Solution

Based on Eq. (3.3.5), the wavelength spacing between the mth mode and the $(m+1)$th modes can be found as $\Delta\lambda \approx \lambda_m^2/(2nL)$. Assume $\lambda_m = 1550$ nm, this mode spacing is $\Delta\lambda = 1.144$ nm, which corresponds to a frequency separation of approximately $\Delta f = 143$ GHz.

The threshold gain condition is that after each roundtrip the amplitude of optical field does not change—that is, $\sqrt{R_1}\sqrt{R_2}\exp\{(\Gamma g_{th} - \alpha)2L\} = 1$, where g_{th} is the optical field gain at threshold. Therefore, in order to achieve the lasing threshold, the optical gain has to be high enough to compensate for both the material attenuation and the optical loss at the mirrors,

$$\Gamma g_{th} = \alpha - \frac{1}{4L}\ln(R_1R_2) \tag{3.3.6}$$

Inside a semiconductor laser cavity, the optical field gain coefficient is a function of the carrier density, which depends on the rate of carrier injection,

$$g(N) = a(N - N_0) \tag{3.3.7}$$

In this expression, N is the carrier density in $[cm^{-3}]$ and N_0 is the carrier density required to achieve material transparency. a is the differential gain coefficient in $[cm^2]$; it indicates the gain per unit length along the laser cavity per unit carrier density.

In addition, due to the Fermi-Dirac distribution of carriers and the limited width of the energy bands, the differential gain is a function of the wavelength, which can be approximated as parabolic,

$$a(\lambda) = a_0 \left\{ 1 - \left(\frac{\lambda - \lambda_0}{\Delta \lambda_g} \right)^2 \right\} \tag{3.3.8}$$

for $|\lambda - \lambda_0| << \Delta \lambda_g$, where a_0 is the differential gain at the central wavelength $\lambda = \lambda_0$ and $\Delta \lambda_g$ is the spectral bandwidth of the material gain.

It must be noted that material gain coefficient g is not equal to the actual gain of the optical field. When an optical field travels along an active waveguide, the field propagation is described as

$$E(z, t) = E_0 e^{(\Gamma g - \alpha)z} e^{-j(\omega t - \beta z)} \tag{3.3.9}$$

where E_0 is the optical field at $z = 0$, ω is the optical frequency, and $\beta = 2\pi n/\lambda$ is the propagation constant. The envelop of optical power along the waveguide is then

$$P(z) = |E(z)|^2 = P_0 e^{2(\Gamma g - \alpha)z} \tag{3.3.10}$$

Combine Eqs. (3.3.7), (3.3.8), and (3.3.10), we have

$$P(z, \lambda) = |E(z)|^2 = P(z, \lambda_0) \exp \left\{ -2\Gamma g_0 \left(\frac{\lambda - \lambda_0}{\Delta \lambda_g} \right)^2 z \right\} \tag{3.3.11}$$

where $P(z, \lambda_0) = P(0, \lambda_0) e^{2(\Gamma g_0 - \alpha)z}$ is the peak optical power at z, and $g_0 = a_0(N - N_0)$ is the peak gain coefficient.

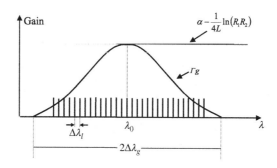

FIG. 3.3.2

Gain and loss profile. Vertical bars show the wavelengths of longitudinal modes.

As shown in Fig. 3.3.2, although there are many potential longitudinal modes that all satisfy the phase condition, the threshold gain condition can be reached only by a limited number of modes near the center of the gain peak.

3.3.2 RATE EQUATIONS

Rate equations describe the nature of interactions between photons and carriers in the active region of a semiconductor laser. Useful characteristics such as output optical power vs. injection current, modulation response, and spontaneous emission noise can be found by solving the rate equations.

$$\frac{dN(t)}{dt} = \frac{J}{qd} - \frac{N(t)}{\tau} - 2\Gamma v_g a(N - N_0)P(t) \tag{3.3.12}$$

$$\frac{dP(t)}{dt} = 2\Gamma v_g a(N - N_0)P(t) - \frac{P(t)}{\tau_{ph}} + R_{sp} \tag{3.3.13}$$

where $N(t)$ is the carrier density and $P(t)$ is the photon density within the laser cavity, they have the same unit [cm^{-3}]. J is the injection current density in [C/cm^2], d is the thickness of the active layer, v_g is the group velocity of the lightwave in [cm/s], and τ and τ_{ph} are the electron and photon lifetimes, respectively. R_{sp} is the rate of spontaneous emission; it represents the density of spontaneously generated photons per second that are coupled into the lasing mode, so the unit of R_{sp} is [cm^{-3}s^{-1}].

On the right-hand side of Eq. (3.3.12), the first term is the number of electrons injected into each cubic meter within each second time window; the second term is the electron density reduction per second due to spontaneous recombination; and the third term represents electron density reduction rate due to stimulated recombination, which is proportional to both material gain and the photon density.

The same term $2\Gamma v_g a(N - N_0)P(t)$ also appears in Eq. (3.3.13) due to the fact that each stimulated recombination event will generate a photon, and therefore, the first term on the right-hand side of Eq. (3.3.13) is the rate of photon density increase due to stimulated emission. The second term on the right-hand side of Eq. (3.3.13) is the photon density decay rate due to both material absorption and photon leakage from the two mirrors. The escape of photons through mirrors can be treated equivalently as a distributed loss in the cavity with the same unit, [cm^{-1}], as the material attenuation. The equivalent mirror loss coefficient is defined as

$$\alpha_m = \frac{-\ln(R_1 R_2)}{4L} \tag{3.3.14}$$

In this way, the photon lifetime can be expressed as

$$\tau_{ph} = \frac{1}{2v_g(\alpha + \alpha_m)} \tag{3.3.15}$$

where α is the material attenuation coefficient. Using this photon lifetime expression, the photon density rate Eq. (3.3.13) can be simplified as

$$\frac{dP(t)}{dt} = 2v_g[\Gamma g - (\alpha + \alpha_m)]P(t) + R_{sp} \tag{3.3.16}$$

where g is the material gain as defined in Eq. (3.3.7).

Rate Eqs. (3.3.12) and (3.3.13) are coupled differential equations describing the interaction between the carrier density and photon density, generally they can be solved numerically to predict static as well as dynamic behaviors of a semiconductor laser.

3.3.3 STEADY STATE SOLUTIONS OF RATE EQUATIONS

In the steady state, $d/dt = 0$, rate Eqs. (3.3.12) and (3.3.13) can be simplified as

$$\frac{J}{qd} - \frac{N}{\tau} - 2\Gamma v_g a(N - N_0)P = 0 \tag{3.3.17}$$

$$2\Gamma v_g a(N - N_0)P - \frac{P}{\tau_{ph}} + R_{sp} = 0 \tag{3.3.18}$$

With this simplification, the equations can be solved analytically, which will help understand some basic characteristics of semiconductor lasers.

(A) Threshold carrier density and current density

Assume that R_{sp}, τ_{ph}, and α are constants. Eq. (3.3.18) can be expressed as

$$P = \frac{R_{sp}}{1/\tau_{ph} - 2\Gamma v_g a(N - N_0)} \tag{3.3.19}$$

Eq. (3.3.19) indicates that when $2\Gamma v_g a(N - N_0)$ approaches the value of $1/\tau_{ph}$, the photon density would approach infinite, and this operation point is defined as the *threshold*. Therefore, the threshold carrier density can be found as

$$N_{th} = N_0 + \frac{1}{2\Gamma v_g a \tau_{ph}} \tag{3.3.20}$$

In order to ensure a positive value of the photon density, $2\Gamma v_g a(N - N_0) < 1/\tau_{ph}$ is necessary, which requires $N < N_{th}$. Practically, carrier density N can be increased to approach the threshold carrier density N_{th} by increasing the injection current density. However, the level of threshold carrier density, N_{th}, can never be reached. With the increase of carrier density, photon density will be increased. Especially when the carrier density approaches the threshold level, photon density will be increased dramatically and the stimulated recombination becomes significant, which, in turn, reduces the carrier density. Fig. 3.3.3 illustrates the relationships among carrier density N, photon density P, and the injection current density J.

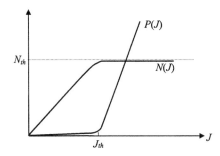

FIG. 3.3.3

Photon density $P(J)$ and carrier density $N(J)$ as functions of injection current density J. J_{th} is the threshold current density and N_{th} is the threshold carrier density.

As shown in Fig. 3.3.3, for a semiconductor laser, carrier density linearly increases with the increase of injection current density to a certain level. After that level, the carrier density increase is quickly saturated due to the significant contribution of stimulated recombination. The current density corresponding to that saturation point is called *threshold current density*, above which the laser output is dominated by stimulated emission. However, below that threshold point spontaneous recombination and emission is the dominant mechanism similar to that in an LED, and the output optical power is usually small because of the low external efficiency. For a LD operating below threshold, stimulated recombination is negligible, and Eq. (3.3.17) can be simplified as $J/qd = N/\tau$. Assume this relation is still valid at the threshold point, $N = N_{th}$, the threshold current density can be found, by considering the threshold carrier density definition (3.3.20), as

$$J_{th} = \frac{qd}{\tau} N_{th} = \frac{qd}{\tau}\left(N_0 + \frac{1}{2\Gamma v_g a \tau_{ph}}\right) \tag{3.3.21}$$

(B) P~J relationship about threshold

In general, the desired operation region of a LD is well above the threshold, where high-power coherent light is generated by stimulated emission. Combining Eqs. (3.3.17) and (3.3.18), we have

$$\frac{J}{qd} = \frac{N}{\tau} + \frac{P}{\tau_{ph}} - R_{sp} \tag{3.3.22}$$

As shown in Fig. 3.3.3, in the operation region well above threshold, carrier density is approximately equal to its threshold value ($N \approx N_{th}$). In addition, since $N_{th}/\tau = J_{th}/qd$, we have

$$\frac{J}{qd} = \frac{J_{th}}{qd} + \frac{P}{\tau_{ph}} - R_{sp} \tag{3.3.23}$$

Thus, the relationship between photon density and current density above threshold is

$$P = \frac{\tau_{ph}}{qd}(J - J_{th}) + \tau_{ph}R_{sp} \tag{3.3.24}$$

Apart from a spontaneous emission contribution term $\tau_{ph}R_{sp}$, which is usually very small, the photon density is linearly proportional to the injection current density for $J > J_{th}$ and the slop is $dP/dJ = \tau_{ph}/qd$.

Then, a question is how to relate the photon density inside the laser cavity to the output optical power of the laser? Assume that the waveguide of the laser cavity has a length l, width w, and thickness d, as shown in Fig. 3.3.4.

FIG. 3.3.4

Illustration of the dimension of a laser cavity.

The output optical power is the flow of photons through the facet, which can be expressed as

$$P_{opt} = P \cdot (lwd)hv2\alpha_m v_g \tag{3.3.25}$$

where $P \cdot (lwd)$ is the total photon number and $P \cdot (lwd)hv$ is the total photon energy within the cavity. α_m is the mirror loss in $[\text{cm}^{-1}]$, which is the percentage of photons that escape from each mirror, and $\alpha_m v_g$ represents the percentage of photon escape per second. The factor 2 indicates that photons travel in both directions along the cavity and escape through two end mirrors.

Neglecting the contribution of spontaneous emission in Eq. (3.3.24), combining Eqs. (3.3.25) and (3.3.15), with Eq. (3.3.24), and considering that the injection current is related to current density by $I = J \cdot wl$, we have

$$P_{opt} = \frac{(I - I_{th})hv}{q} \cdot \frac{\alpha_m}{\alpha_m + \alpha} \tag{3.3.26}$$

This is the total output optical power exiting from both laser end facets. Since α is the rate of material absorption and α_m is the photon escape rate through facet mirrors, $\alpha_m/(\alpha + \alpha_m)$ represents the external efficiency.

3.3.4 SIDE-MODE SUPPRESSION RATIO

As illustrated in Fig. 3.3.2, the phase condition in a LD can be satisfied by multiple wavelengths, which are commonly referred to as *multiple longitudinal modes*. The gain profile has a parabolic shape with a maximum in the middle, and one of the longitudinal modes closest to the material gain peak usually has the highest power, which is the *main mode*. However, the power in the modes adjacent to the main mode may not be negligible for many applications that require single-mode operation. To take into account the multimodal effect in a LD, the photon density rate equation of the mth longitudinal mode can be written as

$$\frac{dP_m(t)}{dt} = 2v_g \Gamma g_m(N) P_m(t) - \frac{P_m(t)}{\tau_{ph}} + R_{sp} \tag{3.3.27}$$

where $g_m(N) = a(N - N_0)$ is the optical field gain for the mth mode. Since all the longitudinal modes share the same pool of carrier density, the rate equation for the carrier density is

$$\frac{dN(t)}{dt} = \frac{J}{qd} - \frac{N(t)}{\tau} - \sum_k 2\Gamma_k v_g g_k(N) P_k(t) \tag{3.3.28}$$

Using a parabolic approximation for the material gain,

$$g(N, \lambda) = g_0(N) \left\{ 1 - \left[(\lambda - \lambda_0)/\Delta\lambda_g \right]^2 \right\}$$

and letting $\lambda_m = \lambda_0 + m\Delta\lambda_l$, where $\Delta\lambda_l$ is the mode spacing as shown in Fig. 3.3.2, there will be approximately $2M + 1$ modes if there are M modes on each side of the main mode $(-M < m < M)$, where $M \approx \Delta\lambda_g/\Delta\lambda_l$.

Therefore, the optical field gain for the mth mode can be expressed as a function of the mode index m as

$$g_m(N) = g_0(N)\left\{1 - \left(\frac{m}{M}\right)^2\right\} \tag{3.3.29}$$

The steady state solution of the photon density rate equation of the mth mode is

$$P_m = \frac{R_{sp}}{1/\tau_{ph} - 2\Gamma v_g g_m(N)} \tag{3.3.30}$$

and the photon density of the main mode ($m=0$) is

$$P_0 = \frac{R_{sp}}{1/\tau_{ph} - 2\Gamma v_g g_0(N)} \tag{3.3.31}$$

The power ratio between the main mode and the mth mode can then be found as

$$SMR = \frac{P_0}{P_M} = 1 + \frac{P_0}{R_{sp}} 2\Gamma v_g g_0(N)\left(m^2/M^2\right) \tag{3.3.32}$$

Eq. (3.3.32) indicates that the side-mode suppression ratio (SMSR) is proportional to m^2 because high-index modes are far away from the main mode and the gain is much lower than the threshold gain. In addition, the SMSR is proportional to the photon density of the main mode. The reason is that at a high photon density level, stimulated emission is predominantly higher than the spontaneous emission; thus side modes that benefited from spontaneous emission become weaker compared to the main mode.

Although SMSR introduced here is from LDs with Fabry-Perot (FP) cavities, the definition is valid also for other types of diode lasers such as DFB and DBR cavities.

3.3.5 MODULATION RESPONSE

Electro-optic modulation is an important functionality in an optical transmitter, which translates electrical signals into the optical domain. The optical power of a LD is a function of the injection current, and thus a convenient way to convert an electrical signal into an optical signal is through the direct modulation of the injection current. Both on-off modulation and linear modulation can be performed. The characteristic of LD under direct current modulation is a practical issue for applications in optical communication systems.

Fig. 3.3.5 illustrates the operating principle of direct intensity modulation of a semiconductor laser. To ensure that the LD operates above threshold, a DC bias current I_B is usually required. A bias-Tee combines the electrical current signal with the DC bias current to modulate the LD. The modulation efficiency is then determined by the slope of the LD P–I curve. Obviously, if the P–I curve is ideally linear, the output optical power is linearly proportional to the modulating current by

$$P_{opt}(t) \approx R_C(I_B - I_{th}) + R_C I(t) \tag{3.3.33}$$

where I_{th} is the threshold current of the laser and $R_C = \Delta P_{opt}/\Delta I$ is the slope of the LD P–I curve. Here, we have neglected the optical power level at threshold.

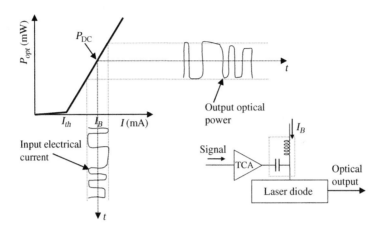

FIG. 3.3.5

Direct intensity modulation of a laser diode. *TCA*: transconductance amplifier.

Frequency response of direct modulation mainly depends on the carrier dynamics of the LD and >20 GHz modulation bandwidth has been demonstrated. However, further increasing the modulation bandwidth to 40 GHz appears to be quite challenging, mainly limited by the carrier lifetime as well as the parasitic effect of the electrode. Another well-known property of direct modulation in a semiconductor laser is the associated frequency modulation, commonly referred to as *frequency chirp*, as discussed later in this section.

(A) Turn-on delay

In directly modulated LDs, when the injection current is suddenly switched on from below to above the threshold, there is a time delay between the signal electrical pulse and the generated optical pulse. This is commonly referred to as *turn-on delay*, which is mainly caused by the slow response of the carrier density below threshold. It needs a certain period of time for the carrier density to build up and to reach the threshold level.

To analyze this process, we have to start from the rate equation at the low injection level J_B below threshold, where photon density is very small and the stimulated recombination term is negligible in Eq. (3.3.12):

$$\frac{dN}{dt} = \frac{J(t)}{qd} - \frac{N(t)}{\tau} \tag{3.3.34}$$

Suppose $J(t)$ switches on from J_B to J_2 at time $t=0$. If J_B is below and J_2 is above the threshold current level, the carrier density is supposed to be switched from a level N_1 far below the threshold to the threshold level N_{th}, as shown in Fig. 3.3.6. Eq. (3.3.34) can be integrated to find the time required for the carrier density to increase from N_B to N_{th}:

$$t_d = \int_{N_B}^{N_{th}} \left[\frac{J(t)}{qd} - \frac{N(t)}{\tau} \right]^{-1} dN = \tau \ln\left[\frac{J - J_B}{J - J_{th}} \right] = \tau \ln\left[\frac{I - I_B}{I - I_{th}} \right] \tag{3.3.35}$$

where $J_B = qdN_B/\tau$ and $J_{th} = qdN_{th}/\tau$. Because laser operates above threshold only for $t \geq t_d$, the actual starting time of the laser output optical pulse is at $t = t_d$. In practical applications, this time delay t_d may limit the speed of digital optical modulation in optical systems. Since t_d is proportional to τ, a LD with shorter spontaneous emission carrier lifetime may help reduce the turn-on delay. Another way to reduce turn-on delay is to bias the low-level injection current J_B very close to the threshold. However, this may result in poor extinction ratio of the output optical pulse.

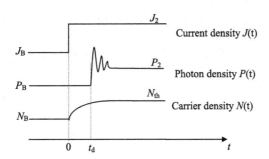

FIG. 3.3.6

Illustration of the laser turn-delay. Injection current is turned on at $t=0$ from J_B to J_2, but both photon density and carrier density will require a certain time delay to build up toward their final values.

(B) Small-signal modulation response

In Section 3.2.2, we show that the modulation speed of an LED is inversely proportional to the carrier lifetime. For a LD operating above threshold, the modulation speed is expected to be much faster than that of an LED thanks to the contribution of stimulated recombination. In fact when a LD is modulated by a small current signal $\delta J(t)$ around a bias point J_B: $J = J_B + \delta J(t)$, the carrier density will be $N = N_B + \delta N(t)$, where N_B and $\delta N(t)$ are the static and small-signal response, respectively, of the carrier density. Rate Eq. (3.3.12) can be linearized for the small-signal response as

$$\frac{d\delta N(t)}{dt} = \frac{\delta J(t)}{qd} - \frac{\delta N(t)}{\tau} - 2\Gamma v_g a P \delta N(t) \qquad (3.3.36)$$

Here for simplicity, we have assumed that the impact of photon density modulation is negligible. Eq. (3.3.36) can be easily solved in frequency domain as

$$\delta \widetilde{N}(\omega) = \frac{1}{qd \left(j\omega + 1/\tau + 2\Gamma v_g a P\right)} \delta \widetilde{J}(\omega) \qquad (3.3.37)$$

where $\delta \widetilde{J}(\omega)$ and $\delta \widetilde{N}(\omega)$ are the Fourier transforms of $\delta J(t)$ and $\delta N(t)$, respectively. If we define an effective carrier lifetime as

$$\tau_{eff} = \left(\frac{1}{\tau} + 2\Gamma v_g a P\right)^{-1} \qquad (3.3.38)$$

the 3-dB modulation bandwidth of the laser will be $B_{3dB} = 1/\tau_{eff}$. For a LD operating well above threshold, stimulated recombination is much stronger than spontaneous recombination, that is, $2\Gamma v_g a P \gg 1/\tau$, and therefore, $\tau_{eff} < < \tau$. This is the major reason that the modulation speed of a LD can be much faster than that of an LED.

In this simplified modulation response analysis, we have assumed that photon density is a constant, and thus there is no coupling between the carrier density rate equation and the photon density rate equation. A more precise analysis has to solve coupled rate Eqs. (3.3.12) and (3.3.13). A direct consequence of coupling between the carrier density and the photon density is that for an increase of injection current, the carrier density will first increase, which will increase the photon density. But the photon density increase tends to reduce carrier density through stimulated recombination. Therefore, there could be an oscillation of both carrier density and photon density immediately after a change of the injection current. This is commonly referred to as *relaxation oscillation*. Detailed analysis of laser modulation can be found in (Agrawal and Dutta, 1986).

(C) Modulation chirp

For semiconductor materials, the refractive index is usually a function of the carrier density, and thus the emission wavelength of a semiconductor laser often depends on the injection current. In addition, a direct current modulation on a LD may introduce both an intensity modulation and a phase modulation of the optical signal, where the phase modulation is originated from the carrier density-dependent refractive index of the material within the laser cavity. The ratio between the phase modulation and the intensity modulation is referred to as the modulation *chirp*.

Carrier-dependent refractive index in a semiconductor laser is the origin of both adiabatic chirp and transient chirp. The adiabatic chirp is caused by the change of phase condition of laser cavity caused by the refractive index change of semiconductor material inside the laser cavity. As refractive index is a parameter of the laser phase condition in Eq. (3.3.5), any change of refractive index will change the resonance wavelength of the laser cavity. Based on Eq. (3.3.5), the wavelength of the mth mode is $\lambda_m = 2nL/m$, and a small index change $n \Rightarrow n + \delta n$ will introduce a wavelength change $\lambda_m \Rightarrow \lambda_m + \delta\lambda$. The linearized relation between $\delta\lambda$ and δn is $\delta\lambda/\lambda_m = \delta n/n$. In a practical LD based on InGaAs, a 1-mA change in the injection current would introduce an optical frequency change on the order of 1 GHz, or 8 pm in 1550 nm wavelength window. This effect is often utilized for laser frequency adjustment and stabilization through feedback control of the injection current.

Transient chirp, on the other hand, is a dynamic process of carrier-induced optical phase modulation. As the optical field is related to the photon density and the optical phase,

$$E(t) = \sqrt{P(t)}e^{j\phi(t)} = \exp\left[j\phi(t) + 0.5 \ln P(t)\right] \qquad (3.3.39)$$

The ratio between phase modulation and amplitude modulation around a static bias point P_B can be defined as

$$\alpha_{lw} = \frac{d\phi(t)/dt}{d[0.5 \ln P(t)]/dt} = 2P_B \frac{d\phi(t)/dt}{dP(t)/dt} \qquad (3.3.40)$$

α_{lw} is the chirp parameter of the laser, which is identical to a well-known linewidth enhancement factor first introduced by Henry (1982), as the spectral linewidth of the LD is also related to α_{lw} which will be discussed in the following section.

The derivative of optical phase modulation through transient chirping is equivalent to an optical frequency modulation. This optical frequency modulation is linearly proportional to the chirp parameter α_{lw} as

$$\delta f = \frac{1}{2\pi}\frac{d\phi(t)}{dt} = \frac{1}{2\pi}\frac{\alpha_{lw}}{2P_B}\frac{dP(t)}{dt} \tag{3.3.41}$$

α_{lw} is an important parameter of a LD, which is determined both by the semiconductor material and laser cavity structure. For intensity modulation-based optical systems, lasers with smaller chirp parameters are desired to minimize the spectral width of the modulated optical signal. On the other hand, for optical frequency modulation-based systems such as frequency-shift key (FSK), lasers with large chirp will be beneficial for an increased optical frequency modulation efficiency.

3.3.6 LASER NOISES

When biased by a constant current, an ideal LD would produce a constant optical power at a single optical frequency. However, a practical LD is not an ideal optical oscillator, which has both intensity noise and phase noise even when driven by an ideally constant injection current. Both intensity noise and phase noise may significantly affect the performance of an optical communication system which uses LD as the optical source.

(A) Relative intensity noise

For a semiconductor laser operating above threshold, although stimulated emission dominates the emission process, there are still a small percentage of photons that are generated by spontaneous emission. The optical intensity noise in a LD is primarily caused by the incoherent nature of these spontaneous photons. As a result, even when a LD is biased by a constant injection current, the effect of spontaneous emission makes both carrier density and photon density fluctuate around their equilibrium values (Agrawal and Dutta 1986).

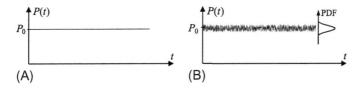

FIG. 3.3.7

Illustration of optical power emitted a CW-operated laser diode without (A), and with relative intensity noise (B)

In general, the intensity noise in a LD caused by spontaneous emission is random in time as illustrated in Fig. 3.3.7, and thus wideband in the frequency domain. Intensity noise is an important measure of a LD quality, which is often a limiting factor for its application in optical communications. Instead of the absolute value of the intensity noise, the ratio between the intensity noise power and the average optical power is a commonly used parameter for laser specification, known as the relative intensity noise (RIN), which is defined as

$$RIN(\omega) = F\left\{\frac{\langle (P(t) - P_{ave})^2 \rangle}{P_{ave}^2}\right\} \qquad (3.3.42)$$

where P_{ave} is the average optical power, $\langle (P(t) - P_{ave})^2 \rangle$ is the mean square intensity fluctuation (variance), and F denotes the Fourier transform. Note that this definition is based on the electrical domain parameters measured by a photodiode in a setup shown in Fig. 3.3.8. As will be discussed in the following chapter, the photocurrent of a photodiode is proportional to the input optical power, the electrical power produced by the photodiode should be proportional to the square of the optical power. The Fourier transform of $\langle (P(t) - P_{ave})^2 \rangle$ is the electrical noise power spectral density in [W/Hz], and P_{ave}^2 is proportional to the average electrical power in [W] created in the photodiode corresponding to the spectral density at DC. Thus, 20 dB of RIN defined in Eq. (3.3.42) is equivalent to 10 dB of actual ratio in the optical domain. The square-law photodetection will be discussed in Chapter 4.

In a practical measurement system, optical loss between the laser and the photodiode, and the gain of the electrical amplifier after the photodiode may not be calibrated. But this does not affect the accuracy of RIN measurement as a ratio between the noise and the signal. This makes RIN a commonly accepted parameter for laser quality specification which is independent of the optical power that comes into the receiver. The unit of RIN is [Hz^{-1}], [dB/Hz].

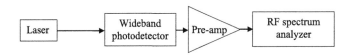

FIG. 3.3.8

System block diagram to measure laser RIN.

For a Gaussian statics of the RIN, the standard deviation of the laser power fluctuation can be found by integrating the RIN spectrum over the electrical bandwidth. Assume a flat spectral density of $RIN = -120$ dB/Hz over the electrical bandwidth $B - 10$ GHz, the variance of power fluctuation is $\sigma^2 = 10^{-2} P_{ave}^2$. In this case, the optical power fluctuation (standard deviation) is roughly 10% of the average power.

In a practical diode laser, the intensity noise can be amplified by the resonance of the relaxation oscillation in the laser cavity, which modifies the noise spectral density. Fig. 3.3.9 shows that each single event of spontaneous emission will increase the photon density in the laser cavity. Meanwhile this increased photon density will consume more carriers in the cavity and create a gain saturation in the medium. As a result, the photon density will tend to be decreased due to the reduced optical gain. This, in turn, will increase the carrier density due to the reduced saturation effect. This resonance process is strong near a specific frequency Ω_p which is determined by the optical gain G, the differential gain dG/dN, and the photon density P in the laser cavity as

$$\Omega_R^2 = G \cdot P \cdot \frac{dG}{dN} \qquad (3.3.43)$$

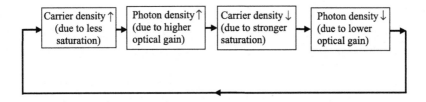

FIG. 3.3.9

Explanation of a relaxation oscillation process in a semiconductor laser.

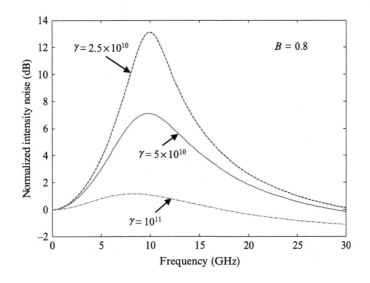

FIG. 3.3.10

Normalized intensity noise spectral density with 10 GHz relaxation oscillation frequency.

Due to relaxation oscillation, the intensity noise of a semiconductor laser becomes frequency dependent, and the normalized intensity noise spectral density can be fit by

$$H(\Omega) \propto \left| \frac{\Omega_R^2 + B\Omega^2}{j\Omega(j\Omega + \gamma) + \Omega_R^2} \right|^2 \tag{3.3.44}$$

where B and γ are damping parameters depending on the specific laser structure and the bias condition. Fig. 3.3.10 shows examples of the normalized intensity noise spectral density with three different damping parameters. The relaxation oscillation frequency used in this figure is $\Omega_R = 2\pi \times 10$ GHz. At frequencies much higher than the relaxation oscillation frequency, the dynamic coupling between the photon density and the carrier density is weak and therefore, the intensity noise becomes less dependent on the frequency.

(B) Phase noise

Phase noise is a measure of random phase variation of an optical signal which determines the spectral purity of the laser beam. Phase noise is an important issue in diode lasers, especially when they are used in coherent optical communication systems where the signal optical phase is utilized to carry information. Although stimulated emission dominates in a LD operating above threshold, a small amount of spontaneous emission still exists. While photons of stimulated emission are mutually coherent, spontaneous emission is not coherent. Fig. 3.3.11 shows that a spontaneous emission event not only generates intensity variation but also produces phase variation. To make things worse, the material refractive index inside a semiconductor laser cavity is a function of the photon density, so that the phase noise is further enhanced by the intensity noise, which makes the phase noise of a semiconductor laser much higher than other types of lasers (Henry, 1986).

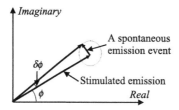

FIG. 3.3.11

Optical field vector diagram. Illustration of optical phase noise generated due to spontaneous emission events.

Assume that the optical field in a laser cavity is

$$E(t) = \sqrt{P(t)} \exp\{-j[\omega_0 t + \varphi(t)]\} \tag{3.3.45}$$

where $P(t)$ is the photon density inside the laser cavity, ω_0 is the central optical frequency, and $\phi(t)$ is the time-varying part of optical phase. A differential rate equation that describes the phase variation in the time domain is (Henry, 1983)

$$\frac{d\varphi(t)}{dt} = F_\varphi(t) - \frac{\alpha_{lw}}{2P} F_P(t) \tag{3.3.46}$$

where α_{lw} is the well-known linewidth enhancement factor of a semiconductor laser, which accounts for the coupling between intensity and phase variations. $F_\phi(t)$ and $F_P(t)$ are the Langevin noise terms for phase and intensity. They are random and their statistic measures are

$$\left\langle F_P(t)^2 \right\rangle = 2R_{sp}P \tag{3.3.47}$$

and

$$\left\langle F_\varphi(t)^2 \right\rangle = \frac{R_{sp}}{2P} \tag{3.3.48}$$

R_{sp} is the spontaneous emission factor of the laser.

Although we directly use Eq. (3.3.46) without derivation, the physical meanings of the two terms on the right-hand side of the equation are very clear. The first term is created directly due to spontaneous emission contribution to the phase variation. Each spontaneous emission event randomly emits a photon which changes the optical phase as illustrated in Fig. 3.3.11. The second term in Eq. (3.3.46) shows that each spontaneous emission event randomly emits a photon that changes the carrier density, and this carrier density variation, in turn, changes the refractive index of the material. Then, this index change will alter the resonance condition of the laser cavity and thus will introduce a phase change of the emitting optical field. Eq. (3.3.46) can be solved by integration:

$$\varphi(t) = \int_0^t \frac{d\varphi(t)}{dt} dt = \int_0^t F_\varphi(t) dt - \frac{\alpha_{lw}}{2P} \int_0^t F_P(t) dt \qquad (3.3.49)$$

If we take an ensemble average, the variance of phase noise can be expressed as

$$\left\langle \varphi(t)^2 \right\rangle = \left\langle \left[\int_0^t F_\varphi(t) dt - \frac{\alpha_{lw}}{2P} \int_0^t F_P(t) dt \right]^2 \right\rangle = \int_0^t \left\langle F_\varphi(t)^2 \right\rangle dt - \left(\frac{\alpha_{lw}}{2P} \right)^2 \int_0^t \left\langle F_P(t)^2 \right\rangle dt$$

$$= \left[\frac{R_{sp}}{2P} + \left(\frac{\alpha_{lw}}{2P} \right)^2 2R_{sp}P \right] |t| = \frac{R_{sp}}{2P} \left(1 + \alpha_{lw}^2 \right) |t| \qquad (3.3.50)$$

Since $\varphi(t)$ is a random Gaussian process

$$\left\langle e^{j\varphi(t)} \right\rangle = e^{-\frac{1}{2}\left\langle \varphi(t)^2 \right\rangle} \qquad (3.3.51)$$

According to the Wiener-Khintchine theorem, the power spectral density of the optical field is proportional to the square of the modulus of the Fourier transformation of the optical field,

$$S_{op}(\omega) = \left| \int_{-\infty}^{\infty} \langle E(t) \rangle e^{-j\omega t} dt \right|^2 = P \left| \int_{-\infty}^{\infty} e^{-j(\omega-\omega_0)t} e^{-\frac{1}{2}\langle \varphi(t)^2 \rangle} dt \right|^2$$

$$= P \left| \int_{-\infty}^{\infty} \exp\left[-j(\omega-\omega_0)t - \frac{R_{sp}}{4P}\left(1+\alpha_{lw}^2\right)|t| \right] dt \right|^2 \qquad (3.3.52)$$

Insert the expression of $\langle \varphi(t)^2 \rangle$ in Eq. (3.3.50) into Eq. (3.3.52), the normalized optical power spectral density can be found as

$$S_{op}(\omega) = \frac{\left[\frac{R_{sp}}{4P}\left(1+\alpha_{lw}^2\right) \right]^2}{\left[\frac{R_{sp}}{4P}\left(1+\alpha_{lw}^2\right) \right]^2 + (\omega-\omega_0)^2} \qquad (3.3.53)$$

This spectrum has a Lorentzian shape with a FWHM spectral linewidth

$$\Delta v = \frac{\Delta \omega}{2\pi} = \frac{R_{sp}}{4\pi P}\left(1+\alpha_{lw}^2\right) \qquad (3.3.54)$$

As the well-known Scholow-Towns formula which describes the laser spectral linewidth is $\Delta v = R_{sp}/(4\pi P)$, Eq. (3.3.34) is commonly referred to as the modified Scholow-Towns formula because of the introduction of linewidth enhancement factor α_{lw}.

Spectral linewidth of a non-modulated laser is determined by the phase noise, which is a measure of coherence of an optical beam. Other measures such as coherence time and coherence length can all be related to the spectral linewidth. Coherence time is usually defined as

$$t_{coh} = \frac{1}{\pi \Delta \nu} \tag{3.3.55}$$

It is the time over which a lightwave may still be considered coherent. Or, in other words, it is the time interval within which the phase of a lightwave is still predictable. Similarly, coherence length is defined by

$$L_{coh} = t_{coh} v_g = \frac{v_g}{\pi \Delta \nu} \tag{3.3.56}$$

It is the propagation distance over which a lightwave signal maintains its coherence, where v_g is the group velocity of the optical signal. As a simple example, for a lightwave signal with 1 MHz linewidth, the coherence time is approximately 320 ns and the coherence length is about 95 m in free space.

(C) Mode partition noise

As discussed in Section 3.3.4, the phase condition in a laser cavity can be simultaneously satisfied by multiple wavelengths, and thus the output from a LD can have multiple longitudinal modes if the material gain profile is broad enough as illustrated in Fig. 3.3.2. All these longitudinal modes compete for the carrier density from a common pool. Those modes near the peak of the material gain profile have competition advantages which consume most of the carrier density, while the power of other modes further away from the material gain peak will be mostly suppressed. In most FP type of diode lasers without extra mode selection mechanisms, the optical gain seen by different FP modes near the material gain peak is usually very small. Perturbations due to spontaneous emission noise, external reflection, or temperature change may introduce random switches of optical power between modes. This random mode hopping is usually associated with the total power fluctuation of the laser output, which is known as *mode partition noise*. This is why RIN in a LD with multiple longitudinal modes is much higher than that with only a single mode. In addition, if the external optical system has wavelength-dependent loss, this power hopping between modes with different wavelength will inevitably introduce additional intensity noise for the system.

3.4 SINGLE-FREQUENCY SEMICONDUCTOR LASERS

So far we have only considered the LD where the resonator consists of two parallel mirrors. This simple structure is called a *Fabry-Perot resonator* and the lasers made with this structure are usually called *Fabry-Perot lasers*, or simply *FP lasers*. An FP LD usually operates with multiple longitudinal modes because a phase condition can be met by a large number of wavelengths and the reflectivity of the mirrors is not wavelength selective. In addition to mode partition noise, multiple longitudinal modes occupy wide optical bandwidth, which results in poor bandwidth efficiency and low tolerance to chromatic dispersion of the optical system.

The definition of a single-frequency laser can be confusing. An absolute single-frequency laser does not exist because of phase noise and frequency noise. A single-frequency LD may simply be a LD with a single longitudinal mode. A more precise definition of single-frequency laser is a laser that not only has a single mode but that mode also has very narrow spectral linewidth. To achieve single-mode operation, the laser cavity has to have a special wavelength selection mechanism. One way to introduce wavelength selectivity is to add a grating along the active layer, which is called distributed feedback. The other way is to add an additional mirror outside the laser cavity, which is referred to as the *external cavity*.

3.4.1 DFB AND DBR LASER STRUCTURES

DFB LDs are widely used as single-wavelength sources in optical communication systems and optical sensors. Fig. 3.4.1 shows the structure of a DFB laser, where a corrugating grating is written just outside the active layer, providing a periodic refractive index perturbation (Kogelink and Shank, 1972). Similarly to what happens in an FP laser, the lightwave resonating within the cavity is composed of two counter-propagating waves, as shown in Fig. 3.4.1B. However, in the DFB structure, the index grating creates a mutual coupling between the two waves propagating in the opposite directions, and therefore, mirrors on the laser surface are no longer needed to provide the optical feedback.

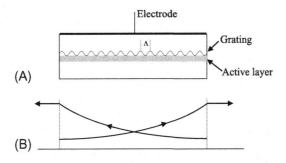

FIG. 3.4.1

(A) Structure of a DFB laser with a corrugating grating just outside the active layer, and (B) an illustration of two counter-propagating waves in the cavity (the ordinate represents optical field amplitude).

The coupling efficiency between the two waves propagating in the opposite directions is frequency dependent, which provides a mechanism for frequency selection of optical feedback in the laser cavity. This frequency-dependent coupling can be modeled by coupled-mode equations which will be discussed in more detail in Chapter 6 where we discuss fiber Bragg grating filters. Briefly, because the refractive index along the grating is varying periodically, constructive interference between the two waves happens only at certain wavelengths. To satisfy the resonance condition, the wavelength has to match the grating period, and the resonance wavelength of a DFB structure is thus,

$$\lambda_B = 2n\Lambda/m \tag{3.4.1}$$

This is called the *Bragg wavelength*, where Λ is the grating pitch, n is the effective refractive index of the optical waveguide, and m is the grating order. For wavelengths away from the Bragg wavelength, the two counter-propagated waves do not reinforce each other along the way; therefore, self-oscillation cannot be sustained for these wavelengths. For example, consider a DFB laser operating at 1550 nm wavelength, and the refractive index of InGaAsP semiconductor material is approximately $n = 3.4$. Operating as a first-order grating ($m = 1$), the grating period is $\Lambda = \lambda_B/(2n) = 228$ nm. With this grating period, the second-order Bragg wavelength would be at around $\lambda_B = n\Lambda = 775$ nm, which would be far outside of the bandwidth of the material gain.

Another way to understand this distributed feedback is to treat the grating as an effective mirror. As shown in Fig. 3.4.2, for a grating of length L, the frequency dependent reflectivity is

$$R(f) \propto \left| \frac{\sin[2\pi(f-f_B)\tau]}{2\pi(f-f_B)\tau} \right|^2 \qquad (3.4.2)$$

where $\tau = 2L/v_g$ is the roundtrip time of the grating length L, v_g is the group velocity, and $f_B = c/\lambda_B$ is the Bragg frequency. This was simply obtained by a Fourier transform of the step-function reflectivity in the spatial domain. At the Bragg frequency $f = f_B$, the power reflectivity is $R(f) = 1$ and the transmissivity is $T(f) = 0$. At frequencies $f = f_B \pm 1/2\tau$, the field reflectivity drop to 0, and the separation between these two frequencies gives the full bandwidth of the Bragg reflectivity, which is $\Delta f = 1/\tau$, or in wavelength, $\Delta\lambda = \lambda_B^2/(2nL)$. This tells that sharper frequency selectivity can be provided by a grating with a longer length.

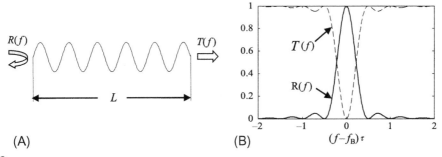

(A)   (B)

FIG. 3.4.2

(A) A Bragg grating with length L, and (B) normalized reflectivity $R(f)$ and transmissivity $T(f)$ of the Bragg grating.

A DFB structure can be regarded as a cavity formed between two equivalent mirrors. As shown in Fig. 3.4.3, from the reference point in the middle of the cavity looking left and right, the effect of the grating on each side can be treated as an equivalent mirror, similar to that in a FP laser cavity. The effective transmissivity of each mirror is $T_{eff} \propto 1 - |(\sin x)/x|^2$, where $x = 2\pi Lc(\lambda - \lambda_B)/(v_g\lambda_B^2)$, v_g is

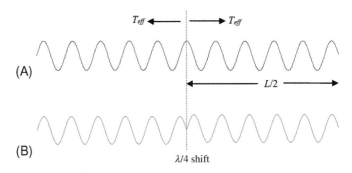

FIG. 3.4.3

(A) A uniform DFB grating and (B) a DFB grating with a quarter-wave shift in the middle.

the group velocity, c is the speed of light, and L is the cavity length. The output optical power spectrum is selected by the frequency-dependent transmissivity of these effective mirrors.

If the grating is uniform, this effective transmissivity has two major resonance peaks separated by a deep stop band. As a result, a conventional DFB laser intrinsically has two degenerate longitudinal modes and the wavelength separation between these two modes is equal to the width of the stop band $\Delta\lambda = \lambda_B^2/(nL)$, as illustrated in Fig. 3.4.4A. Single-mode operation can be obtained with an additional mechanism to breakup the precise symmetry of the transfer function so that the one of the two degenerate modes can be suppress. This extra differentiation mechanism is usually introduced by the residual reflectivities from cleaved laser facets, or by the nonuniformity of the grating structure and current density. Another technique is to etch the Bragg grating inside the active region of the waveguide in the laser cavity so that both the refractive index and the gain (or loss) of the material vary periodically along the waveguide. This complex grating structure yields an asymmetric transfer function, so that single-mode operation can be obtained in a more controllable manor.

A most popular technique to create single-mode operation is to add a quarter-wave shift in the middle of the Bragg grating, as shown in Fig. 3.4.3B. This $\lambda/4$ phase shift introduces a phase discontinuity in the grating and results in a strong transmission peak at the middle of the stop band, as shown in Fig. 3.4.4B. This ensures single-longitudinal mode operation in the LD at the Bragg wavelength.

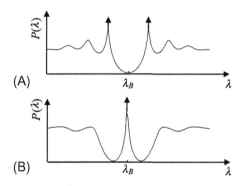

FIG. 3.4.4

(A) Spectrum of a DFB laser with a pair of degenerate modes, and (B) spectrum of a $\lambda/4$-shifted DFB laser with a dominant single mode.

DBR is another structure of single-wavelength laser sources commonly used in optical communication systems. As shown in Fig. 3.4.5, a DBR laser usually has three sections; one of them is the active section which provides the optical gain, another one is a grating section which provides a wavelength selective reflection, and there is a phase control section placed in the middle which is used to adjust the optical phase of the Bragg reflection from the grating section. Both the phase control section and the grating sections are usually passive because no optical gain is required for these sections, and thus they are usually biased by reverse voltages only to adjust the index of refractions through the electro-optic effect.

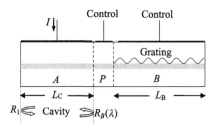

FIG. 3.4.5

Structure of a DBR laser with an active section A, phase control section P, and Bragg reflector section B. The laser cavity is formed between the left-end facet with reflectivity R_1 and the effective reflection from the Bragg section.

Assume the length of the active cavity is L_C, the FP mode spacing is $\Delta\lambda_{FP} \approx \lambda_B^2/(2nL_C)$, while the wavelength-selective Bragg reflectivity has the spectral width $\Delta\lambda_{Bragg} = \lambda_B^2/(2nL_B)$. Thus, to avoid multi-longitudinal mode operation $\Delta\lambda_{Bragg} < \Delta\lambda_{FP}$ is required. Suppose the same semiconductor material is used in all sections of the laser with the same index of refraction n, the Bragg grating section has to be longer than the active section. DBR lasers can be made wavelength tunable through the index tuning in the Bragg reflector section to control the Bragg wavelength. The tuning of phase control section is necessary to satisfy the phase condition of lasing across a continuous region of wavelength.

3.4.2 EXTERNAL CAVITY LDs

The operation of a semiconductor is sensitive to external feedback (Lang and Kobayashi, 1980). Even a $-40\,dB$ optical feedback is enough to bring a laser from single-frequency operation into chaos. Therefore, an optical isolator has to be used at the output of a LD to prevent optical feedback from external optical interfaces. On the other hand, precisely controlled external optical feedback can be used to create wavelength-tunable lasers with very narrow spectral linewidth.

The configuration of a grating-based external cavity laser is shown in Fig. 3.4.6, where laser facet reflectivities are R_1 and R_2 and the external grating has a wavelength-dependent reflectivity of $R_3(\omega)$. In this complex cavity configuration, the reflectivity R_2 of the facet facing the external cavity has to be replaced by an effective reflectivity R_{eff} as shown in Fig. 3.4.6, as

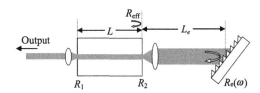

FIG. 3.4.6

Configuration of an external cavity semiconductor laser, where the external feedback is provided by a reflective grating.

$$R_{eff}(\omega) = \left\{ \sqrt{R_2} + (1-R_2)\sqrt{R_3(\omega)} \sum_{m=1}^{\infty} (R_2 R_3(\omega))^{\frac{m-1}{2}} e^{jm\omega\tau_e} \right\}^2 \qquad (3.4.3)$$

If external feedback is small enough ($R_3 \ll 1$), only one roundtrip needs to be considered in the external cavity. Then, Eq. (3.4.3) can be simplified as

$$R_{eff}(\omega) \approx R_2 \left\{ 1 + \frac{(1-R_2)\sqrt{R_3(\omega)}}{\sqrt{R_2}} \right\}^2 \qquad (3.4.4)$$

Then, the mirror loss α_m shown in Eq. (3.3.14) can be modified by replacing R_2 with R_{eff}. Fig. 3.4.7 illustrates the contributions of various loss terms: α_1 is the reflection loss of the grating, which is wavelength selective with only one reflection peak (corresponding to a valley in α_1), and α_2 and α_3 are resonance losses between R_1 and R_2 and between R_2 and R_3, respectively. Combining these three contributions, the total wavelength-dependent loss α_m has only one strong low loss wavelength, which determines the lasing wavelength. In practical external cavity laser applications, an antireflection coating is used on the laser facet facing the external cavity to reduce R_2, and the wavelength dependency of both α_1 and α_2 can be made very small compared to that of the grating; therefore, a large wavelength tuning range can be achieved by rotating the angle of the grating while maintaining single longitudinal mode operation.

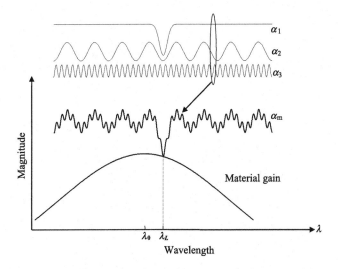

FIG. 3.4.7

Illustration of resonance losses between R_1 and R_2 (α_2) and between R_2 and R_3 (α_3). α_1 is the reflection loss of the grating and α_m is the combined mirror loss. Lasing threshold is reached only by one mode at λ_L.

External optical feedback not only helps to obtain wavelength tuning, it also changes the spectral linewidth of the emission. The linewidth of an external cavity laser can be expressed as

$$\Delta v = \frac{\Delta v_0}{1 + k\cos\left(\omega_0\tau_e + \tan^{-1}\alpha_{lw}\right)} \tag{3.4.5}$$

where Δv_0 is the linewidth of the LD without external optical feedback, ω_0 is the oscillation angular frequency, α_{lw} is the linewidth enhancement factor, and k represents the strength of the optical feedback. When the feedback is not very strong, this feedback parameter can be expressed as

$$k = \frac{\tau_e}{\tau}\frac{(1-R_2)\sqrt{R_3}}{\sqrt{R_2}}\sqrt{1+\alpha_{lw}^2} \tag{3.4.6}$$

$\tau = 2nL/c$, $\tau_e = 2n_eL_e/c$ are roundtrip delays of the laser cavity and the external cavity, respectively, with L and L_e the lengths of the laser cavity and the external cavity. n and n_e are refractive indices of the two cavities.

Eq. (3.4.5) shows that the linewidth of the external cavity laser depends on the phase of the external feedback. To obtain narrow linewidth, precise control of the external cavity length is critical; a mere $\lambda/2$ variation in the length of external cavity can change the impact of optical feedback from linewidth narrowing to linewidth enhancement. This is why an external cavity has to have a very stringent mechanical stability requirement.

An important observation from Eq. (3.4.6) is that the maximum linewidth reduction is proportional to the ratio of the cavity length L_e/L. This is because there is no optical propagation loss in the external cavity and the photon lifetime is increased by increasing the external cavity length. In addition, when photons travel in the external cavity, there is no power-dependent refractive index; this is the reason for including the factor $\sqrt{1+\alpha_{lw}^2}$ in Eq. (3.4.6).

In fact, if the antireflection coating is perfect such that $R_2 = 0$, this ideal external cavity laser can be defined as an *extended-cavity* laser because it becomes a two-section one, with one of the sections passive. With $R_2 = 0$, α_2 and α_3 in Fig. 3.4.7 will be wavelength independent and in this case, the laser operation will become very stable and the linewidth is no longer a function of the phase of the external optical feedback. The extended-cavity LD linewidth can simply be expressed as (Hui and Tao, 1989)

$$\Delta v = \frac{\Delta v_0}{1 + \dfrac{\tau_e}{\tau}\sqrt{1+\alpha_{lw}^2}} \tag{3.4.7}$$

Grating-based external cavity lasers are commercially available and they are able to provide >60 nm continuous wavelength tuning range in the 1550-nm wavelength window with <100 kHz spectral linewidth. Rapid technological advancement in micro-optics, precision mechanics, and micro-mechanical machining has enabled miniaturized and low-cost external cavity lasers with sub-100 kHz linewidth and high reliability. This in turn revived the interest in the research and development of coherent optical communication systems which utilize optical phase as an information carrier. More details of coherent communication will be discussed in Chapter 9.

In practical optical systems using semiconductor lasers, external optical back reflections often exist from fiber connectors and interfaces of optical components. The strengths and phases of these unwanted optical feedbacks vary randomly, affected by the temperature and mechanical stress of the optical fiber. Thus, laser linewidth can be made very unstable, especially when the optical feedback is originated from a distance far away from the LD. This can be understood through Eq. (3.4.6): a long external cavity length results in a longer delay τ_e and a larger feedback parameter k. Sufficiently high optical feedback level can result in the coherence collapse (Schunk and Petermann, 1988) of the laser emission and also significantly enhanced intensity noise. In order to avoid this performance degradation, a single-frequency LD used in high-speed optical communication system usually has to have an optical isolator inserted between the LD and the output optical fiber pigtail, which allows the optical signal to travel only in one direction (see Fig. 3.5.4).

3.5 VCSELs AND ARRAYS

As discussed in Section 3.3, the basic requirement of laser operation requires optical gain of the active material and an optical cavity to initiate self-sustained oscillation and to create stimulated emission. An edge-emitting laser cavity, illustrated in Fig. 3.3.1 is usually formed by an optical waveguide which provides optical confinement and a partial reflection mirror on each end of the waveguide to form the cavity. Lasing threshold I_{th} can be reached when the roundtrip optical gain inside the cavity is equal to the overall loss which includes material loss and mirror loss. As the waveguide is made on planar substrate through epilayer growth, photolithographic patterning and etching, its thickness and width can be well controlled, but the cross section is typically rectangle with the thickness much smaller than the width. Thus, the transversal mode field pattern emitted by a diode laser does not fit well with that of an optical fiber that has a circular cross section. Although one-dimensional (1D) diode laser array can be made on the same wafer with multiple waveguides in parallel, it is not feasible to make a two-dimensional (2D) diode laser array on a single wafer, as it would essentially require a three-dimensional (3D) photonic integration process.

As illustrated in Fig. 3.5.1A, a VCSEL is based on surface emission in which the direction of light emission is perpendicular to the wafer plane. The active layer PN junction of the VCESL is sandwiched between two parallel reflectors formed by multilayer DBR. The DBRs are highly reflective, to form a vertical cavity. The active layer thickness d is typically less than $0.5\,\mu m$, and thus the accumulated optical gain over a roundtrip of the extremely short optical cavity is small. The reflectivity of each DBR has to be high enough, typically more than 99.9%, to minimize the mirror loss so that the lasing threshold gain condition can be achieved.

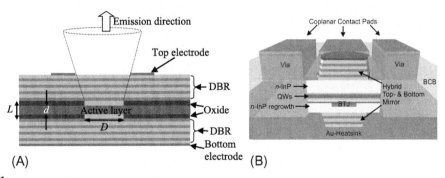

FIG. 3.5.1

(A) Illustration of VCSEL structure, and (B) InP-based VCSEL structure reported in Rodes et al. (2013) operating in 1550 nm wavelength. *BTJ*, buried-tunnel-junction; *BCB*, BenzoCycloButhene.

Lateral confinement of light emission is accomplished by oxide layers which restrict current injection only into a small diameter, D, of the active region. The reflectivity of Bragg gratings is wavelength selective which determines the emission wavelength of the VCSEL device within the bandwidth defined by the semiconductor bandgap structure. Fig. 3.5.1B is an example of an InP-based buried-tunnel-junction (BTJ) VCSEL structure (Rodes et al., 2013) operating in 1550 nm wavelength window designed for optical communication. Benzocyclobuthene (BCB) spacer layers are used to allow for the reduction of parasitic effects so that the device can be used for high-speed modulation. More detailed description of this VCSEL structure can be found in Mueller et al. (2011).

In comparison to an edge-emitting diode laser, a VCSEL has a much smaller active cavity volume. For example, the waveguide of a typical edge-emitting diode laser has a length of $L=300\,\mu m$, a width of $W=3\,\mu m$, and an active layer thickness $d=0.01–0.1\,\mu m$, and thus the active cavity volume is between 9 and 90 μm^3. Whereas for a VCSEL, the typical diameter is on the order of 5 μm, with the actively layer thickness d of 0.01–0.5 μm, and thus the active cavity volume is roughly between 0.02 and 10 μm^3. The extremely short cavity length, and thus small cavity volume, has several important implications on the laser performance.

First, the FP cavity formed between the two DBRs has very wide free-spectral range (FSR) determined by $FSR=\lambda^2/(2n_{eff}L)$, where n_{eff} is the effective index of the material at the lasing wavelength λ, and d is the cavity length. For a $L=10\,\mu m$, $\lambda=1.55\,\mu m$, and $n_{eff}=3.4$, the FSR is approximately 35 nm, which is two orders of magnitude wider than a typical edge-emitting diode laser. Note that the optical cavity length L of a VCSEL is larger than the thickness d of the active layer because of the dielectric buffer layers on both sides of the active layer within the cavity. This wide FSR makes it easier to select a single longitudinal mode with the wavelength-dependent reflectivity of the DBR.

Second, the small active cavity volume results in a low threshold current. In the steady state, Eq. (3.3.21) shows that the threshold current density J_{th} is related to the threshold carrier density N_{th} by $J_{th}=qdN_{th}/\tau$, where q is the electron charge, d is the active layer thickness, and t is the carrier lifetime. Use the cavity volume V as a parameter, the threshold current I_{th} should be linearly related to V by $I_{th}=qVN_{th}/\tau$, where $V=\pi(D/2)^2$ with D the VCSEL active area diameter. We can further express the relationship between threshold current and the mirror reflectivity by using Eqs. (3.3.14), (3.3.15), and (3.3.20),

$$I_{th}=\frac{qV}{\tau}\left[N_0+\frac{\alpha-\ln(R_tR_b)/(2L)}{\Gamma a}\right]$$ (3.5.1)

where N_0 is the transparency carrier density, a is the differential gain, Γ is the confinement factor, R_t and R_b are power reflectivities of the top and the bottom DBR mirrors at the lasing wavelength, α is the material absorption per unit length, and L is the cavity length. Eq. (3.5.1) also shows the important impact of mirror loss on the threshold current. As the mirror loss, defined by $\alpha_m=-\ln(R_tR_b)/(2L)$, is inversely proportional to the cavity length L, DBR reflectivities R_1 and R_2 need to be very close to 100% for a VCSEL with very short cavity.

Third, because of the small cavity volume, although the threshold current is small, the output power of each VCSEL can also be small, which is limited by the maximum allowable current density into the device active region. Based on Eq. (3.3.26), when operating above threshold, the output optical power is linearly proportional to the injection current density J and the cross section area A as

$$P_{opt} = \eta_{ext} \frac{(J - J_{th})Ah\nu}{q} \qquad (3.5.2)$$

where $h\nu$ is the photon energy, and $\eta_{ext} = \alpha_m/(\alpha_m + \alpha)$ represents the external efficiency. In practice the maximum allowable current density is determined by the quality of the semiconductor material as well as thermal dissipation of the device. Currently, a single VCSEL of $25\,\mu m^2$ emitting area can have sub-milliamp threshold current and is able to provide up to a few mW optical power.

Because of the high photon density in a small cavity volume, the effective carrier lifetime in a VCSEL can be very short, and the relaxation oscillation frequency is typically higher than that of an edge emission diode laser. Thus, a VCSEL can have high modulation bandwidth for optical communication applications.

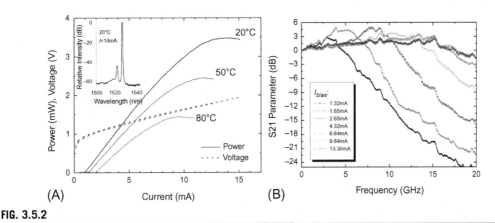

FIG. 3.5.2

Performance of a 1550 nm wavelength VCSEL with $D = 5\,\mu m$ active area aperture diameter and $L = 2.5\,\mu m$ vertical cavity length. (A) P–I curve for device operated at different temperature and the inset shows the emission optical spectrum. Dashed lines show current-voltage relations. (B) Modulation response as the function of the modulation frequency at different bias currents and at 20°C temperature (Rodes et al., 2013).

Fig. 3.5.2 shows the performance of a VCSEL designed for operating at 1550 nm wavelength with the structure shown in Fig. 3.5.1B. The active area aperture diameter is $D = 5\,\mu m$, and the vertical cavity length is $L = 2.5\,\mu m$. P–I curve shown in Fig. 3.5.2A indicates that at room temperature, approximately 3 mW output power can be obtained before saturation at 10 mA injection current. At this operation point, the forward biasing voltage is approximately 1.5 V, so that the electric power consumption is 15 mW, and the electric-to-optical power-conversion efficiency is about 25%. The inset in Fig. 3.5.2A shows that the VCSEL operates in a single mode with a side-mode suppression of >40 dB. Fig. 3.5.2B shows the modulation response of the VCSEL as the function of the modulation frequency at different bias currents. S21 is the network analyzer transfer function representing the power ratio between the output optical signal and the modulating electric signal as the function of modulation frequency. A 3 dB bandwidth of 20 GHz was obtained with this VCSEL.

Because of the unique device structure and fabrication process, the most promising application of VCSELs is the used of VCSEL array. As illustrated in Fig. 3.5.3, light emitted from a 2D VCSEL array fabricated on a planar wafer can be coupled into an array of optical fibers through a planar microlens array. Each VCSEL as a pixel can be individually encoded with data through direct modulation of the injection current. This spatial division multiplexing can significantly increase the overall capacity of the system, which is especially useful for optical interconnection inside datacenters where ultra-high-speed data delivery is needed but for relatively short distances.

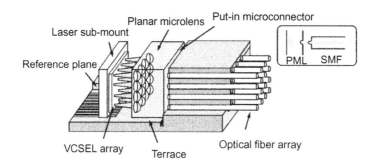

FIG. 3.5.3

Illustration of coupling light from a VCSEL array to a fiber array (Iga, 2000).

So far, VCSELs have been produced for emitting at various wavelength windows from visible to infrared for different applications using different semiconductor materials and photonic structures. The output power of each VCSEL can be increased by increasing the active area. For example, for short-distance interconnection using plastic fiber with a large core diameter of $>100\,\mu$m, large area VCSELs can be used with more than 10 mW optical power.

In addition to the application in optical communications, high-power VCSELs are also prominent candidates as energy-efficient light sources for illumination. For such applications, electrical to optical power-conversion efficiency is the most important parameter. This power-conversion efficiency can be expressed as

$$\eta_{PC} = \frac{P_{opt}}{IV_b} = \eta_{ext}\frac{hv}{qV_b}\left(1 - \frac{I_{th}}{I}\right)$$
(3.5.3)

where V_b is the bias voltage so that IV_b is the electric power. Because the threshold current I_{th} is low and the external efficiency η_{ext} is high in a VCSEL, the power-conversion efficiency can reach to almost 50% in commercial VCSEL array devices. Fig. 3.5.4A is a scanning electron microscope (SEM) image of an actual high-power VCSEL. The active layer is made of multiple quantum wells structure. Electrons and holes are injected via the conductive DBRs and the current is confined to the emission area by a high Al-content oxide insulation layer (Moench et al., 2015). The active emission area has a size of typically 4–20 μm in diameter, which emits approximately 1–30 mW optical power. Fig. 3.5.4B shows a top view image of a VCSEL array. With 8 μm diameter active emitting area of each pixel and 40 μm separation between adjacent pixels, approximately 2500 VCSELs can be made on a $2 \times 2\,\text{mm}^2$ wafer area with 8 W total optical power (Moench et al., 2015). This device can be used for a variety of applications such as illumination and imaging.

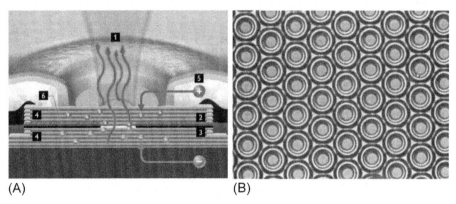

(A)   (B)

FIG. 3.5.4

(A) Illustration of a cross section through a single top-emitting VCSEL superimposed to a SEM picture. (1) vertical light emission, (2) active layer area without current injection, (3) active layer area with current injection, (4) conductive DBRs, (5) metallic ring contact, and (6) oxide layer. (B) Plan view image of VCSEL array (Moench et al., 2015).

Recall that an LED, discussed in Section 3.2, is based on the spontaneous emission with light propagation in all directions. A large portion of the optical power generated inside the active region is trapped inside the device because of the total internal reflection, and thus the external efficiency η_d is very small. In comparison, a VCSEL, based on the stimulated emission, has much better external efficiency because the light is emitted only in the vertical direction, and no total internal reflection is expected. Therefore, high-power VCSEL arrays can eventually replace LEDs for illumination because of their superior power efficiency.

3.6 **LD BIASING AND PACKAGING**

From a practical application point of view, as the optical power of a diode laser is linearly proportional to the injection current when biased above the threshold, the driving electronic circuit has to be a current source instead of a voltage source. The equivalent electrical circuit of a diode laser is an ideal diode with a series resistance as shown in Fig. 3.6.1A.

It is well known that the relation between the current, I_D, and the voltage, V_D, of a diode is

$$I_D = I_s \exp\left[qV_D/(nk_BT)\right] \tag{3.6.1}$$

where I_s is the saturation current, q is the electron charge, T is the absolute temperature, k_B is the Boltzmann's constant, and n is an ideality factor of the diode. When a series resistance R_s is introduced, the external current (I_D) voltage (V_B) relation is then,

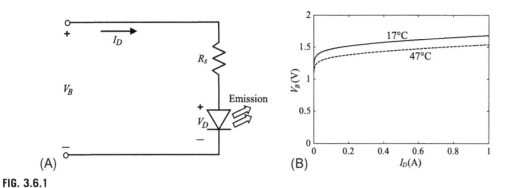

FIG. 3.6.1

(A) Equivalent electrical circuit of a laser diode and (B) voltage vs current relation of a diode laser with $0.1\,\Omega$ series resistance.

$$I_D = I_s \exp\left(q\frac{V_B - R_s I_D}{nk_B T}\right) \tag{3.6.2}$$

The series resistance R_S is primarily caused by the Ohmic contact of the electrode, which is usually less than $1\,\Omega$. For a practical LD used for telecommunication transmitter, the voltage V_B is on the order of $1.5\,V$ at the operation point with a reasonable output optical power. For example, assume a saturation current $I_s = 10^{-13}\,A$, an ideality factor $n=2$, and a series resistance $R_s = 0.1\,\Omega$, current-voltage relation is shown in Fig. 3.6.1B. An injection current increase from 0.1 to 1 A corresponds to a very small voltage increase of less than $0.2\,V$, which is also sensitive to the operation temperature of the LD. As the optical power of a LD is proportional to the injection current, the electrical driver needs to be a current source instead of a voltage source.

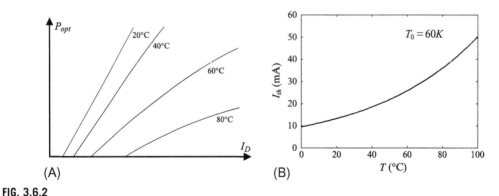

FIG. 3.6.2

(A) Illustration of diode laser P–I curve at different junction temperatures and (B) threshold current as a function of junction temperature.

Threshold current and optical power efficiency are two key parameters of a diode laser, both of them are sensitive to the junction temperature T. The threshold current I_{th} increases with the junction temperature exponentially as $I_{th} = A\exp(T/T_0)$, where A is a proportionality factor and T_0 is the characteristic temperature ranging from 40 to 120 K depending on the material and laser structure. Fig. 3.6.2A illustrates the laser $P\text{--}I$ curves at different junction temperatures, and Fig. 3.6.2B shows the threshold current as the function of the temperature, assuming $A = 0.1$ mA, and $T_0 = 60$ K were assumed.

Fig. 3.6.3 shows an example of LD biasing circuit based on a high-power bipolar junction transistor (BJT), and two operational amplifiers. A straightforward circuit analysis indicates that the DC bias current flowing through the LD is $(V_{cc} - V_1)/R_e$, where, V_{cc} is the DC power supply voltage and V_1 is the voltage at the DC biasing circuit input as marked in Fig. 3.6.3. Here, the LD driving current is linearly proportional to the voltage difference $V_{cc} - V_1$, and $V_1 = V_{cc}R_4/(R_3 + R_4)$ can be adjusted by changing the resistance R_4. The AC signal to be modulated on the LD can be amplified by an inverting voltage amplifier, and added to the DC bias through a capacitor. The overall current ID is then a superposition of DC and AC components as

$$I_D = V_{cc}\frac{1 - R_4/(R_3 + R_4)}{R_e} + \frac{v_s R_2/R_1}{R_e} \tag{3.6.3}$$

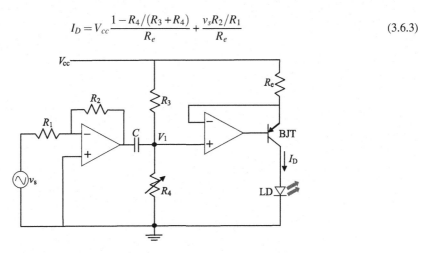

FIG. 3.6.3

Example of laser diode-biasing circuit based on a trans-conductive amplifier stage.

Fig. 3.6.4A shows the picture of a LD chip on mounted on a heat sink. As the size of the electrode on top of the LD chip is very small and fragile, it is usually connected to a staging metal pad through very thin gold wires, and the wire for external connection to the driver electrical circuit is soldered to this metal

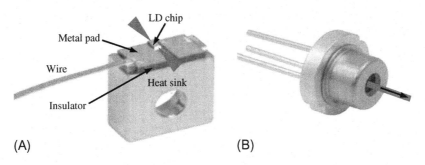

(A)   (B)

FIG. 3.6.4

(A) Picture of diode laser on heat sink and (B) picture of cylindrical packaged diode laser.

pad. The heat sink can also be mounted on a larger heat sink which helps dissipate the heat generated by the diode laser chip when driven by a large injection current. Fig. 3.6.4B shows the picture of a sealed cylindrical package for a diode laser. A collimating lens can be added in front of the LD so that the laser beam emitted from the exit window is collimated.

For lasers used for high-speed optical communication systems, the requirement of temperature stabilization is more stringent than for other applications. The reason is that the junction temperature of a LD not only affects the threshold current and power efficiency, but also significantly affects the emission wavelength through the temperature sensitivity of the refractive index. As a rule of thumb, for an InGaAsP-based DFB laser operating in the 1550 nm wavelength window, each one °C change of junction temperature may introduce as much as 0.1 nm change of the emission wavelength. Thus, a temperature stability of better than 0.1°C is usually required for lasers diodes used for wavelength division multiplexed optical systems, in which each channel has its assigned narrow wavelength window.

Fig. 3.6.5A shows the configuration of a fiber pigtailed LD typically used in fiber-optic communication systems. In this configuration, the laser chip is mounted on a heat sink, and the emission is collimated and passes through an optical isolator to avoid the impact of external optical reflection

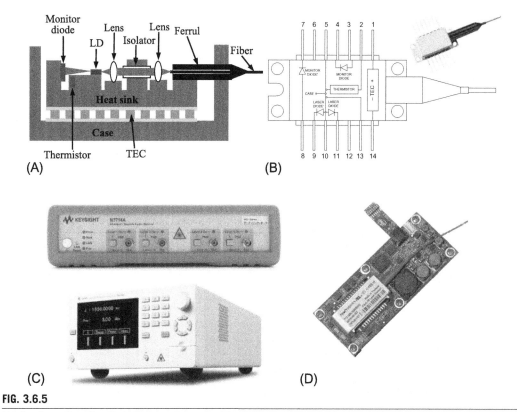

FIG. 3.6.5

(A) Configuration of a diode laser package used for fiber-optic communication systems, (B) pin assignment of a standard 14-pin butterfly package of a laser diode, (C) bench-top external cavity tunable laser instrument, and (D) integrated tunable laser assembly (ITLA) for optical communication systems. ((C) Used with permission from Keysight and Santec. (D) Used with permission from Neophotonics.)

on the LD performance. After the isolator, the collimated optical beam is refocused and coupled into an optical fiber through another lens. A photodiode mounted on the left side of the laser chip is used to monitor the optical power emitted from the back facet of the LD, which is in fact linearly proportional to the power emitted from the right-side laser facet.

The temperature of the heat sink is monitored by a thermistor buried inside the heat sink, and the reading from the thermistor is used to control the thermoelectric cooler (TEC), known as a Peltier, through an electronic feedback circuit. All the elements are hermetically sealed inside a metal case to ensure the mechanical and optoelectronic stability, and reliability for commercial applications. The pin assignment of the standard 14-pin butterfly package is shown in Fig. 3.6.5B.

More sophisticated wavelength tunable semiconductor lasers have been developed with narrow spectral linewidths in the external-cavity configurations. Fig. 3.6.5C shows examples of narrow line-width tunable lasers made as bench-top equipment which are mainly used in laboratories. Because of the very stringent stability requirement of external cavity, mechanical structure has to be well designed with temperature stabilization and motion control to provide continuous tuning of lasing wavelength and to guarantee the narrow spectral linewidth at each wavelength. Integrated Tunable Laser Assembly (ITLA) shown in Fig. 3.6.5D is another format of external cavity LD but with a much smaller footprint. ITLA is usually made with micro-electro-mechanical system (MEMS) for external cavity tuning and stabilization, and is commonly used in fiber-optic systems requiring coherent detection. An ITLA usually guarantees $<100\,kHz$ spectral linewidth, but wavelength tuning may not be continuous. Instead, it may only provide discrete wavelengths specified by the International Telecommunication Union (ITU) wavelength grid for dense wavelength division multiplexing (DWDM) systems.

Although the fabrication of diode laser chips can be cost effective and each processed 6-in. semi-conductor wafer can be cleaved into a large number of LDs, packaging appears even more challenging for cost reduction. The packaging including fiber coupling, optical isolation, temperature control, and power monitoring usually constitutes more than 80% of the total cost of a LD used in the transmitter of a fiber-optic communication system.

3.7 SUMMARY

In this chapter, we have discussed semiconductor-based light sources often used for optical commu-nications, primarily LEDs and LDs. In order to make light sources, direct-bandgap semiconductor ma-terials have to be used with appropriate bandgap energy to provide the desired emission wavelength. Photons can be generated when electrons and holes recombine inside a forward-biased pn junction with proper carrier confinement. We have explained a few important concepts in semiconductor light sources including: direct bandgap vs. indirect bandgap semiconductors, radiative recombination vs. nonradiative recombination, spontaneous emission vs. stimulated emission, and homogeneous broad-ening vs. inhomogeneous broadening.

For electrically pumped light sources such as LEDs and LDs, the ideal electron to photon-conversion efficiency is $h\nu/q$ [W/A] with $h\nu$ the photon energy and q the electron charge. That means every electron is used to generate a photon. However, in practical devices, this conversion efficiency is reduced by both quantum efficiency and external efficiency, where the former is related to the ratio between radiative recombination and nonradiative recombination, and the later is defined by the per-centage of generated photons that can emit out of the device. An LED is based on the spontaneous emission which does not have any preference of direction, and thus only a small percentage of the generated photons can escape while others are trapped inside the high index material by total internal reflection. This results in a low external efficiency. An LED also has broad spectral width and the spontaneously emitted light is considered incoherent. In comparison, a LD is based on highly directional stimulated emission, so that the external efficiency can be much higher when operating above threshold. Generation of stimulated emission in a LD requires gain and phase conditions to be met for each roundtrip in the cavity, and the emission is highly coherent.

Major parameters such as threshold current, slope of the *P–I* curve above threshold, as well as the modulation response of a LD can be predicted by rate equation analysis which describes dynamic interactions between carrier density and photon density inside the laser cavity. We have discussed the importance of linewidth enhancement factor in a LD which is originated from the carrier density-dependent refractive index of the semiconductor material. Linewidth enhancement factor is responsible for the relatively broad spectral linewidth as well as modulation chirp which are both unique for diode lasers.

Our analysis of longitudinal modes started with resonance in a FP cavity in which the roundtrip phase condition is satisfied by many equally spaced frequencies. Although the material gain is band limited, the gain bandwidth is usually not narrow enough to select a single longitudinal mode. Thus, a LD based on the FP cavity generally has multiple longitudinal modes. Readers must not be confused between multimode fiber and multi-longitudinal mode laser: the former refers to multiple spatial modes while the later normally refers to multiple discrete wavelength components. Because all the longitudinal modes in a LD share the same pool of carrier density inside the cavity, they compete with each other dynamically. As a consequence, the power of each individual mode fluctuates as the function of time, known as mode partition noise.

A number of cavity structures have been developed for mode selection, and the most popular ones are DFB, DBR, and external cavity. While DFB and DBR are good in realizing high SMSR, external cavity structure is most effective to narrow the spectral linewidth by the significantly increasing photon life in the passive external cavity. However, fabrication of lasers with long external cavities for practical applications is not trivial because of the extremely stringent requirements in optical alignment accuracy and stability. Recent development in MEMS-based external cavity lasers avoided bulky mechanical apparatus. This technology enabled narrow linewidth Micro Integrable Tunable Laser Assembly (μ-ITLA), which is the key component for fiber-optic systems with coherent detection.

We have also introduced VCSELs in this chapter. Because of the unique cavity structure and fabrication process, VCSELs can be easily made into 2D arrays for low-cost optical interconnection, as well as for display and energy-efficient illumination.

Practical applications of LEDs and diode lasers require appropriate packing and biasing electronic circuits which have been discussed in the last section of this chapter. As the terminal electric characteristic of either an LED or a diode laser is just like a junction diode, a constant current source has to be used for biasing, and a trans-conductive electrical amplifier has to be used to convert a voltage signal into a current signal to drive an LED or a diode laser.

LEDs and LDs are major light sources for fiber-optic communications. Physical mechanisms, device operation principles, and major parameters of semiconductor-based light sources discussed in this chapter are critically important in the design, performance characterization, and maintenance of fiber-optic communication systems and networks.

PROBLEMS

1. The bandgaps of InGaN, GaAs, and InGaAsP are 3.06, 1.24, and 0.8 eV, respectively. These are all direct bandgap semiconductor materials. What would be the emission wavelength if each of them is used to make semiconductor laser?

2. Consider an LED operating in 800 nm wavelength window with 3% external efficiency. The radiative and nonradiative recombination carrier lifetimes are 2 and 8 ns, respectively.
   (a) In order to have 2 mW output optical power, what is the required electrical current?
   (b) What is the 3 dB optical modulation bandwidth of this LED?
   (c) What is the 3 dB modulation bandwidth in electrical domain?

3. An LED operating in 1310 nm wavelength window has 5% external efficiency and 80% internal efficiency. An electrical signal used to modulate this LED is $I(t) = I_B + I_m \cos(\omega t)$, where $I_m = 20$ mA. What is the minimum bias current I_B so that the output optical power waveform has no distortion? What is the average optical power, and what is the peak optical power?

4. Assume the gain parameter of a semiconductor material has a parabolic shape $g(\lambda) = g_0\{1 - [(\lambda - \lambda_0)/\Delta\lambda_g]^2\}$, where $\lambda_0 = 1550$ nm and $\Delta\lambda_g = 15$ nm. The length of the laser cavity is $L = 400$ μm, and refractive index of the material is $n = 3.6$. Please find the number wavelengths which satisfy the phase condition and with $g \geq 0.5g_0$.

5. A laser cavity is formed by the Fresnel reflections between two cleaved facets. The refractive index of the laser cavity is $n = 3.5$, and the absorption parameter is $\alpha = 30\,\mathrm{cm}^{-1}$. At which cavity length, L, the mirror loss is equal to the absorption loss in the laser cavity?

6. A semiconductor laser emits in 1550 nm wavelength band, the cavity length is 350 μm, the refractive index inside the laser cavity is $n = 3.43$ and the confinement factor is $\Gamma = 0.1$.
   (1) What is the wavelength spacing between adjacent lasing modes?
   (2) If the material absorption coefficient for the field is $\alpha = 100\,\mathrm{cm}^{-1}$, find the threshold gain coefficient $g_{th} = ?$ (here consider optical feedback is only caused by the Fresnel reflections of normal incidence)
   (3) Operating above threshold, what is the external efficiency?
   (4) For 10 mA injection current increase, what is the corresponding output optical power change?
   (5) Find the ratio between the optical modulation bandwidth and the corresponding electrical bandwidth: $\frac{f_{3dB}(optical)}{f_{3dB}(electrical)} = ?$

7. A semiconductor laser has two sections as shown in the following figure. The semiconductor section has the length $L_1 = 200$ μm, refractive index $n_1 = 3.4$, power absorption coefficient $\alpha_1 = 50\,\mathrm{cm}^{-1}$. For the air section, the length is $L_2 = 500$ μm, the refractive index is $n_0 = 1$, and both the gain and the absorption coefficients are zero in the air ($\alpha_0 = 0$, $g_0 = 0$). Assume the confinement factor is $\Gamma = 1$, everywhere.

   The laser operates in the 1550 nm wavelength window. (For simplicity, assume the light is perfectly collimated in the air section as illustrated in the figure.)

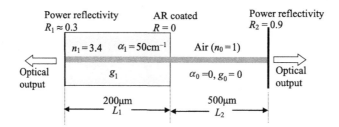

(a) Write the phase condition of this laser and find the wavelength separation $\Delta\lambda$ between adjacent modes (around the lasing wavelength of 1550 nm).

(b) Write the threshold gain condition of this laser (symbolically), and find the threshold power gain coefficient g_{1th}.

(c) Assume the internal quantum efficiency is 100%, find the external efficiency η_{ext} of this laser above threshold.

8. Consider a laser cavity with volume $V=dlb$, as shown in the following figure, the static rate Eqs. (3.3.17) and (3.3.18) for carrier and photon densities can be converted into rate equations for carrier and photon populations.

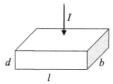

$$\frac{I}{q} = \frac{n}{\tau} + G_N(n-n_0)p$$

$$G_N(n-n_0)p - \frac{p}{\tau_{ph}} + r_{sp} = 0$$

where $I=J\cdot l\cdot b$ is the current in [A], $n=N\cdot V$ is the total carrier population which is unitless, $p=P\cdot V$ is the total photon population which is also unitless, $G_N=2\Gamma v_g a/V$ in [s^{-1}] is the volume differential gain, and $r_{sp}=R_{sp}V$ is the volume spontaneous emission rate in [s^{-1}].

Given: $\tau_{ph}=3ps$, $G_N=10^4[s^{-1}]$, $r_{sp}=0.7\times 10^{12}[s^{-1}]$, $\tau=2ns$ and $n_0=10^8$

(a) Numerically solve the following equations and plot out p vs. I curve.

(b) Plot out carrier population n vs. current I.

(c) Find the threshold carrier population n_{th} and threshold current I_{th}, and compare these values with your numerical solution in (b).

9. Consider a semiconductor laser operating in 1550 nm wavelength with the following parameters: cavity length $L=400\,\mu m$, reflective index $n=3.6$ (assume facet reflectivity is only caused by the Fresnel reflection between semiconductor and air), internal efficiency: 100%, material attenuation coefficient $\alpha_{material}=20\,cm^{-1}$, threshold current: $I_{th}=10\,mA$.

(a) What is the slope of the $P_{opt} \sim I$ curve above threshold in $[mW/mA]$? (P_{opt} is the optical power and I is the injection current.)

(b) What is the photon lifetime τ_{ph}?

(c) If the electric signal of the injection current is $I(t) = I_1 \cos(\omega_1 t) + I_2 \cos(\omega_2 t)$ with $I_1 = 20\,mA$ and $I_2 = 10\,mA$, what is the minimum bias current such that the signal is not distorted?

10. A laser emitting at 1550 nm wavelength has cavity length $l = 400\,\mu m$, width $w = 8\,\mu m$, and thickness $d = 0.2\,\mu m$, so that the volume is $V = 6.4 \times 10^{-16}$ m^3. The refractive index is $n = 3.6$ so that the facet reflectivity is $R = 32\%$. Assume the spontaneous emission factor is $R_{sp} = 2 \times 10^{27} s^{-1} m^{-3}$ and the linewidth enhancement factor is $\alpha_{lw} = 5$. For the optical power from one laser facet is $P_{opt} = 20\,mW$, (a) what is the photon density inside the laser cavity, and (b) based on the modified Scholow-Towns formula what is the spectral linewidth?

11. For a LD with spontaneous emission factor $R_{sp} = 10^{27} s^{-1} m^{-3}$, photon lifetime $\tau_{ph} = 2ps$, differential gain $a = 1.5 \times 10^{-19} m^2$, confinement factor $\Gamma = 0.2$, transparency carrier density $N_0 = 2 \times 10^{23} m^3$, and the group velocity $v_g = 8.3 \times 10^7 m/s$. When the laser is operating above threshold with the photon density inside the cavity is $P = 5 \times 10^{20} m^{-3}$, (a) what is the normalized carrier density difference with respect to the threshold value? [i.e., $(N - N_{th})/N_{th} = ?$], please pay attention to the sign of $(N - N_{th})$ and explain the physical meaning, and (b) if the spontaneous emission carrier lifetime is $\tau = 2 \times 10^{-9} s$, what is the effective carrier lifetime τ_{eff} considering stimulated emission?

12. The spectral density of laser emission has a Lorentzian shape with 10 MHz spectral width at $-3\,dB$ (known as FWHM). (a) What is the spectral width at $-20\,dB$ level? and (b) what is the coherence length of this laser in free space?

13. A single-frequency LD operating in 1550 nm wavelength has the linewidth enhancement factor $\alpha_{lw} = 6$. The optical power of the laser is modulated by a sinusoid such that $P(t) = 10 + 5\cos(2\pi f t)$ in $[mW]$ with $f = 5\,GHz$. (a) What is the maximum optical frequency deviation caused by this modulation? (b) If the modulated laser output is launched into a standard single-mode fiber with $L = 80\,km$ length and $D = 16\,ps/nm/km$ dispersion parameter, what is the maximum differential delay of the optical signal caused by the modulation chirp? Please discuss if this system is limited by the modulation chirp and why?

14. A LD emits 10 mW constant optical power with a flat RIN spectral density of $-130\,dB/Hz$. What is the electrical bandwidth so that the standard deviation of optical power fluctuation is less than 0.1 mW?

15. For a DFB laser based on first-order Bragg grating, and the refractive index of the semiconductor material is $n = 3.6$, (a) in order for the Bragg wavelength to 1540.8 nm, what is the grating period? and (b) if the cavity length is 400 μm, what is the width of the stop band?

16. A LD has a cavity length of $L = 300\,\mu m$, refractive index $n = 3.5$, linewidth enhancement factor $\alpha_{lw} = 5$, and the spectral linewidth of 100 MHz. A mirror is placed at one side of the laser to form an external cavity with the effective reflectivity $R_3 = 0.1$, and the refractive index in the external cavity is $n_e = 1$. The facet reflectivity of the LD facing the external cavity is $R_2 = 0.05$. In order to reduce the spectral linewidth to 1 MHz, what should be the length of the external cavity? (Assume that the phase of the external feedback can be precisely controlled.)

17. Consider the same LD as described in problem 16, (cavity length $L=300\,\mu$m, refractive index $n=3.5$, linewidth enhancement factor $\alpha_{lw}=5$, and the spectral linewidth of 100 MHz). But both of the two laser facets have the same reflectivity $R_2=0.3$. The laser output is coupled into a single-mode fiber, and there is a connector which is 1 m away from the LD with -80 dB optical power reflectivity ($R_3=10^{-8}$) (assume refractive index of the fiber is 1.5). The phase of the optical feedback cannot be controlled, what will be the worst-case (the widest) spectral linewidth of this laser?

18. A semiconductor laser emitting at 850 nm wavelength has the current-voltage relation following the typical diode equation of (3.5.2), where $V_B=1.5$ V, $I_{s=}10^{-13}$ A, $n=2$, and $T=300$ K and assume the internal resistance is negligible ($R_s=0$) for simplicity. This laser has an internal material loss $\alpha=10$ cm^{-1}, mirror loss $\alpha_m=15$ cm^{-1}, and a threshold current $I_{th}=30$ mA.
   (a) What is the applied voltage to reach the threshold?
   (b) What is wall-plug efficiency (the ratio between optical power and electrical power) at the optical power of 100 mW?
   (c) Please plot the what relation between the optical power and the applied voltage in the region between threshold and the operating point ($P_{opt}=100$ mW). If the applied voltage is increase for 10% from the operating point, what is the percentage of the optical power increase corresponding? Compare that to the case when the injection current is increased by 10% and discuss why a current source is usually required for the driving electrical circuit (instead of a voltage source).

REFERENCES

Agrawal, G.P., Dutta, N.K., 1986. Long-Wavelength Semiconductor Lasers. John Wiley & Sons, New York.

Henry, C., 1982. Theory of the linewidth of semiconductor lasers. IEEE J. Quantum Electron. 18 (2), 259–264.

Henry, C.H., 1983. Theory of the linewidth of semiconductor lasers. IEEE J. Quantum Electron. 18, 259.

Henry, C., 1986. Theory of spontaneous emission noise in open resonators and its application to lasers and optical amplifiers. J. Lightwave Technol. 4 (3), 288–297.

Hui, R., Tao, S., 1989. Improved rate-equation for external cavity semiconductor lasers. IEEE J. Quantum Electron. 25 (6), 1580.

Iga, K., 2000. Surface-emitting laser—its birth and generation of new optoelectronics field. IEEE J. Sel. Top. Quantum Electron. 6 (6), 1201–1215.

Kogelink, H., Shank, C.V., 1972. Coupled-wave theory of distributed feedback lasers. J. Appl. Phys. 43 (5), 2327.

Lang, R., Kobayashi, K., 1980. External optical feedback effects on semiconductor injection laser properties. IEEE J. Quantum Electron. 16 (3), 347–355.

Moench, H., Conrads, R., Deppe, C., Derra, G., Gronenborn, S., Gu, X., Heusler, G., Kolb, J., Miller, M., Pekarski, P., Pollman-Retsch, J., Pruijmboom, A., Weichmann, U., 2015. High power VCSEL systems and applications. Proc. of SPIE 9348, 93480W.

Mueller, M., Hofmann, W., Grundl, T., Horn, M., Wolf, P., Nagel, R.D., Ronneberg, E., Bohm, G., Bimberg, D., Amann, M.-C., 2011. 1550 nm highspeed short-cavity VCSELs. IEEE J. Sel. Top. Quantum Electron. 17 (5), 1158–1166.

Rodes, R., et al., 2013. High-speed 1550 nm VCSEL data transmission link employing 25 GBd 4-PAM modulation and hard decision forward error correction. J. Lightwave Technol. 31 (4), 689–695.

Schawlow, A.L., Townes, H., 1958. Infrared and optical masers. Phys. Rev. 112 (6), 1940.

Schunk, N., Petermann, K., 1988. Numerical analysis of the feedback regimes for a single-mode semiconductor laser with external feedback. IEEE J. Quantum Electron. 24 (7), 1242–1247.

PHOTODETECTORS

CHAPTER OUTLINE

Introduction 125
4.1 PN and PIN photodiodes 126
4.2 Responsivity and electric bandwidth 131
    4.2.1 Quantum efficiency and responsivity 131
    4.2.2 Speed of photodetection response 133
    4.2.3 Electrical characteristics of a photodiode 135
4.3 Photodetector noise and SNR 137
    4.3.1 Sources of photodetection noise 137
    4.3.2 SNR of an optical receiver 139
    4.3.3 Noise-equivalent power 140
4.4 Avalanche photodiodes 141
    4.4.1 APD used as a linear detector 142
    4.4.2 APD used as a single-photon detector 144
4.5 Other types of photodetectors 145
    4.5.1 Photovoltaic 145
    4.5.2 Charge-coupled devices 148
4.6 Summary 150
Problems 151
References 154
Further reading 154

INTRODUCTION

Photodetector is the key device in the front end of an optical receiver that converts the incoming optical signal into an electrical signal, known as O/E convertor. Semiconductor photodetectors, commonly referred to as *photodiodes*, are the predominant types of photodetectors used in optical communication systems because of their small size, fast detection speed, and high detection efficiency. Similar to the structures of laser diodes, photodiodes are also based on the PN junctions. However, unlike a laser

Introduction to Fiber-Optic Communications. https://doi.org/10.1016/B978-0-12-805345-4.00004-4

diode in which the PN junction is forward biased, the PN junction of a photodetector is eversely biased so that only a very small reverse saturation current flows through the diode without an input optical signal. Although the basic structure of a photodiode can be a simple PN junction, practical photodiodes can have various device structures to enhance quantum efficiency. For example, the popular PIN structure has an intrinsic layer sandwiched between the *p*- and *n*-type layers, and that is why a semiconductor photodetector is also known as a PIN diode. An avalanche photodiode (*APD*) is another type of often used detector that can introduce significant photon amplification through avalanche gain when the bias voltage is high enough.

Ideally, a photodiode translates each photon of the received optical signal into a free electron instantaneously so that the photocurrent is linearly proportional to the power of the optical signal. However, for a practical semiconductor material, not every incoming photon is able to create an electron, which can be caused by non-efficient photon absorption and carrier collection. The speed of photodetection can also be limited by carrier transient effect and the RC parasitic of the electric structures. Thus, the responsivity and the detection speed of a photodiode depend on a number of factors including the bandgap structure of the semiconductor, material quality, device photonic structure, and electrode design.

In addition, a practical photodiode also generates noises in the photodetection process, which is especially detrimental to the performance of an optical communication system by introducing signal-to-noise ratio (SNR) degradation. Major noise sources include shot noise, thermal noise, and dark-current noise. Thermal noise can be reduced by increasing the load resistance, and dark-current noise can be reduced by decreasing the reverse saturation current through material improvement and junction structure optimization. However, shot noise cannot be reduced because it is intrinsically associated with photo-detection process, which sets the fundamental limit of the optical system performance known as the quantum limit.

Because photocurrent is proportional to the received signal optical power which is the square of the signal optical field, photodiode is also known as a square-law detector. This squaring process can generate mixing products between different frequency components of the received optical field, as well as the mixing between optical signal and the broadband optical noise which introduces beating noises in the electric domain. The properties of these additional noises will be discussed in Chapter 5 where optical amplifiers are introduced.

The last section of this chapter introduces the basic structure and applications of photovoltaic (PV) and charge-coupled devices (CCDs). Although PV and CCD are both based semiconductor PN or PIN structures converting photons into electrons, their main parameters and specifications are quite different from those of photodiodes for communications. For example, PV, as energy-harvesting device, does not require high detection speed so that reverse electric biasing is not necessary. On the other hand, CCD, as a primary imaging acquisition device, requires programmable charge transfer and storage, low dark charge, and large dynamic range are most desirable.

This chapter is intended to introduce the basic concept of photodetection as well as key parameters and definitions of photodiodes relevant for various applications including fiber-optic communications.

4.1 PN AND PIN PHOTODIODES

The homojunction between a *p*-type and an *n*-type semiconductor shown in Fig. 4.1.1 is similar to that in Fig. 3.1.1. The Fermi levels for the *p*- and *n*-type sections are, respectively,

$$E_{F,p} = E_{Fi}^{(p)} - kT \ln\left(\frac{N_a}{n_i^2}\right) \tag{4.4.1a}$$

$$E_{F,n} = E_{Fi}^{(n)} + kT \ln\left(\frac{N_d}{n_i^2}\right) \tag{4.1.1b}$$

where k is the Boltzmann's constant, T is the absolute temperature, N_a is the electron doping densities on the p-type side and N_d is the hole doping density on the n-type side, and n_i is the intrinsic carrier density. $E_{Fi}^{(p)}$ and $E_{Fi}^{(n)}$ are intrinsic Fermi levels of the p- and n-type sections. When the p- and n-type doped sections are put together with intimate contact, their Fermi levels will align under thermal equilibrium so that $E_{F,p} = E_{F,n} = E_F$, while their intrinsic Fermi levels will separate, and the energy level offset is

$$E_{Fi}^{(p)} - E_{Fi}^{(n)} = kT \ln\left(\frac{N_d N_a}{n_i^2}\right) \tag{4.1.2}$$

This is equivalent to an electrical potential barrier across the pn junction (Neamen, 2003),

$$\Delta V_{pn} = \frac{E_{Fi}^{(p)} - E_{Fi}^{(n)}}{q} = \frac{kT}{q} \ln\left(\frac{N_d N_a}{n_i^2}\right) \tag{4.1.3}$$

where q is the electron charge.

Near the p–n interface, the p-type section becomes negatively charged because the loss of holes, and the n-type section becomes positively charged because the loss of electrons. The density of the trapped charges is $-N_a$ and N_d, respectively, determined by the doping density as shown in Fig. 4.1.1C. The space-charged regions extend from $-x_p$ in the p-side to x_n in the n-side. The Gauss's Law of electromagnetics tells that the divergence of the electrical field is proportional to the charge density and thus, the electrical field E_x inside the space-charged region can be found through an integration as

$$\mathbf{E}_x(x) = \begin{cases} -\dfrac{qN_a}{\varepsilon_s \varepsilon_0}(x + x_p) & \text{for} \quad -x_p < x \leq 0 \\ \dfrac{qN_d}{\varepsilon_s \varepsilon_0}(x - x_n) & \text{for} \quad 0 \leq x < x_n \end{cases} \tag{4.1.4}$$

where ε_s is the relative dielectric constant of the semiconductor material and ε_0 is the dielectric constant of vacuum. The maximum value of the built-in electrical field is at the p–n interface ($x=0$), where $\mathbf{E}_{max} = qN_d x_n/(\varepsilon_s \varepsilon_0) = qN_a x_p/(\varepsilon_s \varepsilon_0)$. Due to charge conversion constrain, $N_d x_n = N_a x_p$ is always true. The electrical potential barrier across the pn junction is equal to the integration of the electrical field, which can be obtained through Eq. (4.1.4)

$$\Delta V_{pn} = \int_{-\infty}^{\infty} \mathbf{E}_x(x)dx = \frac{q}{2\varepsilon_s \varepsilon_0}\left(N_d x_n^2 + N_a x_p^2\right) \tag{4.1.5}$$

Consider the charge conversion constrain, $N_d x_n = N_a x_p$, we can find

$$x_n = \sqrt{\frac{2\varepsilon_s \varepsilon_0 \Delta V_{pn}}{q}\left(\frac{N_a}{N_d}\right)\frac{1}{N_d + N_a}} \tag{4.1.6a}$$

$$x_p = \sqrt{\frac{2\varepsilon_s \varepsilon_0 \Delta V_{pn}}{q}\left(\frac{N_d}{N_a}\right)\frac{1}{N_d + N_a}} \tag{4.1.6b}$$

The total width of the space-charged region (also known as the *depletion* region because all free electrons and holes are depleted) is thus,

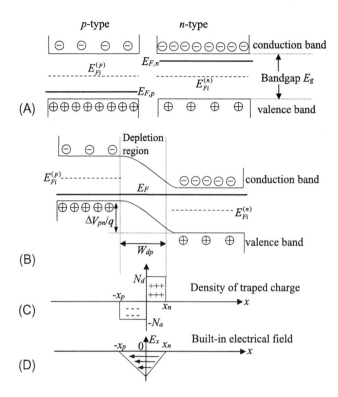

FIG. 4.1.1

(A) Energy band diagram of separate n- and p-type semiconductors, (B) energy diagram of a pn junction under equilibrium, (C) density of space change in the junction region, and (D) electrical field distribution inside the space-charged region.

$$W_{dp} = x_n + x_p - \sqrt{\frac{2\varepsilon_s\varepsilon_0 \Delta V_{pn}}{q}\left(\frac{N_a + N_d}{N_a N_d}\right)} \qquad (4.1.7)$$

Unlike a laser diode where the pn junction is forward biased, a photodiode requires a reverse bias of the pn junction. As illustrated in Fig. 4.1.2, consider a reverse bias voltage V_B applied on a pn junction, the potential barrier across the junction is increased from ΔV_{pn} to $\Delta V_{pn} + V_B$. As the consequence, the width of the depletion region is increased to (Neamen, 2003)

$$W_{dp} = \sqrt{\frac{2\varepsilon_s\varepsilon_0 \left(\Delta V_{pn} + V_B\right)}{q}\left(\frac{N_a + N_d}{N_a N_d}\right)} \qquad (4.1.8)$$

and the maximum electrical field inside the depletion region will be increased to

$$|\mathbf{E}_{max}| = \frac{2\left(\Delta V_{pn} + V_B\right)}{W_{dp}} \qquad (4.1.9)$$

As an example, assume in a silicon photodetector with the dielectric constant $\varepsilon_s = 11.7$, intrinsic carrier density at the room temperature $n_i = 1.5 \times 10^{10} \text{cm}^{-3}$, electron and hole-doping densities at n- and p-type sides are equal, $N_a = N_d = 10^{18} \text{cm}^{-3}$, the built-in potential barrier is $\Delta V_{pn} = 0.865$ V, and the width of the depletion region is approximately 4.73 μm. With a reverse bias of $V_B = 5$ V, the potential barrier is increased to $\Delta V_{pn} = 10.865$ V, the width of the depletion region will be approximately 12.3 μm, and the maximum electrical field inside the depletion region will be about 9520 V/cm.

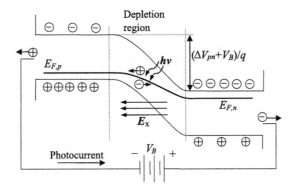

FIG. 4.1.2

Energy band diagram of a pn junction with reverse bias and photocurrent generation.

Because of the strong electrical field, there are no free electrons and holes inside the depletion region. At this time, when photons are launched into this region and if the photon energy is higher than the material bandgap ($h\nu > E_g$), they may breakup initially neutral electron-hole pairs into free electrons and holes. Then, under the influence of the electric field, the holes (electrons) will move to the right (left) and create an electrical current flow, called a *photocurrent*.

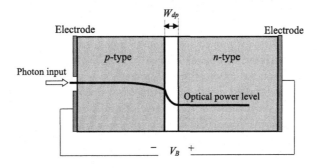

FIG. 4.1.3

Illustration of photon absorption in PN photodiode.

Fig. 4.1.3 illustrates the structure of a PN junction photodiode, in which the optical signal is injected from the p-type side. The p- and n-regions are both highly conductive and the electrical field is built-up only within the depletion region where the photons are absorbed to generate photocurrent. Photodetection efficiency, η, is proportional to the photon absorption within the depletion region as $\eta \propto 1 - \exp(-\alpha W_{dp})$, where α and W_{dp} are the absorption coefficient and the width of the depletion layer. Thus, it is desirable to have a wide depletion layer for efficient optical absorption. For example with $\alpha = 500$ cm^{-1} and $W_{dp} = 12$ μm, the absorption efficiency is on the order of 45%.

A well-known technique to increase photon absorption efficiency is to insert an undoped (or very lightly doped) intrinsic layer between the p- and n-type layers to form the so-called PIN structure as illustrated in Fig. 4.1.4.

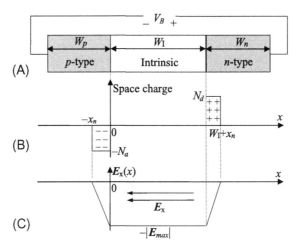

FIG. 4.1.4

(A) Illustration of PIN structure, (B) distribution of space charge, and (C) built-in electrical field distribution.

Since the intrinsic section is not doped, the space charges are only built in the p- and n-type sections near the interface with the intrinsic region as shown in Fig. 4.1.4. The maximum amplitude of the electrical field $|\mathbf{E}_{max}|$ produced by the space charges can be evaluated in the same way as in a PN junction, which is, $|\mathbf{E}_{max}| = qN_d x_n/(\varepsilon_s \varepsilon_0) = qN_a x_p/(\varepsilon_s \varepsilon_0)$. Since the charge density inside the intrinsic region is negligible in comparison to those in the doped regions, the maximum electrical field $|\mathbf{E}_{max}|$ is maintained as a constant within the intrinsic region. The major advantages of a PIN photodiode structure are that the width of the depletion region can be significantly increased and determined in the fabrication, and the built-in electrical field is uniform in the intrinsic region. Typically, the thickness of the intrinsic layer of a PIN photodiode is on the order of 50–100 μm.

Assume that each photon absorbed within the intrinsic layer produces an electrical carrier, whereas the photons absorbed outside the intrinsic layer are lost; the quantum efficiency η can then be defined by the ratio between the number of electrical carriers generated and the number of photons injected:

$$\eta = (1-R)\exp(-\alpha_p W_P)[1 - \exp(-\alpha_i W_I)] \tag{4.1.10}$$

where W_P and W_I are the widths of p-type and the intrinsic layers and α_p and α_i are the absorption coefficients of these two layers. R is the surface reflectivity of the p-type layer which faces the input light. Since photon absorption in the n-type layer does not contribute to the photocurrent, its width and absorption coefficient are not considered in Eq. (4.1.10). For a comparison with a PN junction, if we still assume the absorption coefficient of $\alpha = 500$ cm^{-1} in the depletion region, the increase of the depletion layer thickness from 12 μm in a PN structure to 50 μm in a PIN structure will increase the absorption efficiency from 45% to 92%. To obtain high detection efficiency, a PIN photodiode should have a thin p-type region to minimize the attenuation there, and a thick intrinsic region to completely absorb the incoming signal photons and converts them into photocurrent. An antireflection coating is also necessary to reduce the reflection of the front surface.

Another desired property of a photodiode is a sufficiently large optical window to accept the incoming photons. An image of a packaged large area photodiode is shown in Fig. 4.1.5A, where the optical signal comes from the p-type side of the wafer. Because the doped semiconductor layers are highly conductive, the reverse bias can be applied across the intrinsic layer uniformly without the metal contacts covering the entire surface of the device as illustrated in Fig. 4.1.5B, and thus the optical window size can be made large. However, large detection area and thick intrinsic layer will inevitably increase the response time of photodetection due to the increased RC constant and longer photon/electron transient time (Donati, 2000), which will be discussed in the next section.

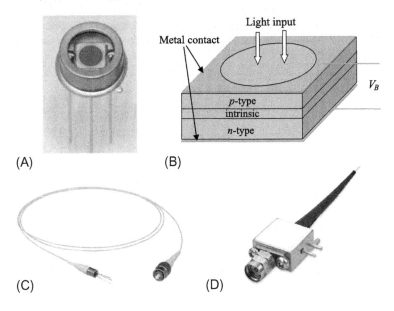

(A)   (B)

(C)   (D)

FIG. 4.1.5

(A) Image of a packaged large area PIN photodetector, (B) illustration of the layer structure and electrodes, (C) image of a fiber pigtailed photodiode of relatively low speed, and (D) image of a high-speed fiber pigtailed photodiode with an RF connector.

4.2 RESPONSIVITY AND ELECTRIC BANDWIDTH
4.2.1 QUANTUM EFFICIENCY AND RESPONSIVITY

As we have discussed previously that quantum efficiency, η, is defined by the number of electrons generated for every incoming photon, which is determined by both the semiconductor material and the structure of the photodiode. Whereas, responsivity, $\mathfrak{R}$, is a more engineering measure of the photodiode performance, which tells how many milliamps of photocurrent can be generated for every milliwatts of input signal optical power. Within a time window Δt, if the photodiode receives N photons, the optical power is $P = N \cdot h\nu / \Delta t$, where $h\nu$ is the photon energy. If each photon generates an electron, that is, $\eta = 1$, the photocurrent will be $I = N \cdot q / \Delta t$, so that the responsivity is $\mathfrak{R} = I/P = q/h\nu$, where q is the electron charge. Considering the nonideal quantum efficiency η as explained in Eq. (4.1.5), the photodiode responsivity will be

$$\mathfrak{R} = \frac{I(mA)}{P(mW)} = \eta \frac{q}{h\nu} = \eta \frac{q\lambda}{hc} \tag{4.2.1}$$

where λ is the wavelength and c is the speed of light. It is interesting to note that the responsivity is linearly proportional to the wavelength of the optical signal. With the increase of wavelength, the energy per photon becomes smaller, and each photon is still able to generate a carrier but with a lower energy. Therefore the responsivity becomes higher at longer wavelengths. However, when the wavelength is too long and the photon energy is too low, the responsivity will suddenly drop to zero because the necessary condition $hv > E_g$ is not satisfied anymore, as illustrated in Fig. 4.2.1A. The longest wavelength to which a photodiode can still have nonzero responsivity is called *cutoff wavelength*, which is $\lambda_c = hc/E_g$, where E_g is the bandgap of the semiconductor material used to make the photodiode. Typically, the spectral response of a silicon-based photodiode covers the entire visible wavelength window with a cutoff wavelength at about 1000 nm. An InGaAs-based photodiode can extend the wavelength to approximately 1700 nm as shown in Fig. 4.2.1B. In general, silicon technology is more mature thanks to the wide application of CMOS and CCD in computer systems and consumer electronics, and the price of silicon-based photodetectors can be orders of magnitude lower than those based on the compound semiconductors such as InGaAs or InP. However, because the responsivity of silicon-based photodetectors cutoff at 1000 nm, they cannot be used for optical communication systems operating in 1310 or 1550 nm wavelength windows.

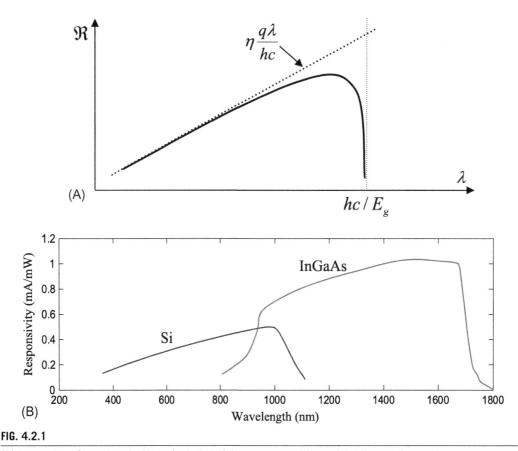

FIG. 4.2.1

(A) Illustration of wavelength-dependent photodetector responsivity and the impact due to photon energy and material bandgap and (B) responsivity of practical photodiodes based on Si and InGaAS.

As an example, for a photodiode operating at 1550 nm wavelength, if the quantum efficiency is $\eta = 0.85$, the responsivity can be easily found with Eq. (4.2.1), as $\Re \approx 1.06$ [mA/mW]. As a rule of thumb, for a typical photodiode, 1 mW of optical power should produce about 1 mA photocurrent.

4.2.2 SPEED OF PHOTODETECTION RESPONSE

The responsivity discussed in the previous section describes the efficiency of photodetection, which can be improved by increasing the thickness of the intrinsic layer so that fewer photons can escape into the n-type layer, as shown in Fig. 4.1.2. Larger area of photodetectors may also be desired to minimize the difficulty of optical alignment in the receiver. However, both of these can result in a slower response speed of the photodiode (Anderson and McMurtry, 1966). Now let us see why.

Suppose the input optical signal is amplitude modulated as

$$P(t) = P_0(1 + k_m \cos \omega t) \tag{4.2.2}$$

where P_0 is the average optical power, ω is the modulation angular frequency, and k_m is the modulation index. The free electrons generated through photodetection are distributed along the z-direction in various locations within the intrinsic region proportional to the optical power distribution as illustrated in Fig. 4.1.2. Assume that carrier drift velocity is v_n under the electrical field; the photocurrent density distribution will be a function of z as

$$J(z, \omega) = J_0(1 + k_m \cos \omega z / v_n) \tag{4.2.3}$$

where J_0 is the average current density. The total electrical current will be the collection of contributions across the entire intrinsic layer. Assume all carriers are generated at the middle of the intrinsic region, the photo-current is approximately,

$$I(\omega) = \int_{-W_I/2}^{W_I/2} J(z, \omega) dz = J_0 W_I \left[1 + k_m \text{sinc} \left(\frac{\omega \tau_i}{2\pi} \right) \right] \tag{4.2.4}$$

where $\tau_t = W_I/v_n$ is the carrier drift time across the intrinsic region. Neglecting the *DC* parts in Eqs. (4.2.2) and (4.2.4), the photodiode response can be obtained as

$$H(\omega) = \frac{\eta}{j\omega \tau_t} \left(e^{j\omega \tau_t} - 1 \right) = \eta \exp \left(\frac{j\omega \tau_t}{2} \right) \cdot \text{sinc} \left(\frac{\omega \tau_t}{2\pi} \right) \tag{4.2.5}$$

where $\eta = J_0/P_0$ is the average responsivity, and $\text{sinc}(x) = \sin(x)/x$. The 3-dB bandwidth of $|H(\omega)|^2$ can be easily found as approximately (Liu, 1996)

$$\omega_c \approx \frac{2.8}{\tau_t} = 2.8 \frac{v_n}{W_I} \tag{4.2.6}$$

Clearly, increasing the thickness of the intrinsic layer will reduce the response speed of the photodiode, and this effect is known as *carrier transient*. The velocity of carrier drift v_n increases with the increase of external bias voltage, but it saturates at a speed of approximately 8×10^6cm/sin silicon, which corresponds to the field strength of about 2×10^4 V/cm. As an example, for a silicon-based photodiode with 50 µm intrinsic layer thickness, according to Eq. (4.2.6) the 3-dB electrical bandwidth is approximately $f_c = \omega_c/2\pi = 710$ MHz. As the carrier mobilities in InGaAs and InP are much higher than that in silicon, carrier drift can be much faster in these compound semiconductors, allowing much high speed of photodetection. In fact, 100 GHz electrical bandwidth has been reported for InGaAs-based photodiodes for application in 1550 nm wavelength high-speed optical communication systems.

Ideally, all the electrical carriers are generated inside the intrinsic region where the electrical field is high and carrier drift is fast. But in a practical photodiode, incoming photons have to penetrate through the p-type region before reaching to the intrinsic layer as shown in Fig. 4.1.2, and a small fraction of photons could be absorbed to generate electrical carriers. As the doped region is highly conductive and thus the electrical field is practically zero, there is no drift for the carriers generated there. The only mechanism of carrier transport in the doped region is diffusion, which is usually much slower than the carrier drift in the intrinsic region. Therefore, the diffusion of carriers generated outside the intrinsic region may introduce a long tail after the falling edge of the impulse response.

Another important parameter affecting the speed of a photodiode is the junction capacitance, which can be expressed as

$$C_j = \frac{\varepsilon_s \varepsilon_0 A}{W_I} \tag{4.2.7}$$

where ε_s is the dielectric constant and ε_s is the free space permittivity of the semiconductor, and A is the junction cross section area. For example, for a silicon photodiode with intrinsic layer thickness $W_I = 500$ µm, and $A = 3 \times 3$ mm^2 junction area, the capacitance is approximately 19 pF. Considering a $R = 50 \Omega$ load resistance, the RC constant is on the order of $t_{RC} \approx 1 ns$, which limits the electrical bandwidth to 1 GHz. For a photodiode requiring high detection speed, the junction area has to be made small enough. For a photodiode used in a high-speed fiber-optic receiver, the cross section area can be made as small as ~ 100 µm^2 so that the RC constant can be small. But the optical coupling from optical fiber to the photodiode has to be precisely aligned to avoid coupling loss.

The overall detection speed of a photodiode is determined by the combination of all the effects discussed above. The parameter of electrical bandwidth alone may not be enough to describe the characteristics of a photodiode, and a measure of time domain response may provide more details. Fig. 4.2.2 shows the photocurrent $I(t)$ generated in a photodiode from a square optical pulse. Ideally, the rise time t_r and the fall time t_f are equal $t_r = t_f = (t_t + t_{RC})$ based on the contributions from carrier transient effect and circuit RC parasitics. However, in a practical photodiode, the rise time may not be the same as the fall time due to the impact of carrier accumulation and diffusion. Because of the relatively slow diffusion time, carriers generated outside the intrinsic region may accumulate along with the pulse, shown as a slightly positive slope in the photocurrent. The slow diffusion of carriers generated outside the intrinsic region is also responsible for a long tail after the pulse as shown in Fig. 4.2.2, which may introduce inter-symbol interference in a digital optical system.

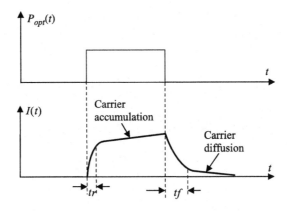

FIG. 4.2.2

Illustration of time-domain impulse response of a photodiode.

4.2.3 ELECTRICAL CHARACTERISTICS OF A PHOTODIODE

The terminal electrical characteristic of a photodiode is similar to that of a conventional diode; its current-voltage relationship is shown in Fig. 4.2.3. The diode equation is

$$I_j = I_D \left[\exp\left(\frac{V}{xV_T}\right) - 1 \right]$$

(4.2.8)

where $V_T = kT/q$ is the thermal voltage ($V_T \approx 25$ mV at the room temperature), $2 > x > 1$ is a device structure-related parameter, and I_D is the reverse saturation current, which may range from pico-ampere to nano-ampere, depending on the structure of the device. When a photodiode is forward biased (please do not try this; it could easily damage the photodiode), current flows in the forward direction, which is exponentially proportional to the bias voltage. On the other hand, when the photodiode is reversely biased, the reverse current is approximately equal to the reverse saturation current I_D when there is no optical signal received by the photodiode. Thus, I_D is also called dark current which will

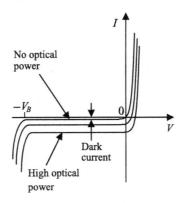

FIG. 4.2.3

Photodiode current-voltage relationship.

be discussed in the next section. With the increase of the signal optical power, the reverse current linearly increases as described by Eq. (4.2.1). Reverse bias also helps increase the detection speed as described in the last section. This is the normal operation region of a photodiode. When the reverse bias is too strong ($V \leq -V_B$), the diode may breakdown, where V_B is the breakdown voltage.

To construct an optical receiver, a photodiode has to be reversely biased and the photocurrent has to be amplified. Figs. 4.2.4 and 4.2.5 show two typical electrical circuit examples of optical receivers. The small-signal electrical signal model of a photodiode consists of a current source I_d representing photocurrent generation, a capacitance C_d with the contributions from the junction capacitance C_j of the photodiode and the parasitic capacitance of packaging, a parallel resistance R_p representing the slope of the current/voltage relation near the bias point, and a series resistance R_s arises from the resistance of the electrical contact. For a typical photodiode, R_p is on the order of mega-Ohms and R_s should be much less than a kilo-Ohm.

Fig. 4.2.4 is a voltage amplifier, where a high-load resistance R_L is usually used to achieve a high conversion gain, because a small photocurrent would convert into a large signal voltage. We will see in the following section that a high-load resistance also helps reducing the contribution of thermal noise. R_s can be neglected in this case because it is much lower than the load resistance R_L. Then, the transimpedance gain can be easily found as

$$G_v(\omega) = \frac{V_0(\omega)}{I_d(\omega)} = \frac{AR_{PL}}{1 + j\omega C_d R_{PL}} \tag{4.2.9}$$

where A is the voltage gain of the amplifier and $R_{PL} = (R_p R_L)/(R_p + R_L)$ is the parallel combination of R_p and R_L. The 3-dB electrical bandwidth of this circuit is $\omega_C = 1/(C_d R_{PL})$. At low frequencies, the transimpedance gain $G_v(0) = AR_{PL}$ can be made very high by using a mega-Ohm load resistance, but the electrical bandwidth of this circuit is usually narrow which is inversely proportional to the load resistance. For this reason, the voltage amplifier circuit shown in Fig. 4.2.4 is typically used in optical sensors which require high detection sensitivity but relatively low speed.

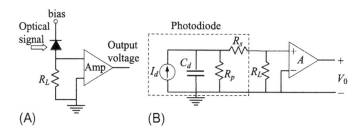

(A)   (B)

FIG. 4.2.4

(A) Photodiode preamplifier based on a voltage amplifier and (B) small-signal equivalent circuit.

Fig. 4.2.5 shows another photodiode preamplifier circuit known as the transimpedance amplifier (TIA). The equivalent load resistance seen by the photodiode is usually low enough so that the impact of parallel resistance R_p can be neglected. The transimpedance gain is

$$G_{TIA}(\omega) = \frac{V_0(\omega)}{I_d(\omega)} = \frac{-A}{1+A} \left\{ \frac{R_F}{1 + j\omega C_d [R_s + R_F/(1+A)]} \right\} \tag{4.2.10}$$

For $A \gg 1$ and $R_F/(1+A) < \, <R_s$, the transimpedance gain at low frequency is $G_{TIA}(0) = R_F$, and the electrical bandwidth is $\omega_c \approx 1/(R_s C_d)$. As an example, for a fiber-coupled photodiode with $80\,\mu m^2$ detection area, the device capacitance is $C_d = 0.4\,pF$ and the series resistance is less than $10\,\Omega$, the electrical bandwidth can be higher than $40\,GHz$. In a high-speed fiber-optic receiver, the photodiode is typically integrated with the TIA circuit in the same package, known as PIN-TIA module, with carefully matched impedance between the PIN output and the TIA input to minimize the reflection.

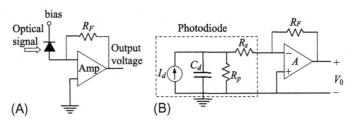

FIG. 4.2.5

(A) Photodiode preamplifier based on a transimpedance amplifier (TIA) and (B) small-signal equivalent circuit.

4.3 PHOTODETECTOR NOISE AND SNR

In an optical communication receiver, SNR is the most important parameter which determines the quality of the signal system. To achieve a high SNR, the photodiode must have a high responsivity and a low noise level. We have already discussed the quantum efficiency and the responsivity of a photodiode previously; in this section, we discuss major noise sources associated with photodetection.

Based on the responsivity defined in Eq. (4.2.1), it is straightforward to find the expression of a signal photocurrent as

$$I_s(t) = \mathfrak{R} P_s(t) \tag{4.3.1}$$

where the signal photocurrent $I_s(t)$ is linearly proportional to the received optical power $P_s(t)$ and the responsivity $\mathfrak{R}$ represents the conversion efficiency.

4.3.1 SOURCES OF PHOTODETECTION NOISE

Major noise sources in a photodiode can be categorized as thermal noise, shot noise, and dark current noise. Because of the random nature of the noises, the best way to specify them is to use their statistical values, such as spectral density, power, and bandwidth.

Thermal noise, also known as Johnson noise, is generated by the load resistor. The spectral density of the thermal noise is independent of the frequency, commonly referred to as a white noise. Thermal noise follows a Gaussian statistics and can be most conveniently characterized by its power spectral density

$$\sigma_{th}^2 = 4kT/R_L \tag{4.3.2}$$

where R_L is the load resistance, k is the Boltzmann's constant, and T is the absolute temperature. The unit of σ_{th}^2 is [W/Hz]. For a receiver with a spectral bandwidth B, the mean-square noise current representing the total thermal noise power can be expressed as $\langle i_{th}^2 \rangle = \sigma_{th}^2 B$. As the thermal noise power spectral density σ_{th}^2 is inversely proportional to the load resistance R_L, a large load resistance helps reduce thermal noise; however, as we discussed in the last section, an increased load resistance would reduce the receiver bandwidth because the increase of RC constant. In most high-speed optical receivers, $R_L = 50\,\Omega$ is usually used as a standard load resistance.

Shot noise, also known as quantum noise, arises from the statistic nature of photodetection. For example, assume 1 µW optical power at 1550 nm wavelength is received by a photodiode, it means that statistically about 7.8 trillion photons hit the photodiode within each second. These photons are not synchronized and they come randomly. Imaging that you walk on a windy beach with sand blown onto your face, although the average number of sand grains hit your face every second may be relatively constant, you may still fell the tickling of individual sand grains that shot on your face. In the case of photodetection, the generated photocurrent will fluctuate as the result of the random nature of photo arrival. Shot noise is also a white noise and the noise power spectral density can be expressed as

$$\sigma_{sh}^2 = 2qI_s = 2q\Re P_s \qquad (4.3.3)$$

which is proportional to the signal photocurrent detected by the photodiode, where q is the electron charge. Under an approximation of Gaussian statics, the mean-square noise current, or equivalently the shot noise power, can be represented by the variance of the of the shot noise current $\langle i_{sh}^2 \rangle = \sigma_{sh}^2 B$, where B is the receiver electrical bandwidth.

It is important to note that in comparison to Gaussian, Poisson distribution is a more accurate description of shot noise statistics. Assuming the average number of carriers generated by an optical signal is N_{ave} over a unit time window, the probability of having N carriers within that time window is

$$P[N] = \frac{N_{ave}^N}{N!} e^{-N_{ave}} \qquad (4.3.4)$$

where $N \geq 0$. Poisson distribution is asymmetric with respect to N_{ave} when N_{ave} is small enough, while it approaches a Gaussian distribution for N near N_{ave} when $N_{ave} \gg 1$. Thus, Gaussian statistics is a good approximation for shot noise when photocurrent is not too small, and Gaussian approximation greatly simplifies the process of receiver SNR analysis in comparison to the use of a Poisson statistics.

Dark current noise is the constant current that exists when no light is incident on the photodiode. This is the reason it is called *dark* current. As shown in Eq. (4.2.8), this dark current is the same thing as the reverse saturation current I_s because the photodiode is always reversely biased. Because of the statistical nature of the carrier generation process similar to that has been discussed for the shot noise, the dark current noise can also be treated as a white noise with the power spectral density

$$\sigma_{dk}^2 = 2qI_D \qquad (4.3.5)$$

where I_D is the dark current of the photodiode which depends on the structure of the pn junction, the doping levels of the material, and the temperature of the photodiode. For an electric bandwidth B of the receiver, the mean-square noise current, or equivalently the dark current noise power is $\langle i_{dk}^2 \rangle = \sigma_{dk}^2 B$.

In an optical receiver, an electrical preamplifier has to be used immediately after the photodiode as shown in Figs. 4.2.4 and 4.2.5 to amplify the photocurrent signal and convert it into an electrical voltage. Noise generated by the preamplifier also has to be considered in the receiver SNR analysis. This preamplifier noise is commonly expressed as an "input-referred rms noise" $\langle i_{amp}^2 \rangle$ so that it can be used together with $\langle i_{th}^2 \rangle$, $\langle i_{sh}^2 \rangle$, and $\langle i_{dk}^2 \rangle$. For example, assume a TIA is used as the preamplifier with the transimpedance gain $Z_{TIA} = v_{out}/i_{in}$, where i_{in} is the input signal current and v_{out} is the output signal voltage. Due to the noise contribution of the amplifier, an output noise power $\langle v_{amp}^2 \rangle$ is measured when the input signal current is turned off. The input-referred rms noise of the TIA can be obtained as $\langle i_{amp}^2 \rangle = \langle v_{amp}^2 \rangle / Z_{TIA}^2$.

4.3.2 SNR OF AN OPTICAL RECEIVER

Since all the noise components discussed so far are referred to the input of the preamplifier, and assume they all have Gaussian statistics, the addition of their powers represents the variance of the electrical signal. In an optical receiver, SNR is usually defined as the ratio of signal electrical power and noise electrical power. This is equivalent to the ratio of their mean-square currents,

$$SNR = \frac{\langle I_s^2(t) \rangle}{\langle i_{th}^2 \rangle + \langle i_{sh}^2 \rangle + \langle i_{dk}^2 \rangle + \langle i_{amp}^2 \rangle} \tag{4.3.6}$$

For the four noise sources of photodetection considered above, shot noise is a signal-dependent noise, while others are independent of the signal level. For the purpose of optical system performance evaluation and optimization, it is useful to distinguish the relative contributions of different noise sources to the overall SNR of the receiver.

As an example, we consider an InGaAs photodiode which operates in the 1550 nm wavelength window with 100 μm diameter of the active area. The dark current of this photodiode is $I_{dk}=5$ nA at a 5 V reverse bias voltage, the load resistance is $R_L=50\,\Omega$, the electrical bandwidth is $B=1$ GHz, and the photodiode operates in the room temperature so that $T=300$ K. The quantum efficiency of this photodiode is $\eta=0.85$, corresponding to a responsivity of $\Re=1.06$[A/W].

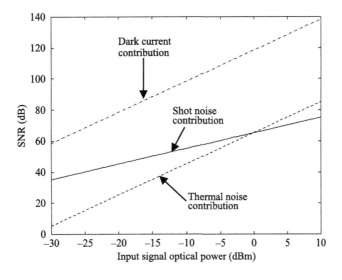

FIG. 4.3.1

Decomposing SNR by its three contributing factors.

Fig. 4.3.1 shows the calculated SNR as the function of the input optical power. Contributions due to the thermal noise, shot noise, and dark current noise are separately displayed for comparison. We have neglected the electrical preamplifier noise for simplicity.

When the received optical signal is very weak, which is the case for most optical receivers, thermal noise is often the dominant noise. In fact, in this particular receiver, the thermal noise power is $\langle i_{th}^2 \rangle = 4kTB/R_L \approx 3.3 \times 10^{-13}$ W which is independent of the signal optical power P_s. If we only consider thermal noise and neglect the contribution due to other noises, the SNR will be simplified as

$$SNR_{th} = \frac{\Re^2 P_s^2}{4kTB/R_L} = 3.2 \times 10^{12} P_s^2 \tag{4.3.7}$$

Here because the signal electrical power is proportional to the square of the input optical power, and the thermal noise is constant, the slope of the SNR vs. the signal optical power curve shows a 2 dB per dB characteristic in Fig. 4.3.1.

In terms of the dark current noise, it is also independent of the optical signal. Thus, the SNR caused by dark current noise also has the 2 dB/dB characteristic similar to that due to the thermal noise. Specifically, the dark current noise power in this example is $\langle i_{dk}^2 \rangle = 2qI_DB = 1.6 \times 10^{-18}$ W which is more than five orders of magnitude lower than the thermal noise. Therefore, the SNR caused by the dark current noise is more than 50 dB higher than that due to the thermal noise as shown in Fig. 4.3.1. In general, dark current noise is only relevant in low-speed optical receivers where the load resistance R_L can be very high, in the Megaohm level, so that the thermal noise can be lower than the dark current noise.

Unlike thermal noise and dark current noise which are independent of the signal optical power, shot noise electrical power is linearly proportional to the signal optical power as described in Eq. (4.3.3). If only shot noise is considered, the SNR will be

$$SNR_{sh} = \frac{\Re^2 P_s^2}{2q\Re P_s B} = 3.3 \times 10^9 P_s \tag{4.3.8}$$

In this case, SNR is linearly proportional to the incident optical power, so that the SNR vs signal optical power curve has a slope of 1 dB/dB. Fig. 4.3.1 indicates that shot noise becomes the limiting factor for the overall SNR only when the signal optical power is high enough and P_s approaches 0 dBm.

Theoretically in an optical receiver, thermal noise can be reduced by increasing the load resistance R_L, and dark current noise can be reduced by improving material quality and junction structural design. Shot noise, caused by the quantum nature of photons, cannot be reduced, so that it sets a fundamental limit for optical detection. In fact, quantum-limited photodetection refers to the receiver in which shot noise is the dominant noise source. This subject will be revisited later when we discuss optical systems based on the coherent detection.

4.3.3 NOISE-EQUIVALENT POWER

Noise-equivalent power (NEP) is another useful parameter commonly used to specify a photodetector. *NEP* is defined as the minimum optical power required to obtain a unity SNR for a 1 Hz electrical bandwidth. Only thermal noise is considered in the definition of NEP. From Eq. (4.3.7), if we let $\Re^2 R_L P_s^2/(4kTB) = 1$, it requires

$$NEP = \frac{P_s}{\sqrt{B}} = \sqrt{\frac{4kT}{\mathcal{R}^2 R_L}} \qquad (4.3.9)$$

According to this definition, the unit of NEP is $[W/\sqrt{Hz}]$. Some manufacturers specify NEP for their photodiode products. Obviously, small NEP is desired for high-quality photodiodes. As an example at the room temperature, for a photodetector operating in the 1550 nm wavelength window with quantum efficiency $\eta = 0.85$ and load resistance $R_L = 50\,\Omega$, the NEP value is approximately $NEP = 17.2\,[pW/\sqrt{Hz}]$. In a high-sensitivity optical receiver operating with low levels of optical signals, the NEP parameter has to be low enough to guarantee the required SNR.

4.4 AVALANCHE PHOTODIODES

The typical responsivity of a PIN photodiode is limited to the level of approximately 1 mA/mW because the quantum efficiency η cannot be higher than 100%. To further increase the detection responsivity, avalanche photodiodes (APDs) were introduced in which photocurrent is internally amplified through carrier multiplication inside the junction structure before going to the electric load. This requires a section in the junction structure to have very high electrical field so that the photon-generated carriers can acquire sufficiently high kinetic energy and initiate the carrier multiplication process (Smith and Forrest, 1982; Brain, 1982). As illustrated in Fig. 4.4.1A, an APD consists of a highly doped p-type section (p^+) with the acceptor doping density N_a^+, a highly doped n-type section (n^+) with the acceptor doping density N_a^+, an intrinsic section, and a regularly doped p-type region (p) which interfaces with the n^+ region. As has been discussed in Section 4.1, the derivative of the built-in electrical field, $|E_x|$, is proportional to the density of the space charge, a sharp increase of electrical field can be created at the

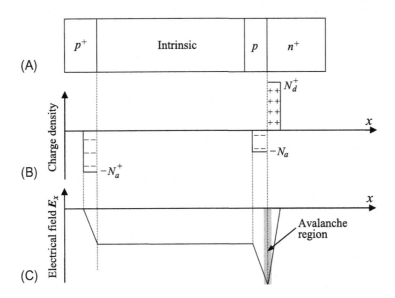

FIG. 4.4.1

(A) APD layer structure, (B) charge density distribution, and (C) electrical field density profile.

pn^+ interface as shown in Fig. 4.4.1C. This highly nonuniform field distribution allows the creation of a very high electrical field in a relatively short section, known as the avalanche region.

When a reverse biasing electrical voltage is applied on an APD, a significant portion of the biasing voltage is dropped across the avalanche region to create an extremely high electrical field locally. Electrons are generated in the intrinsic region when signal photons are injected, and these electrons will drift through the avalanche region. When the biasing voltage is high enough, each photon-generated electron is able to gain sufficient kinetic energy inside the avalanche region and knock several electrons loose from neutral electron-hole pairs; this is called *impact ionization*. These newly generated free electrons will also be able to gain sufficient kinetic energy under the same electrical field inside the avalanche region and to create free carriers known as higher-order carriers; this is commonly referred to as the *avalanche effect*. Fig. 4.4.2 illustrates this carrier multiplication process, where one input electron generates ten output electrons and four holes.

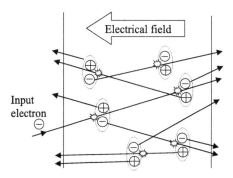

FIG. 4.4.2

Illustration of the carrier multiplication process.

4.4.1 APD USED AS A LINEAR DETECTOR

For application in optical communication systems, an APD is usually used as a linear detector in which the photocurrent is linearly proportional to the received signal optical power. The contribution of the avalanche process is that each input photon may generate multiple free electrons and holes. Thus, the responsivity expression of an APD has to include the effect of carrier multiplication,

$$\mathfrak{R}_{APD} = M_{APD}\mathfrak{R} = M_{APD}\eta\frac{q\lambda}{hc} \tag{4.4.1}$$

where M_{APD} is defined as the APD gain, and $\mathfrak{R}$ is the responsivity of a PIN photodiode defined by Eq. (4.2.1). Since the avalanche process depends on the electrical field in the avalanche region, the APD gain strongly depends on the voltage of the reverse bias. A simplified expression commonly used for APD gain is

$$M_{APD} = \frac{1}{1 - (V_B/V_{BD})^{n_B}} \tag{4.4.2}$$

where n_B is a parameter that depends on the device structure and the material. V_B is the applied reverse bias voltage and V_{BD} is defined as the breakdown voltage of the APD, and Eq. (4.4.2) is valid only for

$V_B < V_{BD}$. Obviously, when the reverse bias voltage approaches the breakdown voltage V_{BD}, the APD gain approaches infinity.

In addition to APD gain, another important parameter in an APD is its frequency response. In general, the avalanche process increases the response time of the APD and reduces the electrical bandwidth. This bandwidth reduction is proportional to the APD gain. A simplified equation describing the frequency response of APD gain is

$$M_{APD}(\omega) = \frac{M_{APD,0}}{\sqrt{1 + (\omega \tau_e M_{APD,0})^2}} \tag{4.4.3}$$

where $M_{APD,0} = M_{APD}(0)$ is the APD gain at DC as shown in Eq. (4.4.2) and τ_e is an effective transient time, which depends on the thickness of the avalanche region and the speed of the carrier drift. Therefore, the 3-dB bandwidth of APD gain is

$$f_c = \frac{1}{2\pi \tau_e M_{APD,0}} \tag{4.4.4}$$

In practical applications, the frequency bandwidth requirement has to be taken into account when choosing APD gain.

Due to the effect of carrier multiplication in an APD, signal photocurrent is

$$I_{s,APD}(t) = \Re M_{APD} P_s(t) \tag{4.4.5}$$

As far as the noises are concerned, since the thermal noise is generated in the load resistor R_L, it is not affected by the APD gain. However, both shot noise and the dark current noise are generated within the photodiode, and they will be enhanced by the APD gain (Fyath and O'Reilly, 1989). Within a receiver bandwidth B, the mean-square shot noise current in an APD is

$$\langle i^2_{sh,APD} \rangle = 2q \Re P_s B M^2_{APD} F(M_{APD}) \tag{4.4.6}$$

The dark current noise in an APD is

$$\langle i^2_{dk,APD} \rangle = 2q I_D B M^2_{APD} F(M_{APD}) \tag{4.4.7}$$

In both Eqs. (4.4.6) and (4.4.7), $F(M_{APD})$ is a noise figure associated with the random nature of carrier multiplication process in the APD. This noise figure is proportional to the APD gain M_{APD}. The following simple expression is found to fit well with measured data for most practical APDs:

$$F(M) = (M_{APD})^x \tag{4.4.8}$$

where $0 \leq x \leq 1$, depending on the material. For often used semiconductor materials, $x = 0.3$ for Si, $x = 0.7$ for InGaAs, and $x = 1$ for Ge APDs.

From a practical application point of view, APD has advantages compared to conventional PIN when the received optical signal is very weak and the receiver SNR is limited by thermal noise. In quantum-noise-limited optical receivers, such as coherent detection receivers, APD should, in general, not be used, because it would only increase noise level and introduce extra limitations in the electrical bandwidth.

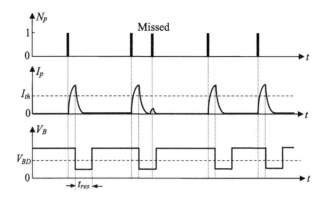

FIG. 4.4.3

Operation principle of a single-photon detector. N_p: number of photons received, I_p: photocurrent, V_B: applied biasing voltage.

4.4.2 APD USED AS A SINGLE-PHOTON DETECTOR

For an APD used as a linear photodetector, the APD is biased slightly below the breakdown threshold V_{BK}. In this case, the photocurrent is linearly proportional to the signal optical power, and the APD gain is typically less than 25 dB. On the other hand, an APD can also be used as a single-photon detector to detect very weak optical signals (Jiang et al., 2007). Fig. 4.4.3 explains the operation principle of a single-photon detector, in which N_p is the number of arriving photons, I_p is the photocurrent, and V_B is the reverse bias voltage. For the application as a single-photon counter, the reverse bias voltage of the APD is set to be higher than the breakdown voltage ($V_B > V_{BK}$) so that the detector operates in an unstable condition. Because the strong impact ionization effect above breakdown voltage, a single photon arriving is enough to trigger an avalanche breakdown of the APD and generate large electric current in the milliamper level. In order to detect another photon which may arrive later, the APD has to be reset to the pre-breakdown state by temporarily reducing the bias to the level below than the breakdown voltage. This operation is commonly referred to as the Geiger mode (Donnelly et al., 2006). The number of photons can be measured by counting the number of photocurrent pulses. The reset window has to be placed immediately after each photo-induced breakdown event, and the width of the reset window, t_{res}, has to be small enough to allow high-speed photon counting. This bias reset is commonly referred to as quenching. In the example shown in Fig. 4.4.3, the third photon is not counted because its arrival time is within the bias reset window. In practical applications, the resetting of electric biasing has to be triggered by each photodetection event, which can be achieved by either passive quenching, or active quenching as shown in Fig. 4.4.4. Passive quenching is accomplished in the component level using a series load resistor to provide a negative feedback. As shown in Fig. 4.4.4A, upon each avalanche event, a large photocurrent pulse is created which results in a large voltage drop V_R on the load resistor. Thus, the voltage applied on the APD is reduced to $V_S - V_R$, which is lower than the breakdown voltage V_{BD} of the APD. On the other hand, active quenching, shown in Fig. 4.4.4B, is a circuit-level approach which uses an amplified negative feedback circuit to switch off the electric bias after each avalanche event. Active quenching usually allows better control in the reset pulse width and depth.

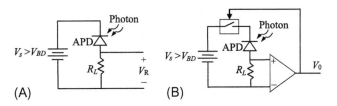

FIG. 4.4.4

Biasing circuits for passive (A) and active (B) quenching of a Geiger mode APD.

4.5 OTHER TYPES OF PHOTODETECTORS

Although photodiodes are most commonly used in optical communication receivers, there are a number of other types of photodetectors including photovoltaic (PV) for solar energy harvesting and charge-coupled devices (CCDs) for imaging. These photodetectors may also be used in optical communication systems for power supplying, sensing, and spectrum analyzing. In this section, we briefly discuss the basic operation principles of PV and CCD.

4.5.1 PHOTOVOLTAIC

PV, or commonly referred to as solar cell, has the similar structure as a photodiode made of a pn junction (Neamen, 2003). Instead of reversely biased in the operation, a PV is not electrically biased.

As discussed in Section 4.1, for an unbiased pn junction, there exists a built-in electric field inside the space-charge region ($-x_p < x < x_n$), which creates a potential barrier across the junction as

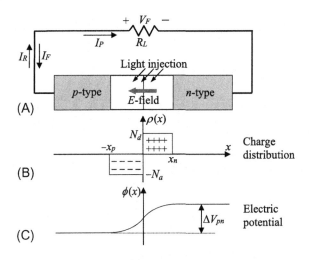

FIG. 4.5.1

(A) Illustration of a photovoltaic based on a pn junction, (B) charge distribution, and (C) potential distribution

$$\Delta V_{pn} = V_T \ln \left(\frac{N_d N_a}{n_i^2} \right) \qquad (4.5.1)$$

where N_d and N_a are the donor densities in the n-type section and the acceptor density in the p-type section, respectively. n_i is the intrinsic carrier density, and V_T is the thermal voltage which is approximately $25\,mV$ at the room temperature.

Incident photons into the space charge region with the photon energy higher than the bandgap can create electron-hole pairs. These electrons and holes will drift to the right and left of the space charge region, respectively, due to the built-in electric field, and creating a photocurrent I_R in the reverse direction of the pn junction as shown in Fig. 4.5.1. Based on Eq. (4.2.1), this photocurrent can be calculated as

$$I_R = \mathfrak{R} P_{opt} = \eta \frac{q\lambda}{hc} P_{opt} \qquad (4.5.2)$$

where P_{opt} is the received optical power, $\mathfrak{R}$ is the responsivity, η is the quantum efficiency, h is the Planck's constant, c is the speed of light, and λ is the photon wavelength. Meanwhile, a forward biasing voltage V_F is induced when this reverse photocurrent flows through the load resistor R_L. This forward biasing voltage will create a forward current through the diode due to the increased diffusion, which can be described by a standard diode equation $I_F = I_S[\exp(V_F/V_T) - 1]$ with I_S the reverse saturation current. The combination of the reverse photocurrent I_R and the forward diode current I_F results in the net photocurrent I_P flowing through the load resistor, which is

$$I_P = I_R - I_S \left[\exp\left(\frac{V_F}{V_T} \right) - 1 \right] \qquad (4.5.3)$$

The property of a PV is commonly represented by an open-circuit voltage and a short-circuit current. The short-circuit current I_{SC} is equal to I_R, as it can be obtained with $R_L = 0$, and thus $V_F = 0$. The open-circuit voltage V_{OC} is obtained with $R_L = \infty$, and $I_P = 0$. That is, $I_R = I_S[\exp(V_{OC}/V_T) - 1]$, so that

$$V_{OC} = V_T \ln(1 + I_R/I_S) \qquad (4.5.4)$$

Therefore, the photocurrent, I_P, can be expressed as the function of V_F and V_{OC} as

$$I_P = I_S \left[\exp\left(\frac{V_{OC}}{V_T} \right) - \exp\left(\frac{V_F}{V_T} \right) \right] \qquad (4.5.5)$$

The electric power delivered from the PV to the load resistor R_L is

$$P_L = I_P V_F = I_R V_F - I_S \left[\exp\left(\frac{V_F}{V_T} \right) - 1 \right] V_F \qquad (4.5.6)$$

Fig. 4.5.2 shows the relations photocurrent and the electric power as the function of the load voltage V_F. Both photocurrent and electric power increase with the open-circuit voltage V_{OC}, which is, in turn, related to the received optical power as described by Eqs. (4.5.4) and (4.5.2). Fig. 4.5.2B indicates that there is an optimum voltage V_F which results in a maximum electric power at the load for each V_{OC}. This corresponds to an optimum load resistance which matches to the equivalent output resistance of the PV.

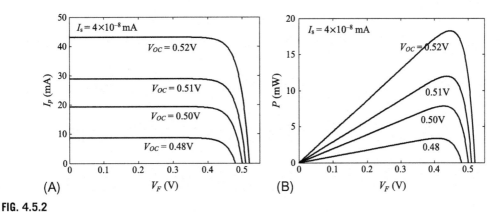

FIG. 4.5.2

(A) Photocurrent as the function of voltage on the load and (B) electric power delivered to the load resistor as the function of voltage on the load. $I_s = 4 \times 10^{-8}$ mA is assumed in the calculation.

Fig. 4.5.3A shows the equivalent circuit of a PV. The optimum load can be found by maximizing the electrical power from Eq. (4.5.6),

$$\frac{d}{dV_F}\left\{I_R V_F - I_S\left[\exp\left(\frac{V_F}{V_T}\right) - 1\right]V_F\right\} = 0 \tag{4.5.7}$$

This results in

$$I_R = I_S\left[\exp\left(\frac{V_F}{V_T}\right) - 1\right] + I_S\frac{V_F}{V_T}\exp\left(\frac{V_F}{V_T}\right) \approx I_S\exp\left(\frac{V_F}{V_T}\right)\left[1 + \frac{V_F}{V_T}\right] \tag{4.5.8}$$

As $I_R/I_S \approx \exp(V_{OC}/V_T)$, the optimum voltage $V_F = V_m$ can be found as

$$V_m = V_{OC} - V_T \ln\left[1 + \frac{V_m}{V_T}\right] \tag{4.5.9}$$

Because the thermal voltage is $V_T \approx 0.025$ V at the room temperature, $V_T\ln(1 + V_m/V_T) < < V_{OC}$ can be assumed in Eq. (4.5.9). So that $V_m \approx V_{OC}$, and Eq. (4.5.9) can be approximated as

$$V_m = V_{OC} - V_T \ln\left[1 + \frac{V_{OC}}{V_T}\right] \tag{4.5.10}$$

Using the numerical example shown in Fig. 4.5.2 with $I_s = 10^{-8}$ mA and $V_{OC} = 0.52$ V, the optimum voltage is approximately $V_m = 0.443$ V as illustrated in Fig. 4.5.3B. This corresponds to an optimum current $I_p = 41.2$ mA, and thus an optimum load resistance of $R_L = 10.75\,\Omega$.

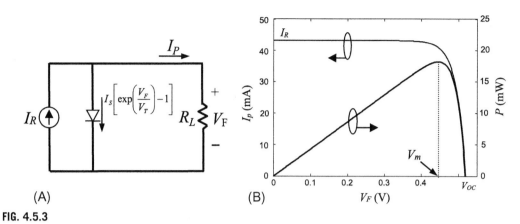

(A)   (B)

FIG. 4.5.3

(A) Equivalent electric circuit of a photovoltaic and (B) photocurrent *(left vertical axis)* and electric power delivered to the load *(right vertical axis)* as the function of load voltage. $I_s = 4 \times 10^{-8}$ mA and $V_{OC} = 0.52$ V are assumed.

4.5.2 CHARGE-COUPLED DEVICES

Charge-coupled device, known as CCD, is another very popular photodetector commonly used for digital imaging and video (Theuwissen, 1995). Unlike a pn junction-based photodiode, CCD is based on a metal-oxide-semiconductor (MOS) structure.

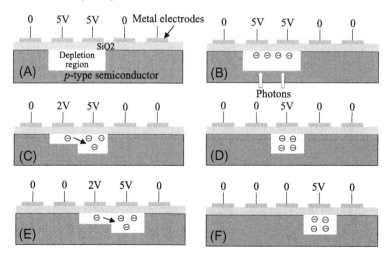

FIG. 4.5.4

(A): Cross-sectional structure of CCD and (B)–(F): illustration of programmable charge shift in the readout process.

Fig. 4.5.4A shows an example of CCD structure with a p-type semiconductor body, a thin silicon dioxide insulating layer and an array of gating electrodes. A positive bias voltage applied on a gate electrode repels holes away from the area underneath the electrode, creating a depletion region.

Incoming photons are able to generate photoelectrons in the depletion region as illustrated in Fig. 4.5.4B. These photon-induced charges are then shifted programmably in the horizontal direction to one side of the array so that they can be electrically amplified and collected. Charge shifting can be accomplished by progressively shifting gate voltage along the array as shown in Fig. 4.5.4C–F. The gating electrodes of CCD made for imaging are usually arranged in a two-dimensional (2D) array as illustrated in Fig. 4.5.5. As the imaging sensor, CCD is usually mounted on the focal plane of a camera. After each exposure, a charge distribution pattern is created on the 2D plane of the CCD, which is proportional to the intensity distribution of the image. The readout circuits of CCD performs a parallel to series conversion. The row-shifter circuit shifts the charges recorded by each row downward in the vertical direction into horizontal registers. Corresponding to each step of row shift, the pixel shifter circuit moves the charge stored in the horizontal registers pixel-by-pixel in the horizontal direction into a preamplifier. This process effectively translates the 2D image array into a waveform in the time domain, which can then be digitized, processed, and recorded.

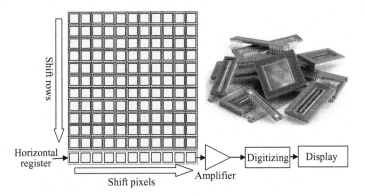

FIG. 4.5.5

Illustration of row and pixel shift for converting a 2D imaging array into a time-domain waveform and a picture of 1D and 2D image sensor products.

Used with permission from Teledyne DALSA.

The quality of CCD is determined by several parameters, including quantum efficiency, dark charge level, and dynamic range. Here, quantum efficiency is defined as the number of photoelectrons produced divided by the number of impinging photons, which is the same as that defined for a photodiode. The dark charge is defined as the number of charge electrons that leaks into a pixel during the exposure time in the absence of light. The dark charge level limits the minimum light intensity that the CCD can detect. In fact, the number of the received photons times the quantum efficiency has to be greater than the dark charge to ensure the quality of imaging. The dynamic range is related to the well depth defined as the maximum number of charged electrons a pixel can store without overflowing into nearby pixels. The saturation of a CCD camera often seen when the light intensity is too high is caused by insufficient dynamic range of the CCD device.

Because of the relatively large capacitance of pixels and the time needed for charge accumulation and sequential readout, the speed of CCD is usually not fast enough for receivers in optical communication systems. But CCDs are often used as a part of optical sensors and spectrometers for performance monitoring in optical systems and networks.

4.6 SUMMARY

In this chapter, we have discussed device structures, operation principles, and key parameter definitions and specifications of photodiodes as an indispensable basic building block of a fiber-optic system. Basically, a photodiode is a reversely biased PN (or PIN) junction which converts the incoming signal optical power into an electrical current.

Important parameters of a photodiode include responsivity, detection speed, and the noises introduced in the detection process. Responsivity in [A/W] is a measure of detection efficiency which is primarily related to the quantum efficiency. The responsivity of a photodiode is generally wavelength dependent, which is determined by the bandgap of the semiconductor material as well as the photon energy of the optical signal. Detection speed is another important property of a photodiode which is often represented by the electrical bandwidth. Detection speed is determined by the speed of carrier drift and the electric RC constant. Increase the reverse biasing voltage helps increasing the electric field in the active region, while decreasing the detection area helps reducing the junction capacitance. Practically, the maximum speed of carrier drift is eventually limited by the carrier mobility of the semiconductor material, and the minimum detection area is limited by the minimally achievable spot size of the optical signal which is usually the core size of the optical fiber. In general, a photodiode with large detection area is easy to use, but usually has a narrow detection electric bandwidth.

APD is a special type of photodiode which has significantly higher responsivity than a conventional photodiode. This increased responsivity is achieved by introducing a section in the photodiode with very high electric field, so that carriers can be accelerated to very high speed when passing through that section. These carriers with high kinetic energy can collide with the material lattice structure to create new carriers in a process like an avalanche, which effectively amplifies the photocurrent. From an application point of view, an APD requires a high reverse biasing voltage to provide the required electric field within the avalanche region. Although the avalanche gain can significantly increase the responsivity, it does not increase the electric bandwidth of the photodetection because the process of carrier colliding and regeneration of new carriers may take time.

Noise generated in the photodetection process in a photodiode is another important aspect discussed in this chapter. Major noise sources include shot noise, thermal noise, and dark-current noise. In essence, thermal noise is created in the load resistor which can be reduced by increasing the load resistance. However, the increase of load resistance R will directly increase the RC constant, resulting in the reduction of the electric bandwidth of the receiver. Indeed, 50 Ω is the standard load resistance for a high-speed receiver. Noise equivalent power (NEP) is a useful parameter which is defined as the signal optical power required to achieve electrical SNR $= 1$ within 1 Hz bandwidth. Thermal noise is the only noise source that is considered in the definition of NEP. Dark current, on the other hand, is the reverse saturation current of the pn junction when the photodiode is placed in dark. The value of dark current depends on the semiconductor material, operation temperature, and the area of the pn junction. A reduced junction area could result in smaller dark current, but would certainly impose stringent requirement on signal optical alignment and focusing into the photodiode. While both thermal noise and dark-current noise are independent of the signal optical power, shot noise is directly proportional to the signal optical power. Shot noise is intrinsically associated with the statistical nature of photodetection process, which imposes the fundamental limit on the electric SNR of the optical receiver. It is important to note that even though the avalanche gain of an APD can significantly increase the signal photocurrent, it also creates a higher level of noise related to signal optical power known as the excess APD noise. In a situation where optical signal is very small so that SNR is limited by thermal noise and dark-current noise, APD gain can help improve SNR by increasing the signal photocurrent. However, when signal optical power is high enough so that SNR is limited by the shot noise, the introduction of APD gain will likely deteriorate SNR because of the excess APD noise.

To amplify the weak photocurrent into a voltage signal in an optical receiver, a low-noise preamplifier has to be used. For a low-speed optical receiver usually used in optical sensors, a high-load resistance of photodiode can be used which translates the photocurrent into a relatively high-voltage signal which is then amplified by a voltage amplifier. Although the electric bandwidth is narrow, the high-load resistance in the mega-ohm level helps significantly reducing the impact of thermal noise. Otherwise for high-speed optical receivers required for fiber-optic communications, 50 W standard load resistance has to be used with transimpedance preamplifier to maximize the electric bandwidth.

Because photocurrent generated in a photodiode is proportional to the signal optical power which is the square of the optical field, photodiode is known to follow the square-law detection rule. This square-law detection can generate mixing products between different frequency components of the received optical field, and this property is utilized for coherent detection which mixes the received optical signal with an optical local oscillator at the photodiode which will be discussed later in Chapter 9. Similarly, optical signal can also mix with broadband optical noise in the photodiode which translates the optical noise into electric baseband. This will be discussed in Chapter 5 where optical amplifiers are discussed which generate broad optical noise.

As both PV and CCDs are also photodetectors which convert photons into electrons, we have briefly discussed the structures and applications of PV and CCD at the end of this chapter. As the major application of PV is to generate electric energy from sunlight, it has to be made with large area, low cost, and high photon conversion efficiency. The load resistance of a PV panel is selected based on maximizing the output power. CCD on the other hand is the dominant device used for imaging acquisition in cameras and cell phones. Programmable charge transfer and storage, low dark charge and large dynamic range are most desirable. Instead of using pn junction structure, a CCD is based on the MOS structure which can be made in the form of 2D array with large number pixels. Pixel size, and programmable charge translation and storage are among main design considerations. Low dark charge will improve detection sensitivity and large dynamic range will allow acquiring images with large contrast.

Basic structures, operation principles, and key parameter definitions and specifications discussed in this chapter are essential knowledge in fiber-optic system design, performance evaluation, and trouble shooting. Understanding and the ability to characterize different types of noises in an optical receiver is especially important in optical communication system performance design and optimization.

PROBLEMS

1. Consider a silicon pn junction diode operating at temperature $T = 300\,\mathrm{K}$, and its intrinsic carrier density is $n_i = 1.5 \times 10^{10}\mathrm{cm}^{-3}$. The dopant concentrations are $N_a = 2.25 \times 10^{16}\mathrm{cm}^{-3}$ on the p-type side and $N_d = 10^{16}\mathrm{cm}^{-3}$ on the n-type side. For silicon, the relative dielectric constant is $\varepsilon_s = 11.7$, and dielectric constant of vacuum $\varepsilon_0 = 8.85 \times 10^{-14}\mathrm{F/cm}$.

Find
(a) the built-in electric voltage ΔV_{pn} across the pn junction,
(b) the width of depletion region for the bias voltage $V_B=0$ and $V_B=10\,V$, and
(c) the maximum electric field in the space charge region for the bias voltage $V_B=0$ and $V_B=10\,V$.

2. Consider a silicon PIN structure with the doping profile shown in the following figure and the temperature is $T=300\,K$. The intrinsic carrier density is $n_i=1.5 \times 10^{10} cm^{-3}$, the relative dielectric constant is $\varepsilon_s=11.7$, and dielectric constant of vacuum $\varepsilon_0=8.85 \times 10^{-14} F/cm$.

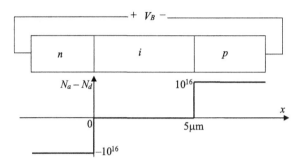

(a) For $V_B=0\,V$, find the widths of depletion region x_n on the n-type side and x_p on the p-type.
(b) For $V_B=10\,V$, draw and label the field distribution across the junction region.

3. Bandgaps of silicon, germanium, and InGaAs are 1.13, 0.78, and 0.73 eV, respectively, what are their cutoff wavelengths when used as photodiodes?
4. Silicon has a bandgap of 1.13 eV. For a silicon photodiode with a quantum efficiency of 0.85, what are the responsivities of this photodiode at signal wavelengths of 700, 1000, and 1500 nm?
5. For a PIN photodiode shown in the following figure, the photon absorption coefficient rates are $\alpha=400\,cm^{-1}$ for the p- and n-doped regions, and $\alpha=1000\,cm^{-1}$ for the intrinsic region. Assume both p- and n-doped regions are highly conductive so that the depletion region is only restricted within the intrinsic layer. Layer thicknesses are $W_p=2\,\mu m$ and $W_n=4\,\mu m$, and the thickness of the intrinsic layer is a design variable. Neglect all surface reflections, and assume the electrodes are all transparent.

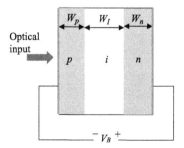

(a) Plot the quantum efficiency of this photodiode as the function of intrinsic layer thickness for $2\,\mu m < W_I < 20\,\mu m$.

(b) Assume the cross section area of this photodiode is $A=100\,\mu m^2$, the relative dielectric constant is $\varepsilon_s=11.7$, dielectric constant of vacuum is $\varepsilon_0=8.85\times10^{-14}F/cm$, and the load resistance is 50Ω, plot the cutoff frequency determined by the RC constant $[f_c=1/(2\pi RC)]$ as the function of intrinsic layer thickness for $2\,\mu m<W_I<20\,\mu m$.

(c) Assume that carrier velocity inside the intrinsic region is $v_n=2\times10^7cm/s$, plot the cutoff frequency f_c determined by the carrier transient time as the function of intrinsic layer thickness for $2\,\mu m<W_I<20\,\mu m$. To optimize the frequency bandwidth, what is the optimum W_I?

6. An optical signal $P(t)=P_{ave}[1+\cos(2\pi f_0 t)]$ is detected by a photodiode with the bandwidth much higher than f_0. The photodiode has a responsivity $\mathfrak{R}=0.9A/W$, and there is a DC block (which removes the DC component from the photocurrent) at the photodiode output circuit).

(a) For $P_{ave}=-20\,dBm$, what is the mean square of the photocurrent at the photodiode output?

(b) For every dB increase of the P_{ave}, what is the corresponding increase (in dB) of the electrical signal power at the photodiode output?

7. A PIN photodiode operates in the 1550 nm wavelength window with responsivity $\mathfrak{R}=0.5A/W$. The operating temperature is $T=300\,K$, load resistance is $R_L=50\Omega$, and the dark current is 1 nA.

(a) Please find the total power of thermal noise, shot noise, and dark current noise within 10 GHz electric bandwidth for each of the following input signal optical power levels: -20, -10, and 0 dBm,

(b) At which input signal optical power shot noise is equal to thermal noise?

8. A PIN photodiode operates in the 1550 nm wavelength window with responsivity with 100% quantum efficiency. The operating temperature is $T=300\,K$, load resistance is $R_L=50\Omega$.

(a) If only shot noise is considered (neglect all other noises), and assume that the receiver bandwidth is 1 Hz, what is the signal optical power required to reach SNR$=1$? This signal optical power is equivalent to how many photons per second?

(b) Repeat question (a) but only thermal noise is considered.

9. An APD operating in the 850 nm wavelength has a quantum efficiency $\eta=0.85$, an APD gain of $M_{APD}=100$ and a noise figure of $F(M)=M_{APD}^{0.3}$. The operating temperature is $T=300\,K$, and load resistance is $R_L=50\Omega$. Dark current is neglected for simplicity.

(a) If the input optical power is $-30\,dBm$, what is the signal photocurrent?

(b) If the input optical power is $-30\,dBm$ and the receiver bandwidth is 1 GHz, what is the SNR?

(c) At which signal optical power that shot noise is equal to thermal noise?

(d) If the APD gain can be varied by change the bias voltage, for the input optical power of $-30\,dBm$ and the receiver bandwidth of 1 GHz, please plot the SNR as the function of APD gain M_{APD} within the range of $1<M_{APD}<200$.

10. Same APD and operation conditions as used in problem 9, but now the signal optical power is 0 dBm. Only consider thermal noise and shot noise,

(a) Compare the SNR for $M_{APD}=100$ and $M_{APD}=1$.

(b) At which APD gain M_{APD} shot noise is equal to thermal noise?

(c) Explain why higher APD gain in this case may result in a lower SNR, and discuss the general rule on how to optimize APD gain.

11. Consider a PV with thermal voltage $V_T=0.025\,\text{V}$, reverse saturation current $I_S=10\,\text{pA}$, and the sun light produces an open-circuit voltage $V_{OC}=0.6\,\text{V}$.

   (a) if the load resistance is $R_L=8\,\Omega$, use a numerical method please find the photocurrent I_p that flows through the load resistor? (Note: search within the range of $50 < I_P < 85\,\text{mA}$.)

   (b) In order to obtain the maximum electric power at the load in this situation, what is the optimum load resistance?

REFERENCES

Anderson, L.K., McMurtry, B.J., 1966. High-speed photodetectors. Appl. Opt. 5 (10), 1573–1587.

Brain, M.C., 1982. Comparison of available detectors for digital optical fiber systems for 1.2-1.5 pm wavelength range, IEEE J. Quantum Electron. QE-18 (2), 219–224.

Donati, S., 2000. Photodetectors: Devices, Circuits and Applications. Prentice Hall.

Donnelly, J.P., Duerr, E.K., McIntosh, K.A., et al., 2006. Design considerations for 1.06-μm InGaAsP-InP Geiger-mode avalanche photodiodes. IEEE J. Quantum Electron. 42, 797–809.

Fyath, R.S., O'Reilly, J.J., 1989. Performance degradation of APD-optical receivers due to dark current generated within the multiplication region. J. Lightwave Technol. 7 (1), 62–67.

Jiang, X., Itzler, M.A., Ben-Michael, R., Slomkowski, K., 2007. InGaAsP-InP avalanche photodiodes for single photon detection. IEEE Sel. Top. Quantum Electron. 13, 895–905.

Neamen, D., 2003. Semiconductor Physics and Devices, third ed. McGraw Hill.

Smith, R.G., Forrest, S.R., 1982. Sensitivity of avalanche photodetector receivers for long-wavelength optical communications. Bell Syst. Tech. J. 61 (10), 2929–2946.

Theuwissen, A.J., 1995. Solid-State Imaging with Charge-Coupled Devices (Solid-State Science and Technology Library), 1995th ed., Springer.

Liu, M-K., 1996. Principles and Applications of Optical Communications (Eq. 5.25). IRWIN.

FURTHER READING

Beling, A., Campbell, J.C., 2009. InP-based high-speed photodetectors. J. Lightwave Technol. 27 (3), 343–355.

OPTICAL AMPLIFIERS

5

CHAPTER OUTLINE

Introduction 155
5.1 Optical gain, gain bandwidth, and saturation 157
5.2 Optical noise and noise figure 160
    5.2.1 Optical noise power spectral density 160
    5.2.2 Impact of ASE noise in the electrical domain 161
    5.2.3 Noise figure 163
5.3 Semiconductor optical amplifiers 165
    5.3.1 Steady-state analysis 165
    5.3.2 Gain dynamics of SOA 169
    5.3.3 All-optical signal processing based on the cross-gain and cross-phase modulation 172
    5.3.4 Wavelength conversion based on four-wave mixing in an SOA 176
5.4 Erbium-doped fiber amplifiers 179
    5.4.1 Absorption and emission cross sections 181
    5.4.2 Rate equations 183
    5.4.3 Numerical solutions of Rate equations 186
    5.4.4 Additional considerations in EDFA design 190
5.5 Raman amplifiers 195
5.6 Summary 201
Problems 203
References 206
Further reading 207

INTRODUCTION

The introduction of the optical amplifier has been one of the most important advances in optical fiber communications. Linear optical amplifiers are often used to compensate losses in optical communication systems and networks due to fiber attenuation, connecting loss, optical power splitting, and other loss factors. Optical amplifiers can also be used to perform nonlinear optical signal processing and waveform shaping when they are used in a nonlinear regime.

Fig. 5.0.1 shows a typical multi-span point-to-point optical transmission system in which optical amplifiers are used to perform various functions. A post-amplifier (post-amp) usually refers to the

Introduction to Fiber-Optic Communications. https://doi.org/10.1016/B978-0-12-805345-4.00005-6

amplifier used at the transmitter to boost the optical power (sometimes also referred to as boost amplifier). Sometimes the post-amp is integrated with the transmitter to form a high-power optical transmitter. It is especially useful if the optical signal from a transmitter is intended to be divided into a number of broadcasting outputs in an optical network, where a post-amp can help compensate the splitting loss. The basic requirement for a post-amp is to be able to supply high enough output optical power.

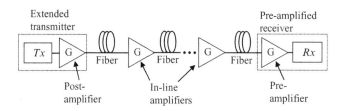

FIG. 5.0.1

Illustration of a point-to-point optical transmission system and the functions of optical amplifiers. *Tx*: transmitter and *Rx*: receiver.

In-line optical amplifiers (line-amps) are used along the transmission system to compensate for the attenuation caused by optical fibers. In high-speed optical transmission systems, line-amps are often spaced periodically along the fiber link, one for approximately every 80 km of transmission distance. Although the basic requirement of a line-amp is to provide high enough optical gain, it also needs to have wide optical bandwidth and flat optical gain within the bandwidth to support wavelength-division multiplexed (WDM) optical systems. In addition, the gain of a line-amp has to be linear to prevent nonlinear cross talk between different wavelength channels.

A preamplifier is often used immediately before the photodiode in an optical receiver to form the so-called preamplified optical receiver. In addition to the requirement of high optical gain, the most important qualification of a preamplifier is that the noise should be low. The sensitivity of a preamplified optical receiver is largely dependent on the noise characteristic of the preamplifier. Because of their different applications, various types of optical amplifiers will generally have to be designed and optimized differently to achieve the best performance.

The most popular optical amplifiers used for optical communications and other electro-optic systems are semiconductor optical amplifiers (SOAs) and erbium (Er)-doped fiber amplifiers (EDFAs). Raman amplifiers based on the stimulated Raman scattering (SRS) in optical fibers can also be used in fiber-optic systems to provide the required optical gain. SOA is electrically pumped with miniature footprint and can be monolithically integrated with other photonic devices. However, an SOA can have very fast gain dynamics determined by the short carrier lifetime on the sub-nanosecond level, which may introduce interchannel cross talk through dynamic gain saturation in a wavelength multiplexed (WDM) optical system. Whereas, an EDFA has long carrier lifetime on the order of milliseconds so that the gain dynamics is much slower than the time scale of data bits carried by the optical signal. This slow gain dynamics thus allows EDFAs to be used in WDM systems without interchannel cross talk. In addition to erbium, doping or co-doping of other rare earth elements such as ytterbium (Yb), neodymium (Nd), and thulium (Tm) can also be found in fiber amplifiers in order to cover different wavelength windows and to improve the power efficiency. Fig. 5.0.2A and B shows examples of packaged fiber amplifiers as a desktop equipment and as a module, respectively. Fig. 5.0.2C is an SOA in a butterfly package.

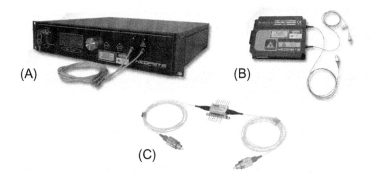

FIG. 5.0.2

(A) and (B) Packaged fiber amplifiers as a desktop equipment and as a module. (C) SOA in a butterfly package.

(A) and (B) Used with permission from Keopsys. (C) Used with permission from Thorlabs.

The peak gain wavelengths and gain bandwidths of SOA and EDFA depend on the band structures of semiconductor and rare earth materials used, while a Raman amplifier has the peak gain at an optical frequency approximately 13 THz lower than that of the optical pump when silica optical fiber is used. Thus, the gain spectrum of a Raman amplifier can be flexible by selecting the pump wavelength(s). Distributed Raman amplifier is another possible choice by utilizing part of the transmission fiber as the gain medium, which makes Raman amplification more flexible.

This chapter discusses common characteristics of optical amplifiers, such as optical gain, gain bandwidth, and noise figure. Specific characteristics of SOA, EDFA, and Raman amplifiers will also be discussed.

5.1 OPTICAL GAIN, GAIN BANDWIDTH, AND SATURATION

In an optical amplifier, the optical signal is amplified through the stimulated emission process in the gain medium where carrier density is inverted. This is similar to that required for the laser operation discussed in Chapter 3. However, an optical amplifier differs from a laser in that it does not require optical feedback, and the optical signal travels through the gain medium only once as illustrated in Fig. 5.1.2. In addition to the stimulated emission, an optical gain medium with population inversion also produces spontaneous emission, creating optical noise which is added to the amplified optical signal (Olsson, 1992; Connelly, 2002).

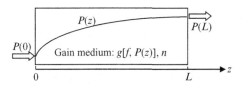

FIG. 5.1.2

Illustration of optical amplifier with a one-dimensional gain medium.

In continuous wave (CW) operation and neglect the spontaneous emission noise, the propagation equation of the optical field E in the gain medium along the longitudinal direction of the gain medium can be expressed as

$$\frac{dE}{dz} = \left[\frac{g-\alpha}{2} + jn\right]E \approx \left[\frac{g}{2} + jn\right]E \tag{5.1.1}$$

where g is the optical power gain coefficient of the material at position z, α is the optical loss, and n is the refractive index of the optical medium. In general for the application of optical amplifier, $g \gg \alpha$, so that the loss coefficient is neglected in Eq. (5.1.1) for simplicity.

Because the semiconductor gain medium has limited optical bandwidth, the local gain coefficient g is a function of optical wavelength. g is also a function of optical power, P, because of the saturation effect. For homogeneously broadened optical systems, gain saturation does not affect the wavelength dependency of the gain, a parabolic approximation can be used for the gain spectrum near the central frequency of the gain medium f_0,

$$g(f, P, z) = \frac{g_0}{1 + 4\left(\frac{f-f_0}{\Delta f}\right)^2 + \frac{P(z)}{P_{sat}}} \tag{5.1.2}$$

where g_0 is the linear gain, Δf is the FWHM bandwidth, and P_{sat} is the saturation optical power.

Neglecting the impact of spontaneous emission, optical power amplification along the gain medium can be easily evaluated from Eq. (5.1.1) as

$$\frac{dP(z)}{dz} = g(f, P)P(z) \tag{5.1.3}$$

When signal optical power is much smaller than the saturation power ($P \ll P_{sat}$), the system is considered linear, and the gain coefficient is independent of z, so that the solution of Eq. (5.1.3) is

$$P(L) = P(0)G_0 \tag{5.1.4}$$

where L is the amplifier length, $P(0)$ and $P(L)$ are the input and the output optical power, respectively, and G_0 is the small-signal optical power gain.

$$G_0 = \frac{P(L)}{P(0)} = \exp\left[\frac{g_0 L}{1 + 4(f-f_0)^2/\Delta f^2}\right] \tag{5.1.5}$$

It is worth noting that the FWHM bandwidth of optical power gain G_0 can be found as

$$B_0 = \Delta f \sqrt{\frac{\ln 2}{g_0 L - \ln 2}} \tag{5.1.6}$$

which is different from the material gain bandwidth Δf. Although Δf is a constant, the bandwidth B_0 optical power gain is inversely proportional to the square root of peak optical gain $g_0 L$.

On the other hand, if the signal optical power is high enough, the optical amplifier is said to operate in the nonlinear regime where nonlinear saturation has to be considered. In such a case, the nonlinear propagation Eq. (5.1.1) may not have a simple analytical solution, and it has to be solved numerically. To simplify the problem, we make a narrowband approximation by consider only the peak optical gain at $f=f_0$, where optical power evolution along the amplifier can be described by

$$\frac{dP(z)}{dz} = \frac{g_0 P(z)}{1 + P(z)/P_{sat}} \tag{5.1.7}$$

This equation can be converted into

$$\int_{P(0)}^{P(L)} \left(\frac{1}{P} + \frac{1}{P_{sat}}\right) dP = g_0 L \tag{5.1.8}$$

And its solution is

$$\ln\left[\frac{P(L)}{P(0)}\right] + \frac{P(L) - P(0)}{P_{sat}} = g_0 L \tag{5.1.9}$$

We know that $G_0 = \exp(g_0 L)$ is the small-signal optical gain. If we define $G = P(L)/P(0)$ as the large-signal optical gain, Eq. (5.1.9) can be written as

$$\frac{\ln(G/G_0)}{1 - G} = \frac{P(0)}{P_{sat}} \tag{5.1.10}$$

With the increase of the signal optical power, the large-signal optical gain G will decrease monotonically. Fig. 5.1.3 shows the large-signal gain and output optical power as the function of the input optical signal power, where the small-signal optical gain is 30 dB and the saturation optical power is 3 dBm.

The 3-dB saturation input power $P_{3dB}(0)$ is defined as the input power at which the large-signal optical gain is reduced by 3 dB. From Eq. (5.1.10), $P_{3dB}(0)$ can be found as

$$P_{3dB}(0) = P_{sat} \frac{2 \ln(2)}{G_0 - 2} \tag{5.1.11}$$

In the example shown in Fig. 5.1.3, $P_{3dB}(0)$ can be found as -25.6 dBm, which depends on the small-signal optical gain G_0. Increasing the value of G_0 will reduce $P_{3dB}(0)$ because of the increase of the output power and the operation is more nonlinear.

In a similar way, we can also define a 3-dB saturation output power $P_{3dB}(L)$, which will be less dependent on the small-signal optical gain. This can be found from $P_{3dB}(L) = GP_{3dB}(0)$, so that,

$$P_{3dB}(L) \approx P_{sat} \ln(2) \tag{5.1.12}$$

where we have assumed that $G_0 \gg 2$. In the example shown in Fig. 5.1.3, $P_{3dB}(L)$ is approximately 1.4 dBm.

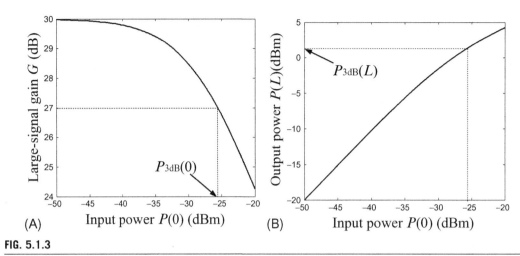

FIG. 5.1.3

Optical gain (A) and output optical power (B) as the function of the input power calculated with Eq. (5.1.10). $G_0 = 30\,\text{dB}$ and $P_{sat} = 3\,\text{dBm}$.

The bandwidth reduction and gain saturation are general characteristics of optical amplifiers when they operate in the high gain and high-power regime. Whereas, the gain bandwidth, the spectral shape of the optical gain, and the saturation power depends on the specific type of the optical amplifier, and the operation condition, which will be discussed in the following sections.

5.2 OPTICAL NOISE AND NOISE FIGURE

Similar to electronic amplifiers, an optical amplifier not only provides optical gain, but also introduces optical noise which degrades the optical signal-to-noise ratio (OSNR). Carrier inversion in the gain medium of an optical amplifier is essential to produce stimulated emission which is responsible for the coherent process of optical amplification. At the same time, the inverted carriers in the gain medium also spontaneously recombine to generate spontaneous emission photons. These spontaneously emitted photons are not coherent with the input signal and they constitute optical noise. To make things worse, the spontaneous emission photons are also amplified by the gain medium while they travel through the optical amplifier; thus the optical noise generated by an optical amplifier is commonly known as the *amplified spontaneous emission* (ASE). By its nature, ASE noise is random in wavelength, phase, and the state of polarization.

5.2.1 OPTICAL NOISE POWER SPECTRAL DENSITY

Generally, the level of ASE noise produced by an optical amplifier depends on both the optical gain and the level of carrier inversion in the gain medium. ASE noise power spectral density (PSD) with the unit of [W/Hz] can be expressed as (Desurvire, 1994)

$$\rho_{ASE}(f) = 2n_{sp}hf[G(f) - 1] \tag{5.2.1}$$

In this expression, $G(f)$ is the gain of the amplifier at the optical frequency f. h is the Planck's constant, c is the speed of light, and the factor 2 indicates two orthogonal polarization states. $n_{sp} > 1$ is a unitless spontaneous emission factor of the gain medium, which is, in general, weakly dependent on the wavelength.

As both the optical gain and the spontaneous emission factor are generally wavelength dependent, the ASE noise spectral density is also a function of wavelength. In a broadband optical amplifier, optical noise PSD $\rho_{ASE}(f)$ can be simply measured by an optical spectrum analyzer in terms of dBm/nm. Within a certain optical bandwidth B_0, the total noise optical power can be obtained by an integration,

$$P_{ASE} = \int_{-B_0/2}^{B_0/2} \rho_{ASE}(f)df \tag{5.2.2}$$

If the optical bandwidth is narrow enough and the ASE noise spectral density is flat within this bandwidth, the noise power will be directly proportional to the optical bandwidth and Eq. (5.2.2) can be simplified as

$$P_{ASE} \approx 2n_{sp}hf[G(f)-1]B_0 \tag{5.2.3}$$

In practical applications, both optical power and OSNR are often used to characterize an optical signal. Although an optical amplifier can enhance the power of an optical signal, it cannot improve the OSNR.

5.2.2 IMPACT OF ASE NOISE IN THE ELECTRICAL DOMAIN

Although the noise performance of an optical amplifier can be characterized in the optical domain as described in the last section, its ultimate impact in the performance of an optical communication system is the signal-to-noise ratio (SNR) in the electrical domain after the optical signal and the ASE noise are detected by a photodetector. Understanding the relationship between optical noise spectral density and the corresponding electrical SNR is essential in the design and performance characterization of optical communication systems.

As discussed in Chapter 4, a photodiode performs square-law detection and the photocurrent is linearly proportional to the input optical power,

$$I = \Re P = \Re \left| E_{sig} + E_{noise} \right|^2 \tag{5.2.4}$$

where E_{sig} and E_{noise} are the fields of the amplified optical signal and optical noise, respectively. $\Re = \eta q \lambda/(hc)$ is the responsivity and η is the quantum efficiency, q is the electron charge, h is the Planck's constant, and λ is the wavelength.

The signal photocurrent is $I_{sig} = \Re |E_{sig}|^2 = \Re P_{sig}$, where $P_{sig} = |E_{sig}|^2$ is the signal optical power. The shot noise produced in the photodetection process is linearly proportional to the total optical power which includes both the optical signal and the ASE noise,

$$\sigma_{sh}^2 = 2q\Re \left(P_{sig} + P_{ASE} \right) \tag{5.2.5}$$

where P_{ASE} is the ASE noise power with the receiver optical bandwidth bandwidth as defined by Eq. (5.2.2). Nevertheless, the most significant impact of optical noise on the electrical SNR often comes from the mixing between the optical signal and the ASE noise, known as signal-ASE beat noise, and the mixing among ASE noise components, known as ASE-ASE beat noise.

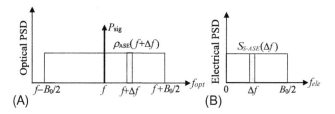

FIG. 5.2.1

Illustration of optical (A) and electrical (B) power spectral densities (PSD) and signal-ASE beat noise generation. f_{opt} and f_{ele} represent optical and electrical frequencies, respectively.

As illustrated in Fig. 5.2.1, assume $E_{sig}(f) = \sqrt{P_{sig}}e^{j2\pi ft}$, is the field of the optical signal at a central optical frequency f, and $E_{noise}(f + \Delta f) = \sqrt{\rho_{ASE}(f + \Delta f)}e^{j2\pi(f + \Delta f)t}$ is the broadband ASE noise with the PSD $\rho_{ASE}(f + \Delta f)$, where Δf represents a frequency deviation between the signal and the noise. The mixing between $E_{sig}(f)$ and the noise component $E_{noise}(f + \Delta f)$ in the photodiode produces an electric noise at the frequency Δf,

$$I(\Delta f) = 2\Re\sqrt{P_{sig}\rho_{ASE}(\Delta f)}\cos(2\pi\Delta ft) \tag{5.2.6}$$

Thus, the radio frequency (RF) PSD at frequency Δf can be found as

$$S_{S-ASE}(f_{ele}) = 2\Re^2 P_{sig}\rho_{ASE}(f_{ele}) \tag{5.2.7}$$

where Δf has been replaced by f_{ele} to represent the electrical-domain frequency. Note that this is a double-sideband RF PSD because $\rho_{ASE}(\Delta f)$ and $\rho_{ASE}(-\Delta f)$ produce the same amount of RF beating noise; thus the single-sideband PSD should be twice as high. However, consider that the optical signal is usually polarized while the ASE noise is unpolarized; only half the ASE power can coherently mix with the optical signal. The combination of these two effects makes Eq. (5.2.7) the correct expression of the single-sideband PSD of signal-ASE beat noise.

Although signal-ASE beat noise is generated by the mixing between the optical signal and the ASE noise, ASE-ASE beat noise is generated by the beating between different optical frequency components of the ASE noise at the photodiode. Like signal-ASE beat noise, ASE-ASE beat noise exists only in the electrical domain as well. Fig. 5.2.2 illustrates that the beating between the two frequency components $\rho_{ASE}(f)$ and $\rho_{ASE}(f + \Delta f)$ in a photodetector generates an RF component at Δf.

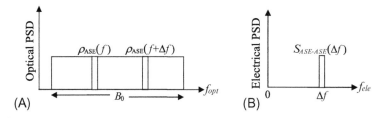

FIG. 5.2.2

Illustration of optical (A) and electrical (B) power spectral densities (PSD) and ASE-ASE beat noise generation. f_{opt} and f_{ele} represent optical and electrical frequencies, respectively.

Assuming that the optical bandwidth of the ASE noise is B_0 and the ASE noise spectral density is constant within the optical bandwidth, the ASE-ASE beat noise generation process can be described by a self-convolution of the ASE noise optical spectrum as illustrated in Fig. 5.2.3A. The size of the shaded area is a function of the frequency separation Δf and the result of correlation is a triangle function vs. Δf, as shown in Fig. 5.2.3B.

Another important consideration in calculating ASE-ASE beat noise is the unpolarized nature of the ASE noise. Statistically it has half-energy horizontally polarized and the other half-vertical polarized, whereas optical mixing happens only between noise components in the same polarization. Therefore, the single-sideband ASE-ASE beat noise PSD in the electrical domain can be found as

$$S_{ASE-ASE}(f_{ele}) = \frac{\Re^2 \rho_{ASE}^2}{2}(B_0 - f_{ele})$$ (5.2.8)

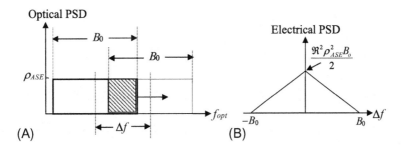

FIG. 5.2.3

Illustration of optical (A) and electrical (B) power spectral densities in the process of ASE-ASE beat noise generation.

Again, the unit of ASE-ASE beat noise is in $[A^2/Hz]$. In most cases, the electrical bandwidth is much less than the optical bandwidth ($\Delta f \equiv f_{ele} < < B_0$); therefore, Eq. (5.2.8) is often expressed as

$$S_{ASE-ASE} \approx \Re^2 \rho_{ASE}^2 B_0 / 2$$ (5.2.9)

which can be treated as a white noise within the receiver RF bandwidth. It is important to notice that in some textbooks, the RF spectral density of spontaneous-spontaneous beat noise is expressed as $S_{ASE-ASE} = 2\Re^2 \rho_{ASE}^2 B_0$, which seems to be four times higher than that predicted by Eq. (5.2.9). The reason for this discrepancy is due to their use of single-polarized ASE noise PSD $\rho_{ASE} = n_{sp}hf$ $[G(f) - 1]$, which is half of the unpolarized ASE noise spectral density given by Eq. (5.2.1).

5.2.3 NOISE FIGURE

In electronic amplifiers, the noise figure is defined as the ratio between the input electrical SNR and the output electrical SNR. For an optical amplifier, both the input and the output are optical signals. To use the similar noise figure definition, a photodetector has to be used both in the input and the output, and therefore (Haus, 2000)

$$F = \frac{SNR_{in}}{SNR_{out}} \qquad (5.2.10)$$

where SNR_{in} and SNR_{out} are the input and the output signal-to-noise power ratios in the electrical domain, which can be measured using the setup shown in Fig. 5.2.4, where the optical amplifier has an input optical power P_{in} and an optical gain G.

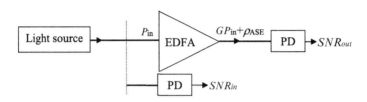

FIG. 5.2.4

Measurement of optical amplifier noise figure. PD: photodetector.

It is important to note that at the input side of the optical amplifier there is essentially no optical noise. However, since there is a photodetector used to convert the input optical signal into the electrical domain, shot noise will be introduced even if the photodetector is ideal. Therefore, the SNR_{in} in the amplifier input side is determined by the shot noise generated in the photodetector. Assume that the photodetector has an electrical bandwidth B_e; the total shot noise power is $P_{shot} = 2q\Re P_{in}B_e$, where P_s is the input optical signal power. At the same time, the detected signal electrical power at the optical amplifier input side is $(\Re P_{in})^2$, so the input SNR is

$$SNR_{in} = \frac{(\Re P_{in})^2}{2q\Re P_{in}B_e} = \frac{\Re P_{in}}{2qB_e} \qquad (5.2.11)$$

At the optical amplifier output, in addition to the amplified optical signal, ASE noise is also generated by the amplifier. After photodetection, the major noise sources are shot noise, signal-ASE beat noise, and ASE-ASE beat noise. Since the ASE-ASE beat noise can be significantly reduced by using a narrowband optical filter in front of the photodetector, in the calculation of SNR_{out} only shot noise and signal-ASE beat noise are included. Considering that the signal electrical power is $(\Re GP_{in})^2$, the signal-ASE beat noise electrical power is

$$P_{S-ASE} = 2\Re^2 GP_{in}\rho_{ASE}B_e = 4\Re^2 GP_{in}n_{sp}hf(G-1)B_e \qquad (5.2.12)$$

where Eq. (5.2.1) has been used.

As the optical bandwidth B_0 is considered narrow enough, ASE contribution to the shot noise is negligible and thus the shot noise power is only created by the optical signal so that $P_{shot} = 2q\Re GP_{in}B_e$. The electrical SNR at optical amplifier output is

$$SNR_{out} = \frac{(\Re GP_{in})^2}{P_{shot} + P_{S-ASE}} = \frac{\Re GP_{in}}{2qB_e + 4\Re n_{sp}hf(G-1)B_e} \qquad (5.2.13)$$

Based on the definition in Eq. (5.2.10), the optical amplifier noise figure is

$$F = \frac{1}{2q}\frac{4\Re n_{sp}hf(G-1)+2q}{G} = 2n_{sp}\frac{(G-1)}{G} + \frac{1}{G} \qquad (5.2.14)$$

where $\Re = q/(hf)$ has been used. In fact, the noise figure is the property of the optical amplifier, which should be independent of the property of the photodetector used for the measurement. Thus, an ideal photodetector with 100% quantum efficiency is used ($\eta = 1$). In most practical applications, the amplifier optical gain is large enough ($G \gg 1$) and therefore, the noise figure can be further simplified as

$$F \approx 2n_{sp} \tag{5.2.15}$$

As the spontaneous emission factor $n_{sp} \geq 1$ with the minimum value of $n_{sp} = 1$ happens with complete carrier inversion, the minimum noise figure of an optical amplifier is 2, which is 3 dB.

Although the definition of the noise figure is straightforward and the expression is simple, there is no simple way to predict the spontaneous emission factor n_{sp} in an optical amplifier, and therefore, actual measurements have to be performed (Baney and Gallion, 2000).

5.3 SEMICONDUCTOR OPTICAL AMPLIFIERS

A SOA is similar to a semiconductor laser operating below threshold. It requires an optical gain medium and an optical waveguide, but it does not require an optical cavity (Connelly, 2002; Shimada and Ishio, 1992; Olsson, 1992). An SOA can be made by a laser diode with antireflection (AR) coating on each facet; an optical signal passes through the gain medium only once and therefore, it is also known as a *traveling-wave optical amplifier*. Because of the close similarity between an SOA and a laser diode, the analysis of SOA can be based on the rate Eq. (3.3.12) we have already discussed in Section 3.3.

$$\frac{dN(z,t)}{dt} = \frac{J}{qd} - \frac{N(z,t)}{\tau} - 2\Gamma v_g a(N - N_0)P(z,t) \tag{5.3.1}$$

where $N(z, t)$ is the carrier density and $P(z, t)$ is the photon density, and both of them are position dependent along the SOA optical waveguide. J is the injection current density, d is the thickness of the active layer, v_g is the group velocity of the lightwave, τ is the spontaneous emission carrier lifetime, a is the differential optical field gain, N_0 is the transparency carrier density, and Γ is the optical confinement factor of the waveguide.

5.3.1 STEADY-STATE ANALYSIS

Let's start with a discussion of steady-state characteristics of an SOA. In the steady state, $d/dt = 0$ and the rate equation is simplified as

$$\frac{J}{qd} - \frac{N(z,t)}{\tau} - 2\Gamma v_g a(N - N_0)P(z,t) = 0 \tag{5.3.2}$$

If the power of the optical signal is low enough ($P(z,t) \approx 0$), the small-signal approximation is justified and the carrier density can be found as $N = J\tau/qd$. Thus, the small-signal material gain can be found as

$$g_0 = 2a(N - N_0) = 2a\left(\frac{J\tau}{qd} - N_0\right) \tag{5.3.3}$$

We can then use this small-signal gain g_0 as a parameter to investigate the general case where the signal optical power may be large. Eq. (5.3.2) can be written as

$$\frac{g_0}{2a} - (N - N_0) - 2\Gamma v_g \tau a (N - N_0) P(z, t) = 0 \tag{5.3.4}$$

In the large-signal case, if we define $g = 2a(N - N_0)$ as the material gain, Eq. (5.3.4) yields

$$g = \frac{g_0}{1 + P(z, t)/P'_{sat}} \tag{5.3.5}$$

where

$$P'_{sat} = \frac{1}{2\Gamma v_g \tau a} \tag{5.3.6}$$

is the saturation photon density, and it is an SOA parameter that is independent of the actual signal optical power. Similar to that discussed in Eq. (3.3.25), the saturation photon density can also be found related to the saturation optical power by

$$P_{sat} = P'_{sat} w d v_g hf \tag{5.3.7}$$

where w is the waveguide width (wd is the area of cross-section) and hf is the photon energy. Therefore, the corresponding saturation optical power is

$$P_{sat} = \frac{wdhf}{2\Gamma \tau a} \tag{5.3.8}$$

As an example, for an SOA with waveguide width $w = 2\,\mu m$ and thickness $d = 0.25\,\mu m$, optical confinement factor $\Gamma = 0.3$, differential field gain $a = 5 \times 10^{-20} m^2$, and carrier lifetime $\tau = 1\,ns$, the saturation optical power is approximately 22 mW.

In the analysis of semiconductor lasers, we often assume that the optical power along the laser cavity is uniform; this is known as *mean-field approximation*. For an SOA, on the other hand, the signal optical power at the output side is much higher than that at the input side, and they differ by the amount of the SOA optical gain. In this case, mean-field application is no longer accurate. The wavelength dependency of the optical gain in an SOA fits to a parabolic reasonably well as described in Eq. (5.1.2), so that the steady-state gain saturation and bandwidth reduction in the nonlinear operation regime discussed in Section 5.1 are readily applicable.

Although an ideal SOA has no reflection from the end facets through AR coating, a practical SOA always has residual reflectivity at each facet which forms a weak resonance cavity and introduces unwanted ripple in the gain spectrum. Fig. 5.3.1 shows an SOA structure with cavity length L, single-pass optical gain G_s, facet reflectivity R, and refractive index n, so that the propagation constant is $\beta = 2\pi fn/c$, with f the optical frequency.

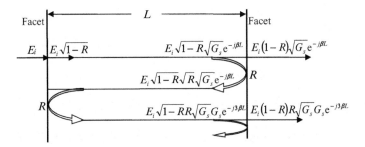

FIG. 5.3.1

Illustration of SOA with cavity length L, single-pass optical gain G_s, facet reflectivity R, and propagation constant β. The colinear optical beam is purposely displaced after each reflection for better display.

Fig. 5.3.1 shows the evolution of the optical field after each reflection and transmission, where the input optical field is E_i. As R is the facet power reflectivity, the optical field reflectivity and transmission are $\sqrt{R}$ and $\sqrt{1-R}$, respectively, and the single-pass complex optical field gain is $\sqrt{G_s}e^{-j\beta L}$. After each roundtrip inside the SOA, the optical field experiences a gain of $G_s e^{-j2\beta L}$, and a reflection loss of R. The overall output optical field is then,

$$E_0 = E_i(1-R)\sqrt{G_s}e^{-j\beta L}\left(\sum_{m=0}^{\infty} G_s^m R^m e^{-j2m\beta L}\right) = \frac{E_i(1-R)\sqrt{G_s}e^{-j\beta L}}{1-RG_s e^{-j2\beta L}} \tag{5.3.9}$$

The power transfer function is thus,

$$T = \left|\frac{E_0}{E_i}\right|^2 = \frac{(1-R)^2 G_s}{(1-RG_s)^2 + 4RG_s \sin^2(2\pi f\tau)} \tag{5.3.10}$$

where $\phi = \beta L = 2\pi\tau_d f$ is the single-trip phase shift, and $\tau_d = nL/c$ is the single-trip propagation delay. Neglect the effect of saturation, the optical gain G_s is a function of optical frequency,

$$G_s(f) = \exp\left\{\frac{\Gamma g(f_0)L}{1+4(f-f_0)^2/\Delta f^2}\right\} \tag{5.3.11}$$

where $g(f_0)$ is the peak material gain and Δf is the FWHM material gain bandwidth. The peak optical gain is $G_s(f_0) = \exp\{\Gamma g(f_0)L\}$.

For an ideal SOA with perfect AR coating so that $R=0$, the power transfer function is $T(f) = G_s(f)$. But even a weak facet reflection would introduce strong ripple in the power transfer function.

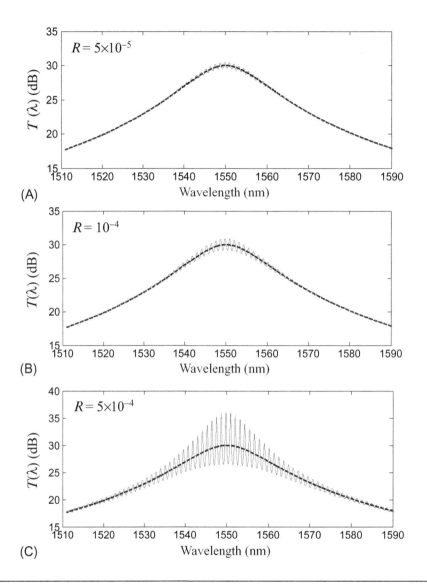

FIG. 5.3.2

SOA power transfer functions calculated with peak gain $G_s(f_0)=30$-dB peak gain, cavity length $L=300\,\mu m$, refractive index $n=3.5$, and material gain bandwidth $\Delta f=2.5\,THz$ (equivalent to 20 nm in the 1550 nm wavelength window). The dashed lines show the power transfer function with $R=0$, and the thin solid lines show the power transfer functions with $R=5\times10^{-5}$ (A), 10^{-4} (B), and 5×10^{-4} (C).

As an example, Fig. 5.3.2 shows SOA power transfer functions with a peak optical gain $G_s(f_0)=$ 30 dB, SOA length $L=300\,\mu m$, material refractive index $n=3.5$, and material gain bandwidth $\Delta f=2.5\,THz$ (equivalently 20 nm in the 1550 nm wavelength window). The dashed line in Fig. 5.3.2 shows the power transfer function with $R=0$, and the thin dashed lines show the power transfer function with $R=5 \times 10^{-5}$, 10^{-4}, and 5×10^{-4} for Fig. 5.3.2A–C, respectively. In general, $G_sR \ll 1$ is required for a high quality, and thus R has to be extremely low for an SOA with high optical gain.

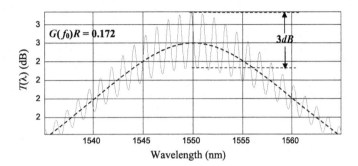

FIG. 5.3.3

SOA power transfer function calculated with peak gain $G_s(f_0)=30\,dB$, $R=0$ (dashed line) and $R=0.172 \times 10^{-3}$ so that the ripple amplitude is exactly 3 dB.

Because of the ripple in the SOA power transfer function, the determination of 3 dB bandwidth may become complicated. In fact, without the cavity effect ($R=0$), the 3 dB bandwidth can be found to be $B_0 = \Delta f \sqrt{\ln 2/(g_0L - \ln 2)}$ as defined in Eq. (5.1.6). Whereas when the cavity resonance is strong enough so that the ripple strength near f_0 is higher than 3 dB, the 3 dB bandwidth will become much narrower and determined only by the Febry-Perot resonance. Based on Eq. (5.3.10) and assume the free-spectral range (FSR) of the Febry-Perot cavity is much smaller than Δf (so that G_s is regarded as wavelength independent), the 3 dB bandwidth can be found as

$$B_0 = \frac{2 \cdot FSR}{\pi} \cdot \sin^{-1}\left(\frac{1-RG_s}{2\sqrt{RG_s}}\right)$$

where $FSR = 1/(2\tau_d)$ represents the period of the Febry-Perot resonance in the frequency domain. Note that for this bandwidth expression to be valid, $1 - RG_s \leq 2\sqrt{RG_s}$ has to be satisfied, or $G_s(f_0)R \geq 0.172$. This is equivalent to the ripple amplitude near $f=f_0$ to be larger than, or as least equal to, 3 dB as shown in Fig. 5.3.3.

5.3.2 GAIN DYNAMICS OF SOA

Due to its short carrier lifetime, optical gain in an SOA can change quickly. Fast-gain dynamics is one of the unique properties of semiconductor-based optical amplifiers as opposed to fiber amplifiers, which are discussed later in this chapter. The consequence of fast-gain dynamics is twofold: (1) It introduces cross talk between different wavelength channels through cross-gain saturation and (2) it can be used to accomplish all-optical switch based on the same cross-gain saturation mechanism. The fast carrier dynamics of SOA not only introduces fast gain saturation, but also creates fast index modulation which is responsible for cross-phase modulation. The application perspective of SOA in all-optical wavelength conversion and pulse shaping is discussed in this section.

Using the same carrier density rate equation as Eq. (5.3.1) but converting photon density P into optical power P_{opt},

$$\frac{dN(z,t)}{dt} = \frac{J}{qd} - \frac{N(z,t)}{\tau} - \frac{2\Gamma a}{hfA}(N-N_0)P_{opt}(z,t) \tag{5.3.12}$$

where A is the waveguide cross-section area. To focus on the time-domain dynamics, it is helpful to eliminate the spatial variable z. This can be achieved by investigating the dynamics of the total carrier population N_{tot} defined as

$$N_{tot}(t) = \int_0^L N(z,t)dz$$

where L is the length of the SOA. Thus, Eq. (5.3.12) becomes

$$\frac{dN_{tot}(t)}{dt} = \frac{I}{qA} - \frac{N_{tot}(t)}{\tau} - \frac{2\Gamma a}{hfA}\int_0^L [N(z,t)-N_0]P_{opt}(z,t)dz \tag{5.3.13}$$

where $I=JL$ is the total injection current. Considering that relationship between the optical power and the optical gain as shown in Eq. (5.1.3), the last term of Eq. (5.3.13) which is an integration along the SOA can be expressed by the output-input power difference,

$$2\Gamma a\int_0^L [N(z,t)-N_0]P_{opt}(z,t)dz = P_{opt}(L,t) - P_{opt}(0,t) \tag{5.3.14}$$

Here, $P_{opt}(0,t)$ and $P_{opt}(L,t)$ are the input and the output optical signal waveforms, respectively. We have assumed that the power distribution along the amplifier does not change during a single pass through the SOA, which implies that the rate of optical power change is much slower than the transit time of the amplifier. For example, if the length of an SOA is $L=300\mu m$ and the refractive index of the waveguide is $n=3.5$, a single pass of the optical pulse through the SOA takes only approximately 3.5 ps. Under this adiabatic approximation, Eq. (5.3.13) can be simplified as

$$\frac{dN_{tot}(t)}{dt} = \frac{I}{qA} - \frac{N_{tot}(t)}{\tau} - \frac{P_{opt}(0,t)}{hfA}[G(t)-1] \tag{5.3.15}$$

where $G(t)=P_{opt}(L,t)/P_{opt}(0,t)$ is the optical gain. On the other hand, since the net optical gain of the amplifier can be expressed as

$$G(t) = \exp\left\{2\Gamma a\int_0^L (N(z,t)-N_0)dz\right\} = \exp\{2\Gamma a(N_{tot}(t)-N_0)L\} \tag{5.3.16}$$

That is, $N_{tot}(t)=\ln G(t)/(2\Gamma a)+N_0 L$, or,

$$\frac{dN_{tot}(t)}{dt} = \frac{1}{2\Gamma aG(t)}\frac{dG(t)}{dt} \tag{5.3.17}$$

Eqs. (5.3.15) and (5.3.17) can be used to investigate how an SOA responds to a short optical pulse. Consider a short and intense optical pulse injected into an SOA with the pulse width T much shorter than the carrier lifetime τ. Within this short time interval of optical pulse duration, the contribution of both current injection I/qd and spontaneous recombination N_{tot}/τ are very small and negligible. Thus Eq. (5.3.15) can be simplified as

$$\frac{dN_{tot}(z)}{dt} \approx -\frac{P_{opt}(0,t)}{hfA}[G(t)-1] \tag{5.3.18}$$

Eqs. (5.3.17) and (5.3.18) can be combined to eliminate dN_{tot}/dt, and then the equation can be integrated over the pulse duration from $t=0$ to $t=T$,

$$\ln\left[\frac{1-1/G(T)}{1-1/G(0)}\right] + \frac{1}{\tau P_{sat}}\int_0^T P_{opt}(0,t)dt = 0 \tag{5.3.19}$$

where P_{sat} is the saturation optical power as defined in Eq. (5.3.8). $G(0)$ and $G(T)$ are SOA optical gains immediately before and immediately after the optical pulse, respectively.

Eq. (5.3.19) can be written as

$$G(T) = \frac{1}{1-\left(1-\dfrac{1}{G_0}\right)\exp\left[-\dfrac{W_{in}(T)}{\tau P_{sat}}\right]} \tag{5.3.20}$$

where $W_{in}(T) = \int_0^T P_{opt}(t,0)dt$ is the optical pulse energy at the SOA input port, and τP_{sat} can be considered as the saturation pulse energy. Eq. (5.3.20) indicates that when the normalized pulse energy $W_{in}/(\tau P_{sat})$ is small, the optical gain is equal to the small-signal gain G_0. Whereas when the normalized pulse energy is very high, the optical gain will be suppressed to $G(T)\approx 1$ immediately after the pulse. As illustrated in Fig. 5.3.4, this fast gain dynamics originates from the fast depletion of the carrier population N_{tot} by the intense optical pulse, but the recovery of the carrier population is dependent on the carrier lifetime which might be on the sub-nanosecond level.

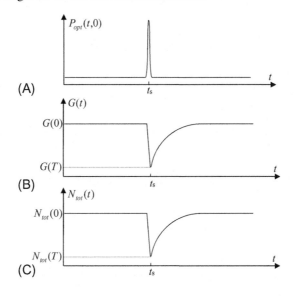

FIG. 5.3.4

Illustration of short pulse response of optical gain and carrier population of an SOA: (A) short optical pulse, (B) optical gain of the SOA, and (C) carrier population.

5.3.3 ALL-OPTICAL SIGNAL PROCESSING BASED ON THE CROSS-GAIN AND CROSS-PHASE MODULATION

All-optical signal processing is the modulation of an optical signal directly by another optical signal, which circumvents the need of optical to electrical (O/E) and electrical to optical (E/O) conversions. SOA operating in the nonlinear regime is an ideal device for all-optic signal processing thanks to its fast carrier dynamics. All-optical switching and wavelength conversion can be accomplished by an SOA through cross-gain and cross-phase modulation (Contestabile et al., 2007; Hui et al., 1994).

Fig. 5.3.5A shows the block diagram of all-optical wavelength conversion based on the cross-gain modulation of an SOA. In this application, a high-power intensity-modulated optical signal at wavelength λ_2 is combined with a relatively weak continuous-wave (CW) signal at λ_1 before they are both injected into an SOA. The optical gain of the SOA is then modulated by the signal at λ_2 through SOA gain saturation effect as illustrated in Fig. 5.3.5B. At the same time, the weak CW optical signal at wavelength λ_1 is amplified by the time-varying gain of the SOA, and therefore, its output waveform is complementary to that of the optical signal at λ_2. At the output of the SOA, an optical bandpass filter is used to select the wavelength component at λ_1 and reject that at λ_2. In this way, the wavelength of the optical carrier is converted from λ_2 to λ_1. This process is all-optical and without the need for electric modulation.

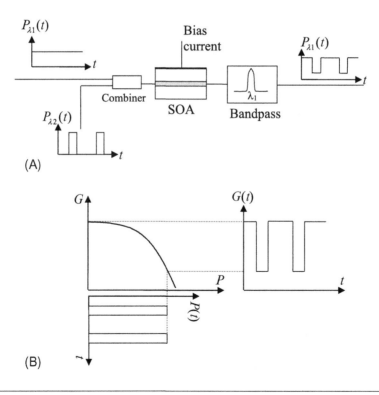

FIG. 5.3.5

All-optical wavelength conversion using cross-gain saturation effect of SOA: (A) system configuration and (B) illustration of principle.

For practical applications, all-optical wavelength conversion based on the cross-gain saturation requires low saturation power P_{sat} of the SOA so that the nonlinearity is strong. The on/off ratio of the converted optical signal at wavelength λ_1 is determined by the ratio of SOA gain suppression introduced by the signal at wavelength λ_2. Assume the input optical pulse is much shorter than the carrier lifetime and thus Eq. (5.3.20) is valid. As an example, for an SOA with a saturation power $P_{sat} = 110$mW and a carrier lifetime $\tau = 1$ns so that the saturation energy is $P_{ast}\tau = 110$pJ, the optical gain suppression ratio vs. input pulse energy is shown in Fig. 5.3.6 for three different small-signal gains G_0 of the SOA. With a high small-signal optical gain, it is relatively easy to achieve a high gain suppression ratio because the output pulse energy is high. For example, if the input optical pulse energy is 10 pJ, to achieve 10-dB optical gain suppression, the small signal gain has to be at least 20 dB, as shown in Fig. 5.3.6.

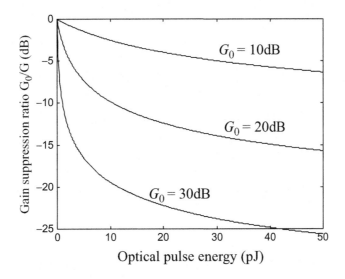

FIG. 5.3.6

Gain suppression ratio vs. input optical pulse energy in an SOA with 110-mW saturation optical power and 1 ns carrier lifetime.

Another important issue in all-optical signal processing is the response speed. Although the optical gain suppression is fast in an SOA, which to some extent is only limited by the width of the input optical pulse as illustrated in Fig. 5.3.4, the gain recovery after the optical pulse can be quite long; this gain recovery process depends on the carrier lifetime τ of the SOA.

After an optical pulse passes through the SOA, the gain of the SOA starts to recover toward its small-signal value G_0 due to the constant carrier injection into the SOA. Since the photon density inside the SOA is low after the event of optical pulse, stimulated emission can be neglected in the gain recovery process, and the carrier population rate Eq. (5.3.15) becomes

$$\frac{dN_{tot}(t)}{dt} = \frac{I}{qA} - \frac{N_{tot}(t)}{\tau} \qquad (5.3.21)$$

The solution of Eq. (5.3.21) is

$$N_{tot}(t) = [N_{tot}(t_s) - N_{tot}(0)] \exp\left(-\frac{t-t_s}{\tau}\right) + N_{tot}(0) \tag{5.3.22}$$

where $N_{tot}(0) = N_{tot}(\infty)$ is the small-signal carrier population of the SOA, and $N_{tot}(t_s)$ is the minimum carrier population corresponding at the falling edge of the pulse as shown in Fig. 5.3.4. Since optical gain is exponentially proportional to the carrier population, as indicated in Eq. (5.3.16), the normalized optical gain during the recovery process is

$$\ln\left(\frac{G(t)}{G_0}\right) = \ln\left(\frac{G(t_s)}{G_0}\right) \exp\left(-\frac{t-t_s}{\tau}\right) \tag{5.3.23}$$

Eq. (5.3.23) clearly demonstrates that the dynamics of optical gain recovery primarily depend on the carrier lifetime τ of the SOA.

In addition to cross-gain modulation, all-optical wavelength conversion can also be accomplished with cross-phase modulation. As we discussed in Chapter 3, in a semiconductor laser the carrier density change not only changes the optical gain, it also changes the refractive index of the material, and thus introduces an optical phase modulation. This is also true in SOAs. We know that photon density P of an optical amplifier can be related to the material gain g as, $P = P_{in}\exp(gL)$, where P_{in} is the input photon density and L is the SOA length. In the case of small-signal modulation, an infinitesimal change in the carrier density induces an infinitesimal change of the material gain by Δg. The corresponding change in the photon density is $\Delta P = P \cdot \Delta gL$. At the same time, the optical phase will be changed by $\Delta\phi = \omega\Delta nL/c$, where Δn is the change of the carrier density and ω is the signal optical frequency. According to the definition of linewidth enhancement factor α_{lw} in Eq. (3.3.40), the gain change and the phase change are related by

$$\alpha_{lw} = 2P\frac{\Delta\phi}{\Delta P} = \frac{2\Delta\phi}{\Delta gL} \tag{5.3.24}$$

This can be extended to large-signal modulation. In such a case, if the phase is switched from ϕ_0 to ϕ_1, the corresponding material gain is changed from g_0 to g_1, and the corresponding optical gain will change from $G_0 = \exp(g_0L)$ to $G_1 = \exp(g_1L)$. Eq. (5.3.24) can be used to find the relationship between phase change and gain change as

$$\phi_1 - \phi_0 = \frac{\alpha_{lw}}{2}\ln\left(\frac{G_1}{G_0}\right) \tag{5.3.25}$$

Fig. 5.3.7A shows an example of SOA gain spectra without (solid circles) and with (open circles) gain saturation. The SOA operating in the 1300 nm wavelength window with an approximately 23-dB small-signal peak gain. The gain saturation was introduced by a pump laser with -8.2-dBm optical power at the SOA input. The gain saturation is stronger in the short wavelength side of the peak gain, which is caused by a red shift of the gain peak when the carrier density is suppressed by the optical pump. This implies that the saturation power $P_{sat}(\lambda)$ is wavelength dependent, and thus converting an optical modulation from a long-wavelength carrier to a shorter-wavelength carrier is more efficient then the opposite. Fig. 5.3.7B shows that the gain extinction ratio is reduced by approximately 5 dB with the wavelength increase from 1270 to 1340 nm. However, for this 70 nm wavelength increase, the optical phase change created by the saturation effect only reduces from 0.64π to 0.4π, which is approximately 2 dB.

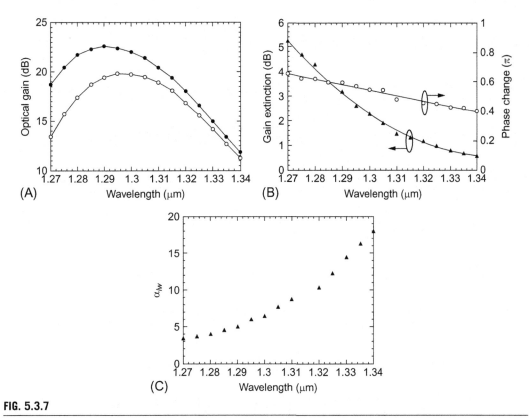

FIG. 5.3.7

(A) SOA gain spectra without (solid circles) and with (open circles) gain saturation. (B) Gain suppression ratio and optical phase change as the function of wavelength. (C) Linewidth enhancement factor a_{lw} as the function of wavelength (Hui et al., 1994).

Fig. 5.3.7B reveals that the ratio between the gain extinction and the phase change varies with the wavelength, which means that the linewidth enhancement factor a_{lw} is wavelength dependent. Using Eq. (5.3.25), a_{lw} can be calculated based on the measured G_1/G_0 and $\phi_1 - \phi_0$ as the function of the wavelength, and the results are shown in Fig. 5.3.7C. Note that in a laser diode, typical value of a_{lw} ranges from 3 to 8 because the lasing wavelength is usually near the wavelength of peak gain. Fig. 5.3.7C indicates that for wavelengths longer than the gain peak, the a_{lw} value can potentially be much higher.

All-optical conversion of signal waveform from an intensity modulated pump to an optical phase modulation of the probe through an SOA allows the generation of phase-modulated optical signal. Meanwhile, this cross-phase modulation can also be translated into an all-optical intensity modulation with the help of a phase discrimination circuit. A typical system configuration utilizing cross-phase modulation of SOA for all-optical intensity modulation is shown in Fig. 5.3.8, where two identical SOAs are used in a Mach-Zehnder interferometer (MZI) setting (Stubkjaer, 2000; Wang et al., 2008). A control optical beam at wavelength λ_2 is injected into one of the two SOAs to create an imbalance of the interferometer through optical power-induced phase change. A $\pi/2$ phase change in an interferometer arm will change the interferometer operation from constructive to destructive interference, thus switching off the optical signal at wavelength λ_1.

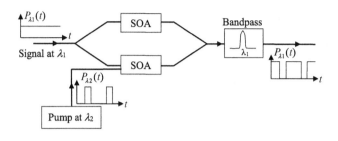

FIG. 5.3.8

All-optical switch using SOAs in a MZI configuration.

The major advantage of an all-optical switch using phase modulation in SOA is that the extinction ration can be made very high by carefully balancing the amplitude and the phase of the two interferometer arms. While in gain saturation-based optical switch, a very strong pump power is required to achieve a reasonable level of gain saturation, as illustrated in Fig. 5.3.5B. In addition, phase modulation is usually less wavelength sensitive compared to gain saturation as shown in Fig. 5.3.7B, which allows a wider wavelength separation between the signal and the pump.

5.3.4 WAVELENGTH CONVERSION BASED ON FOUR-WAVE MIXING IN AN SOA

Four-wave mixing (FWM) is a nonlinear phenomenon in which the mixing among three optical carriers of different frequencies, f_j, f_k, and f_l, creates new frequency components at $f_{jkl} = f_j + f_k - f_l$. FWM can also be created with two original optical carriers, which is a special case with $f_j = f_k$, known as degenerate FWM. The mechanism of FWM in an optical fiber has been discussed in Section 2.6, where dynamic refractive index gratings along the fiber are created from the mixing among optical carriers, and new frequency components are generated from the interaction between optical carriers and the dynamic index gratings. The concept of FWM in an SOA is similar to that in an optical fiber except for a different nonlinear mechanism. In an SOA, the mixing between two optical carriers generates a modulation of carrier population N_{tot} and therefore, the complex optical gain is modulated at the difference frequency between the two carriers. Each original optical carrier is then modulated by this beating frequency, producing sidebands. FWM in an SOA can be much more efficient than that in a passive optical fiber because the length of an SOA is typically less than a millimeter, while the fiber length has to be on the order of kilometers. Rigorous analysis of FWM in an SOA can be performed by numerically solving the rate equations. However, to understand the physical mechanism of FWM, we can simplify the analysis using a mean-field approximation. With this approximation, the carrier population rate Eq. (5.3.13) can be written as

$$\frac{dN_{tot}(t)}{dt} = \frac{I}{qA} - \frac{N_{tot}(t)}{\tau} - \frac{2\Gamma a}{hfA}[N_{tot}(t) - N_{0,tot}]P_{ave}(t) \qquad (5.3.26)$$

where P_{ave} is the spatially averaged optical power and $N_{0,tot}$ is the transparency carrier population. Since there are two optical carriers that coexist in the SOA, the total optical power inside the SOA is the combination of these two carriers,

$$P_{ave}(t) = \left| E_1 e^{-j2\pi f_1 t} + E_2 e^{-j2\pi f_2 t} \right|^2 \tag{5.3.27}$$

where f_1 and f_2 are the optical frequencies and E_1 and E_2 are the field amplitudes of the two carriers, respectively.

Eq. (5.3.27) can be expanded as

$$P_{ave}(t) = P_{DC} + E_1 E_2^* e^{j2\pi \Delta f t} + E_1^* E_2 e^{-j2\pi \Delta f t} \tag{5.3.28}$$

where $P_{DC} = E_1^2 + E_2^2$ is the constant part of the power and $\Delta f = f_2 - f_1$ is the frequency difference between the two carriers. As a consequence of this optical power modulation at frequency Δf as shown in Eq. (5.3.28), the carrier population will also be modulated at the same frequency due to gain saturation:

$$N_{tot}(t) = N_{DC} + \Delta N e^{j2\pi \Delta f t} + \Delta N^* e^{-j2\pi \Delta f t} \tag{5.3.29}$$

where N_{DC} is the constant part of carrier population and ΔN is the magnitude of carrier population modulation. Substituting Eqs. (5.3.28) and (5.3.29) into the carrier population rate Eq. (5.3.26) and separating the DC and time-varying terms, we can determine the values of ΔN and N_{DC} as

$$\Delta N = \frac{(N_{DC} - N_{0,tot})E_1 E_2^*}{P_{sat,opt}\left(1 + P_{DC}/P_{sat,opt} + j2\pi \Delta f \tau\right)} \tag{5.3.30a}$$

$$\Delta N^* = \frac{(N_{DC} - N_{0,tot})E_1^* E_2}{P_{sat,opt}\left(1 + P_{DC}/P_{sat,opt} - j2\pi \Delta f \tau\right)} \tag{5.3.30b}$$

so that,

$$N_{tot}(t) = N_{DC} + \frac{(N_{DC} - N_{0,tot})|E_1 E_2|}{P_{sat,opt}\sqrt{\left(1 + P_{DC}/P_{sat,opt}\right)^2 + (2\pi \Delta f \tau)^2}}\cos(2\pi \Delta f t + \Phi) \tag{5.3.31}$$

where, the phase angle is $\Phi = \tan^{-1}\left(\frac{2\pi \Delta f \tau}{1 + P_{DC}/P_{sat,opt}}\right)$, and

$$N_{DC} = \frac{I\tau/(qA) + N_{0,tot}P_{DC}/P_{sat,opt}}{1 + P_{DC}/P_{sat,opt}} \tag{5.3.32}$$

Since the optical gain of the amplifier is exponentially proportional to the carrier population as $G \propto \exp(N_{tot}L)$, the magnitude of carrier population modulation reflects the efficiency of FWM. Eq. (5.3.31) indicates that FWM efficiency is low when the carrier frequency separation Δf is high and the effective FWM bandwidth is mainly determined by the carrier lifetime τ of the SOA. In a typical SOA, the carrier lifetime is on the order of nanosecond and therefore, this limits the effective FWM bandwidth to the order of gigahertz. However, experimentally, the effect of FWM in SOA has been observed even when the frequency separation between the two optical carriers was as wide as in the Terahertz range, as illustrated in Fig. 5.3.9.

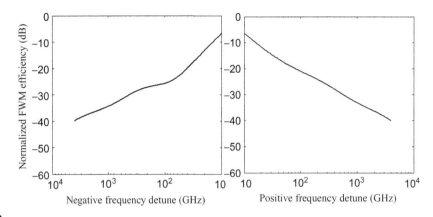

FIG. 5.3.9

Normalized FWM signal power vs. frequency detune (Zhou et al., 1993).

The wide bandwidth of FWM measured in SOAs indicated that the gain saturation is not only determined by the lifetime of interband carrier recombination τ; other effects, such as intraband carrier relaxation and carrier heating, have to be considered to explain the ultrafast carrier recombination mechanism. In Fig. 5.3.9, the normalized FWM efficiency, which is defined as the ratio between the power of the FWM component and that of the initial optical carrier, can be estimated as a multiple time constant system (Zhou et al., 1993):

$$\eta_{FWM} \propto \left| \sum_n \frac{k_i}{1 + i2\pi\Delta f \tau_n} \right|^2 \tag{5.3.33}$$

where τ_n and k_n are the time constant and the importance of each carrier recombination mechanism. To fit the measured curve, three time constants had to be used, with $k_1 = 5.65e^{-1.3i}$, $k_2 = 0.0642e^{1.3i}$, $k_3 = 0.0113e^{1.53i}$, $\tau_1 = 0.2$ns, $\tau_2 = 650$fs, and $\tau_3 = 50$fs (Zhou et al., 1993).

The most important application of FWM in SOA is frequency conversion, as illustrated in Fig. 5.3.10. A high-power CW optical carrier at frequency f_1, which is usually referred to as *pump*, is injected into an SOA together with an optical signal at frequency f_2, which is known as the *probe*. Due to FWM in SOA, a new frequency component is created at $f_{FWM} = f_2 - 2f_1$. A bandpass optical filter (BPF) is then used to select this FWM frequency component. In addition to performing frequency conversion from f_2 to f_{FWM}, an important advantage of this frequency conversion is frequency conjugation, as shown in Fig. 5.3.10B. The upper (lower) modulation side of the probe signal is translated into the lower (upper) sideband of the wavelength converted FWM component. This frequency conjugation effect has been used to combat chromatic dispersion in fiber-optical systems.

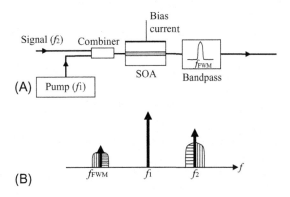

FIG. 5.3.10

Wavelength conversion using FWM in an SOA: (A) system configuration and (B) illustration of frequency relationship of FWM, where $f_{FWM} = f_2 - 2f_1$.

Suppose the optical fiber has an anomalous dispersion. High-frequency components travel faster than low-frequency components. If an SOA is added in the middle of the fiber system, performing wavelength conversion and frequency conjugation, the fast-traveling upper sideband over the first half of the system will be translated into the lower band, which will travel more slowly over the second half of the fiber system. Therefore, signal waveform distortion due to fiber chromatic dispersion can be canceled by this mid-span frequency conjugation.

To summarize, an SOA has a structure similar to that of a semiconductor laser except it lacks optical feedback. The device is small and can be integrated with other optical devices such as laser diodes and detectors in a common platform. This is especially attractive for photonic integration to reduce per-device cost and increase the reliability. Because of its fast carrier dynamics, an SOA can often be used for all-optical switches and other nonlinear optical signal processing purposes. However, for the same reason, an SOA is usually not suitable to be used as a line amplifier in a multiwavelength optical system. In such an application, cross-gain and cross-phase modulations in the SOA would create significant cross talk between different wavelength channels and introducing performance degradation. For that purpose, optical fiber amplifiers would be a better choice.

5.4 ERBIUM-DOPED FIBER AMPLIFIERS

An EDFA is one of the most popular optical devices in modern fiber-optic communication systems (Giles and Desurvire, 1991). EDFAs provide many advantages over SOAs in terms of high gain, high optical power, low cross talk between wavelength channels, and easy optical coupling from and to optical fibers (Desurvire, 1994). The basic structure of an EDFA shown in Fig. 5.4.1 is composed of an erbium-doped fiber (EDF), a pump laser, an optical isolator, and a wavelength-division multiplexing (WDM) coupler.

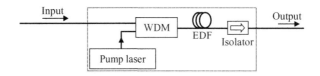

FIG. 5.4.1

Configuration of an EDFA. *EDF:* erbium-doped fiber and *WDM:* wavelength-division multiplexing coupler.

In contrast to SOAs, an EDFA is optically pumped and therefore, it requires a pump source, which is usually a high-power semiconductor laser. The WDM coupler is used to combine the short-wavelength pump with the longer-wavelength signal. The reason to use a WDM combiner instead of a simple optical coupler is to avoid the combination loss, discussed in detail in the next chapter. The optical isolator is used to minimize the impact of optical reflections from interfaces of optical components, and avoid Febry-Perot resonance discussed in Section 5.3. Since an EDFA may provide a significant amount of optical gain, even a small amount of optical reflection may be able to cause oscillation and therefore, degrade EDFA performance. Some high-gain EDFAs may also employ another optical isolator at the input side.

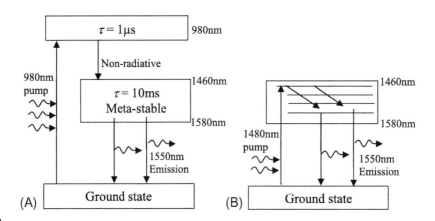

FIG. 5.4.2

Simplified energy band diagrams of erbium ions in silica: (A) three-level system and (B) two-level system.

An optical pumping process in an EDF is usually described by a simplified three-level energy system as illustrated in Fig. 5.4.2A. The bandgap between the ground state and the excited state is approximately 1.268 eV; therefore, pump photons at 980 nm wavelength are able to excite ground-state carriers to the excited state and create population inversion. The carriers stay in the excited state for only about 1 μs, and after that they decay into a metastable state through a nonradiative transition. In this process, the energy loss is turned into mechanical vibrations in the fiber. The energy band of the metastable state extends roughly from 0.78 to 0.85 eV, which correspond to a wavelength window ranging approximately from 1460 to 1580 nm. Under thermal equilibrium, carrier density within the metastable energy band follows the Fermi-Dirac distribution.

Finally, radiative recombination happens when carriers step-down from the metastable state to the ground state and emit photons in the 1550 nm wavelength region. The carrier lifetime in the metastable state is on the order of 10 ms, which is four orders of magnitude longer than the carrier lifetime in the excited state. Therefore, with constant optical pumping at 980 nm, almost all the carriers are accumulated in the metastable state. Therefore, the three-level system can be simplified into two levels for most of the practical applications.

As illustrated in Fig. 5.4.2B, an EDF can also be pumped at 1480 nm wavelength, which corresponds to the bandgap between the top of the metastable state and the ground state. In this case, 1480 nm pump photons excite carriers from ground state to the top of metastable state directly. Then, these carriers relax down through an intraband relaxation process to form the Fermi-Dirac distribution across the metastable energy band. Typically, 1480 nm pumping is more efficient than 980 nm pumping because it does not involve the nonradiative transition from the excited state to the metastable state. Therefore, 1480 nm pumping is often used for high-power optical amplifiers where energy efficiency is the most important concern. However, amplifiers with 1480 nm pumps usually have higher noise figures than 980 nm pumps, which will be discussed later.

5.4.1 ABSORPTION AND EMISSION CROSS SECTIONS

Absorption and emission cross sections are two very important properties of EDFs. Although the name *cross section* may seem to represent a geometric size of the fiber, it does not. The physical meanings of absorption and emission cross sections in an EDF are absorption and emission efficiencies at each wavelength (Desurvire, 1994).

Now let's start with the pump absorption efficiency W_p, which is defined as the probability that a pump photon is absorbed within each second to promote carrier transition. It is equivalent to the probability that a ground-state carrier been pumped to the metastable state within each second. If the pump optical power is P_p, within each second the number of pump photons is P_p/hf_p, where f_p is the optical frequency of the pump. Then, pump absorption efficiency is

$$W_p = \frac{\sigma_a P_p}{hf_p A} \tag{5.4.1}$$

where A is the effective fiber core cross-section area and σ_a is the absorption cross section. In another words, the absorption cross section is defined as the ratio of pump absorption efficiency and the density of pump photon flow rate:

$$\sigma_a = W_p \left/ \left(\frac{P_p}{hf_p A} \right) \right. \tag{5.4.2}$$

Since the unit of W_p is $[s^{-1}]$ and the unit of the density of the pump photon flow rate $P_p/(hf_p A)$ is $[s^{-1} \, m^2]$, the unit of absorption cross section is $[m^2]$. This is probably where the term absorption *cross section* comes from. As we mentioned, it is the property of the doped erbium ions, which has nothing to do with the geometric cross section of the fiber. We also need to note that the *absorption* cross section does not mean attenuation; it indicates the efficiency of energy conversion from photons to the excited carriers.

Similarly, we can define an emission cross section as

$$\sigma_e = W_s \left/ \left(\frac{P_s}{hf_s A} \right) \right. \tag{5.4.3}$$

where P_s and f_s are the emitted signal optical power and frequency, respectively. W_s is the stimulated emission efficiency, which is defined as the probability that a carrier in the metastable state drops down to the ground state to produce a signal photon within each second. It has to be emphasized that both absorption and emission cross sections are properties of the EDF itself. They are independent of the operation conditions of the fiber such as pump and signal optical power levels.

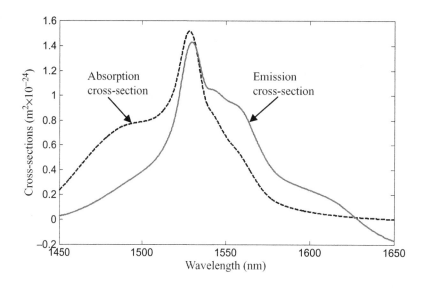

FIG. 5.4.3

Example of absorption and emission cross sections of Lucent HE980 erbium-doped fiber.

At each wavelength both absorption and emission exist because photons can be absorbed to generate carriers through a stimulated absorption process; at the same time, new photons can be generated through a stimulated emission process. Fig. 5.4.3 shows an example of absorption and emission cross sections; both of them are functions of wavelength.

If the carrier densities in the ground state and the metastable state are N_1 and N_2, respectively, the net stimulated emission rate per cubic meter is $R_e = W_s(\lambda_s)N_2 - W_p(\lambda_s)N_1$, where $W_s(\lambda_s)$ and $W_p(\lambda_s)$ are the emission and absorption efficiencies of the EDF at the signal wavelength λ_s. Considering the definitions of absorption and emission cross sections, this net emission rate can be expressed as

$$R_e = \frac{\Gamma(\lambda_s)\sigma_e(\lambda_s)P_s}{hf_sA}\left(N_2 - \frac{\sigma_a(\lambda_s)}{\sigma_e(\lambda_s)}N_1\right) \tag{5.4.4}$$

where a field confinement factor Γ is introduced to take into account the overlap fact between the signal/pump optical field and the erbium-doped area in the fiber core. The physical meaning of this net emission rate is the number of photons that are generated per second per cubic meter. In the ideal case without attenuation, this is equal to the pump absorption rate.

5.4.2 **RATE EQUATIONS**

The rate equation for the carrier density N_2 in the metastable state is

$$\frac{dN_2}{dt} = \frac{\Gamma(\lambda_p)\sigma_a(\lambda_p)P_p}{hf_pA}N_1 - \frac{\Gamma(\lambda_s)\sigma_e(\lambda_s)P_s}{hf_sA}\left(N_2 - \frac{\sigma_a(\lambda_s)}{\sigma_e(\lambda_s)}N_1\right) - \frac{N_2}{\tau} \qquad (5.4.5)$$

On the right side of this equation, the first term is the contribution of carrier generation by stimulated absorption of the pump (Γ_p is the overlap factor at pump wavelength). The second term represents net stimulated recombination which consumes the upper-level carrier density. The last term is spontaneous recombination, in which upper-level carriers spontaneously recombine to generate spontaneous emission. τ is the spontaneous emission carrier lifetime. For simplicity we neglected the emission term at pump wavelength because usually $\sigma_e(\lambda_p) \ll \sigma_a(\lambda_s)$.

In Eq. (5.4.5), both lower (N_1) and upper (N_2) state carrier densities are involved. However, these two carrier densities are not independent and they are related by

$$N_1 + N_2 = N_T \qquad (5.4.6)$$

where N_T is the total erbium ion density that is doped into the fiber. The energy state of each erbium ion stays either in the ground level or in the metastable level.

In the steady state ($d/dt = 0$), if only one signal wavelength and one pump wavelength are involved, Eq. (5.4.5) can be easily solved with the help of Eq. (5.4.6):

$$N_2 = \frac{\Gamma(\lambda_p)\sigma_a(\lambda_p)f_sP_p + \Gamma(\lambda_s)\sigma_a(\lambda_s)f_pP_s}{\Gamma(\lambda_p)\sigma_a(\lambda_p)f_sP_p + \Gamma(\lambda_s)f_pP_s(\sigma_e(\lambda_s) + \sigma_a(\lambda_s)) + hf_sf_pA/\tau}N_T \qquad (5.4.7)$$

Note that since the optical powers of both the pump and the signal change significantly along the length of the fiber, the upper-level carrier density is also a function of the location parameter z along the fiber: $N_2 = N_2(z)$. To find the z-dependent optical powers for the pump and the signal, propagation equations need to be established for both them.

The propagation equations for the optical signal at wavelength λ_s and for the pump at wavelength λ_p are

$$\frac{dP_s(\lambda_s)}{dz} = g(z, \lambda_s)P_s(z) \qquad (5.4.8)$$

$$\frac{dP_p(\lambda_p)}{dz} = \alpha(z, \lambda_p)P_p(z) \qquad (5.4.9)$$

where

$$g(z, \lambda_s) = \Gamma(\lambda_s)[\sigma_e(\lambda_s)N_2(z) - \sigma_a(\lambda_s)N_1(z)] \qquad (5.4.10)$$

is the effective gain coefficient at the signal wavelength and

$$\alpha(z, \lambda_p) = \Gamma(\lambda_p)[\sigma_e(\lambda_p)N_2(z) - \sigma_a(\lambda_p)N_1(z)] \qquad (5.4.11)$$

is the effective absorption coefficient at the pump wavelength.

These propagation equations can be easily understood by looking at, for example, the term $\sigma_e(\lambda_s)$ $N_2(z)P_s(z) = W_shf_sAN_2(z)$. It is the optical power generated per unit length at the position z. Similarly,

$\sigma_a(\lambda_s)N_1(z)P_s(z) = W_a hf_s A N_1(z)$ is the optical power absorbed per unit length at position z. Note that, similar to the material gain and absorption coefficients discussed for SOAs, the unit of both g and α is in $[\text{m}^{-1}]$.

Since the confinement factor $\Gamma(\lambda)$ is only weakly dependent on the wavelength, the wavelength dependence of g and α is primarily determined by the emission and the absorption cross sections. For that reason, in manufacturers data sheets, emission and absorption cross-section values are often presented in terms of the maximum gain and absorption rates along the fiber defined as $\exp[\Gamma\sigma_e(\lambda)N_T]$ and $\exp[\Gamma\sigma_a(\lambda)N_T]$, with the unit of $[\text{dB/m}]$. Fig. 5.4.4 shows the maximum gain and absorption rates based on the emission and absorption cross sections shown in Fig. 5.4.3, and assume a doping density of $N_T = 5.2 \times 10^{24} \text{ m}^{-3}$, and a constant confinement factor of $\Gamma = 0.1$.

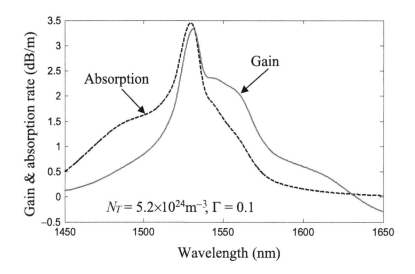

FIG. 5.4.4

Maximum gain and absorption rates along the Lucent HE980 erbium-doped fiber.

The two propagation Eqs. (5.4.8) and (5.4.9) are mutually coupled through carrier density $N_2(z)$ and $N_1(z)$ [which is equal to $N_T - N_2(z)$]. Once the position-dependent carrier density $N_2(z)$ is found, the performance of the EDFA will be known and the overall signal optical gain of the amplifier can be found as

$$G(\lambda_s) = \exp\left\{\Gamma(\lambda_s)\left[[\sigma_e(\lambda_s) + \sigma_a(\lambda_s)]\int_0^L N_2(z)dz - \sigma_a(\lambda_s)N_T L\right]\right\} \qquad (5.4.12)$$

which depends on the accumulated carrier density along the fiber length.

In practice, since both the pump and the signal are externally injected into the EDF and they differ only by the wavelength. There could be multiple pump channels and multiple signal channels. Therefore, Eqs. (5.4.8) and (5.4.9) can be generalized as

$$\frac{dP_k(z)}{dz} = g_k(z, \lambda_k)P_k(z) \qquad (5.4.13)$$

where the subscript k indicates the kth channel with the wavelength λ_k, and the gain coefficient for the kth channel is

$$g_k(z) = \Gamma(\lambda_k)[\sigma_e(\lambda_k)N_2(z) - \sigma_a(\lambda_k)N_1(z)] \tag{5.4.14}$$

In terms of the impact on the carrier density, the only difference between the pump and the signal is that at signal wavelength $\sigma_e > \sigma_a$, whereas at the pump wavelength we usually have $\sigma_a \gg \sigma_e$. Strictly speaking, both signal and pump participate in the emission and the absorption processes. In a more generalized approach, the upper-level carrier density depends on the optical power of all the wavelength channels, including pump channel(s), and the rate equation of N_2 is

$$\frac{dN_2}{dt} = -\sum_j \left\{ \frac{P_j(z)}{Ahf_j}\Gamma(\lambda_j)\left[\sigma_e(\lambda_j)N_2 - \sigma_a(\lambda_j)N_1\right] \right\} - \frac{N_2}{\tau} \tag{5.4.15}$$

where j is the channel index. Considering expressions in Eqs. (5.4.13) and (5.4.14), Eq. (5.4.15) can be simplified as

$$\frac{dN_2(t,z)}{dt} = -\frac{N_2}{\tau} - \sum_j \left\{ \frac{1}{Ahf_j}\frac{dP_j(z)}{dz} \right\} \tag{5.4.16}$$

In the steady state, $d/dt = 0$, Eq. (5.4.16) can be written as

$$N_2(z) = -\tau\sum_j \left\{ \frac{1}{Ahf_j}\frac{dP_j(z)}{dz} \right\} \tag{5.4.17}$$

In addition, Eq. (5.4.13) can be integrated on both sides,

$$\int_{P_{k,in}}^{P_{k,out}} \frac{1}{P_k(z)}dP_k(z) = \ln\left(\frac{P_{k,out}}{P_{k,in}}\right) = \int_0^L g_k(z)dz \tag{5.4.18}$$

Since $g_k(z) = \Gamma(\lambda_k)\{[\sigma_e(\lambda_k) + \sigma_a(\lambda_k)]N_2(z) - \sigma_a(\lambda_k)N_T\}$, based on Eq. (5.4.17), we have

$$g_k(z) = -\Gamma(\lambda_k)\left\{ [\sigma_e(\lambda_k) + \sigma_a(\lambda_k)]\left[\sum_j \frac{\tau}{Ahf_j}\frac{dP_j(z)}{dz}\right] + \sigma_a(\lambda_k)N_T \right\} \tag{5.4.19}$$

$g_k(z)$ can be integrated over the EDF length L:

$$\int_0^L g_k(z)dz = -\Gamma(\lambda_k)\left\{ [\sigma_e(\lambda_k) + \sigma_a(\lambda_k)]\left[\sum_j \frac{\tau}{Ahf_j}\int_0^L \frac{dP_j(z)}{dz}dz\right] + \sigma_a(\lambda_k)N_T L \right\} =$$

$$-\Gamma(\lambda_k)\left\{ [\sigma_e(\lambda_k) + \sigma_a(\lambda_k)]\left[\tau\sum_j \frac{(P_{j,out} - P_{j,in})}{Ahf_j}\right] + \sigma_a(\lambda_k)N_T L \right\}$$

Define

$$a_k = \Gamma(\lambda_k)\sigma_a(\lambda_k)N_T \tag{5.4.20}$$

$$S_k = \Gamma(\lambda_k)\tau[\sigma_e(\lambda_k) + \sigma_a(\lambda_k)]/A \tag{5.4.21}$$

which are EDF properties and independent of the operation condition. We can have a simpler form of the integration,

$$\int_0^L g_k(z)dz = -S_k[Q_{OUT} - Q_{IN}] - a_k L \tag{5.4.22}$$

where $Q_{OUT} = \sum_j P_{j,out}/hf_j$ and $Q_{IN} = \sum_j P_{j,in}/hf_j$ are total photon flow rates at the output and the input of the EDFA, respectively (including both signal channels and pump channels).

Combining Eqs. (5.4.18) and (5.4.22), we have (Saleh et al., 1990)

$$P_{k,out} = P_{k,in} e^{-a_k L} \cdot e^{-S_k(Q_{OUT} - Q_{IN})} \tag{5.4.23}$$

In this equation, both a_k and S_k are properties of the EDF, Q_{IN} is the total photon flow rate at the input, which is known. However, Q_{OUT} is the total photon flow rate at the output, which is unknown. To find Q_{OUT}, one can solve the following equation iteratively:

$$Q_{OUT} = \sum_{k=1}^{N} \left\{ \frac{1}{hf_k} P_{k,in} \exp(S_K Q_{IN} - a_k L) \exp(-S_k Q_{OUT}) \right\} \tag{5.4.24}$$

In this equation, everything is known except Q_{OUT}. A simple numerical algorithm should be sufficient to solve this iterative equation and find Q_{OUT}. Once Q_{OUT} is obtained using Eq. (5.4.24), optical gain for each channel can be found using Eq. (5.4.23).

5.4.3 NUMERICAL SOLUTIONS OF RATE EQUATIONS

Although Eqs. (5.4.23) and (5.4.24) together provides a semi-analytical formulation that can be used to investigate the performance of an EDFA, it has several limitations: (1) in this calculation, we have neglected the carrier saturation effect caused by ASE noise. In fact, when the optical gain is high enough, ASE noise may be significant and it contributes to saturate the EDFA gain. Therefore, the semi-analytical formulation is only valid when the EDFA has a relatively low gain. (2) It only predicts the relationship between the input and the output optical power levels; power evolution along the EDF, $P(z)$, is not given. (3) It only calculates the accumulated carrier population $\int_0^L N_2(z)dz$ in the EDF, while the actual carrier density distribution along z is not provided. Since ASE noise is generated and amplified in a distributive manor along the EDF, we are not able to accurately calculate the ASE noise at the EDFA output using this simple method. Therefore, although this semi-analytical method serves as a quick evaluation tool to find the optical gain spectrum, it does not give accurate enough results when the optical gain is sufficiently high. Precise modeling of EDFA performance has to involve numerical simulations. The model has to consider optical pumps and optical signals of various wavelengths. The spontaneous emission noise generated along the EDF propagates both in the forward direction and in the backward direction, and their contribution to the depletion of carrier density $N_2(z)$ also needs to be considered.

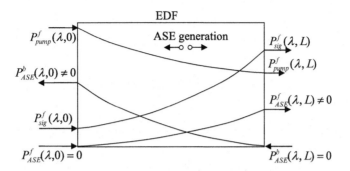

FIG. 5.4.5

Illustration of signal and ASE noise in an EDFA: $P^f_{sig}(\lambda, 0)$ and $P^f_{sig}(\lambda, L)$ represent input and output optical signal in the forward direction, $P^f_{pump}(\lambda, 0)$ is the input pump, $P^f_{ASE}(\lambda)$ and $P^b_{ASE}(\lambda)$ represent forward and backward propagated ASE power spectral densities.

Fig. 5.4.5 illustrates the evolution of optical signals and ASE noise along the EDF. $P^f_s(\lambda, 0) = P^f_{sig}(\lambda, 0) + P^f_{pump}(\lambda, 0)$ is the combination of the input optical signal and the input pump, all in the forward direction. After propagating through the fiber, the optical signal is amplified and the optical pump is depleted. $P^f_s(\lambda, L) = P^f_{sig}(\lambda, L) + P^f_{pump}(\lambda, L)$ is the combination of optical signal and the remnant pump at the EDF output. Because of carrier inversion, spontaneous emission is generated along the fiber, and they are amplified in both directions. $P^f_{ASE}(\lambda)$ is the ASE noise PSD propagating in the forward direction, which is zero at the fiber input side ($z=0$), and becomes nonzero at the fiber output ($z=L$). Similarly, $P^b_{ASE}(\lambda)$ is the ASE noise PSD propagating in the backward direction, which is zero at the fiber output side ($z=L$), and nonzero at the fiber input ($z=0$). In the steady state, the propagation equations of the optical signal (including the pump), the forward ASE and the backward ASE are, respectively,

$$\frac{dP^f_s(\lambda, z)}{dz} = P^f_s(\lambda, z)g(\lambda, z) \tag{5.4.25}$$

$$\frac{dP^f_{ASE}(\lambda, z)}{dz} = P^f_{ASE}(\lambda, z)g(\lambda, z) + 2n_{sp}(z)g(\lambda, z)hc/\lambda \tag{5.4.26}$$

$$\frac{dP^b_{ASE}(\lambda, z)}{dz} = -P^b_{ASE}(\lambda, z)g(\lambda, z) - 2n_{sp}(z)g(\lambda, z)hc/\lambda \tag{5.4.27}$$

where n_{sp} is the spontaneous emission factor, which depends on the carrier inversion level N_2 as

$$n_{sp}(z) = \frac{N_2(z)}{N_2(z) - N_1(z)} = \frac{N_2(z)}{2N_2(z) - N_T} \tag{5.4.28}$$

The carrier density N_2 is described by a steady-state carrier density rate equation,

$$\frac{N_2}{\tau} + \int_\lambda \left[\frac{P^f_s(\lambda, z)}{Ahc/\lambda} + \frac{P^f_{SP}(\lambda, z)}{Ahc/\lambda} + \frac{P^b_{SP}(\lambda, z)}{Ahc/\lambda} \right] g(\lambda, z)d\lambda = 0 \tag{5.4.29}$$

where

$$g(\lambda, z) = \Gamma(\lambda)\{[\sigma_e(\lambda) + \sigma_a(\lambda)]N_2(z) - \sigma_a(\lambda)N_T\} \tag{5.4.30}$$

is the gain coefficient and $\Gamma(\lambda)$ is the wavelength-dependent confinement factor.

Eqs. (5.4.25) through (5.4.30) are coupled through $N_2(z)$ and can be solved numerically. A complication in this numerical analysis is that ASE noise has a continuous spectrum that varies with the wavelength. In the numerical analysis, the ASE optical spectrum has to be divided into narrow slices, and within each wavelength slice, the noise PSD can be assumed to be a constant. Therefore, if the ASE spectrum is divided into m wavelength slices, Eqs. (5.4.26) and (5.4.27) each must be split into m equations with P_{SP}^f and P_{SP}^b representing the noise powers within the each spectral slice. Since the carrier density $N_2(z)$ is a function of z, the calculation usually has to divide the EDF into many short sections, and within each short section N_2 can be assumed to be a constant.

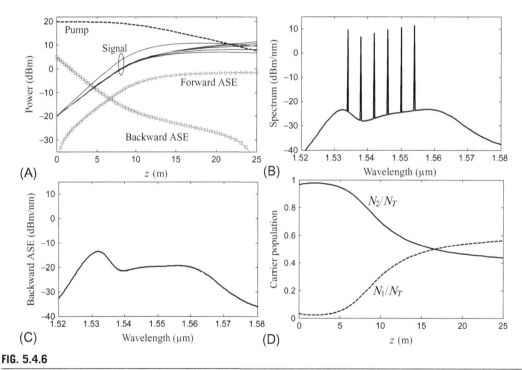

FIG. 5.4.6

Numerically calculated performance of an EDFA with 100-mW forward pump, and six signal channels each with −20 dBm power at the EDFA and 4-nm channel spacing. The length of the EDF is $L=25$ m. (A) Evolution of pump, ASE and signal powers along the EDF, (B) optical PSD at EDFA output, (C) backward ASE PSD at the input side of EDF, and (D) upper level carrier density distribution along the EDF.

As shown in Fig. 5.4.5, if the calculation starts from the input side of the EDF, the power levels of the input optical signal and the pump are known, and we also know that the forward ASE noise at the fiber input ($z=0$) is zero. But we do not know the level of the backward ASE noise at $z=0$. To solve the rate equations, we have to assume a spectrum of $P_{ASE}^b(\lambda,0)$ at $z=0$. Then, rate equations can be solved section by section along the fiber. At the end of the fiber, an important boundary condition has to be met—that is, the backward ASE noise has to be zero ($P_{SP}^b(\lambda,L)=0$). If this condition is not satisfied, we have to perform the calculation again using a modified guess of $P_{SP}^b(\lambda,0)$. This process usually has to be repeated several times until the backward ASE noise at the fiber output is lower than a certain level (say, $-40\,$dBm); then the results can be considered accurate.

Fig. 5.4.6 shows an example of the EDFA performance numerically calculated. The length of the EDF is $L=25\,$m. A forward pump at 980 nm is used with 100-mW optical power. There are six signal channels each with $-20\,$dBm input optical power at wavelengths of 1533, 1538, 1542, 1546, 1550, and 1554 nm. Fig. 5.4.6A shows the power evolution of the pump and the six signal channels along the EDF. It also shows the spectrally integrated total ASE powers in the forward and the backward directions, respectively. The gain of the six signal channels are on the order of 30 dB. However, the optical gain is wavelength dependent because of the wavelength-dependent nature of emission and absorption cross sections. Fig. 5.4.6B shows the PSD at the EDFA output, which consists of amplified optical signal channels and a broadband ASE noise. The integrated (forward) ASE total power is approximately $-1\,$dBm which is indicated in Fig. 5.4.6A. The spectrum of backward propagated ASE emerging from the input side of the EDF is shown in Fig. 5.4.6C with the integrated total power of about 5 dBm, which is 6 dB higher than the forward ASE. This is due to the high gain co-efficient near the input side of the EDF where pump power is high. Fig. 5.4.6D shows the distribution of carrier densities $N_1(z)$ and $N_2(z)$ normalized by the doping density N_T. Almost complete carrier inversion is achieved near the input of the EDF because of the high pump power there. As the pump power decreases along the fiber, the population inversion reduces and the gain coefficient also diminishes.

At a certain pump power, if the EDF is too long, carrier inversion near the end of the EDF will be too low to maintain carrier inversion, and the overall signal gain may be reduced. Whereas if the EDF is too short to absorb the majority of the pump power, the pump efficiency will be low and the signal optical gain will be low. In the EDFA design, the length of EDF has to be optimized based on the available pump power, the range of optical signal power, and the optical gain requirement.

Because of the wavelength-dependent absorption and emission cross sections determined by the property of the erbium ions, the shape of EDFA gain spectrum cannot be represented by a simple parabolic function like in an SOA. In addition, the shape of the EDFA gain spectrum may also change significantly with the operation condition. Fig. 5.4.7 shows an example of EDFA optical gain vs. wavelength at different input optical signal power levels. This example was obtained using a 35 m long, EDF and a 75-mW forward pump power at 980 nm wavelength. Only one signal wavelength was used at 1550 nm. In the EDFA gain spectrum, there is typically a narrow gain peak around 1530 nm and a relatively flat region between 1540 and 1560 nm. Fig. 5.4.7 shows that with the 1550 nm input signal optical power increased from -30 to $-10\,$dBm, the optical gain at 1550 nm decreases from 36 to 27 dB, which is caused by gain saturation. In addition, the spectral shape of the optical gain also changes with the signal optical power. With the increase of signal optical power, the gain tends to tilt toward the longer-wavelength side and the short-wavelength peak around 1530 nm is suppressed. This phenomenon is commonly referred to as *dynamic gain tilt*. In an optical system with multiple signal wavelengths, this dynamic gain tilt is a major concern for channel performance equalization and optimization, which needs to be compensated.

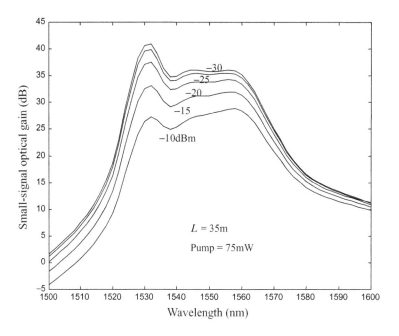

FIG. 5.4.7

Example of EDFA optical gain vs. wavelength at different input optical signal power levels.

5.4.4 ADDITIONAL CONSIDERATIONS IN EDFA DESIGN

EDFA performance depends on the spectral shapes of emission and absorption cross sections, pumping configuration and power, and EDF length. In general, an EDFA can have either a forward pump or a backward pump, or both of them as illustrated in Fig. 5.4.8. Forward pumping refers to the pump that propagates in the same direction as the signals, whereas in backward pumping scheme the pump travels in the opposite direction of the optical signal.

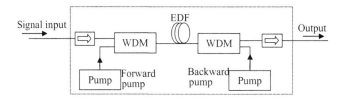

FIG. 5.4.8

Configuration of an EDFA with both a forward pump and a backward pump.

An EDFA with a forward pump alone usually has relatively low signal output optical power, but at the same time the forward ASE noise level is also relatively low. However, the backward propagated ASE noise level at the input side of the EDF may become quite high. The reason is that although the pump power can be strong at the fiber input, its level is significantly reduced at the fiber output because the pump energy is converted to the amplified signal along the fiber. Because of the weak pump at the EDF output, the carrier inversion level is low as well; therefore, the differential gain reduces monotonically along the EDF and the output optical signal power is limited. On the other hand, since the carrier inversion level is very strong near the fiber input side, the backward propagated spontaneous emission noise meets high amplification in that region, and thus the backward ASE noise power is strong at the fiber input side. An example of the EDFA performance using only a forward bump is shown in Fig. 5.4.6.

With backward pumping, the pump power is strong near the fiber output side and its level is significantly reduced when it reaches to the input side. Because of the strong pump power and the relatively high carrier inversion at the EDF output, the output optical signal power can be higher than the forward pumping configuration at the same pump power. But the forward ASE noise level can also be high because of the high differential gain near the output port. Bidirectional pumping can be used to increase both optical gain and output signal optical power for an EDFA. In such a case, pump power distribution along the EDF is more uniform compared to unidirectional pumping. Fig. 5.4.9 shows the power levels of pump and signal channels along the EDF in a bidirectional pumping scheme with 100 mW pump in each direction, and the output PSD. The variation of the combined pump power is less than 5 dB along the EDF. Because of the increase of the total pump power, the optical gain and the total output optical signal power are also increased in comparison to the case of forward pumping as shown in Fig. 5.4.6.

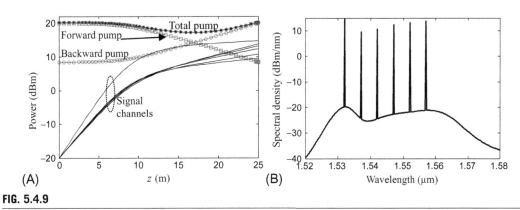

FIG. 5.4.9

Performance of EDFA with bidirectional pumping and 25 m EDF length. 100-mW pump power is used in each direction, and there are six signal wavelength channels each with −20 dBm input power. (A) Power evolution along the EDF for forward pump (squares), backward pump (circles), total pump (stars), and optical signal channels (solid lines). (B) Output optical power spectral density.

Because the optical gain and the shape of the gain spectrum of an EDFA can be affected by the variation of the signal optical power through gain saturation, optical gain has to be stabilized to ensure the stability of the optical system. Automatic gain control (AGC) and automatic power control (APC) are important features in practical EDFAs that are used in optical communication systems and networks. AGC and APC are usually used in in-line optical amplifiers to regulate the optical gain and the output signal optical power of an EDFA. Because both the optical gain and the output signal optical power are dependent on the power of the pump, the automatic control can be accomplished by adjusting the pump power.

Fig. 5.4.10 shows an EDFA design with AGC in which two photodetectors, PD_I and PD_O, are used to detect the input and the output signal optical powers. An electronic circuit compares these two power levels, calculates the optical gain of the EDFA, and generates an error signal to control the injection current of the pump laser if the EDFA gain is different from the target value. If the EDFA carries multiple WDM channels and we assume that the responsivity of each photodetector is wavelength independent within the EDFA bandwidth, this AGC configuration controls the overall gain of the total optical power such that

$$G = \frac{\sum_{j=1}^{N} P_{j,out}}{\sum_{j=1}^{N} P_{j,in}} = \text{constant} \tag{5.4.31}$$

where N is the number of wavelength channels. In this case, although the gain of the total optical power is fixed by the feedback control, the optical gain of each individual wavelength channel may vary, depending on the spectral shape of the optical gain over the EDFA bandwidth.

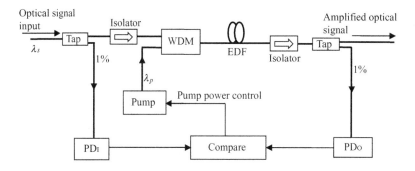

FIG. 5.4.10

Configuration of an EDFA with both automatic gain control. PD_I: input photodiode and PD_O: output photodiode.

In long-distance optical transmission, many in-line optical amplifiers may be concatenated along a system. If the optical gain of each EDFA is automatically controlled to a fixed level, any signal power fluctuation, for example, through dynamic wavelength channel add/drop, along the fiber system will make the output optical power variation throughout the system.

APC, on the other hand, regulates the total output optical power of an EDFA, as shown in Fig. 5.4.11. In this configuration, only one photodetector is used to detect the total output signal optical power. The injection current of the pump laser is controlled to ensure that this measured power is equal to the desired level such that

$$\sum_{j=1}^{N} P_{j,out} = \text{constant} \tag{5.4.32}$$

The advantage of APC is that it isolates signal optical power fluctuations along the system because loss change in one span does not propagate to other spans. However, since the total power is regulated, channel add/drop in WDM systems will affect the optical power of each individual channel. Therefore, in advanced optical communication systems, EDFAs are controlled by intelligent systems taking into account the number of wavelength channels, OSNR, and other important system parameters.

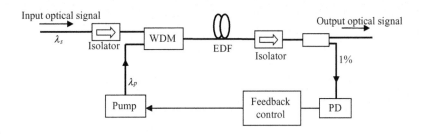

FIG. 5.4.11

Configuration of an EDFA with automatic power control.

In a multiwavelength optical system, the wavelength-dependent gain characteristic of EDFAs shown in Fig. 5.4.7 makes transmission performance different from channel to channel and thus greatly complicates system design and provisioning. In addition, the shape of the optical gain spectrum also changes with the signal optical power as illustrated in Fig. 5.4.7 which is known as dynamic gain tilt. Considering signal power level variations and wavelength add/drop in optical networks, the dynamic gain tilt may also significantly affect transmission performance, and needs to be taking into EDFA design.

In high-end optical amplifiers for long distance and multiwavelength optical systems, gain variation vs. wavelength can be compensated by passive optical filters, as shown in Fig. 5.4.12A. To flatten the gain spectrum, an optical filter with the transfer function complementing to the original EDFA gain spectrum is used with the EDFA so that the overall optical gain does not vary with the wavelength, and this is illustrated in Fig. 5.4.13. However, the passive gain flattening is effective only with one particular optical gain of the amplifier. Any change of EDFA operation condition, such as signal optical power, would change the spectral shape of the gain spectrum and therefore, require a different filter.

To address the dynamic gain tilt problem, a spatial light modulator (SLM) can be used as schematically shown in Fig. 5.4.12B. In this configuration, the output of the EDFA is demultiplexed so that different spectral components are dispersed into the spatial domain. Then, a one-dimensional (1D) SLM is used to attenuate each wavelength component individually to equalize the optical gain across the EDFA bandwidth. Finally, all spectral components are combined through a multiplexer at the output. With the rapid advance of liquid crystal technology, SLM is commercially available. Since each wavelength component is addressed independently and dynamically in this configuration, the effect of dynamic gain tilt can be minimized.

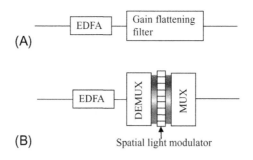

FIG. 5.4.12

EDFA gain flattening using (A) a passive filter and (B) a spatial light modulator.

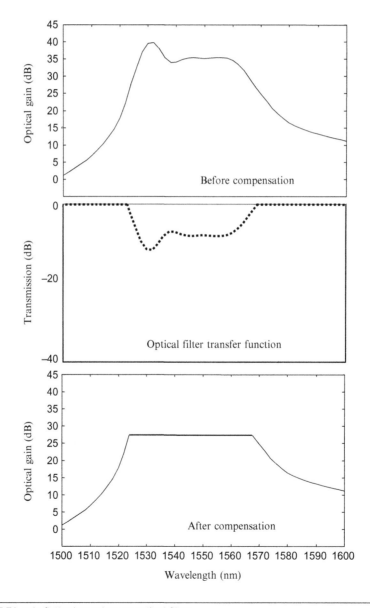

FIG. 5.4.13

Illustration of EDFA gain flattening using an optical filter.

5.5 RAMAN AMPLIFIERS

In Chapter 2, we have briefly discussed the mechanism of SRS in an optical fiber, which is caused by the interaction between the optical signal photons and the energy states of the silica molecules. In this scattering process, a photon spends part of its energy to excite a molecule from a lower to a higher vibrational or rotational state known as optical phonon. The energy ΔE_e of the optical phonon determines the frequency difference between the input photons and the scattered photons commonly referred to as a Stokes wave,

$$f_p - f_s = \frac{\Delta E_e}{h} \tag{5.5.1}$$

where f_p and f_s are frequencies of the input and the scattered photons, respectively, and h is the Planck's constant. If the material structure has a regular crystal lattice, the vibrational or rotational energy levels of the molecules are specific; in such a case, the spectrum of the scattered photons through the SRS process would be relatively narrow. For amorphous materials, such as glasses, the SRS spectrum is quite wide because the electron energy levels are not very well regulated. Fig. 2.6.1 shows the normalized SRS spectrum as a function of the frequency shift $f = f_p - f_s$ in a silica fiber. The frequency shift peaks at about 13 THz and the FWHM width of the spectrum is approximately 7 THz. Although the spectral shape of SRS is not Lorentzian, it provides a mechanism for optical amplification, in which a high-power laser beam is used as the pump to amplify optical signals at frequencies about 7 THz lower than that of the pump. Because the SRS process is effective for both forward and backward directions in an optical fiber, the Raman amplifier can be bidirectional (Headley and Agrawal, 2005; Bromage, 2004).

When a pump wave with an optical power density of $I_p(z)$ propagates along an optical fiber, a Stokes wave at a longer wavelength can be generated and amplified along the fiber in both forward and backward directions with a power density of $I_s(f, z)$. Energy transfer between the pump and the Stokes can be described by the following coupled wave equations:

$$\frac{dI_s(f,z)}{dz} = \mp \alpha_s I_s(f,z) \pm g_R(f,z)I_p(z)I_s(f,z) \tag{5.5.2}$$

$$\frac{dI_p(f,z)}{dz} = \mp \alpha_p I_p(f,z) \mp g_R(f,z)I_p(z)I_s(f,z) \tag{5.5.3}$$

where the $\pm$ sign represents forward (+) and backward (−) propagating Stokes waves and the pump. $g_R(f, z)$ is the SRS gain coefficient, which is generally dependent on the frequency difference f between the pump and the Stokes waves, as shown in Fig. 2.6.1. α_s and α_p are fiber attenuation coefficients at the Stokes and the pump wavelengths, respectively, and hf_s is the Stokes photon energy. We have assumed that the SRS gain coefficient may be nonuniform along the fiber, and thus is a function of z. In general, Eqs. (5.5.2) and (5.5.3) are mutually coupled, however, in most of the cases of a Raman amplifier, the pump power is much higher than that of the Stokes, and the pump depletion can be neglected for simplicity. In such a case, the second term on the right-hand side of Eq. (5.5.3) can be neglected, so that the pump power along the fiber can be simply described by $I_{FP}(z) = I_{FP,\,in}e^{-\alpha_p z}$ for forward pumping with $I_{FP,\,in}$ the input pump power at $z = 0$, and $I_{BP}(z) = I_{BP,\,in}e^{-\alpha_p(L-z)}$ for backward pumping with $I_{BP,\,in}$ the input pump power at $z = L$. Assume that an optical signal at the Stokes frequency, $P_s(f, 0)$, is injected at the input of the fiber $z = 0$, and propagates in the forward direction, neglecting the saturation effect due to spontaneous emission, the signal optical power along the fiber can be found by integrating Eq. (5.5.2) as

$$P_s(f,z) = P_s(f,0)\exp\left[\frac{g_R(f)}{A_{eff}}P_{FP,in}\left(\frac{1-e^{-\alpha_p z}}{\alpha_p}\right) - \alpha_s z\right]$$

(5.5.4)

for forward pumping, and

$$P_s(f,z) = P_s(f,z)\exp\left[\frac{g_R P_{BP,in}}{A_{eff}}\left(\frac{e^{\alpha_p(z-L)} - e^{-\alpha_p L}}{\alpha_p}\right) - \alpha_s z\right]$$

(5.5.5)

for backward pumping. Where $P_{FP,\ BP} = I_{FP,\ BP}A_{eff}$ and $P_s = I_s A_{eff}$ are the pump and the signal optical power, respectively, and A_{eff} is the effective cross-section area of the fiber, and L is the fiber length. We have also assumed that the Raman gain coefficient $g_R(f)$ remains constant along the fiber. Because pump depletion is neglected, the signal optical gain over the entire fiber of length, $z=L$, will be the same for both forward pump and the backward pump schemes, which is

$$G(f,L) = \exp\left[\frac{g_R(f)}{A_{eff}}P_{P,in}\left(\frac{1-e^{-\alpha_p L}}{\alpha_p}\right) - \alpha_s L\right]$$

(5.5.6)

where $P_{P,in}$ represents $P_{FP,in}$ or $P_{BP,in}$ depending on the pump scheme that is used. The Raman gain coefficient is a material property that is proportional to the resonant contribution associated with molecular vibrations. In a silica fiber operating in 1550 nm wavelength window, the peak Raman gain coefficient is on the order of $g_R = 2 \sim 10 \times 10^{-14}$ m/W depending on the fabrication process and doping.

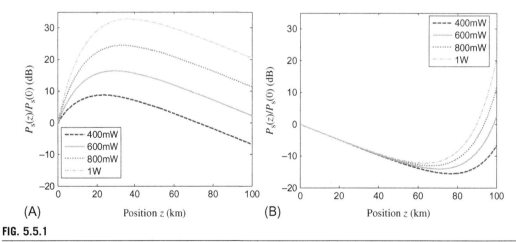

FIG. 5.5.1

Normalized signal optical power as the function of position along the 100 km fiber with forward pump (A) and backward pump (B) for different pump power levels. Parameters used are: Raman gain coefficient $g_R = 5.8 \times 10^{-14}$ m/W, fiber effective core area $A_{eff} = 80\,\mu m^2$, spontaneous scattering factor $n_{sp} = 1.2$, and fiber attenuation parameters $\alpha_p = 0.3$ dB/km and $\alpha_s = 0.25$ dB/km at the pump and the Stokes wavelengths, respectively.

Fig. 5.5.1 shows the normalized signal optical power as the function of the fiber length for different input pump power levels. For forward pump, as shown in Fig. 5.5.1A, signal optical power increases near the input side of the fiber where pump power is relatively high, and reduces after a certain distance when the pump power becomes weak and Raman gain effect diminishes, and eventually the optical signal only sees the fiber attenuation. By increasing the pump power, the Raman gain is increased, and the peak position of the signal power extends further into the fiber. For backward pump, the optical signal injected from $z=0$ primarily experiences fiber attenuation at the beginning, and the Raman gain is effective only near the final section of the fiber, as shown in Fig. 5.5.1B, where the pump power is high. Although the signal optical power emerging from the end of the fiber [at $L=100$ km in Fig. 5.5.1A and B] is the same for both forward pump and backward pump at the same pump power level, the power profiles are quite different.

Note that in the calculation of Raman gain in Eq. (5.5.6), we have used a small-signal approximation which neglected the saturation effect caused by the power $P_s(f)$ in the Stokes frequency. We have also neglected the impact of spontaneous emission in the wave-propagation Eqs. (5.5.2) and (5.5.3). In fact, the spontaneous emission generated along the fiber in the Stokes frequency can be amplified by the Raman effect similar to the ASE in an EDFA. The ASE noise PSD at the Stokes wavelength can be calculated from the differential equation similar to Eq. (5.5.2), but with an additional noise source term. Without losing generality, we use forward propagated ASE noise PSD as an example, which is relevant for most of the applications,

$$\frac{d\rho_{ASE}(z)}{dz} = \left(\frac{g_R P_p(z)}{A_{eff}} - \alpha_s\right)\rho_{ASE}(z) + 2n_{sp}hf_s g_R \frac{P_p(z)}{A_{eff}} \tag{5.5.7}$$

where $P_p(z)$ is the pump power density along the fiber, $n_{sp}=1/\exp(hf/k_BT)$ is a temperature-dependent spontaneous scattering factor with k_B the Boltzmann's constant, T is the absolute temperature, and f is the pump-Stokes frequency difference. The ASE noise PSD at the fiber output can be found as

$$\rho_{ASE,FP}(L) = 2n_{sp}hf_s G(L) g_R \frac{P_{FP,in}}{A_{eff}} \int_0^L \frac{e^{-\alpha_p z}}{G_{FP}(z)} dz \tag{5.5.8}$$

for forward pumping, and

$$\rho_{ASE,BP}(L) = \frac{2n_{sp}hf_s g_R}{A_{eff}} G(L) P_{BP,in} \int_0^L \frac{e^{-\alpha_p(L-z)}}{G_{BP}(z)} dz \tag{5.5.9}$$

where $G(L)$ is the Raman gain over the fiber length L, as defined in Eq. (5.5.6), which is the same for forward pump and backward pump, but $G_{FP}(z)$ and $G_{BP}(z)$ are position-dependent gains for forward and backward pumping schemes defined as

$$G_{FP}(z) = \frac{P_s(f_s,z)}{P_s(f_s,0)} = \exp\left[\frac{g_R P_{FP,in}}{A_{eff}}\left(\frac{1-e^{-\alpha_p z}}{\alpha_p}\right) - \alpha_s z\right] \tag{5.5.10}$$

$$G_{BP}(z) = \frac{P_s(f_s,z)}{P_s(f_s,0)} = \exp\left[\frac{g_R P_{BP,in}}{A_{eff}}\left(\frac{e^{\alpha_p(z-L)} - e^{-\alpha_p L}}{\alpha_p}\right) - \alpha_s z\right] \tag{5.5.11}$$

The noise figure of a fiber Raman amplifier can be calculated based on the procedure outlined in Section 5.2, for forward pump and backward pump, respectively,

$$NF_{FP} = \frac{\rho_{ASE,FP}}{G(L)hf_s} + \frac{1}{G(L)} = 2n_{sp}g_R \frac{P_{FP,in}}{A_{eff}} \int_0^L \frac{e^{-\alpha_p z}}{G_{FP}(z)} dz + \frac{1}{G(L)} \tag{5.5.12}$$

$$NF_{BP} = \frac{\rho_{ASE,BP}}{G(L)hf_s} + \frac{1}{G(L)} = 2n_{sp}g_R \frac{P_{BP,in}}{A_{eff}} \int_0^L \frac{e^{-\alpha_p(L-z)}}{G_{BP}(z)} dz + \frac{1}{G(L)} \tag{5.5.13}$$

In addition to the dependence of pumping scheme (forward or backward), the noise figure of a fiber Raman amplifier also depends on the Raman gain coefficient, fiber length, fiber loss at the pump and the Stokes wavelength, and the pump power. In general, both the gain and the noise figure also depend on the optical signal power at the Stokes wavelength f_s which may saturate the optical gain. But this signal-induced pump depletion has not been considered in Eqs. (5.5.4)–(5.5.13) for simplicity.

Fig. 5.5.2 shows the noise figures calculated from Eqs. (5.5.12) and (5.5.13). For the case of forward pump shown in Fig. 5.5.2A, ASE noise is generated primarily near the input end of the fiber, and significantly attenuated by the fiber while propagating to the output, and thus the noise figure is relatively low. Noise figure becomes almost independent of the fiber length when the pump power is high enough. This is because both signal amplification and ASE noise generation happen near the input side of the fiber and they are equally attenuated by the rest part of the fiber. While having the same signal optical gain as the forward pumping, backward pumping has a much higher noise figure as shown in Fig. 5.5.2B. This is because, with backward pump, the ASE noise is primarily generated near the output side of the fiber where the pump power is the highest, and fiber attenuation has much more impact in the optical signal than for the generated ASE noise. For this reason, the noise figure shown in Fig. 5.5.2B increases with the increase of the fiber length.

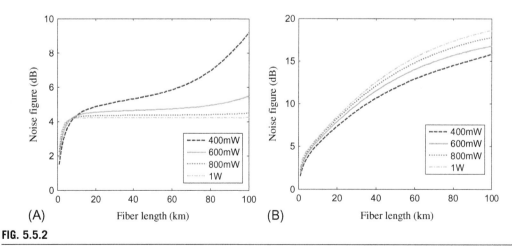

FIG. 5.5.2

Noise figure as the function of the fiber length for forward pump (A) and backward pump (B) at different pump power levels. Parameters used are the same as those in Fig. 5.5.1.

For a forward pumped fiber Raman amplifier with spontaneous scattering factor $n_{sp}=1.2$, the noise figure is slightly higher than 4 dB when the fiber is long enough and the pump power is sufficiently high, as shown in Fig. 5.5.2A. This is similar to the intrinsic (without considering input and output connection losses) noise figure, 3.8 dB, of an EDFA with $n_{sp}=1.2$. Whereas for a backward pumped fiber Raman amplifier, the noise figure can be much higher, as shown in Fig. 5.5.2B. But this does not mean backward pumped Raman amplifier is not useful. In the contrary, backward-pumped Raman amplification is widely used to extend the reach of fiber-optic system beyond what can be accomplished with EDFAs. Here we explain why.

For a localized fiber Raman amplifier as illustrated in Fig. 5.5.3A, the length of the fiber can be chosen to maximize the Raman gain depending on the available pump power. In that case, a fiber Raman amplifier is packaged into a box, and it has no advantage in terms of noise figure compared to an EDFA. However, a popular application of fiber-based Raman amplifier is the distributed Raman amplification in which the transmission fiber is used as the amplification medium as illustrated in Fig. 5.5.3B. In such a case, the length of the fiber is usually longer than 50 km and the overall amplification $G(L)$ can even be lower than 0 dB. The noise figure may also be quite high, especially when $G(L)$ is low. It is important to note that in a distributed Raman amplifier, since the transmission fiber is used as the amplification medium, the attenuation of the transmission fiber has already been counted against the overall Raman gain (Bristiel et al., 2004). For a fair comparison with localized optical amplifiers such as SOA and EDFA, we can separate the attenuation of the transmission fiber which is $e^{-\alpha_s L}$ from an equivalent localized Raman amplifier as shown in Fig. 5.5.2C. The effective optical gain of this localized amplifier should be $G_{eff}=G(L)e^{\alpha_s L}$, and thus the effective noise figure of this localized amplifier should be $NF_{eff}=NF \cdot e^{-\alpha_s L}$.

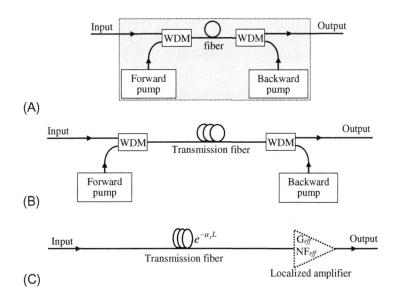

FIG. 5.5.3

(A) A localized Raman amplifier with both forward and backward pumps, (B) a distributed Raman amplifier using transmission fiber as the gain medium, and (C) a model separates a distributed Raman amplifier into a transmission fiber span and a localized amplifier.

For example, in Fig. 5.5.1, the signal optical gain is about 2 dB with 600-mW pump power and 100-km fiber length. Consider the 25-dB fiber loss without the Raman pump, the effective optical gain is equivalent to approximately 27 dB when used as a distributed amplifier. Fig. 5.5.4 shows the effective noise figure of the equivalent localized amplifier as the function of the fiber length, for forward pumping (A) and backward pumping, respectively. All parameters used are the same as those in Figs. 5.5.1 and 5.5.2. The negative values of effective noise figure in Fig. 5.5.4 seem to suggest that the output SNR is lower than the input SNR, which would clearly violate the fundamental rules of physics. But we have to remember that this is an "equivalent" noise figure which equivalently sets the fiber attenuation to be zero at the signal wavelength. In fact, even for the "distributed" Raman amplifier, the overall SNR at the output is still higher than that at the input, and the physically relevant noise figure is NF rather then NF_{eff}.

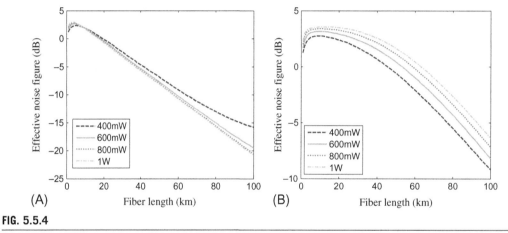

FIG. 5.5.4

Effective noise figure NF_{eff} vs. fiber length for forward pump (A) and backward pump (B) for different forward pump power levels. All parameters are the same as those used to obtain Figs. 5.5.1 and 5.5.2.

For distributed fiber Raman amplification with forward pumping, although it has much smaller noise figure compared to backward pumping, there are a number other issues need to be considered in optical system applications. For forward pumping, the optical signal co-propagates with the pump in the same direction along the fiber, relative intensity noise (RIN) of the pump laser can be transferred into both the intensity noise (Bromage, 2004) and the phase noise (Martinelli et al., 2006; Xu et al., 2016) of the optical signal. This will degrade system performance especially when high-power pump lasers are used which usually have high levels of RIN. For backward pumped fiber Raman amplification, the counter propagation of the pump and the signal makes the efficiency of RIN transfer much lower than the forward pump scheme where the walk-off between the pump and the signal is significantly slower.

Another concern of using forward pumping is the impact of fiber nonlinearity due to the increased signal optical power along the fiber. The accumulated nonlinear phase shift can be calculated through the integration of the signal optical power along the fiber as $\Phi_{NL} = \gamma \int_0^L P_s(z)dz$, where γ is the nonlinear parameter of the fiber defined by Eq. (2.6.4). Consider Raman amplification with forward pumping, there is a large portion along the fiber where the signal optical power is higher than the input level as shown in Fig. 5.5.1A, and thus the induced Φ_{NL} can be significant. Whereas, for backward pump, the signal optical power is kept at low level for most part of the fiber, except the very last portion where the Raman gain is high. Therefore, backward Raman pump does not introduce significant nonlinear phase shift.

The best system performance can be obtained by the combination of EDFA, and distributed Raman amplification with both backward and forward pumps (Cai et al., 2015), but at the expense of increased system complexity and additional amplifier cost. The reduced effective noise figure through distributed Raman amplification allows the use of high-order modulation formats to increase the signal spectral efficiency, as well as to extend the transmission distance (Pelouch, 2016).

5.6 SUMMARY

In this chapter, we have discussed configurations, important parameters, and major applications of optical amplifiers. Optical amplification is one of the most important functions required in most fiber-optic systems and networks to compensate for transmission and splitting losses. An good optical amplifier should have sufficiently high optical gain, wide enough gain bandwidth which covers the desired wavelength window. An optical amplifier can either be electrically pumped such as a SOA, or optically pumped such as an EDFA. The optical gain and the gain spectrum of an optical amplifier are determined by the material properties and the pump levels. The gain of an optical amplifier also depends on the power of the optical signal which amplifies, and the gain can be reduced if the signal optical power is too enough, known as gain saturation. Gain saturation is a nonlinear phenomenon, which not only limits the maximally achievable optical gain but also is responsible for cross talk between different wavelength channels through cross-gain modulation.

An optical amplifier not only amplifies the optical signal, but also it introduces broadband optical noise, and the PSD of the optical noise increases with the increase of the optical gain. This optical noise is usually referred to as ASE. Thus, although an optical amplifier can magnify the signal optical power, it cannot improve the OSNR. This broadband optical noise generated by optical amplifier can have profound impact in the electrical domain SNR after the optical signal is detected by a photodiode in an optical receiver. In the square-law photodetection process, the optical signal translates the optical noise of the same wavelength into the baseband through mixing, which results in signal-ASE beat noise in the electric domain. Similarly, broadband ASE noise also mixes with itself in the photodetector generating ASE-ASE beat noise in the electric domain. Noise figure, defined as the ratio of the input electric SNR and the output electric SNR, is a terminology most commonly used to specify the noise characteristic of an optical amplifier. Note that in order to make the noise figure definition unique and independent of the optical receiver, the photodiode has to be ideal with 100% quantum efficiency, only shot noise is used in the calculation of input SNR, and both shot noise and signal-ASE noise are used in the calculation of output SNR.

We have introduced three types of optical amplifiers which are most commonly used in fiber optical communication systems, namely SOA, EDFA, and Raman amplifier. The structure of an SOA is very similar to a Fabry-Perot-type diode laser, except that optical reflections from the two facets are minimized by AR coating. Optical gain along the optical waveguide in the active layer is introduced by electric current injection similar to that in a diode laser, and therefore, it is categorized as electrical pumping. Optical signal is amplified by traveling through the active optical waveguide in only a single pass, and thus an SOA it is also known as a traveling-wave amplifier (TWA). On the other hand, an EDFA is based on the silica optical fiber which is doped with erbium due to its useful bandgap structure. Optical pumping, at either 980 or 1480 nm wavelength, can elevate erbium ions from the ground state to a metastable state, known as pump-induced carrier inversion. Optical gain in 1550 nm wavelength window is introduced when carriers in the metastable state fall down to the ground state through the stimulated recombination process.

An SOA has miniature size with the length typically less than a millimeter, and can be monolithically integrated with other photonic components. The maximum optical gain of an SOA is usually limited by the quality of AR coating on the end facets so that gain ripple caused by the Fabry-Perot effect is not too high. The gain spectrum of an SOA can be well described by a parabolic shape, and a commercial SOA can have the gain bandwidth well exceed 30 nm in the 1550 nm wavelength window. As the gain spectrum of an SOA is determined by the band structure of semiconductor material used, SOAs operating at a variety of wavelength windows have been made through bandgap engineering of semiconductor materials. Although a large number of SOAs can be fabricated from a single wafer, input and output optical coupling with optical fibers have to be made for each SOA, so that packaging is quite a challenging issue. The coupling loss between fiber and SOA chip can also have significant impact degrading the noise figure of the amplifier. Another important issue in the application of SOA is the fast gain dynamics originated from the short carrier lifetime, typically on the order of 100 ps. This fast gain dynamics makes an SOA susceptible to cross-gain saturation between waveforms carried by different wavelengths, which introduces cross talk among these waveforms. On the other hand, the fast gain dynamics can be utilized to perform all-optical signal processing in which an optical signal can be controlled by another optical signal through cross-gain and cross-phase modulations. All-optical wavelength conversion has been reported based on the cross-gain modulation, cross-phase modulation, and FWM in SOAs.

In comparison, the length of erbium fiber used to make an EDFA is usually on the order of meters, and an EDFA has to be optically pumped by one or two high-power diode laser(s). The gain spectrum of an EDFA is determined by the emission and absorption cross sections of erbium fiber, which has multiple peaks and cannot be represented by a simple parabolic function. However, as EDFA is based on the optical fiber, its input and output ports can be simply fusion-spliced into a fiber-optic system with minimum coupling loss. In addition, the carrier lifetime of erbium ions on the metastable state is on the order of milliseconds, and thus the gain dynamics in EDFA is at least several orders of magnitude slower than bit lengths of data carried by the optical signal. Therefore, gain saturation in an EDFA only depends on average signal optical power but is independent of signal optical waveforms, and thus interchannel cross talk due to cross-gain saturation can be avoided.

We have also discussed the modeling of SOA and EDFA based on the rate equations. Although approximations can be made to simplify the analysis which helps understand the impact of some key parameters, numerical solutions of coupled rate equations are necessary in the amplifier design and performance prediction. Fast carrier dynamics has to be considered in the operation of an SOA, which can be challenging. While for an EDFA, the carrier dynamics is slow but the "crooked" shapes of emission and absorption cross sections, which cannot be represented by simple analytic functions, can make the modeling difficult.

In addition to SOA and EDFA, Raman amplifier is another type of optical amplifiers which are often used in long-distance fiber-optic systems. Based on SRS in optical fibers, the operating wavelength of a Raman amplifier is only determined by the pump laser wavelength which is much more flexible than an EDFA. Broadband Raman amplification can be achieved by using a group of wavelength division multiplexed pump lasers, and the gain flatness within the band can be controlled by optimizing the wavelengths and powers of the pump lasers. Especially, distributed fiber Raman amplifier has a unique advantage of using transmission fiber as the gain medium, instead of transmission loss, the fiber provides the optical gain. Thus, the equivalent noise figure of a distributed fiber Raman amplifier can be negative without violating the law of physics. Because of the relatively low efficiency of SRS compared to direct band-to-band transitions used in EDFA and SOA, a Raman amplifier requires high-power pump lasers which can be expensive.

As an enabling technology, optical amplifiers made long-distance fiber-optic communication possible without electrical domain signal regeneration along the way. Good understanding of optical gain, gain bandwidth, gain saturation, and noise figure of optical amplifiers discussed in this chapter are essential in the design and performance analysis of fiber optic systems and networks.

PROBLEMS

1. The material gain of an optical amplifier has a parabolic shape with the FWHM bandwidth of $\Delta f = 30$ nm. Neglect gain saturation effect for simplicity, please find the optical bandwidths when the peak optical gains of the amplifier are 10, 20, and 30 dB, respectively.

2. For an optical amplifier with the small-signal optical gain $G_0 = 30$ dB at the peak gain wavelength, and the saturation optical power is $P_{sat} = 10$ mW. Find optical gain of this amplifier for the input signal optical power of -30, -20, and 0 dBm at the peak gain wavelength. (Note: numerical method has to be used.)

3. An optical amplifier operates in the 1550 nm wavelength window with a spontaneous emission factor of $n_{sp} = 2.5$ and an optical gain of 30 dB. What is the optical noise power within 0.1 nm optical bandwidth?

4. Consider an optical amplifier in the 1550 nm wavelength window with 6-dB noise figure and optical gain $G \gg 1$. The optical signal at the out of this amplifier is detected by a photodiode as shown in the following figure. If there is a BPF with the bandwidth $B_o = 1$ nm before the photodiode, what is the signal optical power at which the signal-ASE beat noise is equal to the ASE-ASE beat noise? (Assume that electric bandwidth is much less than the optical bandwidth.)

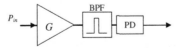

5. There is an optical preamplifier in an optical receiver before the photodiode. The input optical power to the optical amplifier is $P_{in} = -20$ dBm, and the optical amplifier has a 6-dB noise figure. Other parameters are, operation temperature $T = 300$ K, load resistance $R_L = 50\,\Omega$, photodiode responsivity $\mathfrak{R} = 0.9$ A/W, and the operation wavelength $\lambda = 1550$ nm.
   (a) If the gain of the optical amplifier is $G = 30$ dB, find the noise power spectral densities of thermal noise, shot noise, and signal-ASE beat noise.
   (b) What is the required optical gain of the amplifier so that signal-ASE beat noise is 10 dB higher than the thermal noise after photodetection?

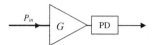

6. In a quantum-limited optical receiver, both thermal noise and dark current are negligible, and assume that the quantum efficiency of the photodiode is 100%.
   (a) Based on the Gaussian noise model, what is the number of photons required per second to achieve SNR$=1$ within 1 Hz receiver bandwidth?
   (b) Now an ideal optical preamplifier is used in front of the photodiode (referring to the figure of problem 5). Please discuss and explain whether or not this will reduce the required number of photons per second to achieve SNR$=1$ within 1 Hz receiver bandwidth.

7. For a SOA with 25-dB peak optical gain, if both end surface have the same power reflectivity R, what is the maximum R allowed so that the maximum gain ripple is less than 1 dB near the peak gain wavelength?

8. Consider an SOA with a spontaneous emission carrier lifetime $\tau=0.2$ ns, saturation power $P_{sat}=10$ mW, and small-signal optical gain $G_0=30$ dB. When a single short optical pulse with <1 ps temporal width and 1-pJ pulse energy is injected into the SOA,
   (a) Find optical gain of the SOA at the time immediately after the optical pulse.
   (b) How long it needs for the optical gain to recover to the level of $0.9G_0$?

9. Same as problem 8, but with some additional information: the linewidth enhancement factor is $\alpha_{lw}=5$, signal wavelength is 1550 nm, and the SOA length is $L=500\,\mu m$.
   (a) What is the maximum refractive index change Δn that the optical pulse produces?
   (b) Now if the optical signal is a pulse train with a repetition time only much shorter than the spontaneous emission carrier lime time τ, what will be the impact on the optical gain? Please discuss.

10. For the MZI configuration shown below with two identical SOAs, one in each arm. Both SOAs have saturation power $P_{sat}=10$ mW, spontaneous emission carrier lifetime $\tau=0.1$ ns, small-signal gain $G_0=20$ dB, and linewidth enhancement factor $\alpha_{lw}=6$.

   Using steady-state analysis, please find the pump power that is required to switch the MZI from constructive to destructive interference. (Since the wavelengths of the pump and the optical signal are very close, we assume that they are the same for simplicity.)

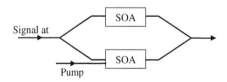

11. At $\lambda=1548$ nm wavelength, the emission and absorption cross sections of an EDF are $\sigma_e=1\times10^{-24}m^2$ and $\sigma_a=0.7\times10^{-24}m^2$, respectively. Erbium-doping density of this EDF is $N_T=7.5\times10^{24}m^{-3}$, and the confinement factor is $\Gamma=0.1$.
   (a) Please find the emission and absorption rate in dB/m for this EDF.
   (b) If the length of the EDF is $L=2$ m, and the pump power is strong enough so that the carrier inversion is complete (i.e., $N_2=N_T$, $N_1=0$), what is the small-signal optical gain at 1548 nm wavelength?

12. At $\lambda=1548\,\text{nm}$ wavelength, the emission and absorption cross sections of an EDF are $\sigma_e=1\times10^{-24}\text{m}^2$ and $\sigma_a=0.7\times10^{-24}\text{m}^2$, respectively. Erbium-doping density of this EDF is $N_T=7.5\times10^{24}\text{m}^{-3}$, and the confinement factor is $\Gamma=0.1$ at both the signal and the pump wavelengths.

Assume that the upper-level carrier density varies along the EDF as $N_2(z)=N_T\exp\{-\Gamma\sigma_a(\lambda_p)N_{Tz}\}$, where $\sigma_a(\lambda_p)=0.1\times10^{-24}\text{m}^2$ is the absorption cross section at the pump wavelength.

(a) Please find the optical gain of this EDFA at 1548 nm wavelength for the EDF lengths of $L=2$, 10, and 20 m, respectively.

(b) Please explain the reason why longer EDF may not provide higher optical gain.

13. An EDFA originally has an optical gain of $G=25\,\text{dB}$ and a noise figure $F=5\,\text{dB}$ at a certain wavelength. Neglect gain saturation effect.

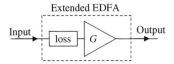

(a) If there is an addition 3 dB loss at the input side of the EDFA, then what is the noise figure of the extended EDFA? Please explain.

(b) If that 3 dB loss is at the output side of the EDFA, what is the noise figure of the extended EDFA?

14. Based on Eq. (5.4.26), assume that the inversion carrier density N_2 is a constant along the EDF, show that the ASE noise PSD at the EDFA output can be expressed as Eq. (5.2.1), where $G=e^{gL}$ with g the gain coefficient and L is the EDF length.

15. To design a localized fiber-based Raman amplifier with forward pump, if your pump laser has 300-mW optical power and the fiber has a Raman gain coefficient $g_R=5\times10^{-14}\text{m/W}$. The attenuation coefficient for both the optical signal and the pump is $\alpha=\alpha_p=\alpha_s=0.25\,\text{dB/km}$, and the effective cross-section area is $55\,\mu\text{m}^2$. Neglect power saturation effect, please find the optimum length of the fiber to achieve the maximum optical gain.

16. A fiber link using standard single-mode fiber with the following parameters: fiber length $L=150\,\text{km}$, Raman gain coefficient $g_R=5\times10^{-14}\text{m/W}$, attenuation coefficient for both the optical signal and the pump are the same $\alpha=\alpha_p=\alpha_s=0.25\,\text{dB/km}$, and the effective cross-section area $A_{eff}=80\,\mu\text{m}^2$.

(a) If forward Raman pump is used, use numerical simulation please find the pump power at which the *effective noise figure* is $NF_{eff}=-10\,\text{dB}$?

(b) Does this negative value of effective noise figure mean that no noise is generated by the Raman amplification? Please explain.

17. Please derive Eqs. (5.5.12) and (5.5.13) for the noise figure with forward pump and backward pump.

18. A fiber link using standard single-mode fiber with the following parameters: fiber length $L=100\,\text{km}$, Raman gain coefficient $g_R=5\times10^{-14}\text{m/W}$, attenuation coefficient for both the optical signal and the pump are the same $\alpha = \alpha_p = \alpha_s = 0.25\,\text{dB/km}$, and the effective cross-section area $A_{eff} = 80\,\mu\text{m}^2$.

If backward Raman pump is used with 700-mW pump power, please find

(a) optical signal gain $G(L)$ at $L = 100\,\text{km}$

(b) the ratio of output signal optical power levels with and without the pump turned on?

(c) the effective noise figure NF_{eff} (use numerical solution)

REFERENCES

Baney, D.M., Gallion, P., 2000. Theory and measurement techniques for the noise figure of optical amplifiers. Opt. Fiber Technol. 6, 122–154.

Bristiel, B., Gallion, P., Jaouën, Y., Pincemin, E., 2004. Intrinsic noise figure derivation for fiber Raman amplifiers from equivalent noise figure measurement. In: IEEE LTIMC 2004—Lightwave Technologies in Instrumentation & Measurement Conference, Palisades, New York, USA, pp. 135–140.

Bromage, J., 2004. Raman amplification for fiber communications systems. J. Lightwave Technol. 22 (1), 79.

Cai, J., et al., 2015. 49.3 Tb/s transmission over 9100 km using C+L EDFA and 54 Tb/s transmission over 9150 km using hybrid-Raman EDFA. J. Lightwave Technol. 33 (13), 2724–2734.

Connelly, M.J., 2002. Semiconductor Optical Amplifiers. Kluwer Academic Publishers.

Contestabile, G., Proietti, R., Calabretta, N., Ciaramella, E., 2007. Cross-gain compression in semiconductor optical amplifiers. J. Lightwave Technol. 25 (3), 915–921.

Desurvire, E., 1994. Erbium-Doped Fiber Amplifiers: Principles and Applications. John Wiley.

Giles, C.R., Desurvire, E., 1991. Propagation of signal and noise in concatenated erbium-doped fiber optical amplifiers. J. Lightwave Technol. 9 (2), 147–154.

Haus, H.A., 2000. Noise figure definition valid from RF to optical frequencies. IEEE J. Sel. Top. Quantum Electron. 6 (2), 240–247.

Headley, C., Agrawal, G., 2005. Raman Amplification in Fiber Optical Communication Systems. Academic Press.

Hui, R., Jiang, Q., Kavehrad, M., Makino, T., 1994. All-optical phase modulation in a traveling wave semiconductor laser amplifier. IEEE Photon. Technol. Lett. 6 (7), 808.

Martinelli, C., Lorcy, L., Durecu-Legrand, A., Mongardien, D., Borne, S., 2006. Influence of polarization on pump-signal RIN transfer and cross-phase modulation in copumped Raman amplifiers. J. Lightwave Technol. 24 (9), 3490–3505.

Olsson, N.A., 1992. Semiconductor optical amplifiers. IEEE Proc. 80, 375–382.

Pelouch, W.S., 2016. Raman amplification: an enabling technology for long-haul coherent transmission systems. J. Lightwave Technol. 34 (1), 6–19.

Saleh, A.A.M., Jopson, R.M., Evankow, J.D., Aspell, J., 1990. Accurate modeling of gain in erbium-doped fiber amplifiers. IEEE Photon. Technol. Lett. 2, 714–717.

Shimada, S., Ishio, H. (Eds.), 1992. Optical Amplifiers and Their Applications. John Wiley.

Stubkjaer, K.E., 2000. Semiconductor optical amplifier-based all-opticalgates for high-speed optical processing. IEEE J. Sel. Top. Quantum Electron. 6 (6), 1428–1435.

Wang, J.P., et al., 2008. Efficient performance optimization of SOA-MZI devices. Opt. Express 16 (5), 3288–3292.

Xu, L., et al., 2016. Experimental verification of relative phase noise in Raman amplified coherent optical communication system. J. Lightwave Technol. 34 (16), 3711–3716.

Zhou, J., et al., 1993. Terahertz four-wave mixing spectroscopy for study of ultrafast dynamics in a semiconductor optical amplifier. Appl. Phys. Lett. 63 (9), 1179–1181.

FURTHER READING

Ju, H., Zhang, S., Lenstra, D., de Waardt, H., Tangdiongga, E., Khoe, G.D., Dorren, H.J.S., 2005. SOA-based all-optical switch with subpicosecond full recovery. Opt. Express 13 (3), 942–947.

Olsson, N.A., 1989. Lightwave systems with optical amplifiers. J. Lightwave Technol. 7 (7), 1071–1082.

PASSIVE OPTICAL COMPONENTS

6

CHAPTER OUTLINE

Introduction 210
6.1 Fiber-optic directional couplers 211
    6.1.1 Basic parameters of a fiber-optic directional coupler 213
    6.1.2 Transfer matrix of a 2×2 optical coupler 217
6.2 Optical interferometer based on two-beam interference 219
6.3 Interferometers based on multi-pass interference 223
    6.3.1 Fabry-Perot Interferometers 223
    6.3.2 Optical ring resonators 230
6.4 Fiber Bragg gratings 234
6.5 WDM multiplexers and demultiplexers 239
    6.5.1 Thin-film-based interference filters 239
    6.5.2 Arrayed waveguide gratings 243
    6.5.3 Acousto-optic filters 245
6.6 Optical isolators and circulators 249
    6.6.1 Optical isolators 250
    6.6.2 Optical circulators 252
6.7 PLCs and silicon photonics 255
    6.7.1 Slab optical waveguides 256
    6.7.2 Rectangle optical waveguides 263
    6.7.3 Directional couplers 267
    6.7.4 Silicon photonics 270
6.8 Optical switches 274
    6.8.1 Types of optical switches 274
6.9 Summary 290
Problems 291
References 296
Further reading 297

Introduction to Fiber-Optic Communications. https://doi.org/10.1016/B978-0-12-805345-4.00006-8

209

INTRODUCTION

Passive optical components are essential in optical systems for the connection, manipulation, and processing of optical signals. Passive optical components are basic building blocks of optic systems which generally include directional optic couplers, optical filters, wavelength-division multiplexers and demultiplexers, and optical isolators and circulators. As opposite to active components such as lasers and photodetectors, passive optical components process signals in the optical domain without electro-optic conversion and high-speed control. For an optical coupler, it can have multiple input and output ports, in general, to split or combine optical signals with a desired ratio. Ideally, a wideband optical coupler is independent of both wavelength and modulation format of the optical signal. On the other hand, optical filters and wavelength-division multiplexers and demultiplexers are intentionally made wavelength dependent so that the spectrum of optical signal can be reshaped, and optical power carried by different wavelength components can be selected, rejected, or selectively combined or split. Various mechanisms can be used to make optical filters such as Mach-Zehnder interferometers (MZIs), Fabry-Perot interferometers (FPIs), array-waveguide gratings, fiber-Bragg gratings, and spatial optical interference. Similar to the specification of radio frequency (RF) filters, pass-band flatness, stop-band rejection ratio, and the sharpness of transition region are important parameters of an optical filter. An optical isolator is designed to prevent optical reflection in the optical path, while an optical circulator redirects the reflected optical signal to a different path. Both optical isolators and circulators are nonreciprocal devices based on the Faraday rotation of polarization of the optical signal. Polarization-sensitive isolators and circulators are generally used for free-space optical setups, whereas in fiber-optic systems, they have to be polarization insensitive because the state of polarization (SOP) of optical signal in a fiber can often be random and vary with time.

Passive optical components can be made by free-space optical assembly. Examples of free-space optical assemblies include optical filters, wavelength-division multiplexers and demultiplexers, optical isolators, and circulators. Because optical coating is a very mature technology, high-quality multilayer optical thin films can be created to allow sharp frequency selectivity and low loss for optical filters and wavelength division multiplexing (WDM) couplers. Birefringence optical materials such as YVO_4 crystal and materials with strong Faraday effect such as yttrium iron garnet (YIG) are often used to make optical nonreciprocal isolators and circulators. Fig. 6.0.1 shows images of a 16-channel WDM multiplexer, a manually tunable optical filter, and an optical isolator. Over the last few decades, the technology of micro-optic assembly and packaging has been improved significantly and optical devices based on free-space optics can be made reasonably small and mechanically stable. However, packaging of free-space optical components can be quite labor intensive and their sizes cannot be very small.

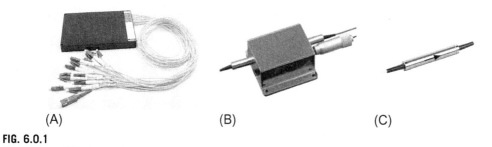

(A)   (B)   (C)

FIG. 6.0.1

Images of (A) a 16-channel WDM multiplexer, (B) a manually tunable optical filter, and (C) an optical isolator.

Another category of passive optical components is based on the optical fibers. Good examples of such devices include fiber Bragg gratings (FBG) and fused fiber directional couplers (also known as splitters and combiners). Because these devices are made with optical fibers, input and output optical coupling are straightforward with minimum loss. In addition, the interaction length can be made very long in fiber-based devices because of the guided-wave structure and low propagation loss. Thus, FBG-based optical filters can have excellent frequency selectivity.

Optical components based on the free-space optics and fiber optics are usually discrete, and each device is dedicated to perform a specific task. These components as building blocks can be interconnected through optical fibers to construct complex optical circuits to perform more sophisticated functionalities. Photonic integration is a promising technology that has the potential to integrate a large number of optical components on a "photonic chip." In photonic integration, optical components are usually made on a planar substrate, and thus, it is also known as planar lightwave circuits (PLCs). Complex guided-wave optical structures can be made in PLC through the process of photolithographic patterning and itching. A number of material platforms can be used for PLC, but the most popular materials are silicon dioxide (SiO_2, also known as silica), indium-phosphate (InP), and silicon-on-insulator (SOI). Silica-based PLC has the low propagation loss and with the best index match to the optical fiber, but it is difficult to integrate active components such as lasers and photodiodes on the silica platform. On the other hand, InP is a III–V semiconductor material, which can be monolithically integrated with lasers and photodiodes. However, InP material has relatively high propagation loss, and material growth and fabrication are much more complex than silica, and thus the cost of InP-based PLC is usually high. Photonic integration on SOI platform, also known as silicon photonics, has attracted much interest in the recent years mainly because fabrication process can be CMOS compatible. This allows the integration of photonic devices with CMOS electronic devices on the same platform.

At this point, except laser sources, almost all photonic components can be made on SOI including modulators, WDM multiplexers and demultiplexers, filters, splitters, and combiners. Complex photonic structures can be made on the SOI platform through photolithography and etching as illustrated in Fig. 6.0.2. However, as silicon is an indirect-bandgap semiconductor material, it cannot be used to make lasers and light-emitting diodes (LEDs). Monolithic integration of silicon photonic circuits with lasers and LEDs is also generally prohibitive because of crystal lattice mismatch between silicon and III–V semiconductor materials. Laser sources have been added into integrated silicon photonic circuits through hybrid integration and flip-chip bonding. Because of the miniaturized footprint, automated large scale fabrication process, and the compatibility with CMOS electronics, integrated silicon photonic circuit has the potential to revolutionize the photonics industry.

A good understanding of passive optical components is very important in the design andcharacterization of optical circuits, as well as trouble shooting of optical systems when problems occur.

6.1 FIBER-OPTIC DIRECTIONAL COUPLERS

An optical directional coupler is one of the most basic inline fiber-optic components, often used to split and combine optical signals, or tap-off a small portion of the optical power for monitoring. For example, a simple 2×2 optical directional coupler can be used to construct a Michelson interferometer.

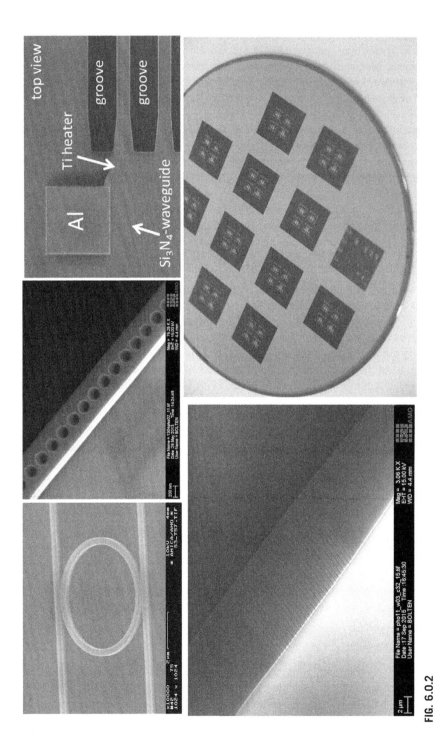

FIG. 6.0.2

Examples of silicon photonic structures fabricated on SOI platform.

Used with permission from AMO web page.

A MZI can be made by simply concatenating two optical directional couplers. The well-known passive optical networks (PON) are basically constructed with directional optical couplers so that a large number of users can share the broadcasted optical signal from the central office through passive splitting. An $N \times N$ optical coupler, known as a star coupler, can also be used to make a wavelength-switched broadcasting optical network.

If a directional optical coupler is used for broadband applications, the dependence on wavelength has to be minimized through proper design of the coupler structure. But in practice the coupling coefficient of an optical coupler is often wavelength dependent to some extent, so that wavelength sensitivity of a directional coupler has to be considered in the system design. For other applications, wavelength-dependent coefficient of an optical coupler can be utilized to make wavelength-selective devices such as wavelength interleavers and wavelength-division multiplexers (MUX) or demultiplexers (DEMUX).

A directional optical coupler can be made by simply fusing fibers together for a certain length known as fused fiber coupler, or using coupled ridge optical waveguides on a PLC. A fused fiber coupler uses all fibers so that optical connections are convenient with input and output fibers. On the other hand, a PLC-based optical coupler made through photolithography and etching can be flexible in design especially for a large number of input and output ports, but connection between rectangle-core ridge waveguides and input/output optical fibers may introduce significant insertion loss because of the mode field mismatch. In the following, we use fused fiber coupler as an example to explain the general characteristics and basic parameters.

6.1.1 BASIC PARAMETERS OF A FIBER-OPTIC DIRECTIONAL COUPLER

A simple fiber directional coupler can be made by fusing two parallel fibers together on the tip of a flame. With a proper tension applied along the longitudinal direction, the diameters of both fibers are reduced in the fused region, so that the evanescent waves into the fiber cladding are enhanced. When the cores of the two fibers are brought together sufficiently close to each other laterally, their evanescent mode fields start to overlap and interfere, and the optical power can be transferred between the two fibers. Because of the interference nature, the power transfer coefficient can generally be periodic along the coupling region as illustrated in Fig. 6.1.1.

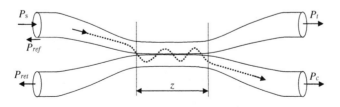

FIG. 6.1.1

Illustration of a fused fiber directional coupler. z is the length of the coupling region.

Fused fiber directional couplers are easier to fabricate compared to many other optical devices, and their fabrication can be automated by online monitoring of input and output optical powers from different ports. Because the interference between evanescent waves of the two fibers is coherent and often wavelength dependent, power coupling coefficient of a fused fiber coupler is generally wavelength dependent. The degree of wavelength dependency can be controlled to some extent through the selection of coupling length z, so that by modifying the fabrication process, the same fabrication equipment can also be used to make wavelength-selective optical devices, such as wavelength-division multiplexers, demultiplexers, and WDM channel interleavers.

As shown in Fig. 6.1.1, the power-splitting ratio of an optical directional coupler is defined as

$$\alpha = \frac{P_c}{P_t + P_c} \tag{6.1.1}$$

This is the ratio between the output of the opposite fiber P_c and the total output power which is $P_c + P_t$. In an ideal fiber coupler without insertion loss, the total output power is equal to the input power P_s; therefore, $\alpha \approx P_c/P_s$. In practical fiber couplers, absorption and scattering loss always exist in the coupling region, which add up to an excess loss defined as

$$\eta = \frac{P_c + P_t}{P_s} \tag{6.1.2}$$

This is another important parameter of the fiber coupler, which is a quality measure of the device. In a high-quality 2×2 fiber coupler, the excess loss is generally lower than a fraction of a dB. From an application point of view, *insertion loss* of a fiber coupler is often used as a system design parameter. The insertion loss is defined as the transmission loss between the input and the designated output:

$$T_{c,t} = \frac{P_{c,t}}{P_s} \tag{6.1.3}$$

The insertion loss is, in fact, affected by both the splitting ratio and the excess loss of the fiber coupler, that is, $T_c = P_c/P_s = \alpha\eta$ and $T_t = P_t/P_s = (1 - \alpha)\eta$. In a 3-dB coupler, although the intrinsic loss due to power splitting is 3 dB, if the excess loss is, for example, 0.4 dB, the actual insertion loss should be 3.4 dB.

In practical fiber couplers, due to the back scattering caused by imperfections in the fused region, there might be optical reflections back to the input ports. Directionality is defined as

$$D_r = \frac{P_{ret}}{P_s} \tag{6.1.4}$$

and reflection is defined as

$$R_{ref} = \frac{P_{ref}}{P_s} \tag{6.1.5}$$

For a high-quality fiber coupler, both the directionality and the reflection should be typically lower than -55 dB.

From a coupler design point of view, the coupling ratio is a periodic function of the coupling region length z (Saleh and Teich, 1991) as indicated in Fig. 6.1.1:

$$\alpha = F^2 \sin^2\left(\frac{K}{F}(z - z_0)\right) \tag{6.1.6}$$

where $F \leq 1$ is the maximum coupling ratio, which depends on the core separation between the two fiber cores and the uniformity of the core diameter in the coupling region; z_0 is the length of the fused region of the fiber before stretching; and K is the parameter that determines the periodicity of the coupling ratio. An experience-based formula of K is often used for coupler design based on standard single mode fibers (Hunsperger, 2009):

$$K \approx \frac{21\lambda^{5/2}}{a^{7/2}} \tag{6.1.7}$$

where a is the radius of the fiber cladding within the fused coupling region and λ is the wavelength of the optical signal. Since the radius of the fiber reduces when the fiber is fused and stretched, the parameter K is a function of z. For example, if the radius of the fiber is a_0 before stretching, the relationship between a and z should be approximately $a \approx \sqrt{a_0^2 z_0 / z}$.

Fig. 6.1.2 shows the calculated splitting ratio α based on Eqs. (6.1.6) and (6.1.7), where $z_0 = 6\,\text{mm}$, $a_0 = 62.5\,\mu\text{m}$, $\lambda = 1550\,\text{nm}$, and $F = 1$ were assumed. Practically, any splitting ratio can be obtained by gradually stretching the length of the fused fiber section. Therefore, in the fabrication process, an *in situ* monitoring of the splitting ratio is usually used. This can be accomplished by simply launching an optical signal P_s at the input port while measuring the output powers P_t and P_c during the fusing and stretching process.

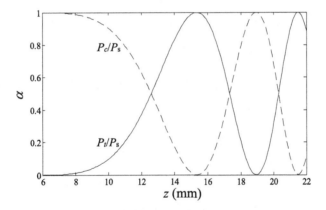

FIG. 6.1.2

Calculated splitting ratio as a function of the fused section length, with $z_0 = 6\,\text{mm}$, $a_0 = 62.5\,\mu\text{m}$, and $\lambda = 1550\,\text{nm}$.

Eqs. (6.1.6) and (6.1.7) also indicate that for a certain fused section length z, the splitting ratio is also a function of the signal wavelength λ. This means that the splitting ratio of a fiber directional coupler is, in general, wavelength dependent. The design of a wideband fused fiber coupler is challenging. For a commercial 3-dB fiber directional coupler, the variation of the splitting ratio is typically 0.5 dB across the telecommunication wavelength C-band (1530–1570 nm). Fig. 6.1.3A shows the calculated splitting ratio of a 3-dB fiber coupler with the same parameters as for Fig. 6.1.2 except that the fused section length was chosen as $z = 12.59\,\text{mm}$. In this case, the variation of splitting ratio within the 1530- and 1570-nm wavelength band is less than ± 0.2 dB.

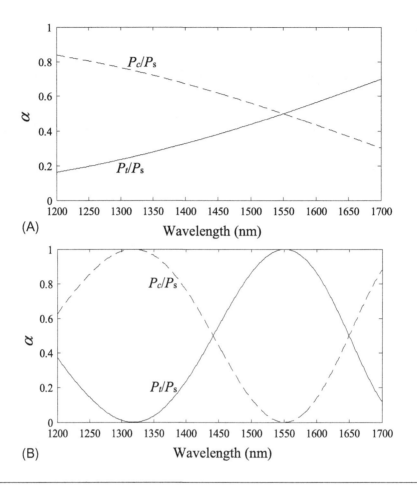

FIG. 6.1.3

Calculated splitting ratio as a function of wavelength with $z_0 = 6$ mm and $a_0 = 62.5$ μm. The fused section lengths are $z =$ (A) 12.59 mm and (B) 21.5 mm.

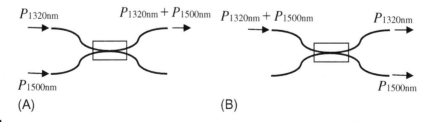

FIG. 6.1.4

A wavelength-dependent fiber coupler used as (A) a mux and (B) a demux for optical signals at 1320 and 1550 nm wavelengths.

On the other hand, the wavelength-dependent power-splitting ratio of a fiber coupler can be utilized to make a WDM multiplexer or interleaver. Fig. 6.1.3B shows the splitting ratio vs. wavelength for the same fiber coupler used for Fig. 6.1.2. But here the fused section length was chosen as $z = 21.5\,mm$, where $P_t/P_s = 100\%$ at 1550 nm wavelength, as can be seen from Fig. 6.1.2. It is interesting that in this case, the coupling ratio for 1320 nm is $P_t/P_s = 0\%$, which means $P_c/P_s = 100\%$ at this wavelength. As illustrated in Fig. 6.1.4, this fiber coupler can be used as a multiplex to combine optical signals of 1550 and 1320 nm wavelengths or as a demultiplexer to split the 1550- and 1320-nm wavelength components from the incoming optical signal. Compared to using wavelength-independent 2×2 couplers, this wavelength-division multiplexing and demultiplexing are only affected by the excess loss of the coupler, but do not suffer from intrinsic combining and splitting losses. For system design and characterization, the wavelength dependency of power splitting ratio and excess loss of fiber directional couplers are critical for many applications; therefore, these parameters have to be measured carefully before being used in the systems.

6.1.2 TRANSFER MATRIX OF A 2×2 OPTICAL COUPLER

The previous section used a 2×2 optical coupler as an example for the discussion of design principles and basic parameters. This section will present the general optical field transfer function of a 2×2 optical coupler. To be general, a 2×2 optical coupler is a 4-terminal device which can be either free space or made by all-fiber, both having two input ports and two output ports as illustrated in Fig. 6.1.5.

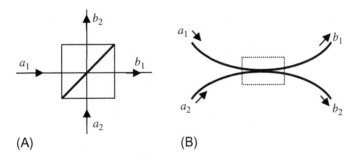

(A)   (B)

FIG. 6.1.5

2×2 optical couplers with (A) free-space optics and (B) fiber optics.

The input/output relationship of a 2×2 coupler can be represented as a transfer matrix (Pietzsch, 1989; Green, 1991):

$$\begin{bmatrix} b_1 \\ b_2 \end{bmatrix} = \begin{bmatrix} s_{11} & s_{12} \\ s_{21} & s_{22} \end{bmatrix} \begin{bmatrix} a_1 \\ a_2 \end{bmatrix} \qquad (6.1.8)$$

where a_1, a_2 and b_1, b_2 are electric fields at the two input and output ports, respectively. This device is reciprocal, which means that the transfer function will be identical if the input and the output ports are exchanged; therefore,

$$s_{12} = s_{21} \qquad (6.1.9)$$

We also assume that the coupler has no excess loss so that energy conservation applies; therefore, the total output power is equal to the total input power:

$$|b_1|^2 + |b_2|^2 = |a_1|^2 + |a_2|^2 \tag{6.1.10}$$

Eq. (6.1.8) can be expanded into

$$|b_1|^2 = |s_{11}|^2|a_1|^2 + |s_{12}|^2|a_2|^2 + s_{11}s_{12}^*a_1a_2^* + s_{11}^*s_{12}a_1^*a_2 \tag{6.1.11}$$

and

$$|b_2|^2 = |s_{21}|^2|a_1|^2 + |s_{22}|^2|a_2|^2 + s_{21}s_{22}^*a_1a_2^* + s_{21}^*s_{22}a_1^*a_2 \tag{6.1.12}$$

where * indicates complex conjugate. Using the energy conservation condition described in Eq. (6.1.10), Eqs. (6.1.11) and (6.1.12) yield

$$|s_{11}|^2 + |s_{21}|^2 = 1 \tag{6.1.13}$$

$$|s_{12}|^2 + |s_{22}|^2 = 1 \tag{6.1.14}$$

and

$$s_{11}s_{12}^* + s_{21}s_{22}^* = 0 \tag{6.1.15}$$

For a coupler, if ε is the fraction of the optical power coupled from input port 1 to output port 2, the same coupling ratio will be from input port 2 to output port 1. That is, $|s_{12}| = |s_{21}|$. For the optical field, this coupling ration is $\sqrt{\varepsilon}$. Since $|s_{12}| = |s_{21}|$, Eqs. (6.1.13) and (6.1.14) give

$$|s_{11}|^2 = |s_{22}|^2 \tag{6.1.16}$$

This means that the optical field coupling rate from input port 1 to output port 1 is equal to that from input port 2 to output port 2, which is $\sqrt{1-\varepsilon}$. If we assume $s_{11} = \sqrt{1-\varepsilon}$ is real, $s_{22} = \sqrt{1-\varepsilon}$ should also be real because of the symmetry assumption. Then, we can assume there is a phase shift for the cross-coupling term, $s_{12} = s_{21} = \sqrt{\varepsilon}e^{j\Phi}$. The transfer matrix is then

$$\begin{bmatrix} b_1 \\ b_2 \end{bmatrix} = \begin{bmatrix} \sqrt{1-\varepsilon} & e^{j\Phi}\sqrt{\varepsilon} \\ e^{j\Phi}\sqrt{\varepsilon} & \sqrt{1-\varepsilon} \end{bmatrix} \begin{bmatrix} a_1 \\ a_2 \end{bmatrix} \tag{6.1.17}$$

Using Eq. (6.1.15), we have

$$e^{-j\Phi} + e^{j\Phi} = 0 \tag{6.1.18}$$

The solution of Eq. (6.1.18) is $\Phi = \pi/2$, that is $e^{j\Phi} = j$. Thus, the general transfer matrix of a 2 × 2 optical coupler is

$$\begin{bmatrix} b_1 \\ b_2 \end{bmatrix} = \begin{bmatrix} \sqrt{1-\varepsilon} & j\sqrt{\varepsilon} \\ j\sqrt{\varepsilon} & \sqrt{1-\varepsilon} \end{bmatrix} \begin{bmatrix} a_1 \\ a_2 \end{bmatrix} \tag{6.1.19}$$

For a 3 dB coupler, the power coupling coefficient is $\varepsilon = 0.5$, so that

$$\begin{bmatrix} b_1 \\ b_2 \end{bmatrix} = \frac{1}{\sqrt{2}} \begin{bmatrix} 1 & j \\ j & 1 \end{bmatrix} \begin{bmatrix} a_1 \\ a_2 \end{bmatrix} \tag{6.1.20}$$

An important conclusion of Eq. (6.1.19) is that there is a 90-degree relative phase shift between the direct pass (s_{11}, s_{22}) and the cross-coupling (s_{12}, s_{21}). This property is important for several applications including phase-balanced detection in coherent receivers. 2 × 2 optical couplers are basic building blocks also for a number of optical devices including Michelson and MZIs.

6.2 OPTICAL INTERFEROMETER BASED ON TWO-BEAM INTERFERENCE

In the previous section, we have discussed the basic parameters and field transfer function of a 2×2 optical coupler. We now introduce a few optical interferometers based on 2×2 optical couplers, which include MZIs, Michelson interferometers, and fiber-optic ring resonators.

The MZI is one of the oldest optical structures, in use for more than a century. The basic free-space configuration of an MZI is shown in Fig. 6.2.1, which consists of a beam splitter, a beam combiner, and two mirrors to alter beam directions.

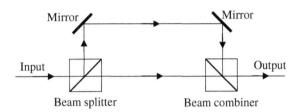

FIG. 6.2.1

Free-space configuration of a Mach-Zehnder interferometer.

The beam splitter splits the incoming optical signal into two equal parts. After traveling through two separate arms, these two beams recombine at the beam combiner. The nature of beam interference at the combiner depends, to a large extent, on the coherence length of the optical signal. If the path length difference between these two arms is shorter than the coherent length of the optical signal, the two beams coherently interfere with each other at the combiner, constructive, or destructive interference may happen depending on the relative phase of the two optical signals at the combiner. On the other hand, if the path length difference is much longer than the coherence length of the optical signal, there will be no deterministic phase relation between optical signals emerging from the two paths, and thus there will be no coherent interference between them. In fiber-optic systems, the beam splitter and the combiner can be replaced by fiber couplers; therefore, all-fiber MZIs can be made of two 2×2 directional fiber couplers and a fiber-optic delay line as shown in Fig. 6.2.2.

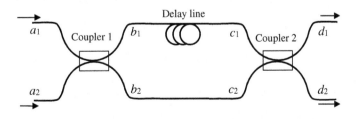

FIG. 6.2.2

Illustration of an all-fiber MZI with two inputs and two outputs.

The transfer function of an MZI can be obtained by cascading the transfer functions of two optical couplers and that of the optical delay line. Suppose the optical lengths of the delay lines in arm 1 and arm 2 are n_1L_1 and n_2L_2, respectively, the transfer matrix of the two delay lines is simply

$$\begin{bmatrix} c_1 \\ c_2 \end{bmatrix} = \begin{bmatrix} \exp(-j\varphi_1) & 0 \\ 0 & \exp(-j\varphi_2) \end{bmatrix} \begin{bmatrix} b_1 \\ b_2 \end{bmatrix} \tag{6.2.1}$$

where $\varphi_1 = (2\pi/\lambda)n_1L_1$ and $\varphi_2 = (2\pi/\lambda)n_2L_2$ are the phase delays of the two delay lines. If we further assume that the two optical couplers have their power splitting coefficients of ε_1 and ε_2, respectively, we have,

$$\begin{bmatrix} d_1 \\ d_2 \end{bmatrix} = \begin{bmatrix} \sqrt{1-\varepsilon_2} & j\sqrt{\varepsilon_2} \\ j\sqrt{\varepsilon_2} & \sqrt{1-\varepsilon_2} \end{bmatrix} \begin{bmatrix} e^{-j\varphi_1} & 0 \\ 0 & e^{-j\varphi_2} \end{bmatrix} \begin{bmatrix} \sqrt{1-\varepsilon_1} & j\sqrt{\varepsilon_1} \\ j\sqrt{\varepsilon_1} & \sqrt{1-\varepsilon_1} \end{bmatrix} \begin{bmatrix} a_1 \\ a_2 \end{bmatrix} \tag{6.2.2}$$

To achieve the highest extinction ratio, most MZIs use 50% power splitting ratio for both optical couplers ($\varepsilon_1 = \varepsilon_2 = 0.5$). In that case, the equation can be simplified as

$$\begin{bmatrix} d_1 \\ d_2 \end{bmatrix} = \frac{1}{2} \begin{bmatrix} (e^{-j\varphi_1} - e^{-j\varphi_2}) & j(e^{-j\varphi_1} + e^{-j\varphi_2}) \\ j(e^{-j\varphi_1} + e^{-j\varphi_2}) & -(e^{-j\varphi_1} - e^{-j\varphi_2}) \end{bmatrix} \begin{bmatrix} a_1 \\ a_2 \end{bmatrix} \tag{6.2.3}$$

If the input optical signal is only at the input port 1, while the input port 2 is disconnected ($a_2 = 0$), the optical field at the two output ports of the MZI will be

$$d_1 = \frac{1}{2}(e^{-j\varphi_1} - e^{-j\varphi_2})a_1 = -e^{-j\varphi_0}\sin\left(\frac{\Delta\varphi}{2}\right)a_1 \tag{6.2.4}$$

and

$$d_2 = \frac{j(e^{-j\varphi_1} + e^{-j\varphi_2})}{2}a_1 = je^{-j\varphi_0}\cos\left(\frac{\Delta\varphi}{2}\right)a_1 \tag{6.2.5}$$

where $\varphi_0 = (\varphi_1 + \varphi_2)/2$ is the average phase delay, and $\Delta\varphi = (\varphi_1 - \varphi_2)$ is the differential phase shift of the two MZI arms. Then, the optical power transfer function from input port 1 to output port 1 is

$$T_{11} = \frac{d_1}{a_1}\bigg|_{a_2=0} = \sin^2\left[\frac{\pi f}{c}(n_2L_2 - n_1L_1)\right] \tag{6.2.6}$$

and the optical power transfer function from input port 1 to output port 2 is

$$T_{12} = \frac{d_2}{a_1}\bigg|_{a_2=0} = \cos^2\left[\frac{\pi f}{c}(n_2L_2 - n_1L_1)\right] \tag{6.2.7}$$

where $f = c/\lambda$ is the signal optical frequency, and c is the speed of light. Obviously, the optical powers coming out of the two output ports are complementary such that $T_{11} + T_{12} = 1$, which results from energy conservation as we assumed that there is no loss.

The transmission efficiencies T_{11} and T_{12} of an MZI are the functions of both the wavelength λ and the differential optical length $\Delta l = n_2L_2 - n_1L_1$ between the two arms. Therefore, an MZI can be used for two different types applications: optical filter and electro-optical modulator.

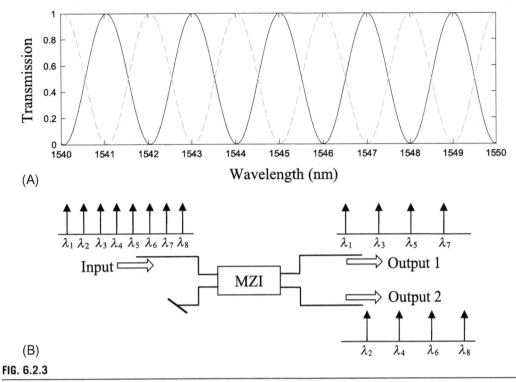

(A)

(B)

FIG. 6.2.3

(A) MZI power transfer functions T_{11} (solid line) and T_{12} (dashed line) with $n_1=n_2=2.387$, $L_1-L_2=0.5$ mm.
(B) An MZI is used as a wavelength interleaver.

The application of MZI as electro-optic modulator will be detailed in Chapter 7, where the differential arm length can be modulated to modulate the optical transfer function. Here we only discuss the application of MZI as an optical filter in which wavelength-dependent characteristics of $T_{11}(\lambda)$ and $T_{12}(\lambda)$ are used, as illustrated in Fig. 6.2.3A. For example, if we assume that the two arms have the same refractive index $n_1=n_2=1.5016$, the physical length difference between the two arms is $L_1-L_2=0.8$ mm, and the operation wavelength is in the 1550 nm window, the wavelength spacing between adjacent transmission peaks of each output arm is

$$\Delta\lambda \approx \frac{\lambda^2}{n_2L_2 - n_1L_1} = 2\text{nm} \tag{6.2.8}$$

In this case, the wavelength difference of the transmission peaks between the two output ports is equal to $\Delta\lambda/2=1$ nm shown as the difference between the solid and the dashed lines in Fig. 6.2.3A. As these transfer functions are strictly periodical, application as band-pass filters is usually not convenient because of the multiple transmission peaks. A quite elegant application of MZI is to make a wavelength interleaver, as shown in Fig. 2.3.4B (Cao et al., 2004; Oguma et al., 2004). In this application, a WDM optical signal with narrow channel spacing can be interleaved into two groups of optical signals each with a doubled channel spacing compared to the input.

The optical configuration of a *Michelson interferometer* is a fold-back version of an MZI where only one 2×2 optical coupler is used bidirectionally. Optical signals are reflected back by two total reflection mirrors at the end of the interferometer arms and recombine in the same optical coupler. In this configuration, there is only one input port and one output port. As shown in Fig. 6.2.4, a Michelson interferometer can be made by free-space optics as well as fiber-optic components.

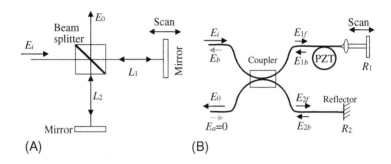

FIG. 6.2.4

Basic configurations of (A) free-space optics and (B) a fiber-optic Michelson interferometer. *PZT*, Piezo-electric transducer.

Using notations in Fig. 6.2.4B, assume an optical field E_i is input into the top left port of the 2×2 coupler with the power coupling coefficient ε.

$$\begin{bmatrix} E_{1f} \\ E_{2f} \end{bmatrix} = \begin{bmatrix} \sqrt{1-\varepsilon} & j\sqrt{\varepsilon} \\ j\sqrt{\varepsilon} & \sqrt{1-\varepsilon} \end{bmatrix} \begin{bmatrix} E_i \\ 0 \end{bmatrix} \tag{6.2.9}$$

The round-trip field transfer matrix of the two delay lines is

$$\begin{bmatrix} E_{1b} \\ E_{2b} \end{bmatrix} = \begin{bmatrix} \sqrt{R_1}e^{-j\phi_1} & 0 \\ j\sqrt{\varepsilon} & \sqrt{R_2}e^{-j\phi_2} \end{bmatrix} \begin{bmatrix} E_{1f} \\ E_{2f} \end{bmatrix} \tag{6.2.10}$$

where R_1 and R_2 are power reflectivities of the two mirrors and $\phi_1 = (2\pi/\lambda)2n_1 L_1$ and $\phi_2 = (2\pi/\lambda)2n_2 L_2$ are round-trip phase delays with n_1, n_2 and L_1, L_2 refractive indices and lengths of the two optical delay lines, respectively. The reflected optical fields E_{1b} and E_{2b} reenter the optical coupler from the right side, so that the overall transfer matrix is

$$\begin{bmatrix} E_{1b} \\ E_0 \end{bmatrix} = \begin{bmatrix} \sqrt{1-\varepsilon} & j\sqrt{\varepsilon} \\ j\sqrt{\varepsilon} & \sqrt{1-\varepsilon} \end{bmatrix} \begin{bmatrix} \sqrt{R_1}e^{-j\phi_1} & 0 \\ 0 & \sqrt{R_2}e^{-j\phi_2} \end{bmatrix} \begin{bmatrix} \sqrt{1-\varepsilon} & j\sqrt{\varepsilon} \\ j\sqrt{\varepsilon} & \sqrt{1-\varepsilon} \end{bmatrix} \begin{bmatrix} E_i \\ 0 \end{bmatrix}$$

This results in a power transfer function of

$$T = \left| \frac{E_0}{E_i} \right|^2 = \left| j\sqrt{\varepsilon(1-\varepsilon)} \left(\sqrt{R_1}e^{-j\phi_1} + \sqrt{R_2}e^{-j\phi_2} \right) \right|^2$$
$$= \varepsilon(1-\varepsilon) \left[R_1 + R_2 + 2\sqrt{R_1 R_2} \cos(\Delta\phi) \right] \tag{6.2.11}$$

where $\Delta\phi = \phi_1 - \phi_1 = \frac{4\pi f}{c}(n_1 L_1 - n_2 L_2)$ is the phase delay difference between the two arms of the interferometer. Ideally, if a 3 dB coupler is used with $\varepsilon = 0.5$, and assume $R_1 = R_2 = 1$, the power transfer function will be

$$T = \frac{1 + \cos(\Delta\phi)}{2} = \cos^2\left[\frac{2\pi f}{c}(n_1 L_1 - n_2 L_2)\right] \qquad (6.2.12)$$

which is very similar to the MZI transfer function in Eq. (6.2.7) except for the fact of 2 in the phase term because the optical signal travels round-trip along each arm.

An important issue in the implementation of an optical interferometer is the extinction ratio, which is defined as the ratio between the maximum and the minimum transmissions. For a Michelson interferometer, based on Eq. (6.2.11), the extinction ratio is equal to $(R_1 + R_2)/(R_1 - R_2)$, which approaches infinity when $R_1 = R_2$. Although the coupling coefficient ε does not affect the extinction ratio of a Michelson interferometer, $\varepsilon = 0.5$ is always desired because it introduces the minimum transmission loss.

6.3 INTERFEROMETERS BASED ON MULTI-PASS INTERFERENCE

The interferometers discussed in the last section, such as Mach-Zehnder and Michelson interferometers, are based on two beam interference so that the transfer functions are primarily cosine shaped. For many applications as optical filters, the sharpness of wavelength selectivity is an important consideration, so that two-beam interference may not be sufficient. Another category of interferometers is based on the optical interference among multiple light beams so that spectral shapes of the transfer functions can be much sharper. Two such examples are FPIs and ring resonators which will be discussed in this section.

6.3.1 FABRY-PEROT INTERFEROMETERS

A FPI is based on the interference among multiple reflections between two parallel mirrors so that sharp frequency selectivity can be expected. The traditional configuration of a free-space FPI is shown in Fig. 6.3.1, where two parallel mirrors, both having power reflectivity R, are separated by a distance d (Hernandez, 1986; Vaughan, 1989). If a light beam is launched onto the mirrors at an incidence angle α, a part of the light will penetrate through the left mirror and propagates to the right mirror at point A. At this point, part of the light will pass through the mirror and the other part will be reflected back to the left mirror at point B. This process will be repeated many times until the amplitude is significantly reduced due to the multiple-reflection losses.

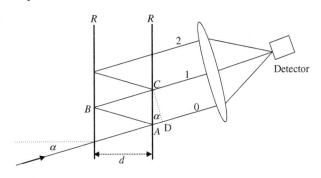

FIG. 6.3.1

Illustration of a Febry-Perot Interferometer with two partially reflecting mirrors separated by a distance d.

The propagation phase delay can be found after each round-trip between the two mirrors. The path length difference between the adjacent output traces, as illustrated in Fig. 6.3.1, is

$$\Delta L = AB + BC - AD = 2d\cos\alpha \tag{6.3.1}$$

After passing through the second mirror, the optical field amplitudes of light rays 0, 1, and 2 can be expressed as, respectively,

$$E_0(f) = E_i(f)\sqrt{1-R}\exp(j\beta nx_0)\sqrt{1-R}$$

$$E_1(f) = E_i(f)\sqrt{1-R}\exp(j\beta nx_0)\exp(j\beta n\Delta L)R\sqrt{1-R}$$

and

$$E_2(f) = E_i(f)\sqrt{1-R}\exp(j\beta nx_0)\exp(j2\beta n\Delta L)R^3\sqrt{1-R}$$

where $E_i(f)$ is the input optical field, f is the optical frequency, x_0 is the propagation delay of the beam 0 shown in Fig. 6.3.1, $\beta = \frac{2\pi}{\lambda} = \frac{2\pi f}{c}$ is the propagation constant, and n is the refractive index of the material between the two mirrors.

The general expression of the optical field, which experienced N round-trips between the two mirrors, can be expressed as

$$E_N(f) = E_i(f)\sqrt{1-R}\exp(j\beta nx_0)\exp(jN\beta n\Delta L)R^{2N+1}\sqrt{1-R} \tag{6.3.2}$$

Adding all the components together at the FPI output, the total output electrical field is

$$E_{out}(f) = E_i(f)(1-R)\exp(j\beta nx_0)\sum_{m=0}^{\infty}\exp(jm\beta n\Delta L)R^m$$

$$= E_i(f)\frac{(1-R)\exp(j\beta nx_0)}{1-\exp(j\beta n\Delta L)R} \tag{6.3.3}$$

Because of the coherent interference between a large number of output ray traces, the transfer function of an FPI becomes frequency dependent. The field transfer function of the FPI is then

$$H(f) = \frac{E_{out}(f)}{E_i(f)} = \frac{(1-R)\exp\left(j\frac{2\pi f}{c}nx_0\right)}{1-\exp\left(j\frac{2\pi f}{c}n\Delta L\right)R} \tag{6.3.4}$$

The power transfer function is the square of the field transfer function:

$$T_{FP}(f) = |H(f)|^2 = \frac{(1-R)^2}{(1-R)^2 + 4R\sin^2\left(\frac{2\pi f}{2c}n\Delta L\right)} \tag{6.3.5}$$

Equivalently, this power transfer function can also be expressed using signal wavelength, instead of optical frequency, as the variable:

$$T_{FP}(\lambda) = \frac{(1-R)^2}{(1-R)^2 + 4R\sin^2\left(\frac{2\pi nd\cos\alpha}{\lambda}\right)} \tag{6.3.6}$$

For a fixed signal wavelength, Eq. (6.3.6) is a periodic transfer function of the incidence angle α. For example, if a point light source is illuminated on an FPI, as shown in Fig. 6.3.2, a group of bright rings will appear on the screen behind the FPI. The diameters of the rings depend on the thickness of the FPI as well as on the signal wavelength. To provide a quantitative demonstration, Fig. 6.3.3 shows the FPI power transfer function vs. the incidence angle at the input. To obtain Fig. 6.3.3, the power reflectivity of the mirror $R=0.5$, mirror separation $d=300\,\mu$m, and media refractive index $n=1$ were used. The solid line shows the transfer function with the signal wavelength at $\lambda=1540.4\,$nm. At this wavelength, the center of the screen is dark. When the signal wavelength is changed to 1538.5 nm, as shown by the dashed line, the center of the screen becomes bright. Note too that with the increase of the mirror separation d, the angular separation between transmission peaks will become smaller and the rings on the screen will become more crowded.

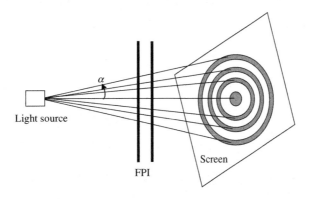

FIG. 6.3.2

Illustration of a circular fringes pattern when a noncollimated light source is launched onto a screen through an FPI.

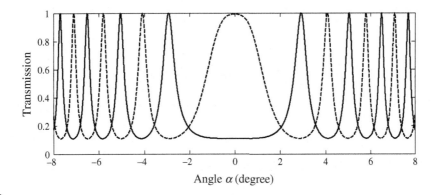

FIG. 6.3.3

Transmission vs. the beam incident angle to the FPI with mirror reflectivity $R=0.5$, mirror separation $d=300\,\mu$m, and media refractive index $n=1$. Calculations were made at two wavelengths: $\lambda=1538.5\,$nm *(dashed line)* and $\lambda=1540.4\,$nm *(solid line)*.

In fiber-optic FPI applications, because of the small numerical aperture of the single-mode fiber, collimated light is usually used, and one can assume that the incidence angle is approximately $\alpha=0$. This is known as a *collinear configuration*. With $\alpha=0$, Eq. (6.3.6) can be simplified to

$$T_{FP}(\lambda) = \frac{(1-R)^2}{(1-R)^2 + 4R\sin^2(2\pi nd/\lambda)} \tag{6.3.7}$$

In this case, power transmission is a periodic function of the signal optical frequency. Fig. 6.3.4 shows two examples of power transfer functions in a collinear FPI configuration where the mirror separation is $d=5$ mm, the media refractive index is $n=1$, and the mirror reflectivity is $R=0.95$ for Fig. 6.3.4A, and $R=0.8$ for Fig. 6.3.4B. With a higher mirror reflectivity, the transmission peaks become narrower and the transmission minima become lower. Therefore, the FPI has better frequency selectivity with high mirror reflectivity.

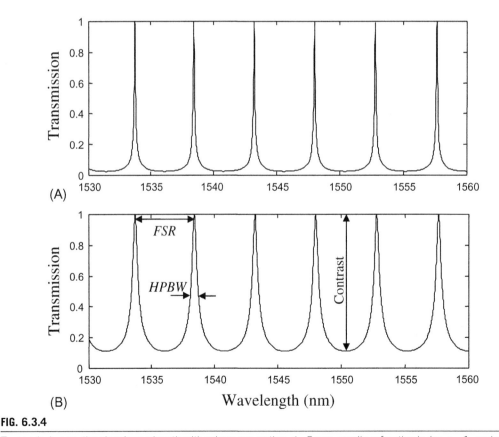

FIG. 6.3.4

Transmission vs. the signal wavelength with mirror separation $d=5$ mm, media refractive index $n=1$, and mirror reflectivity $R=0.95$ (A) and $R=0.8$ (B).

In the derivation of FPI transfer function, we have assumed that there is no loss within the cavity. But in practice optical loss always exists. After each round-trip in the cavity, the reduction in the optical field amplitude should be ηR instead of just R. The extra loss factor $\eta < 1$ may be introduced by cavity material absorption and beam misalignment. The major cause of beam misalignment is that the two mirrors that form the FPI are not exactly parallel. As a consequence, after each round-trip the beam exit from the FPI only partially overlaps with the previous beam; therefore, the strength of interference between them is reduced. For simplicity, we still use R to represent the FPI mirror reflectivity; however, we have to bear in mind that this is an effective reflectivity which includes the effect of cavity loss. The following are a few parameters that are often used to describe the properties of an FPI.

Free-spectral range (FSR) is an important parameter of FPI defined as the frequency separation Δf between adjacent peaks in the power transfer function. For a certain incidence angle α, the FSR can be found from transfer function (6.3.6) as

$$FSR = \Delta f = \frac{c}{2nd\cos\alpha} \tag{6.3.8}$$

FSR is inversely proportional to the cavity optical length nd. For a fiber-based FPI, where $\alpha = 0$, $FSR = c/(2nd)$. For a simple Febry-Perot type laser diode, if the cavity length is $d = 300\,\mu m$ and the refractive index is $n = 3.5$, the FSR of this laser diode is approximately $143\,GHz$. If the laser operates in a 1550-nm wavelength window, this FSR is equivalent to $1.14\,nm$. This is the mode spacing of the laser, as we discussed in Section 3.3.

Half-power bandwidth (HPBW) is the width of each transmission peak of the FPI power transfer function, which indicates the frequency selectivity of the FPI. From Eq. (6.3.6), if we assume at $f = f_{1/2}$ the transfer function is reduced to ½ its peak value, then

$$4R\sin^2\left(\frac{2\pi f_{1/2}nd\cos\alpha}{c}\right) = (1-R)^2$$

Assuming that the transmission peak is narrow enough, which is the case for most of the FPIs, $\sin(2\pi f_{1/2}nd\cos\alpha/c) \approx (2\pi f_{1/2}nd\cos\alpha/c)$ and $f_{1/2}$ can be found as

$$f_{1/2} = \frac{(1-R)c}{4\pi nd\sqrt{R}\cos\alpha}$$

Therefore, the full width of the transmission peak is

$$HPBW = 2f_{1/2} = \frac{(1-R)c}{2\pi nd\sqrt{R}\cos\alpha} \tag{6.3.9}$$

In most applications, large FSR and small HPBW are desired for good frequency selectivity. However, these two parameters are related by a FPI quality parameter known as *finesse*.

Finesse of an FPI is related to the percentage of the transmission window within the FSR. It is defined by the ratio between the FSR and HPBW:

$$F = \frac{FSR}{HPBW} = \frac{\pi\sqrt{R}}{1-R} \tag{6.3.10}$$

Finesse is a quality measure of FPI that depends only on the effective mirror reflectivity R. Technically, very high R is hard to obtain, because the effective reflectivity not only depends on the quality of mirror itself, but also it depends on the mechanical alignment between the two mirrors. Current state-of-the-art technology can provide finesse of up to a few thousand.

The ratio between the transmission maximum and the transmission minimum of the FPI power transfer function is known as the *contrast*, or the extinction ratio. The contrast specifies the ability of wavelength discrimination if the FPI is used as an optical filter. Again, from the transfer function (6.3.6), the highest transmission is $T_{max}=1$ and the minimum transmission is $T_{min}=(1-R)^2/[(1-R)^2+4R]$. Therefore, the contrast of the FPI is

$$C=\frac{T_{max}}{T_{min}}=1+\frac{4R}{(1-R)^2}=1+\left(\frac{2F}{\pi}\right)^2 \tag{6.3.11}$$

The physical meanings of FSR, HPBW, and contrast are illustrated in Fig. 6.3.4B.

EXAMPLE 6.1

The reflection characteristics of a Fabry-Perot (FP) filter can be used as a narrowband notch filter, as illustrated in Fig. 6.3.5. To achieve a power rejection ratio of 20 dB, what is the maximally allowed loss of the mirrors?

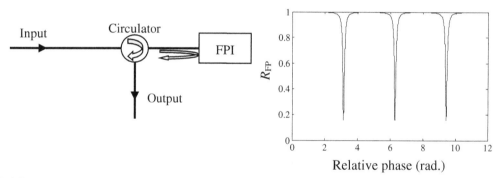

FIG. 6.3.5

Using FPI as a notch filter.

Solution

If the FPI is ideal and the mirror has no reflection loss, the wavelength-dependent power transmission of an FP filter is given by Eq. (6.3.7) as

$$T(\lambda)=\frac{(1-R)^2}{(1-R)^2+4R\sin^2(2\pi nd/\lambda)}$$

and the reflection of the FPI is

$$R_{FP}(\lambda)=1-T(\lambda)=\frac{4R\sin^2(2\pi nd/\lambda)}{(1-R)^2+4R\sin^2(2\pi nd/\lambda)}$$

If this is true, the notch filter would be ideal because $R_{FP}(\lambda)=0$ whenever $(2nd/\lambda)$ is an integer. However, in real devices, absorption, scattering, and non-ideal beam collimation would contribute to reflection losses; therefore, the wavelength-dependent power transmission of an FP filter becomes

$$T(\lambda) = \frac{(1-R)^2}{(1-R\eta)^2 + 4R\eta \sin^2(2\pi nd/\lambda)}$$

where $\eta < 1$ is the power loss of each reflection on the mirror. Then, the FPI power reflectivity is

$$R_{FP}(\lambda) = \frac{(1-R\eta)^2 - (1-R)^2 + 4R\eta \sin^2(2\pi nd/\lambda)}{(1-R\eta)^2 + 4R\eta \sin^2(2\pi nd/\lambda)}$$

Obviously, the minimum reflection of the FPI happens at wavelengths where $\sin^2(2\pi nd/\lambda) = 0$; therefore, the value of minimum reflection is

$$R_{FP}(\lambda) = 1 - \frac{(1-R)^2}{(1-R\eta)^2}$$

In order for the minimum reflectivity to be $-20\,$dB (0.01), the reflection loss has to be

$$\eta \geq \frac{1}{R}\left[1 - \frac{(1-R)}{\sqrt{0.99}}\right]$$

Suppose that $R = 0.9$. The requirement for the excess reflection loss of the mirror is $\eta > 0.9994$. That means it allows for only 0.12% power loss in each round-trip in the FP cavity, which is not usually easy to achieve.

An FPI is usually used as a tunable band-pass optical filter. Wavelength tuning of transmission peaks can be accomplished by either changing the length, d, of the cavity or the refractive index of material inside the cavity. The power transfer function of an FPI is exactly periodic as the function of optical frequency. From Eq. (6.3.7), the wavelength λ_m corresponding to the mth transmission peak can be found as

$$\lambda_m = \frac{2nd}{m} \tag{6.3.12}$$

By changing the length of the cavity, this peak transmission wavelength will move. When the cavity length is scanned by an amount of $\delta d = \lambda_m/(2n)$, the mth transmission peak frequency will be scanned across an entire FSR. That means in order to tune the transmission peak over a wavelength band corresponding to an FSR, one only need to sweep the cavity length, nd, for approximately half the wavelength. This mechanical scanning can usually be accomplished using a voltage-controlled piezo-electrical transducer (PZT); therefore, the mirror displacement (or equivalently, the change in cavity length d) is linearly proportional to the applied voltage.

For an FPI using free-space optics, beam collimation and mirror alignments are usually not easy tasks. Even a slight misalignment may result in significant reduction of the cavity finesse. For applications in fiber-optic systems, an FPI based on all-fiber configuration is attractive for its simplicity and compactness. An example of an all-fiber FPI is shown in Fig. 6.3.6 (Clayton et al., 1991). In this configuration, micro-mirrors are made by high-reflective (HR) coating on the fiber surfaces. The gap between two mirrors can be partially filled by short pieces of fibers to reduce the divergence of the light beam. Only a small air gap between fibers is enough to scan the cavity length for the transmission peaks to scan across the entire FSR. Antireflection (AR) coating is applied on each side of the air gap to eliminate the unwanted reflection. The entire package of an all-fiber FPI can be made small enough to mount onto a circuit board.

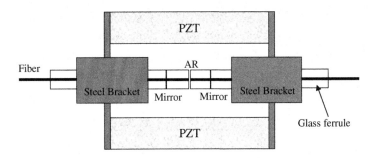

FIG. 6.3.6

Configuration if an all-fiber FPI.

Because the optical signal within the cavity is mostly confined inside single-mode optical fiber and it does not need additional optical alignment, an all-fiber FPI can provide much higher finesse compared to free-space PFIs. A state-of-the-art all-fiber FPI can provide a finesse of as high as several thousand.

6.3.2 OPTICAL RING RESONATORS

Another simple interferometer based on the multi-pass optical interference is a ring resonator. The basic configuration of a fiber-optic ring resonator is shown in Fig. 6.3.7A in which an optical ring is connected with fiber input/output through a 2×2 optical coupler. The power transfer function of this 2-port optical device can be calculated based on the transfer function of a 2×2 coupler as shown in Eq. (6.1.19).

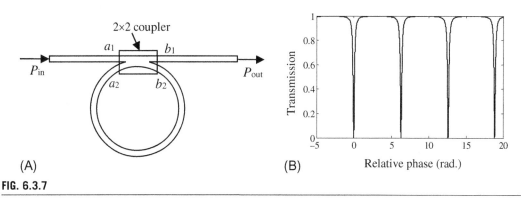

(A)   (B)   Relative phase (rad.)

FIG. 6.3.7

Fiber-based ring resonator (A) and transfer function (B).

Assume that the power-splitting ratio of the fiber coupler is ε, the length of the fiber ring is L, the refractive index of the fiber is n, and based on the notation in Fig. 6.3.7A,

$$b_1 = a_1\sqrt{1-\varepsilon} + ja_2\sqrt{\varepsilon} \tag{6.3.13a}$$

$$b_2 = ja_1\sqrt{\varepsilon} + a_2\sqrt{1-\varepsilon} \tag{6.3.13b}$$

The fiber ring then connects the output port b_2 to the input port a_2, and thus

$$a_2 = \sqrt{\eta}b_2 e^{-j\phi} \tag{6.3.14}$$

where η is the power transmission coefficient of the fiber, $\phi = 2\pi f\tau$ is the phase delay introduced by the fiber ring, and $\tau = nL/c$ is the time delay of the ring. Combine these three equations:

$$T(\phi) = \left|\frac{b_1}{a_1}\right|^2 = \left|\sqrt{1-\varepsilon} - \frac{\varepsilon\sqrt{\eta}e^{-j\phi}}{1-e^{-j\phi}\sqrt{\eta}\sqrt{1-\varepsilon}}\right|^2 \tag{6.3.15}$$

It is easy to find that if the fiber loss in the ring is neglected, $\eta = 1$, the power transfer function will be independent of the phase delay of the ring so that $T(\phi) \equiv 1$. In practice, any transmission media would have loss so that $\eta < 1$, even though it can be very close to unity.

Fig. 6.3.7B shows the transfer function with fiber coupler power splitting ratio $\varepsilon=0.1$ and the transmission coefficient $\eta=0.998$. Obviously, this is a periodic notch filter. The transmission stays near to 100% for most of the phases except at the resonance frequencies of the ring, where the power transmission is minimized. These transmission minima correspond to the phase delay of $\varphi=2m\pi$ along the ring, where m are integers and therefore the FSR of this ring resonator filter is $\Delta f=c/nL$, where c is the speed of light. The value of the minimum transmission can be found from Eq. (6.3.15) as

$$T_{min} = \frac{1-\varepsilon-2\sqrt{\eta}\sqrt{1-\varepsilon}+\eta}{1-2\sqrt{\eta}\sqrt{1-\varepsilon}+\eta(1-\varepsilon)} \tag{6.3.16}$$

From a notch filter application point of view, a high extinction ratio requires the minimum transmission T_{min} to be as small as possible. Fig. 6.3.8 shows the minimum power transmission of a ring resonator-based filter as the function of the power-splitting ratio of the fiber coupler. Three different loss values are used for the fiber ring in Fig. 6.3.8. Obviously, the optimum power-splitting ratio depends on the loss of the fiber ring. In fact, in order for the minimum transmission to be zero, $\eta=1-\varepsilon$ is required.

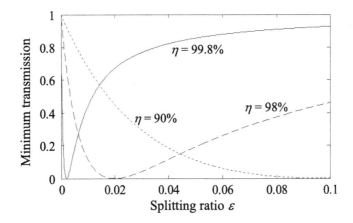

FIG. 6.3.8

Minimum power transmission of a ring resonator-based filter.

In addition to the extinction ratio, *finesse* is also an important parameter to specify a ring resonator. The finesse of a ring resonator is defined by the FSR divided by the full-width at half-maximum (FWHM) of the transmission notch. If we use the optimized value of splitting ratio such that $\eta=1-\varepsilon$, the ring resonator transfer function will be simplified to

$$T(\phi) = \left|\frac{\sqrt{\eta}(1-e^{-j\phi})}{1-\eta e^{-j\phi}}\right|^2 \tag{6.3.17}$$

Assuming that at a certain phase, ϕ_Δ, this power transfer function is $T(\phi_\Delta)=0.5$, we can find that

$$\phi_\Delta = \cos^{-1}\left(\frac{1-4\eta+\eta^2}{-2\eta}\right) \tag{6.3.18}$$

Since the FSR corresponds to 2π in phase, the finesse of the resonator can be expressed as

$$\text{Finesse} = \frac{2\pi}{2\phi_\Delta} = \frac{\pi}{\cos^{-1}\left(\dfrac{1-4\eta+\eta^2}{-2\eta}\right)} \tag{6.3.19}$$

Fig. 6.3.9 shows the calculated finesse as the function of the round-trip loss, $1-\eta$, of the fiber ring. Apparently high finesse requires low loss of the optical ring. In practice, this loss can be introduced either by the fiber loss or by the excess loss of the directional coupler.

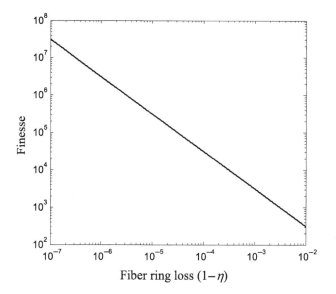

FIG. 6.3.9

Calculated finesse as a function of the fiber ring loss.

Similar to Finesse, there is also a Q-value which defines the quality of a ring resonator,

$$Q = \frac{\lambda_m}{\delta\lambda_m} \tag{6.3.20}$$

where λ_m is the resonance wavelength at which the transmission is minimum, that is at $\lambda_m = \tau c/m$, $\phi = 2m\pi$ with m an integer. $\delta\lambda_m$ is the FWHM of the absorption peak around λ_m in the wavelength domain. Because the round-trip phase is related to the wavelength by $\phi = 2\pi\tau c/\lambda$, its derivative is $\delta\phi = -2\pi\tau c\delta\lambda/\lambda^2$, thus $\delta\lambda_m$ at λ_m can be obtained from Eq. (6.3.18) as

$$\delta\lambda_m = \frac{\lambda_m^2}{2\pi\tau c}(2\phi_\Delta) \tag{6.3.21}$$

where the negative sign is removed because $\delta\lambda_m$ is a measure of linewidth. We can than find the simple relation between Q-value and the finesse defined in Eq. (6.3.19) by

$$Q = m \cdot Finesse \qquad (6.3.22)$$

where $m = \tau c / \lambda_m$.

From an application point of view, the biggest problem with the fiber-based ring resonator is probably the birefringence in the fiber. If there is a polarization rotation of optical signal in the fiber ring, the interference effect after each round-trip will be reduced and thus the overall transfer function will be affected. Therefore, PLC technology might be more appropriate to make high-Q ring resonators compared to all-fiber devices. Ring resonators have been used in evanescent wave biosensors, in which attachment of biomolecules on the surface of the waveguide change the effective propagation delay of the optical signal due to the interaction of biomolecules with the evanescent wave. Thus, the concentration of certain biomolecules can be quantified by monitoring the wavelength shift of the transmission notches of the optical transfer function (Sun and Fan, 2011). Ring resonators with $Q > 10^7$ have been reported, which can potentially make ring-based biosensors very attractive.

The configuration of optical ring resonator has also been used to make electro-optic modulators. In this application, an electrode is applied on the ring so that the optical delay can be modulated by the applied electric voltage. For an optical signal at a certain wavelength near the transmission notch, intensity modulation can be accomplished by modulating the notch wavelength on-and-off the optical signal wavelength (Xu et al., 2006). More details of ring modulators can be found in Chapter 7.

EXAMPLE 6.2

A fiber-optic Sagnac loop is shown in Fig. 6.3.10 in which the two output ports of a 2×2 coupler are connected. The coupler has a power coupling coefficient α. The length of the fiber in the loop is L and the refractive index is n.

Derive the power reflectivity $R = \left| \dfrac{E_a'}{E_a} \right|^2$ and power transmissivity $T = \left| \dfrac{E_b'}{E_a} \right|^2$.

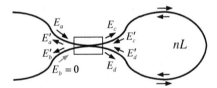

FIG. 6.3.10

Configuration of a fiber-optic Sagnac loop.

Solution

The field transfer function of the 2×2 coupler is

$$\begin{bmatrix} E_c \\ E_d \end{bmatrix} = \begin{bmatrix} \sqrt{1-\alpha} & j\sqrt{\alpha} \\ j\sqrt{\alpha} & \sqrt{1-\alpha} \end{bmatrix} \begin{bmatrix} E_a \\ 0 \end{bmatrix}$$

that is, $E_c = \sqrt{1-\alpha} E_a$, $E_d = j\sqrt{\alpha} E_a$

After traveling through the fiber delay line in the opposite directions, the optical fields reentering the coupler are, $E_c' = E_d e^{-j\phi} = j\sqrt{\alpha} E_a e^{-j\phi}$ and $E_d' = E_c e^{-j\phi} = \sqrt{1-\alpha} E_a e^{-j\phi}$, where $\phi = (2\pi/\lambda)nL$ is the propagation phase shift. Thus,

$$\begin{bmatrix} E_a' \\ E_b' \end{bmatrix} = \begin{bmatrix} \sqrt{1-\alpha} & j\sqrt{\alpha} \\ j\sqrt{\alpha} & \sqrt{1-\alpha} \end{bmatrix} \begin{bmatrix} j\sqrt{\alpha} \\ \sqrt{1-\alpha} \end{bmatrix} E_a e^{-j\phi}$$

so that, $E_a' = j2\sqrt{\alpha(1-\alpha)} E_a e^{-j\phi}$, and $E_b' = (1-2\alpha)E_a e^{-j\phi}$

Therefore, the power reflectivity is, $R = |E_a'/E_a|^2 = 4\alpha(1-\alpha)$, and the power transmissivity is $T = |E_b'/E_a|^2 = (1-2\alpha)^2$. Note that when a 3 dB coupler is used with $\alpha = 0.5$, the Sagnac loop has $R = 1$ and $T = 0$. That is equivalent to a perfect reflector or an ideal mirror.

To verify that the results satisfy energy conservation constrain $T + R = 1$, we have,

$$T + R = (1 - 2\alpha)^2 + 4\alpha(1 - \alpha) = 1 - 4\alpha + 4\alpha^2 + 4\alpha - 4\alpha^2 = 1$$

6.4 FIBER BRAGG GRATINGS

FBG is an all-fiber device which can be used to make low-cost, low-loss, and compact optical filters and demultiplexers. In an FBG, the Bragg grating is written into the fiber core to create a periodic refractive index perturbation along the axial direction of the fiber, as illustrated in Fig. 6.4.1. The periodic grating can be made in various shapes, such as sinusoid, square, or triangle; however, the most important parameters are the grating period Λ, the length of the grating region L, and the strength of the index perturbation δn. Although the details of the grating shape may contribute to the higher-order harmonics, the characteristic of the fiber grating is mainly determined by the fundamental periodicity of the grating. Therefore, the index profile along the longitudinal direction z, which is most relevant to the performance of the fiber grating can be approximated with (Kashyap, 2010)

$$n(z) = n_{core} + \delta n \left\{ 1 + \cos\left(\frac{2\pi}{\Lambda}z\right) \right\}$$ (6.4.1)

where n_{core} is the refractive index of the fiber core.

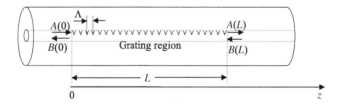

FIG. 6.4.1

Illustration of fiber Bragg grating. Λ: grating period and L: grating length.

The frequency selectivity of FBG is originated from the multiple reflections from the index perturbations and their coherent interference. Obviously, the highest reflection of an FBG happens when the signal wavelength matches the spatial period of the grating, $\lambda_{Bragg} = 2n_{core}\Lambda$, which is defined as the Bragg wavelength. The actual transfer function around the Bragg wavelength can be calculated using coupled-mode equations (McCall, 2000). Assume a forward-propagating wave $A(z) = |A(z)| \exp(j\Delta\beta z)$ and a backward-propagating wave $B(z) = |B(z)| \exp(-j\Delta\beta z)$, where $\Delta\beta = \beta - \pi/(\Lambda n_{core})$ is the wave number detuning around the Bragg wavelength. These two waves couple with each other due to the index perturbation of the grating along the fiber, and the coupled-wave equations are

$$\frac{dA(z)}{dz} = -j\Delta\beta A(z) - j\kappa B(z)$$ (6.4.2a)

$$\frac{dB(z)}{dz} = j\Delta\beta B(z) + j\kappa A(z) \tag{6.4.2b}$$

where κ is the coupling coefficient between the two waves. $\kappa \approx \pi(1 - V^{-2})\delta n / \lambda_{Bragg}$ is linearly proportional to the strength of the index perturbation δn of the grating, where V is the V-number of the fiber as discussed in Chapter 2. To solve the coupled-wave Eq. (6.4.2), we assume

$$A(z) = a_1 e^{-\gamma z} + a_2 e^{\gamma z} \tag{6.4.3a}$$

$$B(z) = b_1 e^{-\gamma z} + b_2 e^{\gamma z} \tag{6.4.3b}$$

where a_1, a_2, b_1, and b_2 are constants. Substituting Eq. (6.4.3) into Eq. (6.4.2), the coupled-wave equations become

$$(j\Delta\beta - \gamma)a_1 = -j\kappa b_1 \tag{6.4.4a}$$

$$(j\Delta\beta + \gamma)b_1 = -j\kappa a_1 \tag{6.4.4b}$$

$$(j\Delta\beta + \gamma)a_2 = -j\kappa b_2 \tag{6.4.4c}$$

$$(j\Delta\beta - \gamma)b_2 = -j\kappa a_2 \tag{6.4.4d}$$

To have nontrivial solutions for a_1, a_2, b_1, and b_2, we must have

$$\begin{vmatrix} j\Delta\beta - \gamma & 0 & j\kappa & 0 \\ 0 & j\Delta\beta + \gamma & 0 & j\kappa \\ j\kappa & 0 & j\Delta\beta + \gamma & 0 \\ 0 & j\kappa & 0 & j\Delta\beta - \gamma \end{vmatrix} = 0$$

This leads to $-\Delta\beta^2 - \gamma^2 + \kappa^2 = 0$ and therefore,

$$\gamma = \pm\sqrt{\kappa^2 - \Delta\beta^2} \tag{6.4.5}$$

Now, we define a new parameter:

$$\rho = \frac{j(j\Delta\beta + \gamma)}{\kappa} \equiv \frac{\kappa}{j(j\Delta\beta - \gamma)} \tag{6.4.6}$$

The relationships between a_1, b_1, a_2, and b_2 in Eq. (6.4.4) become $b_1 = a_1/\rho$ and $b_2 = a_2\rho$. If we know the input and the reflected fields A and B, the boundary conditions at $z = 0$ are $A(0) = a_1 + a_2$ and $B(0) = a_1/\rho + a_2\rho$. Equivalently, we can rewrite the coefficients a_1 and a_2 in terms of $A(0)$ and $B(0)$ as

$$a_1 = \frac{\rho A(0) - B(0)}{\rho - 1/\rho} \tag{6.4.7a}$$

$$a_2 = \frac{A(0) - \rho B(0)}{1 - \rho^2} \tag{6.4.7b}$$

Therefore, Eq. (6.4.3) can be written as

$$A(L) = \frac{\rho[B(0) - \rho A(0)]}{1 - \rho^2} e^{-\gamma L} + \frac{[A(0) + \rho B(0)]}{1 - \rho^2} e^{\gamma L}$$

$$B(L) = \frac{[B(0) - \rho A(0)]}{1 - \rho^2} e^{-\gamma L} + \frac{\rho[A(0) + \rho B(0)]}{1 - \rho^2} e^{\gamma L}$$

where L is the length of the grating. This is equivalent to a transfer matrix expression,

$$\begin{bmatrix} A(L) \\ B(L) \end{bmatrix} = \begin{bmatrix} S_{11} & S_{12} \\ S_{21} & S_{22} \end{bmatrix} \begin{bmatrix} A(0) \\ B(0) \end{bmatrix} \tag{6.4.8}$$

where the matrix elements are

$$S_{11} = \frac{e^{\gamma L} - \rho^2 e^{-\gamma L}}{1 - \rho^2} \tag{6.4.9a}$$

$$S_{22} = \frac{e^{-\gamma L} - \rho^2 e^{\gamma L}}{1 - \rho^2} \tag{6.4.9b}$$

$$S_{12} = -S_{21} = \frac{\rho(e^{-\gamma L} - e^{\gamma L})}{1 - \rho^2} \tag{6.4.9c}$$

Since there is no backward-propagating optical signal at the fiber-grating output, $B(L) = 0$, the complex field reflectivity of the fiber grating can be easily found as

$$R = \frac{B(0)}{A(0)} = -\frac{S_{21}}{S_{22}} = \rho \left(\frac{e^{-2\gamma L} - 1}{e^{-2\gamma L} - \rho^2} \right) \tag{6.4.10}$$

Fig. 6.4.2 shows the calculated power reflectivity of grating as the function of frequency detune from the Bragg wavelength. Since the reflectivity is a complex function of the frequency detune, Fig. 6.4.2 shows both the power reflectivity $|R|^2$ and the phase angle of R. The power reflectivity clearly shows a band-pass characteristic and the phase shift is quasi-linear near the center of the passband.

The peak reflectivity is at zero-detune which corresponds to the Bragg wavelength $\lambda_{Bragg} = 2n_{core}\Lambda$. The first reflectivity notch at each side of the peak reflection wavelength corresponds to $e^{-2\gamma L} = 1$, that is, $\gamma = j\pi/L$. Considering the definition of γ in Eq. (6.4.5), this is equivalent to $\Delta\beta = \sqrt{(\pi/L)^2 + \kappa^2}$. When the filter bandwidth is narrow enough,

$$\Delta\beta = \frac{2\pi n_{core}}{\lambda_{Bragg}} - \frac{2\pi n_{core}}{\lambda_{notch}} \approx \frac{2\pi n_{core}}{\lambda_{Bragg}^2} \left(\lambda_{notch} - \lambda_{Bragg} \right)$$

where λ_{notch} is the wavelength of the nearest notch of reflectivity adjacent to the peak reflection. Then, we can find that the full width of the band-pass filter (wavelength separation between the two notches one on each side of the reflection peak) is

$$\Delta\lambda = 2|\lambda_{notch} - \lambda_{Bragg}| = \frac{\lambda_{Bragg}^2}{2\pi n_{core} L} \sqrt{\pi^2 + (\kappa L)^2} \tag{6.4.11}$$

The peak reflectivity of the grating at the Bragg wavelength can also be obtained from Eq. (6.4.10). At the Bragg wavelength ($\Delta\beta = 0$) $\gamma = \kappa$ and $\rho = j$, Eq. (6.4.10) becomes

$$R(\lambda_{Bragg}) = j \left(\frac{e^{-\kappa L} - e^{\kappa L}}{e^{-\kappa L} + e^{\kappa L}} \right) = -j\tanh(\kappa L) \tag{6.4.12}$$

Thus, the peak power reflectivity at the Bragg wavelength is $R^2(\lambda_{Bragg}) = \tanh^2(\kappa L)$.

The dashed line in Fig. 6.4.2 shows the reflectivity in a grating with uniform coupling coefficient $\kappa = 1.5\,\mathrm{m}^{-1}$, and in this case, the stopband rejection ratio is only approximately 15 dB. This poor stopband rejection ratio is mainly caused by the edge effect because the grating abruptly starts at $z = 0$ and suddenly stops at $z = L$. To minimize the edge effect, apodization can be used in which the coupling coefficient κ is nonuniform and is a function of z. The solid line in Fig. 6.4.2 shows the reflectivity of a fiber grating with $\kappa(z) = \kappa_0 \exp\{-5(z - L/2)^2\}$, as shown in Fig. 6.4.3A, where $\kappa_0 = 1.5\,\mathrm{m}^{-1}$ is the peak coupling coefficient of the grating and $L = 10\,\mathrm{mm}$ is the grating length. The apodizing obviously increases the stopband rejection ratio by an additional 10 dB compared to the uniform grating. In general, more sophisticated apodizing techniques utilize both z-dependent coupling coefficient $\kappa(z)$ and z-dependent grating period $\Lambda(z)$, which help both to improve the FBG performance including the stopband rejection, the flatness of the passband, and the phase of the transfer function (Mihailov et al., 2000; Erdogan, 1997).

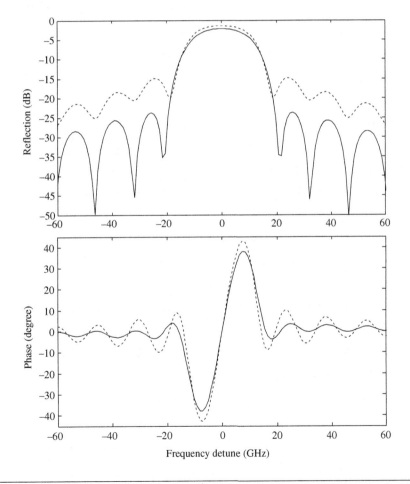

FIG. 6.4.2

Calculated reflectivities of fiber gratings with $\kappa = 1.5\,\mathrm{m}^{-1}$, $L = 10\,\mathrm{mm}$, and $\lambda_{bragg} = 1557\,\mathrm{nm}$. Dashed lines: uniform grating; solid line: apodized grating.

For nonuniform coupling coefficient $\kappa = \kappa(z)$ and grating period $\Lambda(z)$, calculation can be performed by dividing the grating region into a large number of short sections, as illustrated in Fig. 6.4.3B, and assuming both κ and Λ are constant within each short section. Thus, the input/output relation of each section can be described by Eq. (6.4.8). The overall grating transfer function can be found by multiplying the transfer matrices of all short sections:

$$\begin{bmatrix} A(L) \\ B(L) \end{bmatrix} = \left\{ \prod_{m=1}^{N} \begin{bmatrix} S_{11}^{(m)} & S_{12}^{(m)} \\ S_{21}^{(m)} & S_{22}^{(m)} \end{bmatrix} \right\} \begin{bmatrix} A(0) \\ B(0) \end{bmatrix} \tag{6.4.13}$$

where N is the total number of sections and $S_{i,j}^{(m)}$, $(i = 1, 2, \ j = 1, 2)$ are transfer matrix elements of the mth short section that use the κ and Λ values of that particular section.

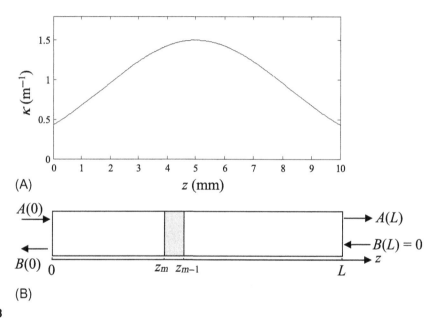

FIG. 6.4.3

(A) Nonuniform coupling coefficient $\kappa(z)$ and (B) dividing FBG into short sections for calculation using transfer matrices.

From an application point of view, the band-pass characteristic of the FBG reflection is often used for optical signal demultiplexing. Since FBG attenuation in regions away from the Bragg wavelength is very small, multiple FBGs, each having a different Bragg wavelength, can be concatenated, as illustrated in Fig. 6.4.4, to make multiwavelength demultiplexers. With special design, the phase shift introduced by an FBG can also be used for chromatic dispersion compensation in optical transmission systems. In the recent years, FBGs are often used to make distributed sensors, which utilize the temperature or mechanical sensitivities of FBG transfer functions. Another note is that although an FBG can be made low cost, an optical circulator has to be used to redirect the reflection from an FBG, which significantly increases the cost. Although a 3-dB fiber directional coupler can be used to replace the circulator, it will introduce a 6-dB intrinsic loss for the round-trip of the optical signal.

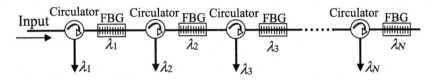

FIG. 6.4.4

Configuration of a WDM demux based on FBGs.

6.5 WDM MULTIPLEXERS AND DEMULTIPLEXERS

In WDM optical systems, multiple wavelengths are used to carry wideband optical signals; therefore, precisely filtering, multiplexing, and demultiplexing these optical channels are very important tasks. Similar to RF filters, the specification of an optical filter includes bandwidth, flatness of passband, stopband rejection ratio, transition slope from passband to stopband, and phase distortion. Simple optical filters based on the Mach-Zehnder, Michelson, and FP configurations are too simple to provide desired transfer functions for many applications. Multilayer thin films are flexible enough to create high-quality optical filters with desired specifications and excellent wavelength selectivity. Especially with advanced thin-film deposition technology, hundreds of thin-film layers with precisely controlled thickness and index profiles can be deposited on a substrate. Multichannel WDM multiplexers (MUXs), demultiplexers (DEMUXs), and wavelength add/drop couplers have been built based on the thin-film technology. In terms of disadvantages, since thin-films filters are free-space devices, precise collimation of optical beams is required and the difficulty becomes significant when the channel count is high. On the other hand, arrayed waveguide grating (AWG) is another configuration to make MUX, DEMUX, and add/drop couplers.

AWG is based on the PLC technology, in which multipath interference is utilized through multiple waveguide delay lines. PLC is an integrated optics technology that uses photolithography and etching; very complex optical circuit configurations can be made with submicrometer-level precision. WDM MUX and DEMUX with very large channel counts have been demonstrated by AWG.

6.5.1 THIN-FILM-BASED INTERFERENCE FILTERS

Fresnel reflection and refraction on an optical interface have been discussed in Section 2.2. If we consider normal incidence on the interface between materials of refractive indices n_0 and n_1, as shown in Fig. 6.5.1A, the Fresnel reflectivity is simply

$$\rho = \frac{E_{0-}}{E_{0+}} = \frac{n_1 - n_0}{n_1 + n_0} \tag{6.5.1}$$

Further, if we consider a thin film with thickness d_1 and refractive index n_1 is sandwiched between two bulk materials with refractive indices n_0 and n_2, as shown in Fig. 6.5.1B, a transfer matrix will be useful to describe the relationship between input and output optical fields,

$$\begin{bmatrix} E_{0+} \\ E_{0-} \end{bmatrix} = \begin{bmatrix} 1 & \rho_{0,1} \\ \rho_{0,1} & 1 \end{bmatrix} \begin{bmatrix} e^{j\beta_1 d_1} & 0 \\ 0 & e^{-j\beta_1 d_1} \end{bmatrix} \begin{bmatrix} 1 & \rho_{1,2} \\ \rho_{1,2} & 1 \end{bmatrix} \begin{bmatrix} E_{2+} \\ E_{2-} \end{bmatrix} \tag{6.5.2}$$

where $\rho_{0,1} = \frac{n_1 - n_0}{n_1 + n_0}$ and $\rho_{1,2} = \frac{n_2 - n_1}{n_2 + n_1}$ are interface Fresnel reflectivity, and $\beta_1 = \frac{2\pi}{\lambda} n_1$ is the propagation constant of the thin film.

The field reflectivity can be found as

$$r_1 = \frac{E_{0-}}{E_{0+}} = \frac{\rho_{0,1} e^{j\beta_1 d_1} + \rho_{1,2} e^{-j\beta_1 d_1}}{e^{j\beta_1 d_1} + \rho_{0,1}\rho_{1,2} e^{-j\beta_1 d_1}} \tag{6.5.3}$$

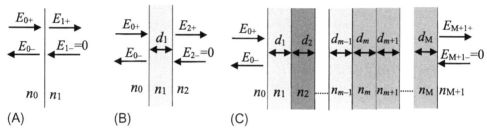

FIG. 6.5.1

Reflection and refraction of (A) single optical interface, (B) three interfaces, and (C) multilayer thin-film stacks.

For a general multilayer thin-film structure shown in Fig. 6.5.1C with M layers, the overall transfer function can be found by multiplying the transfer matrices of all layers,

$$\begin{bmatrix} E_{0+} \\ E_{0-} \end{bmatrix} = \begin{bmatrix} A & B \\ C & D \end{bmatrix} \begin{bmatrix} 1 & \rho_{M-1,M} \\ \rho_{M-1,M} & 1 \end{bmatrix} \begin{bmatrix} E_{M+1+} \\ E_{M+1-} \end{bmatrix} \tag{6.5.4}$$

with

$$\begin{bmatrix} A & B \\ C & D \end{bmatrix} = \prod_{m=1}^{M} \begin{bmatrix} e^{j\beta_m d_m} & \rho_{m,m-1} e^{j\beta_m d_m} \\ \rho_{m,m-1} e^{-j\beta_m d_m} & e^{-j\beta_m d_m} \end{bmatrix} \tag{6.5.5}$$

where, $\rho_{m,m-1} = \frac{n_m - n_{m-1}}{n_m + n_{m-1}}$, $\rho_{M-1,M} = \frac{n_M - n_{M-1}}{n_M + n_{M-1}}$, and $\beta_m = \frac{2\pi}{\lambda} n_m$.

For a thin film with M dielectric layers, if $n_0 = 1$ (air) and $n_{M+1} = 1.47$ (silica substrate) are fixed, there are $2M$ free parameters (thickness and refractive index of each layer) that can be used for the design of the desired transfer function. When the number of layers is large enough, numerical simulation has to be used (Macleod, 2010) to find filter transfer function based on Eqs. (6.5.4) and (6.5.5).

The field transfer function of a multilayer film is complex even for the simplest structure of $M = 1$, as can be easily seen from Eq. (6.5.3), in which both the intensity transmissivity and the optical phase are functions of the wavelength. While the intensity transfer function determines the selectivity and rejection ratio of different wavelength components in the optical signal, phase transfer function represents the chromatic dispersion the filter introduces. Fig. 6.5.2 show examples of intensity and phase transfer functions of multilayer thin-film filters with 100 and 200 GHz bandwidths. These filters were made with four cavities each containing two quarter-wave stacks spaced by one half-wave layer, and each stack contains alternative layers of SiO_2 and Ta_2O_5.

Multilayer thin films can be used to make optical filters both in the transmission mode and in the reflection mode as illustrated in Fig. 6.5.3. Fig. 6.5.3A shows the simplest thin-film fiber-optic filter in

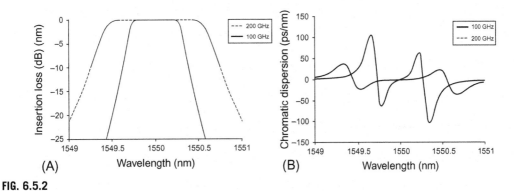

FIG. 6.5.2

(A) Intensity transfer function and (B) chromatic dispersion of thin-film band-pass filters (Zhang et al., 2002).

which the optical signal is collimated and passes through the optical interference film in the transmission mode. At the output, another collimator couples the signal light beam into the output fiber. With a large number of layers of thin films deposited on transparent substrates, high-quality edge filters and band-pass filters can be made with very sharp transition edges, as illustrated in Fig. 6.5.3C. If the transmitted and reflected optical signals are separately collected by two output fibers, a simple two-port WDM coupler can be made as shown in Fig. 6.5.3B. Typical transfer function of a band-pass filter is shown in Fig. 6.5.3D. Note that in practical applications of WDM couplers based on the multilayer films, beam incidence may not be normal with respect to the film surface, and the transfer function is a function of the beam incidence angle. The angular dependence of field transfer function of multilayer thin films can be calculated by taking into account the incidence angle in the Fresnel reflection described in Eqs. (2.2.4) and (2.2.6), and the modification of phase shift in each layer.

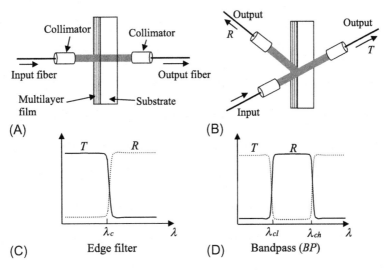

FIG. 6.5.3

Thin-film-based optical filters. T: transmission, R: reflection. (A) Example of an edge filter in simple transmission mode with transfer function shown in (C). (B) Example of a bandpass filter with transfer function shown in (D).

Fig. 6.5.4A shows the configuration of a four-port WDM DEMUX that uses four interference films, each having a different edge frequency of the transmission spectrum. Fig. 6.5.4B shows an alternative arrangement of the thin-film filters to construct an eight-channel WDM DEMUX in which only edge filter characteristics are required for the thin films. Fig. 6.5.4C is an example of the typical transfer function of a four-port WDM DEMUX, which shows the maximum insertion loss of about 1.5 dB, very flat response within the passband, and more than 40 dB attenuation in the stop-band. It is also evident that the insertion loss increases at longer wavelengths because long-wavelength channels pass through larger numbers of thin-film filters.

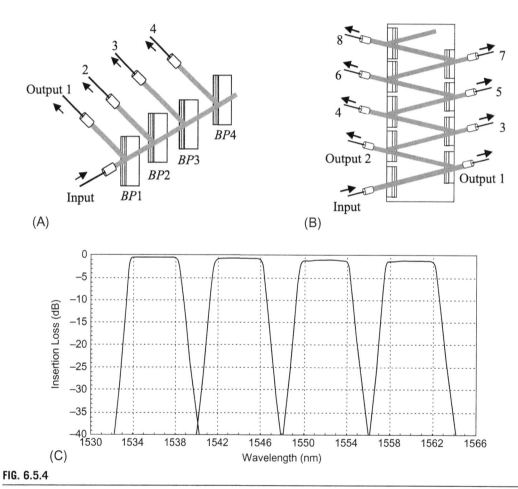

FIG. 6.5.4

Thin-film-based WDM DEMUX. (A) and (B) optical circuit configuration and (C) transfer function of a four-channel DEMUX.

Thin-film technology has been used for many years; sophisticated structural design, material selection, and fabrication techniques have been well studied. However, thin-film-based WDM MUX and DEMUX are based on the discrete thin-film units. The required number of cascaded thin-film units is equal to the number of wavelength channels, as shown in Fig. 6.5.4; and the overall insertion loss linearly increases with the number of ports. In addition, the optical alignment accuracy requirement becomes more stringent when the number of channels is high and thus the fabrication yield becomes low. For this reason, the channel counts of commercial thin-film filter-based MUX and DEMUX rarely go beyond 16. For applications involving large numbers of channels, such as 64 and 128, AWGs are more appropriate.

6.5.2 ARRAYED WAVEGUIDE GRATINGS

The wavelength selectivity of an AWG is based on the multipath optical interference. Unlike transmission or reflection gratings or thin-film filters, an AWG is composed of integrated waveguides deposited on a planar substrate, commonly referred to as *PLCs*. As shown in Fig. 6.5.5, the basic design of an AWG consists of input and output waveguides, two star couplers, and an array of waveguides bridging the two star couplers. Within the array, each waveguide has a slightly different optical length, and therefore, interference happens when the signals combine at the output. The wavelength-dependent interference condition also depends on the design of the star couplers. For the second star coupler, as detailed in Fig. 6.5.6, the input and the output waveguides are positioned at the opposite sides of a Roland sphere with a radius of $L_f/2$, where L_f is the focus length of the sphere. In an AWG operation, the optical signal is first distributed into all the arrayed waveguides by the input star coupler, and then at the output star coupler, each wavelength component of the optical field emerging from the waveguide array is constructively added up at the entrance of an appropriate output waveguide. The phase condition of this constructive interference is determined by the following equation (Takahashi et al., 1995):

$$n_c \Delta L + n_s d \sin\theta = m\lambda \tag{6.5.6}$$

where $\theta = kb/L_f$ is the diffraction angles in the output star coupler, b is the separation between adjacent output waveguides, k indicates the particular output waveguide number, ΔL is the length difference between two adjacent waveguides in the waveguide array, n_s and n_c are the effective refractive indices in the star coupler and waveguides, m is the diffraction order of the grating and is an integer, and λ is the wavelength.

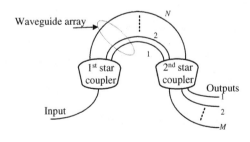

FIG. 6.5.5

Illustration of an arrayed waveguide-grating structure.

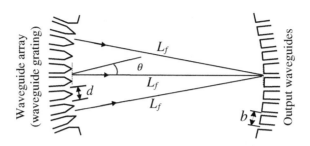

FIG. 6.5.6

Configuration of the star coupler used in AWG.

Due to the concentric configuration of the star coupler, the path length from the end of each grating waveguide to the center output waveguide is the same, and thus the condition for a constructive interference at the center output waveguide ($\theta=0$) at wavelength λ_0 is determined only by the differential length ΔL of the waveguide array:

$$\lambda_0 = \frac{n_c \cdot \Delta L}{m} \tag{6.5.7}$$

The wavelength of constructive interference depends on the angular position of the output waveguide, which is regarded as an angular dispersion of the star coupler. Based on Eq. (6.5.6), this angular dispersion is

$$\frac{d\lambda}{d\theta} = -\frac{n_s d}{m}\cos\theta \tag{6.5.8}$$

When d is a constant, the dispersion is slightly lower for waveguides further away from the center. Based on this expression, the separation of constructive interference wavelength between adjacent output waveguides can be found as

$$\Delta\lambda = \frac{b}{L_f}\left(\frac{d\lambda}{d\theta}\right) = \frac{b}{L_f}\frac{n_s d}{m}\cos\theta \tag{6.5.9}$$

I the design of AWG, ΔL is usually a constant which determines the central wavelength λ_0 through Eq. (6.5.7), and thus, $m = n_c\Delta L/\lambda_0$. Also based on the small angle ($\theta\approx0$) approximation, Eq. (6.5.9) can be simplified as

$$\Delta\lambda = \frac{b}{L_f}\frac{n_s d}{n_c\Delta L}\lambda_0 \tag{6.5.10}$$

Note that the central wavelength λ_0 shown in Eq. (6.5.7) depends on the grating order m. Wavelength difference between successive grating orders is known as the FSR. In AWG, the FSR can be found from Eq. (6.5.7) as

$$\Delta\lambda_{FSR} = \frac{\lambda_0^2}{n_c\Delta L} \tag{6.5.11}$$

The maximum number of wavelength channels that an AWG can support within a FSR can be determined by

$$N_{max} = \frac{\Delta\lambda_{FSR}}{\Delta\lambda} = \frac{L_f\lambda_0}{bn_s d} \qquad (6.5.12)$$

N_{max} can be increased by increasing the radius of the Roland sphere and decreasing waveguide separation between waveguides on both the input and the output sides. Because AWG is based on the multibeam interference, spectral resolution is primarily determined by the number of waveguides in the waveguide array between the two star couplers. Larger number of waveguides will provide better spectral resolution.

As an example, Fig. 6.5.7 shows the calculated insertion loss spectrum of a 32-channel AWG, in which 100 grating waveguides are used between the two star couplers. In this example, the pass band of each channel has a Gaussian-shaped transfer function. With advanced design and apodizing, flat-top transfer function can also be designed, which are commercially available.

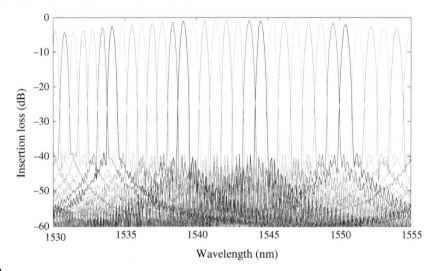

FIG. 6.5.7

Transfer function of an AWG with 32 wavelength channels.

Both thin-film filters and AWGs are often used to make MUX, DEMUX, and wavelength add/drop devices for WDM applications. To characterize these devices, important parameters include: width of the passband, adjacent channel isolation, nonadjacent channel isolation, passband amplitude and phase ripple, return loss, and polarization-dependent loss (PDL).

6.5.3 ACOUSTO-OPTIC FILTERS

Another category of optical filters is made by utilizing the acousto-optic (AO) effect. Acousto-optic filters (AOFs) are based on the Bragg gratings but which are created by acoustic waves propagating in AO crystal materials. AO effect for optical device applications is primarily based on the photoelastic property which is introduced by the pressure-dependent refractive index. Popular AO materials used for AOF include lead molybdate ($PbMoO_4$), tellurium dioxide (TeO_2), and quartz (SiO_2).

As illustrated in Fig. 6.5.8, an acoustic wave is generated from a piezoelectric transducer driven by an RF oscillator, and propagates as a plane wave inside a solid transparent material. This acoustic wave creates a moving periodical pressure pattern along the propagation direction z, and the period is

$$\Lambda = v_a/F \tag{6.5.13}$$

where F is the modulation frequency of the driving RF source and v_a is the velocity of the acoustic wave along the propagation direction in the solid material. For example, for an RF frequency of $F = 100\,\text{MHz}$, and a velocity of acoustic wave of $v_a = 4000\,\text{m/s}$ inside an AO crystal, this grating constant is $\Lambda = 40\,\mu\text{m}$.

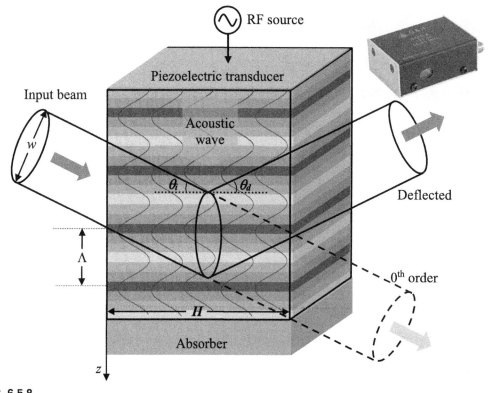

FIG. 6.5.8

Illustration of an acousto-optic modulator/filter. The inset is a picture of acousto-optic deflector from Gooch & Housego (used with permission)

This periodical pressure pattern further creates a periodic index perturbation along the z-direction through photoelastic effect of the solid material, which is equivalent to a moving Bragg grating with a Bragg constant Λ. Then, if an optical beam is launched onto the Bragg grating, it will be diffracted, and the diffraction angle depends both on the wavelength of the optical signal and on the frequency of the RF driving source.

Assume the angle between the incidence optical beam and the plane of the acoustic wave is θ_i, constructive interference of the diffracted wave from the pressure wave induced Bragg grating will be at θ_d to satisfy the grating condition, $\Lambda(\sin\beta_d+\sin\beta_i)=m\lambda/n$ with m the grating order, λ the optical signal wavelength, and n the refractive index of the AO crystal. For $\theta_d=\theta_i$ where the diffraction efficiency is the highest, the Bragg angle for the incident light beam is defined as

$$\theta_B = \sin^{-1}\left(\frac{m\lambda}{2n\Lambda}\right) \tag{6.5.14}$$

Assume that the θ_B is small enough, the first-order Bragg angle ($m=1$) is then,

$$\theta_B \approx \lambda\frac{F}{2nv_a} \tag{6.5.15}$$

Fig. 6.5.8 also shows the remaining optical beam that directly passes through without diffracted by the Bragg grating, which is defined as the zeroth-order beam ($m=0$). The zeroth-order and the first-order beams are separated by an angle of $2\theta_B$. As an example, for the TeO_2 AO crystal with $n=2.26$ at $\lambda=1550$ nm wavelength, and $v_a=4260$ m/s, if the RF modulation frequency is $F=100$ MHz, the grating constant is $\Lambda=42.6$ μm, and the Bragg angle will be approximately $\theta_B\approx7.4\times10^{-3}$ rad, or $0.425°$. This is indeed a very small angle.

As the Bragg angle θ_B depends on both the optical signal wavelength and the frequency of the driving RF source, this device can be used to make a tunable optical filter. For the filter application, the diffracted beam is collected around the Bragg angle $\theta_d=\theta_B$, and the RF frequency F is used as the tuning mechanism. The spectral resolution depends on the beam size and the width of the acoustic wave that interacts with the light beam.

Based on the first-order grating condition at the Bragg angle $2n\Lambda\sin\theta_B=\lambda_0$, where λ_0 is the center wavelength of the optical signal. Consider that there are N grating periods that interact with the light beam, and they are all constructive at the output, this is equivalent to $2Nn\Lambda\sin\theta_B=N\lambda_0$. The interference will become destructive at the output if the signal wavelength is changed from λ_0 to $\lambda_0+\delta\lambda$, that is, $2Nn\Lambda\sin\theta_B=(N-1/2)(\lambda_0+\delta\lambda)$. Thus,

$$\delta\lambda=\frac{\lambda_0}{2(N+1)}\approx\frac{\lambda_0}{2w/\Lambda} \tag{6.5.16}$$

Because θ_B is very small, $N\approx w/\Lambda$ is assumed as the total number of layers that interact with the input light beam, with w the beam diameter as indicated in Fig. 6.5.8. Another limitation for the spectral resolution is the diffraction-limit which causes the uncertainty of the diffraction angle $\delta\theta_B=1.22\lambda_0/(nw)$. Since $\delta\lambda/\delta\theta_B\approx2nv_a/F$ which can be derived from the first-order grating Eq. (6.5.15), we have

$$\delta\lambda=2.44\frac{\lambda_0 v_a}{wF}=2.44\frac{\lambda_0}{w/\Lambda} \tag{6.5.17}$$

This is almost five times bigger than that given in Eq. (6.5.16), and thus, diffraction limit is often the major reason that limits the spectral resolution of an AOF. As an example, for $\Lambda=42.6\mu m$ and $w=2$ mm, the normalized spectral resolution is approximately $\delta\lambda/\lambda_0\approx0.02$.

An AOF can also be used as a wavelength-selective modulator. The modulation speed depends on the incident optical beam diameter w. The switch on time τ is the time in which the acoustic wave passes through the diameter of the optical beam, that is simply, $\tau = w/v_a$, and switch off needs the same time. Therefore, the maximum attainable modulation frequency is

$$f_{modulation} = \frac{v_a}{2w} \tag{6.5.18}$$

For application as an optical filter, in order to achieve fine spectral resolution the acoustic wave frequency F has to be high, and the optical beam diameter w needs to be wide. On the other hand, for application as an optical modulator with high modulation speed, tightly focused beam with small diameter needs to be used. For example for TeO_2 AO crystal with $v_a = 4260\,m/s$, in order to achieve 1 MHz modulation speed, the optical beam diameter has to be $w < 2.3$ mm.

For the filter application of acoustic-optics, the most useful configuration is the colinear propagation between the optical wave and the acoustic wave as illustrated in Fig. 6.5.9.

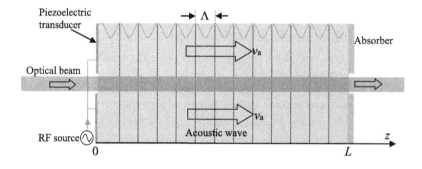

FIG. 6.5.9

Acousto-optic filter based on the colinear configuration.

In this case, both the optical signal and the acoustic wave are propagating in the z-direction, and the index perturbation is

$$n(z, t) = n_0 + \delta n \cos\left(\Omega t - \frac{2\pi}{\Lambda} z\right) \tag{6.5.19}$$

where n_0 is the average refractive index of the AO crystal, δn is the amplitude of index perturbation caused by the acoustic wave, Ω is the angular frequency of the acoustic wave, and $\Lambda = F/v_a$ is the grating constant. This is similar to that described in Eq. (7.5.1) for the FBG, except that the grating created by the acoustic wave is propagating in the z-direction at a speed v_a.

At each moment of time, the grating pattern is periodic with a Bragg wavelength of $\lambda_{Bragg} = 2n_0\Lambda$. The interaction length between the optical and the acoustic beams is L, as indicated in Fig. 6.5.9. In comparison to the incidence at the Bragg angle where the interaction length is approximately equal to the beam width w of the optical wave, the interaction length of this colinear configuration is much longer, and the spectral resolution can be significantly improved. Since diffraction limit is no longer an issue here, the spectral resolution can be roughly estimated from Eq. (6.5.16), but $N = L/\Lambda$ is used here so that $\delta\lambda \approx \lambda_0\Lambda/(2L)$. For example, for an AOF made with TeO_2 with $v_a = 4260\,m/s$, if the device length is $L = 5$ cm and the RF modulation frequency is 100 MHz, the normalized spectral resolution is

$\delta\lambda/\lambda_0 \approx 4.26 \times 10^{-4}$. The actual transfer functions of the filter, both used in the transmission mode and in the reflection mode, can be calculated based on the coupled-mode equation discussed in Section 6.4, but here the index perturbation δn is proportional to the amplitude of the acoustic wave, and with the Bragg wavelength $\lambda_{Bragg} = 2n_0 v_a/F$ is determined by the RF frequency and the velocity of the acoustic wave.

Note that in this colinear configuration, the optical wave interacts with the acoustic wave which propagates in the same direction. Thus, the optical signal will experience a Doppler effect, so that frequency of the optical signal at the filter output becomes, $f_{out} = f_0 - F$, where $f_0 = c/\lambda_0$ is the input optical frequency and c is the speed of light. This is equivalent to a red shift of optical wavelength by

$$\Delta\lambda_{Doppler} = \frac{\lambda_0^2}{c}F \qquad (6.5.20)$$

It is also possible for the optical wave and the acoustic wave to propagate in the opposite direction, known as counter propagation. In that case, the optical filtering property is the same as that of co-propagation, except that the optical frequency is increased due to the Doppler effect, $f_{out} = f_0 + F$, and the optical wavelength is blue shifted.

Frequency shift caused by Doppler effect also exists for diffractive AO modulators shown in Fig. 6.5.8. This is similar to an optical beam reflected by a moving mirror, and the Doppler effect is determined by the velocity of the acoustic wave projected on the direction of the incident optical beam. Thus, the frequency of the diffracted optical signal is $f_{out} = f_0 \pm F\sin\theta_i$. AO modulators without optical frequency shift can be obtained by cascading two sections of crystals driven by acoustic waves in the opposite directions, so that the Doppler-induced frequency shift can be canceled out. Another approach is to apply acoustic waves in the opposite directions along a single AO crystal. In this way, a standing wave pattern is generated due to their interference, and the induced Bragg grating is not moving. Thus, Doppler frequency shift can be avoided.

6.6 OPTICAL ISOLATORS AND CIRCULATORS

An optical isolator is a device that only allows unidirectional transmission of the optical signal. It is often used in optical systems to avoid unwanted optical reflections. The purpose of an optical circulator is to redirect the reflected optical signal into a different destination. Both optical isolators and circulators are classic nonreciprocal devices which are designed based on the polarization rotation induced by Faraday effect. For free-space optical applications with relatively large beam sizes, isolators and circulators with are often polarization dependent. As the SOP of beam propagation in free space can be easily controllable and maintained, polarization dependence of optical devices is acceptable. In a fiber-optic system, on the other hand, the SOP of an optical signal can vary randomly when propagating along a fiber, so that polarization-independent isolators and circulators have been developed for fiber-optic systems. Optical configurations and operation principles of optical circulators and circulators will be discussed in this section. Definitions of key parameters and specifications will also be presented.

6.6.1 OPTICAL ISOLATORS

An optical isolator is a device that is designed to allow the optical signal travel in the forward direction while block reflections that would travel in the backward direction. Optical isolators are critically important in many applications in optical systems. For example, a single-frequency semiconductor laser is very susceptive to external optical feedback. Even a very low level of optical reflection from an external optical circuit, on the order of $-50\,dB$, is sufficient to cause a significant increase in laser phase noise, intensity noise, and wavelength instability. Thus, an optical isolator is usually required at the output of each laser diode in applications that require low optical noise and stable single optical frequency. Another example is in an optical amplifier where unidirectional optical amplification is required. In this case, the bidirectional nature of optical amplification of the optical gain medium would cause self-oscillation if the external optical reflection from, for example, connectors and other optical components is strong enough.

A traditional optical isolator is based on a Faraday rotator sandwiched between two polarizers, as shown in Fig. 6.6.1. In this configuration, the optical signal coming from the left side passes through the first polarizer whose optical axis is in the vertical direction, which matches the polarization orientation of the input optical signal. Then, a Faraday rotator rotates the polarization of the optical signal by 45 degree in a clockwise direction. The optical axis of the second polarizer is oriented 45 degree with respect to the first polarizer, which allows the optical signal to pass through with little attenuation. If there is a reflection from the optical circuit at the right side, the reflected optical signal has to pass through the Faraday rotator from right to left. Since the Faraday rotator is a nonreciprocal device, the polarization state of the reflected optical signal will rotate for an additional 45 degree in the same direction as the input signal, thus becoming horizontal, which is perpendicular to the optical axis of the first polarizer. In this way, the first polarizer effectively blocks the reflected optical signal and assures the unidirectional transmission of the optical isolator.

For the Bi:YIG-based Faraday rotator of a certain thickness, the angle of polarization rotation is proportional to the magnitude of the applied magnetic field, but is generally not linear. It saturates at around 1000 Gauss. Therefore, a magnetic field of >1500 Gauss will guarantee the stable rotation angle for a certain Bi:YIG thickness. The value of isolation is, to a large extent, determined by the accuracy of the polarization rotation angle and thus by the thickness of the YIG Faraday rotator.

The Faraday rotator is the key component of an optical isolator. In long-wavelength applications in 1.3 and 1.5 μm windows for optical communications, Bismuth-substitute yttrium iron garnet (Bi:YIG) crystals are often used; they have high Verdet constant and relatively low attenuation at these wavelength windows. Packaged optical isolators with <1-dB insertion loss and $>30\,dB$ isolation are commercially available. If a higher isolation level is required, multistage isolators can be used to provide optical isolation of $>50\,dB$.

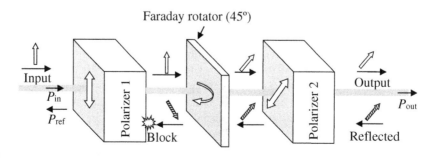

FIG. 6.6.1

Optical configuration of a polarization-sensitive optical isolator.

Although the configuration of this type of isolators is simple, the SOP of the optical signal has to match the orientation of the polarizer's principle axis of the isolator. Any polarization state mismatch would result in additional optical attenuation. An isolator of this type is usually used in free-space optics where the SOP of the optical signal is often stable at the input of the isolator. It is often used immediately after a laser diode chip within the same package to prevent optical reflection that goes back into the laser diode. In this case, the optical signal from the laser diode is launched into the isolator through free-space optical coupling so that the SOP is deterministic. The optical signal output from the isolator can be coupled into an optical fiber pigtail. On the other hand, for many inline fiber-optic applications where the SOP of the optical signal at the input of the isolator may vary randomly, polarization-insensitive optical isolators are required.

Fig. 6.6.2 shows the configuration of a polarization-independent optical isolator made by two birefringence beam displacers and a Faraday rotator (Jameson, 1989). In the 1550-nm wavelength window, a YVO_4 crystal is often used for the birefringence beam displacers due to its high birefringence, low loss, and relatively low cost. The operation principle of this polarization-independent optical isolator can be explained using Fig. 6.6.3. The vertical and horizontal polarized components (or o beam and e beam) of the incoming light beam is first split into two separated beams by the first YVO_4 birefringence beam displacer; they are shown as solid and dashed lines, respectively, in Fig. 6.6.3. The distance of beam separation is determined by the birefringence and the thickness of the crystal. Then, both of these two beams are polarization rotated for 45 degree by a Bi:YIG Faraday rotator without changing their spatial beam positions. The second YVO_4 crystal has the same birefringence and thickness as the first one. By arranging the orientation of the birefringence axis of the second YVO_4 crystal, the o beam and e beam in the first YVO_4 beam displacer become the e beam and o beam, respectively, in the second YVO_4 beam displacer. Therefore, the two beams converge at the end of the second YVO_4 crystal.

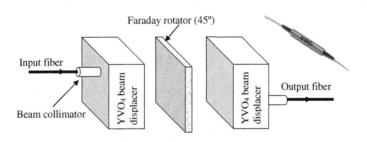

FIG. 6.6.2

Configuration of polarization-independent optical isolator.

In the backward direction, the light beams pass through the second YVO_4 crystal along the same routes as for the forward direction. However, due to its nonreciprocal characteristic, the Faraday rotator rotates the polarization of the backward-propagated light beams by an additional 45 degree (in the same rotation direction as the forward-propagated light beams). The total polarization rotation of each beam is 90 degree after a round-trip through the Faraday rotator. Thus, the initially $\perp$ (//) polarized light beam in the forward direction becomes $\perp$(//) polarized in the backward propagation. The spatial separation between these two beams will be further increased when they pass through the first YVO_4 crystal in the backward direction and will not be captured by the optical collimator in the input side. Fig. 6.6.3 also illustrates the polarization orientation of each light beam at various interfaces. In the design of an optical isolator, the thickness of YVO_4 has to be chosen such that the separation of o and e beams is larger than the beam cross-section size.

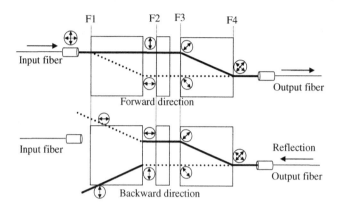

FIG. 6.6.3

Illustration of the operating principle of a polarization-independent optical isolator.

To specify the performance of an optical isolator, the important parameters include isolation, insertion loss, PDL, polarization-mode dispersion (PMD), and return loss. *Insertion loss* is defined as the output power divided by the input power. *Isolation* is defined as the reflected power divided by the input power when the output is connected to a total reflector. *Return loss* is defined as the reflected power divided by the input when the output side of the isolator has no reflection; thus return loss is the measure of the reflection caused by optical interfaces inside the isolator itself. By definition, return loss should always be lower than the isolation. Practically, AR coating can never be perfect which is also wavelength dependent. The Verdet constant of Faraday rotator may also be wavelength dependent. These effects usually limit the width of wavelength window in which an isolator can operate within the specification. For a single stage isolator within a ± 15 nm window around 1550 nm wavelength, the insertion loss is typically less than 0.5 dB, the isolation should be about -30 dB, and the PDL should be less than 0.05 dB. With good AR coating, the return loss of an isolator is on the order of -60 dB.

6.6.2 OPTICAL CIRCULATORS

Similar to an optical isolator that has just been discussed; an optical circulator is device that is also based on the nonreciprocal polarization rotation of an optical signal by the Faraday effect. Different from an isolator, an optical circulator is a three-terminal device as illustrated in Fig. 6.6.4, where terminal 1 is the input port and terminal 2 is the output port, while the reflected signal back into terminal 2 will be redirected to terminal 3 instead of terminal 1.

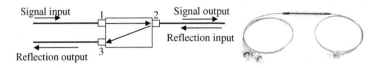

FIG. 6.6.4

Basic function of a three-terminal optical circulator.

Optical circulators have many applications in optical circuits and optical communication systems for redirecting bidirectional optical signals into different ports. One example is the use with FBGs, as shown in Fig. 6.6.5A. Since the wavelength-selective reflection of a FBG can be used either as a band-pass optical filter or as a dispersion compensator, an optical circulator has to be used to redirect the reflected optical signal into the output. Although a 3-dB fiber directional coupler can also be used to accomplish this job, as shown in Fig. 6.6.5B, there will be a 6-dB intrinsic insertion loss for the optical signal going through a round-trip in the fiber coupler.

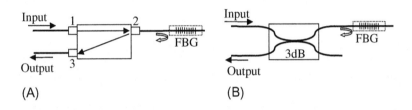

FIG. 6.6.5

Redirect FBG reflection using (A) a circulator and (B) a 3-dB fiber directional coupler.

Fig. 6.6.6 illustrates the configuration of a polarization-independent optical circulator (Van Delden, 1991). Similar to a polarization-independent optical isolator discussed previously, an optical circulator also uses YVO_4 birefringence material as beam displacers and Bi:YIG for the Faraday rotators. However, configuration of an optical circulator is much more complex than an isolator because the backward-propagated light has also to be collected at port 3 shown in Fig. 6.6.6.

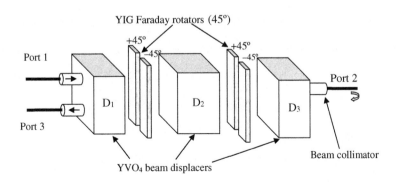

FIG. 6.6.6

Configuration of a polarization-insensitive optical circulator.

The operating principle of the optical circulator can be explained using Fig. 6.6.7. In the forward-propagation direction, the incoming light beam into port 1 is first split into o and e beams by the first YVO_4 beam displacer, and these two beams are shown as the solid and the dashed lines, respectively. The polarization state is also labeled near each light beam. These two beams are separated in the horizontal direction after passing through the first YVO_4 displacer (D_1) and then they pass through a pair of separate Bi:YIG Faraday rotators. The left Faraday rotator (a_1) rotates the o beam by +45 degree and the right Faraday rotator (b_1) rotates the e beam by −45 degree without shifting their beam spatial positions. In fact, after passing through the first pair of Faraday rotators, the two beams become copolarized and they are both o beams in the second YVO_4 displacer (D_1). Since these two separate beams now have the same polarization state, they will pass the second displacer D_2 without further divergence. At the second set of Faraday rotators, the left beam will rotate an additional +45 degree at (a_2) and the right beam will rotate an additional −45 degree at (b_2), then their polarization states become orthogonal with each other. The third beam displacer D_3 will then combine these two separate beams into one at the output, which recreates the input optical signal but with a 90-degree polarization rotation.

For the reflected optical signal into port 2, which propagates in the backward direction, the light beam is separated into two after passing through D_3. However, because of the nonreciprocal characteristic, the Faraday rotator a_2 rotates the reflected beam by +45 degree and b_2 rotates the reflected beam by −45 degree, all in the same direction as rotating the forward-propagated light beams. The total polarization rotation is then 90 degree after a round-trip through the second pair of Faraday rotators. Thus, in the second beam displacer D_2, the backward-propagated beams are again copolarized, but their polarization orientations are both e beams (recall that the two beams are both o beams in the forward-propagating direction). Because of this 90 degree difference in the polarization orientation, the backward-propagated beams will not follow the same routes as the forward-propagated beams in the second beam displacer D_2, as shown in Fig. 6.6.7. After passing through the first set of Faraday rotators and the first beam displacer D_1, the backward-propagated beams are eventually recombined at port 3 at the input side, which is in a different spatial location than port 1.

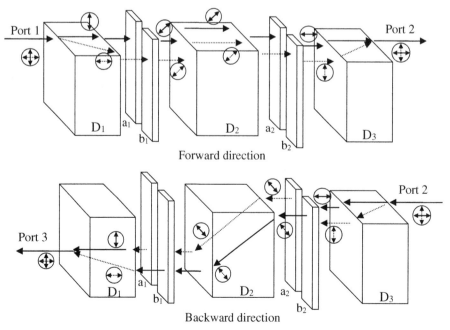

FIG. 6.6.7

Illustration of the operating principle of an optical circulator.

From an application point of view, both optical isolators and circulators need to operate with minimum wavelength dependence for all parameters. Similar to an optical isolator, important specifications for an optical circulator also include insertion loss, isolation, PDL, and return loss. In addition, since a circulator has more than two terminals, directionality is also an important measure. For a three-terminal circulator, as illustrated in Fig. 6.6.4, the insertion loss includes the losses from port 1 to port 2 and from port 2 to port 3. For example, if the optical power launched into port 1 is P_1, port 2 is connected to a total reflector, and the output power measured from port 3 is P_3, the insertion loss is $10\log_{10}(P_1/P_3)$ in dB. Likewise, the isolation also includes the isolation from port 2 to port 1 and from port 3 to port 2. In this case, the isolation between port 2 and port 1 is $10\log_{10}(P_{1+}/P_{1-})$, where $P_{1\pm}$ are forward and reflected powers into and from port 1, with port 2 connected to a total reflector. The isolation between port 3 and port 2 is $10\log_{10}(P_{2-}/P_{2+})$, where $P_{2\pm}$ are backward launched and reflected (in the forward direction) powers into and from port 2, with port 3 connected to a total reflector. As the device has more than two ports, directionality is another parameter for the device specification, which is defined by the transmission loss from port 1 to port 3 when port 2 is terminated without reflection. A typical circulator operating in 1550 ± 15-nm wavelength window usually has insertion loss of about 0.8 dB, which is slightly higher than an isolator due to its more complicated optical structure. The isolation value is on the order of -40 dB and the directionality is usually better than -50 dB.

6.7 PLCs AND SILICON PHOTONICS

PLC is a technology based on the optical waveguides fabricated on planar substrates (Okamoto, 2005). Because of the layered structure of the substrates, waveguides of PLC usually have rectangular cross sections. Sophisticated photonic structures can be fabricated through photolithography and etching on planar substrates, and a large number of devices, such as couplers, arrayed-waveguide gratings (AWGs), MZIs, and ring resonators, can be created on the same substrate, which allows batch fabrication of photonic chips with sophisticated functionalities. Miniature sizes of photonic structures and functional devices allow photonic integration which provides the possibility to significantly reduce the per-device cost and improve the reliability. Many different materials have been utilized for PLC, but the most popular materials are SiO_2, Si, InP, and optical polymer.

Silica-based PLC is often used to make AWG for wavelength-division multiplexer and demultiplexer in WDM systems and networks. Because the refractive index of silica, $n \approx 1.5$, matches that of optical fiber, optical coupling between optical fiber and PLC is straightforward because of the close match of their fundamental mode cross sections. The typical cross-section size of a silica-based PLC waveguide is on the order of $6 \times 6 \, \mu m$ (Okamoto, 2005), and thus precision tolerance of photolithography is not as stringent as those using smaller waveguide sizes based on higher index materials such as silicon. The intrinsic loss of silica-based PLC is also small over a very wide spectral window. Although silica-based PLC is still the dominant technique for AWG currently used in commercial fiber-optic systems, the low index and relatively large waveguide cross-section size as well as the large bending radius limit the achievable device density of photonic integration. For example, a typical WDM DEMUX based on the silica AWG takes an area on the order of 1×1 cm on the substrate. In addition, the platform of silica is not compatible with materials for active photonic and electronic devices, which further limit the prospective of large-scale photonic/electronic integration.

Silicon-based PLC is gaining momentum in the recent years, which provides an ideal integrated photonic circuit platform known as *silicon photonics* (Okamoto, 2014). Silicon is semiconductor with a bandgap energy of $E_g = 1.12$ eV, corresponding to a cutoff wavelength of approximately $\lambda_c = 1.13 \, \mu m$. Thus, silicon is transparent in the optical communications wavelength window of 1550 nm. The refractive index of silicon is approximately $n = 3.5$ at 1550 nm wavelength which is much higher than that of silica. As the result, the cross section of a single-mode silicon waveguide is on the order of $0.2 \times 0.4 \, \mu m$ at the 1550-nm operating wavelength. In addition, carrier injection can be used to dynamically change the refractive index of silicon, so that high-speed modulation can be realized in silicon photonic devices such as ring resonators and MZIs. The small feature size of silicon PLC requires much higher fabrication precision for photolithograph and etching. But at the same time, the small feature size results in reduced footprint of photonic integrated circuits and allowing large-scale photonic integration of a large number of functional devices on the same chip. After all, the most important advantage of silicon photonic circuits is in its material platform and fabrication process which are CMOS compatible: that means the same standard CMOS fabrication process for digital electronic circuits can be used to produce photonic circuits on the same substrate. This further allows hybrid electro-optic integration. It needs to point out that silicon is an indirect semiconductor which, in principle, cannot be used to make light-emitting devices such as lasers and LEDs. Thus, silicon photonic circuits have to rely on external light sources.

InP is a III–V compound semiconductor material with the bandgap at $E_g = 1.34 \, eV$, corresponding to a cutoff wavelength of approximately $\lambda_c = 870 \, nm$. Thus, InP is also transparent in the optical communications wavelength window of 1550 nm. The refractive index of InP is $n = 3.17$ at 1550 nm wavelength, which is slightly lower than that of silicon, but still significantly higher than the refractive index of silica. Because the electron mobility of InP is much higher than that of silicon, faster electro-optic modulation can be achieved in InP-based photonic circuits through carrier injection. A distinct advantage of InP-based photonic circuits is that it can be monolithically integrated with light sources and near-infrared wavelength photodiodes made of III–V semiconductor materials such as GaAs and InGaAsP. Optical couplers, WDM multiplexers and demultiplexers in the AWG configuration made on the InP platform have been reported with monolithically integrated laser diodes and photodiodes. However, PLC fabrication with III–V materials has been shown much more challenging with low yield compared to silicon. Thus, photonic devices based on III–V semiconductor materials are generally much more expensive than their silicon counterparts.

Polymer is a plastic material starting in the liquid form, which can be spin coated onto a substrate and cured into a solid transparent film after baking. Refractive index of polymer is on the order of $n = 1.45$, and can be modified to some extent by changing the material composition. Photonic circuit structures can be made through photolithography and wet etching on the polymer. Multiple layers of polymers each with a different refractive index can be made on the substrate to make complex photonic structures. The major advantage of polymer-based PLC is the low cost, simple fabrication process, and the flexibility and variety of substrates that can be used to support polymer PLC. There are also polymers with very high electro-optic coefficient which can be used to make high efficiency electro-optic modulators. On the other hand, major disadvantages of polymer PLC include the high-temperature sensitivity and the potential long-term reliability issue due to material property degradation overtime common to most plastic products caused by oxidation and aging.

6.7.1 SLAB OPTICAL WAVEGUIDES

A slab waveguide is a two-dimensional (2D) structure formed by a high-index wave-guiding layer sandwiched between a top and a bottom cladding layer with lower refractive indices as illustrated in Fig. 6.7.1.

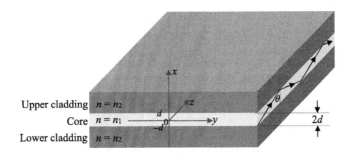

FIG. 6.7.1

Structure of a slab optical waveguide with a core layer thickness $2d$ sandwiched between the upper and lower cladding layers.

In Fig. 6.7.1, the thickness of the core layer is d with the refractive index n_1, and the dimension is $-\infty < y < \infty$, $-d < x < d$, and $-\infty < z < \infty$. To simplify the analysis, we assume that both the upper and the lower claddings have the same refractive index n_2 which is slightly lower than n_1. Because the y-direction is uniform and infinite, this slab waveguide is a 2D problem where the optical field propagates in the z-direction and its amplitude varies over the transversal x-direction. Similar to those discussed in Section 2.3.1, the propagation constant can be decomposed into the propagation direction z and the transversal direction x. Although in practice the waveguide may not be symmetric, we use the symmetric structure to simplify the analysis as shown in Fig. 6.7.1 in which the upper and the lower cladding layers have the same refractive index. The technique can be straightforwardly extended to general cases of asymmetric slab waveguides.

Based on the rule of Fresnel reflection discussed in Chapter 2, the critical angle for total reflection at the core-cladding interface is $\theta_c = \sin^{-1}(n_2/n_1)$. Because of the phase continuity constrain, the propagation constant of the guided mode in the z-direction β_z has to be the same for the core and the claddings, and it is limited within $\beta_1 \geq \beta_z \geq \beta_2$, where $\beta_1 = 2\pi n_1/\lambda$ and $\beta_2 = 2\pi n_2/\lambda$. As the optical field is guided by the core, the field distribution in the transversal x-direction inside the core layer should be a standing wave, whereas in the cladding layers, field distribution should be evanescent. Transverse electric (TE) and transverse magnetic (TM) modes are usually used to categorize optical fields in the waveguide which have no E_z or H_z components, respectively (Lizuka, 2002).

Now, let us use the TE mode as an example. The major electric field component of the TE mode is E_y, which has standing wave patterns along the transversal x-direction inside the core layer. These patterns can either be symmetric or antisymmetric as

$$E_y = A\cos(\beta_{1x}x)e^{j\beta_z z} \quad \text{for}(|x| \leq d) \tag{6.7.1}$$

or

$$E_y = A\sin(\beta_{1x}x)e^{j\beta_z z} \quad \text{for}(|x| \leq d) \tag{6.7.2}$$

where A is the peak amplitude of the field. The optical field in the upper and lower cladding regions should be evanescent with the amplitudes decaying in the $\pm x$ directions away from the core, that is,

$$E_y = \begin{cases} B\exp[j\beta_{2x}(x-d)]e^{j\beta_z z} & \text{for } (x \geq d) \\ \pm B\exp[-j\beta_{2x}(x+d)]e^{j\beta_z z} & \text{for } (x \leq -d) \end{cases} \quad (6.7.3)$$

where B is the peak amplitude at $|x|=d$, and the $\pm$ sign depends on the symmetric (*cos*) or antisymmetric (*sin*) field distribution in the core. β_{1x} and β_{2x} are projections of β_1 and β_2 on the x-direction, and they are defined as, $\beta_{1x}^2 = \beta_1^2 - \beta_z^2$ and $\beta_{2x}^2 = \beta_2^2 - \beta_z^2$ with β_z the propagation constant along the z-direction. For the guided wave modes β_{1x} is real, but β_{2x} has to be imaginary so that $\beta_{2x} = j\alpha_2 = j\sqrt{\beta_z^2 - \beta_2^2}$, where α_2 represents a decay constant. Thus, optical field in the cladding layers are

$$E_y = \begin{cases} B\exp[-\alpha_2(x-d)]e^{j\beta_z z} & \text{for } (x \geq d) \\ \pm B\exp[\alpha_2(x+d)]e^{j\beta_z z} & \text{for } (x \leq -d) \end{cases} \quad (6.7.4)$$

The vector magnetic field can be found through Faraday's law,

$$\vec{H} = \frac{-1}{j\omega\mu_0}\nabla \times \vec{E} \quad (6.7.5)$$

The transversal x-components of magnetic fields are

$$H_x = \frac{A}{\omega\mu_0}\beta_z \begin{cases} (\sin\beta_{1x})e^{j\beta_z z} & \text{symmetric} \\ (\cos\beta_{1x})e^{j\beta_z z} & \text{antisymmetric} \end{cases} \quad (6.7.6)$$

for the core and,

$$H_x = \begin{cases} \dfrac{B}{\omega\mu_0}\beta_z\exp[-\alpha_2(x-d)]e^{j\beta_z z} & \text{for } (x \geq d) \\ \dfrac{\pm B}{\omega\mu_0}\beta_z\exp[\alpha_2(x+d)]e^{j\beta_z z} & \text{for } (x \leq -d) \end{cases} \quad (6.7.7)$$

for the claddings.

The longitudinal z-components of magnetic fields in the core are

$$H_z = \frac{jA}{\omega\mu_0}\beta_{1x} \begin{cases} -(\sin\beta_{1x}x)e^{j\beta_z z} & \text{symmetric} \\ (\cos\beta_{1x}x)e^{j\beta_z z} & \text{antisymmetric} \end{cases} \quad (6.7.8)$$

for the core and,

$$H_z = \begin{cases} \dfrac{-B}{\omega\mu_0}j\alpha_2\exp[-\alpha_2(x-d)]e^{j\beta_z z} & \text{for } (x \geq d) \\ \dfrac{\mp B}{\omega\mu_0}j\alpha_2\exp[\alpha_2(x+d)]e^{j\beta_z z} & \text{for } (x \leq -d) \end{cases} \quad (6.7.9)$$

for the claddings. Now, we need to determine the relation between the proportionality constants A and B, as well as constrains imposed by propagation constants β_{1x} and α_2. First, let us consider the case of symmetric field distribution. The E_y component has to be continuous at the core/cladding interface $x=d$. For the symmetric field distribution this is

$$A\cos(\beta_{1x}d) = B\exp[-\alpha_2(x-d)] \quad (6.7.10)$$

The magnetic field H_z also has to be continuous at $x=d$, that is,

$$A\beta_{1x}\sin(\beta_{1x}d)=B\alpha_2\exp[-\alpha_2(x-d)] \tag{6.7.11}$$

Combining Eqs. (6.7.10) and (6.7.11), the relation between α_2 and β_{1x} for the case of symmetric field distribution is

$$\alpha_2=\beta_{1x}\tan(\beta_{1x}d) \tag{6.7.12}$$

Because of the field symmetry, the continuity condition of E_y at $x=-d$ would arrive at the same relation between α_2 and β_{1x} as that described by Eq. (6.7.12).

Based on the similar continuity constrains of E_y and H_z at the core/cladding interface, the relation between α_2 and β_{1x} for the case of antisymmetric field distribution can be found as

$$\alpha_2=-\beta_{1x}\cot(\beta_{1x}d) \tag{6.7.13}$$

Considering that $\beta_{1x}^2=\beta_1^2-\beta_z^2$ and $\alpha_2^2=\beta_z^2-\beta_2^2$, for both symmetric and antisymmetric field distributions, the relation between α_2 and β_{1x} can also be expressed as

$$\alpha_2^2=\beta_1^2-\beta_2^2-\beta_{1x}^2=\left(\frac{2\pi}{\lambda}\right)^2\left(n_1^2-n_2^2\right)-\beta_{1x}^2 \tag{6.7.14}$$

which is known as the dispersion relation. Based on the combination of Eqs. (6.7.12) through (6.7.14) the coefficients of electric and magnetic field distributions shown in Eqs. (6.7.1)–(6.7.9) can be determined.

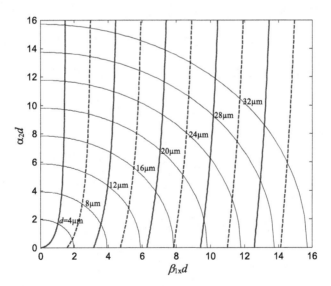

FIG. 6.7.2

Solutions of Eqs. (6.7.12) (bold solid lines) and (6.7.13) (dashed lines) for relationship between normalized propagation constants α_2d and $\beta_{1x}d$ for symmetric and antisymmetric field profiles, respectively. Thin solid lines (quarter circles) represent dispersion relation of Eq. (6.7.14) for $n_1=1.47$, $n_2=1.465$, $\lambda=1550\,\text{nm}$, and $d=4$, 8, and $32\,\mu\text{m}$.

Bold lines in Fig. 6.7.2 show mode-cutoff conditions which are numerical solutions of Eqs. (6.7.12) (solid lines) and (6.7.13) (dashed lines). These curves represent relationship between transversal direction propagation constants α_2 and β_{1x} of different modes in the core and cladding corresponding to symmetric and antisymmetric field profiles, respectively. Normalized values $\alpha_2 d$ and $\beta_{1x}d$ are used in Fig. 6.7.2 for convenience because they are unit-less, and modes exist only in regions of $\beta_{1x}d$, where $\alpha_2 d$ are nonzero. Quarter circles shown in Fig. 6.7.2 show dispersion relations of Eq. (6.7.14) with $n_1 = 1.47$, $n_2 = 1.465$, $\lambda = 1550\,\text{nm}$, and the half-thickness of the core values d changes from 4 to $32\,\mu\text{m}$ in a 4-μm increment. Actual propagation constants, β_{1x}, of discrete modes can be found at intersects of the two groups of curves between mode-cutoff conditions and dispersion relations.

The mode-cutoff condition is $\alpha_2 \leq 0$ when optical field is no longer guided by the core layer because there is no decay of evanescent field in the cladding. Based on Eqs. (6.7.12) and (6.7.13), guided waves exist in the core only when $\alpha_2 > 0$ so that $\tan(\beta_{1x}d) \geq 0$ and $\cot(\beta_{1x}d) \leq 0$ for symmetric and antisymmetric field profiles, respectively. These are equivalent to $(m-1)\pi/2 \leq \beta_{1x}d \leq m\pi/2$ for symmetric modes and $m\pi/2 \leq \beta_{1x}d \leq (m+1)\pi/2$ for antisymmetric modes, respectively, with $m = 1, 3, 5, 7 \dots$.

The mode of the lowest order is the one with symmetric field profile and $m = 1$, and the cutoff condition is $0 \leq \beta_{1x}d \leq \pi/2$. Further considering the dispersion relation of Eq. (6.7.14) at $\alpha_2 = 0$, $0 \leq (2\pi d/\lambda)\sqrt{n_1^2 - n_2^2} \leq \pi/2$. Therefore, this lowest-order mode supports optical signal in the wavelength region $4d\sqrt{n_1^2 - n_2^2} \leq \lambda \leq \infty$, and the cutoff wavelength of the lowers order mode is thus defined as $\lambda_c = 4d\sqrt{n_1^2 - n_2^2}$.

Although mode-cutoff condition discussed above helps understand parameter regions in which guide modes exist, each propagation mode has a unique propagation constant. Substitute α_2 of Eq. (6.7.14) into Eqs. (6.7.12) and (6.7.13), a value of β_{1x} can be found for each guided mode through,

$$\sqrt{\left(\frac{2\pi}{\lambda}\right)^2 \left(\frac{n_1^2 - n_2^2}{\beta_{1x}^2}\right) - 1} = \tan(\beta_{1x}d) \tag{6.7.15}$$

and

$$\sqrt{\left(\frac{2\pi}{\lambda}\right)^2 \left(\frac{n_1^2 - n_2^2}{\beta_{1x}^2}\right) - 1} = -\cot(\beta_{1x}d) \tag{6.7.16}$$

for symmetric and antisymmetric field profiles, respectively. Both Eqs. (6.7.15) and (6.7.16) have multiple discrete solutions for β_{1x} corresponding to discrete modes.

EXAMPLE 6.7.1

Consider a slab waveguide with the following parameters: $n_1 = 1.47$, $n_2 = 1.465$, and $\lambda = 1550\,\text{nm}$ (same as those used to obtain Fig. 6.7.2). Assume the thickness of the core layer is $8\,\mu\text{m}$, that is, $d = 4\,\mu\text{m}$. Find the propagation constants β_z in the longitudinal z-direction for the symmetric and the antisymmetric field modes, and the field distributions of these two modes.

Solution

Based on Fig. 6.7.2, for $d = 4\,\mu\text{m}$, there is only one solution for symmetric field mode and one solution for antisymmetric field mode. Numerically solving Eqs. (6.7.15) and (6.7.16) we can find the normalized propagation constants in the x-direction in the core as $\beta_{1x,s}d = 1.023$ and $\beta_{1x,a}d = 1.874$ for symmetric and antisymmetric modes, respectively. The corresponding propagation constants of this mode along the longitudinal z-direction can be found through $\beta_z = \sqrt{(2\pi n_1/\lambda)^2 - \beta_{1x}^2}$. This results in $\beta_{z,s} = 5.9534 \times 10^6\,\text{m}^{-1}$ and $\beta_{z,s} = 5.9404 \times 10^6\,\text{m}^{-1}$. The corresponding decay constants in the cladding in the transversal x-direction are $\alpha_{2,s} = 4.19215 \times 10^5\,\text{m}^{-1}$, and $\alpha_{2,a} = 1.4657 \times 10^5\,\text{m}^{-1}$, which can be obtained from Eqs. (6.7.12) and (6.7.13). The relation between the proportionality constants A and B can be determined as $B/A = \cos(\beta_{1x,s}d) = 0.5208$ and $B/A = \sin(\beta_{1x,a}d) = 0.9544$ for symmetric and antisymmetric modes, respectively. Based on Eqs. (6.7.1), (6.7.2), (6.7.4), and the parameters calculated above for this example, E_y field distributions of the lowest-order symmetric (dotted line) and antisymmetric (solid line) modes are shown in Fig. 6.7.3. Red and black colors of the curves represent field inside and outside of the slab core layer, respectively. As the antisymmetric mode is near cutoff, more energy is outside of the core region and the decay rate $\alpha_{2,a}$ in the cladding is also smaller than that of the symmetric mode.

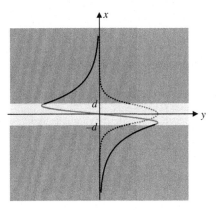

FIG. 6.7.3

Field distributions of lowest-order symmetric *(dashed line)* and antisymmetric *(solid line)* modes on the cross section of a slab waveguide. *Red* and *black* colors represent field inside and outside of the core layer, respectively.

Note that β_z is the projection of $\beta_1 = 2\pi n_1/\lambda$ in the wave propagation direction z, so that $\beta_z \leq \beta_1$. Now, we introduce the concept of *effective index* n_{eff}: when a wave is propagating in the core of a waveguide in the z-direction, it can be treated as if the wave is propagating in free space filled with a material of refractive index n_{eff}. That is, $n_{eff} = \beta_z/(2\pi/\lambda)$. In Example 6.7.1 discussed above, $\beta_{z,s} = 5.9534 \times 10^6 \mathrm{m}^{-1}$ and $\beta_{z,\,s} = 5.9404 \times 10^6 \mathrm{m}^{-1}$ for symmetric and antisymmetric modes, respectively. Thus, we can find $n_{eff,\,s} = 1.4686$ and $n_{eff,a} = 1.4654$ for these two modes. This can be understood as a part of the optical field is outside the core layer so that $n_2 \leq n_{eff} \leq n_1$. As the antisymmetric mode has more energy in the cladding than the symmetric mode, its effective index is smaller.

As the field distribution of each mode is partitioned between the core and the cladding of the waveguide, effective index can be explained as the weighted average of the refractive indices of the core and the cladding with the mode field distribution as the weighting factor. For a low-order mode, the mode field is mostly concentrated inside the core so that the effective index is very close to n_1, whereas for a high-order mode, there is more spreading of mode field into the cladding and thus the effective index will be relatively low.

Another group of modes in the slab waveguide is the TM mode in which $H_z = 0$. H_y component is used for TM mode analysis using continuity conditions on the core/cladding interfaces. Similar to the analysis of TE modes, H_y component which should have standing wave patterns along the transversal x-direction inside the core layer, and these patterns can either be symmetric or antisymmetric. Field in the cladding is an evanescent wave, so that the amplitude has to decay with the increase of $|x|$. Thus, for symmetric modes

$$H_y = \begin{cases} A\cos(\beta_{1x}d)\exp[-\alpha_2(x-d)]e^{j\beta_z z} & x \geq d \\ A\cos(\beta_{1x}x)e^{j\beta_z z} & |x| \leq d \\ A\cos(\beta_{1x}d)\exp[\alpha_2(x+d)]e^{j\beta_z z} & x \leq d \end{cases} \tag{6.7.17}$$

and for antisymmetric modes,

$$H_y = \begin{cases} A\sin(\beta_{1x}d)\exp[-\alpha_2(x-d)]e^{j\beta_z z} & x \geq d \\ A\sin(\beta_{1x}x)e^{j\beta_z z} & |x| \leq d \\ -A\sin(\beta_{1x}d)\exp[\alpha_2(x+d)]e^{j\beta_z z} & x \leq d \end{cases} \tag{6.7.18}$$

where A is the peak amplitudes of the field in the core. Continuity of amplitude across the core/cladding interface at $x = \pm d$ has been used. Electric field components E_x and E_z can then be found through Ampere's law.

For symmetric modes,

$$E_x = \frac{\beta_z}{\omega \varepsilon} H_y = \frac{A\beta_z}{\omega \varepsilon_0} \begin{cases} \frac{1}{n_2^2} \cos(\beta_{1x}d) \exp[-\alpha_2(x-d)]e^{j\beta_z z} & x \geq d \\ \frac{1}{n_1^2} \cos(\beta_{1x}x)e^{j\beta_z z} & |x| \leq d \\ \frac{1}{n_2^2} \cos(\beta_{1x}d) \exp[\alpha_2(x+d)]e^{j\beta_z z} & x \leq d \end{cases} \tag{6.7.19}$$

$$E_z = \frac{j}{\omega \varepsilon} \frac{\partial H_y}{\partial x} = \frac{jA}{\omega \varepsilon_0} \begin{cases} \frac{-\alpha_2}{n_2^2} \cos(\beta_{1x}d) \exp[-\alpha_2(x-d)]e^{j\beta_z z} & x \geq d \\ \frac{-\beta_{1x}}{n_1^2} \sin(\beta_{1x}x)e^{j\beta_z z} & |x| \leq d \\ \frac{\alpha_2}{n_2^2} \cos(\beta_{1x}d) \exp[\alpha_2(x+d)]e^{j\beta_z z} & x \leq d \end{cases} \tag{6.7.20}$$

and for antisymmetric modes,

$$E_x = \frac{A\beta_z}{\omega \varepsilon_0} \begin{cases} \frac{1}{n_2^2} \sin(\beta_{1x}d) \exp[-\alpha_2(x-d)]e^{j\beta_z z} & x \geq d \\ \frac{1}{n_1^2} \sin(\beta_{1x}x)e^{j\beta_z z} & |x| \leq d \\ \frac{-1}{n_2^2} \sin(\beta_{1x}d) \exp[\alpha_2(x+d)]e^{j\beta_z z} & x \leq d \end{cases} \tag{6.7.21}$$

$$E_z = \frac{j}{\omega \varepsilon} \frac{\partial H_y}{\partial x} = \frac{jA}{\omega \varepsilon_0} \begin{cases} \frac{-\alpha_2}{n_2^2} \sin(\beta_{1x}d) \exp[-\alpha_2(x-d)]e^{j\beta_z z} & x \geq d \\ \frac{\beta_{1x}}{n_1^2} \cos(\beta_{1x}x)e^{j\beta_z z} & |x| \leq d \\ \frac{\alpha_2}{n_2^2} \sin(\beta_{1x}d) \exp[\alpha_2(x+d)]e^{j\beta_z z} & x \leq d \end{cases} \tag{6.7.22}$$

The definition of β_{1x} and α_2 are the same as those in the analysis of TE modes. The magnetic field H_z has to be continuous at $x = d$, so that for symmetric modes,

$$\alpha_2 = \frac{n_2^2}{n_1^2} \beta_{1x} \tan(\beta_{1x}x) \tag{6.7.23}$$

and for antisymmetric modes,

$$\alpha_2 = -\frac{n_2^2}{n_1^2} \beta_{1x} \cot(\beta_{1x}x) \tag{6.7.24}$$

Note that the relation between β_{1x} and α_2 for TM modes defined by Eqs. (6.7.23) and (6.7.24) are similar to that for TE modes defined in Eqs. (6.7.12) and (6.7.13), and they are commonly referred to as characteristic equations of modes. A propagation mode exists only when $\tan(\beta_{1x}d) \geq 0$ and $\cot(\beta_{1x}d) \leq 0$ for symmetric and antisymmetric field profiles, respectively. Dispersion Eq. (6.7.14) is applicable for both TE and TM modes, and similar to the V-number of a cylindrical fiber defined in Eq. (2.3.25), a V-number can also be defined for a slab waveguide as

$$V = \frac{2\pi d}{\lambda} \sqrt{n_1^2 - n_2^2} \tag{6.7.25}$$

Based on the characteristic equations, TE and TM modes have the same cutoff condition, $V = m\pi/2$ for symmetric modes and $V = (m+1)\pi/2$ for antisymmetric modes, respectively, with $m = 1, 3, 5, 7 \ldots$. The number of modes increases for every $\pi/2$ radians increase in the V-number so that the total number of modes in a slab waveguide is

$$N = \frac{2V}{\pi} = \frac{4d}{\lambda} \sqrt{n_1^2 - n_2^2} \tag{6.7.26}$$

Note that since characteristic equations of TM modes shown in Eqs. (6.7.23) and (6.7.24) are slightly different from those for the TE modes by a coefficient $(n_2/n_1)^2$, the effective indices of TE and TM modes of the same order are thus slightly different.

Another, probably more straightforward, way to find mode characteristic equation is to use geometric analysis technique based on the round-trip phase matching of a light ray in the transversal x-direction as illustrated in Fig. 6.7.4.

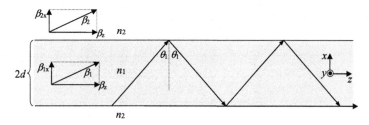

FIG. 6.7.4

Illustration of ray traces in a slab waveguide with a thickness of $2d$, and the decomposition of propagation constant into transversal x and longitudinal z-directions.

In Fig. 6.7.4, the wave vectors β_1 and β_2 for the core and the cladding, respectively, are decomposed into the longitudinal and the transversal components. For a guided mode inside the waveguide core, the round-trip phase change of light ray in the transversal direction has to be the multiple of 2π, that is,

$$4\beta_{1x}d + 2\Delta\Phi = 2m\pi \tag{6.7.27}$$

where m is an integer and $\Delta\Phi$ is the phase shift of reflection at the core/cladding interface which is defined by Eqs. (2.2.17) and (2.2.18) for the electric fields parallel and perpendicular to the incidence plane (which is x/z plane in this case), respectively. As an example, for a TM mode the major electric field components are E_x and E_z which are both parallel to the incidence plane, and thus,

$$\Delta\Phi = \Delta\Phi_{//} = -2\tan^{-1}\left(\frac{n_1\sqrt{n_1^2\sin^2\theta_1 - n_2^2}}{n_2^2\cos\theta_1}\right) = -2\tan^{-1}\left(\frac{n_1^2\,\alpha_2}{n_2^2\,\beta_{1x}}\right) \tag{6.7.28}$$

where θ_1 is the incidence angle and we have used the decay factor in the cladding as $\alpha_2^2 = (j\beta_{2x})^2 = \beta_z^2 - \beta_2^2$ with $\beta_z = \beta_1\sin\theta_1$, $\beta_1 = 2\pi n_1/\lambda$, and $\beta_{1x} = \beta_1\cos\theta_1$. Eq. (6.7.28) can then be expressed as

$$\alpha_2 = \frac{n_2^2\beta_{1x}}{n_1^2}\tan\left(\beta_{1x}d - \frac{m\pi}{2}\right) \tag{6.7.29}$$

If m is an even number, $\tan(\beta_{1x}d - m\pi/2) = \tan(\beta_{1x}d)$ so that,

$$\alpha_2 = \frac{n_2^2\beta_{1x}}{n_1^2}\tan(\beta_{1x}d) \tag{6.7.30}$$

which is identical to the characteristic Eq. (6.7.23) for symmetric optical field distribution. Similarly for a odd number of m, $\tan(\beta_{1x}d - m\pi/2) = -\cot(\beta_{1x}d)$, so that,

$$\alpha_2 = -\frac{n_2^2\beta_{1x}}{n_1^2}\cot(\beta_{1x}d) \tag{6.7.31}$$

which is identical to characteristic Eq. (6.7.24) for antisymmetric optical field distribution. For TE modes, on the other hand, the major electric field component is E_y which is perpendicular to the incidence plane. Substitute the expression of $\Delta\Phi_\perp$ from Eq. (2.2.18) into Eq. (6.7.27), characteristic equations of TE modes can be found.

6.7.2 RECTANGLE OPTICAL WAVEGUIDES

Although slab waveguide is simple, it only provides one-dimensional (1D) optical field confinement, and thus its application in photonic circuits is limited. Instead, the most useful structure in PLC is the rectangular waveguide. Fig. 6.7.5 shows the cross section of a rectangular dielectric optical waveguide, which can be divided into nine regions with the orientation of coordination based on Marcatili (1969). The core has a high refractive index n_1, while refractive indices in the cladding regions are n_2, n_3, n_4, and n_5, respectively, which are lower than that in the core. The height and the width of the waveguide core are $2d$ and $2w$, respectively. Optical fields in the four corner regions (regions 6, 7, 8, and 9) are considered small enough so that they can be neglected for simplicity.

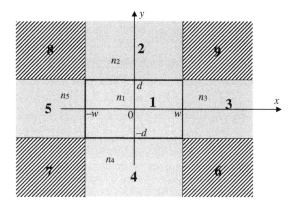

FIG. 6.7.5

Cross section of a rectangle optical waveguide with a thickness $2d$ and a width $2w$.

As the guided-wave optical field is confined in both x- and y-directions in this 3D configuration, the propagation constant β has to be decomposed into three orthogonal components β_x, β_y, and β_z. In the waveguide core (region 1), this decomposition is

$$\left(\frac{2\pi n_1}{\lambda}\right)^2 = \beta_{1x}^2 + \beta_{1y}^2 + \beta_z^2 \tag{6.7.32}$$

Whereas, in the cladding regions 2 and 4, the decay constants are

$$\alpha_2 = \sqrt{\beta_z^2 + \beta_{1x}^2 - \left(\frac{2\pi n_2}{\lambda}\right)^2} \tag{6.7.33}$$

$$\alpha_4 = \sqrt{\beta_z^2 + \beta_{1x}^2 - \left(\frac{2\pi n_4}{\lambda}\right)^2} \tag{6.7.34}$$

Note that because of the field continuity, propagation constant in the x-direction β_{1x} has to be continuous across n_2/n_1 and n_4/n_1 boundaries. Similarly in cladding regions 3 and 5, the decay constants are

$$\alpha_3 = \sqrt{\beta_z^2 + \beta_{1y}^2 - \left(\frac{2\pi n_3}{\lambda}\right)^2} \tag{6.7.35}$$

$$\alpha_5 = \sqrt{\beta_z^2 + \beta_{1y}^2 - \left(\frac{2\pi n_5}{\lambda}\right)^2} \tag{6.7.36}$$

where propagation constant in the y-direction β_{1y} is continuous across n_3/n_1 and n_5/n_1 boundaries. β_z is the same for all five regions in which electric fields are considered. Now, let's consider a TM-like modes which has the major electric field of E_x. Based on the geometric analysis, the round-trip phase match condition in the x-direction is

$$4\beta_{1x}w - 2\tan^{-1}\left(\frac{n_1^2\alpha_3}{n_3^2\beta_{1x}}\right) - 2\tan^{-1}\left(\frac{n_1^2\alpha_5}{n_5^2\beta_{1x}}\right) = 2m_x\pi \qquad (6.7.37)$$

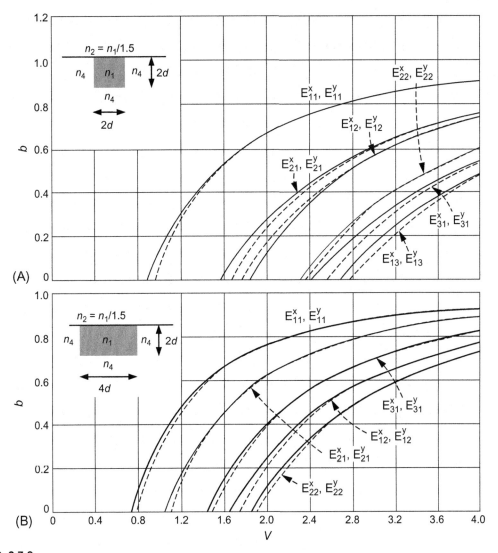

FIG. 6.7.6

Normalized propagation constant for different modes for two different waveguides. These are numerical solutions of Eqs. (6.7.32)–(6.7.38) (Marcatili, 1969). (A) A square waveguide, and (B) a rectangular waveguide with the width twice the length of the thickness.

In this case for the reflections on the interfaces of n_1/n_3 and n_1/n_5, the electric field is parallel to the incidence plane. While for the reflections on the interfaces of n_1/n_2 and n_1/n_4, the electric field is perpendicular to the incidence plane so that the round-trip phase match condition in the y-direction is

$$4\beta_{1y}d - 2\tan^{-1}\left(\frac{\alpha_2}{\beta_{1y}}\right) - 2\tan^{-1}\left(\frac{\alpha_4}{\beta_{1y}}\right) = 2m_y\pi \qquad (6.7.38)$$

where m_x and m_y are integers. Now the seven Eqs. (6.7.32)–(6.7.38), are sufficient to solve for the seven variables, β_{1x}, β_{1y}, β_z, α_2, α_3, α_4, and α_5. Since there is no analytical solution to these equations, they have to be solved numerically.

Fig. 6.7.6 show the normalized propagation constant b as the function of the V-number for various different modes calculated by numerically solving Eqs. (6.7.32)–(6.7.38). Results shown in Fig. 6.7.6 are restricted to relatively weak-guiding waveguides with $n_4 < n_1 < 1.05n_4$. The insets show waveguide cross sections with the core thickness $2d$ and widths $w = 2d$ and $w = 4d$, respectively, for the top and the bottom parts of Fig. 6.7.6. Since the refractive index of the core is n_1, the top cladding layer has an index of n_2, and all other regions have the same index of n_4, the definitions of b and V in Fig. 6.7.6 are, respectively,

$$b = \frac{\beta_z^2 - \beta_4^2}{\beta_1^2 - \beta_4^2} \qquad (6.7.39)$$

with $\beta_1 = 2\pi n_1/\lambda$ and $\beta_4 = 2\pi n_4/\lambda$, and

$$V = \frac{4\pi d}{\lambda}\sqrt{n_1^2 - n_4^2} \qquad (6.7.40)$$

Fig. 6.7.6 indirectly shows the relationship between the z-directional propagation constant β_z and the bottom cladding refractive index n_4 because V is the function of n_4 and b is a function of β_z. The notations of modes in Fig. 6.7.6 are expressed by E_{ij}^x for TM modes and E_{ij}^y for TE modes, where i and j represent the number of extrema in the x- and y-directions inside the waveguide core. A propagation mode exists only when $0 < b < 1$. The lowest-order TE and TM modes are E_{11}^y and E_{11}^x, respectively. Field distributions of E_{ij}^x with $i,j = 1$ and 2,

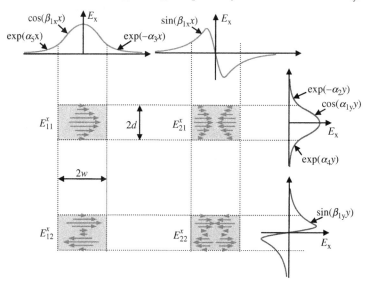

FIG. 6.7.7

Illustration of electric field E_x distribution of low-order TM modes on the waveguide cross section.

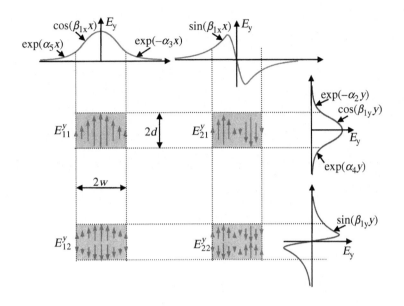

FIG. 6.7.8

Illustration of electric field E_y distribution of low-order TE modes on the waveguide cross section.

on the waveguide cross section are shown in Fig. 6.7.7 for the TM modes in which E_x and H_y are major field components.

Similarly, field distributions of E_{ij}^y with $i, j = 1$ and 2 on the waveguide cross section are shown in Fig. 6.7.8 for the TE modes in which E_y and H_x are major field components.

6.7.3 DIRECTIONAL COUPLERS

Directional coupler is a basic function in an integrated photonic circuit, in which energy of the optical signal is coupled between adjacent optical waveguides. As illustrated in Fig. 6.7.9, energy transfer can happen between two waveguides parallel to each other when they are close enough. This energy coupling is accomplished through evanescent waves of the two waveguide modes outside their cores.

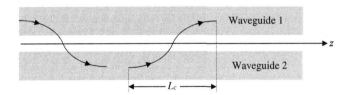

FIG. 6.7.9

Illustration of energy transfer between two parallel waveguides.

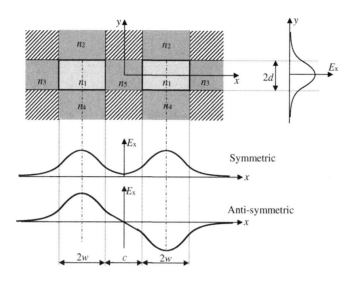

FIG. 6.7.10

Cross section of two identical rectangular optical waveguides parallel to each other, and field distributions of the symmetric and antisymmetric modes.

Fig. 6.7.10 shows the cross section of two identical rectangular optical waveguides parallel to each other. Each waveguide core has a refractive index n_1, a width $2w$, and a thickness $2d$. Separation between these two waveguides is c. Refractive indices of cladding regions are indicated in the figure. Considering only the electric field E_{11}^y of the lowest-order TE mode in the coupled waveguide configuration, E_{11}^y can be either symmetric or antisymmetric as illustrated in Fig. 6.7.10 as both of these two mode structures are allowed to exist.

Coupling length is a parameter that is defined as the length at which optical signal power is completely transferred from one waveguide to the other. It can be found that (Marcatili, 1969) the coupling length is

$$L_p = \frac{\pi}{2} \frac{\beta_5}{\beta_{1x}^2} \frac{w}{\alpha_5} \frac{1+\beta_{1x}^2 \alpha_5^2}{\exp(-c/\alpha_5)} \tag{6.7.41}$$

where $\beta_5 = 2\pi n_5/\lambda$, α_5 is the decay factor in the cladding region between the two waveguides, and β_{1x} is the propagation constant in the transversal x-direction. Eq. (6.7.41) was derived with the weak coupling approximation such that both the fundamental field profile and the propagation constant of each individual waveguide mode are not significantly changed by the introduction of the other waveguide. Thus, α_5 and β_{1x} can be found by solving Eqs. (6.7.32)–(6.7.38). Eq. (6.7.41) indicates that $L_p \propto e^{c/\alpha_5}$, which means that the coupling length exponentially increases with the separation c between the two waveguides.

Analytic solutions based on the approximations can be found in the literature for various different guided-wave photonic structures, which is useful to understand wave propagation, mode formation, and coupling mechanisms. However, complexity of analytic formulations can often be overwhelming for even relatively simple photonic structures. Thus, numerical simulations are commonly used for photonic circuit design and performance evaluation. Various numerical simulation packages are commercially available based on the finite element method (FEM), or finite-difference time-domain method (FDTD).

As an example, Fig. 6.7.11A shows the cross section of a ridge waveguide and the refractive index of each layer is color coded with the color bar on the right side of Fig. 6.7.11A. Fig. 6.7.11B shows mode field distributions numerically solved by FDTD software [synopsys webpage] for all five modes that the waveguide can support. The effective index n_{eff} is also shown on top of each mode field profile, which indicates the extent of mode field confinement in the core.

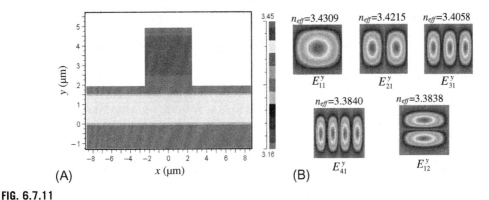

FIG. 6.7.11

(A) Index profile of a ridge waveguide and (B) mode profiles of y-polarized optical field and effective index of each mode (Synopsys simulation software generated).

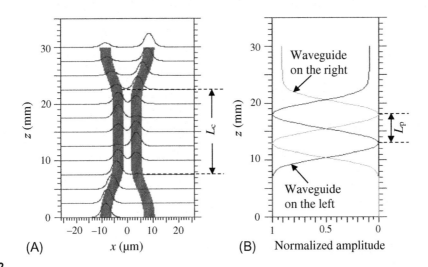

FIG. 6.7.12

(A) Top view of a 2×2 waveguide coupler (red) and optical field distribution at different positions along the waveguides. At the input ($z = 0$), optical field is launched into the waveguide on the left side. (b) Optical field amplitudes carried in the two waveguides at different positions along the waveguides.

Fig. 6.7.12A shows the top view of a directional coupler composed of two optical waveguides with varying separation along the wave propagation direction z. Only the fundamental mode is considered in each waveguide. At the input ($z=0$), the fundamental mode field is launched into the waveguide on the left. At the location where the separation between these two waveguides is close enough, energy coupling between them happens. In the region of $7 < z < 20$ mm optical signal amplitude oscillates between these two waveguides with a periodicity of approximately 10 mm which is known as the coupling region. The length L_p required for a complete energy transfer from one waveguide to the other is determined by the mode structure and the separation between these two waveguides in the coupling region L_p in Fig. 6.7.12B is approximately 5 mm in this example. By choosing an appropriate length of coupling region, L_c in Fig. 6.7.12C, the power ratio between the two outputs of the coupler can be determined.

Fig. 6.7.13 shows an example of a 1×3 interference coupler in which the coupling region is a multimode waveguide, and energy splitting from the input waveguide to a number of output waveguides relies on the modal distribution of the multimode waveguide section. This multimodal field distribution is shown in Fig. 6.7.13B, which is similar to the interference pattern in free-space optics. Similar star couplers based on the multimode interference have been used in AWG to make WDM multiplexers and demultiplexers as previously discussed in Section 6.5.2. More sophisticated PLC structures such as complex waveguide cross sections and plasmonic waveguides based on the metallic properties of materials can also be designed with numerical simulation software (synopsys webpage).

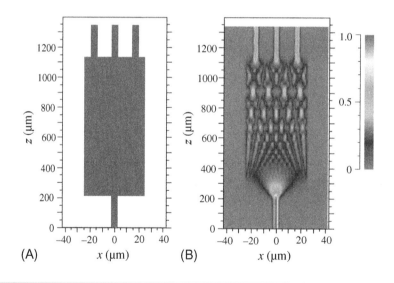

FIG. 6.7.13

(A) Top view of a 1×3 interference coupler and (B) multimodal field distribution.

6.7.4 SILICON PHOTONICS

Integrated silicon photonic circuits known as silicon photonics is now widely accepted as a key technology for the next-generation communication systems and networks (Chrostowski and Hochberg, 2015; Shoji et al., 2002). Compared to other material platforms, a distinctive advantage of silicon photonics is the ability to use CMOS fabrication technology (so-called CMOS compatible) so that photonic circuits can be monolithically integrated with CMOS electronic circuits, and thus be produced in high volume at low cost.

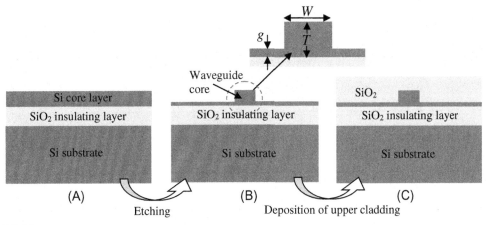

FIG. 6.7.14

(A) SOI wafer layer structure and (B) top layer of the wafer is etched into optical waveguide of width W and thickness T. Silicon-wire waveguide is defined when $g=0$, and silicon-rib waveguide is defined when $T>g>0$.

Integrated silicon photonic circuits are usually fabricated on standard SOI wafer as shown in Fig. 6.7.14A. The SOI wafer has a SiO_2 layer sandwiched between a silicon substrate and a thin silicon core layer on top. The bottom silicon substrate serves as a mechanical support in this structure, and the SiO_2, known as the buried oxide (BOX), layer with a lower index than silicon is a lower cladding which isolates the core layer from the substrate. The thickness of the BOX layer has to be much larger than the signal wavelength so that evanescent wave of the optical signal in the waveguide does not penetrate into the silicon substrate. Optical waveguides can be made by photolithography and etching as shown in Fig. 6.7.14B, and the waveguide thickness T is determined by the thickness of the top silicon layer, which is usually a few hundred nanometers. The most used SOI wafer has the top-layer thickness of 220 nm, which has been suggested in an effort to standardize the fabrication process. Although the thickness of the waveguide, T, cannot be changed across the entire wafer, the width, W, of the waveguide can be much flexible which can be set by optical circuit design and fabrication through photolithography. For the waveguide structure shown in Fig. 6.7.14B, the top silicon layer is not completely etched out on both sides of the waveguide core and the residual thickness is g. For $g=0$, the waveguide is usually referred to as *wire* waveguide, whereas for $T>g>0$ the waveguide is generally referred to as *rib* waveguide. Finally, an upper-cladding layer, also based on the SiO_2 material, is deposited on top of the silicon waveguide as shown in Fig. 6.7.14C.

As the refractive index of silicon in the 1550-nm wavelength region is known, the effective index n_c of a silicon wire waveguide, which depends on the cross-section width and thickness, can be calculated. Fig. 6.7.15A shows the effective index n_c of the fundamental mode E_{11} and the second-order mode E_{21} as the function of core width W for a fixed core thickness $T=220$ nm. It also shows the derivative

dn_c/dW of the fundamental mode on the right y-axis which indicates the sensitivity of effective index against the small change of waveguide width that may be caused due to fabrication errors or waveguide wall roughness. Similarly Fig. 6.7.15B shows the effective index n_c and its derivative as the function of core thickness T for a fixed core width $W = 500$ nm.

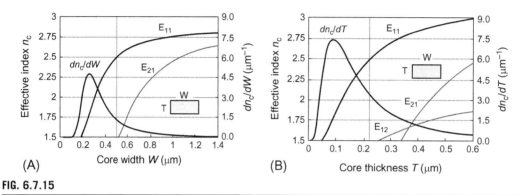

FIG. 6.7.15

Effective index (left y-axis) and its derivative (right y-axis) for a silicon wire waveguide. (A) As the function of core width W with a fixed thickness $T = 220$ nm and (B) as the function of core thickness T with a fixed width $W = 500$ nm (Okamoto, 2014).

As the refractive index of the silica under cladding layer of the waveguide is approximately 1.5, the cutoff condition can be found from Fig. 6.7.15 with $n_c > 1.5$ for the E_{11} mode so that it can be supported, and $n_c \leq 1.5$ for the E_{21} mode so that it is attenuated. For the standard 220 nm thickness of the silicon core layer, the width of the waveguide has to be narrower than 500 nm so that the E_{21} mode is not supported.

Thanks to the high-index difference between the core and the cladding, optical field concentration in the core of the silicon wire waveguide is much tighter than in silica-based optical waveguides. The minimally allowable bending radius, on the order of a few micrometers for silicon wire waveguides, is also much smaller than that of silica waveguides which is typically a few millimeters. Thus, footprint of integrated silicon photonic circuits on the SOI wafer can be several orders of magnitude smaller than silica-based PLC.

However, small waveguide cross sections of silicon wire waveguides pose stringent requirements on the wafer thickness uniformity and fabrication precision. In addition, small waveguide cross section also makes it difficult for optical coupling from and into optical fibers with a 9-μm cross-section diameter. One way to enlarge silicon waveguide cross section while maintaining single-mode condition is to use rib waveguide as shown in Fig. 6.7.14B with $g > 0$. A silicon-rib waveguide can be single mode when $W/T < 0.3 + (g/T)/\sqrt{1 - (g/T)^2}$ and $0.5 \leq g/T < 1$ (Okamoto, 2014). As an example, single-mode operation will be maintained for a silicon-rib waveguide with cross-section parameters $T = W = 3$ μm and $g = 1.8$ mm. This large silicon waveguide cross section helps optical coupling with optical fibers, and greatly relaxes fabrication tolerance, as $dn_c/dW \leq 5.5 \times 10^{-3}$ μm^{-1} which is three orders of magnitude lower than that of a silicon-wire waveguide shown in Fig. 6.7.15. However, the minimally allowed bending radius is also increased by about three orders of magnitude so that the footprint of the optical circuit can become much larger, and thus, silicon-rib waveguides are rarely used in practical silicon photonic circuits.

Mode field mismatch between a single-mode optical fiber and the high-index-contrast silicon wire waveguide makes it difficult to couple light in and out of a silicon photonic chip. A number of techniques have been used to improve the coupling efficiency (Carroll et al., 2016), including tapered fiber, tapered waveguide, and grating coupler, to name a few. Fig. 6.7.16A shows an edge-coupling structure of optical coupling from a tapered fiber to tapered waveguide through a mode converter. In this structure, the taper section of the waveguide is 100–300 μm long which is embedded in an integrated polymer or nitride-based spot-size converter. This spot-size converter increases the effective mode field diameter of the silicon waveguide to approximately 3×3 μm, so that a good modal overlap can be achieved with a tapered optical fiber (Shoji et al., 2002; Pu et al., 2010).

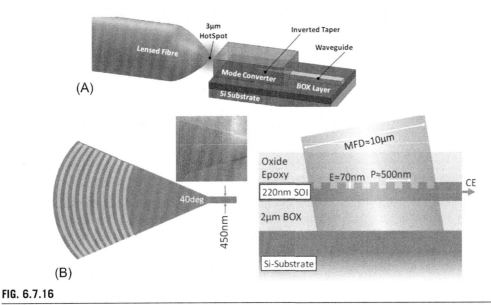

FIG. 6.7.16

(A) Edge-coupling structure of optical coupling from a tapered fiber to tapered waveguide through a mode converter and (B) grating based out-of-plane coupling (Carroll et al., 2016).

Fig. 6.7.16B shows the configuration of grating based out-of-plane coupling, where a sub-μm periodic structure is lithographically etched into the waveguide layer to create a coherent interference condition that diffractively couples the fiber mode into the silicon waveguide (Lee et al., 2016). The peak coupling wavelength λ_p can be found as: $\lambda_p = P(n_e - n_o \cos \theta)$, where P is the pitch of the trenches, n_e is the effective index of the grating-coupler region, n_o is the index of the oxide-layer, and θ is the incidence angle of the fiber. The value of n_e is determined by the etch depth and duty cycle of the trenches, as well as the polarization of the fiber mode (Van Laere et al., 2007; Roelkens et al., 2011). Typically, a near-normal angle-of-incidence $\theta \approx 10°$ is used, to minimize back reflections into the fiber.

6.8 OPTICAL SWITCHES

Optical switching is a basic functionality needed in most optical communication systems and networks. An optical switch can be considered as an optical directional coupler but with binary splitting ratio which is determined by a control signal. Fig. 6.8.1 compares a 1×2 optical directional coupler and a 1×2 optical switch. Ideally without considering excess loss, both a directional coupler and a switch should be energy conserved with $(P_{out,1} + P_{out,2}) = P_{in}$. For a directional coupler, power-splitting ratio $\eta = P_2/P_3$ is often a fixed value determined by coupler design, while a switch operates in one of the two binary states, $P_{out,1} = P_{in}$ (with $P_{out,2} = 0$) and $P_{out,2} = P_{in}$ (with $P_{out,1} = 0$), depending on the control signal. Practically, an optical switch has both insertion loss and cross talk. For a simplest 1×2 optical switch, insertion losses are defined by $\alpha_1 = P_{out, 1}/P_{in}$ and $\alpha_2 = P_{out, 2}/P_{in}$ for the output ports 1 and 2 switched to the *on* state. Correspondingly, the extinction (on/off) ratio, is defined by $\beta_1 = P_{out, 1}/P_{out, 2}$ or $\beta_2 = P_{out, 2}/P_{out, 1}$, respectively, for the two states. Extinction ratio is also often referred to as *cross talk*.

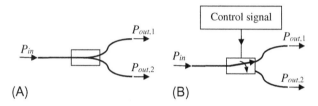

FIG. 6.8.1

Comparison between a 1×2 optical directional coupler (A) and a 1×2 optical switch (B).

For a high-quality switch, the insertion loss should be less than 0.5 dB and the extinction ration should be higher than 30 dB within the specified spectral bandwidth. Other parameters of an optical switch include switching speed, back reflectivity, PMD, PDL, maximum optical power that the switch can handle, and the lifetime which specifies the maximum number of switching cycles the device can be used for. These parameters depend on the type of the switch, the material platform, and the port count of the switch.

6.8.1 TYPES OF OPTICAL SWITCHES

Various technologies have been used to make optical switches for different applications, which include semiconductor, electro-optic, liquid crystal, AO, and micro-electro-mechanical systems (MEMS). Each technology may have its unique advantages and disadvantages, and thus selecting the type of optical switch should fit to the specific application that will be used for. In general, semiconductor or electro-optic switches are used in situations where nanosecond-level switching time is needed, but they are usually expensive and with higher losses. An AO switch can have a switching time on the order of sub-microsecond, but it is relatively bulky. Liquid crystal-based optical switches can be miniaturized and made into arrays with low loss and minimum cross talk, but the speed of a liquid crystal switch is typically on the orders of milliseconds or sub-millisecond. With the rapid advance in micro- and nano-fabrication techniques, MEMS-based optical switches have become highly reliable, with large port counts, and low insertion loss, which enabled large-scale optical circuit switching for network applications. As optical signals are switched by mechanical actuators in MEMS, the switching time is typically on the millisecond level. With that said, faster MEMS switches, with microsecond-level switching time, have been reported by further reducing the masses of the moving parts. Now, let us discuss each of these optical switching devices with more details.

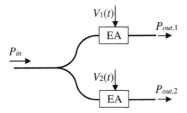

FIG. 6.8.2

Illustration of a 1×2 optical switch based on the EA modulators.

Semiconductor-based optical switches can be made with electro-absorption (EA) or optical interference, similar to the mechanisms of making electro-optic modulators which will be discussed in Chapter 7. In principle, a 1×2 optical switch can be made by using a 3-dB optical coupler and two EA modulators as shown in Fig. 6.8.2. The input optical signal can be switched to the desired output port by minimizing the absorption of the EA modulator at that port and maximizing the EA absorption at that other port through the control of voltage signals $V_1(t)$ and $V_2(t)$. The switching time is determined by the modulation speed of the EA modulators, which can be on the order of sub-nanosecond. However, this optical switch suffers 3-dB splitting loss in addition to the insertion loss of each EA.

A more sophisticated solution is to utilize controllable vertical optical coupling (VOC) between two waveguides, one on top of the other, through evanescent waves. As illustrated in Fig. 6.8.3A, energy

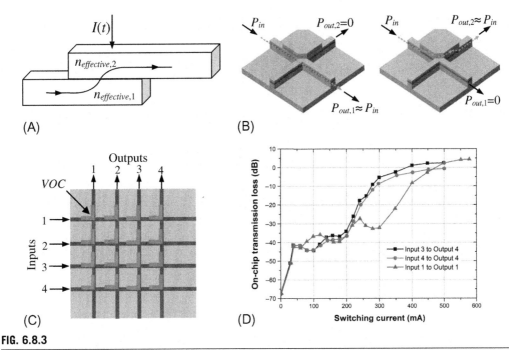

FIG. 6.8.3

(A) Illustration of evanescent-mode coupling between two optical waveguides, (B) optical switching based on the controllable vertical optical coupling (VOC) between two straight waveguides, (C) a 4×4 optical switch fabric based on VOC, and (D) optical loss between different input/output port pairs as the function of injection current into the VOC structure (Chen et al., 2015).

transfer between two parallel optical waveguides through evanescent wave coupling depends on their spatial separation and the mismatch between their effective refractive indices. As both of these waveguides are made by semiconductor materials, their differential refractive index can be controlled by carrier injection into the top waveguide through an electrode, and thus, the efficiency of energy coupling between them can be electrically controlled.

This electrically controlled VOC has been demonstrated to make optical switches with the basic coupling structure shown in Fig. 6.8.3B where the top waveguide is bent by 90 degree. If the top and the bottom waveguides have matched indices, the VOC is able to couple the optical signal from the input waveguide to the waveguide perpendicular to it through the junction. The left part of Fig. 6.8.3B shows that for the *through* state without effective VOC, the input optical power propagates through the straight optical waveguide and exits from output port 1 so that $P_{out,1} \approx P_{in}$. No power is coupled into the perpendicular waveguide, that is, $P_{out,2} \approx 0$. In the *cross* state shown on the right side of Fig. 6.8.3B, the effective VOC redirects the input optical signal to output port 2 so that $P_{out,2} \approx P_{in}$ and $P_{out,1} \approx 0$. Fig. 6.8.3D shows the optical transmission loss between different input/output port pairs as the function of the injection current on the VOC nodes (Chen et al., 2015; Chi et al., 2006). This was measured on a 4×4 optical switch made with VOC as illustrated in Fig. 6.8.3C. The switching time was measured to be less than 10 ns so that this device can be suitable for the applications in packet-switched optical networks.

Optical switching can also be realized by using an interferometer configuration as shown in Fig. 6.8.4. This MZI configuration is similar to that used to make frequency interleavers in Section 6.2, and electro-optic modulators which will be discussed in Chapter 7.

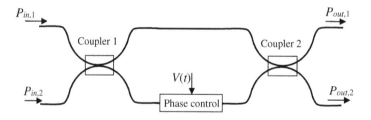

FIG. 6.8.4

2×2 optical switch made by a Mach-Zehnder interferometer structure.

Similar to Eqs. (6.2.6) and (6.2.7), the power transfer functions of the MZI shown in Fig. 6.8.4 are

$$T_{11} = \frac{P_{out,1}}{P_{in,1}} = \sin^2[\varphi_0(\lambda) + \delta\varphi(t)] \tag{6.8.1a}$$

$$T_{12} = \frac{P_{out,2}}{P_{in,1}} = \cos^2[\varphi_0(\lambda) + \delta\varphi(t)] \tag{6.8.1b}$$

$$T_{21} = \frac{P_{out,1}}{P_{in,2}} = \cos^2[\varphi_0(\lambda) + \delta\varphi(t)] \tag{6.8.1c}$$

$$T_{22} = \frac{P_{out,2}}{P_{in,2}} = \sin^2[\varphi_0(\lambda) + \delta\varphi(t)] \tag{6.8.1d}$$

with $\varphi_0(\lambda) = n\pi\Delta L/\lambda$ and $\delta\varphi(t) = \pi V(t)/(2V_\pi)$, where n is the refractive index of both waveguides assuming they are the same, ΔL is the differential length of the two arms, V_π is defined as the voltage required to create a π optical phase change in one of the two MZI arms, which depends on the EO efficiency, and $V(t)$ is the applied voltage.

In general, the transfer functions of the MZI are wavelength dependent, which allows the selection of wavelength components in applications such as frequency interleaver. For the application as a broadband optical switch, $\phi_0(\lambda)$ needs to be minimized by making the interferometer as symmetric as possible so that $\Delta L \approx 0$. The 2×2 optical switch made from an MZI structure can only have two states, *cross* and *bar*, as shown in Fig. 6.8.5A and B, respectively. The only limitation is that the two inputs cannot be routed into the same output port.

Using the 2×2 optical switch as the basic building block, larger non-blocking cross-connect switch fabrics can be built. Base on *Benes* network arrangements, a network with $2N$ input ports can be constructed with $[2\log_2(2N) - 1]N$ units of 2×2 switches arranged in $2\log_2(2N) - 1$ stages with N switches in each stage. Fig. 6.8.5C and D shows 4×4 ($N = 2$) and 8×8 ($N = 4$) cross-connect switches using 6 and 20 2×2 switches, respectively.

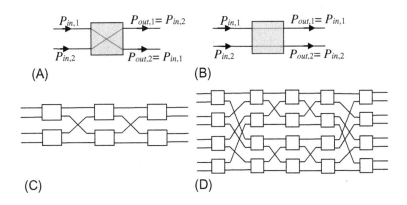

FIG. 6.8.5

Cross (A) and bar (B) states of a 2×2 optical switch. 4×4 (C) and 8×8 (D) non-block switch architecture using 2×2 optical switches as building blocks.

Optical switches based on the MZI structure require optical materials with high enough electro-optic coefficient for efficient switching control with an electrical signal. LiNbO$_3$ is an efficient electro-optic material commonly used for high-speed optical modulators. However, integrating a large number of MZI structures to make switches with large port counts is quite challenging. Integrated silicon photonic circuits have been used to make optical switches based on the MZI structure. A *PIN*-junction built into each arm of the interferometer allows fast electro-optic modulation by changing the control voltage. A switching time of approximately 3 ns has been achieved with this technique (Lu et al., 2016) for a 16×16 cross-connect switch on a single SOI chip.

The major advantage of optical switches built on PLCs, including evanescent coupling in semiconductor waveguides and electro-optic modulation on the MZI structure, is the fast switching time on the nanosecond level, which is suitable for applications in packet-switched optical networks. However, these devices are usually polarization dependent, and with relatively high insertion loss. The extinction ratio is also relatively low which may allow cross talk to build up in a large-scale switch fabric. In addition, switches based on the MZI structure tend to be temperature sensitive and thus monitoring and feedback control is needed to stabilize the operation points (Qiao et al., 2017).

AO switch is based on the interaction between the optical signal and an acoustic wave. AOFs and modulators have been discussed in Section 6.5. An AO switch is similar to an AO modulator but with both of the first-order and the zeroth-order output beams utilized as illustrated in Fig. 6.8.6. In this configuration, the RF source driving the piezoelectric transducer is switched on and off by a binary control signal. When the connection of the RF source is switched off, the input optical beam passes through the AO crystal in the direction of the zeroth-order beam so that $P_{out,1} \approx P_{in}$ and $P_{out,2} = 0$. When the connection of the RF source is switched on, Bragg gratings created by the acoustic wave diffracts the input beam into the direction of the first-order diffraction at the output, so that $P_{out,2} \approx P_{in}$ and $P_{out,1} \approx 0$. In Fig. 6.8.6, the Bragg grating created by the acoustic wave has a grating constant $\Lambda = v_a/F$, with F the frequency of the RF source. As discussed in Section 6.5, the switch time is determined by the optical beam diameter, as well as the velocity v_a of the acoustic wave propagating in the AO crystal, which is typically on the orders of 10s of nanoseconds to sub-microsecond.

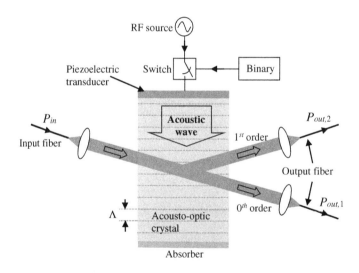

FIG. 6.8.6

Illustration of a 1×2 acousto-optic switch.

The major advantages of AO switch are the relatively fast switching speed, the insensitivity to the SOP of the optical signal, and the high extinction ratio of usually more than 50 dB. However, an AO switch is generally bulky and power inefficient because the need of an RF driving source. Integrating multiple 1×2 AO switches into a large switching matrix is also challenging.

Liquid crystal is another optical material whose optical property can be controlled by an electric voltage signal, and thus liquid crystals have been used to make optical modulators and switches. The most important and unique property of liquid crystal is that the orientation of the long molecules

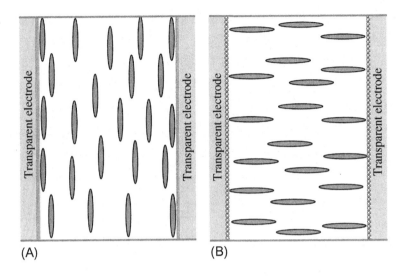

FIG. 6.8.7

Illustration of molecular orientations of nematic-type liquid crystals: (A) homogenous and (B) homeotropic.

can be altered by relatively low amplitudes of electric or magnetic fields. This allows the electronic control of the birefringence which can be utilized to make tunable optical components for a wide range of applications from optical communications to flat screen TV display.

The name of *liquid crystal* suggests that it is a liquid, and thus has to be placed inside containers or cells. Otherwise it would flow away. To a large extent, the property of the cell wall inner surface determines the molecular orientation of the liquid crystal without the application of an external electric or magnetic field. The surface of cell inner wall is usually coated with a thin layer of polymers, such as polyvinyl alcohol, on which microscopic structures can be patterned. Fig. 6.8.7 shows two basic types of molecular orientations of liquid crystal contained in cells. In Fig. 6.8.7A, molecules are aligned parallel to the surface, which is known as *homogenous*. This can be obtained when the cell wall inner surface is printed with fine-structured parallel groves. These parallel groves help anchor the molecules and align them in the same direction. In Fig. 6.8.7B, molecules are aligned perpendicular to the surface, which is known as *homeotropic*. This can be obtained by chemically treating the inner surface of the cell wall with a coupling agent so that one end of the molecules can be anchored on it. This also helps align all other molecules which are not on the wall to the same direction perpendicular to the cell wall.

The alignment of these long molecules in a liquid crystal can easily introduce strong birefringence. Typically when the polarization direction of an optical signal is the same as that of the aligned molecules, the optical signal sees a higher refractive index. Otherwise, if the polarization of the optical signal is perpendicular to the molecular direction, the optical signal sees a lower refractive index. This is commonly known as positive birefringence with $n_e > n_o$.

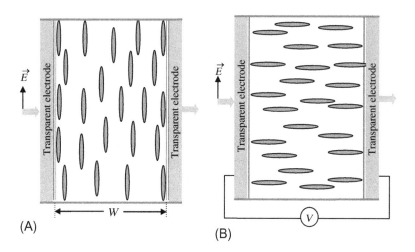

FIG. 6.8.8

(A) A homogenous liquid crystal cell without an applied electric voltage. Light beam is input from the left. (B) A voltage is applied on transparent electrodes across the liquid crystal cell, and molecules are aligned with the electric field.

Now if a an electric field is applied across a liquid crystal cell, the molecular orientation of liquid crystal will be forced to align with the direction of the applied field. This can change the refractive index or birefringence seen by a polarized light. As shown in Fig. 6.8.8A for a homogenous liquid crystal with molecules aligned in the vertical directions, if the input optical beam is also polarized in the vertical direction, the effective refractive index n_e seen by the optical signal is high. When a voltage is applied between the two transparent electrodes, it creates an electric field in the horizontal direction. This electric field forces liquid crystal molecules to rotate to the horizontal direction as shown in Fig. 6.8.8B, and thus effective refractive index seen by the vertically polarized input optical field is reduced to n_o, which is smaller than n_e. This can be used to make an optical phase modulator. If the optical path length inside the liquid crystal cell is W, phase change of the optical signal with wavelength λ caused by the modulation is $\Delta\phi = 2\pi W(n_e - n_o)/\lambda$.

Alternatively, if the input optical beam is not polarized in the same direction as the molecules, the birefringence of this liquid crystal cell will create polarization rotation of the optical signal. For example, a half-wave plate corresponds to the cell thickness of $W = \lambda/2(n_e - n_o)$. When the voltage is applied on the transparent electrodes as shown in Fig. 6.8.8B, the birefringence of the liquid crystal cell will be reduced to zero with the refractive index $n = n_o$. This can be used to make tunable wave plates in optical systems.

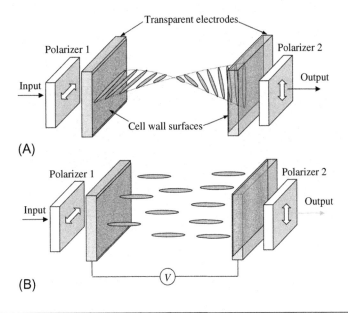

FIG. 6.8.9

(A) Twisted nematic liquid crystal for polarization rotation of optical signal and (B) molecular orientation is aligned with applied electric field so that no polarization rotation is introduced.

Another often used wave plate is based on the twisted nematic (TN) liquid crystal as illustrated in Fig. 6.8.9A. A TN liquid crystal is similar to the homogeneous liquid crystal shown in Fig. 6.8.7A, but surface line structures on the opposite cell inner walls are orientated in different directions. In the example shown in Fig. 6.8.9A, the orientations of liquid crystal molecules twist gradually for 90 degree from one side to the other to meet the orientations of the cell walls. The twist of the molecular orientation can introduce rotation of the optical signal. Consider that the optical signal incident from the left side is linearly polarized in the horizontal direction which matches to the orientation of liquid crystal molecules near the left wall. Traveling inside the liquid crystal cell, the polarization of the optical signal will gradually rotate following the rotation of molecules, and exit from the right side wall vertically polarized as shown in Fig. 6.8.9A. This output signal can pass the polarizer on the right side with minimum loss.

When an electric voltage is applied as shown in Fig. 6.8.9B, all liquid crystal molecules are aligned with the electric field and the input optical signal sees an isotropic medium without birefringence. Thus, the polarization orientation of the incident optical signal remains horizontal and is blocked by the polarizer on the right side.

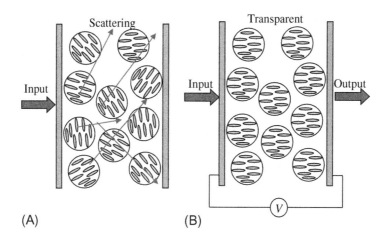

FIG. 6.8.10

(A) Polymer-dispersed liquid crystal (PDLC) cell without electric biasing, and microdroplets introduce strong scattering of input light, and (B) with the application of electric biasing, light scattering is minimized because of the index match between droplets and surrounding polymer material.

The inconvenience of homogenous and twisted liquid crystals shown in Figs. 6.8.8 and 6.8.9 is that they both relay on polarized optical signal at the input. If the input optical signal is unpolarized, a polarizer has to be added in front of the liquid crystal cell. Another technique based on the polymer-dispersed liquid crystal (PDLC) is able to operate with unpolarized optical signal. PDLC is made from liquid crystal microdroplets which are imbedded into polymer. The refractive index n_{pol} of the polymer is equal to n_o of the liquid crystal, but $n_{pol} < n_e$. The sizes of these microdroplets are on the order of 1 μm which is similar to the wavelength of the optical signal. As shown in Fig. 6.8.10A, without electric biasing these liquid crystal microdroplets are randomly oriented with the refractive indices seen by the incident optical signal statistically higher than the surrounding polymer. In this case, the optical signal is scattered by these microdroplets through the well-known *Mie scattering* mechanism, which introduces strong optical attenuation. With an electric biasing voltage applied as shown in Fig. 6.8.10B, liquid crystal molecular orientation of these microdroplets is aligned by the applied electric field in the propagation direction of the optical signal. Thus, the refractive index of these liquid crystal microdroplets seen by the optical signal becomes n_o, which is equal to the refractive index of the surrounding polymer n_{pol}, and the cell suddenly becomes transparent without scattering. This technique has been widely used to make flat TV screens with display pixels made by arrays of small PDLC cells, as shown in Fig. 6.8.11A, where the optical loss of each pixel can be individually addressed electrically so that images can be displayed. For fiber-optic applications, a PDLC cell can be inserted between two fibers, as shown in Fig. 6.8.11B, and the insertion loss can be adjusted by an electric voltage $V(t)$ applied on the transparent electrodes.

Configurations discussed so far for liquid crystal devices include optical phase modulation and intensity modulation on optical signals. A 1×2 optical switch, on the other hand, should be able to redirect the input optical beam to one of the two outputs. There are a number of ways to make optical switches based on liquid crystals. Here, we discuss an example of 1×2 optical switch based on total reflection on the surface of a liquid crystal cell as shown in Fig. 6.8.12. As illustrated in Fig. 6.8.12A, an optical beam is launched onto a homogeneous liquid crystal cell

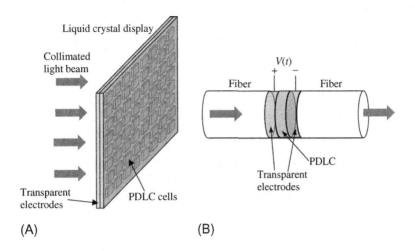

FIG. 6.8.11

Illustration of (A) liquid crystal display and (B) variable fiber-optic attenuator based on the liquid crystal inserted between fibers.

with an angle θ_i with respect to its surface normal. The input optical signal is polarized horizontally which is aligned with the orientation of liquid crystal molecules, and thus the optical signal sees a refractive index $n=n_e$ of the liquid crystal. Assume that the refractive index of the transparent electrodes is n_{ito}, with $n_e \approx n_{ito}$, there is no reflection at the interface, so that $P_T \approx P_{in}$ and $P_R \approx 0$ as illustrated by the inset between (A) and (B) in Fig. 6.8.12. When an electric voltage is applied between the two transparent electrodes, the electric field forces liquid crystal molecules to align in the direction perpendicular to the electrodes, and these molecules are also perpendicular with respect to the polarization of the input optical signal. Then, the optical signal sees a lower index n_o of the liquid crystal with $n_o < n_{ito}$. Total reflection can be achieved at the liquid crystal interface if the incidence angle is larger than the critical angle, as defined by Eq. (2.2.14), that is, $\theta_i \geq \sin^{-1}(n_o/n_{ito})$. In this switching state, $P_R \approx P_{in}$ and $P_T \approx 0$.

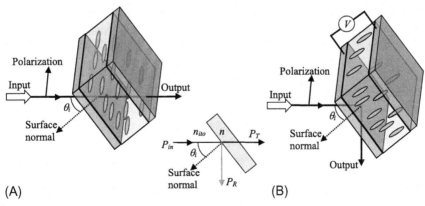

FIG. 6.8.12

1×2 optical switch based on the total internal reflection of a liquid crystal cell without (A) and with (B) electric biasing.

Using the 1×2 optical switches as building blocks, large-scale optical switch fabrics can be constructed. Fig. 6.8.13 shows two examples of multi-port optical switches. Fig. 6.8.13A is an 8×8 optical circuit switch, where optical signals from eight input fibers are collimated and launched onto an array of total-reflection liquid crystal cells. The electric voltage applied on each liquid crystal cell is individually controlled to determine the switching state (pass or reflection) as described in Fig. 6.8.12. Light beam from each input fiber can be switched to output 1 or output 2 depending on the electrical signal from the switch control. This spatial switch is wavelength independent, and each fiber port can carry multiple wavelength channels.

Fig. 6.8.13B shows the configuration of a 1×2 wavelength-selective switch, known as WSS. In this configuration, multiple wavelength channels carried by the input fiber are first spatially separated by a reflective grating and a collimating lens. Then, each wavelength channel is projected onto a liquid crystal cell which performs *pass* or *reflection* operation as described in Fig. 6.8.13. All wavelength channels at the output side are combined and focused into the output fiber. A WSS can selectively route any wavelength channel to either one of the two output fibers depending on the switch control.

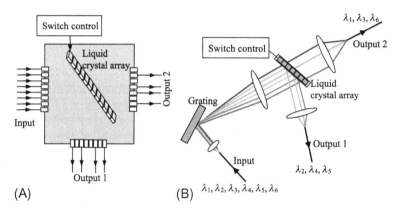

FIG. 6.8.13

(A) Space optical switch and (B) wavelength-selective switch (WSS) based on the liquid crystal arrays.

Optical switches based on the liquid crystals can be low cost and highly efficient. The switching speed is typically on the sub-millisecond level, and there is no mechanically moving part involved for switching so that the devices are quite reliable. However, free-space optics is usually needed for collimating and focusing of optical signals into liquid crystal cells, so that the device size cannot be sufficiency miniaturized, especially when the port count is high. The electric voltage required to control a liquid crystal device is usually a few tens of volts.

Mechanical switch is another technique that has been used to make optical switches for many years. It is well known that the direction of a light beam reflected from a mirror can be changes by steering the angle of the mirror. With rapid advances in material science and micro/nano-fabrication technologies, micro-electro-mechanical systems (MEMS) have been introduced. MEMS technology enabled the fabrication of complex mechanical systems with very small sizes and electrically controllable. A large number of small mirrors can be made into an array on a small chip, and each mirror can be steered by electrical actuation. Fig. 6.8.14A shows an example of a MEMS mirror array made by the Lucent Technologies (Lucent LambdaRouter) in early 2000 (Bishop et al., 2002). Each mirror in the array, as shown in Fig. 6.8.14B, can scan around the two principle axes, so that a light beam reflected by a MEMS mirror can be scanned in both the horizontal and the vertical directions. The scan angle on each axis is about ± 6 degree, and the diameter of each mirror is on the order of a few hundred micrometers. Fig. 6.8.14C shows a configuration of free-space optical switch based on two sets of MEMS mirror arrays. Optical signal from each input optical fiber is collimated onto a micro-mirror in the input mirror array. It can be directed to a target mirror in the output mirror array through adjusting the tilting angle θ. A fixed concave mirror is used between the two sets of MEMS so that light beams launched onto the output mirror array are mostly parallel, which helps reducing the requirement of mirror tilt. Then, the light beam is coupled into the desired output fiber port through an array of output collimators (Kawajiri et al., 2012).

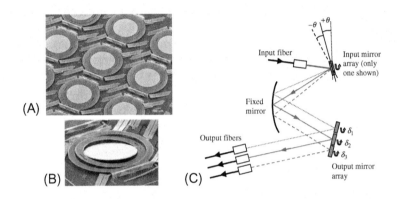

FIG. 6.8.14

(A) Image of MEMS array, (B) image of one mirror in the array with tilt (Bishop et al., 2002), and (C) optical configuration of MEMS switch.

Fig. 6.8.15 shows the three-dimensional (3D) architecture of a 512×512 MEMS optical switch module, which uses four 128×128 MEMS chips to construct each 512×512 mirror array (Kawajiri et al., 2012), and a picture of 128×128 MEMS chip is shown on the right of the figure. This makes MEMS-based optical switches scalable.

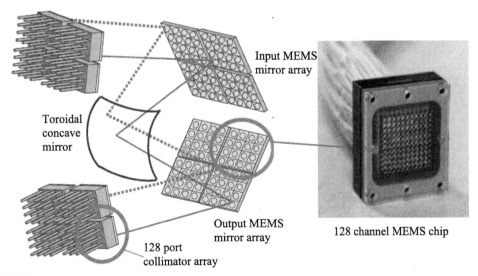

FIG. 6.8.15

Optical configuration of a 512×512 port 3D MEMS optical switch module. Picture on the right is a 128×128 MEMS chip (Kawajiri et al., 2012).

The MEMS chip shown in Fig. 6.8.15 has relatively large mirror size with a mirror diameter on the order of a few hundred micrometers to ensure negligible impacts of diffraction limit and low cross talk in fiber coupling. However, large size mirrors not only make the switch bulkier, but also reduce the switching speed due to the large mass of each mirror. Thus, the switching time of a typical MEMS-based optical switch is on the order of milliseconds.

Another type of MEMS-based mirror arrays, known as Digital Micromirror Device (DMD) (Van Kessel et al., 1998), developed for image projectors have much higher densities and filling factors of mirrors as shown in Fig. 6.8.16A. This is also known as the Digital Light Processing (DLP) technology which was first developed by the Texas Instruments. For example, a standard DMD (TI DLP660TE) made for image projector application has the mirror size on the order of 5.4 μm with less than 0.1 μm gap between adjacent mirrors. The number of mirrors on a MEMS chip is more than 3 million arranged in a 2D 2176×1528 array with an active chip area of $8.25 \times 14.67 \, \text{mm}^2$. The switch operation is binary with the tilt angle of each mirror either 0 or ± 17 degree depending on the control signal state, as shown in Fig. 6.8 16B. Because of the small size and mass of each mirror, the switch time of DMD can be quite fast on the order of microseconds. Inside a projector, each mirror represents an image pixel which can either be "off" or "on" represented by tilt or non-tilt, so that moving pictures can be created when a broad light beam is reflected off the DMD and projected onto a screen.

When DMD is used for fiber-optic systems and networks, interference between light reflected from different mirrors has to be considered because of the small mirror size and the monochromatic nature of the optical signal. As the pitch size of DMD is only a few micrometers which is on the same order of magnitude of an optical wavelength, this 2D array with a large number of mirrors can operate just like an optical grating but with the angle of grooves switchable.

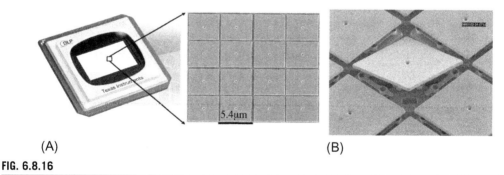

(A)   (B)

FIG. 6.8.16

Images of MEMS-based DMD. (A) A packaged DMD chip with 5.4 μm pitch of micro-mirrors made by Texas Instruments and (B) zoomed-in image of tilt micro-mirror. (B) Used with permission from Dr. Larry Hornbeck.

Assume that a collimated light beam is launched onto the surface of a DMD with an incidence angle θ_i with respect to the plane normal as shown in Fig. 6.8.17, and the mirror is tilt at an angle θ_t with respect to the plane normal. Based on the grating equation, the maximum power is diffracted by the grating at an exit angle θ_r defined by

$$d(\sin\theta_r - \sin\theta_i) = m\lambda \tag{6.8.2}$$

where θ_r is the angle of the diffracted beam with respect to the plane normal, d is the pitch of the mirror array, λ is the signal wavelength, and m is an integer. Meanwhile, the maximum reflectivity of each mirror corresponds to the blaze angle determined by $\theta_r - \theta_t = \theta_i + \theta_t$, at which the input and the output are symmetric with respect to the surface normal of the mirror. When the grating condition overlaps with the blaze condition, the diffractive beam has the maximum intensity.

EXAMPLE

For a DMD with pitch $d = 5.4\,\mu m$ and mirror tilt angle $\theta_t = 17$ degree mirror tilt angle, and assume the signal wavelength is $\lambda = 1550\,nm$. What is the incidence angle at which the first-order ($m = 1$) grating condition overlaps the mirror blaze angle?

Solution

Based on the grating Eq. (6.8.2), and mirror glazing condition $\theta_r = \theta_i + 2\theta_t$, we find the incidence angle which satisfies: $\sin(\theta_i + 2\theta_t) - \sin\theta_i = \lambda/d$, where $m = 1$ is assumed. Insert parameters $d = 5.4\,\mu m$, $\theta_t = 17$ degree, and $\lambda = 1.55\,\mu m$, the solution can be found numerically as $\theta_i \approx 43.6$ degree and $\theta_r \approx 77.6$ degree.

In the example discussed above, if the mirrors tilt angle is switched to $\theta_t = 0$ degree, then the mirror blazing condition will be switched to a diffraction angle of $\theta_r = 43.6$ degree, and the diffraction efficiency at $\theta_r \approx 77.6$ degree will become very small.

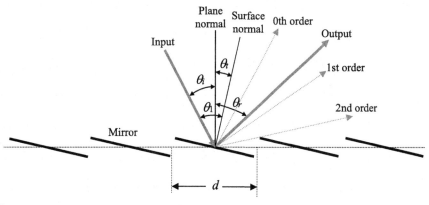

FIG. 6.8.17

Illustration of diffraction angles of a grating.

An optical switch can be made by switching the tilt angle of mirrors, on and off, so that the output beam can be switched to different directions as illustrated in Fig. 6.8.18. Multiple input beams can be switched independently each using a separate area on the DMD surface. Because the mirror switch is binary, each output beam can only be switched between two discrete angles, and this simple design cannot provide any-to-any optical switch capability.

A more sophisticated design of optical switch takes advantage of the high mirror density of DMD, and use the DMD as a high-resolution spatial light modulator. When optical signal from an input fiber is launched onto a DMD, the diffraction pattern is determined by the black and white image (reflectivity profile) on the DMD consisting of a large number of small mirrors. By loading the computer-calculated images, known as holograms, into the DMD, its diffraction patterns can be precisely determined. This allows the input fiber port to be imaged onto a plane with an array of output fiber, and the switch of input optical signal to different output fiber ports can be accomplished be loading different holograms on the DMD (Lynn et al., 2013). A 7×7 DMD-based diffractive switch with fiber pigtails has been demonstrated (Miles et al., 2015).

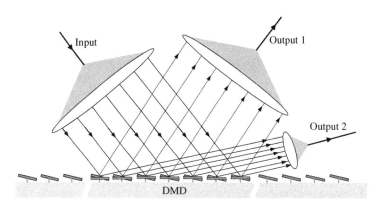

FIG. 6.8.18

Optical 1×2 switch uses an area on a DMD.

MEMS optical switch can also be based on the mechanical switch of optical coupling between waveguides. Because of the improved optical confinement, waveguide-based optical switches can be made much smaller than those made by switching mirrors based on the free-space optics. Because the coupling between waveguide is sensitive to their positions and alignments, a very smaller mechanical change can be sufficient to introduce optical switching with high extinction ratio. There are various different optical switching techniques based on the mechanical control of waveguide alignments (Ollier, 2002). Fig. 6.8.19 shows an example of a MEMS-based guided wave optical crossbar switch fabricated on silicon (Han et al., 2015). In this example, the switch has 50 input ports, 50 through ports, and 50 drop ports to be able to route optical signals to different directions with a total of 2500 MEMS cantilever switches. Fig. 6.8.19A shows the configuration of a single MEMS cantilever and the crossing optical waveguides. When the cantilever is pulled up by an electrostatic force from the plane of the crossing waveguides as shown in Fig. 6.8.19B, there is no optical coupling between the input optical waveguide and the waveguide in the cantilever. Thus, the input optical signal stays in the straight optical waveguide through the crossing point with minimum loss. Note that there is no optical coupling between two orthogonal waveguides at the crossing point. This is defined as the *through* state. When the electrostatic force is removed and the cantilever drops down to the same plane as the crossing waveguides, two directional optical couplers are formed between the curved waveguide in the cantilever and the crossing waveguides as shown in Fig. 6.8.19C. By carefully designing the length of the coupling region of the directional couplers ($a=1$ as defined by Eq. 6.1.1), the input optical signal is completely coupled to the orthogonal waveguide through the two couplers and the cantilever. This is defined as the *drop* state.

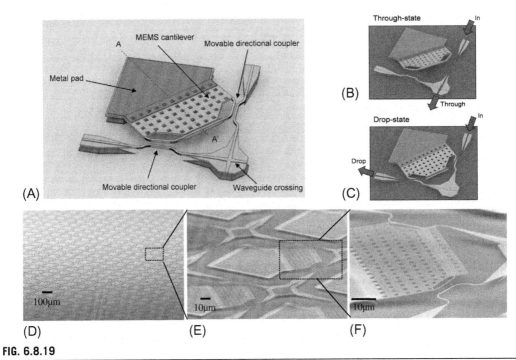

FIG. 6.8.19

Optical switch made with switchable waveguide coupling through a MEMS cantilever (Han et al., 2015). (A) A switch point configuration, (B) and (C) through and drop states, and (D) SEM images of fabricated MEMS switch array.

The waveguide cross-section size on this MEMS switch is 350×220 nm, which is typical of integrated silicon photonic circuits-based SOI platform. A small vertical offset, on the order of 200 nm, of the waveguide in the cantilever can completely switch the coupling coefficient α of the directional couplers from 1 to 0. Because the MEMS-actuated cantilevers are small with very low mass, the switching time of this device can be on the microseconds level which is fast enough for many circuit-switched optical network applications. Fig. 6.8.19D–F shows the scanning electron microscope (SEM) images of the fabricated switching chip and the size of the cantilever.

For the application in optical networks, Fig. 6.8.20 shows the schematic of crossbar switch and add/drop multiplexing with N input and N output ports ($N = 6$ in this figure) arranged in an orthogonal grid. By lifting the cantilever, the corresponding waveguide crossing point does not allow optical coupling between the horizontal and the vertical waveguides, and thus light stays in the straight waveguide and passes through the cross-point. By drop the cantilever down to the waveguide plane, light at the corresponding crossing node can be coupled from the horizontal waveguide to the vertical waveguide and be redirected to the desired output port through a pair of directional couplers. This provides any-to-any crossbar switch capability.

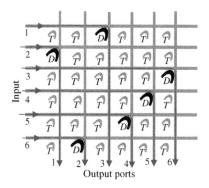

FIG. 6.8.20

Crossbar optical switch architecture. *T*: through state and *D*: drop state.

For the example shown in Fig. 6.8.20, the signals from input ports [1, 2, 3, 4, 5, 6] are switched to output ports [3, 1, 6, 5, 4, 2]. This any-to-any optical cross-connect switch capability is a highly desirable functionality for optical networks.

6.9 SUMMARY

In this chapter, we have discussed a number of basic passive optical components including directional couplers, optical interferometers, interleavers, FP optical filters, FBGs, WDM multiplexers and demultiplexers, and optical isolators and circulators. We have also introduced PLCs as an important platform for photonic integration.

Optical splitting and combining based on the directional couplers are necessary to realize two-beam optical interference in Mach-Zehnder or Michelson configurations. Such interferometers are commonly used to make electro-optic modulators, optical filters, and WDM channel interleavers. A fiber-optic directional coupler can also be used as a tapper in an optical system which taps a small amount of optical signal off for monitoring. Another category of optical interferometer is based on the multi-beam interference, which includes FPI, ring resonator, and grating-based optical interferometers. Because a large number of optical beams are involved in the interference process each having a different relative phase delay, frequency selectivity of the interferometer can be significantly improved compared to a simple two-beam-based Mach-Zehnder configuration. In principle, the larger the number of beams involved in the interference process, the better the frequency selectivity that can be achieved. For FPI and ring resonators, this requires low optical loss of each cavity round-trip. For a planar diffractive grating, this requires either a large optical beam size or a high groove line density on the grating. For a FBG, this requires increased grating section length while maintaining good uniformity of grating period. From an engineering design point of view, a proper trade-off has to be made between the achievable performance and the complexity of the device.

Optical isolators and circulators are in the category of nonreciprocal devices in which the forward and the backward optical paths have different characteristics. The nonreciprocity of these devices is usually realized by the Faraday rotation which changes the SOP between the forward and backward paths. Birefringent optical materials are also used to spatially separate the positions of light beams polarized along the fast and the slow axes. An optical isolator is often necessary to be used at the output of a coherent light source to prevent optical reflection which may compromise the state of laser operation. The function of an optical circulator is to redirect the reflected optical signal to a different direction. A typical example of optical circulator application is to use with FBG. As an FBG operates in the reflection mode to make a narrowband optical filter, an optical circulator is often necessary to redirect the reflected optical signal to the output without suffering from the splitting and recombining loss of using a directional coupler.

Although a variety of high-quality optical components can be made based on the free-space optics, optical systems will benefit from integrating many photonic devices into small functional chips. Similar to large-scale electronic integration which revolutionized the field of electronics and enabled many new applications, photonic integration is a promising technology for photonics and fiber optics. PLC is the major platform for photonic integration. The process of photolithography and etching allows the fabrication of very complex photonic structures on a planar substrate. Several materials have been used in PLC including silica, polymer, gallium-nitride (GaN), InP, and silicon. Silicon photonics has the potential advantage of CMOS compatible. This means silicon photonic circuits can be fabricated on the same substrate with CMOS electronics, and using the same fabrication techniques previously developed for CMOS electronics. So far almost all optical components can be fabricated with silicon photonics except laser sources.

Optical switches are enabling components, which can also be considered as subsystems, for optical networks. Various technologies have been used to make optical switches including semiconductor, electro-optic, liquid crystal, AO, and MEMS. Semiconductor and electro-optic switches are used in situations where nanosecond-level switching time is needed, but these devices are usually expensive and with higher losses. An AO switch can have a switching time on the order of sub-microsecond, but it is relatively bulky. Liquid crystal-based optical switches can be miniaturized and made into arrays with low loss and minimum cross talk, but the speed of a liquid crystal switch is typically on the orders of milliseconds or sub-millisecond. MEMS-based optical switches have become very popular because of their high extinction ratio, wavelength transparency, and the potential of large port counts. Although most MEMS optical switches operate at millisecond switching speed, fast MEMS switches with microsecond-level switching time have been reported by further reducing the masses of the moving parts. Each technology may have its unique advantages and disadvantages, and thus selecting the type of optical switch should fit to the specific application that will be used in.

PROBLEMS

1. The data sheet of a 2×2 fiber directional coupler indicates that the power splitting ratio is 20%, the excess loss is 0.5 dB, and the directionality is -50 dB.

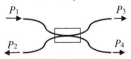

If the input optical power is $P_1 = 0.1$ mW, find the power values of P_2, P_3, and P_5 as indicated in the figure above.

2. An optical directional coupler can be used to redirect a reflection into a transmission as shown in the following figure, so that $P_0(f) = \eta R(f) P_i(f)$. Where $R(f)$ is the frequency-dependent reflectivity of the filter and η is the insertion loss caused by the coupler which is determined by the power splitting ratio α.

   Neglect excess loss of the coupler, what is the optimum value α to minimize the insertion loss? and what is the minimum insertion loss?

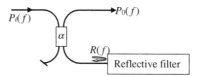

3. Using Eqs. (7.2.6) and (7.2.7), design a 1480-nm/1550-nm WDM coupler based on the single-mode fiber by find the proper values of z and z_0.

4. Optical power launched into a 10-km single-mode fiber through a 2×2 fiber coupler is $P_i = 1\,\text{mW}$ as shown in the following figure. The fiber coupler has 10-dB power coupling ratio ($\alpha = 0.1$ or usually referred to as 10/90 coupler) has no excess loss. The refractive index of the fiber core is $n = 1.5$. The fiber pigtail at port C is AR coated ($R_1 = 0$) and the far end of the 10 km fiber is open in the air (cleaved perpendicular to the fiber axis). (The isolators assure only unidirectional transmission and their losses are negligible.)
   (a) Neglecting all fiber losses, what is the output optical power P_0?
   (b) If the attenuation coefficient of the 10 km fiber is 0.5 dB/km and neglect losses of other fibers, what is the output optical power P_0?

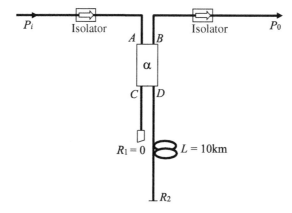

5. A Mach-Zehnder-type optical interleaver is designed to separate even and odd number of equally spaced WDM channels as shown in the following figure.

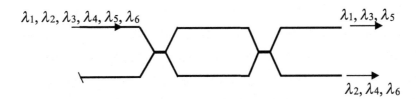

$\lambda_1, \lambda_2, \lambda_3, \lambda_4, \lambda_5, \lambda_6$

$\lambda_1, \lambda_3, \lambda_5$

$\lambda_2, \lambda_4, \lambda_6$

The system works in 1550-nm wavelength band and the input WDM signal channel spacing is 0.8 nm in wavelength (between adjacent channels). Find the length difference between the two arms of the MZ interferometer (assume the refractive index $n=1.5$).

6. A FPI is made by two identical mirrors in the air. The interferometer has a 150 GHz FSR and a 1-GHz FWHM (3-dB bandwidth). Find the finesse of this interferometer and the cavity length of this interferometer

7. For a FPI, if the facet reflectivity of both mirrors is $R=0.98$ and neglecting the absorption loss, find the ratio between the maximum power transmission and the minimum power transmission (i.e., commonly refereed to as extinction ratio).

8. The following figure shows a fiber ring laser. There is a semiconductor optical amplifier (SOA) in the fiber loop to provide the optical gain. The fiber coupler has 90% power coupling coefficient.

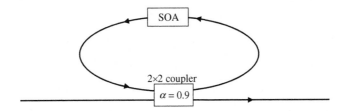

SOA

2×2 coupler
$\alpha = 0.9$

(a) Neglect fiber attenuation, find the SOA optical gain G required to achieve lasing threshold.
(b) If the total fiber length in the ring is $L_1=19$ mm, fiber refractive index is $n_1=1.47$, SOA length is $L_2=550\,\mu$m, and SOA refractive index is $n_2=3.6$, find the frequency spacing between adjacent lasing modes.

9. The following figure is a 3-dB (50%) 2×2 fiber coupler with four ports A, B, C, and D. The two output ports are each connected to a fiber delay line and their lengths are L_1 and L_2, respectively. A total reflector ($R=1$) is added at the end of each delay line (this can be done by coating gold at the fiber-end surface). The refractive index of optical fiber is $n=1.5$.

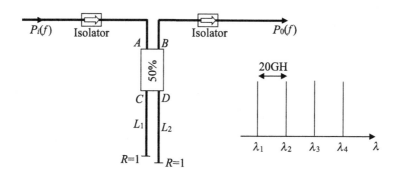

(a) Derive an expression to describe the power transfer function $P_0(f)/P_i(f)$

(b) A 4-wavelength WDM signal with 20 GHz channel spacing is launched into port A. If we only want to have wavelengths at λ_2 and λ_4 to come out from port B, what is the length difference between L_1 and L_2: $(\Delta L = L_1 - L_2 = ?)$.

10. A FP cavity is made by a material which has the refractive index of $n = 3.6$. The cavity length is 0.5 mm. Neglect absorption losses,

(a) What is the spacing in frequency between the adjacent FP resonance peaks?

(b) What is the finesse of this cavity?

(c) Plot out the power transmission and reflection spectrum (in a 5-nm wavelength range from 1550 to 1555 nm).

11. Assume a single-mode fiber core refractive index is $n_{core} = 1.47$. To design a FBG band-pass filter with the peak reflectivity at 1560 nm, what should be the grating period Λ?

12. Consider a FBG with the coupling coefficient $\kappa = 10\,m^{-1}$, fiber core refractive index $n_{core} = 1.47$, and Bragg wavelength $\lambda_{Bragg} = 1550$ nm.

(a) Find the minimum length of coupling region, L, so that the power reflectivity at the Bragg wavelength is -3 dB.

(b) At this L value, what is the full bandwidth of this filter in GHz?

13. In FBG band-pass filter design, assume $\lambda_{Bragg} = 1550$ nm and $n_{core} = 1.47$. If we want to have the power reflectivity at the center of the filter to be 90%, and the full bandwidth to be 0.1 nm, find the coupling coefficient κ and the length of the coupling region L.

14. Based on Eqs. (6.4.8) and (6.4.9), derive the wavelength-dependent transmissivity of FBG, $T = A(L)/A(0)$.

Assume $\lambda_{Bragg} = 1550$ nm, $n_{core} = 1.47$, $\kappa = 2\,cm^{-1}$, and $L = 1$ cm, plot the amplitude (in dB) and phase (in radians) of T as the function of frequency detune.

15. An AWG has the following parameters: central wavelength $\lambda_0 = 1550$ nm, Roland sphere radius $L_f = 20$ mm, effective index of array channels $n_c = 1.457$, refractive index of star coupler $n_s = 1.47$, array waveguide separation $d = 12\,\mu m$, output waveguide separation $b = 12\,\mu m$, and differential waveguide length $\Delta L = 100\,\mu m$. Find

(a) grating order,
(b) channel spacing in [nm], and
(c) FSR in [nm].

16. Consider a four-channel WDM demultiplexer made by optical thin films based on the configuration shown in Fig. 6.5.4A. Each film is a band-pass reflective filter with characteristics shown in Fig. 6.5.3D. Assume each film has 0.2 dB out-of-band reflection loss (due to nonideal AR coating) and 0.4-dB absorption loss, and assume the in-band reflection is ideally 100%. Find the insertion losses of channels 1, 2, 3, and 4. If a 16-channel WDM demultiplexer is constructed this way, what is the maximum insertion loss?

17. An optical beam is perpendicularly incident onto an optical film. Assume the film has a refractive index $n = 3.5$, a thickness of $20 \, \mu m$, and negligible optical absorption.
    Find the ratio between the maximum and the minimum power transmissivity (also known as the visibility). And find the FSR.

18. For a polarization-insensitive optical isolator, the return loss is limited by Fresnel reflections from component interfaces. For the isolator configuration shown in Fig. 6.6.3, how many interfaces are most relevant in introducing reflections which decrease the return loss?

19. For both polarization-sensitive and polarization-insensitive isolators, the isolation depends on the precision of 45 degree Faraday rotation. For the isolator configuration shown in Fig. 6.6.1, the Faraday rotator is thicker then specified, which introduces ± 50 degree polarization rotation (axes of the two polarizers are still 45 degree apart).
    (a) How many dB insertion loss can be caused by this polarization over-rotation?
    (b) What is the isolation value only caused by this polarization over-rotation?

20. A slab waveguide consists of a core layer with index $n_1 = 1.49$, and both the upper and lower cladding layer have refractive index of $n_2 = 1.475$. The optical signal is at 1550 nm wavelength. To ensure only a single-mode operation with symmetric field distribution, what is the maximum thickness of the core layer?

21. Consider a berried optical waveguide with a square cross section as shown in the following figure with $W = d$. The core with index $n_1 = 1.5$, is buried inside a material with index $n_2 = 1.49$, and the top is air so that $n_0 = 1$. Find the maximum cross-section size of the waveguide for single-mode operation (use Fig. 6.7.6).

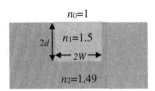

22. A silicon "wire" waveguide fabricated on silicon nitride (Si_3N_4) substrate with refractive index $n_2 = 2.25$ and the cross section is shown in the following figure. The silicon waveguide thickness is $T = 220 \, nm$.

Find the maximum width of the waveguide for single-mode operation (use Fig. 6.7.15).

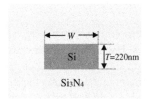

REFERENCES

Bishop, D.J., Giles, C.R., Austin, G.P., 2002. The Lucent LambdaRouter: MEMS technology of the future here today. IEEE Commun. Mag., 75–79.

Cao, S., Chen, J., Damask, J.N., Doerr, C.R., Guiziou, L., Harvey, G., Hibino, Y., Li, H., Suzuki, S., Wu, K.-Y., Xie, P., 2004. Interleaver technology: comparisons and applications requirements. J. Lightwave Technol. 22 (1), 281–289.

Carroll, L., Lee, J.-S., Scarcella, C., Gradkowski, K., Duperron, M., Lu, H., Zhao, Y., Eason, C., Morrissey, P., Rensing, M., Collins, S., Hwang, H.Y., O'Brien, P., 2016. Photonic packaging: transforming silicon photonic integrated circuits into photonic devices. MDPI Appl. Sci. 6. (www.mdpi.com/journal/applsci), paper 426.

Chen, X., et al., 2015. Monolithic InP-based fast optical switch module for optical networks of the future. In: 2015 International Conference on Photonics in Switching (PS), Florence, Italy 2015. paper # 15601425.

Chi, N., Vegas Olmos, J.J., Thakulsukanant, K., Wang, Z., Ansell, O., Yu, S., Huang, D., 2006. Experimental characteristics of optical crosspoint switch matrix and its applications in optical packet switching. J. Lightwave Technol. 24 (10), 3646–3653.

Chrostowski, L., Hochberg, M., 2015. Silicon Photonics Design: From Devices to Systems. Cambridge University Press.

J. B. Clayton, M. A. El, L. J. Freeman, J. Lucius, C. M. Miller, 1991. Tunable Optical Filter, US Patent #5,073,004.

Erdogan, T., 1997. Fiber grating spectra. J. Lightwave Technol. 15 (8), 1277–1294.

Green, P.E., 1991. Fiber Optic Networks. Prentice-Hall.

Han, S., Scok, T.J., Quack, N., Yoo, B.-W., Wu, M.C., 2015. Large-scale silicon photonic switches with movable directional couplers. Optica 2 (4), 370–375.

Hernandez, G., 1986. Fabry-Perot Interferometers. Cambridge University Press.

Hunsperger, R., 2009. Integrated Optics: Theory and Technology, sixth ed. Springer.

R. S. Jameson, 1989. Polarization Independent Optical Isolator. US Patent: US5033830 A.

Kashyap, R., 2010. Fiber Bragg Gratings, second ed. Academic Press.

Kawajiri, Y., Nemoto, N., Hadama, K., Ishii, Y., Makihara, M., Yamaguchi, J., Yamamoto, T., 2012. 512 × 512 port 3D MEMS optical switch module with toroidal concave mirror. NTT Technical Rev. 10 (11), 1–7.

Lee, J.-S., Carroll, L., Scarcella, C., Pavarelli, N., Menezo, S., Bernabe, S., Temporiti, E., O'Brien, P., 2016. Meeting the electrical, optical, and thermal design challenges of photonic-packaging. IEEE J. Sel. Top. Quantum Electron. 22 (6). paper 8200209.

Lizuka, K., 2002. In: Saleh, B.E.A. (Ed.), Elements of Photonics Volume II: For Fiber and Integrated Optics. In: Willey Series in Pure and Applied Optics, John Willey & Sons, Inc.

Lu, L., Zhao, S., Zhou, L., Li, D., Li, Z., Wang, M., Li, X., Chen, J., 2016. 16 × 16 non-blocking silicon optical switch based on electro-optic Mach-Zehnder interferometers. Opt. Express 24 (9), 9295–9307.

Lynn, B., Blanche, P.-A., Miles, A., Wissinger, J., Carothers, D., LaComb Jr., L., Norwood, R.A., Peyghambarian, N., 2013. Design and preliminary implementation of an N × N diffractive all-optical fiber optic switch. J. Lightwave Technol. 31 (24), 4016–4021.

Macleod, H.A., 2010. Thin-Film Optical Filters, fourth ed. Taylor & Francis.

Marcatili, E.A.J., 1969. Dielectric rectangular waveguide and directional coupler for integrated optics. Bell Syst. Tech. J. 48 (7), 2071–2102.

McCall, M., 2000. On the application of coupled mode theory for modeling fiber Bragg gratings. J. Lightwave Technol. 18, 236.

Mihailov, S.J., Bilodeau, F., Hill, K.O., Johnson, D.C., Albert, J., Holmes, A.S., 2000. Apodization technique for fiber grating fabrication with a halftone transmission amplitude mask. Appl. Opt. 39, 3670–3677.

Miles, A., Lynn, B., Blanche, P.-A., Wissinger, J., Carothers, D., LaComb Jr., L., Norwood, R.A., Peyghambarian, N., 2015. 7x7 DMD-based diffractive fiber switch at 1550 nm. Opt. Commun. 334, 41–45.

Oguma, M., Kitoh, T., Inoue, Y., Mizuno, T., Shibata, T., Kohtoku, M., Kibino, Y., 2004. Compact and low-loss interleave filter employing lattice form structure and silica-based waveguide. J. Lightwave Technol. 22 (3), 895–902.

Okamoto, K., 2005. Fundamentals of Optical Waveguides, second ed. Academic Press.

Okamoto, K., 2014. Wavelength-division-multiplexing devices in thin SOI: advances and prospects. IEEE J. Sel. Top. Quantum Electron. 20 (4). paper 8200410.

Ollier, E., 2002. Optical MEMS devices based on moving waveguides. IEEE J.Sel. Top. Quantum Electron. 8, 155–162.

Pietzsch, J., 1989. Scattering matrix analysis of 3×3 fiber couplers. J. Lightwave Technol. 7 (2), 303–307.

Pu, M., Liu, L., Ou, H., Yvind, K., Hvam, J.M., 2010. Ultra-low-loss inverted taper coupler for silicon-on-insulator ridge waveguide. Opt. Commun. 283, 3678–3682.

Qiao, L., Tang, W., Chu1, T., 2017. 32×32 silicon electro-optic switch with built-in monitors and balanced-status units. Sci. Rep. 7 42306.

Roelkens, G., Vermeulen, D., Selvaraja, S., Halir, R., Bogaerts, W., Van Thourhout, D., 2011. Grating-based optical fiber interfaces for silicon-on-insulator photonic integrated circuits. IEEE J. Sel. Top. Quantum Electron. 17, 571–580.

Saleh, B.E.A., Teich, M.C., 1991. Fundamentals of Photonics. John Willey & Sons, Inc.

Shoji, T., Tsuchizawa, T., Watanabe, T., Yamada, K., Morita, H., 2002. Low loss mode size converter from 0.3 µm square Si wire waveguides to single mode fibers. Electron. Lett. 38, 1669–1670.

Sun, Y., Fan, X., 2011. Optical ring resonators for biochemical and chemical sensing. Anal. Bioanal. Chem. 399 (1), 205–211.

Takahashi, H., Oda, K., Toba, H., Inoue, Y., 1995. Transmission characteristics of arrayed waveguide $N \times N$ wavelength multiplexer. J. Lightwave Technol. 13, 447–455.

J. S. Van Delden, 1991. Optical Circulator Having a Simplified Construction. US Patent: US5212586 A.

Van Kessel, P.F., Hornbeck, L.J., Meier, R.E., Douglass, M.R., 1998. A MEMS-based projection display. Proc. IEEE 86 (8), 1687–1704.

Van Laere, F., Claes, T., Schrauwen, J., Scheerlinck, S., Bogaerts, W., Taillaert, D., O'Faolain, L., Van Thourhout, D., Baets, R., 2007. Compact focusing grating couplers for silicon-on-insulator integrated circuits. IEEE Photon. Technol. Lett. 19, 1919–1921.

Vaughan, J.M., 1989. The Fabry-Perot Interferometer: History, Practice and Applications. Adam Hilger, Bristol England.

Xu, Q., Schmidt, B., Shakya, J., Lipson, M., 2006. Cascaded silicon micro-ring modulators for WDM optical interconnection. Opt. Express 14 (20), 9430–9435.

Zhang, K., Wang, J., Schwendeman, E., Dawson-Elli, D., Faber, R., Sharps, R., 2002. Group delay and chromatic dispersion of thin-film-based, narrow bandpass filters used in dense wavelength-division-multiplexed systems. Appl. Opt. 41 (16), 3172–3175.

FURTHER READING

www.amo.de/blog/2017/11/20/amo-joins-silicon-photonics-alliance-epixfab
For example: https://www.synopsys.com/optical-solutions/rsoft.html.

EXTERNAL ELECTRO-OPTIC MODULATORS

7

CHAPTER OUTLINE

Introduction 299
7.1 Basic operation principle of electro-optic modulators 301
    7.1.1 EO coefficient and phase modulator 301
    7.1.2 Electro-optic intensity modulator 307
    7.1.3 Optical intensity modulation vs. field modulation 310
    7.1.4 Frequency doubling and high-order harmonic generation 311
7.2 Optical single-sideband modulation 312
7.3 Optical I/Q modulator 316
7.4 Electro-optic modulator based on ring-resonators 317
7.5 Optical modulators using electro-absorption effect 324
7.6 Summary 329
Problems 330
References 334
Further reading 335

INTRODUCTION

In an optical transmitter, encoding electrical signals into optical domains can be accomplished either by directly modulating the injection current of a laser diode, known as direct modulation, or by electro-optic (EO) modulation using an external modulator, known as external modulation.

The speed of direct modulation is primarily limited by the carrier lifetime of the laser source, which is typically in the sub-nanosecond level. Meanwhile, a more important concern of direct current modulation is that it not only modulates the intensity but also modulates the frequency of the optical signal through the carrier density-dependent refractive index of the semiconductor material (Petermann, 1988). This effect is known as the frequency chirp, which broadens the spectral width of the optical signal and may significantly degrade the transmission performance of a high-speed fiber-optic system when chromatic dispersion is considered. The broadened signal optical spectrum also reduces the bandwidth efficiency when wavelength division multiplexing (WDM) is used.

On the other hand, an external modulator manipulates the optical signal after the laser, and ideally it does not affect the laser operation. Nowadays, a commercial, high-speed external EO modulator can have >100 GHz modulation bandwidth with well-controlled frequency chirp (Wooten et al., 2000).

Introduction to Fiber-Optic Communications. https://doi.org/10.1016/B978-0-12-805345-4.00007-X

Because of their high performances, external modulators have been widely adopted in long-distance and high-speed optical communication systems. Depending on the device configuration and system requirement, an external modulator can perform intensity modulation and optical phase modulation, as well as complex optical field modulation, which is the combination of amplitude and phase modulation. Both the EO effect and electro-absorption (EA) have been used to make external optical modulators. The former relies on the modulation of refractive index by the applied electric field, whereas the latter is the attenuation that is dependent on the applied electric field.

This chapter starts by discussing basic operation principles and configurations of EO modulators, including phase modulators and intensity modulators. Among various EO materials, lithium niobate ($LiNbO_3$) is most often used to make external modulators in fiber-optic systems because of the high EO coefficient and low attenuation in the optical communications wavelength window. While the EO effect is capable of directly introducing optical phase modulation through the applied electric voltage waveform, optical circuit configuration such as a Mach-Zehnder interferometer (MZI) can convert the phase modulation into intensity modulation, so that EO amplitude modulators can be made. The relationship between phase modulation and amplitude modulation in a modulator based on a single MZI can be explored to create single-sideband as well as carrier-suppressed optical modulation. A more sophisticated optical circuit based on the combination of three MZIs, known as the I/Q modulator, can provide more flexibility of modulating the optical phase and optical amplitude independently. This ability of complex optical field modulation is required in many coherent optical communication systems that encode data into both the amplitude and the phase of the optical signal to enhance the transmission capacity. A commercially available $LiNbO_3$ based on I/Q modulator with input and output fiber pigtails is shown in Fig. 7.0.1A.

(A)   (B)

FIG. 7.0.1

Examples of commercial EO modulator based on $LiNbO_3$ (A), and EA modulator integrated with a laser source (B).

(A) Used with permission from Thorlabs. (B) Used with permission from Neophotonics.

Although EO modulators based on $LiNbO_3$ can have excellent performance, they are usually expensive and traditionally cannot be monolithically integrated with diode laser sources. EA modulators, on the other hand, are based on pn-junction structures made of III–V semiconductor materials, which are the same materials as those used to make laser diodes. Instead of forward biasing with injection current such as in-diode lasers, the pn junction of an EA modulator is reversely biased (similar to a photodiode), which absorbs photons and converts them into electrons. The optical absorption of an EA modulator is thus proportional to the applied reverse bias voltage, directly allowing intensity modulation of the optical signal. Because of the use of the same material platform and the similar pn-junction and waveguide structures, an EA modulator can be monolithically integrated with a diode laser to reduce the cost and to miniaturize the footprint of an optical transmitter. A commercially available laser diode integrated with and EA modulator is shown in Fig. 7.0.1B; it only needs a single output fiber pigtail. However, the change of electro-absorption is often associated with a change of signal optical phase in the semiconductor material when modulated with an electric voltage waveform, which often introduces a modulation chirp. In addition, directly integrating an EA modulator with a diode laser source without sufficient isolation between them may cause an optical reflection back into the laser diode, affecting the coherence property of the optical signal.

From an engineer's point of view, the trade-off between cost and performance often dictates the system design and devices selection.

7.1 BASIC OPERATION PRINCIPLE OF ELECTRO-OPTIC MODULATORS

EO modulators rely on EO materials in which the refractive indices are functions of the electrical field applied on them. More specifically, for an EO material, the index change δn is linearly proportional to the applied electric field E as $\delta n = \alpha_{EO} E$, where α_{EO} is the linear EO coefficient, which is a material property. The most popular EO material used so far for an EO modulator is $LiNbO_3$, which has a high EO coefficient and low attenuation in the telecommunications optical wavelength window.

7.1.1 EO COEFFICIENT AND PHASE MODULATOR

The EO coefficient of an optical crystal can usually be represented by a tensor (Heismann et al., 1997; Noguchi et al., 1995). Now let us start with the commonly used index ellipsoid for a crystal without an applied electric field,

$$\left(\frac{x}{n_x}\right)^2 + \left(\frac{y}{n_y}\right)^2 + \left(\frac{z}{n_z}\right)^2 = 1 \tag{7.1.1}$$

where n_x, n_y, and n_z are indices of refraction along the three principal axes in the Cartesian coordinates. Then, with an electric field vector E applied in an arbitrary direction, the expression of the index ellipsoid has to be modified as

$$\left(\frac{1}{n^2}\right)_1 x^2 + \left(\frac{1}{n^2}\right)_2 y^2 + \left(\frac{1}{n^2}\right)_3 z^2 + 2\left(\frac{1}{n^2}\right)_4 yz + 2\left(\frac{1}{n^2}\right)_5 xz + 2\left(\frac{1}{n^2}\right)_6 xy = 1 \tag{7.1.2}$$

where $(1/n^2)_i$, with $i = 1, 2, \ldots, 6$, representing indices along the six directions determined by the possible combinations of x, y, and z. The field-induced index change in each of these six directions is then

$$\Delta\left(\frac{1}{n^2}\right)_i = \left(\frac{1}{n^2}\right)_i - \left(\frac{1}{n^2}\right)_i\Big|_{E=0} = \sum_{j=1}^{3} r_{ij}E_j \tag{7.1.3}$$

where $i = 1, 2, \ldots, 6$, and j represents x, y, and z. In Eq. (7.1.3), r_{ij} are the elements of a 6×3 EO tensor matrix of the crystal, that is,

$$\begin{bmatrix} \Delta(1/n^2)_1 \\ \Delta(1/n^2)_2 \\ \Delta(1/n^2)_3 \\ \Delta(1/n^2)_4 \\ \Delta(1/n^2)_5 \\ \Delta(1/n^2)_6 \end{bmatrix} = \begin{bmatrix} r_{11} & r_{12} & r_{13} \\ r_{21} & r_{22} & r_{23} \\ r_{31} & r_{32} & r_{33} \\ r_{41} & r_{42} & r_{43} \\ r_{51} & r_{52} & r_{53} \\ r_{61} & r_{62} & r_{63} \end{bmatrix} \begin{bmatrix} E_x \\ E_y \\ E_z \end{bmatrix} \tag{7.1.4}$$

The LiNbO$_3$ single crystal has a trigonal lattice structure with nonzero elements r_{13}, r_{22}, r_{33}, and r_{51} in the EO tensor matrix, and some of the elements are symmetric or antisymmetric due to the crystal structure. The EO tensor matrix of LiNbO$_3$ crystal is usually expressed as

$$\begin{bmatrix} \Delta(1/n^2)_1 \\ \Delta(1/n^2)_2 \\ \Delta(1/n^2)_3 \\ \Delta(1/n^2)_4 \\ \Delta(1/n^2)_5 \\ \Delta(1/n^2)_6 \end{bmatrix} = \begin{bmatrix} 0 & -r_{22} & r_{13} \\ 0 & r_{22} & r_{13} \\ 0 & 0 & r_{33} \\ 0 & r_{51} & 0 \\ r_{51} & 0 & 0 \\ -r_{22} & 0 & 0 \end{bmatrix} \begin{bmatrix} E_x \\ E_y \\ E_z \end{bmatrix} \tag{7.1.5}$$

These nonzero matrix elements relevant for EO modulation are $r_{13}=8.6\,\text{pm/V}$, $r_{22}=3.4\,\text{pm/V}$, $r_{33}=30.8\,\text{pm/V}$, and $r_{51}=28.0\,\text{pm/V}$. Fig. 7.1.1 shows a *longitudinal* free-space EO modulator based on the LiNbO$_3$ crystal. In this case, the applied electric field is along the crystal's z-axis.

With the applied electric field existing only in the z-direction, Eq. (7.1.5) indicates that $\Delta(1/n^2)_1=\Delta(1/n^2)_2=r_{13}E_z$ and $\Delta(1/n^2)_3=r_{33}E_z$, and thus the index ellipsoid equation becomes

$$\left(\frac{1}{n_o^2} + r_{13}E_z \right)x^2 + \left(\frac{1}{n_o^2} + r_{13}E_z \right)y^2 + \left(\frac{1}{n_e^2} + r_{33}E_z \right)z^2 = 1 \tag{7.1.6}$$

where n_o and n_e are linear indices of the LiNbO$_3$ birefringence crystal in the x- and y-axes (n_o), and z-axis (n_e). For an optical plane wave propagating in the z-direction, the impact of the applied electric field E_z is to change the original value of $1/n_o^2$ into $1/n_o^2+r_{13}E_z$, which is equivalent to the change of n_o into $n_o+\delta n_o$. Assume $|\delta n_o| \ll n_o$, which is always the case because the EO coefficient is usually very small, the expression $1/n_o^2+r_{13}E_z=1/(n_o+\delta n_o)^2$ is equivalent to

$$\delta n_o \approx -\frac{1}{2}n_o^3 r_{13}E_z \tag{7.1.7}$$

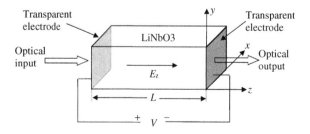

FIG. 7.1.1

Crystal and applied field orientation of a free-space longitudinal EO modulator made with LiNbO$_3$ EO material. The applied electric field is in the same direction of the wave propagation.

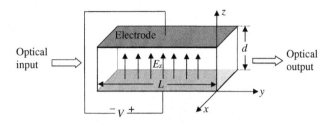

FIG. 7.1.2

Crystal and applied field orientation of a free-space transverse EO modulator made with LiNbO$_3$ EO material. The applied electric field is perpendicular to the direction of wave propagation.

Specifically, for this longitudinal free-space EO modulator, $n_x=n_y=n_o-0.5n_o^3r_{13}E_z$ and $n_z=n_e-0.5$-$n_e^3r_{33}E_z$. Thus, the longitudinal EO modulation only introduces a pure phase modulation of the optical signal, while the birefringence is not modulated as $n_x=n_y$ is always true in this case.

As an alternative, Fig. 7.1.2 shows a free-space *transverse* EO modulator made with LiNbO$_3$ EO material. In this case, the electric field is applied perpendicular to the wave propagation direction, but still along the z-axis of the LiNbO$_3$ crystal. Although the index ellipsoid equation is identical to that expressed in Eq. (7.1.6), the plane wave of the optical signal sees either $n_z=n_e-0.5n_e^3r_{33}E_z$ or $n_x=n_o-0.5n_o^3r_{13}E_z$ depending on the state of polarization (SOP) of the optical signal in the vertical or horizontal directions, respectively.

In comparison to the longitudinal EO modulator, the transverse EO modulator can be more efficient because at the same applied voltage the electric field can be increased by reducing the separation between the electrodes. However, a transverse EO modulator also introduces the modulation of birefringence, and thus the optical signal has to be linearly polarized with the SOP aligned with the z-axis of the LiNbO$_3$ crystal to achieve the highest modulation efficiency as r_{33} is much higher than r_{13}. In addition, the alignment of optical signal SOP with the crystal z-axis is necessary to avoid the SOP modulation by the EO modulator.

EXAMPLE 7.1

For the LiNbO$_3$ EO material operating in the 633 nm wavelength, parameters are: $n_o=2.286$, $n_e=2.2$, $r_{13}=8.6$ pm/V, and $r_{33}=30.8$ pm/V.

  (a) For the longitudinal EO modulator shown in Fig. 7.1.1 with the device length $L=1$cm, in order to produce a π phase shift on a 633-nm optical signal passing through the modulator, what is the required electric voltage?

Solution

To create a π phase shift on a 633-nm optical signal passing through the modulator of length L, the index change has to be

$$\delta n = \frac{\lambda}{2L} = \frac{633 \times 10^{-9}}{2 \times 10^{-2}} = 3.165 \times 10^{-5}$$

Based on Eq. (7.1.7),

$$|E_z| = \frac{\delta n}{0.5n_o^3r_{13}\delta n} = \frac{3.165 \times 10^{-5}}{0.5 \times 2.286^3 \times 8.6 \times 10^{-12} \text{m/V}} = 6.16 \times 10^5 \text{V/m}$$

Thus the required voltage is $V_b=E_zL=6.16 \times 10^5 \times 10^{-2}=6.16$kV

(b) For a transverse EO modulator shown in Fig. 7.1.2 with device length $L=1$ cm and separation between the two electrodes $d=2$ mm (which limits the signal optical beam size), if the input optical signal at 633 nm wavelength is vertically polarized in the z-direction, what is the required electric voltage to create a π phase shift?

Solution

A π phase shift on the 633-nm optical signal requires $\delta n=3.165 \times 10^{-5}$, which is the same as part (a) of this example. But,

$$|E_z| = \frac{\delta n}{0.5 n_e^3 r_{33} \delta n} = \frac{3.165 \times 10^{-5}}{0.5 \times 2.2^3 \times 30.8 \times 10^{-12} \text{m/V}} = 1.93 \times 10^5 \text{V/m}$$

Thus, the required voltage is $V_b = E_z d = 1.93 \times 10^5 \times 2 \times 10^{-3} = 386$V

(c) For part (b) of this example, if the input optical signal is linearly polarized 45° between the z- and the x-axes, what is the differential phase shift between the z- and the x-components caused by a 386-V applied voltage?

Solution

Electric field induced by 386-V applied voltage is

$$E_z = V_b/d = 1.93 \times 10^5 \text{V/m}$$

$$
\begin{aligned}
|\delta n_{zx}|L &= \frac{L}{2}\left(n_e^3 r_{33} - n_o^3 r_{13}\right)E_z \\
&= \frac{10^{-2} \times 10^{-12}}{2}\left(2.2^3 \times 30.8 - 2.286^3 \times 8.6\right) \times 1.93 \times 10^5 = 2.17 \times 10^{-7}
\end{aligned}
$$

The differential phase shift is then

$$\phi_{zx} = 2\pi \frac{|\delta n_{zx}|L}{\lambda} = 2\pi \frac{2.17 \times 10^{-7}}{633 \times 10^{-9}} = 0.69\pi$$

In fiber-optic systems, waveguide-based EO modulators are usually used with optical fiber pigtails, and requiring low enough bias voltage. Waveguides on a LiNbO$_3$ substrate can be created by the diffusion of other materials, such as titanium, along a stripe near the surface, which increases the local refractive index. Thus, the cross section of the created waveguide is often a half-ellipse. The crystal orientation of the waveguide is predetermined by the cut direction of the crystal substrate. Fig. 7.1.3 illustrates the cross sections of LiNbO$_3$ waveguides and the orientations of crystal axes. In both (A) and (B) of Fig. 7.1.3, optical waveguides are formed along the y-axis of the crystal which is also the direction of optical signal propagation (Sanna and Schmidt, 2010). The indices of orientation 1, 2, and 3 correspond to x, y, and z, respectively, are indicated in the figure. In Fig. 7.1.3A the crystal is z-cut, so that the z-axis of the crystal is in the vertical direction, which is perpendicular to the surface of the substrate. To support the modulation of a vertically polarized optical signal propagating along the waveguide, electrodes can be arranged in a way that the electric fields are primarily in the z-direction when a biasing voltage $-V_b$ is applied.

In Fig. 7.1.3B the crystal is x-cut, and the z-axis of the crystal is in the horizontal direction, which is parallel to the surface of the substrate (Minakata et al., 1978). This configuration can support the modulation of a horizontally polarized optical signal propagating along the waveguide by arranging the configuration of the electrodes so that the electric fields are primarily in the horizontal z-direction when a biasing voltage V_b is applied.

In the configurations of both Fig. 7.1.3A and B, the electric field E_z is applied in the z-axis of the crystal similar to the transverse EO modulator described in Fig. 7.1.2. However, the narrow waveguide

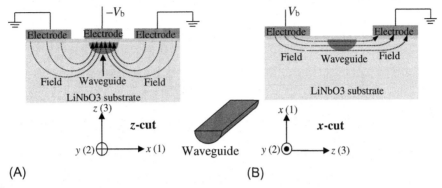

FIG. 7.1.3

Illustration of cross sections of z-cut (A) and x-cut (B) LiNbO$_3$ waveguides and the corresponding orientations of crystal axes. Optical signal propagates in the crystal's y-direction along the waveguide. The indices of orientation 1, 2, and 3 correspond to x, y, and z, respectively.

width allows much smaller separation between electrodes and thus requiring much lower applied voltage than the free-space EO modulator to achieve the same index change. For the optical signal linearly polarized in the z-direction, which is vertical for z-cut shown in Fig. 7.1.3A and horizontal for x-cut shown in Fig. 7.1.3B, the field-induced index modulation is

$$\delta n_e \approx -\frac{1}{2}n_e^3 r_{33} E_z \tag{7.1.8}$$

where n_e is the linear index along the z-axis of the crystal. The EO coefficient of the waveguide can be defined as

$$\alpha_{EO} \approx -\frac{1}{2}n_e^3 r_{33} \tag{7.1.9}$$

so that $\delta n_e = \alpha_{EO} E_z$.

For LiNbO$_3$, although EO coefficients are not particularly sensitive to the wavelength in the visible and near infrared regions, linear refractive indices of the two birefringence axes, n_o and n_e, vary with the wavelength, which can be expressed by a Sellmeier equation,

$$n_i = \sqrt{A_i + \frac{B_i}{\lambda^2 + C_i} + D_i \lambda^2} \tag{7.1.10}$$

where the wavelength λ is in micrometers, $i = o$ or e, with $A_o = 4.9048$, $B_o = 0.11768$, $C_o = -0.0475$, $D_o = -0.02717$, $A_e = 4.582$, $B_e = 0.099169$, $C_e = -0.04443$, and $D_e = -0.02195$.

At $\lambda = 1550$ nm wavelength, $n_o = 2.211$ and $n_e = 2.138$ as shown in Fig. 7.1.4. As an example with $r_{33} = 30.8$ pm/V at 1550 nm wavelength, the EO coefficient of LiNbO$_3$ is approximately $\alpha_{EO} \approx 1.5 \times 10^{-10}$ V/m. With an electrical field $E_z = 10^6$ V/m, which can be created with 10 V voltage across two electrodes separated by 10 μm, the field-induced index change is about $\delta n_e \approx 1.5 \times 10^{-4}$.

EO-phase modulator is the simplest external modulator, which can be made by a LiNbO$_3$ optical waveguide and a pair of electrodes. For simplicity and without considering specific electrode design details, we can assume that the waveguide is sandwiched between two electrodes as shown in Fig. 7.1.5.

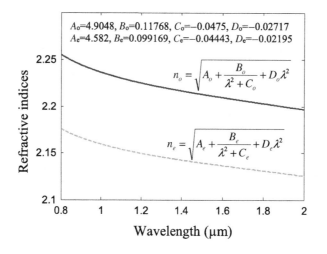

A_o=4.9048, B_o=0.11768, C_o=−0.0475, D_o=−0.02717
A_e=4.582, B_e=0.099169, C_e=−0.04443, D_e=−0.02195

$$n_o = \sqrt{A_o + \frac{B_o}{\lambda^2 + C_o} + D_o \lambda^2}$$

$$n_e = \sqrt{A_e + \frac{B_e}{\lambda^2 + C_e} + D_e \lambda^2}$$

FIG. 7.1.4

Refractive indices of LiNbO$_3$ as the function of wavelength along the directions of the o- and the e-axes of the crystal.

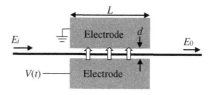

FIG. 7.1.5

Electro-optic phase modulator.

If the length of the electrode is L, the separation between the two electrodes is d, and the applied voltage is V, the optical phase change introduced by the linear EO effect is

$$\phi(V) = \left(\frac{2\pi a_{EO}L}{\lambda d}\right)V \qquad (7.1.11)$$

The modulation efficiency, defined as $d\varphi/dV$, is directly proportional to the length of the electrode L and inversely proportional to the electrodes separation d. Increasing the length and reducing the separation of the electrodes would certainly increase the modulation efficiency, but will inevitably increase the parasitic capacitance and thus reducing the modulation speed. Another speed limitation is due to the device transit time, $t_d=L/v_p$, where $v_p=c/n$ is the propagation speed in the waveguide of refractive index n. For a 1-cm-long LiNbO$_3$ waveguide with n=2.138 at 1550nm wavelength, the transit time is approximately 35 ps, which sets a speed limit to the modulation. Another limitation on the modulation speed is the RC constant. The capacitance between the electrodes $C \propto L/d$ is proportional to the electrode length and inversely proportional separation between them. In very high-speed EO modulators, traveling-wave electrodes are usually used, where the RF modulating signal propagates along the electrodes in the same direction as the optical signal. This results in a phase matching between the RF and the optical signal and thus eliminates the transit time induced speed limitation.

7.1.2 ELECTRO-OPTIC INTENSITY MODULATOR

Although phase modulation can be useful in some optical systems, intensity modulation is more popular. The EO effect used for optical phase modulators discussed above can also be used to make optical intensity modulators. A very popular external intensity modulator is made by planar waveguide circuits designed in a Mach-Zehnder interferometer (MZI) configuration as shown in Fig. 7.1.6.

In this configuration, the input optical signal is equally split into two interferometer arms and then recombined at the output. Similar to the phase modulator, an electrical field is applied across one of the two MZI arms to introduce relative phase delay and thus control the differential phase between the two arms.

If the phase delays of the two MZI arms are ϕ_1 and ϕ_2, respectively, the output optical field is

$$E_0 = \frac{1}{2}\left(e^{j\phi_1} + e^{j\phi_2}\right)E_i \qquad (7.1.12)$$

where E_i is the complex field of the input optical signal. Then, Eq. (7.1.12) can be rewritten into

$$E_0 = \cos\left(\frac{\Delta\phi}{2}\right)e^{j\phi_c/2}E_i \qquad (7.1.13)$$

where $\phi_c = \phi_1 + \phi_2$ is the average (common-mode) phase delay and $\Delta\phi = \phi_1 - \phi_2$ is the differential phase delay of the two arms.

The input-output power relationship of the modulator is then

$$P_0 = \cos^2\left(\frac{\Delta\phi}{2}\right)P_i \qquad (7.1.14)$$

where $P_i = |E_i|^2$ and $P_0 = |E_0|^2$ are the input and the output powers, respectively. Obviously, in this intensity modulator transfer function, the differential phase delay between the two MZI arms plays a major role. Again, if we use L and d for the length and the separation of the electrodes and α_{OE} for the linear EO coefficient, the differential phase shift will be

$$\Delta\phi(V) = \phi_0 + \left(\frac{2\pi}{\lambda}\right)\frac{\alpha_{EO}V}{d}L \qquad (7.1.15)$$

where ϕ_0 is the initial differential phase without the applied electrical signal. Its value may vary from device to device and may change with temperature mainly due to practical fabrication tolerance. In addition, if the driving electric voltage has a DC bias and an AC signal, the DC bias can be used to control this initial phase ϕ_0.

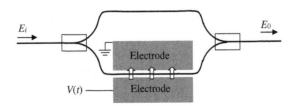

FIG. 7.1.6

Electro-optic modulator based on MZI configuration.

A convenient parameter to specify the efficiency of an EO intensity modulator is V_π, which is defined as the voltage required to change the optical power transfer function from the minimum to the maximum. From Eqs. (7.1.14) and (7.1.15), V_π can be found as

$$V_\pi = \frac{\lambda d}{2\alpha_{EO}L} \qquad (7.1.16)$$

V_π is obviously a device parameter depending on the device structure as well as the material EO coefficient. With the use of V_π, the power transfer function of the modulator can be simplified as

$$T(V) = \frac{P_0}{P_i} = \cos^2\left[\phi_0 + \frac{\pi V}{2V_\pi}\right] = \frac{1}{2}\left[1 + \cos\left(2\phi_0 + \frac{\pi V}{V_\pi}\right)\right] \qquad (7.1.17)$$

Fig. 7.1.7 illustrates the relationship between the input electrical voltage waveform and the corresponding output optical signal waveform. V_{b0} is the DC bias voltage, which determines the initial phase ϕ_0 in Eq. (7.1.17). The DC bias is an important operational parameter because it determines the electrical-to-optical (E/O) conversion efficiency. If the input voltage signal is bipolar, the modulator is usually biased as the quadrature point, as shown in Fig. 7.1.7. This corresponds to the initial phase $\phi_0 = \pm \pi/4$, depending on the selection of the positive or the negative slope of the transfer function. With this DC bias, the E/O transfer function of the modulator has the best linearity and allows the largest swing of the signal voltage, which is $\pm V_\pi/2$. In this case, the input/output optical power relation is

$$P_0(t) = \frac{P_i}{2}\left\{1 \pm \sin\left(\frac{\pi V(t)}{V_\pi}\right)\right\} \qquad (7.1.18)$$

Although this transfer function is nonlinear, it is not a big significant concern for binary digital modulation, where the electrical voltage switches between $-V_\pi/2$ and $+V_\pi/2$ and the output optical power switches between zero and P_i. However, for analog modulation, the nonlinear characteristic of the modulator transfer function may introduce signal waveform distortion. For example, if the modulating

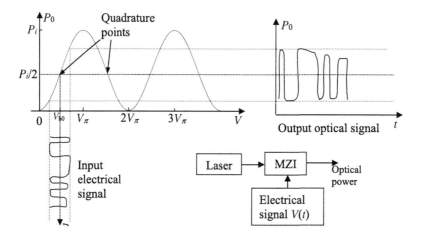

FIG. 7.1.7

Electro-optic modulator transfer function and input (electrical)/output (optical) waveforms.

electric signal is a sinusoidal, then $V(t) = V_m \cos(\Omega t)$, and the modulator is biased at the quadrature point with $\phi_0 = m\pi - \pi/4$, where m is an integer, the modulator output optical power is then $P_0(t) = P_i[1 + \sin(x \cdot \cos \Omega t)]/2$, with $x = \pi V_m/V_\pi$. This output power expression can be expanded in a Bessel series,

$$P_0(t) = \frac{P_i}{2} + P_i J_1(x) \cos(\Omega t) - P_i J_3(x) \cos(3\Omega t) + P_i J_5(x) \cos(5\Omega t) + \ldots \quad (7.1.19)$$

where $J_n(x)$ is the nth-order Bessel function. On the right-hand side of Eq. (7.1.19), the first term is the average output power, the second term is the result of linear modulation, and the third and the fourth terms are high-order harmonics caused by the nonlinear transfer function of the modulator. Because of the antisymmetric nature of power transfer function when biased at the quadrature point, there are only odd order harmonics at frequencies of 3Ω, 5Ω, and so on. To minimize these high-order harmonics, the amplitude of the modulating voltage signal has to be very small, such that $V_m << V_\pi/2$, and therefore $J_1(x)$ is much higher than $J_3(x)$ and $J_5(x)$.

It is noticed that so far we have only discussed the intensity transfer function of the external modulator as given by Eq. (7.1.17), whereas the optical phase information has not been discussed. In fact, an external optical modulator may also have a modulating chirp similar to the direct modulation of semiconductor lasers but originated from a different mechanism and with a much smaller chirp parameter. To investigate the frequency chirp in an external modulator, the input/output optical field relation given by Eq. (7.1.13) has to be used. If the optical phase ϕ_2 is modulated by a small signal $\delta\phi(t)$, which is superimposed on a static value ϕ_{20} so that $\phi_2 = \phi_{20} + \delta\phi(t)$, we can write $\phi_c = \phi_1 + \phi_{20} + \delta\phi(t)$ and $\Delta\phi = \phi_1 - \phi_{20} - \delta\phi(t)$. Eq. (7.1.13) can be modified as

$$E_0 = \cos\left(\frac{\phi_0 - \delta\phi(t)}{2}\right) e^{j(\phi_{c0} + \delta\phi(t))/2} E_i \quad (7.1.20)$$

where $\phi_{c0} = \phi_1 + \phi_{20}$ and $\phi_0 = \phi_1 - \phi_{20}$ are the static values of the common mode and the differential phases. It is evident that the phase of the optical signal is modulated as represented by the factor $e^{j\delta\phi(t)/2}$ in Eq. (7.1.20).

Similar to the linewidth enhancement factor defined in Eq. (3.3.41), a chirp parameter can be defined as the ratio between the phase modulation and the intensity modulation. It can be found from Eq. (7.1.20) that

$$\frac{dP_0}{dt} = \frac{-P_i}{2} \sin(\phi_0 - \delta\phi(t)) \frac{d\delta\phi(t)}{dt} \quad (7.1.21)$$

Thus, the equivalent chirp parameter of this modulator is

$$\alpha_{lw} = 2P_0 \frac{d\delta\phi(t)/dt}{dP_0/dt}\bigg|_{\delta\phi(t)=0} = \frac{1 + \cos(\phi_0)}{\sin(\phi_0)} \quad (7.1.22)$$

Not surprisingly, the chirp parameter is only a function of the DC bias. The reason is that although the phase modulation efficiency $d\delta\phi(t)/dt$ is independent of the bias, the efficiency of the normalized intensity modulation is a function of the bias. At $\phi_0 = (2m+1)\pi$ corresponding to the minima of the power transfer function shown in Fig. 7.1.7, the chirp parameter is $\alpha_{lw} = 0$ because the output optical power is zero, whereas at $\phi_0 = 2m\pi \pm \pi/2$ corresponding to the maxima of the power transfer function, the chirp parameter is $\alpha_{lw} = \infty$ as the small-signal intensity modulation efficiency is zero at these bias points. It is also interesting to note that the sign of the chirp parameter can be positive or negative depending on the

DC bias on the positive or negative slopes of the power transfer function (Cartledge, 1995). This adjustable chirp of external modulator may be utilized in system designs to compensate the effect of chromatic dispersion of the optical fiber.

However, for many applications chirp is not desired, and thus external modulators with zero chirp have been developed. These zero-chirp modulators can be built based on a balanced MZI configuration with antisymmetric driving of the two MZI arms, as shown in Fig. 7.1.8. In this case, the two MZI arms have the same physical length, but the electrical fields are applied in the opposite directions across the two arms, creating an antisymmetric phase modulation.

Recall that in Eq. (7.1.13), the common-mode phase delay, which determines the modulating chirp, is $\phi_c = \phi_1 + \phi_2$. If both ϕ_1 and ϕ_2 are modulated by the same amount $\delta\phi(t)$ but with opposite signs: $\phi_1 = \phi_{10} + \delta\phi$ and $\phi_2 = \phi_{20} - \delta\phi$, then $\phi_c = \phi_{10} + \phi_{20}$ will not be time dependent and therefore no optical phase modulation is introduced. For the differential phase delay, $\Delta\phi = \phi_{10} - \phi_{20} + 2\delta\phi(t)$, which doubles the intensity modulation efficiency and reducing the value of V_π by half.

7.1.3 OPTICAL INTENSITY MODULATION VS. FIELD MODULATION

For optical systems based on intensity modulation and direct detection (IMDD), optical phase information is not utilized, whereas for optical systems with coherent detection, both the amplitude and the phase of the optical field can be used to carry information. It is important to understand the relation between the power transfer function and the optical field transfer function of an external modulator, as illustrated in Fig. 7.1.7. Compare Eqs. (7.1.13) and (7.1.14), the power transfer function is equal to the square of the field transfer function, and the periodicity is doubled in this squaring operation. As a consequence, if the modulator is biased at a minimum power transmission point, the field transfer function has the best linearity because it swings between ±1, while the power transfer function is always positive. As shown in Fig. 7.1.9, for optical field modulation with the modulator biased at a minimum power transmission point, the output optical field E_0 is almost linearly proportional to the driving electric voltage signal, and this electric voltage is allowed to swing over $2V_\pi$ to obtain the maximum modulation index.

With this modulation, the spectrum of the optical field has two sidebands at $\omega_0 \pm \Omega$, while the optical carrier component at ω_0 is suppressed. This bipolar modulated optical field can be detected in a receiver equipped with an optical phase discriminator, such as an interferometer, or by a coherent homodyne detection receiver. However, if this modulated optical signal is directly detected by a

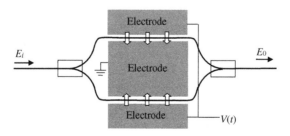

FIG. 7.1.8

Dual-drive EO modulator based on an MZI configuration.

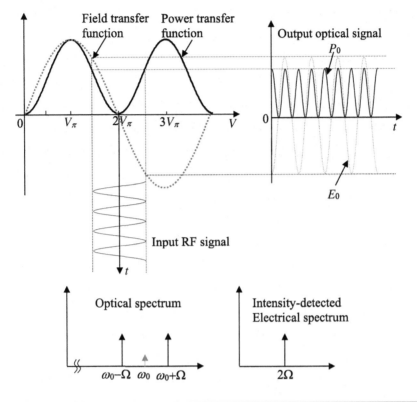

FIG. 7.1.9

Electro-optic modulator power transfer function and field transfer function.

photodiode in which the photocurrent is linearly proportional to the signal optical intensity, the RF spectrum of the photocurrent will have a predominant frequency component at 2Ω, which is twice the modulating frequency. This can be also explained as the mixing between the two-frequency components $\omega_0 + \Omega$ and $\omega_0 - \Omega$ of the optical field at the photodiode.

7.1.4 FREQUENCY DOUBLING AND HIGH-ORDER HARMONIC GENERATION

As illustrated in Fig. 7.1.9, when an EO intensity modulator based on a Mach-Zehnder configuration is biased either at a minimum or a maximum power transmission point, it can be used to perform frequency doubling between the modulating RF signal and the intensity of the optical signal. This can be used to generate high-frequency modulation through a modulator with a relatively low-frequency bandwidth.

Referring to the power transfer function shown in Eq. (7.1.17), the minima or maxima of power transmission correspond to the biasing phase of $\phi_0 = m\pi \pm \pi/2$, so that

$$P_0 = \frac{P_i}{2}\left[1 \pm \cos\left(\frac{\pi V(t)}{V_\pi}\right)\right] \tag{7.1.23}$$

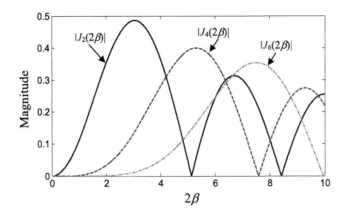

FIG. 7.1.10

Absolute values of Bessel functions $J_2(2\beta)$, $J_4(2\beta)$, and $J_6(2\beta)$.

If the input RF signal is a sinusoid, $V(t) = V_m \cos(\Omega t)$, then the output is

$$P_0 = \frac{P_i}{2}\left[1 \pm \cos\left(\frac{\pi V_m}{V_\pi}\cos(\Omega t)\right)\right] \tag{7.1.24}$$

This can be expanded into a Bessel series as:

$$P_0 = \frac{P_i}{2}[1 \pm J_0(2\beta) \mp 2J_2(2\beta)\cos(2\Omega t) \pm 2J_4(2\beta)\cos(4\Omega t) \mp \ldots] \tag{7.1.25}$$

where $\beta = \pi V_m/(2V_\pi)$ is the modulation index.

In the spectrum of the output optical intensity P_0, the fundamental frequency component Ω of the driving signal is eliminated, and the lowest modulation harmonic is at 2Ω, which doubles the input RF frequency. The absolute values of $J_2(2\beta)$, $J_4(2\beta)$ and $J_6(2\beta)$ are shown in Fig. 7.1.10 for the convenience of indicating the relative amplitudes of the second-, fourth- and six-order harmonics as the function of the modulation index. When the modulating amplitude is small enough so that $\beta \ll 1$, Bessel terms higher than the second order can be neglected, resulting in only a DC and a frequency-doubled component. This technique can also be used to generate quadruple frequency by increasing the amplitude of the modulating RF signal such that $2\beta \approx 5.14$ where the amplitude of the second-order harmonic is zero. In general, a precise control of the modulation index x can help in selecting or eliminating certain orders of harmonics. However, in practice, a large-amplitude RF signal at high frequency is usually difficult to generate; this requires high-speed and high-power electronics.

7.2 OPTICAL SINGLE-SIDEBAND MODULATION

Generally, a real-valued RF signal has a double-sideband spectrum. When modulated by a real-valued RF signal through an EO modulator, the modulated optical signal also has two sidebands—one on each side of the optical carrier, which is usually referred to as *double-sideband modulation*. Since these two sidebands carry redundant information, removing one of them will not affect the information capacity of the optical system. A single sideband optical signal, on the other hand, occupies a narrower spectral bandwidth and

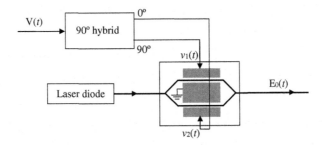

FIG. 7.2.1

Optical single-sideband modulation using a dual-electrode MZI electro-optic modulator.

thus results in better bandwidth efficiency compared to double optical sidebands, which is especially advantageous in multi-wavelength WDM systems. In addition, an optical signal with narrower spectral bandwidth is less susceptible to the effect of chromatic dispersion in optical fiber. These are the major reasons that optical single-sideband (OSSB) modulation is attractive (Smith et al., 1997; Hui et al., 2002).

An straightforward way to generate OSSB is to use a notch optical filter, which directly removes one of the two optical sidebands. However, this technique requires stringent wavelength synchronization between the notch filter and the transmitter. OSSB can also be generated using an EO modulator with a balanced MZI structure and two electrodes, which can be independently modulated with RF signals $v_1(t)$ and $v_2(t)$, as shown in Fig. 7.2.1.

If we assume that the frequency of the optical carrier is ω_0 and the RF modulating signal is a sinusoid at frequency Ω, based on Eq. (7.1.12), the output optical field is

$$E_0(t) = \frac{E_i}{2}[\exp(j\omega_0 t + j\phi_1(t)) + \exp(j\omega_0 t + j\phi_2(t))] \qquad (7.2.1)$$

which includes both the positive and the negative optical sidebands. $\phi_1(t)$ and $\phi_2(t)$ are the phase delays of the two MZI arms. Now we suppose that these two-phase delays can be modulated independently by voltage signals, $v_1(t)$ and $v_2(t)$, respectively. These two RF signals have the same amplitude and the same frequency, but there is a relative RF phase difference θ between them, so that, $v_1(t) = V_b + V_m \cos\Omega t$ and $v_2(t) = V_m(\cos\Omega t + \theta)$, where V_b is a DC bias and V_m is the amplitude. Therefore, the two-phase terms in Eq. (7.2.1) are:

$$\phi_1(t) = \phi_{01} + \beta\cos(\Omega t + \theta)$$

and

$$\phi_2(t) = \beta\cos(\Omega t)$$

where $\beta = \pi V_m/V_\pi$ is the modulation index, and $\phi_{01} = \pi V_b/V_\pi$ is a relative phase due to the DC bias. Then, the output optical field can be expressed as

$$E_0(t) = \frac{E_i e^{j\omega_0 t}}{2}\{\exp[j\phi_{01} + j\beta\cos(\Omega t + \theta)] + \exp[j\beta\cos(\Omega t)]\} \qquad (7.2.2)$$

If the modulator is biased at the quadrature point (e.g., $\phi_{01} = 3\pi/4$ so that $e^{j\phi_{01}} = -j$), Eq. (7.2.2) becomes

$$E_0(t) = \frac{E_i}{2}e^{j\omega_0 t}\{\exp[j\beta\cos(\Omega t)] - j\exp[j\beta\cos(\Omega t + \theta)]\} \tag{7.2.3}$$

Now we can show that the relative RF phase shift between the signals coming to the two electrodes have significant impactin the spectral properties of the modulated optical signal.

In the first case, if the relative RF phase shift between the two RF modulating signals $v_1(t)$ and $v_2(t)$ is 180 degree, that is, $\theta = (2m \pm 1)\pi$ with m an integer, Eq. (7.2.3) becomes

$$E_0(t) = \frac{E_i e^{j\omega_0 t}}{2}[\exp(j\beta\cos\Omega t) - j\exp(-j\beta\cos\Omega t)] \tag{7.2.4}$$

Mathematically, the trigonometric functions can be expanded into a Bessel series, known as the Jacobi-Anger identity,

$$\exp(j\beta\cos\Omega t) = \sum_{k=-\infty}^{\infty} j^k J_k(\beta)e^{jk\Omega t} \tag{7.2.5a}$$

$$\exp(j\beta\sin\Omega t) = \sum_{k=-\infty}^{\infty} J_k(\beta)e^{jk\Omega t} \tag{7.2.5b}$$

Consider that $J_{-k}(\beta) = (-1)^k J_k(\beta)$, Eq. (7.2.4) can be written as

$$E_0(t) = \frac{E_i e^{j\omega_0 t}}{2} \sum_{k=-\infty}^{\infty} j^k J_k(\beta)\left(e^{jk\Omega t} - je^{-jk\Omega t}\right) \tag{7.2.6}$$

For $k=0$, the carrier component is $0.5J_0(\beta)E_i e^{j\omega_0 t}(1-j)$. For $k=\pm 1$, the signal components for the positive and negative sidebands are $0.5(1+j)J_1(\beta)E_i e^{j(\omega_0 \pm \Omega)t}$. The second harmonic on both sideband corresponding to $k=\pm 2$ are $0.5(1+j)J_2(\beta)E_i e^{j(\omega_0 \pm 2\Omega)t}$. This is a typical double-sideband optical spectrum where the major frequency components are ω_0, $\omega_0 \pm \Omega$, $\omega_0 \pm 2\Omega$, and so on. Fig. 7.2.2A shows an example of power spectral density of a double-sideband modulated optical signal. The amplitude of the desired modulation sidebands at $\omega_0 \pm \Omega$, is proportional to $[J_1(\beta)]^2$, while the amplitude of the second-order harmonics at $\omega_0 \pm 2\Omega$, is proportional to $[J_2(\beta)]^2$. In this example, a modulation index of $\beta = 0.1$ is used so that $J_1(\beta)/J_2(\beta) = 40$, so that the second-order harmonic power is about 32 dB lower than the component at the fundamental frequency in the optical spectrum.

The power transfer function of Eq. (7.1.18) can be obtained by squaring on both sides of Eq. (7.2.4) except that the modulation index is doubled because of push-pull driving on the two electrodes. While the output of the power (or envelope) transfer function can be measured by a photodiode after the modulator, the output of the field transfer function is in the optical domain representing the signal optical spectrum which can be measured by an optical spectrum analyzer. The quadrature point of the power transfer function is not the same as the quadrature point of the optical field transfer function. That is why the second-order harmonic does not exist in Eq. (7.1.19) for the envelope transfer function, but is present in the optical spectrum of Eq. (7.2.6).

In the second case, if the relative RF phase shift between the two RF modulating signals is 90°, that is, $\theta = 2m\pi - \pi/2$, Eq. (7.2.3) becomes

$$E_0(t) = \frac{E_i}{2}e^{j\omega_0 t}\{\exp[j\beta\cos(\Omega t)] - j\exp[j\beta\sin(\Omega t)]\} \tag{7.2.7}$$

Again, based on Jacobi-Anger identity shown in Eqs. (7.2.5a) and (7.2.5b), Eq. (7.2.7) can be written as

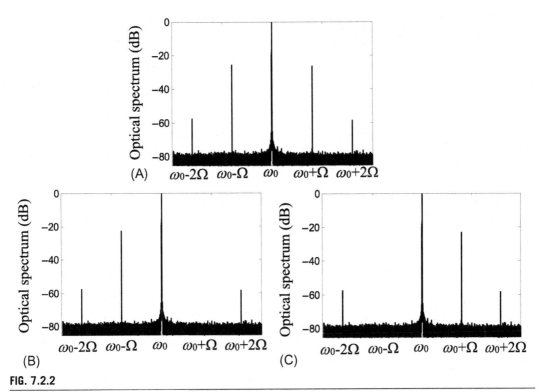

FIG. 7.2.2

Example of calculated power spectral density of modulated optical signal. (A) Double-sideband modulation, (B) single-sideband modulation with upper sideband suppressed, and (C) single-sideband modulation with lower sideband suppressed. ω_0 is the optical frequency, Ω is the modulation frequency, and modulation index is $\beta = 0.1$.

$$E_0(t) = \frac{E_i e^{j\omega_0 t}}{2} \sum_{k=-\infty}^{\infty} J_k(\beta) \left(j^k - j \right) e^{jk\Omega t} \tag{7.2.8}$$

In this case, for $k=0$, the carrier component is still $0.5 J_0(\beta) E_i e^{j\omega_0 t}(1-j)$. However, for $k=\pm 1$, the signal components for the positive and negative sidebands are zero and $J_1(\beta) E_i j e^{j(\omega_0 - \Omega)t}$, respectively. This is an optical single-sideband (OSSB) spectrum with the positive sideband suppressed and the energy in the negative sideband is doubled compared to the optical double sideband spectrum. One can also choose to suppress the negative optical sidebands while keeping the positive sideband either by changing the DC bias to $\phi_{01} = \pi/4$ so that $e^{j\phi_{01}} = j$ in Eq. (7.2.2), or by changing the RF phase shift to $\theta = 2m\pi + \pi/2$ in Eq. (7.2.3).

Fig. 7.2.2B and C show single-sideband modulated optical spectra with the lower and the upper sideband suppressed, respectively. In the single-sideband modulation, the energy of the suppressed sideband is transferred to the opposite sideband, and as a result, the power of the remaining sideband is increased by 3 dB compared to the double-sideband modulation.

7.3 OPTICAL I/Q MODULATOR

Although intensity modulation has been widely used in optical communication systems, complex optical field modulation has been shown much more efficient in terms of transmission capacity as both the intensity and the optical phase can be used to carry information. EO modulators based on the Mach-Zehnder interferometer (MZI) configuration allows both intensity modulation and phase modulation with proper selections of biasing voltage, as well as the polarity and phase of the driving RF signals on the two arms. However, phase modulation and intensity modulation are not independently addressable in a Mach-Zehnder EO modulator, which restricts the degree of freedom in the complex optical field modulation. In this section, we discuss an EO in-phase and quadrature (I/Q) modulator which is capable of complex optical field modulation.

The configuration of an EO I/Q modulator is shown in Fig. 7.3.1A, which is also based on an MZI structure, but an independent MZI (MZIa and MZIb) is built in each arm of a bigger MZI (MZIc). A separate optical phase control section is built in one of the two arms of MZIc to adjust the relative optical phase (Tsukamoto et al., 2006; Cho et al., 2006).

To simplify the analysis, we assume that both MZIa and MZIb are chirp-free, which can be achieved with antisymmetric electric driving of the two electrodes as shown in Fig. 7.1.4 so that common-mode phase modulation is zero. We also assume that both MZIa and MZIb are biased at the minimum power transmission point so that the optical field transfer function of each of them is

$$E_{out1,2} = E_{in1,2} \sin\left(\pi \frac{v_{1,2}(t)}{V_\pi}\right) \tag{7.3.1}$$

where $v_{1,2}(t)$ is the driving electric voltage, $E_{in1,2}$ is the input optical field, and $E_{out1,2}$ is the output optical field of these two MZMs. Combining the transfer functions of MZIa and MZIb, and set a relative optical phase shift between the two arms of MZIc at $\theta = \pi/2$, the overall optical field transfer function of MZIc is

$$E_o = \frac{E_i}{2}\left[\sin\left(\pi\frac{v_1(t)}{V_\pi}\right) + j\sin\left(\pi\frac{v_2(t)}{V_\pi}\right)\right] \tag{7.3.2}$$

The two free parameters $v_1(t)$ and $v_2(t)$ in Eq. (7.3.2) provide the capability of modulating the in-phase and the quadrature components of the optical field independently. This capability of complex optical

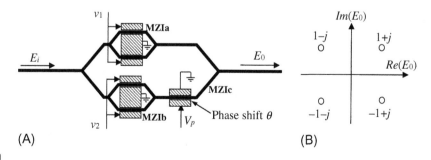

(A)   (B)

FIG. 7.3.1

(A) Configuration of an electro-optic I-Q modulator based on the combination of three MZIs, (B) normalized constellation diagram of complex modulated QPSK optical signal.

field modulation allows a number of applications in optical communication systems. As an example, for QPSK modulation as illustrated in Fig. 7.3.1B, the complex optical field E_0 has four possible positions in the constellation diagram, $E_i(1+j)/2$, $E_i(1-j)/2$, $E_i(-1+j)/2$, and $E_i(-1-j)/2$. This can be obtained by setting four combinations of the driving voltage signals (v_1, v_2) as: $(V_\pi/2, V_\pi/2)$, $(V_\pi/2, -V_\pi/2)$, $(V_\pi/2, V_\pi/2)$, and $(-V_\pi/2, -V_\pi/2)$, respectively.

For analog modulation, a small-signal approach can be used when $v_{1,2}(t) \ll V_\pi$. In such a case, Eq. (7.3.2) can be linearized so that

$$E_0 \approx E_i \left(\frac{\pi}{2V_\pi} \right) [v_1(t) + jv_2(t)] \qquad (7.3.3)$$

In a conventional double-sideband modulation with a real-valued RF signal, the upper and the lower modulation sidebands is a complex conjugate pair which carries redundant information. An I/Q modulation allows single-sideband modulation, as well as double-sideband modulation but with the two sidebands carrying independent information channels. To understand this application, assume $D_1(t)$ and $D_2(t)$ are two different data sequences carrying independent information. We can construct two voltage waveforms as,

$$v_1(t) = [D_1(t) + D_2(t)] \cos(\Delta \omega t) \qquad (7.3.4a)$$

$$v_2(t) = [D_1(t) - D_2(t)] \sin(\Delta \omega t) \qquad (7.3.4b)$$

where $\Delta \omega$ is an RF carrier frequency. Using the linearized transfer function of the I/Q modulator, the output optical field is

$$E_0(t) \approx \sqrt{P_i} e^{j\omega_0 t} \left(\frac{\pi}{2V_\pi} \right) [(D_1 + D_2) \cos(\Delta \omega t) + j(D_1 - D_2) \sin(\Delta \omega t)]$$
$$= \sqrt{P_i} \left(\frac{\pi}{2V_\pi} \right) \{ D_1 \exp[j(\omega_0 + \Delta \omega)t] + D_2 \exp[j(\omega_0 - \Delta \omega)t] \} \qquad (7.3.5)$$

where $E_i = \sqrt{P_i} e^{j\omega_0 t}$ is the input optical field, P_i is the input optical power which is a constant, and ω_0 is the frequency of the optical carrier. The spectrum of the optical field $E_0(t)$ in Eq. (7.3.5) has two sidebands, one on each side of the optical carrier ω_0. However, these two sidebands are not redundant. Instead, the upper and the lower sidebands carry data channels $D_1(t)$ and $D_2(t)$, independently. Fig. 7.3.2 shows a spectrum of complex modulated optical field with the upper and the lower modulation sidebands carrying independent information channels. Eq. (7.3.5) indicates that in this complex modulated optical field, the optical carrier component is suppressed because both MZIa and MZIb are biased at the minimum power transmission point. This helps improve the power efficiency of an optical communication system, as the optical carrier would only be a CW component.

7.4 ELECTRO-OPTIC MODULATOR BASED ON RING-RESONATORS

In addition to Mach-Zehnder configuration, optical ring resonators can also be used to make EO modulators. A ring resonator is frequency selective, and a ring resonator-based modulator can be used to selectively modulate the desired wavelength channel. Because of the compact size of micro-ring resonators and potential for low power consumption, EO modulators based on micro-ring resonators have attracted significant interest. Especially in integrated silicon photonic circuits, modulators based on micro-ring resonators are most common.

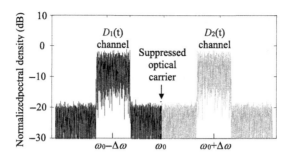

FIG. 7.3.2

Illustration of a modulated optical spectrum with carrier suppression and independent information channels carried by the upper and the lower modulation sidebands.

Although the transfer function of ring resonator based on optical fiber has been discussed previously in Chapter 6, EO modulators are based on planar waveguide rings so that electrodes can be added for modulation. Fig. 7.4.1A illustrates a silicon photonic ring modulator made on silicon-on-insulator (SOI) platform (Xu et al., 2007). The electrodes are made of heavily doped silicon in the substrate layer on both sides of the ring waveguide. This forms a p-i-n junction and carriers can be injected into the silicon waveguide through electric biasing, which allows the electric control of refractive index of the waveguide. The cross section of the silicon waveguide typically has the height of 200 nm and the width of 450 nm, as shown in Fig. 7.4.1B. Assume the radius of the waveguide ring is r. The coupling between the straight waveguide and the ring is determined by the gap between them. The coupling region is a 2×2 waveguide coupler, and its power transfer function has been presented in Chapter 6 as

$$T = \left| \sqrt{1-\varepsilon} - \frac{\varepsilon \sqrt{\eta} e^{j\phi}}{1 - e^{j\phi} \sqrt{\eta} \sqrt{1-\varepsilon}} \right|^2 \tag{7.4.1}$$

where ε is the power-splitting ratio of the coupler, η represents the waveguide loss, $\phi = 2\pi f \tau$ is the phase delay, and $\tau = nL/c$ is the time delay of the ring. $L = 2\pi r$ is the circumference of the ring.

As discussed in Section 6.3, the Q-value of a ring is defined by $Q = \lambda/\delta\lambda$, which is only dependent on the loss of the ring, where λ is the wavelength of the absorption peak, and $\delta\lambda$ is the FWHM of the power

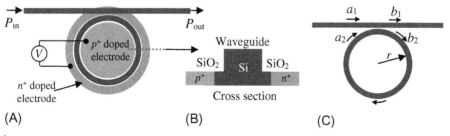

FIG. 7.4.1

(A) configuration of a waveguide ring modulator made on the SOI platform, (B) cross section of the waveguide, and (C) coupling between the straight waveguide and the ring resonator.

transmission notch. The Q-value of a ring resonator can be measured by an optical spectrometer and the propagation loss of the ring η can be calculated based on Eqs. (6.3.19) and (6.3.22) as

$$Q = \frac{2\pi r n_g}{\lambda} \frac{\pi}{\cos^{-1}\left(\frac{1-4\eta+\eta^2}{-2\eta}\right)} \tag{7.4.2}$$

where n_g is the group index of the waveguide, r is the radius of the ring, and η is the waveguide loss of the ring from b_2 to a_2 as shown in Fig. 7.4.1C similar to that shown in Fig. 6.3.7A.

Fig. 7.4.2 shows the power transfer function of a ring resonator near a resonance wavelength λ_0. Based on Eq. (7.4.1), if the waveguide loss η is known, the optimum splitting ratio ε of the coupler should satisfy $\eta = 1 - \varepsilon$ so that the power transmission at the resonance wavelength λ_0 is $T(\lambda_0) = 0$. For a silicon micro-ring modulator, the effective index n_g can be reduced by carrier injection, which provides the mechanism of phase modulation on the ring. The phase modulation modulates the resonance wavelength of the ring so that the transmission loss of an optical signal at λ_0 will be reduced as illustrated in Fig. 7.4.2.

A resonance wavelength change by an amount of FWHM linewidth $\delta\lambda$ is equivalent to a phase change of ϕ_Δ as defined by Eqs. (6.3.18) and (6.3.21). Based on the relation between the finesse and the Q-value defined by Eq. (6.3.22), it can be found that the relative change of refractive index to introduce the resonance wavelength change by a linewidth $\delta\lambda$ is

$$\delta n_g \equiv \left(\frac{n_{g,0} - n_g}{n_{g,0}}\right) = \frac{1}{Q} \tag{7.4.3}$$

where $n_{g,0}$ and n_g are the effective indices of waveguide before and after carrier injection, respectively. It usually needs to shift the resonance wavelength by more than a $\delta\lambda$ to ensure that the loss is low enough for the high transmission state. The dotted line in Fig. 7.4.2 shows that when the resonance is shifted by $4\delta\lambda$, the transmission loss at signal wavelength λ_0 is reduced to less than 1 dB. It has been found that the index reduction in a silicon p-i-n junction waveguide is on the order of $\Delta n \approx 2.1 \times 10^{-3}$ by a carrier density increase of $\Delta N = 10^{18} \text{cm}^{-3}$, so that the efficiency is $n_\Delta = \Delta n/\Delta N \approx 2.1 \times 10^{-21} \text{cm}^3$.

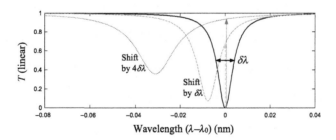

FIG. 7.4.2

Power transfer function of a micro-ring resonator near a resonance wavelength. EO modulation changes both the refractive index and the loss of the ring so that the resonance wavelength can be shifted and the peak absorption is also reduced. Solid line: ideal transfer function without carrier injection, dashed line: resonance wavelength is shifted by a linewidth $\delta\lambda$, and dotted line: resonance wavelength is shifted by a $4\delta\lambda$. λ_0 is the optical signal wavelength.

Thus, the carrier density change required to shift the ring resonance wavelength by $4\delta\lambda$ should be on the order of $\Delta N = (4n_{g,\,0}/Q) \times 4.8 \times 10^{20} \mathrm{cm}^{-3}$.

Meanwhile, carrier injection will also increase the absorption loss of the waveguide (decrease the value of η), this results in the reduction of Q-value and the broadening of the resonance linewidth as also shown in Fig. 7.4.2.

EXAMPLE 7.2

Consider a a silicon ring modulator with waveguide group index $n_g = 4.3$, ring radius $r = 6\,\mu\mathrm{m}$, and the quality factor $Q = 10^5$. The operation wavelength is in the 1550 nm window.
  (1) what is the ring propagation loss in dB from b_2 to a_2 with reference to Fig. 7.4.1C? and what is the optimum coupling ratio ε of the 2×2 coupler?
  (2) What are resonance wavelengths within the wavelength window ranging from 1528 to 1565 nm?
  (3) Choose the signal wavelength at the center resonance wavelength within the 1528 nm to 1565 nm window. The efficiency of carrier induced complex index change for the silicon waveguide is $\Delta n_g/\Delta N = -(2.1 - j0.4) \times 10^{-21}\mathrm{cm}^3$, where the imaginary part of Δn_g represents the loss, and ΔN is the carrier density change. Find the required carrier density change that shifts the resonance wavelength by $4\delta\lambda$ with $\delta\lambda$ the resonance linewidth. What is the loss for the optical signal at this level of carrier injection?

Solution

  (1) the roundtrip loss of the ring is related to the Q-value by Eq. (7.4.2), which can be solved numerically as $\eta \approx 0.9967 = 0.014\mathrm{dB}$ for the Q-value of 10^5. The optimum splittingg ratio of the 2×2 coupler is $\varepsilon = 1 - \eta = 0.0032$.
  (2) resonance condition requires $\lambda = 2\pi r n_g/m$ with m an integer. Resonance wavelengths within the window of 1528–1565 nm are 1529.3, 1543.87, and 1558.71 nm.

In fact the free-spectral range of a ring resonator is

$$\Delta\lambda_{FSR} = \frac{\lambda^2}{2\pi r n_g} \qquad (7.4.4)$$

which is approximately 14.82 nm in the 1550-nm wavelength window in this example.

  (3) choose signal wavelength at $\lambda_0 = 1543.87\mathrm{nm}$. The real part of the group index change has to be $\mathrm{Re}\{\Delta n_g\} = 4n_g/Q = 1.72 \times 10^{-4}$ for resonance wavelength change of $4\delta\lambda$, which corresponds to a carrier density increase of $\Delta N = 8.19 \times 10^{16}\mathrm{cm}^{-3}$. Because of the carrier-induced propagation loss of the silicon waveguide, the complex group index change is $\Delta n_g = 4n_g/Q = (1.72 + j0.1638) \times 10^{-4}$. This corresponds to a phase change of $\Delta\phi = 4\pi^2 r \Delta n_g/\lambda_0 = 0.026 + j0.0025$. Based on the transfer function (7.4.1), the power transfer function at $\phi = \Delta\phi$ is $T(\Delta\phi) = 0.964 = -0.16\mathrm{dB}$ as shown in Fig. 7.4.3.

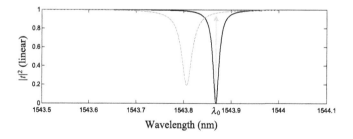

FIG. 7.4.3

Power transfer function of a micro-ring resonator near a resonance wavelength $\lambda_0 = 1543.87\mathrm{nm}$ without *(solid line)* and with carrier injection *(dashed line)*.

Because a ring resonator is largely transparent for wavelengths away from the resonance wavelength, a number of ring resonators can be cascaded to modulate the output of an optical source which emits multiple wavelengths, as shown in Fig. 7.4.4A. In this case, each ring resonator is dedicated to modulate the optical signal at one of the wavelengths without affecting all other wavelength channels. The wavelength window that this composite ring modulator can cover is determined by the FSR of each ring. Practically, for micro-ring-based modulators with the ring diameter on the order of a few micrometers and FSR on the order of tens of nanometers, even a very small fabrication of error could change the resonance wavelength significantly. In order to match the resonance wavelength of a ring resonator to the ITU wavelength grid, thermal tuning can be used to correct any fabrication error in the resonance wavelength, as well as to compensate for environmental changes. This can be done by integrating a resistive heater near the waveguide ring.

In comparison, if Mach-Zehnder modulators are used for the source with multiple wavelengths, a WDM EDMUX has to be used to separate different wavelength channels, and a MUX has to be used to combine them after EO modulation as shown in Fig. 7.4.4B.

For the EO modulators based on micro-rings discussed so far, we have used the static power transfer function as shown in Eq. (6.2.15). We have also shown that a high Q-value is necessary to increase the modulation efficiency and to reduce the required density of carrier injection. However, a high Q-value may become a limiting factor for the modulation speed because of the long photon lifetime. The Q-value is defined by $Q = \lambda_0/\delta\lambda$ with λ_0 and $\delta\lambda$ representing the operation wavelength and the width of the absorption line, respectively. The photon lifetime-limited modulation bandwidth can be estimated from (Li et al., 2011),

$$\delta f = \frac{c}{\lambda_0 Q} \tag{7.4.5}$$

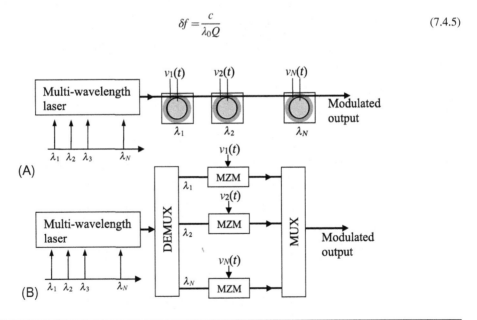

FIG. 7.4.4

Modulation of a light source with multiple wavelength channels by the cascade of multiple ring modulators (A) and by conventional Mach-Zehnder modulators (MZMs) (B).

this is because δf is related to $\delta\lambda$ by $\delta f = \delta\lambda \cdot c/\lambda_0^2$, where c is the speed of light in vacuum.

For a modulator operating in the 1550 nm wavelength window, in order for the photon lifetime limited modulation bandwidth to be wider than 10 GHz ($\delta f > 10$GHz), the Q-value has to be less than 2×10^4.

In addition, factors affecting modulation speed also include parasitic capacitance of the electrodes and carrier response time of the p-i-n junction. There are a number of research papers describing these effects both experimentally and mathematically (Xu et al., 2007; Baba et al., 2013; Sacher and Poon, 2008). Although the carrier response time of a p-i-n junction waveguide is typically of the nanosecond level, as the relationship between optical transmission and carrier density is highly nonlinear, a small change in the group index of the ring waveguide can introduce a big change in the optical transfer function, and thus the speed can be much faster. In fact, based on the ring modulator power transfer function of Eq. (7.4.1), assuming the optimum coupling coefficient $\varepsilon = 1 - \eta$, the transfer function can be written as

$$T(\phi) = 2\eta \frac{1 - \cos\phi}{1 + \eta^2 - 2\eta\cos\phi} \tag{7.4.6}$$

Assume the optical signal is at the resonance wavelength of the ring without carrier injection, and a small phase deviation $\delta\phi$ caused by carrier injection can be linearized with $\cos\delta\phi \approx 1 - \delta\phi^2/2$, so that

$$T(\delta\phi) = 1 - \frac{1}{1 + \frac{\eta}{(1-\eta)^2}\delta\phi^2} = 1 - \frac{1}{1 + \left(\frac{\delta N}{\xi}\right)^2} \tag{7.4.7}$$

where $\xi = \left(\frac{\lambda_0}{4\pi r n_\Delta}\right)\frac{(1-\eta)}{\sqrt{\eta}}$ is a device and material-related parameter with $n_\Delta = \delta n/\delta N$ being the efficiency of carrier induced index change, and δN being the carrier density change. Here we have used $\delta\phi = 4\pi^2 r \delta n/\lambda_0$. Fig. 7.4.5 shows the optical signal transmission loss through the ring modulator as the function of the normalized injection carrier density $\delta N/\xi$. This highly nonlinear transfer function indicates that a sharp reduction of transmission loss can be introduced by a relatively small change of the normalized

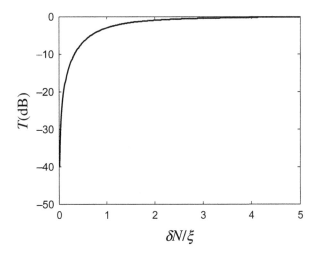

FIG. 7.4.5

Optical signal transmission loss through the ring modulator as the function of normalized injection carrier density.

carrier density. As an example, for a ring modulator operating at $\lambda_0 = 1550$nm with $r = 6\,\mu$m, $Q = 10^5$ (so that $\eta \approx 0.9967$), and $n_\Delta = 2.1 \times 10^{-21}$cm^3, the value of ξ is approximately 10^{16}cm^2.

Assume a carrier lifetime τ for the *p-i-n* junction, with a step function current injection the carrier density increase follows the standard charging function,

$$\delta N(t) = N_B \left(1 - e^{-t/\tau} \right) \tag{7.4.8}$$

where N_B is the final carrier density determined by the injection current level, and at $t = \tau$, the carrier density reaches $N_B(1 - e^{-1}) \approx 0.632 N_B$. Fig. 7.4.6A shows the normalized carrier density $\delta N / N_B$ as the function of the normalized time t/τ for the current turn-on transition.

Because of the nonlinear relation between carrier density and the signal transmission loss, the switch on of the optical signal can be much faster than the carrier lifetime τ depending on the target carrier density N_B. Fig. 7.4.6A shows the signal transmission loss as the function of the normalized time with different N_B levels. When the modulator is overdriving with N_B much higher than ξ, the optical signal switch-on time can be much shorter than the carrier lifetime τ.

Although we have only discussed intensity modulation with micro-ring modulators, it is also possible to realize complex optical I/Q modulation based on these devices (Zhang et al., 2008; Sacher and

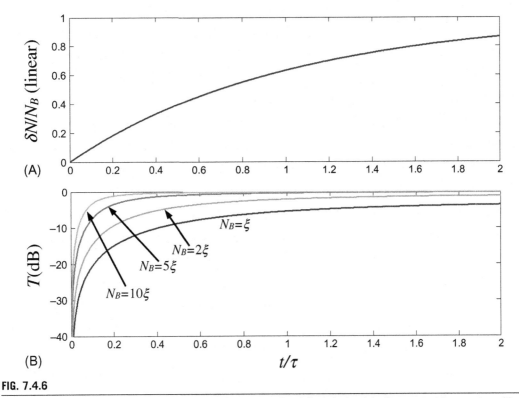

FIG. 7.4.6

(A) normalized carrier density as the function of the normalized time during injection current turn-on transition, (B) signal transmission loss as the function of normalized time for different N_B levels.

Poon, 2009; Dong et al., 2012). This is because the optical field transfer function of a micro-ring is complex, and carrier injection is able to change both the power transmission and the phase of the optical signal. An I/Q modulator has been demonstrated with two micro-rings each on an arm of a Mach-Zehnder interferometer structure.

7.5 OPTICAL MODULATORS USING ELECTRO-ABSORPTION EFFECT

As discussed in the last section, EO modulators made of LiNbO$_3$ in an MZI configuration have good performances in terms of high modulation speed, low-frequency chirp, and the capability of providing complex optical field modulation. However, these modulators cannot be monolithically integrated with semiconductor lasers due to the difference in the material systems. As a consequence, LiNbO$_3$-based external modulators are mostly standalone devices with input and output pigtail fibers, and the insertion loss is typically of the order of 3 to 5 dB. Difficulties in high-speed packaging and input/output optical coupling make LiNbO$_3$ based modulators relatively expensive, and thus they are usually used in high-speed and long-distance optical fiber systems. In addition, LiNbO$_3$-based EO modulators are generally sensitive to the state of polarization of the input optical signal; therefore, a polarization-maintaining fiber has to be used to connect the modulator to the source laser. This prevents LiNbO$_3$ modulators from performing re-modulation in locations far away from the laser source, where the state of polarization can be random.

Electro-absorption (EA) modulation can be made from the same type of semiconductor materials, such as group III–V materials, as are used for semiconductor lasers. By appropriate doping, the bandgap of the material can be engineered such that it does not absorb the signal photons and thus the material is transparent to the input optical signal. However, when an external electrical voltage is reversely biased across the pn junction, its bandgap will be changed to the same level of the signal photon energy, mainly through the well-known Stark effect. Then, the material starts to absorb the signal photons and converts them into photocurrent, similarly to what happens in a photodiode. Therefore, in an EA modulator the optical transmission coefficient through the semiconductor material is a function of the reversely applied voltage. Because EA modulators are made by semiconductor, they can usually be monolithically integrated with semiconductor lasers, as illustrated in Fig. 7.5.1. In this example, the DFB laser section is driven by a constant injection current I_C to produce a constant optical power, whereas the EA section is separately controlled by a reverse-bias voltage $V(t)$ that determines the strength of absorption.

The optical field transfer function of an EA modulator (EAM) can be expressed as

$$E_0(t) = E_i \exp\left\{ -\frac{\Delta\alpha[V(t)]}{2}L - j\Delta\beta[V(t)]L \right\} \qquad (7.5.1)$$

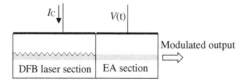

FIG. 7.5.1

EA modulator integrated with a DFB laser source.

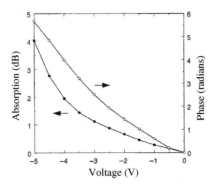

FIG. 7.5.2

Phase and absorption coefficients of an EA modulator (Cartledge, 1998).

where $\Delta\alpha[V(t)]$ is the bias voltage-dependent power attenuation coefficient, which originates from the electro-absorption effect in reverse-biased semiconductor pn junctions. $\Delta\beta[V(t)]$ is the voltage-dependent phase coefficient, which is introduced by the EO effect. L is the length of the EA modulator. In general, both $\Delta\alpha$ and $\Delta\beta$ are strongly nonlinear functions of the applied voltage V. As an example, Fig. 7.5.2 shows the attenuation and phase shift of a 600-μm long straight EA waveguide. Both of these two curves are nonlinear functions of the applied voltage V, and they are determined by the material bandgap structure, as well as the specific waveguide configuration.

In this EA modulator with a 600-μm long EA waveguide, the maximum extinction ratio is approximately 5 dB with 5-V applied voltage. Thus, a longer EA waveguide is required to achieve a higher extinction ratio. From an application point of view, as EA modulators can be monolithically integrated with semiconductor lasers on the same chip, they can be relatively low cost and more compact compared to LiNbO$_3$-based modulator. On the other hand, because both the absorption and the chirp parameters of an EA modulator are nonlinear functions of the bias voltage, the overall performance of optical modulation is generally not as good as using a LiNbO$_3$ external optical modulator based on the MZI configuration. In practical applications, there is generally no optical isolator between the semiconductor laser source and the integrated EA modulator, and thus optical reflection from an EA modulator often affects the wavelength and the phase of the laser to some extent (Hashimoto et al., 1992). Modeling of optical system performance based on an EA modulator has to take into account this residual optical reflection as well as the nonlinear absorption and phase transfer functions.

Because the semiconductor-based waveguide used for the EA modulator introduces both electro-absorption and signal optical phase change due to the applied electrical field, a Mach-Zehnder configuration can also be used to improve the extinction ratio and modify the chirp parameter of the modulator. As the EAM waveguide cannot be forward biased, it is not possible to use the antisymmetric push-pull biasing configuration as that used for dual-electrode LiNbO$_3$ intensity modulator shown in Fig. 7.1.6.

Fig. 7.5.3 shows the configuration of adding an EA-modulator in a Mach-Zehnder interferometer (MZI) configuration, in which the EAM is in one arm and a phase shifter in the other arm of the MZI. As the phase shifter only provides a constant phase control, it can simply be accomplished by a thermal heater. Because the EAM introduces transmission loss with the applied negative voltage, while phase shifter does not, in order to achieve the optimum performance, both of these two directional couplers should not equally split the signal optical power.

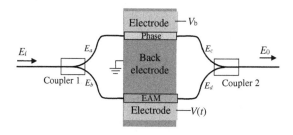

FIG. 7.5.3

Illustration of an intensity modulator (top view) based on an MZI configuration with one of the two arms made by an EAM.

Details of optical directional couplers have been discussed in Chapter 6, but simply based on energy conservation, we can assume that the input-output relations of the first and the second splitters are $|E_a/E_i| = \sqrt{\varepsilon_1}$, $|E_b/E_i| = \sqrt{1-\varepsilon_1}$, $|E_c/E_0| = \sqrt{1-\varepsilon_2}$, and $|E_d/E_0| = \sqrt{\varepsilon_2}$ respectively, where ε_1 and ε_2 are power splitting ratios of the two couplers. This ensures that $|E_a|^2 + |E_b|^2 = |E_i|^2$ and $|E_c|^2 + |E_d|^2 = |E_0|^2$. The optical power transfer function of the modulator shown in Fig. 7.5.3 is then,

$$T(V) = E_0(t)/E_i(t) = \sqrt{A(V)}\sqrt{1-\varepsilon_1}\sqrt{\varepsilon_2}e^{j\Phi_{EAM}(V)+j\Phi_{EA0}} + \sqrt{\varepsilon_1}\sqrt{1-\varepsilon_2}e^{j\Phi_{ph}(V_b)} \qquad (7.5.2)$$

where $A(V) = e^{-\Delta\alpha(V)L}$ is the voltage-dependent power attenuation of EAM, $\Phi_{EAM}(V) = \beta(V)L$ is the voltage-dependent phase delay of the EAM arms, Φ_{EA0} is an initial phase delay without the biasing voltage on the EAM arms, and $\Phi_{ph}(V_b)$ is the phase delay of the phase shifter arm controlled by the biasing voltage V_b. Ideally for the highest power transmission in the "on" state of the modulator with $V=0$, $A(0) = 1$, $\Phi_{EAM}(0) = 0$, and $\Phi_{ph}(V_b) = \Phi_{EA0}$, we should have $\sqrt{1-\varepsilon_1}\sqrt{\varepsilon_2} + \sqrt{1-\varepsilon_2}\sqrt{\varepsilon_1} = 1$. Let $\sqrt{1-\varepsilon_1}\sqrt{\varepsilon_2} = K$, Eq. (7.5.2) can be rewritten as (Ueda et al., 2014)

$$T(V) = E_0(t)/E_i(t) = e^{j\Phi_{EA0}}\left\{\sqrt{A(V)}Ke^{j\Phi_{EAM}(V)} + (1-K)e^{j\Phi_{ph}(V_b)-j\Phi_{EA0}}\right\} \qquad (7.5.3)$$

At the "off" state of the modulator, the ideal power transmission has to be zero. This requires that at the driving voltage $V = V_{off}$,

$$A(V_{off})K^2 = (1-K)^2 \qquad (7.5.4a)$$

that is, $K = 1/\left[1 + \sqrt{A(V_{off})}\right]$, and

$$\Phi_{EAM}(V_{off}) + [\Phi_{ph}(V_b) - \Phi_{EA0}] = \pi \qquad (7.5.4b)$$

Based on the data of attenuation and phase shift for the 0.6-mm long straight AOM waveguide shown in Fig. 7.5.2, $\Delta\alpha$ in [Neper/mm] and $\Delta\beta$ in [rad/mm] can be extracted as the functions of the applied voltage V as shown Fig. 7.5.4.

As an example, assume the EAM section of the waveguide is 1 mm long, and the phase delay of the phase shifter is $\Phi_{ph} = \Phi_{EA0}$ (mod $2n\pi$), in order to satisfy the condition of Eq. (7.5.4b), the bias voltage has to be $V_{off} \approx 2.44$ V. At this voltage, $\Delta\alpha(V_{off}) = 0.317$, and the optimum coupler design should provide $K = 0.5395$. Based on these parameters, the optical power transfer function and phase transfer function of the modulator can be calculated using Eq. (7.5.3), as shown in Fig. 7.5.5. The equivalent V_π value is 2.44 V in this example.

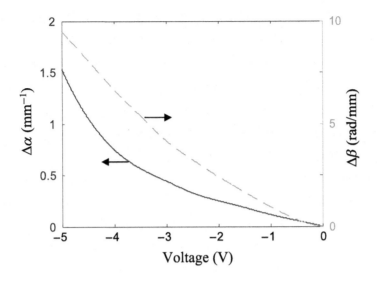

FIG. 7.5.4

Attenuation parameter $\Delta\alpha$ in [Neper/mm] and phase parameter $\Delta\beta$ in [rad/mm] extracted from Fig. 7.5.2.

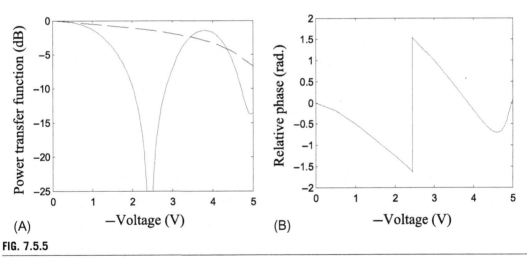

FIG. 7.5.5

Power transfer function (A) and phase transfer function (B) of EAM in a MZI configuration with $\Phi_{ph}=\Phi_{EA0}$. Dashed line in (A) shows the power transfer function of EAM along.

As a biasing mechanism, an adjustable phase shifter on the opposite arm of the MZI is necessary to compensate for the initial phase Φ_{EA0} of the EAM arm and its variation due to temperature change and environmental variations. Fig. 7.5.6 shows the impact of phase error $\delta\Phi=\Phi_{ph}-\Phi_{EA0}$ on the power transfer function and phase.

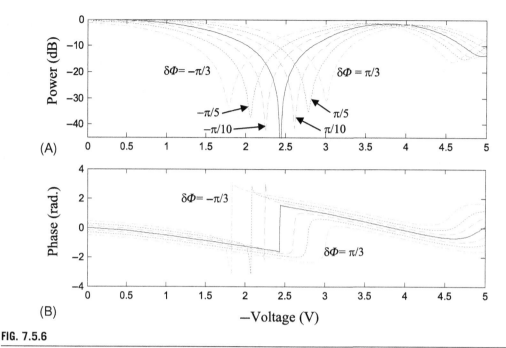

FIG. 7.5.6

Power transfer function (A) and phase transfer function (B) of EAM in an MZI configuration with $\Phi_{ph}=0$ (solid line), $\pm\pi/10$ (dashed line), $\pm\pi/5$ (dotted line), and $\pm\pi/3$ (dash-dotted line).

The ideal performance of the modulator is obtained with $\delta\Phi=0$, which allows the maximum extinction of the optical signal at $V_{off}=2.44\,V$ in this example. For an increase or decrease of $\delta\Phi$, the extinction ratio will be reduced as the voltage V_{off} corresponding to the minimum transmission will have to be changed to satisfy Eq. (7.5.4b), while the requirement of Eq. (7.5.4a) is longer satisfied.

Because both phase modulation and intensity modulation of an EAM are nonlinear functions of the applied voltage as shown in Fig. 7.5.4, the intensity modulator based on the MZI configuration also has a voltage-dependent chirp parameter. Based on the definition of modulation chirp in Eq. (7.1.22), the chirp parameter of the MZI modulator based on EAM can be obtained by

$$\alpha_{lw} = 2|T^2|\frac{d[\text{angle}(T)]/dV}{d|T^2|/dV}$$

where $T(V)$ is the transfer function of the modulator. Based on the intensity and phase transfer functions shown in Fig. 7.5.5, the chirp parameter of the device discussed in this example is shown in Fig. 7.5.7. An obvious singularity of modulation chirp happens at approximately $V=3.8\,V$ where the power transfer function has a local maximum as shown in Fig. 7.5.5A so that the differential power is zero, while the differential phase $d[\text{angle}(T)]/dV$ is nonzero at this voltage.

The device has the highest chirp parameter of approximately 10 at $V_{on}=0\,V$ because $d|T^2|/dV$ is small in the low voltage region. The chirp parameter reduces monotonically and it is approximately zero at $V_{off}=2.44\,V$ because $d|T^2|/dV$ approaches infinity at that voltage.

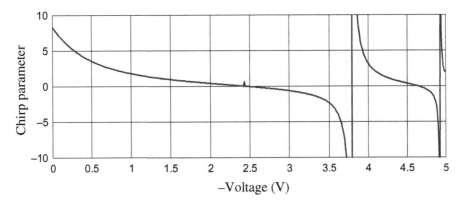

FIG. 7.5.7

Chirp parameter as the function of the applied voltage

7.6 SUMMARY

In this chapter, we have discussed external optical modulators for optical communications, which include phase modulator, intensity modulator and complex optical field modulator. A complex optical field modulator is able to modulate both amplitude and phase of the optical signal independently.

EO modulators based on $LiNbO_3$ are most popular in high-speed fiber-optic communication transmitters because the $LiNbO_3$ crystal has a high EO coefficient and low attenuation in the optical communication's wavelength. The refractive index of an EO material is dependent on the applied electrical field, so that an EO phase modulator can be straightforwardly made by a waveguide of the EO material with appropriate electrodes to introduce the electric field through biasing voltage. An EO intensity modulator can be made by converting optical phase modulation into intensity modulation through an optical interferometer such as a Mach-Zehnder interferometer (MZI).

For an EO intensity modulator, the intensity modulation on the optical signal is often associated with an optical phase modulation, and the ratio between them is known as the modulation chirp. Chirp parameter of an EO modulator depends on the material as well as the optical circuit configuration. Zero-chirp EO intensity modulator can be made with antisymmetric voltage applied on the two arms of a balanced MZI. Optical single-sideband modulation can be achieved by exploiting the relation between amplitude and phase modulation in a dual-drive MZI-based EO intensity modulator. Independent modulation of amplitude and phase of the optical signal is also possible by using a device of three combined MZIs, known as the in-phase/quadrature (I/Q) modulator. An I/Q modulator enables the modulation of the complex optical field as a vector, which is one of the key enabling technologies for coherent optical communication systems where both optical amplitude and phase are used simultaneously to carry the information.

With the rapid advance in silicon photonics for photonic integration, silicon-based micro-ring modulators have been demonstrated. Thanks to the low loss of the silicon waveguides in the 1550 nm optical communications wavelength window, high Q micro-rings have been fabricated on the silicon-on-insulator (SOI) platform. Silicon waveguides can be made into p-i-n junctions through impurity doping, which allows the carrier density modulation through carrier injection. The major advantages of the silicon-based micro-ring modulator include a small size suitable for photonic integration, and low modulation voltage that helps reducing electric power consumption. However, because the high-Q ring resonance wavelength is very sensitive to the temperature, feedback control through thermal tuning is usually required to stabilize the operation condition.

The electro-absorption (EA) modulator is another popular type of modulator used in optic communications. EA modulators are based on the effect of optical absorption when a semiconductor pn-junction structure is reversely biased. Because an EA modulator can be monolithically integrated with a semiconductor laser on the same material platform, both the cost and the footprint of an optical transmitter can be significantly reduced compared to that using a LiNbO$_3$ modulator. However, the extinction ratio of an EA modulator made of a single EA waveguide can be limited by the relatively low unit length absorption. An EA modulator can also be placed in a MZI configuration to increase the extinction ratio. Because of the nonlinear relations of phase delay and attenuation with the applied voltage, the chirp parameter can often be voltage dependent.

In an optical communication system, the selection of optical modulator depends on a number of factors including modulation speed, device size, driving electric signal voltage requirement, as well as the device cost. LiNbO$_3$-based Mach-Zehnder modulators usually have the best performance but are most expensive, and thus they are commonly used in high-speed- and long-distance optical transmission systems. In recent years, microstructure COMS compatible LiNbO$_3$ modulator fabrication technologies have been developed (Mercante et al., 2016; Rao and Fathpour, 2018), which made this material platform even more promising. The electro-absorption modulator (EAM) has cost advantage compared to a LiNbO$_3$-based modulator, but with reduced performance. Silicon micro-ring modulators are most suitable for parallel photonic interconnection, which may require large number of wavelength and spatial channels in a photonic integrated circuit.

PROBLEMS

1. For an EO-phase modulator operating in 1550 nm wave, assume the EO coefficient is $\alpha_{EO} \approx 1.5 \times 10^{-10}$V/m, electrode length is $L = 10$ mm, and the separation between $\pm$ electrodes is $d - 10\,\mu\text{m}$ (LiNbO$_3$ waveguide is sandwiched between the two electrodes), what is the voltage required to change the optical phase by π radians?

2. An EO intensity modulator in a symmetric Mach-Zehnder configuration is shown in the following figure. Only one of its two arms is phase modulated by a microwave source.

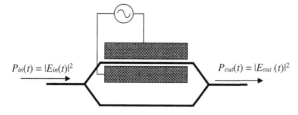

$P_{in}(t) = |E_{in}(t)|^2 \qquad P_{out}(t) = |E_{out}(t)|^2$

If a chirp parameter is defined by $\alpha = 2P_{out}\dfrac{d\Phi_c(t)/dt}{dP_{out}(t)/dt}$, where $P_{out}(t)$ is the output optical power and $\Phi_c(t)$ is the phase of the output optical field.

(a) Express the chirp parameter as a function of ϕ. (ϕ is the phase delay difference between the two arms). If the modulator is biased at $\phi = \pi/2$, what is the small-signal modulation chirp?

(b) Is it possible to change the sign of the chirp with the same device? How it can be accomplished?

3. An EO intensity modulator in a Mach-Zehnder configuration is shown in the following figure. However, one of the two branches of the Mach-Zehnder interferometer has a loss $0 < \eta < 1$, so that the input-output field relation is $E_0 = 0.5(e^{j\phi_1} + \eta e^{j\phi_2})E_i$.

Please derive an expression of the extinction ratio (defined by the ratio of the maximum and the minimum power transmission) of this modulator as the function of η. Please also plot the extinction ratio in [dB] as the function of η for $0.5 < \eta < 1$.

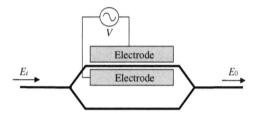

4. An EO intensity modulator in an asymmetric Mach-Zehnder configuration is shown in the following figure. Both of its two arms are phase modulated by the same microwave source but with opposite polarities. The length of the two electrode are not the same; they are L and xL, respectively, with $0 < x < 1$.

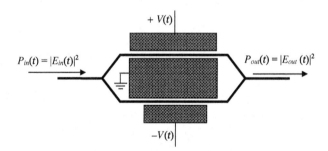

The chirp parameter is defined by $\alpha = 2P_{out}\frac{d\Phi(t)/dt}{dP_{out}(t)/dt}$, where $P_{out}(t)$ is the output optical power and $\Phi(t)$ is the phase of the output optical field. Suppose the modulation index is infinitesimal and the modulator is biased at the quadrature point (where the intensity modulation has the maximum efficiency), find the relationship between the chirp parameter and the parameter x. If $x = 0.5$, what is the chirp parameter?

Note: In the expression $\alpha = 2P_{out}\frac{d\Phi(t)/dt}{dP_{out}(t)/dt}$, P_{out} is the output power at the defined bias point.

5. A Mach-Zehnder EO intensity modulator with $V_\pi = 5\,\text{V}$ is biased at the quadrature point, and the driving RF signal is $V(t) = V_m\cos(\Omega t)$. The modulated signal optical power is detected by a photodiode as shown in the following figure.

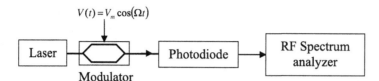

(a) For a modulation index of $m=V_m/V_\pi=0.25$, what are the ratios of the second- and the third-order harmonics with respect to the fundamental frequency component in the RF power spectral density, which can be measured by the RF spectrum analyzer?

(b) Instead of biasing at the quadrature point, the DC biasing is now moved to $\phi_0=\pi/6$, but still with $m=0.25$. Find the RF powers of the second- and the third-order harmonics with respect to that of the fundamental frequency component.

6. Consider an ideal EO modulator based on a MZI configuration with balanced dual-electrode. Assume an RF hybrid coupler ($\theta°$) is used as illustrated in the following figure with the input RF signals $v_1(t)=V_m\cos(\Omega t)$ and $v_2(t)=V_m\cos(\Omega t+\theta)$. The laser diode has 1mW average optical power.

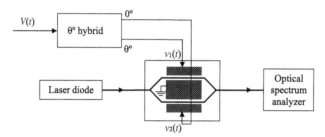

If the modulator is biased at the quadrature point of the power transfer function with $V_m=0.2V_\pi$ and $\theta=\pi/2$, please find the optical powers of the 0th, ±1, ±2, and ±3 harmonics of the optical spectrum.

7. Repeat problem 6, but with the modulator biased at the minimum power transmission point, that is, $\phi_{01}=\pi$ in Eq. (7.2.2).

8. Consider an ideal MZI-based EO I-Q modulator biased at the minimum power transmission point. The input optical signal frequency is ω_0 and the two RF driving waveforms are $v_1(t)=A_1\cos(\Omega t)$ and $v_2(t)=A_2\sin(\Omega t)$. What are the frequencies of major components of the modulated optical spectrum and their relative amplitude?

9. Consider an ideal EO I-Q modulator based on an MZI configuration shown below.

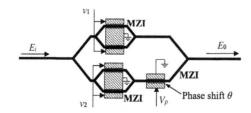

(1) Create two independent binary pseudorandom data sequences at 2-Gb/s data rate in the time domain.

(2) Up-convert these two data sequences onto the upper and the lower sidebands of the output optical spectrum with a 5-GHz RF carrier frequency.

(3) Plot the modulated optical spectrum.

10. Consider an electro-absorption (EA) modulator made of a straight semiconductor waveguide. The optical absorption and phase shift coefficients of the semiconductor material can be expressed in polynomial forms as $\Delta\alpha(V) = -0.015V^3 - 0.05V$ [mm^{-1}] and $\Delta\beta(V) = 0.2V^2 - V$ in [rad/mm], respectively, with reference to Eq. (7.5.1).

   (a) Assume that the waveguide length is 2 mm, for a binary modulation with the two voltage levels of 0 and -5 V, what is the power extinction ratio of the modulated optical signal?

   (b) Plot the chapter factor (also known as the linewidth enhancement factor) defined in Eq. (7.1.22) as the function of the applied voltage in the range of $-5\,\text{V} \le V \le 0$.

11. Consider a Mach-Zehnder modulator with EA waveguide in one of the two arms as shown in the following figure. Both couplers have 50% power splitting ratio. The optical absorption and phase shift coefficients of the semiconductor material can be expressed in polynomial forms as $\Delta\alpha(V) = -0.1V^3$ [mm^{-1}] and $\Delta\beta(V) = -V$ in [rad/mm], respectively, with reference to Eq. (7.5.1). EA waveguide length is $L = 3$ mm.

   (a) By setting a proper DC bias voltage V_b, the differential phase delay between the two arms is zero at $V_a = 0$ so that modulator power transfer function $|E_0/E_i|^2 = 1$. At $\Delta\beta(V_a)L = \pi$, what is the modulator power transfer function?

   (b) Changing the DC bias voltage V_b, so that the differential phase delay between the two arms is π at $V_a = 0$ so that modulator field transfer function $|E_0/E_i|^2 = 0$. At $\Delta\beta(V_a)L = \pi$, what is the modulator power transfer function?

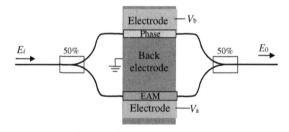

12. Derive Eq. (7.4.3), which relates the required relative refractive index change to introduce a resonance wavelength change by a linewidth $\delta\lambda$.

13. Based on Eqs. (7.4.7) and (7.4.8), determine the required N_B/ξ so that the turn-on time (determined by <3 dB loss) for the optical signal is 0.2τ.

14. Consider a silicon ring modulator with the waveguide group index $n_g = 4.3$, ring radius $r = 8\,\mu$m, and the quality factor $Q = 2 \times 10^4$. The operation wavelength is in the 1550 nm window.

   (a) What are resonance wavelengths within the wavelength window ranging from 1528 to 1565 nm?

   (b) Choose the signal wavelength is at the center resonance wavelength of the window. The efficiency of carrier induced complex index change for the silicon waveguide is $\Delta n_g/\Delta N = -(2.1 - j0.4) \times 10^{-21}\,\text{cm}^3$, where the imaginary part of Δn_g represents the loss, and ΔN is the carrier density change. Find the required carrier density change that shifts the resonance wavelength by $3\delta\lambda$ with $\delta\lambda$ the resonance linewidth. What is the loss for the optical signal at this level of carrier injection?

REFERENCES

Baba, T., Akiyama, S., Imai, M., Hirayama, N., Takahashi, H., Noguchi, Y., Horikawa, T., Usuki, T., 2013. 50-Gb/s ring-resonator-based silicon modulator. Opt. Express 21 (10), 11869–11876.

Cartledge, J.C., 1995. Performance of 10 Gb/s lightwave systems based on lithium niobate Mach-Zehnder modulators with asymmetric Y-branch waveguides. IEEE Photon. Technol. Lett. 7 (9), 1090–1092.

Cartledge, J.C., 1998. Comparison of effective parameters for semiconductor Mach Zehnder optical modulators. J. Lightwave Technol. 16, 372–379.

Cho, P.S., Khurgin, J.B., Shpantzer, I., 2006. Closed-loop bias control of optical quadrature modulator. Photonics Technol. Lett. 18 (21), 2209–2211.

Dong, P., Xie, C., Chen, L., Fontaine, N.K., Chen, Y.-K., 2012. Experimental demonstration of microring quadrature phase-shift keying modulators. Opt. Lett. 37 (7), 1178–1180.

Hashimoto, J.-I., Nakano, Y., Tada, K., 1992. Influence of facet reflection on the performance of a DFB laser integrated with an optical amplifier/modulator. IEEE J. Quantum Electron. 28 (3), 594–603.

Heismann, F., Korotky, S.K., Veselka, J.J., 1997. Lithium niobate integrated optics: selected contemporary devices and system applications. In: Kaminow, I.P., Koch, T.L. (Eds.), Optical Fiber Telecommunications III B. Academic Press, New York, pp. 377–462.

Hui, R., Zhu, B., Huang, R., Allen, C., Demarest, K., Richards, D., 2002. Subcarrier multiplexing for high-speed optical transmission. IEEE J. Lightwave Technol. 20, 417–427.

Li, G., Zheng, X., Yao, J., Thacker, H., Shubin, I., Luo, Y., Raj, K., Cunningham, J.E., Krishnamoorthy, A.V., 2011. 25Gb/s 1V-driving CMOS ring modulator with integrated thermal tuning. Opt. Express 19 (21), 20435–20443.

Mercante, A.J., Yao, P., Shi, S., Schneider, G., Murakowski, J., Prather, D.W., 2016. 110 GHz CMOS compatible thin film LiNbO3 modulator on silicon. Opt. Express 24 (14), 15590–15595.

Minakata, M., Saito, S., Shibata, M., Miyazawa, S., 1978. Precise determination of refractive-index changes in Ti-diffused LiNbO3 optical waveguides. J. Appl. Phys. 49, 4677–4682.

Noguchi, K., Mitomi, O., Miyazawa, H., Seki, S., 1995. A broadband Ti: LiNbO3 optical modulator with a ridge structure. J. Lightwave Technol. 13, 1164–1168.

Petermann, K., 1988. Laser Diode Modulation and Noise. Kluwer Academic Publishers.

Rao, A., Fathpour, S., 2018. Compact lithium niobate electrooptic modulators. IEEE J. Sel. Top. Quantum Electron. 24 (4). paper 3400114.

Sacher, W.D., Poon, J.K.S., 2008. Dynamics of microring resonator modulators. Opt. Express 16 (20), 15741–15753.

Sacher, W.D., Poon, J.K.S., 2009. Microring quadrature modulators. Opt. Lett. 34 (24), 3878–3880.

Sanna, S., Schmidt, W.G., 2010. Lithium niobate X-cut, Y-cut, and Z-cut surfaces from ab initio theory. Phys. Rev. B. 81. paper 214116.

Smith, G.H., Novak, D., Ahmed, Z., 1997. Overcoming chromatic dispersion effects in fiber-wireless systems incorporating external modulators. IEEE Trans. Microwave Technol. 45, 1410–1415.

Tsukamoto, S., Katoh, K., Kikuchi, K., 2006. Coherent demodulation of optical multilevel phase-shift-keying signals using homodyne detection and digital signal processing. IEEE Photon. Technol. Lett. 18 (10), 1131–1133. May 15.

Ueda, Y., Fujisawa, T., Kanazawa, S., Kobayashi, W., Takahata, K., Ishii, H., 2014. Very-low-voltage operation of Mach-Zehnder interferometer-type electroabsorption modulator using asymmetric couplers. Opt. Express 22 (12), 14610–14616.

Wooten, E.L., Kissa, K.M., Yi-Yan, A., Murphy, E.J., Lafaw, D.A., Hallemeier, P.F., Maack, D., Attanasio, D.V., Fritz, D.J., McBrien, G.J., Bossi, D.E., 2000. A review of lithium niobate modulators for fiber-optic communications systems. IEEE J. Sel. Top. Quantum Electron. 6 (1), 69–82.

Xu, Q., Manipatruni, S., Schmidt, B., Shakya, J., Lipson, M., 2007. 12.5 Gbit/s carrier-injection-based silicon microring silicon modulators. Opt. Express 15 (2), 430–436.

L. Zhang, J. Y. Yang, Y. Li, R. G. Beausoleil, and A. E. Willner, Optical Fiber Communication Conference (OFC 2008), (Optical Society of America, 2008), paper OWL5, 2008.

FURTHER READING

Dorrer, C., Kilper, D., Stuart, H., Raybon, G., 2002. Ultra-sensitive optical sampling by coherent-linear detection. In: Tech. Dig. Optical Fiber Communications (OFC 2002), Postdeadline Paper FD5.

Neophotonics 43G EML: https://www.neophotonics.com/product/43g-eml/.

Thorlabs modulator image: https://www.thorlabs.com/newgrouppage9.cfm?objectgroup_id=3948.

OPTICAL TRANSMISSION SYSTEM DESIGN

CHAPTER OUTLINE

Introduction 337
8.1 BER vs. *Q*-value for binary modulated systems 339
    8.1.1 Overview of IMDD optical systems 339
    8.1.2 Receiver *BER* and *Q* 342
8.2 Impacts of noise and waveform distortion on system *Q*-value 347
    8.2.1 *Q*-calculation for optical signals without waveform distortion 347
    8.2.2 *Q*-estimation based on eye diagram parameterization 348
8.3 Receiver sensitivity and required OSNR 352
    8.3.1 Receiver sensitivity 352
    8.3.2 Required OSNR 356
8.4 Concept of wavelength division multiplexing 358
8.5 Sources of optical system performance degradation 361
    8.5.1 Performance degradation due to linear sources 361
    8.5.2 Performance degradation due to fiber nonlinearities 372
    8.5.3 Semi-analytical approaches to evaluate nonlinear crosstalks in fiber-optic systems 376
8.6 Conclusion 408
Problems 409
References 414
Further reading 416

INTRODUCTION

Optical communication is one of the most important applications of fiber-optic technology. The introduction of optical fiber into communications revolutionized the entire telecommunications industry. The wide transmission bandwidth and low propagation loss make optical fiber an ideal medium for transmission. Nowadays, almost 100% of long-distance communication traffic is carried by optical fibers all over the world. Fiber-optic technology is the backbone of the modern internet carried by high-speed communication and data networks including wide area, metro area, and access networks. With the knowledge of optical components discussed in the previous chapters, we discuss how to

Introduction to Fiber-Optic Communications. https://doi.org/10.1016/B978-0-12-805345-4.00008-1

construct optical communication systems in this chapter based on these basic building blocks, and the characterization of system performance.

The physical layer of an optical fiber transmission system comprises a transmitter, a line system, and a receiver. The transmitter provides a means of uploading the electrical signal to be transmitted onto an optical carrier, known as electrical to optical (E/O) conversion. The line system delivers the modulated optical carrier to the receiver, which can be as simple as a length of optical fiber as the transmission medium, or as complex as a multi-span, optically amplified, and switched optical network with wavelength-division multiplexing (WDM). The receiver detects the optical carrier and down-converts the information from the optical carrier back to the electrical domain, known as optical to electric (O/E) conversion. Information on the optical carrier can either be analog or digital. While analog modulation is used for a number of applications including cable TV or radio-over-fiber, digital modulation has clear advantages for high-speed and long-distance transmission. We discuss the fundamentals of binary modulation in this chapter which is the simplest yet the most often used modulation format for digital transmission systems and communication networks, and leave the discussion of high order modulation formats to Chapter 10.

The fidelity of digital transmission is quantified by the bit error rate (BER). The BER is defined as the fraction of transmitted data that is mistakenly decoded by the receiver, which is a function of the system quality factor, Q. The quality factor Q is an electrical domain measure determined by the ratio of the separation of digital states to the noise associated with the states. Both the numerator and the denominator of Q can be partitioned into contributions whose sources are objects of system design. Examples include accumulated optical noise generated by optical amplifiers, signal optical power, polarization-dependent loss (PDL), and polarization mode dispersion (PMD), receiver and transmitter transfer function, accumulated chromatic dispersion, and nonlinear propagation noise and distortion.

Commercial optical systems are designed to operate with a BER lower than a specified maximum value over their lifetime. For example, a maximum BER of 10^{-15} is commonly allowed for fiber links spanning cities, continental, and intercontinental distances. Such links are often designed with forward error correction (FEC) wherein overhead bits encoded with the data payload in such a way as to allow limited correction of errors upon decoding at the receiver (Kumar et al., 2002). There is a wide range of FEC implementations offering a variety of correction capabilities and efficiencies. These FEC algorithms can deliver corrected BER as low as 10^{-15} based on received data with uncorrected (raw) BER as high as 10^{-3}.

Although fundamental communication protocols, modulation formats, and performance evaluation criteria are applicable, optical fiber communication has unique characteristics due to its high data rate and the special system properties due to the use of optical fibers. Understanding basic properties of optical systems and the underline physical mechanisms is very important in the design, development, and installation of fiber-optic transmission systems, subsystems, and networks.

In this chapter we describe basic parameters defining the performance of optical transmission systems. Although the description is focused on intensity-modulated direct detection (IMDD) systems with binary coding, most of the techniques and their fundamental principles are applicable to other types of modulation formats.

Section 8.1 provides an overview of digital optical transmission systems and their performance specifications, such as BER, the quality factor (Q), and their relation for binary modulated systems. Section 8.2 introduces the definitions of receiver sensitivity and the required optical signal-to-noise ratio (ROSNR). In practical applications, receiver sensitivity is an evaluation criterion useful for optical systems whose performance are limited by noise generated in the receiver, whereas ROSNR is a criterion often used for systems with inline optical amplifiers in which performance is mainly determined by the OSNR of the optical signal. Section 8.3 discusses the impact of noise and waveform distortion in the performance of an optical system. While noises, such as thermal noise, shot noise, and signal ASE beat noise, are random, waveform distortion which may be caused by limited bandwidth of the transmitter and receiver, as well as fiber chromatic dispersion, is usually deterministic. Thus, their impacts on the closure of eye diagram and the reduction of *Q*-factor are different. Section 8.4 introduces the concept of WDM, which allows dramatic increase of fiber system capacity by the use of multiple wavelength channels. Finally Section 8.5 discusses various linear and nonlinear sources of system performance degradation, and ways to quantify and model their impacts.

8.1 BER VS. *Q*-VALUE FOR BINARY MODULATED SYSTEMS

BER and quality factor (*Q*-value) are most important parameters describing the quality of digital signals at a telecommunication receiver, and these two parameters are related. In this section, we discuss the definition, the implication, and the limitation of these two parameters.

8.1.1 OVERVIEW OF IMDD OPTICAL SYSTEMS

The simplest yet very often used digital optical transmission systems are based on binary modulation. In the intensity modulation and direct detection (IMDD) mode, data are encoded on the optical power emitted from the transmitter, and the transmitter output has two digital states that are usually chosen to be light-pass (mark) and light-block (space). At the receiver, the signal optical power is converted into a photocurrent by means of a photodiode, in which the photocurrent is linearly proportional to the optical power received. This conversion eliminates wavelength information as well as the phase noise of the optical carrier at the receiver. The digital data can also be encoded as frequency or phase of the optical carrier, such as frequency shift key (FSK) and phase shift key (PSK). However, because the photocurrent of a photodiode is only proportional to the signal optical power, frequency or phase decoders have to be included in the optical receivers before the photodiodes in these systems, so that the data embedded in the phase or frequency of the optical carrier can be recovered.

A block diagram of data flow through the components of an IMDD link is shown in Fig. 8.1.1A. A data sequence to be transmitted is first encoded with FEC algorithm, and the encoded binary data stream, is electrically amplified by a driver and applied on to an O/E converter. Fig. 8.1.1B shows an example of ideal binary data sequence with 10 Gb/s data rate where the bit length is 100 ps. The O/E conversion can be accomplished with a direct-modulated laser diode whose output optical power is linearly proportional to the applied electric current as described in Chapter 3, or through an external electro-optic modulator whose transmission loss is related to the applied electric voltage as described in Chapter 7. The modulated optical signal is then launched into an optical fiber system for transmission. The fiber system can be as simple as a length of optical fiber, or multiple spans of fibers with optical amplifiers to compensate the transmission loss of the fiber. An O/E converter at the receiver detects the received optical signal and converts it into an electric current. The O/E converter can be a simple photodiode for direct detection of the intensity-modulated optical signal. For frequency or phase-modulated optical signals, appropriate frequency, or phase-sensitive optical components have to be employed before the photodiode to extract the information. Because of the impairments throughout the modulation transmission and photodetection, the electric current signal from the O/E converter will be distorted and noisy compared to the transmitted waveform, as illustrated in Fig. 8.1.1C. In the digital receiver, the clock has to be recovered from the corrupted waveform through narrowband filtering and phase-locking, and this recovered clock is used to determine the moment within each bit period when the decision has to be made. Fig. 8.1.1D shows the eye diagram obtained by folding the waveform of Fig. 8.1.1C into a time window of 2 bits, where T_D is the decision time, and v_{th} is the decision threshold.

Within a bit period, if the instantaneous amplitude of the received waveform is higher (or lower) than the threshold v_{th} at the decision time T_D, that bit is then recognized as "1" (or "0"). This decision process converts the analog waveform back into a digital data sequence as shown in Fig. 8.1.1E, and ideally it should be identical to the original binary data sequence shown in Fig. 8.1.1B. But in practice, decision errors may happen by misreading "0" as "1" or vise versa. A proper system design is to minimize these errors under various application scenarios and sources of performance degradation.

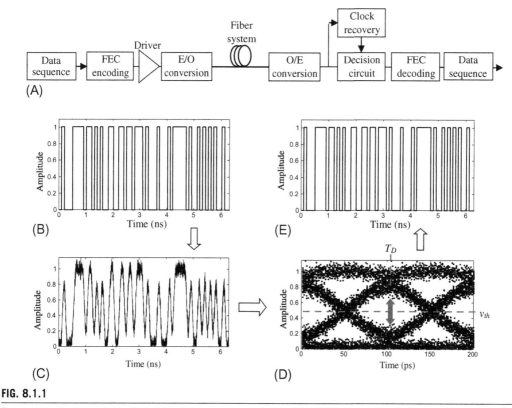

FIG. 8.1.1

Schematic diagram of data flow through the components of an IMDD link. (A) Block diagram of an optical transmission system, (B) ideal binary data sequence, (C) distorted signal waveform after transmission, (D) eye-diagram of the received signal, and (E) recovered data sequence after the decision circuit.

In short-distance and low-speed optical systems without inline optical amplifiers, the system performance is often limited by the signal optical power level that reaches the receiver. Receiver sensitivity is defined as the minimum signal optical power required at the receiver to achieve the targeted BER. If the signal optical power is too low at the receiver, the noise generated by the receiver would make the signal-to-noise ratio (SNR) unacceptable. For high-speed long-distance optical systems employing multiple inline optical amplifiers, signal waveform distortion and accumulated ASE noise throughout the transmission system may become major limitations in the transmission distance. In this case, receiver sensitivity is no longer a relevant parameter to specify an optical receiver. Instead, the receiver should be qualified by its ability to resist the influences of waveform distortion and optical noise.

One of the most important sources of linear performance impairments in high-speed fiber-optic transmission system is the chromatic dispersion of optical fiber. For a standard single-mode fiber, the chromatic dispersion at a 1550 nm wavelength is on the order of 17 ps/nm-km. This limits the transmission distance of a 10 Gb/s IMDD optical system to about 100 km. Beyond which the intersymbol interference (ISI) due to chromatic dispersion will introduce significant waveform distortion shown as the closure of the signal eye diagrams. Dispersion shifted fibers (DSFs) have been developed to minimize this problem, but they were later found unsuitable for WDM systems due to the increased nonlinear crosstalk between different wavelength channels.

Dispersion compensation (DC) has emerged as an effective way to overcome the dispersion-induced waveform distortion and to extend the maximum transmission distance. This is usually accomplished using dispersion-compensating modules (DCMs) that have the opposite sign of dispersion of the transmission fiber. DCM can be made by dispersion-compensating fibers (DCFs) or by passive optical devices such as fiber Bragg gratings. The overall accumulated dispersion in a transmission system can be reduced to an acceptable level with proper system design and dispersion compensation. In WDM systems with large numbers of channels, different wavelengths may experience different levels of dispersion due to the dispersion slope in optical fibers. In this case, slope compensation has also to be applied for high-speed optical transmission to equalize the performance of all WDM channels. Adding a dispersion compensator in each fiber span has become a standard industrial practice for long-distance optical systems. In fact, a dispersion compensator can often be packaged into an inline optical amplifier module to simplify optical system implementation. The disadvantage of DC in optical domains is the increased optical attenuation due to DCM, which requires a higher level of optical amplification in the system to compensate for this additional loss. This increased gain requirement of optical amplifiers will in turn generate more ASE noise, which tends to degrade optical SNR in the receiver.

In an amplified multispan WDM optical system with a large number of wavelength channels, inter-channel crosstalk is another important concern, especially when the system has a large number of spans and the signal optical power level at the begining of each span is high enough. In addition to linear crosstalk that might be caused by leakage from optical filters and switches, nonlinear crosstalk is especially notorious because it cannot be eliminated by improving the qualities of optical components. The major sources of nonlinear crosstalk in high-speed optical transmission systems include cross-phase modulation (XPM), four-wave mixing (FWM), and Raman crosstalk. Understanding the mechanisms of various system performance degradation is essential for the system design, optimization, and performance specification.

An important way to reduce performance degradation due to linear and nonlinear impairments in fiber-optic systems is to use advanced modulation formats. In general, an optical signal with longer pulse duration and (or) narrower spectral width would suffer less from chromatic dispersion. Multilevel modulation (Waklin and Conradi, 1999), phase-shaped binary (PSB) modulation (Penninckx et al., 1997), and digital subcarrier multiplexing (Hui et al., 2002) have been used to reduce the impact of chromatic dispersion because of their reduced spectral width. More recently, electrical domain digital signal processing (DPS) was applied to reduce transmission impairments, which has the potential to completely eliminate the requirement of optical domain DC (McNicol et al., 2005). While advanced modulation formats and DSP-based optical systems will be discussed in Chapters 10 and 11, this chapter focuses on the discussion of most fundamental parameters that specify the quality of an optical transmission system.

In an optical network scenario, data are usually encapsulated in a digital wrapper that can be used to record content partitioning, source and destination, enable synchronization, time-domain partitioning, performance monitoring, fault isolation, internodal communication, and the algorithm of FEC, to name a few. Additional overhead may be added to simplify clock recovery. These essential network functionalities increase the overall line rate (equivalently bandwidth) for a given data rate. Depending on transmission standard and FEC algorithm used, such overhead can possibly add up to 25% to the line rate, but it is typically 3%–7% in practice. Even though the data and overhead are often scrambled to regulate pattern length, in some cases the framing structure (which is not scrambled) can contain long patterns, which place demands on the receiver low-frequency response and clock recovery circuits. For example, long pseudorandom bit sequences (PRBS), such as $2^{31} - 1$, have to be used in transmission experiments to properly exercise the pattern dependence of a link.

8.1.2 RECEIVER *BER* AND *Q*

BER is a fundamental measure of digital communication system quality. BER is essentially an error probability of digital bits in the received signal; it is also known as *bit error probability*. By definition, BER is

$$BER = \frac{Bit_{Error}}{Bit_{Total}} \tag{8.1.1}$$

where Bit_{Error} is the number of misinterpreted bits by the receiver and Bit_{Total} is the total number of received bits. Both the misinterpreted bits and the total received bits are measured within a certain time window ΔT, which is referred to as *gating time*.

A useful alternative to the estimation of BER is the system Q value. It is a quality factor determined by the ratio of separation between implemented digital states and the approximate Gaussian noise associated with those states at the receiver. In an optical receiver, after photodetection and a transimpedance preamplifier (TIA), the time-dependent voltage signal is presented to a decision circuit. This latter is typically a gated threshold device synchronized to the recovered clock. The decision circuit reports a logical *one* for signal voltage above a reference, threshold, value, and a logical zero otherwise. A decision is made at each clock cycle. This scheme is shown in Fig. 8.1.2. An eye diagram, formed by overlapping consecutive segments of the received electrical waveform, shows the site in phase (horizontal axis) and in voltage (vertical axis) of the decision instant and threshold, respectively. Depending on receiver design, the decision instant and threshold might be optimized once at start of life or in a continuous and automatic manner dictated by a performance cost function.

In the eye diagram shown in Fig. 8.1.2, the spread of the voltage values above and below threshold at the sampling instant is attributable to both the waveform distortion caused by ISI and random noises. Sources of ISI include channel memory stemming from receiver and transmitter transfer functions, linear and nonlinear propagation effects such as residual chromatic dispersion, PMD, SPM, XPM, and FWM discussed in Chapter 2, and optical filter transfer functions. Random noises can be caused by photodiode thermal noise and shot noise, as well as signal-ASE beat noise and ASA-ASE beat noise as described in Chapter 5. Phase delay variations within the information bandwidth contribute to spreading at eye crossings, usually located ½ a clock cycle from the decision instant. This spreading is a constituent of timing jitter.

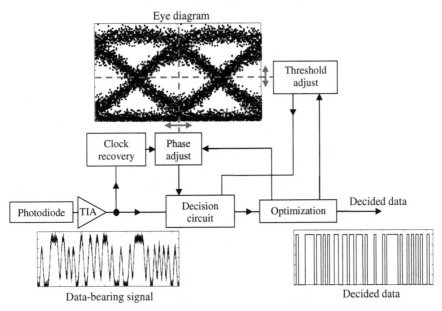

FIG. 8.1.2

Illustration of bit decision in a binary receiver.

Fig. 8.1.3 shows the probability distribution function (PDF) of the eye diagram so that we can derive the fundamentals of BER and *Q*-value calculations. In fact, BER is a conditional probability of receiving signal *y* while the transmitted signal is *x*, $P(y/x)$, where *x* and *y* can each be digital 0 or 1. Since the transmitted signal digital states can be either 0 or 1, we can define $P(y/0)$ and $P(y/1)$ as the PDFs of the received signal at state *y* while the transmitted signals are 0 and 1, respectively. Suppose that the probability of sending digital 0 and 1 are $P(0)$ and $P(1)$ and the decision threshold is v_{th}; the BER of the receiver should be

$$BER = P(0)P(v > v_{th}/0) + P(1)P(v < v_{th}/1) \qquad (8.1.2)$$

where v is the received signal level. In most of the binary transmitters, the probabilities of sending 0 and 1 are the same, $P(0) = P(1) = 0.5$. Also, Gaussian statistics can be applied to most of the noise sources in the receiver as a first-order approximation,

$$P_{Gaussian}(v) = \frac{1}{\sigma\sqrt{2\pi}} \exp\left(-\frac{(v - v_m)^2}{2\sigma^2}\right) \tag{8.1.3}$$

where σ is the standard deviation and v_m is the mean value of the Gaussian probability distribution. Then the probability for the receiver to declare 1 while the transmitter actually sends a 0 is

$$P(v > v_{th}/0) = \frac{1}{\sigma_0\sqrt{2\pi}} \int_{v_{th}}^{\infty} \exp\left(-\frac{(v - v_0)^2}{2\sigma_0^2}\right) dv = \frac{1}{\sqrt{2\pi}} \int_{Q_0}^{\infty} \exp\left(-\frac{\xi^2}{2}\right) d\xi \tag{8.1.4}$$

where σ_0 and v_0 are the standard deviation and the mean value of the received signal photocurrent at digital 0, $\xi = (v - v_0)/\sigma_0$ and

$$Q_0 = \frac{v_{th} - v_0}{\sigma_0} \tag{8.1.5}$$

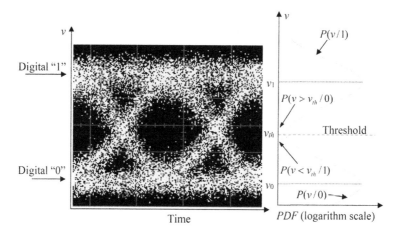

FIG. 8.1.3

Probability distribution function (PDF) of the eye diagram.

Similarly, the probability for the receiver to declare 0 while the transmitter actually sends 1 is

$$P(v < v_{th}/1) = \frac{1}{\sigma_1\sqrt{2\pi}} \int_{-\infty}^{v_{th}} \exp\left(-\frac{(v_1 - v)^2}{2\sigma_1^2}\right) dv = \frac{1}{\sqrt{2\pi}} \int_{Q_1}^{\infty} \exp\left(-\frac{\xi^2}{2}\right) d\xi \tag{8.1.6}$$

where σ_1 and v_1 are the standard deviation and the mean value of the received signal photocurrent at digital 1, $\xi = (v_1 - v)/\sigma_1$ and

$$Q_1 = \frac{v_1 - v_{th}}{\sigma_1} \tag{8.1.7}$$

According to Eq. (8.1.2), the overall error probability is

$$BER = \frac{1}{2}P(v > v_{th}/0) + \frac{1}{2}P(v < v_{th}/1) = \frac{1}{2\sqrt{2\pi}}\left\{ \int_{Q_0}^{\infty} \exp\left(-\frac{\xi^2}{2}\right)d\xi + \int_{Q_1}^{\infty} \exp\left(-\frac{\xi^2}{2}\right)d\xi \right\} \qquad (8.1.8)$$

where $P(0) = P(1) = 0.5$ is assumed.

Mathematically, a widely used special function, the error function, is defined as

$$erf(x) = \frac{2}{\sqrt{\pi}}\int_{0}^{x} \exp\left(-y^2\right)dy \qquad (8.1.9)$$

and a complementary error function is defined as

$$erfc(x) = 1 - erf(x) = \frac{2}{\sqrt{\pi}}\int_{x}^{\infty} \exp\left(-y^2\right)dy \qquad (8.1.10)$$

Therefore Eq. (8.1.8) can be expressed as complementary error functions:

$$BER = \frac{1}{4}\left\{ erfc\left(\frac{Q_0}{\sqrt{2}}\right) + erfc\left(\frac{Q_1}{\sqrt{2}}\right) \right\} \qquad (8.1.11)$$

Since both Q_0 and Q_1 in Eq. (8.1.11) are functions of the decision threshold v_{th}, and usually the lowest BER can be obtained when $P(0)P(v > v_{th}/0) = P(1)P(v < v_{th}/1)$, we can simply set $Q_0 = Q_1$, which is

$$\frac{v_{th} - v_0}{\sigma_0} = \frac{v_1 - v_{th}}{\sigma_1}$$

Or equivalently

$$v_{th} = \frac{v_0\sigma_1 + v_1\sigma_0}{\sigma_0 + \sigma_1} \qquad (8.1.12)$$

Under this "optimum" decision threshold, Eqs. (8.1.5) and (8.1.7) are equal and

$$Q = Q_1 + Q_2 = \frac{v_1 - v_0}{\sigma_1 + \sigma_0} \qquad (8.1.13)$$

The BER function in Eq. (8.1.11) becomes

$$BER = \frac{1}{2}erfc\left(\frac{Q}{\sqrt{2}}\right) \qquad (8.1.14)$$

This is a simple but very important equation that establishes the relationship between BER and the receiver Q-value, as shown in Fig. 8.1.4. As a rule of thumb, $Q = 6$, 7, and 8 correspond to the BER of approximately 10^{-9}, 10^{-12}, and 10^{-15}, respectively.

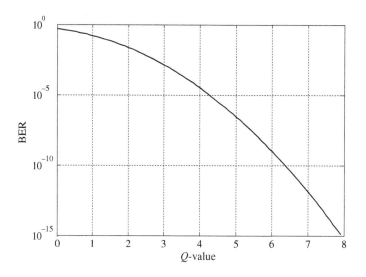

FIG. 8.1.4

BER as a function of receiver Q-value.

Note that the relationship between the BER and the Q-value shown in Eq. (8.1.14) is based on a Gaussian noise assumption. In practical systems, the statistics of noise sources are not always Gaussian. For example, shot noise is a Poisson process whose PDF follows a Poisson distribution (Personick, 1977). Another note is that the received photocurrent at digital 0 should never be negative because the received optical power is always positive. Therefore, the tail of the PDF should be limited to the positive territory and a Rayleigh distribution may be more appropriate to describe the noise statistics associated with digital 0. Nevertheless, Gaussian approximation is widely adopted because of its simplicity.

In the Gaussian approximation discussed so far, we have assumed that the eye diagram only has a single line at the digital 1 level and another single line at the digital 0 level. In practice, the eye diagram may have many lines at each digital level due to pattern-dependent waveform distortion. A normalized eye diagram is recast in Fig. 8.1.5. This diagram plots the normalized signal voltage which is linearly proportional to the optical power waveform at the receiver. Associated with each of the noise-free lines of the eye diagram is an approximately Gaussian noise distribution that is generated from a handful of independent processes. The aggregate noise power distributions at the sampling instant (within the jitter window) are drawn on the right side of the figure. The jitter window is determined by the quality of clock recovery, which is an uncertainty on the decision phase. The noise distribution on transmitted 1 s is typically wider than the distribution on transmitted 0 s due to the signal dependence of some of the noise processes.

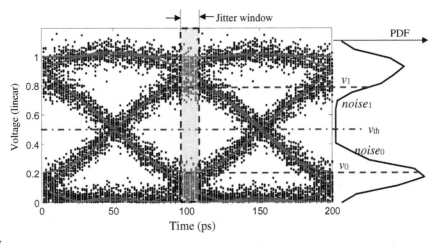

FIG. 8.1.5

Statistic distribution of noise when the eye diagram is distorted. Smooth lines represent noise-free eye diagram in which eye closure is caused by waveform distortion.

Noise processes also depend on receiver optical-to-electrical conversion technology. For example, a receiver based on a PIN photodiode has different noise properties than an avalanche photodiode (APD)-based optical receiver. In optically amplified systems, the presence of ASE noise generated from optical amplifiers makes the receiver performance specification and analysis quite different from unamplified optical systems.

8.2 IMPACTS OF NOISE AND WAVEFORM DISTORTION ON SYSTEM Q-VALUE

Eq. (8.1.13) is a general definition of receiver Q-value under Gaussian noise approximation, and it is the ratio between the signal eye opening and the total noise at the decision time. Although both the waveform distortion and the random noise directly affect the system performance, the ways of their impacts on the receiver Q-value are different.

8.2.1 Q-CALCULATION FOR OPTICAL SIGNALS WITHOUT WAVEFORM DISTORTION

Let us first consider a system without waveform distortion so that the Q-value is only determined by the accumulated noise at the receiver. In such an ideal case, the eye diagram is wide open with P_1 the average optical power corresponding to the signal "1" level, and $P_0 = 0$ the signal "0" level at the decision time of each bit. Thus, the Q-value is

$$Q = \frac{v_1 - v_0}{\sigma_1 + \sigma_0} = \frac{M\Re P_1}{\sqrt{\left(\sigma_{th}^2 + \sigma_{sh}^2 + \sigma_{dk}^2 + \sigma_{S-ASE}^2 + \sigma_{ASE-ASE}^2 + \sigma_{RIN}^2\right)B_e} + \sqrt{\left(\sigma_{th}^2 + \sigma_{dk}^2 + \sigma_{ASE-ASE}^2\right)B_e}} \qquad (8.2.1)$$

where, $\mathfrak{R}$ is the responsivity of photodiode, M is the APD gain if an APD is used ($M=1$ for a PIN photodiode), and B_e is the receiver electric bandwidth.

$$\sigma_{th}^2 = 4kT/R_L \tag{8.2.2}$$

$$\sigma_{sh}^2 = M^2 F_M \cdot 2q\mathfrak{R}P_1 \tag{8.2.3}$$

$$\sigma_{dk}^2 = M^2 F_M \cdot 2qI_{dk} \tag{8.2.4}$$

$$\sigma_{S-ASE}^2 = M^2 F_M \cdot 2\mathfrak{R}^2 P_1 \rho_{ASE} \tag{8.2.5}$$

$$\sigma_{ASE-ASE}^2 = M^2 F_M \cdot \mathfrak{R}^2 \rho_{ASE}^2 B_0/2 \tag{8.2.6}$$

$$\sigma_{RIN}^2 = M^2 F_M \cdot 2\mathfrak{R}P_1 \cdot RIN \tag{8.2.7}$$

These six equations represent single-sided power spectral densities of thermal noise, shot noise, dark current noise, signal-ASE beat noise, ASE-ASE beat noise, and relative intensity noise (RIN), respectively. R_L is the load resistance, k is the Boltzmann's constant, T is the absolute temperature, q is the electron charge, I_{dk} is the photodiode dark current, RIN is the relative intensity noise of the laser source defined by Eq. (3.3.42), B_0 is the optical filter bandwidth, and ρ_{ASE} is the accumulated ASE noise optical power spectral density at the input of the photodiode. F_M is the noise figure of APD which is dependent on the APD gain M. If a photodiode (instead of an APD) is used, $M=1$ and $F_M=1$. Among these noise sources, short noise, signal-ASE beat noise and laser RIN are proportional to the signal optical power, so that they are categorized as *signal-dependent* noise. Whereas, thermal noise, dark current noise, and ASE-ASE beat noise are independent of the signal optical power so that they are referred to as *signal-independent* noise. Note that signal-dependent noise terms do not exist in the σ_0 term in this ideal case without waveform distortion because the associated signal optical power is $P_0=0$.

In the characterization of an optical system, it is always more convenient to measure the average power than measuring the instantaneous power at signal "1"s. For the majority of binary modulated optical signals, the probabilities of "0"s and "1"s in the signal are approximately equal, and they are both 50%. If $P_0=0$, and P_{ave} is the average signal optical power, the "1" level of the optical signal should be $P_1=2P_{ave}$, which can be used in Eqs. (8.2.1)–(8.2.7).

8.2.2 *Q*-ESTIMATION BASED ON EYE DIAGRAM PARAMETERIZATION

In practical optical transmission systems, waveform distortion cannot be avoided which contributes to the degradation of optical system performance. As waveform distortion is generally pattern-dependent, the contributions to eye closure caused by isolated single "1"s and by continuous "1"s can be quite different. For example, if the system transfer function has a low-pass characteristic, isolated "1"s (supported by high-frequency components in the spectrum) will be penalized more than continuous "1"s which are predominately represented by low-frequency components of the spectrum. An accurate calculation of Q-value has to consider all possible bit patterns of the received signal waveform and the probability of their occurrence, as well as the noise associated with each specific bit pattern. This generally pattern-specific nature prohibits analytic estimation of the Q-value even when the signal waveform is pseudorandom, and numerical simulations have to be applied in system performance calculation. Here we discuss a simplified technique for the estimation of system Q-value based on eye-mask parameterization. This technique is based on the separation of waveform distortion (eye closure) and random noise, which is suitable for systems designed to operate at high Q-values ($Q > 7$ or BER $< 10^{-12}$). An optical eye distortion mask parameterization can be made at any reference interface of the system to define the distortion-related link performance contributions, independent of the particular noise characteristic (Hui et al., 1999).

Fig. 8.2.1A is an example of a waveform measured at the output of a dispersive fiber link which is affected by both waveform distortion and random noise. This is a section of a PRBS. As a PRBS waveform repeats itself for each pattern length, an average can be made to remove the random noise so that a much smoother (or deterministic) waveform can be obtained as shown in Fig. 8.2.1B. These waveforms can be converted into eye diagrams with and without the noise contribution as shown in Fig. 8.2.2A and B, respectively.

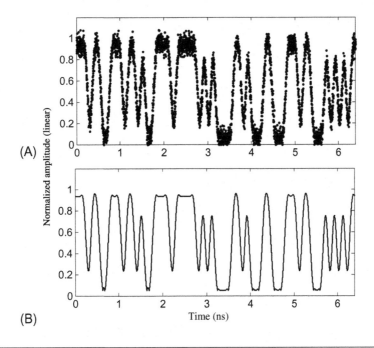

FIG. 8.2.1

An example of photocurrent waveform with (A) and without (B) the random noise.

In Fig. 8.2.2B, the noise-free eye distortion mask is defined by a four-level feature (P_1, P_0, A, B) over a timing window, W, which represents the worst-case phase uncertainty in the sampling instant. W comprises the sum of all bounded uncertainties plus typically 7 times the standard deviation of statistical uncertainty of the decision phase. P_1 and P_0 are the power levels associated with signal long "1"s and long "0"s in the pseudo-random nonreturn-to-zero (NRZ) bit pattern. The dimensionless parameters A and B are the lowest inner upper eye and the highest inner lower eye measured within the phase window W, and they are independent of the noise. According to definitions in Fig. 8.2.2, the average signal optical power is $P_{ave} = (P_1 + P_0)/2$, given that signal "1"s and "0"s have the same probability.

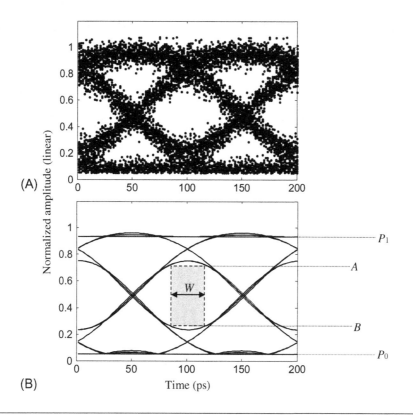

FIG. 8.2.2

Schematic representation of an optical eye distortion mask mapped onto a measured eye diagram with (A) and without (B) the random noise.

In practical systems, waveform distortion, signal-dependent noise, and signal-independent noise are all mixed together at the receiver. According to Eq. (8.1.13), the receiver Q factor can be written as

$$Q = \frac{(A - B)2\Re P_{ave}}{\sqrt{\left(\sigma_{ind}^2 + \zeta A 2 P_{ave}\right) B_e} + \sqrt{\left(\sigma_{ind}^2 + \eta B 2 P_{ave}\right) B_e}} \tag{8.2.8}$$

where $\sigma_{ind}^2 = \sigma_{th}^2 + \sigma_{dk}^2 + \sigma_{ASE-ASE}^2$ is the signal independent noise power spectral density which includes thermal noise, dark current noise and ASE-ASE beat noise. ζ is a system-dependent multiplication factor which represents the impact of signal-dependent noise,

$$\zeta = 2\Re M^2 F_M (q + \Re \rho_{ASE}) \tag{8.2.9}$$

In the absence of distortion, $B = 0$ and $A = 1$, the system Q is determined only by the noise contribution. In this case,

$$Q = Q_0 = \frac{2\Re P_{ave}}{\sqrt{\left(\sigma_{ind}^2 + \zeta 2 P_{ave}\right) B_e} + \sigma_{ind}\sqrt{B_e}} \tag{8.2.10}$$

By this definition of Q_0, the system Q degradation caused only by eye distortion can be written in a general form as

$$D(A,B,x) = \frac{Q}{Q_0} = \frac{A-B}{Y_e} \qquad (8.2.11)$$

In this expression, Y_e is an important factor that shows the effect of interaction between distortion and noise:

$$Y_e(A,B,x) = \frac{\sqrt{1+xA} + \sqrt{1+xB}}{1+\sqrt{1+x}} \qquad (8.2.12)$$

where $x = 2\zeta P_{ave}/\sigma_{int}^2$ is the ratio of signal-dependent noise to signal-independent noise.

In a case where signal-independent noise dominates, $x=0$ and $Y_e=1$, so that $D=A-B$. On the other hand, if signal-dependent noise dominates, $x=\infty$ and $Y_e = \sqrt{A} + \sqrt{B}$, therefore $D = \sqrt{A} - \sqrt{B}$. In general, with $x \in (0, \infty)$, the maximum value of Y_e that corresponds to the worst-case distortion can be expressed as

$$Y_0 = \begin{cases} \sqrt{A} + \sqrt{B} & (\sqrt{A} + \sqrt{B}) \geq 1 \\ 1 & (\sqrt{A} + \sqrt{B}) < 1 \end{cases} \qquad (8.2.13)$$

The two possible maxima $Y_{max} = \sqrt{A} + \sqrt{B}$ and $Y_{max}=1$ correspond to $x=\infty$ and $x=0$, respectively. Using Eq. (8.2.11), the worst-case distortion factor, defined as D_{wc}, can be written as a function of Y_{max}:

$$D_{wc} = (A-B)/Y_{max} \qquad (8.2.14)$$

Obviously, D_{wc} is a global worst-case distortion effect, which is independent of the nature of the noise. To demonstrate the impact of noise characteristic on the system distortion penalty, D, defined by Eq. (8.2.11), is plotted in Fig. 8.2.3 as a function of x. In this plot, two sets of eye-closure parameters were used, corresponding to the conditions for the two solutions of Eq. (8.2.13). In one case, $A=0.7$, $B=0.15$, and $\sqrt{A} + \sqrt{B} > 1$ so that the worst-case distortion happens at $x=\infty$. In the other case, $A=0.4$, $B=0.05$, and $\sqrt{A} + \sqrt{B} < 1$, and the worst-case distortion happens at $x=0$. The dashed lines in Fig. 8.2.3 are $10 \log (\sqrt{A} - \sqrt{B})$, which represents the case where signal-dependent noise dominates ($x=\infty$), whereas the dash-dotted lines are $10 \log(A-B)$, which represents the case of $x=0$. Shown as the solid lines in Fig. 8.2.3, $D(x)$ vs. x characteristics are not monotonic; however, D symptomatically approaches its worst-case D_0 with either $x=0$ or $x \to \infty$, depending on the value of $\sqrt{A} + \sqrt{B}$. It is worthwhile to note that generally, Eqs. (8.2.13) and (8.2.14) overestimate the distortion penalty because $x \in (0, \infty)$ was used to search for the worst case, but in real systems x value can never be infinity. The existence of a worst-case distortion factor D_{wc} implies a possibility to separate distortion from noise in the system link budgeting. Eq. (8.2.14) clearly demonstrates a simple linear relationship between system Q and the worst-case distortion factor D_{wc}. Regardless of the fundamental difference in the origins of noise and distortion, separate counting of these two effects shows a clear picture of system budget allocations. Experimental verification of the linear relationship between Q and D_{wc}, can be found in Hui et al. (1999).

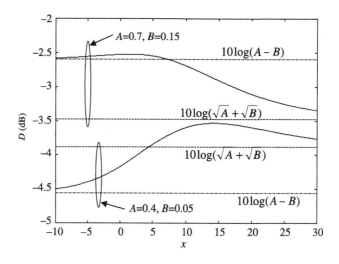

FIG. 8.2.3

Q-degradation parameter D as a function of x (solid line). Dashed line: $10\log(A^{1/2}+B^{1/2})$; dashed-dotted line: $10\log(A-B)$.

8.3 RECEIVER SENSITIVITY AND REQUIRED OSNR

In a fiber-optic transmission system, the receiver has to correctly recover the data carried by the optical carrier. Because the optical signal that reaches the receiver is usually very weak after experiencing attenuation of the optical fiber and other optical components, a high-quality receiver has to be sensitive enough. In addition, high-speed and long-distance optical transmission systems suffer from waveform distortion, linear and nonlinear crosstalk during transmission along the fiber, and the accumulated ASE noise due to the use of inline optical amplifiers. A high-quality receiver also has to tolerate these additional quality degradations of the optical signal. The criterion of optical receiver qualification depends on the configuration of the optical system and the dominant degradation sources. In a relatively short distance optical fiber system without in-line optical amplifiers, optical noise is not a big concern at the receiver. In this case, transmission quality can be guaranteed as long as the signal optical power is high enough, and thus *receiver sensitivity* is the most relevant measure of the system performance. On the other hand, in a long-distance fiber-optic transmission system employing multiple in-line optical amplifiers, accumulated optical noise arriving at the receiver can be overwhelming. In such a case, transmission performance cannot be ensured by simply increasing the signal optical power, and therefore a minimum optical signal-to-noise ratio (OSNR) has to be met. In the following we discuss receiver sensitivity and the required OSNR, two useful receiver specifications.

8.3.1 RECEIVER SENSITIVITY

Receiver sensitivity is one of the most widely used specifications of optical receivers in fiber-optic systems. It is defined as the minimum signal optical power level required at the receiver to achieve a certain BER performance. For example in an optical system, for the BER to be less than 10^{-12} without FEC, the minimum signal optical power reaching the receiver has to be no less than $-35\,$dBm; this means the receiver sensitivity is$-35\,$dBm. Obviously the definition of receiver sensitivity depends on the targeted BER level and the signal data rate. However, signal waveform distortion and optical

SNR are, in general, not clearly specified in the receiver sensitivity definition, but it assumes that the noise originated from the receiver is the major limiting factor of the system performance.

Now we consider two types of optical receivers shown in Fig. 8.3.1, in which configuration (a) is a simple PIN photodiode followed by a transimpedance amplifier (TIA), whereas, configuration (b) has an optical preamplifier added in front of the PIN to boost the optical signal power level.

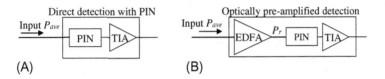

(A)   (B)

FIG. 8.3.1

Direct detection receivers with (A) and without (B) optical preamplifier.

We first consider the simplest receiver configuration shown in Fig. 8.3.1A. Neglecting waveform distortion and the impact of crosstalk in the optical signal for a moment, the Q-value will only depend on the electrical SNR after photodetection. In an intensity-modulated system with direct detection (IMDD), the receiver sensitivity is affected by thermal noise, shot noise, and photodiode dark current noise. Eq. (8.2.1) can be written as

$$Q = \frac{2\Re P_{ave}}{\sqrt{(4kT/R_L + 4q\Re P_{ave} + 2qI_{dk})B_e} + \sqrt{(4kT/R_L + 2qI_{dk})B_e}} \tag{8.3.1}$$

Fig. 8.3.2 shows the calculated receiver Q-value as a function of the received average signal optical power P_{ave}. This is a 10Gb/s binary system with direct detection, and the electrical bandwidth of the receiver is $B_e = 7.5\,\text{GHz}$. Other parameters used are $\Re = 0.85\,mA/mW$, $R_L = 50\,\Omega$, $I_d = 5\,\text{nA}$, and $T = 300\,\text{K}$. Fig. 8.3.2 indicates that to achieve a BER of 10^{-12}, or equivalently $Q = 7$ ($10\log_{10}(Q) = 8.45\,\text{dB}$ on the vertical axis), the received average signal optical power has to be no less than $-19\,\text{dBm}$. Therefore the sensitivity of this 10Gb/s receiver is $-19\,\text{dBm}$. Every dB decrease in signal optical power will result in a dB decrease of the Q-value, as indicated in Fig. 8.3.2.

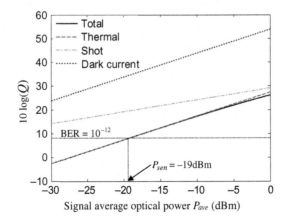

FIG. 8.3.2

Receiver sensitivity plot (continuous line) for a 10-Gb/s system using a PIN photodiode. Dashed, dash-dotted and dotted lines represent Q calculation only consider thermal noise, shot noise, and dark current noise, individually.

Comparing the impacts of thermal noise, shot noise, and dark current noise on the Q-value, it is evident that thermal noise dominates in this type of direct detection receiver in the vicinity of the targeted BER level. Other noise terms can be safely neglected without introducing noticeable errors.

In the thermal noise dominated receiver, if we further consider waveform distortion with $A < 1$ and $B > 0$ [referring to the eye mask defined by Fig. 8.2.2B], the Q-value becomes

$$Q = \sqrt{\frac{R_L}{4kTB_e}} \Re(A - B)P_{ave} \tag{8.3.2}$$

For a targeted performance of $Q = 7$, the receiver sensitivity can then be found as

$$P_{sen} = \frac{7}{\Re(A - B)} \sqrt{\frac{4kTB_e}{R_L}} \tag{8.3.3}$$

which is inversely proportional to the eye closure penalty $(A-B)$.

On the other extreme, if the shot noise is the only noise source while other noises are negligible (for example, if the load resistance R_L is very high), the Q-value is $Q = \sqrt{\Re P_{ave}/(qB_e)}(\sqrt{A} - \sqrt{B})$. Since $\Re = \eta q/(hv)$ with η the quantum efficiency and hv the photo energy, the shot-noise limited Q-value can be expressed as $Q = \sqrt{\eta P_{ave}/(hvB_e)}(\sqrt{A} - \sqrt{B})$. In the ideal case with 100% quantum efficiency and without waveform distortion, $Q = \sqrt{P_{ave}/(hvB_e)}$. This corresponds to a receiver sensitivity of $P_{sen} = 49hvB_e$ for $Q = 7$. Assuming a bandwidth efficiency of 1 bit/Hz, this receiver sensitivity is commonly referred to as quantum-limited detection efficiency, which is of 49 photons per bit in this case. In other words, for each bit of information, 49 photons are required to achieve a Q-value of 7.

In order to improve receiver sensitivity, an optical preamplifier can be added in front of the PIN photodiode as illustrated in Fig. 8.3.1B. As the optical preamplifier is part of the optical receiver, the receiver sensitivity is defined as the minimum optical power that reaches the preamplifier to achieve a targeted Q-value. The optical preamplifier increases the signal optical power before it reaches the photodiode, but it also introduces optical noise in the amplification process. In this case, the level of the ASE noise power spectral density is proportional to the gain of the optical preamplifier. Nevertheless, the receiver Q-value still increases with the increase of the input average signal optical power P_{ave}, but not linearly. If we neglect the waveform distortion so that average signal power is equal to ½ of the instantaneous power at signal digital 1, the Q-value can be calculated by

$$Q = \frac{2\Re P_r}{\sqrt{\left(2q(2\Re P_r + I_d) + \dfrac{4kT}{R_L} + 4\rho_{ASE}\Re^2 P_r + \rho_{ASE}^2\Re^2(2B_o - B_e)\right)B_e} + \sqrt{\left(2qI_d + \dfrac{4kT}{R_L} + \rho_{ASE}^2\Re^2(2B_o - B_e)\right)B_e}} \tag{8.3.4}$$

where $P_r = GP_{ave}$ is the amplified signal average optical power that reaches the PIN photodiode. In preamplified optical receiver, P_r is usually fixed to a reasonably high level around 0 dBm so that thermal noise and dark current contributions become negligible compared to signal-dependent noises. In such a case, the gain of the optical preamplifier becomes a function of the input signal optical power, as does the ASE noise level. For an optical preamplifier with a noise figure of $F = 5$ dB that corresponds to $n_{sp} = 1.58$, and with the signal wavelength of 1550 nm, the ASE noise optical power spectral density is

$$\rho_{ASE} = 2n_{sp}\frac{hc}{\lambda}(G - 1) = 4 \times 10^{-19}\left(\frac{P_r}{P_{ave}} - 1\right) \tag{8.3.5}$$

in the unit of Watt per Hertz. Fig. 8.3.3 (curve marked with "total") shows the calculated receiver Q-value as a function of the received average signal optical power P_{ave} at the input of the EDFA pre-amplifier. The impact of contribution due to each noise term is separately plotted in Fig. 8.3.3. The parameters used in the calculation are $P_r=0\,\mathrm{dBm}$, $R_L=50\,\Omega$, $I_d=5\,\mathrm{nA}$, $T=300\,\mathrm{K}$, $B_0=25\,\mathrm{GHz}$, $B_e=7.5\,\mathrm{GHz}$, and $\lambda=1550\,\mathrm{nm}$. In this case the receiver sensitivity is $P_{sen}=-41.8\,\mathrm{dBm}$ (for $Q=7$), which is approximately 23 dB better than the direct detection receiver without the EDFA preamplifier.

In the optically preamplified PIN receiver, since the signal optical power P_r at the preamplifier output is kept constant, thermal noise, shot noise, and dark current noise are constants and independent of the input optical power P_{ave}. In this example shown in Fig. 8.3.3, Q-values corresponding to thermal, shot, and dark current noises are 27.5, 29.2, and 54 dB, respectively, and their contributions to the overall Q-value are practically negligible. The dominant noise term that limits the receiver Q-value near the targeted BER level of $Q=7$ is the signal-ASE beat noise as can be easily seen from Fig. 8.3.3.

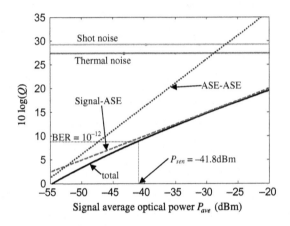

FIG. 8.3.3

Receiver sensitivity plot for a 10-Gb/s system with an optically preamplified PIN receiver.

If we only consider the impact of signal-ASE beat noise, Eq. (8.3.4) can be greatly simplified as

$$Q=\sqrt{\frac{P_r}{\rho_{ASE}B_e}}=\sqrt{\frac{P_r}{(2n_{sp}hc/\lambda)(G-1)B_e}}\approx\sqrt{\frac{P_{ave}}{(2n_{sp}hc/\lambda)B_e}} \tag{8.3.6}$$

Here $G\gg 1$ is assumed so that $G-1\approx G$. In this case, the Q-value is proportional to $\sqrt{P_{ave}}$, instead of P_{ave} as in non-preamplified PIN receiver. This is beneficial because P_{ave} is usually very small so that $\sqrt{P_{ave}}\gg P_{ave}$ is always true. In a preamplified PIN receiver for every dB signal optical power decrease, there is only half a dB decrease in $10\log(Q)$.

With the approximation of only considering signal-ASE beat noise in the Q calculation of preamplified PIN optical receiver, we can consider the impact of waveform distortion assuming $A<1$ and $B>0$ [referring to the eye mask defined by Fig. 8.2.2B]. In this case, the numerator of Eq. (8.1.13) is $2(A-B)\Re P_r$ and terms in the denominator are $\sigma_1=4\rho_{ASE}\Re^2AP_r$ and $\sigma_0=4\rho_{ASE}\Re^2BP_r$, so that the Q-value becomes

$$Q = \sqrt{\frac{P_r}{\rho_{ASE}B_e}}\left(\sqrt{A}-\sqrt{B}\right) = \sqrt{\frac{P_{ave}}{2n_{sp}(hc/\lambda)B_e}}\left(\sqrt{A}-\sqrt{B}\right) \qquad (8.3.7)$$

In comparison to the simple PIN receiver where $Q \propto (A - B)$, optically preamplified PIN receiver has $Q \propto \left(\sqrt{A}-\sqrt{B}\right)$. For an open eye diagram with $A > 0.5$ and $B < 0.5$, $\sqrt{A}-\sqrt{B} < A - B$ is mostly true. This implies that optically preamplified PIN receiver is more susceptible to waveform distortion than a simple PIN receiver.

8.3.2 REQUIRED OSNR

In the systems discussed earlier, optical noise is not accompanied with the optical signal that reaches the optical receiver, and therefore, BER can be reduced by increasing the level of signal optical power. On the other hand, in a long-distance optical transmission system employing multiple inline optical amplifiers, the level of optical power at the receiver can always be increased by increasing the gain of optical amplifiers. But the accumulated ASE noise generated by these optical amplifiers may become significant when the number of inline optical amplifiers is large. When the OSNR at the input of the receiver is too low, increasing optical power at the receiver may not result in the improvement of BER performance. In this type of systems, the performance is no longer limited by the signal optical power that reaches the receiver; rather, it is limited by the OSNR.

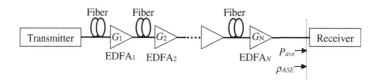

FIG. 8.3.4

Fiber-optic transmission system with N optically amplified fiber spans.

Fig. 8.3.4 illustrates a fiber-optic transmission system with N optically amplified fiber spans. In this system, if the transmission loss of each fiber span is exactly compensated by the gain of the inline optical amplifier in that span, the average signal optical power that reaches the receiver is equal to that emitted from the transmitter. Meanwhile, each EDFA generates an optical noise power spectral density $\rho_{ASE, i} = 2n_{sp}(hc/\lambda)(G_i - 1)$, with $i = 1, 2, ..., N$. As the ASE noise generated by each EDFA is also attenuated by the fibers and amplified by the EDFAs along the following fiber spans, the accumulated ASE noise power spectral density at the input of the receiver will be simply the addition of contributions from all inline EDFAs: $\rho_{ASE} = \sum_{i=1}^{N}\rho_{ASE,i}$. Then, OSNR at the input of the optical receiver is defined as the ratio between the average optical signal power, P_{ave}, and the power spectral density of the accumulated noise ρ_{ASE}, that is

$$\text{OSNR} = \frac{P_{ave}}{\rho_{ASE}} \qquad (8.3.8)$$

Since the unit of the signal average power is [W] and the unit of optical noise power spectral density is [W/Hz], the unit of OSNR should be [Hz], or [dB·Hz]. In practice, the optical noise power spectral density is measured by an optical spectrum analyzer (OSA) with a certain resolution bandwidth, R_B, and the OSA reports the measured noise power within a resolution bandwidth with the unit of [W/R_B]. Whereas, assume the spectral linewidth of the optical signal itself is narrower than the OSA resolution bandwidth, the OSA actually measures the total power of the optical signal. Thus, [$dB \cdot R_B$] is often used as the unit to specify OSNR.

In the system with multiple optical amplifiers, since the level of optical power arriving at the receiver is usually high enough, signal-independent noises such as thermal noise and dark current noise can be neglected in comparison with signal-dependent noises, that include shot noise, signal-ASE beat noise, and ASE-ASE beat noise. If we only consider shot noise, signal-ASE beat noise, and ASE-ASE beat noise, and neglect waveform distortion, Q-value can be calculated by

$$Q = \frac{2\Re P_{ave}}{\sqrt{\left(4\Re(q + \rho_{ASE}\Re)P_{ave} + \rho_{ASE}^2\Re^2(2B_o - B_e)\right)B_e} + \sqrt{\rho_{ASE}^2\Re^2(2B_o - B_e)B_e}} \tag{8.3.9}$$

This can be expressed as the function of OSNR as

$$Q = \frac{2\Re \cdot OSNR}{\sqrt{\left(4\Re q\frac{OSNR^2}{P_{ave}} + 4\Re^2 \cdot OSNR + \Re^2(2B_o - B_e)\right)B_e} + \sqrt{\Re^2(2B_o - B_e)B_e}} \tag{8.3.10}$$

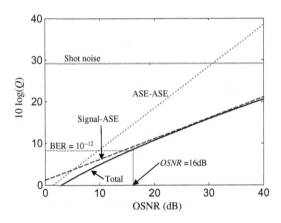

FIG. 8.3.5

Q-value as the function of signal OSNR (curve marked with total) considering contributions from shot noise, ASE-ASE beat noise, and signal-ASE beat noise.

Fig. 8.3.5 shows the calculated receiver Q-value as the function of the signal OSNR for a 10 Gb/s binary system based on Eq. (8.3.10), where waveform distortion is not considered ($A = 1$ and $B = 0$). The OSNR, based on a resolution bandwidth of 0.1 nm, is the signal optical power divided by the noise power within 0.1 nm optical bandwidth. Other system parameters are $P_{ave} = 0$ dBm, $\Re = 0.85 mA/mW$, $B_o = 25$ GHz, $B_e = 7.5$ GHz, and $\lambda = 1550$ nm. Contributions due to shot noise, ASE-ASE beat noise and signal-ASE beat noise are shown in the same figure. In the vicinity of BER $= 10^{-12}$ ($Q = 8.45$ dB), the Q-value is mainly determined by the contribution from signal-ASE beat noise, and the required OSNR (ROSNR) to achieve the targeted BER of 10^{-12} is approximately 16 dB.

Since SAE-ASE beat noise can be reduced by further reducing the bandwidth of the optical filter, the dominant noise in most of the long-distance optical systems employing multiple inline optical amplifiers are signal-ASE beat noise. If we only consider signal-ASE beat noise and also take into account the effect of waveform distortion, the Q-value will be

$$Q = \sqrt{\frac{P_{ave}}{\rho_{ASE}B_e}}\left(\sqrt{A} - \sqrt{B}\right) = \frac{\left(\sqrt{A} - \sqrt{B}\right)}{\sqrt{B_e}}\sqrt{OSNR} \tag{8.3.11}$$

which is similar to that described in Eq. (8.3.7), but here P_{ave} and ρ_{ASE} are input optical signal average power and noise power spectral density, respectively, at the receiver, and their ratio represents the OSNR.

Eq. (8.3.11) indicates that the Q-value of the receiver is not determined by the received signal optical power level, but rather it is linearly proportional to the square root of the OSNR. If we set a target value of $Q = 7$ (for BER $= 10^{-12}$), the required OSNR, or ROSNR is simply

$$ROSNR = \frac{49B_e}{\left(\sqrt{A} - \sqrt{B}\right)^2} \tag{8.3.12}$$

8.4 CONCEPT OF WAVELENGTH DIVISION MULTIPLEXING

Standard single-mode fiber has two low loss wavelength windows. The first window is in the 1310 nm region with about 45 nm width from 1285 to 1330 nm, and the attenuation is on the order of 0.35 ± 0.05 dB/km within this window The second window is in the 1550 nm region with about 50 nm width from 1525 to 1575 nm, and the attenuation is on the order of 0.22 ± 0.05 dB/km. All together this provides an approximately 14 THz optical bandwidth suitable for optical communication. New fibers developed in recent years can also minimize the water absorption peak between the 1310 and 1550-nm wavelength window so that one giant low loss window extending from 1285 nm all the way to 1575 nm has the potential to provide more than 40 THz bandwidth. However, the bandwidth of signal that can be modulated onto an optical carrier is limited by the speed of electronic circuits and the bandwidth of optoelectronic devices. Although wideband RF amplifiers and high-speed electro-optic modulators can provide up to >50 GHz bandwidth, it is still a very small fraction of the available bandwidth of the optical fiber.

Wavelength division multiplexing (WDM) is a technique developed to make efficient use of the vast bandwidth resource provided by the optical fiber. Fig. 8.4.1 shows the block diagram of a point-to-point fiber-optic WDM transmission system which uses multiple transmitters and receivers. Laser diode used in each transmitter is assigned with a unique wavelength so that there is no spectral overlap between different transmitters. Optical signals emitted from all transmitters are combined into a composite multiwavelength signal through a WDM multiplexer (MUX), and launched into a single transmission fiber. At the receiver, the multiwavelength optical signal is first split into individual wavelength channels through a WDM demultiplexer (DEMUX), and each wavelength channel is detected by an optical receiver performing optical to electric conversion. Inline optical amplifiers can be used to compensate the attenuation of the transmission fiber. EDFAs are usually used as inline amplifiers in WDM systems which require wide optical bandwidth and low crosstalk between different wavelength channels.

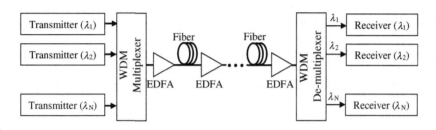

FIG. 8.4.1

Block diagram of point-to-point WDM transmission system.

Determined by the available gain bandwidth of EDFAs, the most used wavelengths window is between 1531 and 1570 nm, known as the C-band. Other wavelength bands include O-band: 1270–1370 nm, E-band: 1371–1470 nm, S-band: 1471–1530 nm, and L-band: 1571–1611 nm. In order to ensure the interoperability of optical communication equipment produced by different companies, a precise wavelength grid is standardized by International Telecommunication Union (ITU) for WDM systems and networks. Use C-band as an example, with 100 GHz grid size the standard WDM optical frequencies are $f_m = (190,000 + 100 \cdot m)$GHz, corresponding to the wavelength of the mth WDM channel of $\lambda_m = c/f_m$, where m is an integer representing channel index. Thus, nearly 50 wavelength channels can be put into the WDM C-band. The number of WDM channels can be increased by reducing channel spacing from 100 to 50 GHz or 25 GHz, which are known as dense wavelength multiplexing (DWDM). Similarly, ITU grid for 25 GHz channel spacing is specified by $f_m = (190,100 + 100 \cdot m)$GHz, and the entire C-band can host about 200 wavelength channels. With binary IMDD, 10 Gb/s dart rate can be loaded on each wavelength channel in a 25 GHz grid DWDM system, so that up to 2 Tb/s total traffic capacity can be carried by a single fiber in the C-band. Nowadays, with complex modulation and coherent detection with polarization multiplexing, 100 Gb/s data rate can be put on each wavelength channel in a 25 GHz DWDM grid, the total traffic capacity in the C-band alone can potentially reach 20 Tb/s.

Basic components enabling DWDM include high-quality WDM MUX and DEMUX, single wavelength semiconductor lasers, and wideband inline optical amplifiers. Characteristics and specifications MUXs and DEMUXs have been discussed in Chapter 6. As far as laser sources are concerned, a DWDM system requires wavelength stabilization of semiconductor lasers so that they can be on the frequency grid defined by ITU. For binary intensity-modulated systems with direct detection, DFB lasers with both temperature and current stabilization are often used with spectral linewidths on the order of Megahertz. Narrower spectral linewidths are required for phase-modulated systems using coherent detection which will be discussed in the next chapter.

As discussed in Chapter 3, the wavelength of a simple DFB laser diode is primarily determined by the period of the Bragg grating used inside the laser cavity. Although the emitting wavelength can be adjusted to some extent by changing the injection current and the operation temperature, the tuning range is realistically not more than 1 nm. Thus, a WDM transmitter has to be specifically designed for each wavelength on the grid. For a WDM system with N wavelength channels, N different transmitters have to be specifically designed each with a different wavelength. For practical system operation, each transmitter also needs to have a backup to avoid the interruption of service when the transmitter is malfunctioning. This requires to maintaining a large inventory which can often be very costly. A good solution to this problem is to use a tunable laser in the transmitter so that it can be provisioned to the required wavelength at the time of system installation, or replacing any transmitter when needed. This requires a wide wavelength tuning range of the tunable laser diode to cover the entire C-band. Although the cost and the complexity of a tunable laser are typically higher than a simple wavelength stabilized DFB laser, it still makes economic sense for developing wavelength tunable WDM transmitters. For the application as the tunable laser source inside a WDM transmitter, the speed of wavelength provisioning does not have to be fast, and thus tunable lasers based on micro-electro-mechanical systems (MEMS) can be used, as illustrated in Fig. 8.4.2.

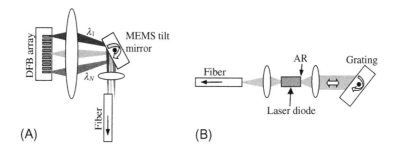

FIG. 8.4.2

Configurations of tunable lasers based on wavelength tuning through MEMS. (A) Use MEMS tilt mirror to select the desired wavelength from an array of DFB laser diodes. (B) Use MEMS tilt grating to control optical feedback condition in an external cavity to select the desired wavelength and minimize spectral linewidth.

Fig. 8.4.2A shows a tunable laser configuration based on a DFB laser diode array which has N components. Each laser diode in the array emits a unique wavelength, and they are selected by a MEMS tilt mirror which decides which wavelength is coupled in the output optical fiber. This array-and-selection configuration does not change the spectral structure of each laser diode in the array, so that the spectral linewidth is on the Megahertz level. For the external cavity configuration shown in Fig. 8.4.2B, a reflective grating is used for wavelength selection in the external cavity, and the angle tilt of the grating is controlled by the MEMS actuating mechanism. In this external cavity configuration, both the angle of grating and the length of the external cavity are important to determine the emission wavelength and the spectral linewidth of the laser. More stringent precision and stability is required in this configuration, and feedback control is also required to optimize the optical feedback condition and minimize the spectral linewidth. 100-kHz spectral linewidth is usually specified for tunable lasers based on external cavity.

8.5 SOURCES OF OPTICAL SYSTEM PERFORMANCE DEGRADATION

WDM is clearly an enabling technology that makes efficient use of the fiber bandwidth and makes transmission of Terabit data traffic possible over a single fiber. But at the same time, it introduces linear and nonlinear crosstalks among different wavelength channels. Understand the nature of these impairments is essential in system design and optimization, and it helps find ways to mitigate and minimize the impact of these crosstalks.

8.5.1 PERFORMANCE DEGRADATION DUE TO LINEAR SOURCES

Linear sources of degradation to the performance of a fiber-optic transmission system include chromatic dispersion, PMD, accumulated ASE noise from inline optical amplifiers, multi-pass interference due to optical reflections in the system, and crosstalk between adjacent wavelength channels due to the imperfect extinction ratio of optical filters and WDM MUX and DEMUX. These degradations are in the *linear* category because they are independent of the signal optical power that is traveling inside the fiber.

8.5.1.1 Eye closure penalty due to chromatic dispersion

Chromatic dispersion, as discussed in Chapter 2, is a frequency-dependent group velocity, which causes different frequency components within a modulated optical signal spectrum to travel in different speeds, resulting in an arrival time difference between them. Fig. 8.5.1A–D show normalized eye diagrams of a 10 Gb/s binary intensity-modulated signal propagating through a standard single-mode fiber with 0, 40, 80, and 120 km lengths, respectively. Only chromatic dispersion is considered in the fiber. At the signal wavelength of 1553 nm, the dispersion parameter of the fiber is 15.8 ps/nm/km, so that the accumulated dispersion values along the system are 0, 632, 1264, and 1896 ps/nm, respectively, for the fiber lengths considered. A 7.5 GHz low-pass filter (fifth-order Bessel) was used in the receiver to eliminate high-frequency components, while maintaining the fundamental frequency component to maximize the eye opening at the decision phase. After 80 km of propagation through the fiber, the eye opening is reduced to approximately 60%. The eye diagrams are further closed at 120 km where differential group delay within the signal spectrum exceeds the 100 ps pulse width of the modulated signal.

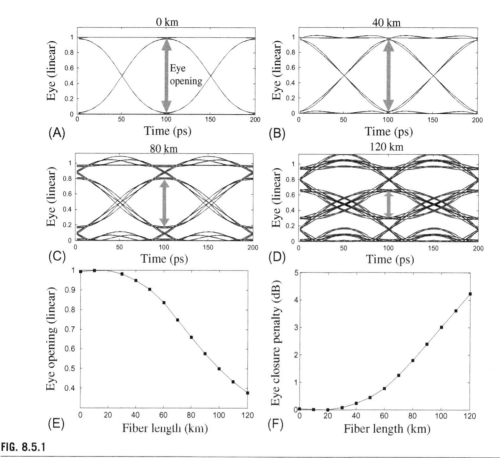

FIG. 8.5.1

(A)–(D) Eye diagrams of received 10 Gb/s binary optical signal propagating through single mode fiber with lengths of 0 km (A), 40 km (B), 80 km (C), and 120 km (D). Fiber dispersion parameter is $D = 15.8$ ps/nm/km at the signal wavelength. (E)–(F): Eye opening (E) and eye closure penalty (F) as the function of fiber length.

Eye diagrams shown in Fig. 8.5.1A–D are obtained through numerical simulations, in which a NRZ pseudo-random bit sequence (PRBS) is used. No phase modulation is associated with the intensity modulation, or equivalently the modulation chirp is zero in the optical signal used in the simulation. Fig. 8.5.1E shows the normalized eye opening as the function of the fiber length, and Fig. 8.5.1F shows the eye closure penalty, $E_{penalty}$, which is related to the eye opening, E_{eye}, by $E_{penalty} = -10\log_{10}(E_{eye})$. In this case without modulation chirp, normalized eye opening reduces monotonically with the increase of the fiber length, and is reduced to about 50% at 100-km fiber transmission distance. When modulation chirp is considered, the eye closure penalty as the function of fiber length may change. Fig. 8.5.2A shows the calculated eye closure penalty for the 10 Gb/s binary modulated optical signal, similar to that shown in Fig. 8.5.1F, but with optical phase modulation associated with the intensity modulation. In this case, we assume optical phase shift $\phi(t)$ is linearly proportional to the normalized instantaneous optical power, $p(t)$, as $\phi(t) = \alpha_c \cdot p(t)$ with $0 < p(t) < 1$, where α_c is the chirp parameter, similar to that defined in Eq. (3.3.40), for the linewidth enhancement factor. The blue (dashed curve in print versions), black, and red (dark gray curve in print versions) curves in Fig. 8.5.2A show eye closure penalties for the chirp parameters of -0.2π, 0, and 0.2π, respectively.

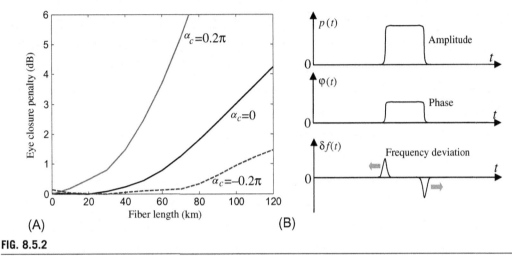

(A)   (B)

FIG. 8.5.2

(A) Calculated eye closure penalty as the function of fiber length using modulated optical signal with chirp parameters of $\alpha_c = -0.2\pi$ (blue, dashed curve in print versions), 0 (black), and 0.2π (red, dark gray curve in print versions), respectively. (B) Illustration of instantaneous frequency deviation $\delta f(t) = d\phi(t)/dt$ for a single pulse with modulation chirp.

When the chirp parameter is positive ($\alpha_c > 0$), the leading edge of a pulse has a positive frequency deviation while the trailing edge of the pulse has a negative frequency deviation as illustrated in Fig. 8.5.2B. For a standard single-mode fiber with anomalous dispersion, $D = d\tau/d\lambda > 0$ and $\beta_2 = d\tau/d\omega < 0$, where τ is the propagation delay, higher-frequency components travel faster (less delay) than the lower-frequency components. Therefore, positive chirp associated with an intensity-modulated pulse introduces additional pulse broadening in this system which accelerateseye closure vs fiber length. On the other hand, a negative chirp can result in pulse squeezing along an anomalous dispersion fiber, which has the effect of increasing eye opening and compensates for the pulse broadening to some extent as shown in Fig. 8.5.2A (blue curve, dashed curve in print versions).

Chromatic dispersion can also be seen as a low-pass filter applied on the modulated optical signal. For a NRZ PRBS pattern, the most vulnerable bit is the isolated "1"s which contains the highest frequency components. As an approximation, if we only consider an isolated digital "1" bit, optical field waveform can be represented by a normalized Gaussian pulse

$$E(t) = \frac{1}{\sigma\sqrt{2\pi}} \exp\left(-\frac{t^2}{2\sigma^2}\right) e^{-j\varphi(t)} \tag{8.5.1}$$

where $\varphi(t) = \alpha_c t^2/(2\sigma^2)$ is the optical phase introduced from modulation chirp. The FWHM pulse width of the optical power is $T_0 = \sigma 2\sqrt{\ln 2}$, and the pulse energy is unity. For a Gaussian pulse in the time domain, its spectrum in the frequency domain, which can be obtained through Fourier transform of $E(t)$, is still Gaussian,

$$\tilde{E}(\omega) = \sqrt{\frac{1}{(1+j\alpha_c)}} \exp\left[\frac{-\omega^2\sigma^2}{2(1+j\alpha_c)}\right] \tag{8.5.2}$$

When this optical pulse propagates through an optical fiber, the impact of chromatic dispersion introduces a frequency-dependent phase into the spectrum. So that at the fiber output, the optical field becomes

$$\tilde{E}(\omega) = \sqrt{\frac{1}{(1+j\alpha_c)}} \exp\left[-\frac{\omega^2\sigma^2}{2(1+j\alpha_c)} + j\frac{\beta_2\omega^2}{2}L\right] \tag{8.5.3}$$

An inverse Fourier transform converts this optical spectrum back to time domain is

$$E(t,L) = \sqrt{\frac{1}{2\pi[\sigma^2 + (\alpha_c - j)\beta_2 L]}} \exp\left[\frac{-(1+j\alpha_c)t^2}{2\sigma^2 + 2(\alpha_c - j)\beta_2 L}\right] \tag{8.5.4}$$

where L is the fiber length and β_2 is the dispersion parameter of the fiber. At the fiber input, the peak power of the normalized Gaussian pulse, defined by Eq. (8.5.1), is $p(0) = 1/(2\pi\sigma^2)$. While at the fiber output, the peak power becomes

$$p(L) = \frac{1}{2\pi\sqrt{(\sigma^2 + \alpha_c\beta_2 L)^2 + (\beta_2 L)^2}} \tag{8.5.5}$$

Thus, the normalized eye opening can be found as

$$E_{eye} = \frac{\sigma^2}{\sqrt{(\sigma^2 + \alpha_c\beta_2 L)^2 + (\beta_2 L)^2}} \tag{8.5.6}$$

For a Gaussian pulse without modulation chirp ($\alpha_c = 0$), the eye opening at the receiver is $E_{eye} = \sqrt{1 + (\beta_2 L/\sigma^2)^2}$. If we further define a dispersion length as

$$L_D = \frac{\sigma^2}{|\beta_2|} \tag{8.5.7}$$

eye opening can have a very simple form,

$$E_{eye} = \frac{1}{\sqrt{1 + (L/L_D)^2}} \approx 1 - \frac{L^2}{2L_D^2} \tag{8.5.8}$$

where linearization can be applied if the closure penalty is small enough. The reason that the dispersion length L_D is proportional to the square of the pulse width is that increasing pulse width linearly reduces the spectral width and reducing dispersion induced differential delay, and at the same time the increased pulse width will also increase the tolerance to the impact of differential delay. For a standard single-mode fiber with a dispersion parameter $D = 15.8\,\text{ps/nm/km}$, or $\beta_2 \approx 2 \times 10^{-26}\,\text{s}^2/\text{m}$, the dispersion length is approximately 180 km for 100 ps FWHM pulse width ($\sigma = 60$ ps). The Gaussian pulse approximation is simple, and it can help find the general rule of dispersion impact, however it underestimates the eye closure penalty compared to simulation using PRBS patterns as shown in Fig. 8.5.2A. PRBS is a combination of pulses of different widths which may cause different rise/fall times after experiencing chromatic dispersion, so that eye closure penalty is higher than using a single Gaussian pulse. In fact, if we consider an isolated digital "0," the 0 level will increase because of the spreading of adjacent "1"s. This can also be seen as the broadening of an inverse Gaussian pulse so that the minimum level increases the same way as the decrease of the amplitude of the "1" pulse. Considering the decrease of the 1 level and the increase of the 0 level, Eq. (8.5.8) should be modified to $E_{eye} \approx 1 - L^2/L_D^2$.

8.5.1.2 Eye closure penalty due to PMD

PMD is a differential delay between the two orthogonally polarized modes in a single-mode fiber, which introduces time jitter in a fiber-optic system, resulting in an eye closure penalty. Unlike chromatic dispersion in the fiber system which is deterministic, the differential delay introduced by PMD is random and largely time-dependent, which makes the modeling of PMD impact in a fiber-optic transmission system challenging. Large number of simulations has to be done to have statistical relevance. As the most vulnerable bit of an NRZ modulated waveform subject to time jitter is the isolated "1," a Gaussian pulse approximation can be used to evaluate the impact of PMD on the eye closure penalty. Assume a normalized Gaussian pulse of the signal optical power with a width of σ_{in} at the input of an optical fiber, corresponding to a FWHM pulse width of $\Delta t_{FWHM} = 2\sigma_{in}\sqrt{2\ln(2)}$,

$$P_{in}(t) = \frac{1}{\sigma_{in}\sqrt{2\pi}} \exp\left(-\frac{t^2}{2\sigma_{in}^2}\right) \tag{8.5.9}$$

Only considering the impact of PMD, and neglecting the polarization-independent propagation delay, the optical pulse at the fiber output is the weighted combination of the powers carried by the two polarization modes, which is

$$P_{out}(t) = \frac{1}{\sigma_{in}\sqrt{2\pi}}\left[\gamma \exp\left(-\frac{t^2}{2\sigma_{in}^2}\right) + (1-\gamma)\exp\left(-\frac{(t-\tau)^2}{2\sigma_{in}^2}\right)\right] \tag{8.5.10}$$

where $0 < \gamma < 1$ is the ratio of the powers carried by the two principle states of polarization (PSP), and τ is the differential group delay (DGD) between them (Poole et al., 1991). The average position of the temporal output pulse can be found as $t_a = \int_{-\infty}^{\infty} t P_{out}(t)dt = (1-\gamma)\tau$. The width σ_{out} of the output pulse can be found through the calculation of the variance

$$\sigma_{out}^2 = \int_{-\infty}^{\infty} (t-t_a)^2 P_{out}(t)dt = \sigma_{in}^2 + \tau^2\gamma(1-\gamma) \tag{8.5.11}$$

As the peak amplitude of a constant-energy Gaussian pulse is inversely proportional to the pulse width, the normalized eye opening can be found as

$$E_{eye} = \frac{\sigma_{in}}{\sigma_{out}} = \frac{1}{\sqrt{1 + \gamma(1-\gamma)\tau^2/\sigma_{in}^2}} \tag{8.5.12}$$

For a NRZ modulated pulse train at a data rate B, assume the FWHM pulse width of an isolated "1" bit is $\Delta t_{FWHM} = 1/B$, we have $\sigma_{in} = 1/(2B\sqrt{2\ln(2)}) = 0.425/B$. Thus, the normalized eye opening can be expressed as the function of the data rate B as

$$E_{eye} = \frac{1}{\sqrt{1+5.6\gamma(1-\gamma)\tau^2 B^2}} \approx 1 - 2.8\gamma(1-\gamma)\tau^2 B^2 \tag{8.5.13}$$

where we consider that eye closure penalty is small enough so that linearization can be used. Fig. 8.5.3A shows a normalized Gaussian pulse with $\sigma_{in} = 1/\sqrt{2}$ ($\Delta t_{FWHM} \approx 1.67$) which is split into two parts (shown as dashed and dash-dotted lines) with a peak power ratio of $\gamma = 0.6$, and a relatively delay $\tau = 1$ between them. The combined pulse shown as the solid line in Fig. 8.5.3A has the average position $t_a = 0.4$ and a normalized width of $\sigma_{out} = 0.8602$ (corresponding to $\Delta t_{FWHM} \approx 2.03$). Obviously the worst-case pulse broadening due to PMD happens when the signal power is equally split into the two polarization modes ($\gamma = 1/2$).

If we further consider an isolated digital "0," the effect of DGD will cause the increase of the 0 level in the similar way as the decrease of the 1 level of digital "1." Thus, the impact on the overall eye closure will be doubled to approximately

$$E_{eye} \approx 1 - 5.6\gamma(1-\gamma)\tau^2 B^2 \tag{8.5.14}$$

Fig. 8.5.3B shows the eye closure penalty as the function of τB, which is the DGD normalized by the FWHM pulse width as $\Delta t_{FWHM} = 1/B$.

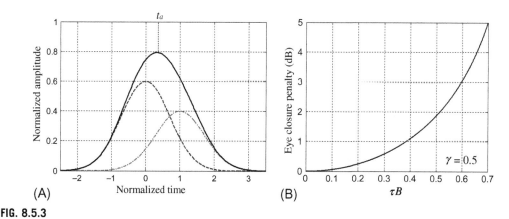

(A)   **(B)**

FIG. 8.5.3

(A) Illustration of a Gaussian pulse which is split into two (dashed and dash-dotted lines) with 0.6/0.4 ratio, and relatively delayed. The recombined pulse (solid line) has a shifted average position and broader width. (B) Eye closure penalty calculated with Eq. (8.5.14).

Since PMD is a random process, both the DGD value τ and the power splitting ratio γ vary with time. The statistics of DGD in long fiber systems has been well studied and the PDF of τ is Maxwellian

$$PDF_{DGD}(\tau) = \sqrt{\frac{6}{\pi}\frac{3\tau^2}{\tau_{rms}^3}}\exp\left(\frac{-3\tau^2}{2\tau_{rms}^2}\right) \tag{8.5.15}$$

where τ_{rms} is the root-mean-square (RMS) value of DGD. Fig. 8.5.4 shows the statistic distribution of t, and the RMS value of DGD is assumed to be $\tau_{rms} = 25$ ps. The average DGD is $\tau_{mean} \approx 27$ ps, which is calculated from the PDF, shown in Eq. (8.5.15), as

$$\tau_{mean} = \int_0^\infty \tau \cdot PDF_{DGD}(\tau)d\tau = \tau_{rms}\sqrt{\frac{3\pi}{8}} \qquad (8.5.16)$$

On the other hand, the ratio γ of power splitting into the two polarized modes has a uniform distribution, so that the average value of $\gamma(1-\gamma)$ is $\gamma_{mean} = \int_0^1 \gamma(1-\gamma)d\gamma = 1/6$, which is less than the worst case of 0.25. Thus, the average eye opening of the received optical signal is $E_{eye,mean} \approx 1 - 0.933\tau_{mean}^2 B^2$

Thus, a mean DGD of $\tau_{mean} = 27$ ps corresponds to an average eye closure penalty of approximately 0.3 dB for a 10 Gb/s NRZ data sequence. However, the Maxwellian distribution has a relatively long tail as shown in Fig. 8.5.4B. For example, the instantaneous DGD can reach 100 ps though at a low probability of 10^{-8}. This may cause short-bursts of system outage over a long period of time. A statistical analysis of system outage probability has to be conducted in the system design to guarantee the system performance.

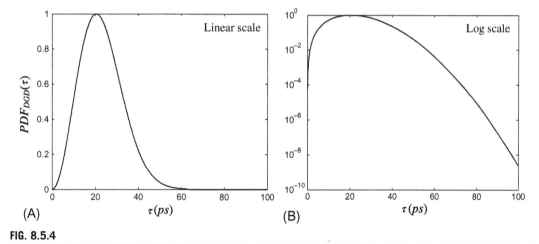

FIG. 8.5.4

Probability of fiber DGD, τ, which has a Maxwellian distribution with $\tau_{rms} = 25$ ps and a mean DGD of $\tau_{mean} = 27$ ps. (A) In linear scale, and (B) in logarithm scale.

The concept of birefringence is relatively straightforward in a short fiber where refractive indices are slightly different in the two orthogonal axes of the transversal cross-section so that DGD is linearly proportional to the length of the fiber. However, if the fiber is long enough, the birefringence axes may rotate along the fiber due to banding, twisting, and fabrication nonuniformity. Meanwhile, there is also energy coupling between the two orthogonally polarized propagation modes in the fiber. In general, both the birefringence axis rotation and the mode coupling are random and unpredictable, which make PMD a complex problem to understand and to solve (Poole and Nagel, 1997).

Despite the random rotation of birefringence axis along a fiber, a concept of *Principal state of polarization* (PSP) is very useful in the analysis of PMD, which indicates two orthogonal polarization states corresponding to the fast and slow axes of the fiber. Under this definition, if the polarization state of the input optical signal is aligned with one of the two PSPs of the fiber, the output optical signal will keep the same SOP. In this case, the PMD has no impact in the optical signal, and the fiber only provides a single propagation delay. It is important to note that PSPs exist not only in "short" fibers but also in "long" fibers. In a long fiber, although the birefringence along the fiber is random and there is energy coupling between the two polarization modes, an equivalent set of PSPs can always be found. Again, PMD has no impact on the optical signal if its polarization state is aligned to one of the two PSPs of the fiber. However, in a practical fiber the orientation of the PSPs usually change with time, especially when the fiber is long. The change of PSP orientation over time can be originated from the random changes in temperature and mechanical perturbations along the fiber.

An important parameter of a single-mode fiber is its *PMD parameter*, which is defined as the mean DGD over a unit length of fiber. For the reasons we have discussed, for a short fiber, mean DGD is $\tau_{mean} \propto L$, where L is the fiber length, so that the unit of PMD parameter is [ps/km]. Whereas for a long fiber, mean DGD is $\tau_{mean} \propto \sqrt{L}$, and thus the unit of PMD parameter is [ps/$\sqrt{km}$]. For example, for a standard single-mode fiber, if the PMD parameter is $PMD = 0.06 ps/\sqrt{km}$, the mean DGD value for a 100-km long fiber will be $\tau_{mean} = 0.6 ps$.

8.5.1.3 ASE noise accumulation in fiber systems with inline optical amplifiers

In a long-distance fiber-optic transmission system, optical amplifiers are used to compensate for the transmission loss as illustrated in Fig. 8.5.5. The optical amplifier at the optical transmitter, commonly referred to as the postamplifier, boosts the signal optical power before it is sent to the transmission fiber. Postamplifier can be packaged inside an optical transmitter to compensate the loss introduced by the electro-optic modulator and other optical components of the transmitter. At a receiver, the optical amplifier in front of the photodiode is commonly referred to as the preamplifier which enhances the signal optical power before photo-detection. The optical preamplifier can be treated as part of the optical receiver to form a preamplified optical receiver. Other optical amplifiers implemented along the fiber transmission line are called inline optical amplifiers. In a WDM optical system, inline optical amplifiers need to have wide gain bandwidth with flat gain spectrum to equally support all wavelength channels in the system.

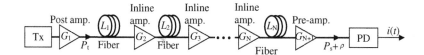

FIG. 8.5.5

Fiber-optic transmission system with multiple optical amplifiers. Tx: optical transmitter, PD: photodiode.

In the optical system shown in Fig. 8.5.5, there are N fiber spans and $N+1$ optical amplifiers including the post and inline amplifiers. Assume the optical gains of the amplifiers are $G_1, G_2, ..., G_N$, and G_{N+1}, and the losses of fiber spans are $L_1, L_2, ..., L_{N-1}$ and L_N, respectively. If the gain of each optical amplifier compensates the loss of each fiber span, the optical signal power P_s that reaches the photodiode is the same as that at the output of the postamplifier which is P_t. The ASE noise generated by each optical amplifier is linearly proportional to its optical gain, so that the accumulated ASE noise at the input of the photodiode is

$$\rho = 2hv \sum_{i=1}^{N} \left[n_{sp,i}(G_i - 1) \left(\prod_{m=i}^{N} G_{m+1}L_m \right) \right] + 2hvn_{sp,N+1}(G_{N+1} - 1) \tag{8.5.17}$$

where $L_i = e^{-\alpha_i l_i}$ is the fiber loss with l_i and α_i the fiber length and attenuation parameter of the ith span. $n_{sp,i}$ is the noise parameter of the ith optical amplifier, as defined in Eq. (5.2.1), and hv is the photon energy. If we further assume that each fiber span has the same loss which is compensated by the gain of the optical amplifier in that span, assume $(G_i - 1) \approx G$, and all amplifiers have the same noise figure $n_{sp,i} = n_{sp}$, the ASE noise power spectral density shown in Eq. (8.5.17) can be simplified as, $\rho \approx hvn_{sp}(N + 1)G$. The optical SNR (OSNR) is

$$OSNR = \frac{P_s}{\rho} \approx \frac{P_t}{2hvn_{sp}(N+1)G} \qquad (8.5.18)$$

which is inversely proportional to the number of EDFAs, $N + 1$. Since the unit of the signal optical power is dBm, and the unit of optical noise power spectral density is dBm/Hz, the unit of OSNR calculated from Eq. (8.5.18) should be [dB·Hz]. In practice, 0.1 nm is often used as the resolution bandwidth of OSA in the process of evaluating OSNR, which results in a unit of [dB·0.1 nm] for OSNR. In the 1550-nm wavelength window, 0.1 nm is equal to 12.5 GHz, so that a factor of 12.5×10^9 has to be used to translate [dB·Hz] into [dB·0.1 nm]. As an example, Fig. 8.5.6 shows the calculated OSNR in [dB·0.1 nm] as the function of the number of fiber spans. In this example, the noise figure of each optical amplifier is 5 dB, equivalent to $n_{sp} = 1.58$. The fiber loss is 0.25 dB/km, and signal optical power is $P_s = 1$ mW. The total fiber lengths of the system are 1000, 3000, and 5000 km, corresponding to the total fiber loss, L_{total}, of 250, 750, and 1250 dB, respectively. To compensate the fiber loss, the gain of each optical amplifier has to be $G^{N+1} = 10^{L_{total}/10}$, where L_{total} is in dB. Thus $G^{N+1} = 10^{L_{total}/10(N+1)}$ and Eq. (8.5.18) can be written as the function of total fiber loss as

$$OSNR = \frac{P_t}{2hvn_{sp}(N+1) \times 10^{\frac{L_{total}}{10(N+1)}}}$$

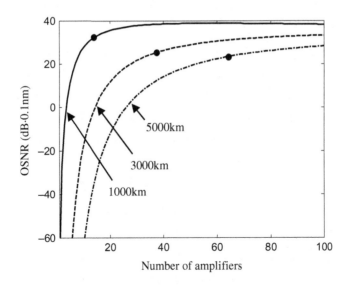

FIG. 8.5.6

OSNR as the function of the number of fiber spans for the total system length of 1000 km (solid line), 3000 km (dashed line), and 5000 km (dash-dotted line). The solid dot on each curve indicates where the length of each fiber span is 80 km.

As shown in Fig. 8.5.6, for a certain total length of the fiber system, OSNR generally increases with the increasing the number of optical amplifiers so that the gain of each optical amplifier is reduced. However, the rate of increase saturates when the length of each fiber span is small enough. From an engineering point of view, small number of optical amplifiers in the system would simply system implementation, maintenance, and reduce the cost, but would reduce the OSNR, and thus a tradeoff has to be made in the determination of the fiber span length. The solid dot on each curve in Fig. 8.5.6 indicates the place where the length of each fiber span is 80 km. This example clearly tells that when the length of each fiber span is shorter than 80 km, further decreasing the span length would not significantly improve the system OSNR. Thus, in practical fiber-optic systems, amplified fiber span lengths are typically between 60 and 100 km.

8.5.1.4 Multipath interference

In practical optical systems, an optical signal may go through different paths before recombining at the receiver. Multipath interference can be caused by beam splitting and combining, as well as reflections along a fiber link with multiple optical connectors and interfaces, as illustrated in Fig. 8.5.7.

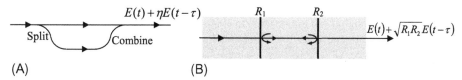

(A)   (B)

FIG. 8.5.7

Linear crosstalk caused by (A) beam splitting and combining, and (B) optical reflections.

In Fig. 8.5.7, η is a crosstalk coefficient, and R_1 and R_2 are reflectivities of optical interfaces along the fiber. $\tau = n\Delta L/c$ is a relative time delay between the main and the delayed paths with n the refractive index, ΔL the difference of path length, and c the speed of light. In most cases, the optical signal going through the delayed path has much lower amplitude than that through the main path, that is $\eta \ll 1$ or $R_1 R_2 \ll 1$, thus multipath interference can be treated as a perturbation. Consider an optical signal $E(t) = A(t)e^{j\varphi(t)}$ with amplitude $A(t)$ and phase $\varphi(t)$, the photocurrent of this optical signal mixes with its delayed and attenuated replica is,

$$\begin{aligned}
I(t) &= \mathfrak{R}\left| A(t)e^{j\varphi(t)} + \zeta A(t-\tau)e^{j\varphi(t-\tau)} \right|^2 \\
&= \mathfrak{R}P(t) + \mathfrak{R}\zeta^2 P(t-\tau) + 2\mathfrak{R}\zeta \sqrt{P(t)P(t-\tau)}\cos\delta\varphi(t)
\end{aligned} \qquad (8.5.19)$$

where $\mathfrak{R}$ is the responsivity of the photodiode, $P(t) = |E(t)|^2$ is the signal optical power, and $\zeta = \eta$ for beam splitting/combining and $\zeta = \sqrt{R_1 R_2}$ for multiple reflections as shown in Fig. 8.5.7A and B, respectively. The optical phase is $\varphi(t) = \omega t + \varphi_0(t)$, where ω is the angular frequency and φ_0 represents a reference phase, the differential phase is

$$\delta\varphi(t) = \varphi(t) - \varphi(t-\tau) = \omega\tau + \varphi_0(t) - \varphi_0(t-\tau) \qquad (8.5.20)$$

The impact of multipath interference can be regarded as a conversion from phase noise of the optical source into an intensity noise. The multipath configuration shown in Fig. 8.5.7A and the multiple-reflection configuration shown in Fig. 8.5.7B have the transfer functions of a Mach-Zehnder interferometer and a Fabry-Perot interferometer, respectively. For a constant wave (CW) optical source with an intensity $P(t) = P_a$, the photocurrent at the detector is $I(t) \approx \Re P_a [1 + 2\zeta \cos \delta\varphi(t)]$, where interference has been assumed small enough so that the ζ^2 term is neglected for simplicity. When the relative delay τ is much shorter than the coherence length, the interference is coherent, and a linear approximation can be used for the differential phase, $\delta\varphi(t) \approx \tau[\omega + \delta\omega(t)]$ where $\delta\omega(t) = d\varphi(t)/dt$ is the optical frequency deviation caused by the phase noise. Maximum conversion efficiency happens when $\omega\tau = m\pi + \pi/2$, with m an integer, and the maximum intensity noise peak-to-peak amplitude on the photocurrent is $4\zeta I_a$, as shown in Fig. 8.5.8, where $I_a \approx \Re P_a$ is the average photocurrent.

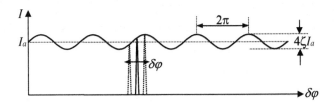

FIG. 8.5.8

Illustration of phase noise to intensity noise conversion through phase-dependent transfer function of multipath interference.

Now, let us look at the impact of coherent multipath interference on a binary intensity-modulated waveform as illustrated in Fig. 8.5.9A. For simplicity, we assume that the waveform is ideal before the multipath interference, so that $P_1 = 2P_{ave}$ and $P_0 = 0$ for optical power levels at digital "0" and "1," respectively, with P_{ave} the average optical power. Based on Eq. (8.5.19), the maximum photocurrent of digital "0" (corresponds to $P(t) = 0$ and $P(t - \tau) = P_1$) is $\Re\zeta^2 P_1$, while the minimum photocurrent of digital "1" (corresponds to $P(t) = P_1$ and $P(t - \tau) = P_1$) is $\Re P_1(1 + \zeta^2 - 2\zeta)$. The worst-case peak-to-peak variation caused by multipath interference at signal digital "1" is $\Delta I = 4\zeta$, and the normalized eye closure penalty is $A - B = (1 - 2\zeta)$. This value has been normalized by the nominal digital "1" photocurrent which is $I_1 = \Re P_1$.

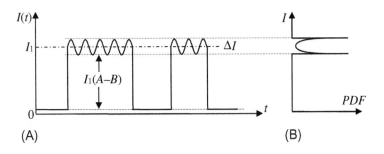

FIG. 8.5.9

(A) illustration of normalized binary waveform with multipath interference, and (B) probability distribution function of the waveform.

Fig. 8.5.9B shows the probability density function (PDF) of the normalized photocurrent assume $P(t-\tau)=P(t)$. For a uniformly distributed differential phase $\delta\varphi$, the PDF of $i(\delta\varphi)=1+\zeta^2+2\zeta\cos(\delta\varphi)$ is

$$PDF = \frac{1}{2\pi\zeta\sqrt{1-\left(\frac{i-1-\zeta^2}{2\zeta}\right)^2}} \tag{8.5.21}$$

This PDF approaches infinity at $i=1+\zeta^2\pm2\zeta$.

The coherent interference happens when the relative delay time τ is much shorter than the coherence time of the laser source defined by Eq. (3.3.55), that is $\tau\ll1/(\pi\Delta\nu\tau)$, where $\Delta\nu$ is the spectral linewidth of the laser source. On the other hand, if the relative delay is very long such that $\tau\gg1/(\pi\Delta\nu\tau)$, statistically, $\delta\varphi(t)=\varphi(t)$ is valid and multipath interference will result in an incoherent addition. In such a case, multipath interference can be seen as a mechanism which converts the phase noise into a RIN, which is (Gimlett and Cheung, 1989)

$$RIN(f) \approx \frac{4\zeta^2}{\pi}\left[\frac{\Delta\nu}{f^2+(\Delta\nu)^2}\right] \tag{8.5.22}$$

which is dependent on the spectral linewidth, $\Delta\nu$, of the laser source, but independent of the relative delay τ. In this case, the impact of multipath interference can be treated simply as an additional noise term in the system performance evaluation as discussed in Section 8.3.

8.5.2 PERFORMANCE DEGRADATION DUE TO FIBER NONLINEARITIES

While sources of linear degradation in a fiber-optic system discussed earlier are independent of the signal optical power level, performance of a fiber-optic system can also be degraded by nonlinear effects in the optical fiber which is sensitive to the level of optical power used. Because of the small diameter of the fiber core, on the order of 9 μm for a standard single-mode fiber, even for a moderate optical power level of 1 mW, the power density is more than 1 kW/cm². Kerr effect nonlinearity originates from the power density-dependent refractive index, which is responsible to a number of well-known effects including SPM, XPM, FWM, and modulation instability (MI). Nonlinear scattering effects such as stimulated Brillouin scattering (SBS) and stimulated Raman scattering (SRS) caused by interaction between the signal photons and the traveling acoustic phonons and photonic phonons, respectively, can cause energy translation from high energy photons to low-energy photons which generates new wavelength components. Mechanisms of both the Kerr effect and the nonlinear scattering have been introduced in Chapter 2. In the following, we discuss the impact of these fiber nonlinearities in the performance of fiber-optical communication systems.

As discussed in Chapter 2, the analysis of Kerr effect nonlinearity in a fiber usually starts from a nonlinear differential equation which describes the envelope of optical field propagating along an optical fiber (Agrawal, 1989):

$$\frac{\partial A(t,z)}{\partial z} + \frac{i\beta_2}{2}\frac{\partial^2 A(t,z)}{\partial t^2} + \frac{\alpha}{2}A(t,z) - i\gamma|A(t,z)|^2A(t,z) = 0 \tag{8.5.23}$$

where $A(t,z)$ is a time and position-dependent complex optical field along the fiber z is the longitudinal coordinate along the fiber, β_2 is the chromatic dispersion parameter, α is the fiber loss parameter, and

$\gamma = n_2\omega/(cA_{eff})$ is the nonlinear parameter with n_2 the nonlinear refractive index, ω the optical frequency, c the speed of light, and A_{eff} the effective cross-section area of the fiber core. We have neglected high-order dispersion terms for simplicity.

Because the complex optical field $A(t, z)$ is both time and position dependent, there is no standard analytical solution to the nonlinear differential equation (8.5.23). Numerical simulation is commonly used for transmission performance estimation. In this section, we first introduce split-step Fourier method (SSFM) of numerical simulation, and then discuss analytical and semi-analytical solutions for the impact of specific nonlinear degradation mechanisms.

SSFM is a popular technique commonly used in numerical simulations for fiber-optic system performance evaluation. It models the dispersion effect in the frequency domain and nonlinear Kerr-Effect in the time domain. Because signal optical power reduces gradually along the fiber due to attenuation, SSFM divides the transmission fiber into many short sections so that an average power can be used for each section.

Based on the nonlinear differential equation (8.5.23), if only linear effects including dispersion and attenuation of the fiber are considered, Eq. (8.5.23) reduces to

$$\frac{\partial A(t, z)}{\partial z} = -\frac{i\beta_2}{2}\frac{\partial^2 A(t, z)}{\partial t^2} - \frac{\alpha}{2}A(t, z) \tag{8.5.24}$$

which has an analytic solution in the frequency domain as

$$\tilde{A}(\omega, L) = \exp\left[\left(\frac{i\beta_2\omega^2}{2} - \frac{\alpha}{2}\right)L\right]\tilde{A}(\omega, 0) \tag{8.5.25}$$

where L is the fiber length. $\tilde{A}(\omega, L)$ and $\tilde{A}(\omega, 0)$ are the Fourier transforms of $A(t, L)$ and $A(t, 0)$ at the output $(z=L)$ and the input $(z=0)$ of the fiber, respectively.

On the other hand, if only the nonlinear Kerr-Effect is considered, and assuming the optical power is constant along the fiber, Eq. (8.5.23) reduces to

$$\frac{\partial A(t, z)}{\partial z} = i\gamma|A(t, z)|^2 A(t, z) \tag{8.5.26}$$

which can be solved analytically in the time domain, and the solution relating the optical field at the fiber input $A(t, 0)$ and that at the fiber output $A(t, L)$ by

$$A(t, L) = A(t, 0)\exp\left(i\gamma|A|^2 L\right) \tag{8.5.27}$$

To generalize, we can separate the frequency domain and the time domain solutions into two operators for the nonlinear differential equation:

$$\frac{\partial A(t, z)}{\partial z} = \left(\hat{D} + \hat{N}\right)A(t, z) \tag{8.5.28}$$

where

$$\hat{D} = -\frac{i\beta_2}{2}\frac{\partial^2}{\partial t^2} - \frac{\alpha}{2} \tag{8.5.29}$$

is a linear differential operator for dispersion and loss, and

$$\hat{N} = i\gamma|A(t, z)|^2 \tag{8.5.30}$$

is a nonlinear operator representing the nonlinear phase shift.

In a practical fiber system, signal optical power $|A(t,z)|^2$ is a function of z, and the dispersion parameter β_2 may also be z-dependent, which makes the solution of Eq. (8.5.28) difficult. In the numerical solution process, one can divide the optical fiber into k sections as illustrated in Fig. 8.5.10, where h_n is the length of the nth section. The purpose of dividing the fiber into many short sections is to make sure that within each section both $|A(t,z)|^2$ and β_2 have negligible changes with z, so that an average value can be used for each of them within the short section.

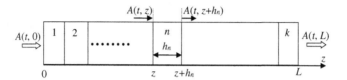

FIG. 8.5.10

Divide a long fiber into k short sections for split-step Fourier simulation.

Assume that over the short fiber section n with length h_n, as shown in Fig. 8.5.10, dispersion and nonlinear effects act independently, the solution of Eq. (8.5.28) can be written as

$$A(t, z+h_n) = \exp\left(\hat{D}h_n\right) \cdot \exp\left(\hat{N}h_n\right)A(t, z) \tag{8.5.31}$$

The linear operator representing dispersion and attenuation can be calculated in the Fourier domain as

$$\exp\left(\hat{D}h_n\right)A(t, z) = F^{-1}\left\{\exp\left[h_n\hat{D}(j\omega)\right] \cdot F[A(t, z)]\right\} \tag{8.5.32}$$

where, $F(x)$ and $F^{-1}(x)$ represent Fourier transform and inverse Fourier transform, respectively, and the frequency domain linear operator is

$$\hat{D}(j\omega) = -i\left(\frac{\lambda_0^2}{4\pi c}\right)\omega^2 D - \frac{\alpha}{2} \tag{8.5.33}$$

where λ_0 is the central wavelength of the optical signal and $D = -2\pi c\beta_2/\lambda_0^2$. The nonlinear effect can be calculated directly in the time domain as

$$\exp\left(\hat{N}h_n\right)A(t, z) = \exp\left(i\gamma|A(z, t)|^2 h_n\right)A(t, z) \tag{8.5.34}$$

Combining these two operators allows us to find the relationship between the input field $A(t, z)$ and the output field $A(t, z+h_n)$ of section n,

$$A(t, z+h_n) = \exp\left(i\gamma|A(z, t)|^2 h_n\right)F^{-1}\left\{\exp\left(\frac{i\beta_2\omega^2}{2}h_n - \frac{\alpha}{2}h_n\right)F[A(z, t)]\right\} \tag{8.5.35}$$

The overall transfer function of a long fiber can then be obtained by repeating the process of Eq. (8.5.35) section by section from the beginning to the end of the fiber. A Fourier transform and an inverse Fourier transform has to be used for each short fiber section.

As a numerical algorithm, the major advantage of SSFM is that the results automatically include both linear chromatic dispersion effect and the nonlinear Kerr effect. Given an input complex optical field in time-domain $A(0, t)$, one can find the output complex optical field in the time-domain, $A(L, t)$, through a section-by-section calculation. The impacts of SPM, XPM, and FWM are all included. Fig. 8.5.11 shows examples of optical spectra and eye diagrams of a three channel WDM system with per channel average optical power of -10, 0, and 5 dBm, respectively. The system has 80-km standard single-mode fiber with 16 ps/nm/km chromatic dispersion at 1550 nm signal wavelength. Fiber cross section area is 80 μm^2 with a nonlinear refractive index $n_2 = 2.35 \times 10^{20} m^2/W$ and a loss coefficient of 0.25 dB/km. NRZ modulation format is used at 10 Gb/s data rate, and WDM channel spacing is 100 GHz. A dispersion compensator is used at the receiver to compensate the impact of chromatic dispersion. Only the middle channel is selected at the receiver to measure the eye diagram.

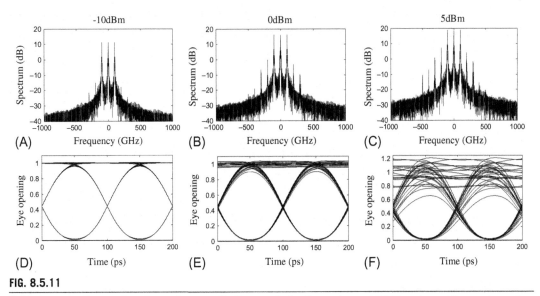

FIG. 8.5.11

Optical spectra (top row) and eye diagrams (bottom row) of a three-channel WDM system with per-channel average powers of -10 dBm [(A) and (D)], 0 dBm [(B) and (E)], and 5 dBm [(C) and (F)] at the input of 80 km standard single-mode fiber.

Because chromatic dispersion is completely compensated in front of the receiver, the eye diagram is widely open when the average signal optical power is low (-10 dBm per channel). With the increase of the launched signal optical power to 0 dBm per channel, Kerr effect nonlinearity starts to show its impact in the eye diagram. Further increasing the per channel signal power to 5 dBm, the effect of nonlinearity becomes significant, primarily due to nonlinear crosstalk introduced through FWM. A number of new frequency components are generated in the optical spectrum as shown in Fig. 8.5.11C. As has been discussed in Section 2.6, the newly generated mixing products coherently interfere with the original optical channel at the same frequency which creates an eye closure penalty.

Examples shown in Fig. 8.5.11 are for a system with only a single fiber span of 80 km. The split step simulation can also be used for WDM fiber systems with multiple fiber spans with inline optical amplifiers. This provides a powerful tool for system design and performance evaluation. Notice that in the simulated results shown in Fig. 8.5.11, we have assumed that all signal optical channels are co-polarized along the fiber so that the effect of FWM is the maximum. In a practical system, the state of polarizations (SOPs) of WDM channels at different wavelengths may walk-off from each other along the fiber due to PMD, and thus the actual efficiency of FWM may be lower especially in long-distance fiber-optical systems with multiple fiber spans.

The numerical method of SSFM is a powerful tool for system design, but on the flip side it is usually time consuming in the computation, which can often be prohibitive especially for systems with wide optical bandwidth and large numbers of WDM channels. In such cases the sizes of Fourier transform and inverse Fourier transform have to be excessively large and the speed of computation can be painfully slow. Optical spectra and eye diagrams obtained through numerical simulation shown in Fig. 8.5.11 automatically include all Kerr effect nonlinearities including SPM, XPM, and FWM. It lacks direct explanation of the physical mechanism behind each individual effect. Therefore, analytic or semi-analytic methods can be more helpful in understanding the impact of each individual nonlinear effect, and avoiding lengthy numerical processes.

8.5.3 SEMI-ANALYTICAL APPROACHES TO EVALUATE NONLINEAR CROSSTALKS IN FIBER-OPTIC SYSTEMS

In an amplified multispan WDM optical system, interchannel nonlinear crosstalk is an important concern, especially when the system has a large number of spans and the signal optical power level is high. Although SSFM discussed in the previous section can be used to evaluate the nonlinear crosstalks through numerical solutions based on the nonlinear wave propagation equation, analytical or semi-analytical solutions based on first principles can be very helpful in understanding physical mechanisms.

In this section we discuss analytic and semi-analytic solutions to evelute the impact of nonlinear crosstalks including XPM, FWM, and MI in multispan WDM fiber-optic systems.

8.5.3.1 XPM-induced intensity modulation in IMDD optical systems

XPM originates from the Kerr effect in optical fibers, in which intensity modulation of one optical carrier can modulate the phases of other co-propagating optical signals in the same fiber [Chraplyvy, 1990; Marcuse et al., 1994 (JLT)]. Unlike phase encoded optical systems, intensity-modulation direct-detection (IMDD) optical systems are not particularly sensitive to signal optical phase fluctuations. Therefore XPM-induced PM alone is not a direct source of performance degradation in IMDD systems. However, due to the chromatic dispersion of optical fibers, phase modulation can be converted into intensity modulation (Wang and Petermann, 1992) and thus can degrade the IMDD system performance. On one hand, nonlinear phase modulation created by XPM is inversely proportional to the signal baseband modulation frequency (Chiang et al., 1996); on the other hand, the efficiency of phase noise to intensity noise conversion through chromatic dispersion increases with the modulation frequency (Wang and Petermann, 1992). Therefore, XPM-induced overall intensity modulation is a complicated function of the signal modulation frequency.

The theoretical analysis of XPM begins with the nonlinear wave propagation equation (8.5.23). As illustrated in Fig. 8.5.12, assume that there are only two wavelength channels copropagating along the fiber,they are defined as the probe and the pump, and their optical fields are denoted as $A_j(t, z)$ and $A_k(t, z)$, respectively. The evolution of the probe wave (a similar equation can be written for the pump wave) is described by

$$\frac{\partial A_j(t,z)}{\partial z} = -\frac{\alpha}{2}A_j(t,z) - \frac{1}{v_j}\frac{\partial A_j(t,z)}{\partial t} - \frac{i\beta_2}{2}\frac{\partial^2 A_j(t,z)}{\partial t^2}$$
$$+ i\gamma_j p_j(t,z)A_j(t,z) + i\gamma_j 2p_k(t - z/v_k, z)A_j(t,z) \tag{8.5.36}$$

where α is the fiber attenuation coefficient, β_2 is the chromatic dispersion parameter, $\gamma_j = 2\pi n_2/(\lambda_j A_{eff})$ is the nonlinear coefficient, n_2 is the nonlinear refractive index, λ_j and λ_k are the probe and the pump wavelengths, A_{eff} is the fiber effective core area, and $p_k = |A_k|^2$ and $p_j = |A_j|^2$ are optical powers of the pump and the probe, respectively. Due to chromatic dispersion, the pump and the probe waves generally travel at different speeds, and this difference must be taken into account in the calculation of XPM because it introduces the walk-off between the two waves. Here we use v_j and v_k to represent the group velocities of these two channels.

On the right-hand side of Eq. (8.5.36), the first term represents the effect of attenuation, the second term is the linear propagation group delay, the third term accounts for chromatic dispersion, the fourth term is responsible for SPM, and the fifth term is the XPM on the probe signal j induced by the pump signal k. The strength of XPM is proportional to the optical power of the pump and the fiber nonlinear coefficient. To simplify our analysis and focus the investigation on the effect of XPM-induced interchannel crosstalk, the interaction between SPM and XPM has to be neglected, assuming that these two act independently. We also assume that the probe is operated in CW, whereas the pump is modulated with a sinusoid at a frequency Ω. Although the effects of SPM for both the probe and the pump channels are neglected in the XPM calculation, a complete system performance evaluation can take into account the effect of SPM and other nonlinear effects separately. This approximation is valid as long as the pump signal waveform is not appreciably changed by the SPM-induced distortion within the nonlinear length of the fiber. Under this approximation the third term on the right hand side of Eq. (8.5.36) can thus be neglected. Using variable substitutions $T = t - z/v_j$ and $A_j(t,z) = E_j(T,z)\exp(-\alpha z/2)$, Eq. (8.5.36) becomes

$$\frac{\partial E_j(T,z)}{\partial z} = -\frac{i\beta_2}{2}\frac{\partial^2 E_j(T,z)}{\partial T^2} + i\gamma_j 2p_k(T - d_{jk}z, 0)\exp(-\alpha z)E_j(T,z) \tag{8.5.37}$$

where $d_{jk} \equiv (1/v_j) - (1/v_k)$ is the relative pump/probe walk-off, which can be linearized as $d_{jk} = D\Delta\lambda_{jk}$ if the channel separation $\Delta\lambda_{jk}$ is not too wide. $D = -2\pi c\beta_2/\lambda^2$ is the fiber dispersion coefficient, λ is the average wavelength, and c is the light velocity. This linear approximation of d_{jk} neglected higher-order dispersion effects.

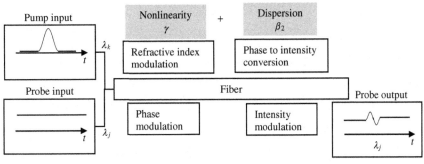

FIG. 8.5.12

Illustration of pump-probe interaction through XPM process.

In general, dispersion and nonlinearity act together along the fiber. However, as illustrated by Fig. 8.5.13, in an infinitesimal fiber section dz, we can assume that the dispersive and the nonlinear effects act independently, the same idea as used in the SSFM discussed in the last section. Let $E_j(T, z) = |E_j| \exp[i\phi_j(T, z)]$, where $|E_j|$ and ϕ_j are the amplitude and the phase of the probe channel optical field. Taking into account the effect of XPM alone, at $z = z'$, the nonlinear phase modulation in the probe signal induced by the pump power in the small fiber section dz can be obtained as

$$d\phi_j(T, z') = \gamma_j 2 p_k (T - d_{jk} z', 0) \exp(-\alpha z') dz$$

The Fourier transformation of this phase variation gives

$$d\widetilde{\phi}_j(\Omega, z') = 2\gamma_j p_k(\Omega, 0) e^{(-\alpha + i\Omega d_{jk})z'} dz \tag{8.5.38}$$

Neglecting the intensity fluctuation of the probe channel, this phase change corresponds to a change in the electrical field, $\overline{E}_j \exp[id\phi_j(T, z')] \approx \overline{E}_j[1 + id\phi_j(T, z')]$, or, in the Fourier domain, $\overline{E}_j[1 + id\widetilde{\phi}_j(\Omega, z')]$, where $d\widetilde{\phi}_j(\Omega, z')$ is the Fourier transform of $d\phi_j(T, z')$, and $\overline{E}_j$ represents the average field amplitude.

Due to chromatic dispersion of the fiber, the phase variation generated at location $z = z'$ is converted into an amplitude variation at the end of the fiber $z = L$. Taking into account a source term of nonlinearity-induced phase perturbation at $z = z'$ and the effect of chromatic dispersion, the Fourier transform of Eq. (8.5.37) becomes

$$\frac{\partial \widetilde{E}_j(\Omega, z)}{dz} = \frac{i\beta_2 \Omega^2}{2} \cdot \widetilde{E}_j(\Omega, z) + \overline{E}_j \left[1 + id\widetilde{\phi}(\Omega, z')\right] \delta(z - z')$$

where the Kronecker delta $\delta(z - z')$ is introduced to take into account the fact that the source term exists only in an infinitesimal fiber section at $z = z'$. Therefore, at the fiber output $z = L$, the probe field is

$$\widetilde{E}_j(\Omega, L) = \overline{E}_j + id\widetilde{\phi}_j(\Omega, z') \overline{E}_j \exp\left[i\beta_2 \Omega^2 (L - z')/2\right]$$

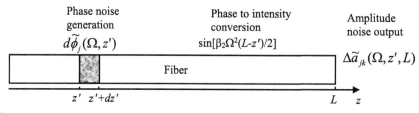

FIG. 8.5.13

Illustration of elementary contribution of XPM from a short fiber section.

The optical power variation caused by the nonlinear phase modulation created in the short section dz at $z = z'$ is thus

$$\Delta \tilde{a}_{jk}(\Omega, z', L) = \left| \tilde{E}_j(\Omega, L) \right|^2 - \overline{E}_j^2 = -2\overline{E}_j^2 d\tilde{\phi}_j(\Omega, z') \sin \left[\beta_2 \Omega^2 (L - z')/2 \right]$$

where a linearization has been made considering that $d\tilde{\phi}_j$ is infinitesimal. Using $E_j(T, z) = A_j(T + z/v_j, z)$ $\exp(\alpha z/2)$ and Eq. (8.5.38), integrating all nonlinear phase contributions along the fiber, the accumulated intensity fluctuation at the end of the fiber can be obtained as

$$\Delta \tilde{s}_{jk}(\Omega, L) = -4\gamma_j p_j(0) e^{-(\alpha - i\Omega/v_j)L} \int_0^L p_k(\Omega, 0) \sin \left[\beta_2 \Omega^2 (L - z')/2 \right] e^{-(\alpha - i\Omega d_{jk})z'} dz' \tag{8.5.39}$$

where $\Delta \tilde{s}_{jk}(\Omega, L) = \Delta \tilde{a}_{jk}(\Omega, L) e^{-\alpha L}$ represents the fluctuation of A_j at frequency Ω. After integration, we have

$$\Delta \tilde{s}_{jk}(\Omega, L) = 2p_j(L)\gamma_j e^{i\Omega/v_j L} p_k(\Omega, 0) \left\{ \frac{\exp\left(i\beta_2 \Omega^2 L/2\right) - \exp\left(-\alpha + i\Omega d_{jk}\right)L}{i\left(\alpha - i\Omega d_{jk} + i\beta_2 \Omega^2/2\right)} \right.$$
$$\left. - \frac{\exp\left(-i\beta_2 \Omega^2 L/2\right) - \exp\left(-\alpha + i\Omega d_{jk}\right)L}{i\left(\alpha - i\Omega d_{jk} - i\beta_2 \Omega^2/2\right)} \right\} \tag{8.5.40}$$

where $p_j(0)$ and $p_j(L)$ are the probe optical powers at the input and the output of the fiber, respectively. If the fiber length is much longer than the nonlinear length, $\exp(-\alpha L) \ll 1$, and the modulation bandwidth is much smaller than the channel spacing, that is, $d_{jk} \gg \beta_2 \Omega/2$, a much simpler frequency domain description of the XPM-induced intensity fluctuation can be derived for the probe channel:

$$\Delta \tilde{s}_{jk}(\Omega, L) = 4\gamma_j p_j(L) p_k(\Omega, 0) \frac{\sin\left(\beta_2 \Omega^2 L/2\right)}{\alpha - i\Omega d_{jk}} e^{i\Omega/v_j L} \tag{8.5.41}$$

Eq. (8.5.41) can be further generalized to analyze multispan optically amplified systems, where the total intensity fluctuation at the receiver is the accumulation of XPM contributions created by all fiber spans, as illustrated in Fig. 8.5.14. For a system with N amplified fiber spans, the nonlinear phase modulation created in the mth span produces an intensity modulation $\Delta \tilde{s}_{jk}^{(m)}(\Omega, L_N)$ at the end of the system. Even though the phase modulation creation depends only on the pump power and the pump/probe walk-off within the mth span, the phase-to-intensity conversion depends on the accumulated dispersion of the fibers from the mth to the Nth fiber spans, and therefore

$$\Delta \tilde{s}_{jk}^{(m)}(\Omega, L_N) = 4\gamma_j p_j(L_N) p_k^{(m)}(\Omega, 0) \exp\left[i\Omega \sum_{n=1}^{m-1} d_{jk}^{(n)} L^{(n)} \right] \frac{\sin\left[\Omega^2 \sum_{n=m}^{N} \beta_2^{(n)} L^{(n)}/2 \right]}{\alpha - i\Omega d_{jk}^{(i)}} \exp\left(i\Omega L_N/v_j \right) \tag{8.5.42}$$

where $L_N = \sum_{n=1}^{N} L^{(n)}$ is the total fiber length in the system, $L^{(m)}$ and $\beta_2^{(m)}$ are fiber length and dispersion of the m-th span (where $L^{(0)} = 0$), $p_k^{(m)}(\Omega, 0)$ is the pump signal input power spectrum in the m-th span, and $d_{jk}^{(m)}$ is the relative walk-off between two channels in the mth span (where $d_{jk}^{(0)} = 0$). To generalize the single-span XPM expression Eqs. (8.5.41) to (8.5.42), which represents a multispan system, the term $\sin(\beta_2 \Omega^2 L/2)$ in Eq. (8.5.41) has to be replaced by $\sin\left[\Omega^2 \sum_{n=m}^{N} \beta_2^{(n)} L^{(n)}/2 \right]$ in Eq. (8.5.42) to take into account the linear accumulation of dispersion. Another important effect that has to be taken into account is the different propagation speeds between the pump and the probe wavelengths. The phase difference between the pump and the probe at the input of the mth span is different from that at the input of the first span. The walk-off-dependent term $\exp\left[i\Omega\sum_{n=1}^{m-1}d_{jk}^{(n)}L^{(n)}\right]$ in Eq. (8.5.42) takes into account the walk-off between the probe and the pump channels before they both enter the m-th fiber span.

Finally, contributions from all fiber spans add up, as illustrated in Fig. 8.5.14, and therefore the intensity fluctuation induced by the XPM of the whole system can be expressed as

$$\Delta\widetilde{S}_{jk}(\Omega, L_N) = \sum_{m=1}^{N} \Delta\widetilde{s}_j^{(m)}(\Omega, L_N) \tag{8.5.43}$$

In the time domain, the output probe optical power with XPM-induced intensity crosstalk is

$$p_{jk}(t, L_N) = p_j(L_N) + \Delta S_{jk}(t, L_N) \tag{8.5.44}$$

where $\Delta S_{jk}(t, L_N)$ is the inverse Fourier transform of $\Delta\widetilde{S}_{jk}(\Omega, L_N)$ and $p_j(L_N)$ is the probe output without XPM. $\Delta S_{jk}(t, L_N)$ has a zero mean.

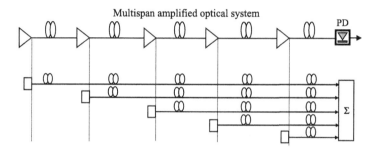

FIG. 8.5.14

Linear superposition of XPM contributions from each amplified span.

When the probe signal reaches an optical receiver, the electrical power spectral density at the output of the photodiode is the Fourier transform of the autocorrelation of the time domain optical intensity waveform. Therefore we have

$$\rho_j(\Omega, L_N) = \eta^2\left\{p_j^2(L_N)\delta(\Omega) + \left|\Delta\widetilde{S}_{jk}(\Omega, L_N)\right|^2\right\} \tag{8.5.45}$$

where δ is the Kronecker delta and η is the photodiode responsivity. For $\Omega > 0$, the XPM induced electrical domain power spectral density in the probe channel, normalized to its power level without an XPM effect, can be expressed as

$$\Delta p_{jk}(\Omega, L_N) = \frac{\eta^2\left|\Delta\widetilde{S}_{jk}(\Omega, L_N)\right|^2}{\eta^2 p_j^2(L_N)} = \left|\sum_{i=1}^{N}\left\{4\gamma_j p_k^{(i)}(\Omega, 0)\exp\left[i\Omega\sum_{n=1}^{i-1}d_{jk}^{(n)}L^{(n)}\right]\frac{\sin\left[\Omega^2\sum_{n=i}^{N}\beta_2^{(n)}L^{(n)}/2\right]}{\alpha - i\Omega d_{jk}^{(i)}}\right\}\right|^2 \tag{8.5.46}$$

$\Delta p_{jk}(\Omega, L_N)$ can be defined as a normalized XPM power transfer function, which can be directly measured by a microwave network analyzer. It is worth noting that in the derivation of Eq. (8.5.46), the waveform distortion of the pump signal has been neglected. This is indeed a small signal approximation, which is valid when the XPM-induced crosstalk is only a weak perturbation to the probe signal (Shtaif and Eiselt, 1997). In fact, if this crosstalk level is less than, for example, 20% of the signal, the second-order effect caused by the small intensity fluctuation through SPM in the pump is considered negligible.

To characterize the XPM-induced interchannel crosstalk and its impact in optical system performance, it is relatively easy to perform a frequency-domain transfer function measurement. A block diagram of the experimental setup is shown in Fig. 8.5.15. Two external-cavity tunable semiconductor lasers (ECL) emitting at λ_j and λ_k, respectively, are used as sources for the probe and the pump. The probe signal is CW and the pump signal is externally modulated by a sinusoid signal from a microwave network analyzer. The two optical signals are combined by a 3 dB coupler and then sent to an EDFA to boost the optical power. A tunable optical filter is used before the receiver to select the probe signal and suppress the pump signal. After passing through an optical preamplifier, the probe signal is detected by a wideband photodiode, amplified by a microwave amplifier, and then sent to the receiver port of the network analyzer. The transfer function measured in this experiment is the relationship between the frequency-swept input pump and its crosstalk into the output probe.

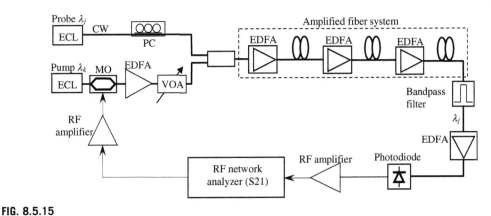

FIG. 8.5.15

Experimental setup for frequency domain XPM measurement.

As an example, Fig. 8.5.16 shows the normalized XPM frequency response measured at the output of a fiber link consisting of a single 114 km span of nonzero dispersion-shifted fiber (NZDSF). The channel spacings used to obtain this figure were 0.8 nm (λ_j=1559 nm, λ_k=1559.8 nm) and 1.6 nm (λ_j=1559 nm, λ_k=1560.6 nm). Corresponding theoretical results obtained from Eq. (8.5.46) are also plotted in the same figure. To have the best fit to the measured results, parameters used in the calculation were chosen to be λ_0=1520.2 nm, S_0=0.075 ps/km/nm^2, n_2=2.35·10^{-20}m^2/W, A_{eff}=5.5·10^{-11}m^2, and α=0.25 dB/km. These values agree with nominal parameter values of the NZDSF used in the experiment. Both the probe and the pump signal input optical powers were 11.5 dBm, and the pump channel modulation frequency was swept from 50 MHz to 10 GHz. To avoid significant higher-order harmonics generated from the LiNbO$_3$ Mach-zehnder intensity modulator, the modulation index is chosen to be approximately 50%. High-pass characteristics are clearly demonstrated in both curves in Fig. 8.5.16. This is qualitatively different from the frequency dependence of phase-modulation described in the following section, where the conversion from phase modulation to intensity modulation through fiber dispersion does not have to be accounted for and the *phase* variation caused by the XPM process has a *low-pass* characteristic (Chiang et al., 1996). In an ideal IMDD system, the phase modulation of the probe signal by itself does not affect the system performance. However, when a nonideal optical filter is involved, it may convert the phase noise to intensity noise. This is significant in the low frequency part where XPM-induced probe PM is high. The discrepancy between theoretical and experimental results in the low-frequency part of Fig. 8.5.16 is most likely caused by electrostriction nonlinearity and the transverse acoustic wave resonating between the center of the fiber core and the circumference of the fiber cladding which causes resonance in the frequency regime lower than 1 GHz (Hui et al., 2015).

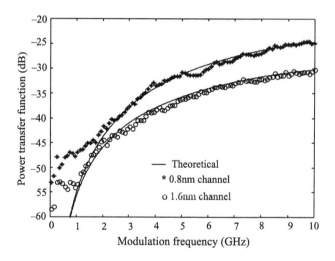

FIG. 8.5.16

XPM frequency response in the system with single span (114 km) nonzero dispersion shifted fiber. Stars: 0.8-nm channel spacing (λ_{probe}=1559 nm and λ_{pump}=1559.8 nm), open circles: 1.6-nm channel spacing (λ_{probe}=1559 nm and λ_{pump}=1560.6 nm). Continuous lines are corresponding theoretical results.

To demonstrate the effect of XPM in multispan systems, Fig. 8.5.17 shows the XPM frequency response of a two-span NZDSF system (114 km for the first span and 116 km for the second span), where, again, two channel spacings of 0.8 and 1.6 nm were used and the optical power launched into each fiber span was 11.5 dBm. In this multispan case, the detailed shape of the XPM frequency response is strongly dependent on the channel spacing, and the ripples in the XPM spectral shown in Fig. 8.5.17 are due to interference between XPM-induced crosstalk created in different fiber spans.

For this two-span system, the notch frequencies in the spectrum can found from Eq. (8.5.46) as approximately

$$1 + e^{i\Omega d_{jk}L_1} = 0 \tag{8.5.47}$$

The frequency difference between adjacent notches in the spectrum is thus, $\Delta f = 1/(d_{jk}L_1)$, where L_1 is the fiber length of the first span.

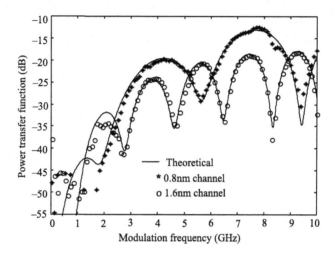

FIG. 8.5.17

XPM frequency response in the system with two spans (114 and 116 km) of NZDSF. Stars: 0.8 nm channel spacing ($\lambda_{probe} = 1559$ nm and $\lambda_{pump} = 1559.8$ nm), open circles: 1.6 nm channel spacing ($\lambda_{probe} = 1559$ nm and $\lambda_{pump} = 1560.6$ nm). Continuous lines are corresponding theoretical results.

Because the resonance structure of the XPM transfer function is caused by interactions between XPM created in various fiber spans, different arrangements of fiber system dispersion maps may cause dramatic changes in the overall XPM transfer function. As another example, the XPM frequency response measured in a three-span system is shown in Fig. 8.5.18, where the first two spans are 114 and 116 km of NZDSF and the third span is 75 km of standard SMF. In this experiment, the EDFAs are adjusted such that the optical power launched into the first two spans of NZDSF is 11.5 dBm and the power launched into the third span is 5 dBm. Taking into account the larger spot size, $A_{eff} = 80\mu m^2$, of the standard SMF (about $55\mu m^2$ for NZDSF) and the lower pump power in the third span, the nonlinear phase modulation generated in the third span is significantly smaller than that generated in the previous two spans.

Comparing Fig. 8.5.18 with Fig. 8.5.17, we see that the level increase in the crosstalk power transfer function in Fig. 8.5.18 is mainly due to the high dispersion in the last standard SMF span. This high dispersion results in a high efficiency of converting the PM, created in the previous two NZDSF spans, into intensity modulation. As a reasonable speculation, if the standard SMF were placed at the first span near the transmitter, the XPM crosstalk level would be much lower.

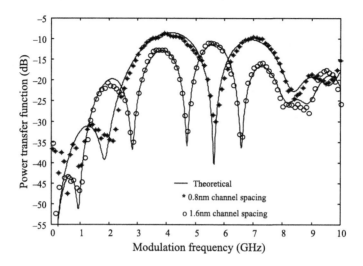

FIG. 8.5.18

XPM frequency response in a system with two spans (114 and 116 km) of NZDSF and one span (75 km) of normal SMF. Stars: 0.8 nm channel spacing ($\lambda_{probe}=1559$ nm and $\lambda_{pump}=1559.8$ nm), open circles: 1.6 nm channel spacing ($\lambda_{probe}=1559$ nm and $\lambda_{pump}=1560.6$ nm).

So far we have discussed the normalized frequency response of XPM-induced intensity crosstalk and the measurement technique. It would also be useful to find its impact on the performance of optical transmission systems. Even though the CW waveform of the probe simulates only the continuous "1"s in an NRZ bit pattern, the results may be generalized to pseudo-random signal waveforms. It is evident in Eq. (8.5.41) that the actual optical power fluctuation of the probe output caused by XPM is directly proportional to the unperturbed optical signal of the probe channel. Taking into account the actual waveforms of both the pump and the probe, XPM-induced crosstalk from the pump to the probe can be obtained as

$$C_{jk}(t) = F^{-1}\left\{ F[m_k(t)]\sqrt{\Delta p_{jk}(\Omega, L)}\sqrt{H_j(\Omega)} \right\}m_j(t) \tag{8.5.48}$$

where $m_j(t)$ is the normalized probe waveform at the receiver and $m_k(t)$ is the normalized pump waveform at the transmitter. For pseudo-random bit patterns, $m_{j,k}(t)=u_{j,k}(t)/2P^{av}_{j,k}$ with $u_{j,k}$, the real waveforms, and $P^{av}_{j,\ k}$, the average optical powers F and F^{-1} indicate Fourier and inverse Fourier transformations. $H_j(\Omega)$ is the receiver electrical power transfer function for the probe channel.

It is important to mention here that the expression of $\Delta p_{jk}(\Omega,L)$ in Eq. (8.5.46) was derived for a CW probe, so that Eq. (8.5.48) is not accurate during probe signal transitions between 0 and 1 s. In fact, XPM during probe signal transitions may introduce an additional time jitter, which is neglected in this analysis. It has been verified experimentally that XPM-induced time jitter due to a probe pattern effect was negligible compared to the XPM-induced eye closure at signal 1 in a system with NRZ coding; therefore the CW probe method might still be an effective approach (Eiselt, 1999 OFC). Another approximation in this analysis is the omission of pump waveform distortion during transmission. This may affect the details of the XPM crosstalk waveforms calculated by Eq. (8.5.48). However, the maximum amplitude of $C_{j,k}(t)$, which indicates the worst-case system penalty, will not be affected as long as there is no significant change in the pump signal optical bandwidth during transmission. In general, the impact of XPM cross-talk on system performance depends on the bit rate of the pump channel, XPM power transfer function of the system, and the baseband filter transfer function of the receiver for the probe channel.

To understand the impact of XPM on the system performance, it is helpful to look at the time-domain waveforms involved in the XPM process. As an example, trace A in Fig. 8.5.19 shows the normalized waveform (optical power) of the pump channel, which is a 10 Gb/s $(2^7 - 1)$ pseudo-random bit pattern, band-limited by a 7.5 GHz raised-cosine filter. Suppose that the probe is launched into the same fiber as a CW wave and its amplitude is normalized to 1. Due to XPM, the probe channel is intensity modulated by the pump, and the waveforms created by the XPM process for two different system configurations are shown by traces (B) and (C) in Fig. 8.5.19. Trace (B) was obtained in a single-span system with 130 km NZDSF, whereas trace (C) shows the XPM crosstalk waveform calculated for a three-span system with 130 km NZDSF + 115 km NZDSF + 75 km standard SMF. Looking at these time-domain traces carefully, we can see that trace (B) clearly exhibits a simple high-pass characteristic that agrees with the similar waveform measured and reported in Rapp (1997) in a single-span fiber system. However, in multispan systems, XPM transfer functions are more complicated. Trace (C) in Fig. 8.5.19 shows that the amplitude of the crosstalk associated with periodic 0 1 0 1 patterns in the pump waveform has been significantly suppressed.

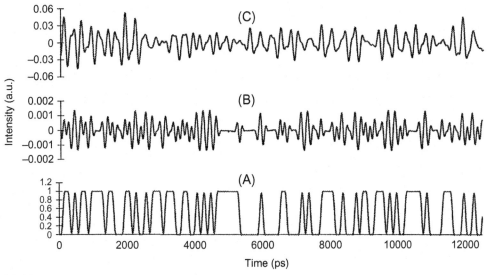

FIG. 8.5.19

Time-domain waveforms. Trace (A): input pump signal (10 Gb/s $(2^7 - 1)$ pseudo-random bit pattern). Trace (B): XPM crosstalk of the probe channel in a single-span 130 km NZDSF system. Trace (C): XPM crosstalk of the probe channel in a three-span system with 130 km NZDSF + 115 km NZDSF + 75 km normal SMF.

To help better understand the features in the time-domain waveforms obtained with various system configurations, Fig. 8.5.20 shows the XPM power transfer functions in the frequency-domain corresponding to trace (B) and trace (C) in Fig. 8.5.19. In the single-span case, the crosstalk indeed has a simple high-pass characteristic. For the three-span system, the XPM power transfer function has a notch at the frequency close to the half bit rate, which suppresses the crosstalk of 0 1 0 1 bit patterns in the time domain.

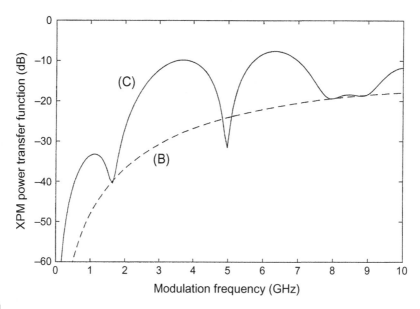

FIG. 8.5.20

XPM power transfer functions: trace (B) corresponds to time domain trace (B) in Fig. 8.5.19 and trace (C) corresponds to time domain trace (C) in Fig. 8.5.19.

It is worth mentioning that the crosstalk waveforms shown in Fig. 8.5.19 were calculated before an optical receiver. In practice, the transfer function and the frequency bandwidth of the receiver will re-shape the crosstalk waveform and may have a strong impact on system performance. After introducing a receiver transfer function, XPM-induced eye closure ECL in the receiver of a system can be evaluated from the amplitude in the crosstalk waveform for the probe channel. The worst-case eye closure happens with $m_j(t) = 1$, where $ECL_{(m_j=1)} = \{\max[C_{jk}(t)] - \min[C_{jk}(t)]\}/2$. It is convenient to define this eye closure as a *normalized XPM crosstalk*. In a complete system performance evaluation, this normalized XPM crosstalk penalty should be added on top of other penalties, such as those caused by dispersion and SPM. Considering the waveform distortion due to transmission impairments, the received probe waveform typically has $m_j(t) \leq 1$, especially for isolated 1 s. Therefore normalized XPM crosstalk gives a conservative measure of system performance.

In WDM optical networks, different WDM channels may have different data rates. The impact of probe channel bit rate on its sensitivity to XPM-induced crosstalk is affected by the receiver bandwidth. Fig. 8.5.21A shows the normalized power crosstalk levels vs the probe channel receiver electrical bandwidth for 2.5, 10, and 40 Gb/s bit rates in the pump channel. This figure was obtained for a single-span system of 100 km with a dispersion of 2.9 ps/nm/km, launched optical power of 11.5 dBm, and a channel spacing of 0.8 nm. In this particular system, we see that for a bit rate of higher than 10 Gb/s, the

XPM-induced crosstalk is less sensitive to the increase in the pump bit rate. This is because the normalized XPM power transfer function peaks at approximately 15 GHz for this system. When the pump spectrum is wider than 15 GHz, the XPM crosstalk efficiency is greatly reduced. This is the reason that the difference in the XPM-induced crosstalk between 40 and 10 Gb/s systems is much smaller than that between 10 and 2.5 Gb/s systems.

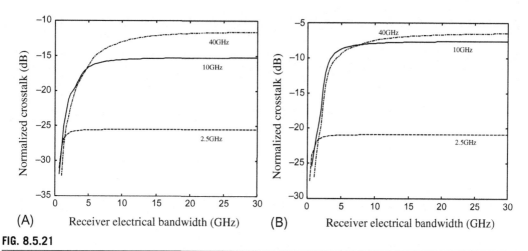

FIG. 8.5.21

Normalized power crosstalk levels vs. the receiver bandwidth for 2.5, 10, and 40 Gb/s bit rates in the pump channel. (A) The system has a 130-km single fiber span with fiber dispersion of 2.9 ps/nm/km and optical channel spacing of 0.8 nm. Launched pump optical power at each span is 11.5 dBm. (B) The system has five fiber spans (100 km/span) with fiber dispersion of 2.9 ps/nm/km and optical channel spacing of 0.8 nm. Launched pump optical power at each span is 8.5 dBm.

Typical receiver bandwidths for 2.5, 10, and 40 Gb/s systems are 1.75, 7.5, and 30 GHz, respectively. Fig. 8.5.21A indicates that when the receiver bandwidth exceeds the bandwidth of the pump channel, there is little increase in the XPM-induced crosstalk level with further increasing of the receiver bandwidth. In principle, the crosstalk between high bit rate and low bit rate channels is comparable to the crosstalk between two low bit rate channels. An important implication of this idea is in hybrid WDM systems with different bit rate *interleaving*; for example, channels 1, 3, and 5 have high bit rates and channels 2, 4, and 6 have low bit rates. The XPM-induced crosstalk levels in both high and low bit rate channels are very similar and are not higher than the crosstalk level in the system with the low bit rate. However, when the channel spacing is too low, XPM crosstalk from channel 3 to channel 1 can be bigger than that from channel 2 with a low bit rate. Fig. 8.5.21B shows the normalized crosstalk levels vs receiver electrical bandwidth in a five-span NZDSF system with a 100 km/span. The fiber dispersion is 2.9 ps/nm/km and the launched optical power at each span is 8.5 dBm. There is little difference in the crosstalk levels for the 10 Gb/s system and the 40 Gb/s system. This is because in systems with higher accumulated dispersion, the XPM power transfer function peaks at a lower frequency and the high-frequency components are strongly attenuated.

Fig. 8.5.22 shows the normalized crosstalk vs fiber dispersion for the same system used to obtain Fig. 8.5.21B. The fixed receiver bandwidths used for 40, 10, and 2.5 Gb/s systems are 30, 7.5, and 1.75 GHz, respectively. The worst-case XPM crosstalk happens at lower dispersion levels with higher signal bit rates. It is worth noting that for the 10 Gb/s system, the worst-case XPM crosstalk happens when the fiber dispersion parameter is 2.5 ps/nm/km, and therefore the total accumulated dispersion of the system is 1250 ps/nm, which is about the same as the dispersion limit for an uncompensated 10 Gb/s system.

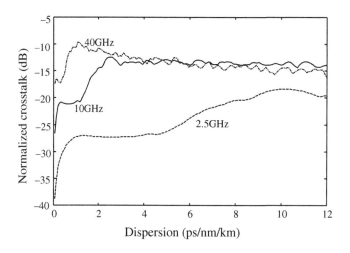

FIG. 8.5.22

Normalized power crosstalk levels vs. the fiber dispersion for 2.5, 10, and 40 Gb/s bit rates. Five cascaded fiber spans (100 km/span). Optical channel spacing 0.8 nm; 8.5 dBm launched pump optical power at each span.

It needs to be pointed out that, for simplicity, in both Fig. 8.5.21A and B, the signal optical powers were chosen to be the same for systems with different bit rates. However, in practice, a higher power level is normally required for a system with a higher bit rate. A generalization of these results to the case with different signal power levels can be made using the simple linear dependence of XPM crosstalk on the launched power level, as shown in Eq. (8.5.41).

Although most people would think that XPM crosstalk was significant only in low dispersion fibers, Fig. 8.5.22 clearly indicates that for uncompensated systems, even before reaching the system dispersion limit, higher dispersion could produce significant XPM crosstalk. On the other hand, in dispersion compensated optical systems, high local dispersion helps reduce the XPM-induced phase modulation and low accumulated system dispersion will reduce the phase noise to intensity noise conversion.

One important way to reduce the impact of XPM-induced crosstalk in a fiber system is to use DC (Saunders et al., 1997). The position of dispersion compensator in the system is also important. The least amount of DC is required if the compensator is placed in front of the receiver. In this position, the dispersion compensator compensates XPM crosstalk created in all fiber spans in the system. The optimum amount of DC for the purpose of XPM crosstalk reduction is about 50% of the accumulated dispersion in the system (Saunders et al., 1997). Although this lumped compensation scheme requires the minimum amount of DC, it does not give the best overall system performance.

Fig. 8.5.23 shows the normalized power crosstalk levels vs the percentage of DC in a 10 Gb/s system with six amplified NZDSF fiber spans of 100 km/span. The dispersion of transmission fiber is 2.9 ps/nm/km and the launched optical power into each fiber span is 8.5 dBm. Nonlinear effects in the DCFs are neglected for simplicity. Various DC schemes are compared in this figure. Trace (1) is obtained with compensation in each span. In this scheme XPM-induced crosstalk created from each span can be precisely compensated, so at 100% of compensation the XPM crosstalk is effectively eliminated. Trace (2) was obtained with the dispersion compensator placed after every other span. In this case, the value of DC can only be optimized for either the first span or the second span but not for both of them. The residual XPM crosstalk level is higher in this case than that with compensation in each span. Similarly, trace (3) in Fig. 8.5.23 was obtained with a dispersion compensator placed after every three spans, and trace (4) is with only one lumped compensator placed in front of the receiver. Obviously, when the number of dispersion compensators is reduced, the level of residual XPM crosstalk is higher and the optimum value of DC is closer to 50% of the total system dispersion. Therefore, in systems where XPM-induced crosstalk is a significant impairment, per-span DC is recommended. However, this will increase the number of dispersion compensators and thus increase the cost.

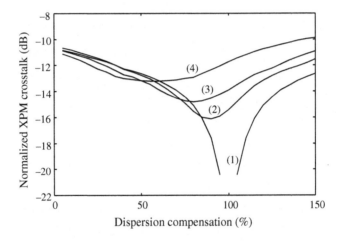

FIG. 8.5.23

Normalized power crosstalk levels vs the percentage of dispersion compensation in a 10 Gb/s, six-span system (100 km/span) with fiber dispersion of 2.9 ps/nm/km. An 8.5-dBm launched pump optical power at each fiber span. (1) Dispersion compensation after each span, (2) dispersion compensation after every two spans, (3) dispersion compensation after every three spans, and (4) one-lumped dispersion compensation in front of the receiver.

8.5.3.2 XPM-induced phase modulation

In the last section, we discussed the intensity modulation introduced by XPM in which the phase modulation is converted into intensity modulation through chromatic dispersion. This XPM-induced intensity crosstalk is a source of performance degradation in IMDD systems and introduces eye closure penalty. On the other hand, if the system is phase modulated, most likely the most relevant impact of XPM is the nonlinear PM itself. In fact, from Eq. (8.5.38), if we directly integrate the elementary contribution of nonlinear phase created over the entire fiber, the overall contribution is

$$\widetilde{\phi}_j(\Omega) = 2\gamma_j p_k(\Omega, 0) \int_0^L e^{(-\alpha + i\Omega d_{jk})z} dz = 2\gamma_j p_k(\Omega, 0) \sqrt{\eta_{XPM}} L_{eff} e^{j\theta} \tag{8.5.49}$$

where $L_{eff} = (1 - e^{-\alpha L})/\alpha$ is the effective nonlinear length, and

$$\eta_{XPM} = \frac{\alpha^2}{\alpha^2 + \Omega^2 d_{jk}^2} \left[1 + \frac{4\sin^2(\Omega d_{jk} L/2) e^{-\alpha L}}{(1 - e^{-\alpha L})^2} \right] \tag{8.5.50}$$

is the XPM-induced phase modulation efficiency, which is obviously a function of the modulation frequency Ω. The phase term in Eq. (8.5.49) is

$$\theta = -\tan^{-1}\left(\frac{\Omega d_{jk}}{\alpha}\right) - \tan^{-1}\left[\frac{e^{-\alpha L}\sin(\Omega d_{jk} L)}{1 - e^{-\alpha L}\cos(\Omega d_{jk} L)}\right] \tag{8.5.51}$$

In time domain, the XPM-induced phase variation of the probe signal can be expressed as (Chiang et al., 1996)

$$\phi_{XPM}(L, t) = \gamma_j\left[\overline{P}_j(0) + 2\overline{P}_k(0)\right] L_{eff} + \left|\widetilde{\phi}_j(\Omega)\right|\cos\left[\Omega\left(t - \frac{L}{v_j}\right) + \theta\right] \tag{8.5.52}$$

where the first term on the right side is a constant nonlinear phase shift with $\overline{P}_j(0)$ and $\overline{P}_k(0)$ the average input powers of the probe and the pump, respectively. In this case the conversion from phase modulation to intensity modulation through chromatic dispersion has been neglected.

If the system has more than one amplified fiber span, the overall effect of XPM is the superposition of contributions from all fiber spans,

$$\phi_{XPM}\left(\sum_{l=1}^N L^{(l)}, t\right) = \sum_{l=1}^N \left|\widetilde{\phi}_j^{(l)}(\Omega)\right|\cos\left[\Omega\left(t - \sum_{n=1}^N \frac{L^{(n)}}{v_j^{(n)}}\right) + \Omega\sum_{n=1}^{l-1} L^{(n)} d_{jk}^{(n)} + \theta^{(l)}\right] \tag{8.5.53}$$

where N is the total number of amplified fiber spans, and each term of summation on the right hand side of Eq. (8.5.53) represents the XPM-induced phase shift created in the corresponding fiber span. The additional phase term $\Omega\sum_{n=1}^{l-1} L^{(n)} d_{jk}^{(n)}$ represents the effect of pump-probe phase walk-off before the lth span.

XPM-induced phase modulation can be measured with phase-sensitive techniques such as coherent detection which will be discussed in details in the following chapter. Fig. 8.5.24 shows examples of the measured XPM-induced phase modulation indices in two-span and three-span amplified fiber-optic systems (Chiang et al., 1996). Similar to that shown in Fig. 8.5.15, in this measurement two tunable lasers are used as the probe and the pump at the wavelengths of λ_j and λ_k, respectively. The pump laser is intensity modulated by a sinusoid wave at frequency Ω through an external modulator. The probe and the pump are combined and sent to an optical system with multiple amplified fiber spans. At the output of the optical system, a narrowband optical filter is used to select the probe wave at λ_j while rejecting the pump. Instead of measuring the intensity modulation created by the XPM process, this measurement measures the optical phase modulation created on the probe through XPM using a phase-sensitive detection technique. As the average pump power level $P_k(0)$ at the input of each amplified fiber span was equal, the XPM index was defined as a normalized phase modulation efficiency $\phi_{XPM}/P_k(0)$. The wavelength spacing between the pump and the probe in this measurement was 3.7 nm.

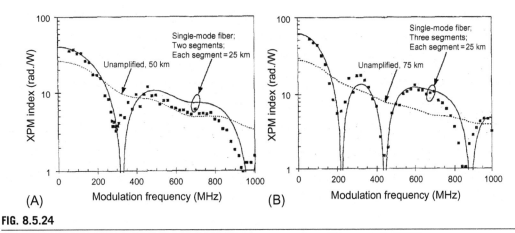

FIG. 8.5.24

XPM index for (A) a two-span system and (B) a three-span system (Chiang et al., 1996).

Fig. 8.5.24A was obtained in a system with two fiber spans, each having 25 km standard SMF. For Fig. 8.5.24B there were three fiber spans in the system, again with 25 km standard SMF in each span. In both cases, optical power at the input of each span was set to 7 dBm. If there is no optical amplifier between fiber spans, the XPM index vs modulation frequency decreases monotonically showing a low-pass characteristic. This is shown in Fig. 8.5.24 as the dotted lines that were calculated with a single 50 km (a) and 75 km (b) fiber span. When an optical amplifier was added at the beginning of each 25 km fiber span, the XPM indices varied significantly over the modulation frequency. Clearly, this is due to coherent interferences between XPM-induced PMs produced in different fiber spans. Compared to the XPM-induced intensity modulation, XPM-induced PM tends to diminish at high modulation frequency, especially when the chromatic dispersion of the fiber is high. From this point of view, XPM-induced intensity modulation discussed in the last section is probably the most damaging effect on high-speed optical systems.

It is important to point out that XPM is the crosstalk originated from the intensity modulation of the pump, which results in the intensity and PMs of the probe. In optical systems based on PM on all wavelength channels, XPM will not exist, in principle, because the pump has a constant optical power. However, the phase-coded optical signals carried by the pump wave can be converted into intensity modulation through fiber chromatic dispersion. Then this intensity modulation will be able to produce an XPM effect in the probe channels, thus causing system performance degradation (Ho, 2003).

8.5.3.3 FWM-induced crosstalk in IMDD optical systems
FWM is a parametric process that results in the generation of signals at new optical frequencies:

$$f_{jkl} = f_j + f_k - f_l \tag{8.5.54}$$

where f_j, f_k, and f_l are the optical frequencies of the contributing signals. There will be system performance degradation if the newly generated FWM frequency component overlaps with a signal channel in a WDM system, and appreciable FWM power is delivered into the receiver. The penalty will be greatest if the frequency difference between the FWM product and the signal, $f_{jkl} - f_i$, lies within the receiver bandwidth. Unfortunately, for signals on the regular ITU frequency grid in a WDM system, this overlapping is quite probable.

Over an optical cable span in which the chromatic dispersion is constant, there is a closed form solution for the FWM product-power-to-signal-power ratio:

$$x_s = \frac{P_{jkl}(L_s)}{P_l(L_s)} = \eta L_{eff}^2 \chi^2 \gamma^2 P_j(0) P_k(0) \tag{8.5.55}$$

where $L_{eff} = (1 - \exp(-\alpha L_s))/\alpha$ is the nonlinear length of the fiber span, L_s is the span length, $P_j(0)$, $P_k(0)$, and $P_l(0)$ are contributing signal optical powers at the fiber input, $\chi = 1, 2$ for nondegenerate and degenerate FWM, respectively, and the efficiency is

$$\eta = \rho \frac{\alpha^2}{\Delta k_{jkl}^2 + \alpha^2} \left[1 + \frac{4e^{-\alpha L_s} \sin\left(\Delta k_{jkl} L_s/2\right)}{(1 - e^{-\alpha L_s})^2} \right] \tag{8.5.56}$$

where the detune is

$$\Delta k_{jkl} = -\frac{2\pi c}{f_m^2} D(\lambda_m) \left[\left(f_j - f_m\right)^2 - \left(f_l - f_m\right)^2 \right] \tag{8.5.57}$$

$0 < \rho < 1$ is the polarization mismatch factor, and λ_m is the central wavelength, corresponding to the frequency of $f_m = \frac{f_j + f_k}{2} = \frac{f_l + f_{jkl}}{2}$.

In practice, in a multispan WDM system, the dispersion varies not only from span to span but also between fiber cabling segments, typically a few kilometers in length, within each span. The contributions from each segment can be calculated using the analytic solution described earlier, with the additional requirement that the relative phases of the signals and FWM product must be taken into account in combining each contribution (Inoue and Toba, 1995). The overall FWM contribution is a superposition of all FWM contributions throughout the system:

$$a_F = \sum_{spans} \sum_{segments} \exp\left(i\Delta\phi_{jkl}\right) \sqrt{\frac{P_{jkl}(z)}{P_l(z)}} \tag{8.5.58}$$

where $\Delta\phi_{jkl}$ is the relative phase of FWM generated at each fiber section. The magnitude of the FWM-to-signal ratio is quite sensitive, not only to the chromatic dispersions of the cable segments but also to their distribution, the segment lengths, and the exact optical frequencies of the contributing signals. Because of the random nature of the relative phase in each fiber section, statistical analysis may be necessary for system performance evaluation.

In the simplest case when the system has only two wavelength channels, degenerate FWM exists and the new frequency components created on both sides of each original optical signal frequency. However, for a WDM system with multiple equally spaced wavelength channels, there will be a large number of FWM components and almost all of them overlap with the original optical signals to create inter-channel crosstalks. For example, in a WDM system with four equally spaced wavelength channels, there are 10 FWM components which overlap with the original signal channels, as illustrated in Fig. 8.5.25A, where $f_1, f_2, f_3,$ and f_4 are the frequencies of the signal optical channels and f_{jkl} ($j, k, l = 1, 2, 3, 4$) are the FWM components created by the interactions among signals at $f_j, f_k,$ and f_l. In system performance evaluation, the power ratio between the original signal channel and the FWM crosstalk created at the same frequency is an important measure of the crosstalk. The spectral overlap between them makes the crosstalk evaluation in the spectral domain difficult. One way to overcome this overlapping problem is to deliberately remove one of the signal channels and observe the power of the FWM components generated at that wavelength, as illustrated in Fig. 8.5.25B. This obviously underestimates the FWM crosstalk because the FWM contribution that would involve that empty channel is not considered. For example, in the four-channel case, if we remove the signal channel at f_3, FWM components at frequency f_3 would only be f_{221} and f_{142}, whereas f_{243} would not exist. But for a WDM system with a large number of signal channels, removing one of the channels does not introduce significant difference in terms of FWM.

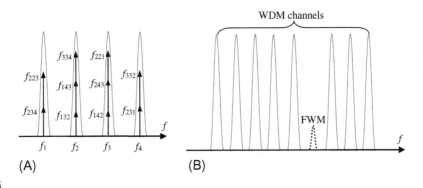

FIG. 8.5.25

(A) Illustration of FWM components in a four-channel system and (B) evaluating FWM crosstalk in a WDM system with one empty channel slot.

Fig. 8.5.26 shows an example of the measured FWM-to-carrier power ratio defined by Eq. (8.5.55). The system consists of five amplified fiber spans with 80 km fiber in each span, and the per-channel optical power level at the beginning of each fiber span ranges from 6 to 8 dBm in a random manner. The fiber used in the system has zero-dispersion wavelength at $\lambda_0 = 1564$ nm, whereas the average signal wavelength is approximately $\lambda = 1558$ nm. In this system, three channels were used with frequencies at $f_1, f_2,$ and f_3, respectively. The horizontal axis in Fig. 8.5.26 is the frequency separation between f_1 and f_3, and the frequency of the FWM component f_{132} is 15 GHz away from f_2, that is, $f_2 = (f_1 + f_3)/2 - 7.5 GHz$. Because there is no frequency overlap between f_{132} and f_2, the FWM-to-carrier power ratio P_{132}/P_2 can be measured. Fig. 8.5.26 clearly demonstrates that FWM efficiency is very sensitive to the frequency detune, and there can be more than 10 dB efficiency variation, with the frequency change only on the order of 5 GHz. This is mainly due to the rapid change in the phase match conditions as well as the interference between FWM components created at different locations in the system.

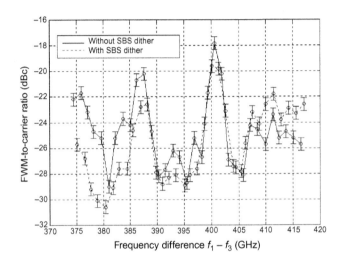

FIG. 8.5.26

FWM-to-carrier power ratio measured in a five-span optical system with three wavelength channels. The horizontal axis is the frequency separation of the two outer channels.

The instantaneous FWM efficiency is also sensitive to the relative polarization states of the contributing signals. In this system, the fiber DGD averaged over the wavelength range 1550–1565 nm is $\tau = 0.056 ps/\sqrt{km}$. From this DGD value, we can estimate the rate at which signals change their relative polarizations (at the system input they are polarization-aligned). For signal launch states that excite both principle states of polarization (PSP) of the fiber, the projection of the output SOP rotates around the PSP vector at a rate of $d\phi/d\omega = \tau$, where τ is the instantaneous DGD and ω is the (radian) optical frequency. In crude terms, we expect the signals to remain substantially aligned within the nonlinear interaction length of around 20 km. There may be a significant loss of SOP alignment between adjacent spans.

In practical long-distance optical transmission systems, frequency dithering on the laser source is often used to reduce the effect of SBS. The SBS dither effectively increases the transmission efficiency of the fiber by decreasing the power loss due to SBS while simultaneously smoothing out the phase match peaks. However, Fig. 8.5.26 indicates that SBS dithering only moderately reduces the level of FWM.

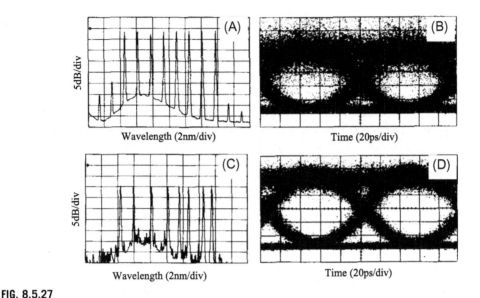

FIG. 8.5.27

(A) and (B) WDM optical spectrum with eight equally spaced channels and the measured eye diagram;
(C) and (D) WDM optical spectrum with eight unequally spaced channels and the corresponding eye diagram
(Forghieri et al., 1995).

An effective way to reduce the FWM crosstalk is to use unequal channel spacing in a WDM system. Fig. 8.5.27 shows a comparison between WDM systems with equal and unequal channel spacing through the measurement of both the optical spectra and the corresponding eye diagrams (Forghieri et al., 1995). This figure was obtained in a system with a single-span 137 km DSF that carries eight optical channels at 10-Gb/s data rate per channel. In an optical system with equally spaced WDM channels, as shown in Fig. 8.5.27A and B, severe degradation on the eye diagram can be observed at the per-channel signal optical power of 3 dBm at the fiber input. This eye closure penalty can be significantly reduced if the wavelengths of the WDM channels are unequally spaced as shown in Fig. 8.5.27C and D, although a higher per-channel signal optical power of 5 dBm was used. This is because the FWM components do not overlap with the signal channels so that they can be spectrally filtered out at the receiver. However, non-equal channel spacing is rarely used in commercial optical transmission systems. Instead, standardized ITU wavelength grid is commonly adopted for interoperability of product from different manufacturers and system operators.

Since FWM crosstalk is generated from nonlinear mixing between optical signals, it behaves more like a coherent crosstalk than a noise. To explain the coherent crosstalk due to FWM, we can consider that a single FWM product is created at the same wavelength as the signal channel f_i. Assuming that the power of the FWM component is p_{fwm}, which coherently interferes with an amplitude-modulated optical signal whose instantaneous power is p_s, the worst-case eye closure occurs when all the contributing signals are at the high level (digital 1). Due to the mixing between the optical signal and the FWM component, the photocurrent at the receiver is proportional to

$$I' \propto \left| \sqrt{p_s} + \sqrt{p_{fwm}} \cos\left(\Delta\phi + 2\pi\left(f_i - f_{jkl}\right)\right) \right|^2 \tag{8.5.59}$$

As a result, the normalized signal 1 level becomes

$$A = \frac{p_s + p_{fwm} \pm 2\sqrt{p_s p_{fwm}}}{p_s} \approx 1 \pm 2\sqrt{\frac{p_{fwm}}{p_s}} \tag{8.5.60}$$

This is a "bounded" crosstalk because the worst-case closure to the normalized eye diagram is $2\sqrt{p_{fwm}/p_s}$.

In a high-density multichannel WDM system there will be more than one FWM product that overlaps with each signal channel. In general, these FWM products will interfere with the signal at independent beating frequencies and phases. The absolute worst-case eye closure can be found by superpositioning the absolute value of each contributing FWM product. However, if the number of FWM products is too high, the chance of reaching the absolute worst case is very low. The overall crosstalk then approaches Gaussian statistics as the number of contributors increases (Eiselt, 1999, JLT).

8.5.3.4 Modulation instability and its impact in WDM optical systems

In addition to nonlinear crosstalk between signal channels such as XPM and FWM discussed above, crosstalk may also happen between optical signal and the broadband ASE noise in the system to cause performance degradations. This can be seen as FWM between the optical signal and the broadband noise, commonly referred to as parametric gain, or MI (Yu et al., 1995). The mechanism for system performance degradation caused by MI depends on system type. In optical phase-coded transmission systems, degradation is mainly caused by the broadening of the optical spectrum (Mecozzi, 1994). On the other hand, for an IMDD, the performance degradation can be introduced by phase noise to intensity noise conversion through chromatic dispersion in the optical fiber. The increased RIN within the receiver baseband is responsible for this degradation.

Since MI is caused by the Kerr effect nonlinearity, its analysis can be based on the nonlinear Schrodinger equation, which was given in Eq. (8.5.23) and we repeat here for your convenience,

$$\frac{\partial A(t, z)}{\partial z} + \frac{i\beta_2}{2}\frac{\partial^2 A(t, z)}{\partial t^2} + \frac{\alpha}{2}A(t, z) - i\gamma|A(t, z)|^2 A(t, z) = 0 \tag{8.5.61}$$

where $A(z,t)$ is the electrical field, $\gamma = \omega_0 n_2/cA_{eff}$ is the nonlinear coefficient of the fiber, ω_0 is the angular frequency, n_2 is the nonlinear refractive index of the fiber, c is the speed of light, A_{eff} is the effective fiber core area, β_2 is the fiber dispersion parameter, and α is the fiber attenuation. High-order dispersions were ignored here. The z-dependent steady-state solution of Eq. (8.5.61) is

$$A_0(z) = \sqrt{P_0} \exp\left(i\gamma|A_0(z)|^2 z\right)\exp\left(-\frac{\alpha}{2}z\right) \tag{8.5.62}$$

Since the signal optical power can vary significantly along the fiber in practical optical systems because of the attenuation, a simple mean-field approximation over the transmission fiber is usually not accurate enough. To obtain a semi-analytical solution, the fiber can be divided into short sections, as illustrated in Fig. 8.5.28, and a mean-field approximation can be applied within each section. For example, in the jth section with length Δz_j, Eq. (8.5.61) becomes

$$\frac{\partial A_j(z, t)}{\partial z} = \frac{-i}{2}\beta_2 \frac{\partial^2 A_j(z, t)}{\partial t^2} + i\gamma_j|A_j(z, t)|^2 A_j(z, t) \tag{8.5.63}$$

where

$$\gamma_j = \frac{1 - \exp\left(-\alpha\Delta z_j\right)}{\alpha\Delta z_j}\gamma \tag{8.5.64}$$

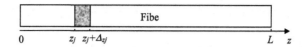

FIG. 8.5.28

Illustration of dividing fiber into short sections for transfer matrix analysis.

With the assumption that noise power at fiber input is much weaker than pump power, the solution of Eq. (8.5.63) can be written as

$$A_j(z,t) = [A_{0j} + \tilde{a}_j(z,t)] \exp\left(i\gamma |A_{0j}|^2 z\right) \tag{8.5.65}$$

where A_{0j} is the steady-state solution of Eq. (8.5.63), $\tilde{a}_j(z,t)$ is a small perturbation and $\tilde{a}_j(z,t) << A_{0j}$ is assumed. With linear approximation of the noise term, the nonlinear Schrodinger Eq. (8.5.63) becomes

$$\frac{\partial \tilde{a}_j(z,t)}{\partial z} = \frac{-i}{2}\beta_2 \frac{\partial^2 \tilde{a}_j(z,t)}{\partial t^2} + i\gamma_j \left[|A_{0j}|^2 \tilde{a}_j(z,t) + A_{0j}^2 \tilde{a}_j^*(z,t)\right] \tag{8.5.66}$$

where the symbol * denotes complex conjugate. We need to emphasize that the linearization used to obtain Eq. (8.5.66) is valid only when the perturbation is small enough such that the effect of pump depletion can be neglected.

By converting Eq. (8.5.66) into frequency domain through Fourier transform [Marcus, 1994 (Electronics letters)], the following two equations can be obtained for the real and the imaginary parts of $\tilde{a}_j(z,t)$, respectively:

$$\frac{\partial a_j(\omega,z)}{\partial z} = \frac{i}{2}\omega^2\beta_2 a_j(\omega,z) + i\gamma_j \left[|A_{0j}|^2 a_j(\omega,z) + A_{0j}^2 a_j^*(-\omega,z)\right] \tag{8.5.67}$$

$$\frac{\partial a_j^*(-\omega,z)}{\partial z} = \frac{-i}{2}\omega^2\beta_2 a_j^*(-\omega,z) - i\gamma_j \left[|A_{0j}|^2 a_j^*(-\omega,z) + A_{0j}^{*2} a_j(\omega,z)\right] \tag{8.5.68}$$

The formal solution of linear differential Eqs. (8.5.67) and (8.5.68) can be expressed in a matrix format:

$$\begin{bmatrix} a_{j+1}(\omega, z_j + \Delta z_j) \\ a_{j+1}^*(-\omega, z_j + \Delta z_j) \end{bmatrix} = \begin{bmatrix} M_{11}^{(j)} & M_{12}^{(j)} \\ M_{21}^{(j)} & M_{22}^{(j)} \end{bmatrix} \begin{bmatrix} a_{j+1}(\omega, z_j) \\ a_{j+1}^*(-\omega, z_j) \end{bmatrix}$$

When we further take into account the linear attenuation of the signal in each section, a factor $\exp(-\alpha\Delta z_j/2)$ has to be added so that

$$\begin{bmatrix} a_{j+1}(\omega, z_j + \Delta z_j) \\ a_{j+1}^*(-\omega, z_j + \Delta z_j) \end{bmatrix} = \begin{bmatrix} M_{11}^{(j)} & M_{12}^{(j)} \\ M_{21}^{(j)} & M_{22}^{(j)} \end{bmatrix} \begin{bmatrix} a_{j+1}(\omega, z_j) \\ a_{j+1}^*(-\omega, z_j) \end{bmatrix} \exp\left(-\frac{\alpha}{2}\Delta z_j\right) \tag{8.5.69}$$

where $\Delta z_j = z_{j+1} - z_j$ and

$$M_{11}^{(j)} = \frac{e^{ik_j z_j} - r_{fj} r_{bj} e^{-ik_j z_j}}{|1 - r_{fj} r_{bj}|} \tag{8.5.70}$$

$$M_{12}^{(j)} = \frac{r_{bj}\left(e^{-ik_j z_j} - e^{ik_j z_j}\right)}{|1 - r_{fj} r_{bj}|} \tag{8.5.71}$$

$$M_{21}^{(j)} = \frac{r_{fj}\left(e^{ik_jz_j} - e^{-ik_jz_j}\right)}{\left|1 - r_{fj}r_{bj}\right|} \tag{8.5.72}$$

$$M_{22}^{(j)} = \frac{e^{-ik_jz_j} - r_{fj}r_{bj}e^{ik_jz_j}}{\left|1 - r_{fj}r_{bj}\right|} \tag{8.5.73}$$

$$r_{fj} = \frac{k_j - \beta\omega^2 - \gamma_j\left|A_{0j}\right|^2}{\gamma_j A_{0j}^2} = \frac{-\gamma_j A_{0j}^{*2}}{k_j + \beta\omega^2 + \gamma_j\left|A_{0j}\right|^2} \tag{8.5.74}$$

$$r_{bj} = \frac{k_j - \beta\omega^2 - \gamma_j\left|A_{0j}\right|^2}{\gamma_j A_{0j}^{*2}} = \frac{-\gamma_j A_{0j}^2}{k_j + \beta\omega^2 + \gamma_j\left|A_{0j}\right|^2} \tag{8.5.75}$$

Eqs. (8.5.74) and (8.5.75) have two eigenmodes whose propagation constants are equal in magnitude and opposite in sign, and they are given by

$$k_j = \pm\sqrt{\left(\frac{\beta_2}{2}\omega^2 + \gamma_j\left|A_{0j}\right|^2\right)^2 - \left(\gamma_j\left|A_{0j}\right|^2\right)^2} \tag{8.5.76}$$

The parameters r_{fj} and r_{bj} can be regarded as effective reflectivities for the two eigenmodes; therefore the sign of k should be chosen such that $\left|r_{fj}\right| \leq 1$ and $\left|r_{bj}\right| \leq 1$.

The evolution of the noise along the fiber can then be calculated simply by matrix multiplication:

$$\begin{bmatrix} a(\omega, L) \\ a^*(-\omega, L) \end{bmatrix} = \begin{bmatrix} B_{11} & B_{12} \\ B_{21} & B_{22} \end{bmatrix} \begin{bmatrix} a(\omega, 0) \\ a^*(-\omega, 0) \end{bmatrix} \exp\left(-\frac{\alpha}{2}L\right) \tag{8.5.77}$$

with

$$\begin{bmatrix} B_{11} & B_{12} \\ B_{21} & B_{22} \end{bmatrix} = \prod_{j=1}^{N} \begin{bmatrix} M_{11}^{(j)} & M_{12}^{(j)} \\ M_{21}^{(j)} & M_{22}^{(j)} \end{bmatrix} \tag{8.5.78}$$

where L is the fiber length and N is the total number of sections.

Let us first look at the power spectral density of the optical field, according to the Wiener-Khintchine theorem it is proportional to the square of the modulus of the Fourier transformation of the complex field amplitude. If the optical field is sampled over a time interval T, this optical power spectral density is

$$S_t(\omega, z) = \left|\left\{\frac{1}{T}\int_{-T/2}^{T/2} A(z, t)\exp(i\omega t)dt\right\}\right|^2\Bigg|_{T\to\infty}$$

Separating the optical field into CW and stochastic components, we have

$$S_t(\omega, z) = \left\langle\left|a(\omega, L)\right|^2\right\rangle + \left|A_0(L)\right|^2\delta(\omega)$$

where $<>$ denotes ensemble average, and normalization by the sample interval. $\delta(\omega)$ is the Kronecker delta function. The noise term has zero mean, so $\langle a(\omega,z)\rangle = 0$ and the cross terms vanish. It is convenient to remove the CW contribution from the power spectrum, so we define

$$S(\omega, L) = S_t(\omega, L) - |A_0(L)|^2\delta(\omega) = \left\langle |a(\omega, L)|^2\right\rangle \tag{8.5.79}$$

Using Eq. (8.5.77) we can find

$$S(\omega, L) = \left\langle |B_{11} \cdot a(\omega, 0) + B_{12} \cdot a^*(-\omega, 0)|^2 \right\rangle \exp(-\alpha L) \tag{8.5.80}$$

Because $a(\omega,0)$ is a random process, amplitudes at distinct frequencies are uncorrelated, so $<a(\omega,0)a^*(-\omega,0)>=0$ and

$$S(\omega, L) = \left[|B_{11}|^2 S(\omega, 0) + |B_{12}|^2 S(-\omega, 0)\right]\exp(-\alpha L) \tag{8.5.81}$$

where $S(\omega,0)=<a(\omega,0)a^*(\omega,0)>$ is the power spectrum of $\tilde{a}(t, 0)$, which is the input noise. To simplify the analysis, we assume that the input noise spectrum is symmetric around the carrier (e.g., white noise): $S(\omega,0)=S(-\omega,0)$. Eq. (8.5.81) becomes

$$S(\omega, L) = \left(|B_{11}|^2 + |B_{12}|^2\right)S(\omega, 0)e^{-\alpha L} \tag{8.5.82}$$

A linear system can be treated as a special case with nonlinear coefficient $\gamma=0$. In this case, $k=\beta_2\omega^2/2$, $r_f=r_b=0$, $|B_{11}|=1$, $B_{12}=0$, and

$$S_L(\omega, L) = S(\omega, 0)e^{-\alpha L} \tag{8.5.83}$$

Using S_L as a normalization factor so that the normalized optical gain or optical noise amplification in the nonlinear system is

$$S_{OG}(\omega, L) = \frac{S(\omega, L)}{S_L(\omega, L)} = |B_{11}|^2 + |B_{12}|^2 \tag{8.5.84}$$

Fig. 8.5.29 shows the normalized optical spectra vs fiber length in a single-span system using DSFs with anomalous (a) and normal (b) dispersions. Fiber parameters used to obtain Fig. 8.5.28 are: loss coefficient $\alpha = 0.22$ dB/km, input signal optical power $P_{in}=13$ dBm, nonlinear coefficient $\gamma = 2.07$ W^{-1} km^{-1} and fiber dispersion $D=2$ ps/nm/km for Fig. 8.5.28A, and $D=-2$ ps/nm/km for Fig. 8.5.28B with D defined as $D=2\pi c\beta_2/\lambda^2$. In both cases of Fig. 8.5.28A and B, optical noise is amplified around the carrier. The difference is that in an anomalous dispersion regime, optical spectrums have two peaks at each side of the carrier, whereas in the normal dispersion regime, spectra are single peaked. The amplification of optical spectra near the carrier can be explained as the spectrum broadening of the carrier caused by the nonlinear PM between the signal and the broadband ASE. Fig. 8.5.28C and D show nonlinear amplifications of ASE in a 100 km fiber with positive (C) and negative (D) chromatic dispersions. Input power level used is $P_{in}=15$ dBm. Three different dispersion values are used in each figure, which are solid line: $D=\pm1$ ps/nm/km, dashed line: $D=\pm0.5$ ps/nm/km, and dashed-dotted line: $D=\pm0.05$ ps/nm/km.

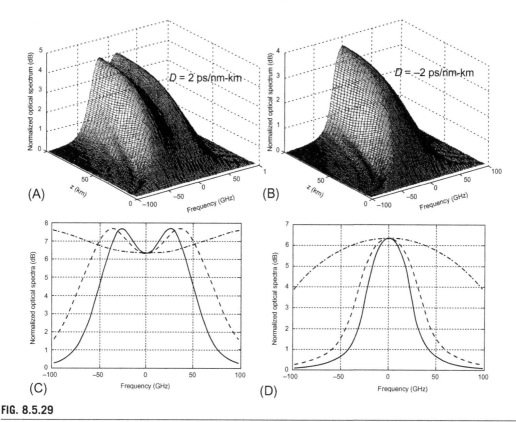

FIG. 8.5.29

(A) and (B) Nonlinear amplification of ASE along the longitudinal direction of a single-span fiber. Input optical power $P_{in} = 13$ dBm, fiber nonlinear coefficient $\gamma = 2.07$ W^{-1} km^{-1}, and fiber loss $\alpha = 0.22$ dB/km. (C) and (D) Nonlinear amplification of ASE in a 100 km fiber for positive (C) and negative (D) dispersions. Input power $P_{in} = 15$ dBm. Solid line: $D = \pm 1$ ps/nm/km, dashed line: $D = \pm 0.5$ ps/nm/km, and dash-dotted line: $D = \pm 0.05$ ps/nm/km.

In the case of coherent optical transmission, the entire optical spectrum is moved to the intermediate frequency after beating with the local oscillator. The frequency components beyond the range of the baseband filter will then be removed, which may cause receiver power reduction. Therefore the broadening of the signal optical spectrum is the major source of degradation in coherent optical transmission systems. For IMDD optical systems, on the other hand, the photodiode detects the total optical power without wavelength discrimination, and the RIN of the optical signal is the major source of degradation related to MI.

Let us then look at the electric noise power spectral density after a direct-detection optical receiver. After the square-law detection of a photodiode, the photo current can be expressed as

$$I(t) = \eta |A_0 + \tilde{a}(t, L)|^2 = \eta \left[|A_0|^2 + A_0 \tilde{a}^*(t, L) + A_0^* \tilde{a}(t, L) \right] \qquad (8.5.85)$$

where η is the photodetector responsivity. For simplicity, second and higher orders of small terms have been omitted in the derivation of Eq. (8.5.85).

In an IMDD system, the receiver performance is sensitive only to the amplitude noise of the photocurrent, which can be obtained from Eq. (8.5.85) as

$$\delta I(t) = I(t) - I_0 = \eta \left[A_0 \tilde{a}^*(t, L) + A_0^* \tilde{a}(t, L) \right] \tag{8.5.86}$$

where $I_0 = \eta |A_0|^2$ is the photocurrent generated by the CW optical signal.

The power spectrum of the noise photocurrent is the Fourier transformation of the autocorrelation of the time-domain noise amplitude:

$$\rho_n(\omega) = \eta^2 \left(|A_0^* B_{11} + A_0 B_{12}|^2 S(\omega, 0) + |A_0^* B_{12} + A_0 B_{22}|^2 S(-\omega, 0) \right) e^{-\alpha L} \tag{8.5.87}$$

where $S(\omega, 0)$ is the power spectral density of $\tilde{a}(t, 0)$.

Under the same approximation as we used in the optical spectrum calculation, the input optical noise spectrum is assumed to be symmetric around zero frequency: $S(\omega, 0) = S(-\omega, 0)$, Eq. (8.5.87) becomes

$$\rho_n(\omega) = \eta^2 \left(|B_{11} + B_{21}|^2 + |B_{12} + B_{22}|^2 \right) |A_0(L)|^2 S(\omega, 0) e^{-\alpha L} \tag{8.5.88}$$

Using Eqs. (8.5.70)–(8.5.76), it is easy to prove that $|B_{11} + B_{21}|^2 = |B_{12} + B_{22}|^2$, and therefore Eq. (8.5.88) can be written as

$$\rho_n(\omega) = 2 P_{in} \eta^2 |B_{11} + B_{21}|^2 S(\omega, 0) e^{-2\alpha L} \tag{8.5.89}$$

where P_{in} is the input signal power such that $|A_0(L)|^2 = P_{in} \exp(-\alpha L)$.

Again, a linear system can be treated as a special case of a nonlinear system with the nonlinear coefficient $\gamma = 0$. In this case, $k = \beta_2 \omega^2 / 2$, $r_f = r_b = 0$, $|B_{11}| = 1$, and $B_{21} = 0$. The electrical noise power spectral density in the linear system is then

$$\rho_0(\omega) = 2 P_{in} \eta^2 S(\omega, 0) e^{-2\alpha L} \tag{8.5.90}$$

Using $\rho_0(\omega)$ as the normalization factor, the normalized power spectral density of the receiver electrical noise, or the normalized RIN, caused by fiber Kerr effect is therefore

$$R(\omega) = |B_{11} + B_{21}|^2 \tag{8.5.91}$$

Comparing Eq. (8.5.91) to Eq. (8.5.84), it is interesting to note that these two spectra are fundamentally different: The relative phase difference between B_{11} and B_{21} has no impact on the amplified optical noise spectrum of Eq. (8.5.84), but this phase difference is important in the electric domain RIN spectrum as given in Eq. (8.5.91).

Fig. 8.5.30 shows the normalized RIN spectra vs fiber length in a single-span system with anomalous dispersion (a) and normal dispersion (b). Fiber parameters used in Fig. 8.5.30 are the same as those used for Fig. 8.5.29. In the anomalous fiber dispersion regime (Fig. 8.5.30A), two main side peaks of noise grow along z, the peak frequencies become closer to the carrier and the widths become narrower in the process of propagating along the fiber. On the other hand, if the fiber dispersion is in the normal regime, as shown in Fig. 8.5.30B, noise power density becomes smaller than in the linear case in the vicinity of the carrier frequency. This implies a noise squeezing (Hui and O'Sullivan, 1996) and a possible system performance improvement. In either regime, the system performance is sensitive to the baseband electrical filter bandwidth. For the purpose of clearer display, Fig. 8.5.30C and D show nonlinear amplifications of RIN over 100 km fiber with positive (C) and negative (D) chromatic dispersions. The input power level is $P_{in} = 15$ dBm. Three different dispersion values are used in both Fig. 8.5.30C and D, they are solid line: $D = \pm 1$ ps/nm/km, dashed line: $D = \pm 0.5$ ps/nm/km and dashed-dotted line: $D = \pm 0.05$ ps/nm/km.

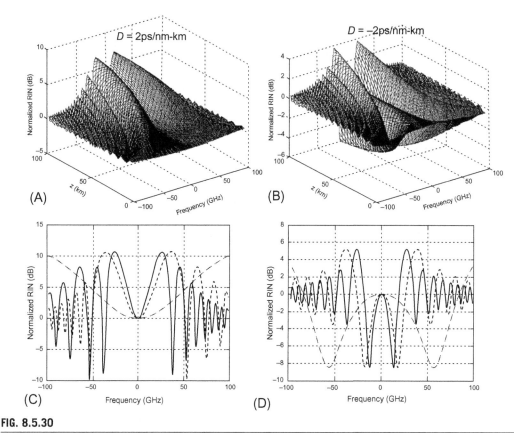

FIG. 8.5.30

(A) and (B) Nonlinear amplification of RIN along the longitudinal direction of a single-span fiber. Input optical signal power $P_{in}=13$ dBm, fiber nonlinear coefficient $\gamma=2.07$ W^{-1} km^{-1} and fiber loss $\alpha=0.22$ dB/km. (C) and (D) Nonlinear amplification of RIN of 100 km fiber for positive (C) and negative (D) dispersions. Input power $P_{in}=15$ dBm. Solid line: $D=\pm1$ ps/nm/km, dashed line: $D=\pm0.5$ ps/nm/km, and dashed-dotted line: $D=\pm0.05$ ps/nm/km.

It is worthwhile to notice the importance of taking into account fiber loss in the calculation. Without considering fiber loss, the calculated RIN spectra would be qualitatively different. For example, in the case of normal fiber dispersion, the normalized RIN spectra were always less than 0 dB in the mean-field approximation [Marcus, 1994 (Electronics Letters)], which is in fact not accurate, as can be seen in Fig. 8.5.30B and D.

After discussing the nonlinearly amplified optical noise spectrum due to MI, and the impact in the electric domain RIN spectrum after a direct-detection receiver, one wonders if DC would affect the impact of MI. It is well known that DC is an important way to reduce eye closure penalty due to chromatic dispersion in fiber systems. It can also reduce XPM-induced nonlinear crosstalk in IMDD system, as discussed in the earlier. The impact of DC on the effect of MI in optical systems is also an important consideration in optical system design and performance evaluation. Neglecting the nonlinear effect of the DC module (DCM), its transfer function can be represented by a conventional Jones Matrix:

$$[C] = \begin{bmatrix} \exp[-i\Phi(\omega)] & 0 \\ 0 & \exp[i\Phi(\omega)] \end{bmatrix} \qquad (8.5.92)$$

If the DC is made by a piece of optical fiber, the phase term is related to the chromatic dispersion, $\Phi(\omega) = \beta_2 \omega^2 z/2$.

The effect of DC on the optical system can be evaluated by simply multiplying the Jones matrix of Eq. (8.5.92) to the MI transfer matrix of Eq. (8.5.77). The RIN spectra at the IMDD optical receiver are sensitive not only to the value of DC but to the position at which the DCM is placed in the system. Let us take two examples to explain the reason. First, if the DCM is positioned after the nonlinear transmission fiber (at the receiver side), the combined transfer function becomes

$$\begin{bmatrix} B_{11} & B_{12} \\ B_{21} & B_{22} \end{bmatrix} = \begin{bmatrix} \exp[-i\Phi(\omega)] & 0 \\ 0 & \exp[i\Phi(\omega)] \end{bmatrix} \begin{bmatrix} B_{11}^f & B_{12}^f \\ B_{21}^f & B_{22}^f \end{bmatrix} = \begin{bmatrix} B_{11}^f \exp(-i\Phi) & B_{12}^f \exp(-i\Phi) \\ B_{21}^f \exp(i\Phi) & B_{22}^f \exp(i\Phi) \end{bmatrix}$$

The normalized RIN spectrum in Eq. (8.5.91) is then

$$R(\omega) = \left| B_{11}^f \exp(-i\Phi) + B_{21}^f \exp(i\Phi) \right|^2$$

where $B_{ij}^f (i=1, 2, j=1, 2)$ are the transfer function elements of the nonlinear transmission fiber only. Obviously, the normalized RIN spectrum is sensitive to the amount of DC represented by the phase shift Φ. Fig. 8.5.31 shows an example of normalized RIN spectra without DC (solid line), with 50% of compensation (dashed-dotted line), and with 100% compensation (dashed line). It is interesting to note that 100% DC does not necessarily bring the RIN spectrum to the linear case.

On the other hand, if the DCM is placed before the nonlinear transmission fiber (at the transmitter side), the total transfer matrix is

$$\begin{bmatrix} B_{11} & B_{12} \\ B_{21} & B_{22} \end{bmatrix} = \begin{bmatrix} B_{11}^f & B_{12}^f \\ B_{21}^f & B_{22}^f \end{bmatrix} \begin{bmatrix} \exp[-i\Phi(\omega)] & 0 \\ 0 & \exp[i\Phi(\omega)] \end{bmatrix} = \begin{bmatrix} B_{11}^f \exp(-i\Phi) & B_{12}^f \exp(i\Phi) \\ B_{21}^f \exp(-i\Phi) & B_{22}^f \exp(i\Phi) \end{bmatrix}$$

The normalized RIN spectrum in Eq. (8.5.91) then becomes

$$R(\omega) = \left| B_{11}^f \exp(-i\Phi) + B_{21}^f \exp(-i\Phi) \right|^2 = \left| B_{11}^f + B_{21}^f \right|^2$$

In this case, it is apparent that the phase term Φ introduced by DC does not bring any difference into the normalized RIN spectrum.

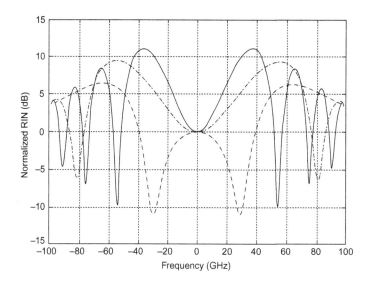

FIG. 8.5.31

Normalized RIN spectra for 0% (solid line), 50% (dash-dotted line), and 100% (dashed line) dispersion compensations. Fiber length $L=100$ km, fiber nonlinear coefficient $\gamma=2.07$ W^{-1} km^{-1}, fiber loss $\alpha=0.22$ dB/km, $P_{in}=15$ dBm, and $D=1$ ps/nm/km (Hui et al., 1997).

Another important observation is that although the RIN spectrum in the electrical domain after photodetection can be affected by the DC, the optical spectrum is not affected by the use of DC, regardless of the position of the DCM. The reason for this can be found in Eq. (8.5.84), where the normalized optical spectrum is related to the absolute values of B_{11} and B_{12}. It does not matter where the DCM is placed in the system; DC has no effect on the normalized optical spectrum.

So far, we have discussed the effect of MI in a single-span fiber-optic system, and it is straightforward to extend the analysis to multi-span fiber systems with in-line optical amplifiers. In an optical fiber system of N spans with N EDFAs (one post-amplifier and $N-1$ line amplifiers), the fiber loss per span is compensated by the optical gain of the EDFA. Suppose that all EDFAs have the same noise figure; the ASE noise power spectral density generated by the ith EDFA is

$$S_i = hv(FG_i - 1)$$
(8.5.93)

where the ASE noise optical spectrum is supposed to be white within the receiver optical bandwidth. After transmission through fibers, amplified by EDFAs and detected by the photodiode, the power spectrum of the detected RIN can be obtained by the multiplication of the transfer function of each span of optical fiber, supposing that ASE noise generated by different EDFAs are uncorrelated:

$$\rho(\omega) = 2hvP_{in}\eta \sum_{m=1}^{N} \left\{ (FG_{N-m+1} - 1) \left| B_{11}^{(N-m+1)} + B_{21}^{(N-m+1)} \right|^2 \right\}$$
(8.5.94)

where $B_{ij}^{(k)}$ $(i, j, =1, 2)$ are matrix elements defined as

$$B^{(k)} = \prod_{m=1}^{k} \begin{bmatrix} B_{11} & B_{12} \\ B_{21} & B_{22} \end{bmatrix}_m \tag{8.5.95}$$

and the matrix

$$\begin{bmatrix} B_{11} & B_{12} \\ B_{21} & B_{22} \end{bmatrix}_m$$

represents the IM transfer function of the mth fiber span.

Setting $\gamma=0$ to obtain the normalization factor, the normalized RIN spectrum is then

$$R(\omega) = \frac{\sum_{m=1}^{N} \left\{ (FG_{N-m+1} - 1) \left| B_{11}^{(N-m+1)} + B_{21}^{(N-m+1)} \right|^2 \right\}}{\sum_{m=1}^{N} (FG_{N-m+1} - 1)} \tag{8.5.96}$$

Assuming Gaussian statistics, the change of the standard deviation of the noise caused by fiber MI can be expressed in a simple way:

$$\delta\sigma = \frac{\sigma^2}{\sigma_0^2} = \int_{-\infty}^{\infty} R(\omega) |f(\omega)|^2 d\omega \tag{8.5.97}$$

where σ and σ_0 are noise standard deviations in the nonlinear and linear cases, respectively, and $f(\omega)$ is the receiver baseband filter transfer function.

Extending from the Q-definition in Eq. (8.1.13), in a direct-detection optical receiver with the effect of MI taken into account, the quality factor Q can be expressed as

$$Q = \frac{\Re(P_1 - P_0)}{\sqrt{\sigma_{sh}^2 + \sigma_{th}^2 + \sigma_{sp-sp}^2 + \sigma_{s-sp}^2 \delta\sigma_1} + \sqrt{\sigma_{sh0}^2 + \sigma_{th0}^2 + \sigma_{sp-sp0}^2 + \sigma_{s-sp0}^2 \delta\sigma_0}} \tag{8.5.98}$$

where P is the signal level, σ_{sh}, σ_{th}, $\sigma_{sp\text{-}sp}$, and $\sigma_{s\text{-}sp}$ are, respectively, the standard deviations of shot noise, thermal noise, ASE-ASE beat noise, and signal-ASE beat noise in the absence of MI. Subscripts 1 and 0 indicate the symbols at signal logical 1 and logical 0, respectively. In practice, MI introduces a signal-dependent noise that affects more on signal during logical 1 than that during logical 0. The change in the RIN spectrum causes system performance degradation primarily due to the increase of signal-ASE beat noise through the ratio $\delta\sigma_1$. However, considering that if the signal extinction ratio is not infinite, signal power at logical 0 may also introduce MI through $\delta\sigma_0$, although this effect should be very relatively small. Meanwhile, optical spectrum change due to MI may also introduce Q degradation through ASE-ASE beat noise, but it is expected to be a second-order small effect in an IMDD receiver.

Eq. (8.5.98) indicates that, in general, system performance degradation due to MI depends on the proportionality of signal-ASE beat noise to other noises. Multispan optical amplified fiber systems, where signal-ASE beat noise predominates, are more sensitive to MI in comparison to unamplified optical systems. For signal-ASE beat noise limited optical receiver, to the first-order approximation, the system Q degradation caused by MI can be expressed in a very simple form:

$$10\log(\delta Q) - 10\log(Q_0) = -5\log(\delta\sigma) \tag{8.5.99}$$

where Q_0 is the receiver Q-value in the linear system without considering the impact of MI, and $\delta\sigma$ is the change of noise standard deviations due to MI.

As the major impact of MI in an IMDD optical system is the increase of the RIN at the receiver due to parametric amplification of ASE noise, it is straightforward to measure MI and its impact in optical transmission systems in frequency domain. To provide experimental evidence and verify the theoretical analysis of MI discussed earlier, Fig. 8.5.32 shows the detailed experimental setup for the MI measurement in a dispersion-compensated multi-span fiber system with optical amplifiers.

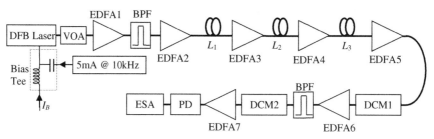

FIG. 8.5.32

Experimental setup. L_1: 84.6 km (1307 ps/nm), L_2: 84.6 km (1291 ps/nm), L_3: 90.4 km (1367 ps/nm). Optical powers at the output of EDFA5, EDFA6, and EDFA7 are less than 0 dBm. *DCM:* dispersion compensation module.

To make sure that the measured receiver RIN is dominated by signal-ASE beat noise, one can increase the broadband ASE noise from EDFAs intentionally. The role of the first optical amplifier in the link was to inject optical noise. It had a noise figure of 7.5 dB and the input optical power of −32 dBm. The narrowband optical filter that followed kept the total ASE power less than that of the signal so as to decrease the ASE-ASE beat noise. The signal power was typically 0.5 dB less than the total power emerging from the EDFA, and this was taken into account when setting the output power of the line amplifiers. The first EDFA was the dominant source of ASE arriving at the PIN detector. The three line amplifiers had output power adjustable by computer control. The fiber spans all had loss coefficient measured by OTDR of 0.2 dB/km. The output power of the next two EDFAs was set below 0 dBm to avoid nonlinear effects in the DCM. The narrowband filter after EDFA suppressed ASE in the 1560 nm region, which would have led to excessive ASE-ASE beat noise. The last optical amplifier was controlled to an output power of 4 dBm, just below the overload level of the PIN detector.

First, a calibration run was made, with an attenuator in place of the system. This measurement was subtracted from all subsequent traces so as to take out the effect of the frequency response of the detection system. The calibration trace was >20 dB higher than the noise floor for the whole 0–18 GHz band.

The RIN spectra at the output of the three-span link are shown in Fig. 8.5.33 with line amplifier output power (signal power) controlled at 8, 10, 12, and 14 dBm, for inverted triangles, triangles, squares, and circles, respectively. In Fig. 8.5.33A, open points represent the measured spectra with the system configuration described in Fig. 8.5.32 except no DC was used. Continuous lines in the same figures were calculated using Eq. (8.5.91). Similarly, Fig. 8.5.33B shows the measured and calculated RIN spectra with the DC of −4070 ps/nm at the receiver side, as shown in Fig. 8.5.32. To obtain the theoretical results shown in Fig. 8.5.33, the fiber nonlinear coefficient used in the calculation was $\gamma = 1.19 \, \text{W}^{-1}\text{km}^{-1}$, and other fiber parameters such as length and dispersion were chosen according to the values of standard single-mode fiber used in the experiment, as shown in the caption of Fig. 8.5.32. Very good agreement between measured and calculated results in the practical power range assures the validity of the two major approximations used in the transfer matrix formulation, namely, the linear approximation to the noise term and the insignificance of pump depletion.

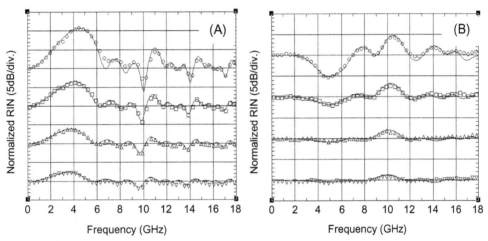

FIG. 8.5.33

Measured (open points) and calculated (solid lines) RIN spectra in the three-span standard SMF as described in Fig. 8.5.32. The optical power at the output of EDFA2, EDFA3, and EDFA4 is 8 dBm (triangles-down), 10 dBm (triangles-up), 12 dBm (squares), and 14 dBm (circles). Curves are shifted for 10 dB between one and another for better display. (A) Without dispersion compensation and (B) with −4070 ps/nm dispersion compensation (Hui et al., 1997).

Although the RIN spectra are independent of the signal data rate, the variance of the noise depends on the bandwidth of the baseband filter as explained in Eq. (8.5.97). Fig. 8.5.34 shows the effect of DC on the ratio of noise standard deviation between nonlinear and linear cases for the three-span fiber system described in Fig. 8.5.32. The optical power was 12 dBm at the input of each fiber span and raised-cosine filters were used with 8 GHz bandwidth. Both theoretical and experimental results demonstrate that $\delta\sigma$ approaches its minimum when the DC is approximately 70% of the total system dispersion. Generally, the optimum level of DC depends on the number of spans, electrical filter bandwidth, optical power levels, and the dispersion in each fiber span.

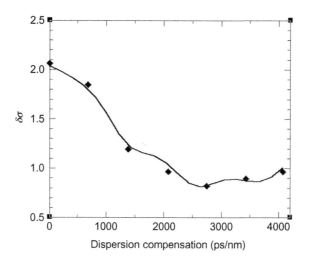

FIG. 8.5.34

Comparison of $\delta\sigma$ between calculation (solid line) and measurement (diamonds) for the three-span system described in Fig. 8.5.32 with optical power $P_{in}=12\,dBm$ (Hui et al., 1997).

As far as MI is concerned, in the anomalous fiber dispersion regime, system performance always becomes worse with increasing signal power. On the other hand, in the normal dispersion regime, system sensitivity may be improved by the nonlinear process. This is an interesting phenomenon that was explained as a noise squeezing (Hui and O'Sullivan, 1996). It can be readily understood from Fig. 8.5.33B, where if the signal optical power is higher than 12 dBm, the noise level is reduced at the low-frequency region. If the receiver bandwidth is less than 7 GHz, the total noise level will be lower compared to the linear system. However, for systems with higher bit rate, sensitivity degradations may also be possible in the normal dispersion regime with high input signal powers. This degradation is caused by the increased noise level at higher frequencies, which happen to be within the receiver baseband.

To conclude this section, MI is in a unique category of fiber nonlinearity. In contract to nonlinear crosstalk due to XPM and FWM, MI is a single-channel process. The high power of the optical signal amplifies the ASE noise through the parametric gain process. The increased RIN due to MI is originated from the nonlinear interaction between the signal and the broadband ASE noise. This effect is also different from SPM because the consequence of MI is the amplification of intensity noise rather than the deterministic waveform distortion.

8.6 CONCLUSION

Optical system design and characterization requires extensive knowledge in both system and component levels. In this chapter we have concentrated in the discussion of basic performance evaluation criteria of binary intensity-modulated optical communication systems with direct detection. High-order modulation and phase-modulated optical systems with coherent detection will be introduced later in

Chapters 9 and 10. BER is an ultimate measure of the signal quality at the receiver, which is often represented by a Q-value. Sources of BER degradation in an optical system can be categorized by waveform distortion and noise. Waveform distortion is usually deterministic which can be caused by chromatic distortion, linear and nonlinear crosstalk between channels, whereas noise is random which is caused by accumulated amplified spontaneous emission (ASE) along the system when optical amplifiers as well as noises created by the photodetector such as thermal noise and shot noise.

In a traditional optical system without inline optical amplifiers, receiver thermal noise and shot noise are major limits to the transmission performance, and *receiver sensitivity*, defined as the minimum signal optical power required to achieve a certain BER, is often used to specify the system. For optical systems employing inline optical amplifiers, on the other hand, signal optical power reaching the photodiode can be easily amplified to a high level. However BER in the optically amplified system is mainly determined by the signal-ASE beat noise in the receiver, and thus required OSNR is a more relevant measure of the system performance.

With the knowledge of the noise statistics, the major impact of noise on BER degradation can often be evaluated analytically. However, waveform distortion depends on specific system configuration as well as and waveforms and optical power levels of modulated optical signals. Especially in WDM optical systems, linear and nonlinear crosstalk can become major limiting factors in the transmission performance. Numerical simulators solving nonlinear Schrödinger equations based on SSFM are power tools for predicting optical system performance, which can be used to guide system design and performance evaluation. Analytical and semi-analytical methods are also indispensible for the understanding of physical origins of eye closure penalties.

The impact of XPM, FWM, and MI are specifically presented at the end of this chapter as examples of nonlinear crosstalks. Semi-analytic methods have been used for the analysis, emphasizing the importance of small-signal perturbation approximation and linearization. Good understanding of various mechanisms introducing system performance degradation helps system design and performance optimization through optimizing dispersion maps, channel spacing, as well as choosing the optimum power levels for optical signals.

PROBLEMS

1. For a binary modulated system without waveform distortion, if the normalized noise standard deviations associated with signal "0" and "1" levels are 0.05 and 0.15, respectively, for the normalized eye diagram,
   (a) Find the normalized optimum decision threshold $(0 < v_{th} < 1)$
   (b) Find the Q-value and the BER
   (c) If the normalized decision threshold is chosen at $v_{th} = 0.5$, what is the BER?

2. A direct detection optical receiver using a PIN photodiode with a responsivity $\Re = 1$A/W, a load resistance $R_L = 50\Omega$, and an electrical bandwidth $B_e = 10$GHz. The optical signal is NRZ modulated. Neglect dark current noise and signal waveform distortion,
   (a) What are the receiver sensitivities for BER $= 10^{-12}$ limited by the thermal noise (only consider thermal noise) and shot noise (only consider shot noise), respectively?
   (b) Approximately what is the receiver sensitivity considering both thermal noise and shot noise?

3. Repeat problem 2 but an APD is used now with a responsivity $\Re=1A/W$, a photo multiplication gain $M=500$ with excess noise factor $F(M)=M^{0.28}$. Neglect dark current noise and signal waveform distortion.

  (a) What are the receiver sensitivities for $BER=10^{-9}$ limited by the thermal noise and shot noise, respectively?

  (b) Approximately what is the receiver sensitivity considering both thermal noise and shot noise? In comparison to a PIN photodiode used in problem 2, what is the benefit introduced by the APD?

  (c) Assume that the average input optical signal power to the optical receiver is $-30\,dBm$, and the APD gain M is adjustable, what is the optimum APD gain in this system to maximize the Q-value? (Numerical solution can be used)

4. Consider an optical transmission system with NRZ intensity modulation at a data rate of $5\,Gb/s$. The average optical signal optical power emitted from the transmitter is $P_{tx}=6\,dBm$, and the loss coefficient of the optical fiber is $\alpha=0.25\,dB/km$. Chromatic dispersion introduces waveform distortion which can be approximated by $v_0=2\times10^{-6}L$ and $v_1=1-2\times10^{-6}L$, where v_0 and v_1 are the upper and lower levels of the normalized eye diagram, and L is the fiber length in meters. In the direct detection receiver a PIN photodiode is used which has a responsivity $\Re=1A/W$, a load resistance $R_L=50\Omega$, and an electrical bandwidth $B_e=5\,GHz$. Thermal noise is considered as the major noise in the receiver (neglect other noise sources).

  (a) Find the maximum fiber length that is allowed for $BER\leq10^{-12}$ in this system. (numerical solution can be used to solve the final equation)

  (b) If chromatic dispersion can be completely compensated so the $v_1=1$ and $v_0=0$, what will be the maximum transmission fiber length?

5. Consider the same optical system as in problem 4, except that an optical preamplifier is added in front of the photodiode with $6\,dB$ noise figure and $30\,dB$ optical gain. Signal-ASE beat noise is considered as the major noise in the receiver (neglect other noise sources). Operating wavelength is $1550\,nm$.

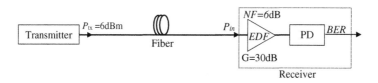

  (a) Find the maximum fiber length that is allowed for $BER\leq10^{-12}$ in this system. (numerical solution can be used to solve the final equation)

  (b) If chromatic dispersion can be completely compensated so the $v_1=1$ and $v_0=0$, what will be the maximum transmission fiber length?

6. An optical system has N in-line optical amplifiers each with $6\,dB$ noise figure, and the gain of each optical amplifier exactly compensates the loss of the optical fiber span immediately before it. Assume all the fiber spans have the same length L and the fiber loss coefficient is $\alpha=0.25\,dB/km$. The signal optical power that enters the first fiber span is $P_t=1\,mW$.

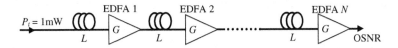

(a) Please derive an expression for optical signal-to-noise ratio (OSNR) at the system output as the function of fiber span length L.

(b) If the total fiber length of the system is $L_T=1500$ km and $L=L_T/N$, please plot OSNR (in logarithmic scale) as the function of the number of fiber spans N for $3<N<50$. (The unit of OSNR in this case is [dB·Hz])

(c) For an engineering design point of view, please discuss how to select the fiber span length between adjacent in-line amplifiers.

7. A 1550-nm wavelength optical system shown below has three spans of single-mode optical fiber, and three optical amplifiers each with 6 dB noise figure. Each amplifier has 20 dB optical gain but the length of each fiber span is different, they are 80, 120, and 50 km, respectively. The fiber attenuation is 0.25 dB/km. The signal optical power emitted from the transmitter is 10 mW.

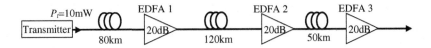

(a) What is the optical noise power spectral density at the output of the laser EDFA (in the unit of [dBm/nm])?

(b) If there is a narrowband filter at the output of the last EDFA at the optical signal wavelength with $B_0=0.1$ nm optical bandwidth, what is the optical signal-to-noise ratio (OSNR) at the system output? (OSNR is defined as the ratio between signal optical power and the optical noise power within the bandwidth of the optical filter)

8. For a 10 Gb/s NRZ intensity-modulated optical system, the receiver is a simple PIN photodiode with a responsivity $\Re=1$ A/W, a load resistance $R_L=50\Omega$, and an electrical bandwidth $B_e=10$ GHz. Assume that there is no waveform distortion, and only consider thermal noise and signal-ASE beat noise. Please find the receiver sensitivity (for $Q=7$) when the OSNR (defined with 0.1-nm optical noise spectral bandwidth) of the input optical signal is:

(a) infinite

(b) 20 dB

(c) 10 dB

(d) Discuss the reason why there is no positive solution for the for receiver sensitivity when OSNR$=10$ dB, and why receiver sensitivity is no longer a relevant parameter in this case.

9. Consider a 40-Gb/s NRZ modulated optical system operating in 1550-nm wavelength window in which the optical receiver has an EDFA preamplifier with a 6-dB noise figure. A band-pass optical filter (BPF) is placed after the preamplifier with 1 nm optical bandwidth. The photodiode has a responsivity $\Re=1$ A/W, a load resistance $R_L=50\Omega$, and an electrical bandwidth $B_e=40 GHz$. The average input optical signal power is $P_{in}=-20$ dBm with an OSNR of 30 dB (OSNR is defined

with 0.1 nm optical noise spectral bandwidth). Signal waveform distortion is negligible, and consider thermal noise, signal-ASE beat noise and ASE-ASE-beat noise in the calculation.

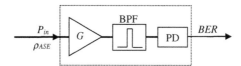

(a) If the gain of the optical preamplifier is $G=30$ dB, what is the OSNR (still use 0.1 nm optical noise spectral bandwidth) immediately after the optical preamplifier?

(b) If the gain of the optical preamplifier is $G=30$ dB, what is the system Q-value at the receiver output?

(c) If the optical gain G of the optical preamplifier is adjustable, plot Q-value as the function of G (for 0 dB $<G<40$ dB) and show that Q-value is independent of G when G is high enough

10. For a high-speed short-reach fiber-optic system with 20 Gb/s NRZ intensity modulation, a PIN photodiode is used in the receiver with a responsivity $\Re=0.8$ A/W, a load resistance $R_L=50\Omega$, and an electrical bandwidth $B_e=20$ GHz. Standard single-mode fiber is used with a dispersion parameter $D=17$ ps/nm/km at the 1550 nm signal wavelength and the loss parameter is $\alpha=0.25$ dB/km. Based on Gaussian pulse approximation, eye closure penalty is described by Eq. (8.5.8), where we assume that the pulse width is $T_0=50$ ps, and the eye closure is equally distributed in the upper and lower levels, that is, $B=L^2/(2L_D^2)$ and $A=1-L^2/(2L_D^2)$.

Transmitter optical power that launched into the fiber is $P_{tx}=10$ dBm. Neglect modulation chirp and consider thermal noise as the dominate noise source.

(a) Please find the maximally allowed fiber length of this system for $Q\geq7$ (Numerical solution can be used to solve the problem).

(b) Is a perfect dispersion compensation is applied so that there is no waveform distortion, what is the maximally allowed fiber length of this system for $Q\geq7$?

11. Same optical system as described in problem 10, but the modulation chirp parameter α_c can be managed. Based on Eq. (8.5.6), what is the optimum chirp parameter α_c to achieve the longest transmission distance?

12. Performance of a long-distance optical fiber system can be affected by polarization mode dispersion (PMD). Consider a 40 Gb/s NRZ intensity-modulated system using standard single-mode fiber. The total fiber length is 500 km and the PMD parameter of the fiber is $PMD=0.1$ps/$\sqrt{km}$. Assume that power splitting ratio γ between two SOP is 50%.

(a) Based on Eq. (8.5.14), what is the DGD value that reduces the eye opening to 50%?

(b) What is the accumulated probability that eye opening is $E_{eye}\leq0.5$(it can be integrated numerically).

13. EDFA in the C-band covers a wavelength window from 1530 to 1560 nm. A WDM system has a channel spacing of 25 GHz.

(a) How many wavelength channels this EDFA can support?

(b) If each channel carries 10 Gb/s data traffic, what is the total data rate that can be carried by this WDM system?

14. In an optical system with coherent multipath interference (MPI) caused by cavity effect due to multiple reflections. Assume $\zeta = \sqrt{R_1 R_2} = R$ is the roundtrip power loss of the cavity with $R = R_1 = R_2$ the effective power reflectivity. Assume $R \ll 1$ so that R^2 term is negligible.
    (a) If the receiver is thermal noise limited, please find the maximally allowed R value so that degradation of receiver sensitivity due to MPI is less than 1 dB
    (b) If the receiver is signal-ASE beat noise limited, please find the maximally allowed R value so that degradation of the required OSNR due to MPI is less than 1 dB

15. In the case of incoherent multipath interference, phase noise of the laser is converted into a relative intensity noise as described by Eq. (8.5.22). If the laser source has a spectral linewidth of $\Delta v = 10 \text{MHz}$ and the optical system has a data rate of 10 Gb/s, could incoherent multipath interference significantly degrade the system performance? Please discuss.

16. Based on split-step Fourier Method, please write a computer program and simulate a single wavelength NRZ modulated optical signal at 10 Gb/s. System configuration: 100 km standard single-mode fiber, chromatic dispersion parameter $D = 17 \text{ps/nm/km}$, loss coefficient $\alpha = 0.23 \text{dB/km}$, fiber effective cross-section area $A = 80 \, \mu\text{m}^2$, and nonlinear index $n_2 = 2.35 \times 10^{20} \text{m}^2/\text{W}$. Plot out optical power waveform at the output of the fiber system for the average input power levels of 0, 10, and 20 dBm, respectively.

17. Repeat problem 16, but use three wavelength channels with 50 GHz channel spacing. Please plot out optical power waveform of the central channel, and plot out optical spectra for the per-channel average input power levels of 0, 10, and 20 dBm, respectively.

18. Please explain the physical mechanism why the XPM crosstalk in a WDM optical fiber system is inversely proportional to the channel spacing.

19. For the following two multi-span optically amplified fiber optical systems, which one has less system penalty due to XPM induced crosstalk? Please explain the reason.

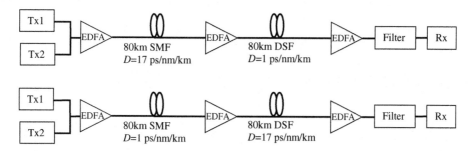

20. In the frequency domain XPM transfer function shown in Eq. (8.5.41), $p_k(\Omega, 0)$ is the power spectral density of the pump channel at the fiber input. Please generate a time-domain 10 Gb/s PRBS NRZ waveform for the pump channel, and plot the crosstalk waveform on the probe channel similar to that shown in Fig. 8.5.19 (the input of the probe channel is CW). System parameters are 100-km standard single-mode fiber, chromatic dispersion parameter $D = 17 \text{ps/nm/km}$, loss coefficient $\alpha = 0.23 \text{dB/km}$, fiber effective cross-section area $A = 80 \, \mu\text{m}^2$, nonlinear index

$n_2 = 2.35 \times 10^{20} m^2/W$, frequency spacing between pump and probe: $\Delta\lambda = 0.4$ nm. Average pump power at the fiber input is 10 dBm.

21. Consider a two-channel WDM system with channel spacing of $\Delta\lambda$ in a single fiber span. Fiber attenuation is α, linear dispersion parameter is D and nonlinear coefficient is γ. Using small-signal approximation and suppose $\exp(-\alpha L) \ll 1$, derive a formula which gives the XPM induced phase modulation efficiency η.

    η is defined by:

$$\eta = \left| \frac{\varphi_{jk}(\Omega, L)}{p_k(\Omega, 0)} \right|$$

    where $\varphi_{jk}(\Omega,L)$ is the phase modulation in the signal channel created by the crosstalk channel, L is the fiber length and $p_k(\Omega,0)$ is the crosstalk channel input optical power spectral. Discuss why η has a low-pass characteristic.

22. In a three channel WDM optical system using dispersion shifted fiber, given $\lambda_1 = 1551$ nm, $\lambda_2 = 1551.8$ nm, and $\lambda_3 = 1552.6$ nm, how many FWM components can be generated, at what wavelengths?

23. Modulation instability (MI) is a parametric gain process. Consider a single-mode fiber with optical attenuation $\alpha = 0.22$ dB/km, and nonlinear coefficient $\gamma\Pi\lambda\alpha\chi\epsilon\eta o\lambda\delta\epsilon\rho$ $T\epsilon\xi\tau = 2.07 \, W^{-1} km^{-1}$. Assume the fiber length is 100 km, and the input pump optical power is 13 dBm at 1550 nm wavelength, plot the MI-induced optical gain as the function of the frequency-detune from the pump wavelength for the following two conditions:
    (a) fiber chromatic dispersion parameter is $D = 2$ ps/nm/km
    (b) fiber chromatic dispersion parameter is $D = -2$ ps/nm/km

24. Same as problem 23. Assume a direct detection receiver is used, the MI-induced optical noise is converted into electrical domain through mixing with the pump in the photodiode. Please also plot the normalized electric noise power spectral density as the function of the frequency. MI-induced optical gain as the function of the frequency-detune from the pump wavelength.

25. In a single-span fiber-optic system, can you use a dispersion compensator (added either before or after the fiber) to change the MI-induced *optical gain spectral*? Use Jones matrix formula to explain why. Then, how about in a multi-span system using distributed dispersion compensator? (use a 2-span system as an example).

REFERENCES

Agrawal, G.P., 1989. Nonlinear Fiber Optics. Academic Press, San Diego, CA.

Chiang, T.-K., Kagi, N., Marhic, M.E., Kazovsky, L., 1996. Cross-phase modulation in fiber links with multiple optical amplifiers and dispersion compensators. IEEE J. Lightwave Technol. 14 (3), 249–260.

Chraplyvy, A.R., 1990. Limitations of lightwave communications imposed by optical fiber nonlinearities. *IEEE J. Lightwave Technol.* 8, 1548–1557.

Eiselt, M., 1999. Limits on WDM systems due to four-wave mixing: a statistical approach. J. Lightwave Technol. 17 (11), 2261–2267.

Forghieri, F., Tkach, R.W., Chraplyvy, A.R., 1995. WDM systems with unequally spaced channels. IEEE J. Lightwave Technol. 13 (5), 889–897.

Gimlett, J.L., Cheung, N., 1989. Effects of phase-to-intensity noise conversion by multiple reflections on gigabit-per-second DFB laser transmission systems. J. Lightwave Technol. 7 (6), 888–895.

Ho, K.-P., 2003. Performance degradation of phase-modulated systems due to nonlinear phase noise. IEEE Photon. Technol. Lett. 15, 1213–1215.

Hui, R., O'Sullivan, M., 1996. Noise squeezing due to Kerr effect nonlinearty in optical fibers with negative dispersion. IEEE Electron. Lett. 32 (21), 2001–2002.

Hui, R., O'Sullivan, M., Robinson, A., Taylor, M., 1997. Modulation instability and its impact in multispan optical amplified systems: theory and experiments. *IEEE J. Lightwave Technol.* 15 (7), 1071–1082.

Hui, R., Vaziri, M., Zhou, J., O'Sullivan, M., 1999. Separation of noise from distortion for high-speed optical fiber system link budgeting. IEEE Photon. Technol. Lett. 11, 910–912.

Hui, R., Zhu, B., Huang, R., Allen, C., Demarest, K., Richards, D., 2002. Subcarrier multiplexing for high-speed optical transmission. J. Lightwave Technol. 20, 417–427.

Hui, R., Laperle, C., Reimer, M., Shiner, A.D., O'Sullivan, M., 2015. Characterization of electrostriction nonlinearity in a standard single-mode fiber based on coherent detection and cross-phase modulation. IEEE/OSA J. Lightwave Technol. 33 (22), 4547–4553.

Inoue, K., Toba, H., 1995. Fiber four-wave mixing in multi-amplifier systems with non-uniform chromatic dispersion. IEEE J. Lightwave Technol. 13 (1), 88–93.

Kumar, P.V., Win, M.Z., Lu, H.F., Georghiades, C.N., 2002. Error–control coding techniques. In: Kaminow, I., Li, T. (Eds.), Optical Fiber Telecommunications IVB. Academic Press.

Marcus, D., 1994. Noise properties of four-wave mixing of signal and noise. Electron. Lett. 30, 1175–1177.

Marcuse, D., Chraplyvy, A.R., Tkach, R.W., 1994. Dependence of cross-phase modulation on channel number in fiber WDM systems. *IEEE J. Lightwave Technol.* 12, 885–890.

J. McNicol, M. O'Sullivan, K. Roberts, A. Comeau, D. McGhan, L. Strawczynski, 2005. Electrical domain compensation of optical dispersion. In Proc. OFC 2005, paper OThJ3.

Mecozzi, A., 1994. Long-distance transmission at zero dispersion: combined effect of the Kerr nonlinearity and the noise of the inline amplifiers. J. Opt. Soc. Am. B 12, 462–465.

Penninckx, D., Chbat, M., Pierre, L., Thiery, J.P., 1997. The Phase-Shaped Binary Transmission (PSBT): a new technique to transmit far beyond the chromatic dispersion limit. IEEE Photon. Technol. Lett. 9, 259–261.

Personick, S.D., 1977. Receiver design for optical fiber systems. Proc. IEEE 65, 1670–1678.

Poole, C.D., Nagel, J., 1997. Polarization effects in lightwave systems. In: Kaminow, I.P., Koch, T. (Eds.), Optical Fiber Telecommunications IIIA. Academic Press.

Poole, C.D., Tkach, R.W., Chraplyvy, A.R., Fishman, D.A., 1991. Fading in lightwave systems due to polarization-mode dispersion. IEEE Photon. Technol. Lett. 3 (1), 68–70.

Rapp, L., 1997. Experimental investigation of signal distortion induced by cross-phase modulation combined with distortion. IEEE Photon. Technol. Lett. 9 (12), 1592–1594.

Saunders, R.A., Patel, B.L., Harvey, H.J., Robinson, A., 1997. Impact of cross-phase modulation seeded modulation instability in 10 Gb/s WDM systems and methods for its suppression. In: Proceedings of Optical Fiber Communication Conference OFC'97, paper WC4, Dallas TX, pp. 116–117.

Shtaif, M., Eiselt, M., 1997. Analysis of intensity interference caused by cross-phase modulation in dispersive optical fibers. IEEE Photon. Technol. Lett. 9 (12), 1592–1594.

Waklin, S., Conradi, J., 1999. Multilevel signaling for increasing the reach of 10 Gb/s lightwave systems. J. Lightwave Technol. 17, 2235–2248.

Wang, J., Petermann, K., 1992. Small signal analysis for dispersive optical fiber communication systems. IEEE J. Lightwave Technol. 10 (1), 96–100.

Yu, M., Agrawal, G.P., McKinstrie, C.J., 1995. Pump-wave effects on the propagation of noisy signals in nonlinear dispersive media. J. Opt. Soc. Am. B 11, 1126–1132.

FURTHER READING

Eiselt, M., Shtaif, M., Garrett, L.D., 1999. Cross-phase modulation distortions in multi-span WDM systems. In: Optical Fiber Communication Conference OFC '99, paper ThC5, San Diego, CA.

COHERENT OPTICAL COMMUNICATION SYSTEMS

9

CHAPTER OUTLINE

Introduction 417
9.1 Basic principles of coherent detection 418
9.2 Receiver signal-to-noise ratio calculation of coherent detection 421
    9.2.1 Heterodyne and homodyne detection 421
    9.2.2 Signal-to-noise-ratio in coherent detection receivers 422
9.3 Balanced coherent detection and polarization diversity 427
    9.3.1 Balanced coherent detection 428
    9.3.2 Polarization diversity 429
9.4 Phase diversity and I/Q detection 430
9.5 Conclusion 435
Problems 436
References 438
Further reading 438

INTRODUCTION

It is well known that an optical field consists of amplitude, frequency, phase, and the state of polarization (SOP). Theoretically all of them, and their combinations, can be encoded to carry information in an optical communication system. As the photocurrent of a photodiode in an optical receiver is proportional to the intensity of the received optical signal, intensity modulation and direct detection (IMDD) provides the simplest mechanism for optical communication in which information is carried only by the intensity of the optical carrier. Wavelength-division multiplexing (WDM) utilizes optical carriers of different frequencies, and the intensity of each optical carrier carries an independent information channel so that they can be separately detected in the receiver after optical filtering through a wavelength-division demultiplexer. In a WDM system, although optical frequency is utilized, it is for multiplexing instead of data encoding. Coherent optical communication systems are designed to make full use of the information capacity provided by the complex optical field. The ability to encode information onto amplitude, frequency, and phase of an optical carrier in the transmitter, and the ability

Introduction to Fiber-Optic Communications. https://doi.org/10.1016/B978-0-12-805345-4.00009-3

to detect the complex optical field of the signal are necessary for the construction of a coherent optical system.

For direct injection current modulation of a laser diode in an optical transmitter, both the amplitude and the phase of the optical carrier are modulated simultaneously, and the relation between phase modulation and intensity modulation is determined by a linewidth enhancement factor, α_{lw}, as defined by Eq. (3.3.40). In this case, the amplitude and the phase of the optical signal cannot be independently modulated as α_{lw} is typically a constant for a laser diode. A more effective way to encode information onto the complex optical field is to use external electro-optic modulators as discussed in Chapter 7. Especially, an in-phase/quadrature (I/Q) electro-optic modulator provides a high degree of flexibility which allows the amplitude and the phase of an optical carrier to be independently modulated.

In order to detect the complex optical field in the receiver, coherent detection usually uses a strong optical local oscillator (LO) which provides a reference for the optical frequency and phase (Betti et al., 1995; Kikuchi, 2016). The mixing between the LO and the received optical signal at the photodiode allows the determination of the amplitude and the phase information carried by the received optical signal. In addition, since the optical power of the LO is much stronger than the received optical signal, it effectively amplifies the weak optical signal in the mixing process, and provides much improved detection sensitivity compared to direct detection. In the coherent detection process by mixing with an LO, the signal optical spectrum is linearly down converted into the electric domain so that electric signal processing can be used to compensate for transmission impairments such as chromatic dispersion of the optical fiber.

Coherent detection technique was investigated extensively in the 1980s primarily for the purpose of improving receiver sensitivity in optical communication systems. However, the introduction of erbium-doped fiber amplifier (EDFA) in the early 1990s made coherent detection less attractive, mainly for two reasons: (1) EDFA provides sufficient optical amplification without the requirement of an expensive low phase noise laser LO in the receiver and (2) EDFA is polarization independent, and has wide gain bandwidth to support multichannel WDM optical systems.

In the recent years, technological advances made tunable laser diodes commercially available with narrow spectral linewidth, low cost, small footprint, and reliable enough to meet telecommunication standards. Coherent transmission systems research and development is revitalized to significantly increase transmission capacity, multiplexing flexibility, and to allow electronic domain compensation of various transmission impairments. In this chapter, we discuss basic operation principles and various optical circuit configurations of coherent optical communication systems.

9.1 BASIC PRINCIPLES OF COHERENT DETECTION

Coherent detection originates from radio communications, where a local carrier mixes with the received radio frequency (RF) signal to generate a product term. As a result, the received RF signal can be frequency translated and demodulated.

A block diagram of coherent detection is shown in Fig. 9.1.1. In this circuit, the received RF signal $m(t)\cos(\omega_{sc}t)$ has an information-carrying amplitude $m(t)$ and an RF carrier at frequency ω_{sc}, whereas the LO has a single frequency at ω_{loc}. The RF signal multiplies with the LO in an RF mixer, generating

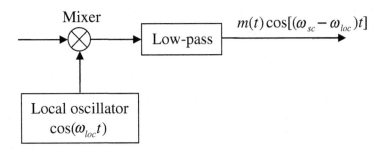

FIG. 9.1.1

Block diagram of coherent detection in radio communications.

the sum and the difference frequencies between the signal and the LO. The process can be described by the following equation:

$$m(t)\cos(\omega_{sc}t) \times \cos(\omega_{loc}t) = \frac{m(t)}{2}\{\cos[(\omega_{sc}+\omega_{loc})t] + \cos[(\omega_{sc}-\omega_{loc})t]\} \qquad (9.1.1)$$

A low-pass filter is usually used to eliminate the sum frequency component and thus the baseband signal can be recovered if the frequency of the LO is equal to that of the signal ($\omega_{loc}=\omega_{sc}$). When the RF signal has multiple and very closely spaced frequency channels, excellent frequency selection in coherent detection can be accomplished by fine tuning the frequency of the LO. This technique has been used in ratio communications for many years, and high-quality RF components such as oscillators, RF mixers, and amplifiers are standardized.

For coherent detection in lightwave systems, although the fundamental principle is similar, its operating frequency is many orders of magnitude higher than the radio frequencies; thus the required components and circuit configurations are quite different. In a lightwave coherent receiver, mixing between the received optical signal and the optical LO is done in a photodiode, which is a square-law detection device. A typical schematic diagram of coherent detection in a lightwave receiver is shown in Fig. 9.1.2, where the incoming optical signal and the optical LO are combined in an optical coupler. The optical coupler can be made by a partial reflection mirror in free space, or more conveniently made by a fiber directional coupler in guided wave optics. To match the SOPs between the input optical signal and the LO, a polarization controller is used; it can be put in the path of either the LO or

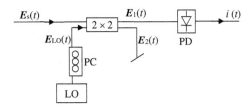

FIG. 9.1.2

Block diagram of coherent detection in a lightwave receiver. *PC*, polarization controller; *PD*, photodiode; *LO*, local oscillator.

the input optical signal. Mixing between the signal and the LO in the photodiode converts the modulated optical signal into a photocurrent in the electric domain.

Consider the complex field vector of the incoming optical signal,

$$\vec{E}_s(t) = \vec{A}_s(t)\exp[-j(\omega_s t + \varphi_s(t))] \tag{9.1.2}$$

and the field vector of the LO,

$$\vec{E}_{LO}(t) = \vec{A}_{LO}\exp[-j(\omega_{LO} t + \varphi_{LO})] \tag{9.1.3}$$

where $\vec{A}_s(t)$ and $\vec{A}_{LO}(t)$ are the real amplitudes of the incoming signal and the LO, respectively, ω_s and ω_{LO} are their optical frequencies, and $\phi_s(t)$ and ϕ_{LO} are their optical phases. The optical transfer function of the 2×2 optical coupler, discussed in Chapter 6, is

$$\begin{bmatrix} \vec{E}_1 \\ \vec{E}_2 \end{bmatrix} = \begin{bmatrix} \sqrt{1-\varepsilon} & j\sqrt{\varepsilon} \\ j\sqrt{\varepsilon} & \sqrt{1-\varepsilon} \end{bmatrix} \begin{bmatrix} \vec{E}_S \\ \vec{E}_{LO} \end{bmatrix} \tag{9.1.4}$$

where ε is the power-coupling coefficient of the optical coupler. In the simplest coherent receiver configuration shown in Fig. 9.1.2, only one of the two output ports of the 2×2 coupler is used, while the other output is nulled. The composite optical signal at the coupler output is

$$\vec{E}_1(t) = \sqrt{1-\varepsilon}\vec{E}_s(t) + j\sqrt{\varepsilon}\vec{E}_{LO}(t) \tag{9.1.5}$$

As discussed in Section 4.2, a photodiode has a square-law detection characteristic and the photocurrent is proportional to the square of the composite optical field, that is,

$$\begin{aligned} i(t) &= \Re\left|\vec{E}_1(t)\right|^2 \\ &= \Re\left\{(1-\varepsilon)|A_s(t)|^2 + \varepsilon|A_{LO}|^2 + 2\sqrt{\varepsilon(1-\varepsilon)}\vec{A}_s(t)\cdot\vec{A}_{LO}\cos(\omega_{IF}t + \Delta\varphi(t))\right\} \end{aligned} \tag{9.1.6}$$

where $\Re$ is the responsivity of the photodiode, $\omega_{IF} = \omega_{LO} - \omega_s$ is the frequency difference between the signal and the LO, which is referred to as the *intermediate frequency* (IF), and $\Delta\varphi(t) = \varphi_{LO} - \varphi_s(t)$ is their phase difference. We have neglected the sum-frequency term in Eq. (9.1.6) because $\omega_s + \omega_{LO}$ will be in the optical domain but at a much shorter wavelength, which will be eliminated by the RF circuit that follows the photodiode.

The first and the second terms on the right side of Eq. (9.1.6) are the direct detection components of the optical signal and the LO, respectively. The last term is the coherent detection term, which is the mixing term between the optical signal and the LO.

Typically, the LO is a laser operating in continuous waves (CW). Ideally, the LO laser has no intensity noise so that $A_{LO} = \sqrt{P_{LO}}$ is a constant with P_{LO} the LO optical power. Since the LO power is, in general, much stronger than the power of the optical signal such that $|A_s(t)|^2 \ll |\vec{A}_s(t)\cdot\vec{A}_{LO}|$, the only significant information-carrying component on the right hand side of Eq. (9.1.6) is the last term, and thus

$$i(t) \approx 2\Re\sqrt{\varepsilon(1-\varepsilon)}\cos\theta\sqrt{P_s(t)\cdot P_{LO}}\cos(\omega_{IF}t + \Delta\varphi(t)) \tag{9.1.7}$$

where $\cos\theta$ results from the dot product of $\vec{A}_s(t)\cdot\vec{A}_{LO}$, which is the angle of SOP mismatch between the optical signal and the LO. $P_s(t) = |A_s(t)|^2$ is the signal optical power. Eq. (9.1.7) shows that the detected electric signal level depends also on the coefficient, ε, of the optical coupler. Since $2\sqrt{\varepsilon(1-\varepsilon)}$ reaches the maximum value of 1 when $\varepsilon = \frac{1}{2}$, a 3-dB coupler is usually used in the coherent detection receiver. If the SOP of the LO ideally matches that of the incoming optical signal, $\cos\theta = 1$, the photocurrent of coherent detection is maximized and can be simplified as

$$i(t) \approx \Re \sqrt{P_s(t) \cdot P_{LO}} \cos(\omega_{IF} t + \Delta\varphi) \tag{9.1.8}$$

Eq. (9.1.8) shows that coherent detection shifts the spectrum of the optical signal from the optical carrier frequency ω_s to an IF ω_{IF}, which can be handled by an electric circuit. In general, coherent detection can be categorized as homodyne detection if $\omega_{IF} = 0$, and heterodyne detection when $\omega_{IF} \neq 0$.

9.2 RECEIVER SIGNAL-TO-NOISE RATIO CALCULATION OF COHERENT DETECTION

Electric-domain signal-to-noise ratio (SNR) after photodetection is an important measure of the signal quality in an optical receiver, which determines the transmission performance of the system. Because the LO has an optical power much higher than the received optical signal, LO-induced shot noise is usually the dominant noise in a coherent receiver. This is fundamentally different from direct detection where receiver thermal noise has the most impact in the SNR because the signal optical power is usually low. Of course, optical amplifiers can also be used in coherent transmission systems to extend the transmission distance. In such systems, accumulated optical noise may become a limiting factor of the system performance, and the electric SNR can be primarily determined by the optical signal-to-noise ratio (OSNR).

9.2.1 HETERODYNE AND HOMODYNE DETECTION

In coherent detection, the photocurrent is proportional to the field of the received optical signal, the phase information of the optical signal can be detected (Yamamoto and Kimura, 1981). In coherent detection, assume an ideally matched SOP between the signal and the LO, the photocurrent is

$$i(t) = \Re \sqrt{P_s(t)P_{LO}} \cos(\omega_{IF} t + \Delta\varphi(t)) \tag{9.2.1}$$

In a heterodyne configuration, the modulated signal optical spectrum centralized at ω_s is down converted to the RF domain centralized at an IF ω_{IF}. In order to avoid spectral aliasing around $\omega = 0$, ω_{IF} is usually chosen to be higher than the modulation bandwidth of the signal. To convert the IF signal further down to the baseband, two basic techniques can be used. One of these techniques, shown in Fig. 9.2.1A, uses an RF bandpass filter and an RF detector. This is usually referred to as *RF envelope detection*. Another technique, shown in Fig. 9.2.1B, uses an RF LO at the IF ω_{IF}, which mixes with the heterodyne signal and down converts it to the baseband. This second technique is called *RF coherent detection*.

Although both of these two techniques recover baseband signals from the IF, each has its unique characteristics. RF envelope detection is relatively insensitive to the IF frequency variation because the width of the bandpass filter can be chosen wider than the signal bandwidth. But on the other hand, a wider filter bandwidth also allows more noise to be included so that the noise power is higher. In comparison, RF coherent detection has higher detection efficiency than RF envelope detection because a strong RF LO effectively amplifies the signal level. In addition, the SNR of RF coherent detection is 3-

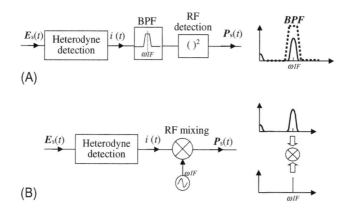

FIG. 9.2.1

(A) Heterodyne optical detection and RF envelope detection and (B) heterodyne optical detection and RF coherent detection. *PC*, polarization controller; *PD*, photodiode; *LO*, local oscillator.

dB higher than the RF envelope detection because only the in-phase noise component is involved and the quadrature noise component can be eliminated. Obviously, RF coherent detection is sensitive to the variation of the IF, and typically a frequency locked loop needs to be used.

In coherent homodyne detection, the IF is set to $\omega_{IF}=0$. This is obtained by setting the frequency of the LO, ω_{LO}, equal to that of the signal optical carrier ω_s. Thus, baseband information carried by the received optical signal is directly obtained in the homodyne detection as

$$i(t) = \Re A_s(t)A_{LO}\cos\Delta\varphi(t) \qquad (9.2.2)$$

Although homodyne detection seems to be simpler than heterodyne detection, it requires both frequency locking and phase locking between the LO and the optical signal because a random phase $\Delta\phi(t)$ in Eq. (9.2.2) would introduce signal fading whenever $\Delta\phi(t)=n\pi+\pi/2$ happens, where n is an integer. On the other hand, as homodyne detection is optical phase sensitive, it provides the capability of optical phase decoding for optical transmission systems. Techniques such as phase diversity, which will be discussed later in this chapter, can be used to retrieve the instantaneous optical phase in a homodyne detection receiver.

9.2.2 SIGNAL-TO-NOISE-RATIO IN COHERENT DETECTION RECEIVERS

In an optical receiver with coherent detection, the power of LO is usually much stronger than the received optical signal; therefore, the SNR is mainly determined by the shot noise in the photodiode caused by the LO.

First, let us consider a coherent system without optical amplifiers so that there is no broadband optical noise associated with the received optical signal. If we consider only thermal noise and shot noise, the receiver SNR is

$$SNR = \frac{\langle i^2(t)\rangle}{\langle i_{th}^2\rangle + \langle i_{sh}^2\rangle} \qquad (9.2.3)$$

where for RF envelope detection usually used in the heterodyne detection scheme, the signal RF power is

$$\langle i^2(t) \rangle = \frac{\mathfrak{R}^2 P_s(t) \cdot P_{LO}}{2} \tag{9.2.4a}$$

where 1/2 is the result of $\langle \cos^2 \varphi(t) \rangle$. While for homodyne detection with optical phase locking between the received optical signal and the LO, the signal RF power is

$$\langle i^2(t) \rangle = \mathfrak{R}^2 P_s(t) \cdot P_{LO} \tag{9.2.4b}$$

where $\cos \varphi(t) = 1$ is assumed.

$$\langle i_{th}^2 \rangle = \frac{4kTB_e}{R_L} \tag{9.2.5}$$

is the thermal noise power, and

$$\langle i_{sh}^2 \rangle = 2q\mathfrak{R}[P_s(t)/2 + P_{LO}]B_e \tag{9.2.6}$$

is the short noise power, where B_e is the signal electric bandwidth, R_L is the load resistance, and q is the electron charge. $P_s(t)/2$ used in the shot noise calculation is due to the 3 dB splitting loss of the optical signal at the LO-signal combining coupler. Then, the SNR for optical heterodyne detection with RF envelope detection can be expressed as

$$SNR_{heterodyne} = \frac{1}{2B_e} \cdot \frac{\mathfrak{R}^2 P_s(t) P_{LO}/2}{4kT/R_L + 2q\mathfrak{R}[P_s(t)/2 + P_{LO}]} \tag{9.2.7a}$$

where $2B_e$ is for the double bandwidth required for heterodyne detection.

For homodyne detection with optical phase-locked LO, the SNR is

$$SNR_{homodyne} = \frac{1}{B_e} \cdot \frac{\mathfrak{R}^2 P_s(t) P_{LO}}{4kT/R_L + 2q\mathfrak{R}[P_s(t)/2 + P_{LO}]} \tag{9.2.7b}$$

which is 6 dB higher than that of heterodyne detection due to the reduced electric bandwidth and phase locking.

Fig. 9.2.2A shows the SNR comparison between direct detection and coherent detection, where we assume that the receiver bandwidth is $B_e = 10\,\mathrm{GHz}$, the load resistance is $R_L = 50\,\Omega$, the photodiode responsivity is $\mathfrak{R} = 0.75\,\mathrm{A/W}$, the temperature is $T = 300\,K$, and the power of the LO is fixed at $P_{LO} = 20\,\mathrm{dBm}$. Obviously, coherent detection significantly improves the SNR compared to direct detection, especially when the signal optical power level is low. In fact, if the LO power is strong enough, the shot noise caused by the LO can be significantly higher than the thermal noise; therefore, Eqs. (9.2.7a) and (9.2.7b) can be simplified into

$$SNR_{heterodyne} \approx \frac{\mathfrak{R}}{8B_e q} P_s(t) \tag{9.2.8a}$$

and

$$SNR_{homodyne} \approx \frac{\mathfrak{R}}{2B_e q} P_s(t) \tag{9.2.8b}$$

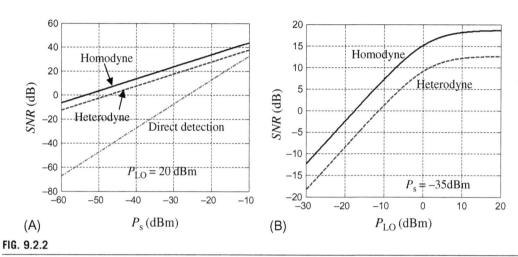

FIG. 9.2.2

(A) Comparison between direct detection and coherent detection when P_{LO} is 20 dBm and (B) effect of LO power when P_s is fixed at −35 dBm. Other parameters: B_e=10 GHz, R_L=50 Ω, $\mathcal{R}$=0.75 A/W, and T=300 K.

In the approximation, we have assumed that $P_{LO} >> P_s$, which is usually true. Therefore, *SNR* in a coherent detection receiver is linearly proportional to the power of the input optical signal P_s. In contrast, in a direct detection receiver, $SNR \propto P_s^2$.

Fig. 9.2.2B shows that SNR in a coherent detection receiver also depends on the power of the LO. If the LO power is not strong enough, the full benefit of coherent detection is not achieved, and the SNR is a function of the LO power. When LO power is strong enough (for $P_{LO} > 10$ dBm in this case), the SNR no longer depends on LO as the strong LO approximation is valid and the SNR can be accurately represented by Eqs. (9.2.8a) and (9.2.8b).

Note that we have used Gaussian statistics for the analysis of shot noise here. However, Poisson distribution is more accurate describing the shot noise process as discussed in Section 4.3. Nevertheless, Gaussian statistics greatly simplifies the performance analysis, and it is a reasonably good approximation when the photocurrent is relatively high, which is indeed the case in a coherent detection receiver where LO is usually strong.

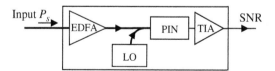

FIG. 9.2.3

Coherent receiver with an EDFA preamplifier.

Now, let us see if an optical preamplifier can help improve system performance in a coherent detection receiver as shown in Fig. 9.2.3. Dominant noise sources in this receiver are shot noise introduced by the LO and signal-ASE beat noise due to the broadband ASE noise introduced by the EDFA preamplifier. The electric signal power is

$$\langle i^2(t)\rangle = \frac{\mathfrak{R}^2 G P_s(t)\cdot P_{LO}}{2} \tag{9.2.9a}$$

for RF envelope detection, and

$$\langle i^2(t)\rangle = \mathfrak{R}^2 G P_s(t)\cdot P_{LO} \tag{9.2.9b}$$

for phase-locked homodyne detection with G the optical gain of the EDFA preamplifier.

The shot noise electric power is

$$\langle i_{sh}^2\rangle = 2q\mathfrak{R}[P_s(t)G + P_{LO}]B_e \tag{9.2.10}$$

and the signal-ASE beat noise power is

$$\langle i_{s-ASE}^2\rangle = 4\mathfrak{R}^2 \left[\frac{P_s(t)G}{2} + P_{LO}\right]\frac{hvn_{sp}(G-1)}{2}B_e \tag{9.2.11}$$

where 3 dB loss of the signal-LO combining coupler is also assumed so that both optical signal and the ASE nose of the preamplifier are attenuated by 3 dB. Single-polarized ASE noise is used because the optical signal is single polarized. The signal-ASE beat noise is usually much higher than the shot noise for reasonable values of EDFA noise figure and with optical gain $G \gg 1$. Thus, signal-ASE beat noise (including LO-ASE beat noise) is most likely the biggest noise source. In fact if we neglecting all other noise contributions, and only considering shot noise and signal ASE-beat noise, the SNR at the receiver output will be,

$$SNR_{\text{heterodyne}} = \frac{\mathfrak{R}^2 G P_s(t)P_{LO}}{4B_e[P_s(t)G/2 + P_{LO}]\left[2q\mathfrak{R} + 2\mathfrak{R}^2 hvn_{sp}(G-1)\right]} \tag{9.2.12a}$$

and

$$SNR_{\text{homodyne}} = \frac{\mathfrak{R}G P_s(t)P_{LO}}{B_e[P_s(t)G/2 + P_{LO}]\left[2q\mathfrak{R} + 2\mathfrak{R}^2 hvn_{sp}(G-1)\right]} \tag{9.2.12b}$$

while for pre-amplified direct detection,

$$SNR_{\text{direct}} = \frac{\mathfrak{R}^2 G^2 P_s^2(t)}{B_e\mathfrak{R}\left[2q + 4\mathfrak{R}hvn_{sp}(G-1)\right]G P_s(t)} = \frac{\mathfrak{R}G P_s(t)}{B_e\left[2q + 4\mathfrak{R}hvn_{sp}(G-1)\right]}$$

Fig. 9.2.4 shows the calculated SNR for coherent homodyne detection and pre-amplified direct detection, respectively, as the function of the input signal optical power. Parameters used in the calculation for coherent detection are: receiver electric bandwidth $B_e = 10\,$GHz, photodiode responsivity $\mathfrak{R} = 0.75\,$A/W, and LO optical power $P_{LO} = 10\,$dBm. In Fig. 9.2.4, the dashed line shows the SNR of coherent homodyne detection without the EDFA preamplifier ($G = 0\,$dB). When an ideal EDFA preamplifier is introduced with 3 dB noise figure ($n_{sp} = 1$) and 20 dB optical gain, the SNR of coherent homodyne detection shown as the dotted line is about 6 dB higher than that without the EDFA preamplifier. As a comparison, for pre-amplified direct detection with the same ideal EDFA preamplifier at 20 dB optical gain, the SNR shown as the solid line which is 3 dB lower than the homodyne detection with the same optical preamplifier. The take-home message from Fig. 9.2.4 is that coherent detection does not significantly improve the SNR compared to an optically preamplified detection optical receiver when the noise figure of the preamplifier is low enough.

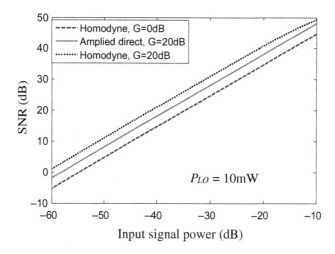

FIG. 9.2.4

Calculated SNR for coherent homodyne detection receiver without an EDFA preamplifier *(dashed line)*, with an EDFA preamplifier with noise figure of 3 dB *(dotted line)*, and preamplified direct detection solid line.

For a long-distance optical transmission system accumulated ASE noise from multiple in-line optical amplifiers can be significant. In such a case, OSNR will be a limiting factor which determines the system performance.

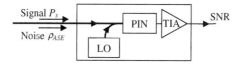

FIG. 9.2.5

Coherent receiver for a long-distance optical-amplified system with accumulated ASE noise.

As heterodyne and homodyne detection techniques are known to have 6 dB SNR difference, we use homodyne detection for the following analysis without losing generality. As illustrated in Fig. 9.2.5, assume that the signal optical power is P_s and the power spectral density of the ASE noise at the signal wavelength is ρ_{ASE}. If the optical bandwidth is sufficiently narrow, the ASE-ASE beat noise, which is linearly proportional to the optical bandwidth, is much lower than the signal-ASE beat noise. In this case, the electric signal power of coherent homodyne detection is: $\langle i^2(t) \rangle = \mathfrak{R}^2 P_s(t) \cdot P_{LO}$, while the signal-ASE beat noise power with bandwidth B_e is $\mathfrak{R}^2(P_{LO}+P_s/2)\rho_{ASE}B_e$, where we have, again, considered a 3 dB attenuation due to the optical coupler for both the signal P_s and the ASE noise spectral density ρ_{ASE}. Note that in a coherent detection receiver, the strong LO also participates in the generation of signal-ASE beat noise. The electric SNR at the receiver output is

$$SNR_{\text{homodyne}} = \frac{\mathfrak{R}^2 P_{LO}P_s}{2\mathfrak{R}^2(P_{LO}+P_s/2)\rho_{ASE}B_e} \approx \frac{P_s}{\rho_{ASE}B_e} \tag{9.2.13}$$

where $P_{LO} \gg P_s$ is assumed, and the unit of SNR is $[\text{Hz}^{-1}]$. It is worthwhile to point out that the electrical *SNR* at the receiver output is related to the *OSNR* at the receiver input by

$$SNR_{\text{homodyne}} = \left(\frac{P_s}{\rho_{ASE}}\right)\frac{1}{B_e} = \left(\frac{OSNR}{B_e}\right) \tag{9.2.14}$$

where $OSNR = P_s/\rho_{ASE}$. Because the unit of ASE noise spectral density of ρ_{ASE} is in [W/Hz], the unit of *OSNR* is [Hz].

Practically, *OSNR* can be measured with an optical spectrum analyzer (OSA), and the value reported by the OSA is in [W/RB], where *RB* stands for resolution bandwidth. A 0.1 nm is commonly used by the industry as the OSA optical resolution bandwidth to define the *OSNR*. As 0.1 nm in wavelength window is equivalent to 12.5 GHz in terms of frequency interval in the 1550 nm telecommunication wavelength, OSNR measured with 0.1 nm OSA resolution bandwidth can be converted to OSNR of 1 Hz bandwidth by $OSNR_{0.1\text{nm}} = OSNR/(1.25 \times 10^{10})$.

In comparison, for a direct detection receiver, if the input optical signal P_s is associated with significantly high ASE noise spectral density ρ_{ASE}, so that signal-ASE beat noise is higher than the thermal noise and shot noise, the receiver SNR is

$$SNR_{dir} = \frac{\mathfrak{R}^2 P_s^2}{2\mathfrak{R}^2 P_s \rho_{ASE} B_e} = \frac{P_s}{2\rho_{ASE} B_e} = \frac{OSNR}{2B_e} \tag{9.2.15}$$

which is only 3 dB lower than coherent homodyne detection. This also indicates that coherent detection receiver in an optically amplified system does not significantly improve the required OSNR compared to direct detection.

It is important to note that in the SNR analysis so far in this chapter, signal optical power is the parameter to be detected. For the detection of signal optical power, coherent detection improves the SNR compared to direct detection when the received signal optical power level is low. A strong LO in a coherent detection receiver effectively amplifies the optical signal, and shot noise caused by LO on the photodiode becomes the dominant noise source, so that quantum limited detection sensitivity can be reached. When the signal optical power level is high enough in systems employing optical amplifiers, the sensitivity advantage of coherent detection diminishes.

Nevertheless, the unique advantage of coherent detection is the ability of recovering the signal optical phase. The information capacity and optical bandwidth efficiency of complex optical field modulation can be much higher than only using intensity modulation.

9.3 BALANCED COHERENT DETECTION AND POLARIZATION DIVERSITY

Coherent detection using the simple configuration shown in Fig. 9.1.2 has two major problems even for the detection of only signal intensity. First, the direct detection terms of input signal and LO are not useful, which may interfere with and cause degradation of the useful coherent detection term. Second, SOP mismatch between the received optical signal and the LO may cause the reduction of coherent mixing efficiency. As the SOP of the received optical signal is often random and unpredictable after a long transmission fiber, random signal fading may happen if the SOP of the LO does not follow that of the received optical signal. In this section, we discuss basic techniques often used to eliminate direct detection term and to avoid signal fading caused by SOP mismatch in a coherent detection receiver.

9.3.1 BALANCED COHERENT DETECTION

A coherent detection receiver based on a single photodiode produces both the mixed-frequency term and the unwanted direct-detection terms shown in Eq. (9.1.6). In the analysis of the last section, we assumed that the LO has no intensity noise, and the direct-detection term of the signal is small enough and negligible. In practice, however, the direct-detection terms may overlap with the IF spectrum of coherent detection introducing significant cross-talk. In addition, since the optical power of the LO is significantly higher than the received optical signal, any relative intensity noise of the LO would introduce excessive noise in coherent detection. Another important concern of coherent detection is that a strong LO may produce large photocurrent in the photodiode which may saturate the electric preamplifier after photodetection. Thus, the first two direct detection terms in Eq. (9.1.6) need to be eliminated in a high-quality coherent detection receiver, and balanced coherent detection configuration is often used for this purpose (Painchaud et al., 2009).

The schematic diagram of a balanced coherent detection is shown in Fig. 9.3.1A. Instead of using a single photodiode as shown in Fig. 9.1.2, two head-to-toe photodiodes are used in this balanced coherent detection configuration and the electric circuit is illustrated in Fig. 9.3.1B.

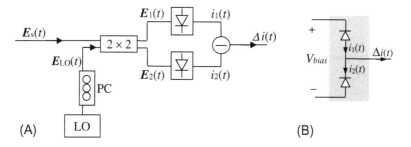

FIG. 9.3.1

(A) Block diagram of balanced coherent detection and (B) electric circuit of a balanced photodiode.

Based on the transfer function of a 2×2 optical coupler shown in Eq. (9.1.4), the optical fields at the two output ports are

$$\vec{E}_1(t) = \left[\vec{E}_s(t) + j\vec{E}_{Lo}(t)\right]/\sqrt{2} \tag{9.3.1}$$

$$\vec{E}_2(t) = j\left[\vec{E}_s(t) - j\vec{E}_{Lo}(t)\right]/\sqrt{2} \tag{9.3.2}$$

and the photocurrents generated by the two photodiodes are

$$i_1(t) = \frac{1}{2}\Re\left[|A_s(t)|^2 + |A_{LO}|^2 + 2A_s(t) \cdot A_{LO} \sin(\omega_{IF}t + \Delta\varphi)\cos\theta\right] \tag{9.3.3}$$

$$i_2(t) = \frac{1}{2}\Re\left[|A_s(t)|^2 + |A_{LO}|^2 - 2A_s(t) \cdot A_{LO} \sin(\omega_{IF}t + \Delta\varphi)\cos\theta\right] \tag{9.3.4}$$

Therefore, the direct-detection components can be eliminated by subtracting these two photocurrents, and we obtain

$$\Delta i(t) = i_1(t) - i_2(t) = 2\Re A_s(t)A_{LO} \sin(\omega_{IF}t + \Delta\varphi)\cos\theta \tag{9.3.5}$$

In the derivation of Eqs. (9.3.1)–(9.3.5), we assumed that the power coupling coefficient of the optical coupler is $\varepsilon = 0.5$ (3-dB coupler) and the two photodetectors have identical responsivities. In system applications, paired photodiodes with identical characteristics are connected in the head-to-toe electric configuration shown in Fig. 9.3.1B, and they are tightly packaged together and commercially available for applications in coherent detection receivers. Practically, more than 30 dB rejection ratio of direct detection terms can be achieved, which is known as the common-mode rejection ratio.

9.3.2 POLARIZATION DIVERSITY

Balanced detection using two photodiodes can eliminate direct detection components in a coherent detection receiver as shown in Eq. (9.3.5). However, the efficiency of coherent mixing between the received optical signal and the LO can still be affected by their SOP mismatch. The impact of polarization mismatch on the coherent detection efficiency is represented by the factor $\cos\theta$ in Eq. (9.3.5), where θ is the angle of SOP mismatch between the received optical signal and the LO. Although the SOP of the LO can be made stable because it is produced locally in the receiver, the SOP of the received optical signal is usually not predictable after the transmission over long distance of optical fiber. As a result, the useful photocurrent term may fluctuate over time causing IF signal fading in the receiver. Polarization diversity is a common and effective technique that overcomes polarization mismatch-induced signal fading in a coherent receiver.

One possible way to correct the polarization mismatch between signal and LO is to use a polarization controller to adjust the SOP of the LO (Kazovsky, 1989). An active feedback control has to be employed to make sure that the LO follows the SOP variation of the received optical signal. Various algorithms have also been developed to optimize the process of the endless polarization control.

However, the most practical technique commonly used to combat signal fading caused by polarization mismatch is the polarization diversity, which does not require active feedback and endless polarization control. The block diagram of polarization diversity in a coherent receiver is shown in Fig. 9.3.2, where two polarization beam splitters (PBSs) are used to separate the input optical signal and LO into horizontal ($E_{s//}$) and vertical ($E_{s\perp}$) polarization components. The polarization state of the LO is aligned midway between the two principle axis of the PBS such that optical power of the LO is equally split between the two outputs of the PBS, that is, $E_{LO//} = E_{LO\perp} = E_{LO}/\sqrt{2}$. Two balanced-photodiode modules are used to detect signals carried by the two orthogonal polarization components. As a result, the photocurrents at the output of the two balanced photodetection branches are

$$\Delta i_1(t) = \sqrt{2} \Re A_s(t) A_{LO} \sin(\omega_{IF} t + \Delta\varphi) \cos\theta \tag{9.3.6}$$

$$\Delta i_2(t) = \sqrt{2} \Re A_s(t) A_{LO} \sin(\omega_{IF} t + \Delta\varphi) \sin\theta \tag{9.3.7}$$

where θ is the angle between the polarization state of the input optical signal and the principle axis of the PBS, $E_{s//} = E_s \cos\theta$ and $E_{s\perp} = E_s \sin\theta$.

Both photocurrents are then squared by the RF power detectors before they combine to produce the RF power of the coherently detected signal,

$$P_{IF}(t) = \Delta i_1^2 + \Delta i_2^2 = 2\Re^2 P_s(t) P_{LO} \sin^2(\omega_{IF} t + \Delta\varphi) \tag{9.3.8}$$

The impact of polarization angle θ is removed by this square-and-add process, and thus the RF power is independent of the SOP of the input optical signal. No polarization tracking based on the feedback control is required in this configuration, so that it is reliable.

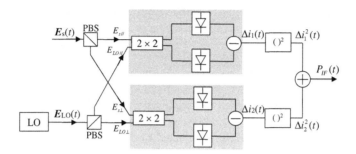

FIG. 9.3.2

Block diagram of balanced coherent detection with polarization diversity.

9.4 PHASE DIVERSITY AND I/Q DETECTION

In an optical system with coherent detection, the photocurrent is linearly proportional to the signal optical field and thus the phase information of the optical signal is preserved. However, phase noise exists in both the LO and the received optical signal, so that the differential phase term $\Delta\phi(t)$ may vary randomly. Optical phase noise may not be detrimental for coherent heterodyne detection with RF power detection, but is critically important for coherent homodyne detection where the useful photocurrent is directly related to the differential phase term, $i(t) = \Re A_s(t) \cdot A_{LO} \sin \Delta\varphi(t)$ as given by Eq. (9.2.2). In a coherent system, phase variation $\Delta\phi(t) = \Delta\phi_s(t) + \Delta\phi_n(t)$ includes contributions from optical phase modulation $\Delta\phi_s(t)$ which carries the useful information and the contribution due to phase noise, $\Delta\phi_n(t)$, of the transmitter and the LO.

In practice, a coherent homodyne system requires narrow linewidth lasers with low phase noise for both the transmitter and the LO so that the variation of $\Delta\phi_n(t)$ is much slower compared to the high speed signal $\Delta\phi_s(t)$. Thus, they can be isolated in the frequency domain and the impact due to phase noise $\Delta\phi_n(t)$ can be eliminated.

Traditionally, a phase-locked loop can be used to track the low-frequency phase noise $\Delta\phi_n(t)$ and to overcome the signal fading problem. Fig. 9.4.1 schematically shows a phase-locked loop, in which a beam splitter taps a small portion of the output RF power for the feedback loop. A low-pass filter averages out high-speed data contributions in the photocurrent due to $\Delta\phi_s(t)$, and allows the slow varying envelope of $\Delta\phi_n(t)$ to be detected by a low-frequency RF power detector which is used as the target for a phase control unit. The optical phase of the LO is adjusted by the feedback signal of the phase control unit to maximize the RF power level, which ensures that $\sin\Delta\varphi_n(t) = 1$.

Phase-locked loops have been used in coherent detection systems for many years. However, its biggest disadvantage is the requirement of an adaptive control system, including expensive electrical and optical devices and the complexity of the control algorithms, as well as the stability concern. With the rapid advance of photonic devices, and the availability of 90 degree optical hybrid, phase diversity becomes more practical than phase-locked loop (Davis et al., 1987).

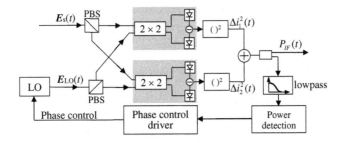

FIG. 9.4.1

Coherent homodyne detection with a phase-locked feedback loop.

As shown in Fig. 9.4.2, phase diversity coherent receiver is based on a 90 degree optical hybrid coupler. It is well known that the transfer matrix of a conventional 3-dB 2×2 optical fiber coupler, that relates the output optical fields $\vec{E}_1$ and $\vec{E}_2$ with the input fields $\vec{E}_s$ and $\vec{E}_{LO}$, is given by

$$\begin{bmatrix} \vec{E}_1 \\ \vec{E}_2 \end{bmatrix} = \frac{1}{\sqrt{2}} \begin{bmatrix} 1 & j \\ j & 1 \end{bmatrix} \begin{bmatrix} \vec{E}_s \\ \vec{E}_{LO} \end{bmatrix} \tag{9.4.1}$$

In the balanced detection coherent receiver shown in Fig. 9.3.1 using a 3-dB, 2×2 fiber coupler, the cross-products are 180 degree apart in the two output arms after photodetectors, as indicated by Eqs. (9.3.3) and (9.3.4). This means that a conventional 2×2 fiber coupler can be, in fact, classified as a 180 degree hybrid.

Theoretically, an ideal 90 degree optical hybrid coupler should have a transfer matrix as

$$\begin{bmatrix} \vec{E}_1 \\ \vec{E}_2 \end{bmatrix} = \frac{1}{\sqrt{2}} \begin{bmatrix} 1 & \exp(j\pi/4) \\ \exp(j\pi/4) & 1 \end{bmatrix} \begin{bmatrix} \vec{E}_s \\ \vec{E}_{LO} \end{bmatrix} \tag{9.4.2}$$

For a simple coherent receiver with balanced detection shown in Fig. 9.3.1A, replacing the 2×2 fiber coupler with a 90 degree hybrid coupler as shown in Fig. 9.4.2, the optical fields at the 90 degree hybrid coupler output will be

$$E_1(t) = E_s(t) + E_{LO}e^{j\pi/4} \tag{9.4.3}$$

$$E_2(t) = E_s(t)e^{j\pi/4} + E_{LO} \tag{9.4.4}$$

Here, we have assumed perfectly matched polarization states between the input optical signal and the LO.

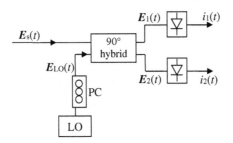

FIG. 9.4.2

Coherent detection based on a 90 degree hybrid coupler.

The corresponding photocurrents are thus,

$$i_1(t) = \Re|E_1|^2 = \Re\left\{|A_s(t)|^2 + |A_{LO}|^2 + A_s(t) \cdot A_{LO} \cos\left(\omega_{IF}t + \Delta\varphi(t) - \frac{\pi}{4}\right)\right\} \qquad (9.4.5)$$

$$i_2(t) = \Re|E_2|^2 = \Re\left\{|A_s(t)|^2 + |A_{LO}|^2 + A_s(t) \cdot A_{LO} \cos\left(\omega_{IF}t + \Delta\varphi(t) + \frac{\pi}{4}\right)\right\} \qquad (9.4.6)$$

where we have assumed that $E_s(t) = A_s(t)e^{j\omega_s t + j\varphi_s(t)}$, $E_{LO}(t) = A_{LO}e^{j\omega_{LO}t + j\varphi_{LO}(t)}$, $\omega_{IF} = \omega_s - \omega_{LO}$, and $\Delta\varphi(t) = \varphi_s(t) - \varphi_{LO}(t)$. Obviously, there is a quadrature relationship between the coherent detection terms in $i_1(t)$ and $i_2(t)$. If we neglect the direct detection terms in Eqs. (9.4.5) and (9.4.6), a square-and-add operation will result in,

$$i_1^2(t) + i_2^2(t) = \Re^2 P(t) \cdot P_{LO} \qquad (9.4.7)$$

which is independent of the differential phase $\Delta\varphi(t)$.

It is clear that in the phase diversity receiver configuration shown in Fig. 9.4.2, the key device is the 90 degree optical hybrid. Unfortunately, the transfer matrix shown in Eq. (9.4.2) cannot be provided by a simple 2×2 fiber directional coupler. In fact, a quick test shows that the transfer function of Eq. (9.4.2) does not even satisfy the energy conservation principle, because $|E_1|^2 + |E_2|^2 \neq |E_s|^2 + |E_{LO}|^2$.

A number of optical structures have been proposed to realize the 90 degree optical hybrid. An example of fiber-optic realization of 90 degree hybrid is a specially designed 3×3 fiber coupler with the following transfer matrix (Epworth, 2005):

$$\begin{bmatrix} E_1 \\ E_2 \\ E_3 \end{bmatrix} = \begin{bmatrix} \sqrt{0.2} & \sqrt{0.4}\exp\left(j\frac{3\pi}{4}\right) & \sqrt{0.4}\exp\left(j\frac{3\pi}{4}\right) \\ \sqrt{0.4}\exp\left(j\frac{3\pi}{4}\right) & \sqrt{0.2} & \sqrt{0.4}\exp\left(j\frac{3\pi}{4}\right) \\ \sqrt{0.4}\exp\left(j\frac{3\pi}{4}\right) & \sqrt{0.4}\exp\left(j\frac{3\pi}{4}\right) & \sqrt{0.2} \end{bmatrix} \begin{bmatrix} E_s \\ E_{LO} \\ 0 \end{bmatrix} \qquad (9.4.8)$$

where only two of the three ports are used at each side of the coupler and therefore, it is used as a 2×2 coupler. On the input side, the selected two ports connect to the input optical signal and the LO, whereas the first two output ports provide

$$E_{01} = \left(\sqrt{0.2}E_s(t) + \sqrt{0.4}\exp\left(j\frac{3\pi}{4}\right)E_{LO}\right) \qquad (9.4.9)$$

$$E_{02} = \left(\sqrt{0.4}\exp\left(j\frac{3\pi}{4}\right)E_s(t) + \sqrt{0.2}E_{LO}\right) \qquad (9.4.10)$$

After photodiodes and neglecting the direct detection components, the AC parts of these two photocurrents are

$$i_1(t) = 2\Re\sqrt{0.08}A_s(t)A_{LO}\cos\left(\Delta\varphi(t) - \frac{3\pi}{4}\right) \qquad (9.4.11)$$

$$i_2(t) = 2\Re\sqrt{0.08}A_s(t)A_{LO}\sin\left(\Delta\varphi(t) - \frac{3\pi}{4}\right) \qquad (9.4.12)$$

Therefore, after squaring and combining, the receiver output is

$$i_1^2(t) + i_2^2(t) = 0.32 \Re^2 P(t) \cdot P_{LO} \tag{9.4.13}$$

Compare this result with Eq. (9.4.2), where an ideal 90 degree hybrid was used. There is a signal RF power reduction of approximately 5 dB using the 3×3 coupler. Because one of the three output ports is not used, obviously a portion of the input optical power is dissipated through this port.

Another way to construct a 90 degree optical hybrid is to use an integrated-optics approach in which the optical phase shift can be precisely controlled. Similar to Eq. (9.4.2), the transfer function of a 90 degree optical hybrid can also be written as

$$\begin{bmatrix} \vec{E}_1 \\ \vec{E}_2 \end{bmatrix} = \frac{1}{\sqrt{2}} \begin{bmatrix} 1 & \exp(j\pi/2) \\ 1 & 1 \end{bmatrix} \begin{bmatrix} \vec{E}_S \\ \vec{E}_{LO} \end{bmatrix} \tag{9.4.14}$$

which will result in the quadrature relation of photocurrents at the two output ports of the coupler after photodetection similar to that shown in Eqs. (9.4.5) and (9.4.6).

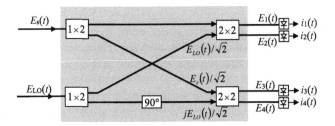

FIG. 9.4.3

A 2×4 90 degree hybrid coupler based on a 90 degree optical phase shifter.

Fig. 9.4.3 shows a 90 degree optical hybrid based on a $\lambda/4$ phase shifter on one of the four branching waveguides between the four couplers. This configuration is not feasible to be implemented by fiber optics because the fiber length control cannot be precise enough to realize a $\lambda/4$ (or equivalently 90 degree) extra optical phase shift in one of the fibers compared to the other three. But planar lightwave circuits (PLCs) made with photolithography patterning and etching can create guided-wave optical circuit structures with a high level of precision (Dong et al., 2014).

For the structure shown in Fig. 9.4.3, the top 2×2 coupler performs a balanced coherent mixing between the signal and the LO the same way as that in Fig. 9.3.1. Assume perfectly matched SOP between the signal and the LO, we have

$$i_1(t) = \frac{1}{4} \Re \left[|A_s(t)|^2 + |A_{LO}|^2 + 2A_s(t) \cdot A_{LO} \sin(\omega_{IF} t + \Delta\varphi) \right] \tag{9.4.15}$$

$$i_2(t) = \frac{1}{4} \Re \left[|A_s(t)|^2 + |A_{LO}|^2 - 2A_s(t) \cdot A_{LO} \sin(\omega_{IF} t + \Delta\varphi) \right] \tag{9.4.16}$$

whereas for the bottom 2×2 coupler in Fig. 9.4.3, the two outputs are

$$E_3(t) = \{E_s(t) + j[jE_{LO}(t)]\}/2,$$

$$E_4(t) = j\{E_s(t) + E_{LO}(t)\}/2.$$

So that,

$$i_3(t) = \frac{1}{4}\Re\left[|A_s(t)|^2 + |A_{LO}|^2 - 2A_s(t) \cdot A_{LO}\cos(\omega_{IF}t + \Delta\varphi)\right] \qquad (9.4.17)$$

$$i_4(t) = \frac{1}{4}\Re\left[|A_s(t)|^2 + |A_{LO}|^2 + 2A_s(t) \cdot A_{LO}\cos(\omega_{IF}t + \Delta\varphi)\right] \qquad (9.4.18)$$

Combining Eqs. (9.4.15) through (9.4.18), we have

$$\Delta i_Q(t) = i_1(t) - i_2(t) = \Re A_s(t) \cdot A_{LO}\sin(\omega_{IF}t + \Delta\varphi) \qquad (9.4.19)$$

$$\Delta i_I(t) = i_4(t) - i_3(t) = \Re A_s(t) \cdot A_{LO}\cos(\omega_{IF}t + \Delta\varphi) \qquad (9.4.20)$$

This 2×4 90 degree hybrid coupler combines balanced coherent detection and phase diversity into the same device.

With the rapid progress of PLC and electro-optic integration technologies, integrated coherent receivers become commercially available and standardized by the industry. As shown in Fig. 9.4.4, an integrated coherent receiver combines polarization diversity, phase diversity, balanced detection, and trans-impedance amplifiers (TIA) into a hermetically sealed small package with fiber pigtails for the connection to the optical signal and the LO. The x- and the y-polarization components of the input optical signal are separately detected through polarization diversity as described in Fig. 9.3.2. Each of the two orthogonal polarization components is coherently mixed with the LO through a 2×4 90 degree hybrid with the optical configuration described in Fig. 9.4.3.

Four photodiodes are used for the detection of each polarization component producing four photocurrent signals defined by Eqs. (9.4.15)–(9.4.18). Direct detection components are eliminated by dif-

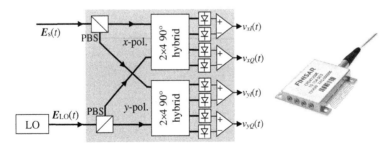

FIG. 9.4.4

Left: coherent optical receiver module including polarization diversity, phase diversity, balanced detection, and TIA for differential amplification. *Right*: photograph of fiber-pigtailed coherent optical receiver packages.

(Right) Used with permission from Finisar.

ferential amplification using TIAs to produce in-phase (I) and quadrature (Q) voltage signals proportional to $A_s(t)A_{LO}\cos[\omega_{IF}t + \Delta\varphi(t)]$ and $A_s(t)A_{LO}\sin[\omega_{IF}t + \Delta\varphi(t)]$, respectively. Information carried by $A_s(t)$ and $\Delta\varphi(t)$ can then be recovered through RF or digital signal processing (DSP) of $v_I(t)$ and $v_Q(t)$ of each polarization. The complex optical field of the optical signal can be recovered by

$$E_s(t) = v_I(t) + jv_Q(t) \tag{9.4.21}$$

In coherent detection, since the complex field of the received optical signal can be obtained through the amplitude and phase information, compensation of chromatic dispersion of the transmission fiber can be performed in the electric domain after photodetection. In addition, because the x- and y-polarized optical signal components are detected simultaneously, instantaneous SOP information of the received optical signal can be obtained, which allows the compensation of polarization mode dispersion (PMD) in the electric domain. This also allows polarization-division multiplexing (PDM) in which each orthogonal polarization state carries an independent information channel to double the overall optical bandwidth efficiency.

Coherent detection enables the complex optical field, including both the intensity and the phase, to be utilized for information delivery. Receiver sensitivity and required OSNR of phase-modulated optical systems can also be improved in comparison to intensity modulated optical systems with coherent detection. More importantly, optical bandwidth efficiency of complex optical field modulation can be much higher than

9.5 CONCLUSION

In this chapter, we have discussed basic principles of coherent detection and various techniques to improve the performance of coherent detection in fiber-optic communication systems. By mixing with a strong optical LO in the coherent detection receiver, the received optical signal is effectively amplified, which improves the receiver sensitivity in comparison to a direct detection receiver. Thanks to the optical phase reference provided by the LO, the phase information of the received optical signal can be detected which allows optical phase modulation to be used in optical transmission systems employing coherent detection. As coherent detection is capable of recovering the complex field of the received optical signal, transmission impairments induced by the optical fibers including chromatic dispersion and PMD can be effectively compensated through electric domain signal processing. Although receiver sensitivity of coherent detection is comparable to a direct-detection receiver employing an optical preamplifier, the flexibility of complex optical field modulation and the increased bandwidth efficiency make coherent optical systems superior to the direct detection counterpart.

Classic techniques used to improve the performance of coherent detection include balanced photodetection, polarization diversity, and phase diversity. Balanced photodetection is effective in eliminating the direct detection components and their interference with the useful coherent detection term. As the SOP of the received optical signal is often unpredictable, and coherent mixing between the signal and the LO is polarization selective, minimizing SOP mismatch between the optical signal and the LO is important for coherent detection. Polarization diversity is a technique that partitions the received optical signal into two orthogonal SOPs through a PBS, and detects them separately. This eliminates the polarization sensitivity of coherent detection when the two photocurrent signals are combined. It also allows the reconstruction of the SOP of the input optical signal so that compensation of PMD is possible through electric processing of the two photocurrents. Phase diversity is a technique that allows the recovery of optical phase information from the received optical signal. Phase diversity utilizes a 90 degree optical hybrid so that the in-phase and the quadrature components can be separately detected to reconstruct the received complex optical field.

Coherent optical communication has been proposed and investigated several decades ago, but only started to gain momentum in the recent years mainly because of the two key technological advances. First, narrow linewidth (<100 kHz) semiconductor lasers became mature enough to meet the telecommunication's standard with small footprint and relatively low cost. Second, advanced PLC technology allowed the creation of coherent optical receiver which integrates polarization diversity, phase diversity, and balanced detection photodiodes into a single package with electronic preamplifiers. In addition, advances in digital electronics based on the silicon integrated circuit technology enabled high-speed DSP, so that advanced modulation formats and electronic domain compensation of various transmission impairments can be performed through DSP. More details of advanced optical modulation formats and phase-modulated coherent transmission performance will be discussed in the next chapter.

PROBLEMS

1. In a coherent detection optical receiver, assume the power coupling coefficient of the 2×2 coupler is ε. Find the optimum coupling coefficient ε to achieve the highest coherent detection efficiency.

   Does this optimum coupling coefficient depend on the power ratio between the LO and the received optical signal?

2. In order for a coherent receiver to operate properly, the power of the LO has to be high enough so that shot noise generated by the LO is much higher than thermal noise. What is the LO power required for the LO-induced shot noise to be 10 times the thermal noise at room temperature? Assume receiver load resistance is $R_L = 50\,\Omega$ and photodiode responsivity is $\mathfrak{R} = 1\text{mA/mW}$.

3. Consider a coherent homodyne receiver with 0 dBm phase-locked LO, and a direct detection receiver shown in the following figure. Both receivers use PIN photodiodes with 0.75 A/W responsivity and a negligible dark current at the room temperature, and 40 GHz electrical bandwidth. The load resistor is 50 Ω.

   (a) Please find the signal optical power required to achieve the electrical SNR of 20 dB for coherent detection and direct detection, respectively.
   (b) What is the signal optical power level below which coherent detection provides better SNR compared to direct detection?

4. All parameters are the same as in problem 2, but now consider an on-off key binary modulated optical signal in which $P_{s,1} = 2P_{s,ave}$, with $P_{s,1}$ and $P_{s,ave}$ signal optical power at signal "1" and average signal optical power, respectively. Assume there is no signal waveform distortion.
   (a) Please find the average signal optical power required to achieve the receiver $Q = 8$ for coherent homodyne detection and direct detection, respectively.

(b) What is the average signal optical power level at which coherent detection has the same Q value as direct detection?

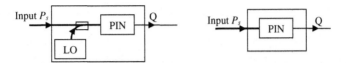

5. Consider an on-off key binary modulated optical system without waveform distortion. Both coherent homodyne receiver and optically preamplified direct detection can be used as shown in the following figure. Operating wavelength is 1550 nm and photodiode has 100% quantum efficiency. For the coherent detection, assume LO is strong enough so that LO-induced shot noise is the dominate noise. For optically preamplified receiver, assume signal ASE beat noise is the dominate noise source, and the optical gain of the amplifier is very high so that $G \gg 1$.

What would be the noise figure of the optical preamplifier for these two receivers to have the same SNR for at the same average signal optical power?

(Note: SNR is evaluated at signal "1," and assume that signal power at "1" is twice the average signal power $P_{s,ave}$)

6. Consider a coherent homodyne receiver used in a binary intensity modulated system with multiple inline optical amplifiers at 1550 nm wavelength. Assume the modulation data rate is 40 Gb/s and the receiver electrical bandwidth is 40 GHz. Neglect signal waveform distortion, and only consider LO-ASE beat noise, what is the required OSNR in [dB · 0.1nm] to achieve a Q value of 7?

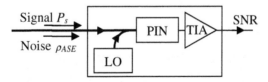

7. Consider an optical system with inline optical amplifiers, the signal optical power is P_s and the optical noise power spectral density is ρ_{ASE}. If a coherent homodyne receiver is used, Eq. (9.2.14) indicates that the electrical SNR at the receiver output is $SNR = OSNR/B_e$. Now an optical preamplifier is inserted as shown in the following figure with an optical gain G. Assume that the LO power is high enough so that the major noise source is due to the mixing between LO and ASE.

Does this optical preamplifier help improve system performance? Please explain the reason by deriving necessary equations.

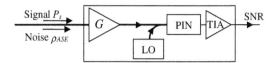

8. For a coherent receiver, the relative intensive noise (RIN) of the LO is $-140\,\text{dB/Hz}$ within the receiver bandwidth. If the photodiode responsivity is $1\,\text{mA/mW}$, what the LO power such that the RIN noise is equal to the shot noise generated by the LO.

9. Consider coherent homodyne detection based on a 2×4 90 degree hybrid coupler as shown in Fig. 9.4.3. If the actual value of the 90 degree optical phase shifter is $90° + \delta$, where $\delta = 10°$ is a fabrication error, what is the maximum percentage of signal power variation $\Delta i_I^2 + \Delta i_Q^2$ with the random variation of $\Delta \varphi$ [reference to Eqs. (9.4.19) and (9.4.20)]?

10. Same coherent homodyne receiver as in problem 9, but the optical signal is QPSK modulated with four constellation points at ($\pi/4$, $3\pi/4$, $5\pi/4$, $7\pi/4$). Because of the $\delta = 10°$ error of the 90° phase shifter, what are the angles of the detected constellation points by this I/Q receiver?

REFERENCES

Betti, S., De Marchis, G., Iannone, E., 1995. Coherent Optical Communications Systems, first ed. Wiley-Interscience.

Davis, A., Pettitt, M., King, J., Wright, S., 1987. Phase diversity techniques for coherent optical receivers. J. Lightwave Technol. 5 (4), 561–572.

Dong, P., Liu, X., Chandrasekhar, S., Buhl, L.L., Aroca, R., Chen, Y.-K., 2014. Monolithic silicon photonic integrated circuits for compact 100^+ Gb/s coherent optical receivers and transmitters. IEEE J. Sel. Top. Quantum Electron. 20 (4) paper 6100108.

R. Epworth, 3 Fibre I and Q Coupler, US patent #6,859,586, 2005

Kazovsky, L.G., 1989. Phase- and polarization-diversity coherent optical techniques. J. Lightwave Technol. 7 (2), 279–292.

Kikuchi, K., 2016. Fundamentals of coherent optical fiber communications. J. Lightwave Technol. 34 (1), 157–178.

Painchaud, Y., Poulin, M., Morin, M., Têtu, M., 2009. Performance of balanced detection in a coherent receiver. Opt. Express 17 (5), 3659–3672.

Yamamoto, Y., Kimura, T., 1981. Coherent optical fiber transmission systems. IEEE J. Quantum Electron. 17 (6), 919–935.

FURTHER READING

Finisar: 40 GHz Integrated High-speed Coherent Photodetector. https://www.finisar.com/communication-components/cpdv1200r

MODULATION FORMATS FOR OPTICAL COMMUNICATIONS

10

CHAPTER OUTLINE

Introduction 439
10.1 Binary NRZ vs. RZ modulation formats 440
10.2 Generation of PRBS patterns and clock recovery 445
10.3 Polybinary, duobinary, and carrier-suppressed RZ modulation 449
    10.3.1. M-ary and Polybinary coding 450
    10.3.2. Duo-binary optical modulation 453
    10.3.3. Carrier-suppressed return-to-zero (CSRZ) 456
10.4 BPSK and DPSK optical systems 460
10.5 High-level PSK and QAM modulation 466
10.6 Analog optical systems and radio over fiber 472
    10.6.1. Analog subcarrier multiplexing and optical single-sideband modulation 473
    10.6.2. Carrier-to-signal ratio, inter-modulation distortion and clipping 478
    10.6.3. Impact of relative intensity noise 481
    10.6.4. Radio-over-fiber technology 483
10.7 Optical system link budgeting 486
    10.7.1. Power budgeting 486
    10.7.2. OSNR budgeting 487
10.8 Summary 489
Problems 490
References 494
Further reading 495

INTRODUCTION

The purpose of an optical communication system is to deliver information from the transmitter to the receiver using optical wave as the carrier. Encoding information onto the optical carrier with high bandwidth efficiency and resistance to transmission impairments is a critical issue which defines the system performance. Digital modulation is commonly used in high-speed transmission systems with superior

Introduction to Fiber-Optic Communications. https://doi.org/10.1016/B978-0-12-805345-4.00010-X

439

performance compared to analog modulation in terms of the tolerance to waveform distortion and noise, and it is suitable for regeneration without the impact of noise accumulation.

Various digital modulation formats have been used for telecommunications. The choice of modulation format depends on the transmission medium, transmission distance, the requirement of the bandwidth efficiency, and the nature to transmission impairments (Winzer and Essiambre, 2006; Lach and Idler, 2011). For example in wireless communication networks, spectrum resource is extremely scarce and expensive, and thus bandwidth efficiency is the most important concern in choosing the modulation format. On the other hand, an optical fiber has wide transparent wavelength window and once considered of having unlimited bandwidth available for optical transmission. Bandwidth efficiency was not a major concern in fiber-optic communication systems for many years. However, with the explosive growth of internet traffic and the rapidly increased demand for the transmission capacity, usable spectral bandwidth in a transmission fiber has become more and more valuable, especially in places where fiber installation is expensive. Thus, bandwidth efficient modulation becomes an important issue in fiber-optic systems and networks. In wireless networks, multipath interference, differential propagation delay, and antenna radiation pattern are major problems to deal with. Whereas in fiber-optic systems, chromatic dispersion, polarization mode dispersion (PMD) and accumulated spontaneous emission noise from in line optical amplifiers are unique concerns in selecting the most appropriate modulation format.

In this chapter, we will discuss basic characteristics of different optical modulation formats. Amplitude modulation formats, including nonreturn-to-zero (NRZ), return-to-zero (RZ), carrier-suppressed return-to-zero (CSRZ), and dual-binary, are often used with direct detection receiver. While complex optical field modulation formats such as quadrature phase shift keying (QPSK), differential QPSK (DQPSK), and quadrature amplitude modulation (QAM) are often used with coherent detection receiver. Bandwidth efficiency, tolerance to chromatic dispersion and PMD, receiver sensitivity, and the techniques of encoding and decoding of each modulation format will be discussed, and pros and cons will be compared.

10.1 BINARY NRZ VS. RZ MODULATION FORMATS

Binary NRZ and RZ are the most popular modulation formats often used in optical fiber systems based on intensity modulation and direct detection (Caspar et al., 1999; Wilson, 1996). In comparison to NRZ, a standard RZ modulation at the same data rate requires wider spectral bandwidth because the pulse width in the time domain only occupies half of the bit slot. As the result, RZ modulated optical signal suffers more from bandwidth dependent impairments such as chromatic dispersion polarization mode dispersion and transfer function ripples in broadband electric amplifiers. On the other hand, RZ modulated optical pulses have higher peak power compared to a NRZ waveform of the same average power, so that the signal-to-noise ratio (SNR) can be better than that of NRZ at the decision points. With the introduction of dispersion compensation which is commonly used in long-distance fiber-optic systems, the impact of chromatic dispersion can be minimized. To some extent, it is more efficient to optimize dispersion compensation for RZ modulated optical signals because of the regular pulse shape. Whereas, for NRZ modulated waveforms, the optimum value of dispersion compensation is often pattern dependent.

Fig. 10.1.1 shows the comparison between NRZ and RZ waveforms of the same data rate and data sequence. Each data bit occupies a time slot of width T, so that the data rate is $B = 1/T$. While digital "1" of a NRZ bit has a length T, digital "1" of a RZ bit returns to zero after $T/2$, so that the pulse width is only $T/2$.

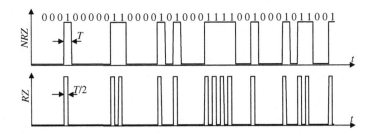

FIG. 10.1.1

Comparison between NRZ *(top)* and RZ *(bottom)* waveforms of the same data sequence and data rate.

The basic component of an ideal NRZ waveform in the time domain is a gate function with a width T. The Fourier transform of this time-domain gate function is a *sinc* function in the frequency domain, and the normalized power spectral density is,

$$S_{NRZ}(f) = \left[\frac{\sin(\pi Tf)}{\pi Tf} \right]^2 = \mathrm{sinc}^2(Tf)$$

with the first null at $f = 1/T$, which is equal to the data rate. Fig. 10.1.2A shows the normalized spectral density of a NRZ waveform at 10 Gb/s data rate so that $T = 10^{-10}$ s. For equal probability of "1"s and

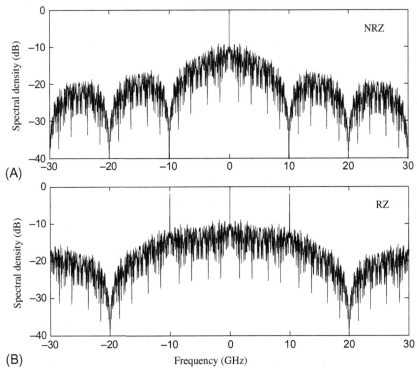

FIG. 10.1.2

Normalized spectral densities of binary NRZ (A) and RZ (B) waveforms with 10 Gb/s data rate.

"0"s in a pseudorandom data sequence, the average level of the signal amplitude is 50% of the peak so that there is a significant DC component shown as a sharp peak in the middle of the spectrum. Note that in the NRZ spectrum the clock component at $f = 1/T$ is completely suppressed because the *sinc* function has a null exactly at the clock frequency. This creates a problem for clock recovery in the receiver.

Fig. 10.1.2B shows the normalized spectral density of a RZ waveform also for 10 Gb/s data rate. Because the pulse width in the time domain is $T/2$, the normalized spectral density is $S_{RZ}(f) = \text{sinc}^2(Tf/2)$, in which the first null is at $f = 2/T$. In addition to the DC component, there is a strong clock component at $f = 1/T$, which can be easily extracted for clock recovery at the receiver. The strong clock component of RZ spectrum is due to the regularly shaped pulses at the fundamental repetition rate $1/T$ for continuous "1"s. In comparison, continuous "1"s in a NRZ waveform is represented by low-frequency components in the spectrum.

Ideal waveforms of NRZ and RZ both have infinitely wide spectral widths because of the rectangle pulse shape. In practical optical systems, these waveforms have to be band-limited through low-pass filtering. In a digital system, the minimally required bandwidth can be determined by having completely open eye diagram only at the decision point which is at the middle of the eye. Fig. 10.1.3 shows eye diagrams of low-pass filtered NRZ and RZ waveforms. For the NRZ eye diagram shown in Fig. 10.1.3A, a fifth-order Bessel filter is used with a 3 dB bandwidth of 6.5 GHz. The eye

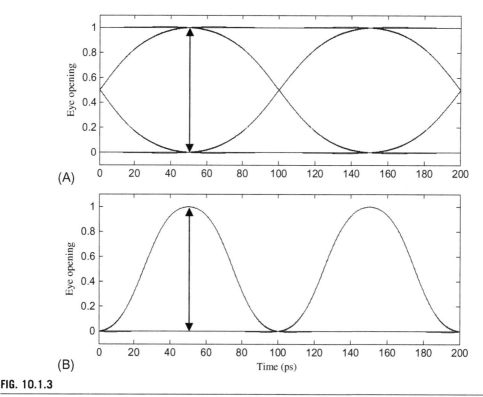

(A)

(B)

FIG. 10.1.3

Eye diagrams of bandwidth-limited NRZ (A) and RZ (B) waveforms at 10 Gb/s data rate. fifth-order Bessel filters are used with 6.5 GHz bandwidth for NRZ waveform and 13 GHz for the RZ waveform.

opening is complete at the middle of the bit, which is sufficient for the decision circuit in the receiver for data recovery.

For the eye diagram of RZ waveform shown in Fig. 10.1.3B, a similar fifth-order Bessel filter is used but with a 3 dB bandwidth of 13 GHz to maintain the completely open eye.

In practice, the eye opening is determined by the overall transfer function of the system including the transmitter, the receiver and the transmission medium. Chromatic dispersion of the fiber can often be treated as a low-pass filter in the system transfer function with the bandwidth determined by the accumulated dispersion and the spectral width of the optical signal. Note that the basic shapes of NRZ and RZ eye diagrams are quite different. As there is no continuous high level ("1" level) in the RZ eye diagram, the energy concentration at the decision phase is higher than that of a NRZ eye diagram.

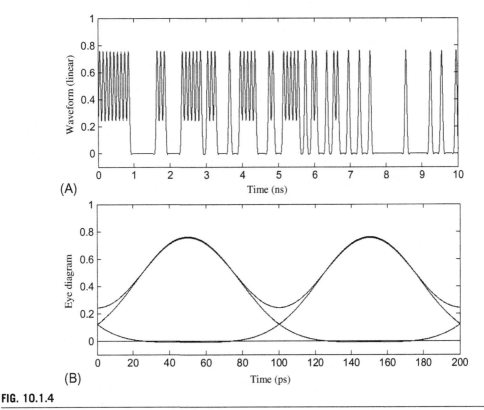

(A)

(B)

FIG. 10.1.4

Band-limited 10 Gb/s RZ waveform (A) and eye diagram (B). fifth-order Bessel filter is used with a 3 dB bandwidth of 6.5 GHz.

Neglect waveform distortion, the peak power P_1 at the signal "1" level is twice the average power P_{ave} for a NRZ waveform, that is $P_1 = 2P_{ave}$. While for a RZwaveform, this relation is $P_1 = 4P_{ave}$. As described in Section 8.3, with a fixed signal level the receiver Q value is inversely proportional to the square-root of the electric bandwidth B_e. If the system is dominated by the signal-independent noise, such as thermal noise, $Q \propto P_1/\sqrt{B_e}$. On the other hand, if the system is dominated by the signal-dependent noises, such as shot noise and signal-ASE beat noise, we have $Q \propto \sqrt{P_1/B_e}$. In comparison to a NRZ waveform with the same average signal power, a RZ waveform has twice the pulse peak power but also twice the electric bandwidth. Thus, RZ has 1.5 dB sensitivity improvement compared to NRZ for a signal-independent noise dominated system, whereas for a signal-dependent noise dominated system, their receiver sensitivities are about the same.

An interesting observation is that if passing through a low-pass filter with an electric bandwidth of much narrower than $2/T$, a RZ waveform can be turned into a waveform more similar to NRZ as the integration time becomes much longer than $T/2$. Fig. 10.1.4 shows an example of normalized 10 Gb/s RZ waveform which is low-pass filtered by a fifth-order Bessel filter with 6.5 GHz bandwidth. Although the same filter bandwidth is wide enough for an NRZ waveform to maintain a completely open eye as shown in Fig. 10.1.3A, it is not sufficient to keep the eye diagram completely open for the RZ waveform. Since the 6.5 GHz bandwidth is equivalent to an integration time of approximately $1.5T$ which is much longer than a RZ pulse length, the eye opening is reduced to about 76%. Nevertheless, although the RZ eye diagram is not completely open, the ratio between the pulse peak power and the average power is $P_1 \approx 3.2 P_{ave}$ which is still higher than that of NRZ even with a perfectly open eye diagram. Fig. 10.1.5A shows the signal peak to average power ratio P_1/P_{ave} as the function of the 3-dB bandwidth B_e of the fifth-order Bessel low-pass filter. Complete eye opening can be achieved with $B_e \geq 6.5$ GHz for NRZ, while $B_e \geq 13$ GHz is required for RZ.

Fig. 10.1.5B compares the normalized Q-values of NRZ and RZ waveforms for *signal-independent* noise dominated systems where $Q_{ind} \propto P_1/(P_{ave}\sqrt{B_e})$. In this figure, Q_{ind} has been normalized by its maximum value for the NRZ waveform obtained at a bandwidth of approximately 5.5 GHz. The maximum Q_{ind} for RZ is obtained at $B_e \approx 8$ GHz, which is about 1.8 dB higher than that of NRZ. For systems dominated by *signal-dependent* noise, the normalized Q-values, denoted as Q_{dep}, are shown in Fig. 10.1.5C based on $Q_{dep} \propto \sqrt{P_1/(P_{ave}B_e)}$. In this case, the Q improvement of using RZ is less than 1.5 dB in comparison to NRZ.

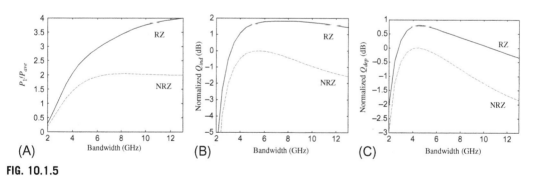

FIG. 10.1.5

(A) pulse peak to average power ratio of 10 Gb/s NRZ and RZ waveforms as the function of the low-pass filter bandwidth. (B) normalized Q-values of systems dominated by signal-independent noise, and (C) normalized Q-values of systems dominated by signal-dependent noise.

Note that in the calculation to obtain Fig. 10.1.5, we have assumed that both the signal and the broadband noise pass through the same filter of bandwidth B_e. But in practice, the signal eye opening is determined by the entire system transfer function including the transmitter, and the receiver. Whereas the noise created after the transmitter is filtered mainly by the receiver transfer function. Therefore, more accurate analysis has to consider specific system configurations.

10.2 GENERATION OF PRBS PATTERNS AND CLOCK RECOVERY

In commercial optical systems and networks, low rate data streams are aggregated into high-speed data rate through several layers of digital multiplexers for optical transmission. However in the development and performance evaluation of optical transmission systems and equipment, pseudorandom binary sequence (PRBS) is commonly used to mimic the actual digital data traffic in the systems. PRBS is a periodic bit pattern, but there are a large number of bit combinations within each period; they compromise between the randomness of the data signal carried in practical systems and the repetitiveness that simplifies the measurements. Major parameters to specify a PRBS bit pattern include sequence length in bits, maximum continuous digital "1"s, and maximum continuous digital "0"s. Spectral density of modulated optical signal not only depends on the specific modulation format but also depends on the PRBS pattern.

By definition, *pseudo*-random implies that the pattern is not really "random" and in fact it is only quasi-random. A PRBS pattern generator generates a random pattern with a certain length and the pattern repeats itself after every pattern length, as illustrated in Fig. 10.2.1. Within each bit pattern, the combination of bits should be as random as possible to simulate actual digital data traffic. This requires the length of the pattern to be long enough. Generally, a long pattern length allows the use of longer continuous "1"s and continuous "0"s. This helps stretch the test to the worst case of the system.

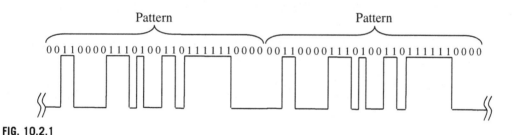

FIG. 10.2.1

Example of a NRZ PRBS pattern.

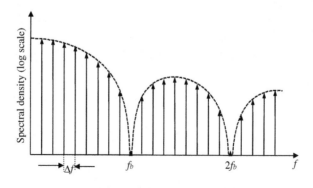

FIG. 10.2.2

Illustration of the spectrum of a PRBS signal with NRZ modulation.

For example, for NRZ modulation, a clock recovery circuit usually works well for a pattern with alternative "0"s and "1"s but not as well for long continuous "0"s or "1"s which only contribute to low-frequency components. In addition, the spectrum of a PRBS in the frequency domain consists of an envelope that is a *sinc* function determined by the time-domain waveform of each individual bit and bit duration (e.g., 100 ps for a 10 Gb/s data-rate), whereas the frequency spacing between adjacent spectral components is determined by the inverse of the pattern length, $\Delta f = f_b/N_b$, as shown in Fig. 10.2.2, where f_b is the bit rate and N_b is the pattern length (the number of bits per pattern). Obviously, a longer pattern length results in a narrower spacing between spectral lines, which is equivalent to a more closely spaced sampling in the frequency domain. From the system transfer function point of view, a probe signal with densely spaced frequency components would be desirable for system performance evaluation. For example, if the data rate is 10 Gb/s, to probe system transfer function at frequencies as low as 100 kHz the number of bits in each PRBS pattern has to be at least 10^5.

PRBS patterns have been standardized by the ITU for testing digital transmission systems. The most commonly used patterns in digital transmission system testing are $(2^N - 1)$ with $N = 7$, 10, 15, 20, 23, and 31. The corresponding pattern length (sometimes referred to as *word length*) is 127, 1023, 32767, 1048575, 8388607, and 2.1475×10^9 bits, respectively, per pattern. In a typical implementation, PRBS patterns are generated using shift registers with feedback as shown in Fig. 10.2.3, where $D_1, D_2, ..., D_N$ are shift registers.

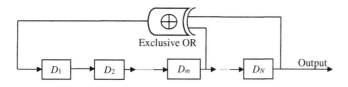

FIG. 10.2.3

PRBS generation using shift registers with feedback.

To explain the operation of PRBS sequence generation, Fig. 10.2.4A shows the simplest case, with $N = 3$ and $m = 1$, where m is the number of shifting shown in Fig. 10.2.3. In this example, suppose that the initial states of the shift registers are $D_1 = 1$, $D_2 = 1$, and $D_3 = 1$. According to the truth table shown in Fig. 10.2.4B, the output of the exclusive OR gate should be $C = 0$. Then in the next time slot, the states become $D_1 = 0$, $D_2 = 1$, and $D_3 = 1$ and the output of the exclusive OR gate becomes $C = 1$. As shown in Table 10.2.1, this process continues until the 7th bit slot, where $D_1 = 1$, $D_2 = 1$, and

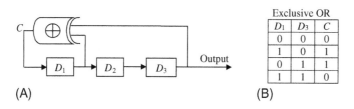

(A)   (B)

FIG. 10.2.4

(A) Simplest PRBS generator with $N = 3$ and (B) truth table of the exclusive OR gate.

$D_3=0$; this makes the exclusive OR gate output $C=1$. After that the states of the shift registers repeat themselves.

In this example, since $N=3$, the total number of bits per pattern is $N_b=2^3-1=7$. This is the main reason that 2^N-1 is used as the word length for PRBS instead of simply 2^N. Another reason for using 2^N-1 as the standard pattern length is that the pattern repetition rate is not harmonically related to the data rate. In a 2^N-1 PRBS bit pattern, the lowest frequency component is

$$f_{start} = \frac{f_b}{N_b} = \frac{f_b}{2^N-1} \qquad (10.2.1)$$

Table 10.2.1 Logical states of shift registers for a PRBS generator with $N=3$

Bit	C	D_1	D_2	D_3	Output $= D_3$
1	1+1 = 0	1	1	1	1
2	0+1 = 1	0	1	1	1
3	1+1 = 0	1	0	1	1
4	0+0 = 0	0	1	0	0
5	0+1 = 1	0	0	1	1
6	1+0 = 1	1	0	0	0
7	1+0 = 1	1	1	0	0
8	0	1	1	1	1
9	1	0	1	1	1
10	0	1	0	1	1
11	0	0	1	0	0
12	1	0	0	1	1
13	1	1	0	0	0
14	1	1	1	0	0

For example, for a 2^7-1 PRBS bit pattern at 10 Gb/s data rate, the lowest frequency component is about 78 MHz. If $2^{31}-1$ PRBS is used at the same data rate, the lowest frequency will be as low as 4.65 Hz. This gives a guideline for choosing RF amplifiers of digital transmission equipment.

In most PRBS pattern generators, in addition to the selection of pattern length, there are usually a number of other choices to further specify the PRBS pattern, such as mark-density-pattern and zero-substitution-pattern. Mark-density-pattern allows varying the ratio of the numbers of marks ("1"s) and spaces ("0"s) in the PRBS pattern. This allows the test of system response to unequal mark and space distributions. Zero-substitution-pattern allows portions within the pattern to be replaced by spaces, which is often used to stretch-test clock recovery circuitry in the receiver. An exceptionally large percentage of spaces in the digital signal usually makes it more difficult to recover the clock.

The performance of clock recovery circuit in the receiver depends on the received bit sequence. Clock recovery from the received PRBS signal involves the selection of the clock frequency component using a narrowband filter and stabilizes the phase with a phase locked loop. As illustrated in Fig. 10.2.5, for a signal with RZ modulation format, the signal waveform returns to 0 within each bit and the continuous "1" pattern has exactly the same periodicity as the clock. Therefore the signal spectrum contains a strong clock frequency component. In this case, a simple narrowband filter at the clock frequency can select the clock component for clock recovery.

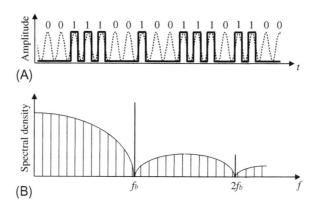

FIG. 10.2.5

(A) Time-domain waveform and (B) its spectrum of an RZ modulated signal. Dashed line in (A) is the clock waveform.

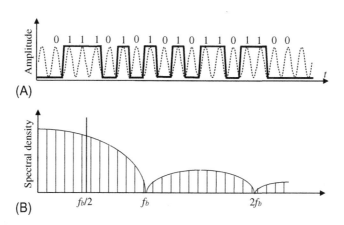

FIG. 10.2.6

(A) Time-domain waveform and (B) its spectrum of an NRZ modulated signal. Dashed line in (A) is the clock waveform.

On the other hand, for signals with NRZ modulation format, as illustrated in Fig. 10.2.6, the signal waveform maintains a constant level over an entire bit length T; therefore the fundamental frequency of an NRZ signal is equal to one-half the clock frequency. This corresponds to the 010101 … bit pattern. The spectrum of an NRZ signal has zero energy at the clock frequency. Therefore, clock extraction for an NRZ signal is more complicated than that for RZ. Typically, to recover the clock from an NRZ data pattern, a narrowband filter has to select the frequency component at half data rate, which is the fundamental frequency of the NRZ signal. Then a nonlinear circuit has to be used to perform frequency doubling to recover the actual clock.

Fig. 10.2.7 shows the block diagram of a clock recovery circuit for an NRZ signal, where a narrowband filter selects the fundamental frequency component at the half bit rate, $f_b/2$, and a nonlinear circuit doubles this frequency to f_b. Then this single-frequency signal is compared with a sinusoid signal generated by a voltage-controlled oscillator (VCO) and their frequency difference is used as the error signal to control the VCO. This ensures that the sinusoid generated by the VCO has exactly the same frequency and phase as the clock extracted from the incoming signal. This also allows the clock to be recovered from PRBS signals with long zeroes.

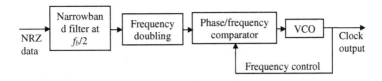

FIG. 10.2.7

Block diagram of a clock recovery circuit for NRZ signal. *VCO*, voltage controlled oscillator.

In practice, although the clock recovery is required in the receiver of commercial optical systems, many laboratory system test-beds do not have clock recovery circuits because the transmitter and the receiver are located in the same laboratory, and the clock can be extracted directly from a master clock generator inside the transmitter. This is commonly regarded as the "cheat" clock because it is not actually recovered from the signal transmitted through the testing system.

10.3 POLYBINARY, DUOBINARY, AND CARRIER-SUPPRESSED RZ MODULATION

Both NRZ and RZ modulation discussed so far are binary with two levels, "0" and "1," and thus bit rate is equal to the symbol rate. RZ modulation improves the receiver sensitivity because of the increased pulse peak power at the decision events, but suffers from reduced tolerance against chromatic dispersion because of the increased spectral width. The introduction of dispersion compensation in the fiber system can help minimize the dispersion-induced waveform distortion, so that RZ modulation is more appropriate in dispersion compensated fiber systems. In a long-distance fiber-optic system employing in-line optical amplifiers, the interaction between fiber nonlinearity and chromatic dispersion may also degrade the transmission system performance, which cannot be eliminated simply by dispersion compensation. High-level modulation and the combination of amplitude and phase coding may help shape the optical spectra, which can have significant impact in the transmission performance.

10.3.1 M-ARY AND POLYBINARY CODING

Increasing the number of digital levels of signal waveform will increase the number of bits per symbol, which allows the increase of the bit rate while maintaining the same symbol rate (also known as baud rate). An M level intensity modulation, known as M-ary, carries $\log_2(M)$ bits per symbol, so that for the same symbol rate, the bit rate increases by $\log_2(M)$. For example, for $M=4$, each symbol has 4 possible levels coded as $0 \Rightarrow 00$, $1 \Rightarrow 01$, $2 \Rightarrow 10$, and $3 \Rightarrow 11$, so that each symbol carries 2 bits of information. Fig. 10.3.1A shows an example of a 4-level amplitude modulated (AM) signal waveform of 10 Gbaud/ s. The eye diagram shown in Fig. 10.3.1B is obtained after the waveform is low-pass filtered by a 5th-order Bessel filter with 6.5 GHz bandwidth. In this example, the symbol rate of 10 Gbaud/s corresponds to a bit rate of 20 Gbit/s. An M-ary modulation with $M=4$ is also known as PAM4, which stands for four-level pulse amplitude modulation.

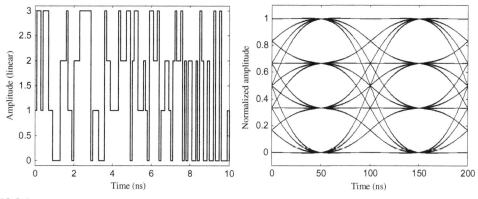

FIG. 10.3.1

(A) ideal waveform of a four-level amplitude modulated signal at 10 Gbaud/s rate and (B) the corresponding eye diagram after a low-pass filter with 6.5 GHz bandwidth.

Fig. 10.3.2A shows that for a fixed data rate, the symbol rate and thus the signal spectral bandwidth is proportional to $1/\log_2(M)$ which decreases with the increase of M. For a chromatic dispersion-limited fiber-optic system, the maximum transmission distance is inversely proportional to the square of the symbol rate, and thus the increase of the dispersion-limited transmission distance is proportional to $[\log_2(M)]^2$, as shown in Fig. 10.3.2B.

Meanwhile as the eye diagram is divided into $M-1$ levels as shown in Fig. 10.3.1B, the opening of each eye level is reduced to $1/(M-1)$ in comparison to the binary eye opening. With the increase of M, although the spectral bandwidth is reduced and the tolerance to chromatic dispersion is increased, the signal becomes more susceptible to both waveform distortion and noise because of the reduced eye opening.

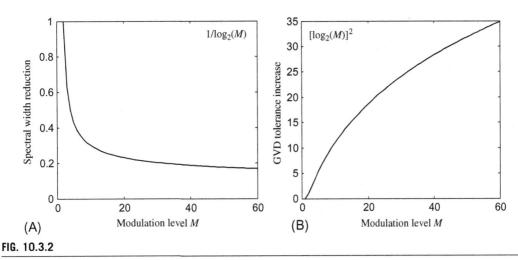

FIG. 10.3.2

(A) reduction of spectral bandwidth, and (B) increase of group-velocity-dispersion (GVD) tolerance as the function of modulation level M.

For a system dominated by signal-independent noise, such as thermal noise, according to Eq. (8.3.3), receiver sensitivity (the lower the better) is inversely proportional to the eye opening and proportional to the square root of the electric bandwidth. Thus, in comparison to a binary system, the power penalty of M-ary is

$$P_{penalty} = \frac{1-M}{\sqrt{\log_2(M)}} \tag{10.3.1}$$

For a system limited by signal-dependent noise, for example, if an optical preamplifier is used in the receiver, according to Eq. (8.3.7), receiver sensitivity is $P_{sen} \propto B_e / \left(\sqrt{A} - \sqrt{B}\right)^2$, where A and B are the upper and the lower levels of the normalized eye, and B_e is the electric bandwidth. Therefore, for an evenly distributed multilevel eye diagram, $A = (M-n)/(M-1)$, and $B = (M-n-1)/(M-1)$ with $n = 1$, 2, ... $(M-1)$. Thus different layers of the eye diagram will have different penalties with respect to an ideal binary system with $A = 1$, and $B = 0$,

$$P_{penalty} = \frac{1}{\left(\sqrt{\dfrac{M-n}{M-1}} - \sqrt{\dfrac{M-n-1}{M-1}}\right)^2 \log_2(M)} \tag{10.3.2}$$

where the top lay of the eye with $n = 1$ has the highest power penalty. For example, for $M = 4$, the power penalties for the three eye levels are 7.4, 6.2, and 2.4 dB, respectively, for $n = 1$, 2, and 3. Thus non-equally spaced eye levels can be used to equalize the power penalty of each level, and help minimize the overall power penalty.

In the multilevel coding described above, data bits are random and there is no correlation between consecutive bits. Thus, the spectral width is dependent only on the symbol rate but independent of the modulation level M. Polybinary modulation is another category of multilevel coding scheme which introduces controlled correlation between adjacent bits so that the signal level change between consecutive bits is not more than one (Lender, 1964; Walklin and Conradi, 1999; Olmos et al., 2013).

The process of polybinary encoding is described in Fig. 10.3.3, where the input NRZ binary data sequence a_k is pre-coded into another binary sequence b_k based on the following rule,

$$b_k = a_k \oplus b_{k-1} \oplus b_{k-2} \cdots \oplus b_{k-(M-2)} \tag{10.3.3}$$

and the polybinary data sequence is

$$c_k = \sum_{i=0}^{M-2} b_{k-i} \tag{10.3.4}$$

where, "$\oplus$" represents an exclusive OR operation, M is the level of coding, and k is the symbol index.

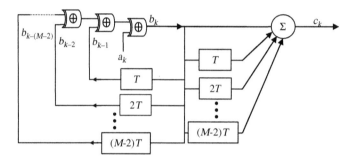

FIG. 10.3.3

Block diagram of polybinary encoding.

Fig. 10.3.4 shows an example of encoding from a binary NRZ data sequence $a_k(t)$ into a five-level ($M=5$) polybinary sequence $c_k(t)$, with $b_k(t)$ the intermediate binary sequence. A key feature of a polybinary sequence is that level change between consecutive symbols is not more than one, which results in a further reduction of the spectral bandwidth compared to an M-ary sequence of the same symbol rate and bit rate.

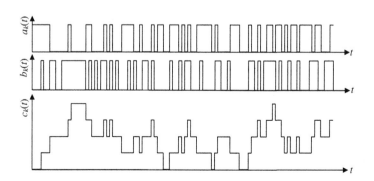

FIG. 10.3.4

Example of polybinary encoding. $a_k(t)$ is the original binary sequence, $b_k(t)$ is the intermediate binary sequence, and $c_k(t)$ is the 5-level polybinary sequence.

In the receiver, the original binary data sequence can be recovered simply by

$$a_k = c_k \, mod \, 2 \tag{10.3.5}$$

Fig. 10.3.5 shows a comparison between the spectra of a binary NRZ waveform, a_k, and a five-level polybinary waveform, c_k, both at 10 Gbaud/s symbol rate. Because of the controlled inter-symbol-interference (ISI) introduced in the polybinary encoding process, the spectral width is significantly reduced. In fact the spectral density of a polybinary waveform is

$$S(f) = \frac{(M-1)^2 T}{4} \, sinc^2 [(M-1)fT] \tag{10.3.6}$$

where T is the symbol period, and M is the level of modulation. Red solid curves in Fig. 10.3.5 are calculated from Eq. (10.3.6), and the spectra in black are the Fourier transforms of $a_k(t)$ and $c_k(t)$ shown in Fig. 10.3.4 for $M=5$.

It is important to clarify the difference between M-ary and polybinary coding formats. For M-ary coding, there is no correlation between adjacent data symbols in the data sequence and the data rate is thus $log_2(M)$ times the binary date sequence of the same symbol rate. On the other hand, because of the correlation introduced in the polybinary coding process, a polybinary data sequence of level M carries the same bit rate of the binary data sequence of the same symbol rate, but with with spectral bandwidth reduced by $1/(M-1)$ as illustrated by Fig. 10.3.5. As an example, for a bit rate of 10 Gbit/s with $M=4$, the symbol rate of M-ary will be 5 Gbaud/s and the spectral bandwidth (1st null in the spectrum) is close to 5 GHz. For the same bit rate but with polybinary coding, the spectral bandwidth (also 1st null in the spectrum) is only 3.3 GHz. The ratio of spectral widths between polybinary and M-ary of the same bit rate is $[log_2(M)]/(M-1)$.

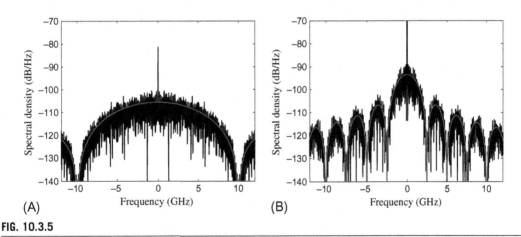

(A)

(B)

FIG. 10.3.5

Red spectra: predicted by Eq. (10.3.4) *(red; gray solid lines in print versions)* for M=2 (A) and M=5 (B). Black spectra: Fourier transforms of $a_k(t)$ (A) and $c_k(t)$ (B) indicated in this figure, both of 10 Gb/s bit rate.

10.3.2 DUO-BINARY OPTICAL MODULATION

Among various levels of polybinary modulation, $M=3$ is most often used in fiber-optic systems, known as dual-binary. Fig. 10.3.6 shows an example of duo-binary encoding with the original binary

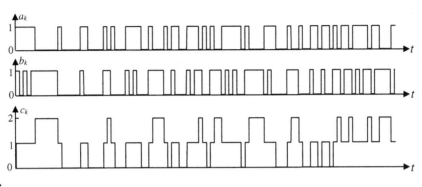

FIG. 10.3.6

Example of duo-binary encoding. a_k: original binary NRZ data sequence, b_k: intermediate data sequence obtained through pre-coding, and c_k: duo-binary data waveform.

NRZ data sequence a_k, the intermediate data sequence b_k obtained through pre-coding, and the duo-binary data waveform c_k. This duo-binary encoded waveform c_k can be converted back to the original binary data in the receiver by $a_k = c_k \, mod \, 2$ operation. In fact, for the waveform c_k shown in Fig. 10.3.6, if the level 2 is folded back to level 0, the original NRZ data sequence can be recovered directly (Penninckx et al., 1997).

For the duo-binary coding process, after pre-coding, the delay-and-add operation is equivalent to time-domain integration over two symbols, which can be accomplished by a low-pass filter. Fig. 10.3.7A shows the 10 Gb/s duo-binary waveforms obtained through the delay-and-add operation, and through a fifth-order Bessel low-pass filter with 2.5 GHz bandwidth. These two waveforms are very similar, except for the unfiltered ripples corresponding to continuous "1"s of the original signal. Because the use of a narrowband filter, high-frequency components of the band-limited duo-binary waveform is much lower than that obtained through the delay-and-add operation, as shown in Fig. 10.3.7B, and the first null at 5 GHz is smoothed out in the band-limited duo-binary spectrum. Fig. 10.3.7C shows the comparison between the band-limited duo-binary spectrum and the spectrum of the original binary NRZ waveform, and the bandwidth reduction is obvious even though their 1st nulls are at the same frequency.

In a fiber-optic transmitter, the simplest way to generate an optical duo-binary waveform is to utilize the complex transfer function of an electro-optic Mach-Zehnder modulator (MZM) as illustrated in the block diagram of Fig. 10.3.8. The original NRZ binary waveform $a_k(t)$ is first pre-coded into $b_k(t)$ based on $b_k - a_k \oplus b_{k-1}$. Then a band-limiting low-pass filter with a 3 dB bandwidth of approximately 25% of the data rate is used to mimic the relation of Eq. (10.3.4) which converts $b_k(t)$ into $c_k(t)$. As discussed in Chapter 7, the relation between the input voltage waveform V and the output optical field waveform E of an electro-optic MZM is a sine function. Fig. 10.3.8 shows that by biasing the MZM at the power transmission null, and applying the DC-blocked duo-binary electric voltage waveform $c_k(t)$ with a peak-to-peak swing of $2V_\pi$, an optical duo-binary waveform can be generated. The corresponding eye diagram of the output signal optical power $P(t)$ is also shown in Fig. 10.3.8, which is equivalent to the eye diagram after a direct-detection optical receiver, in which the photocurrent is linearly proportional to the signal optical power.

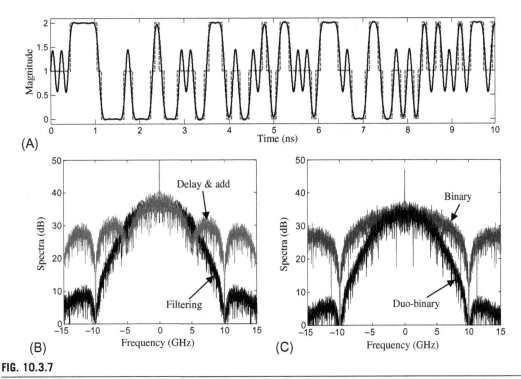

FIG. 10.3.7

(A) 10 Gb/s duo-binary waveforms generated by delay-and-add operation *(dashed line)* and by low-pass filtering *(solid line)* with a bandwidth of 3 GHz. (B) Corresponding spectra of duo-binary waveforms generated by delay-and-add operation *(red; light gray in print versions)* and by low-pass filtering *(black)*. (C) Comparison between the spectra of duo-binary waveform generated by low-pass filtering *(black)* and of the original binary NRZ waveform *(blue; light gray in printed version)*.

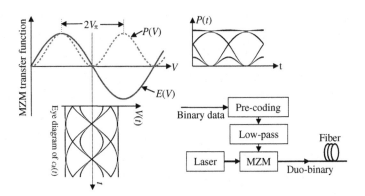

FIG. 10.3.8

Illustration and block diagram of optical duo-binary waveform generation through an electro-optic Mach-Zehnder modulator (MZM). $V(t)$ and $P(t)$ represent input voltage and output optical power eye diagrams.

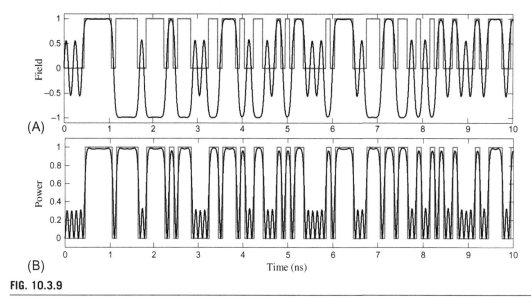

FIG. 10.3.9

Waveforms of optical field [*black* curve in (A)] and optical power [*black* curve in (B)] and the comparison with the inverted original NRZ waveform *(red; light gray curves in print versions)*.

Fig. 10.3.9 compares the time-domain optical field and optical power of the duo-binary waveform created through the MZM and the inverted original binary NRZ waveform. Note that the optical field flips the sign at the middle of each signal "0" bit through the modulator. This optical phase transition helps improve the resistance against the impact of chromatic dispersion. When energy from adjacent signal "1"s is spread into the "0" bit slot due to fiber dispersion, they tend to cancel each other because of their opposite phases, so that signal "0" can be less sensitive to corruption caused by energy leak from adjacent "1"s. Fig. 10.3.10 shows the calculated eye closure penalties of binary NRZ modulated 10 Gb/s optical signal and the converted duo-binary optical signal of the same data rate as the function of the fiber length. The fiber used in this calculation is standard single mode fiber with 16 ps/nm/km chromatic dispersion parameter at 1550-nm signal wavelength. No dispersion compensation is used in this system. Because of the reduced optical spectral width, and the antisymmetric optical phase transition across each signal "0" bit, the dispersion tolerance of duo-binary modulated optical signal is more than 3 times higher compared to the binary modulated optical signal at the same eye closure penalty level of 3 dB. Examples of eye diagrams are also shown in the inset of Fig. 10.3.10 at 120 km for NRZ binary waveform and 150 km for duo-binary waveform.

10.3.3 CARRIER-SUPPRESSED RETURN-TO-ZERO (CSRZ)

Another optical modulation format utilizing inter-symbol phase transition is the CSRZ modulation (Miyamoto et al., 2001; Bosco et al., 2002; Yu et al., 2013). As we have discussed in Section 10.3.1, RZ modulation has improved receiver sensitivity but occupies broader electrical bandwidth due to the reduced pulse width compared to NRZ.

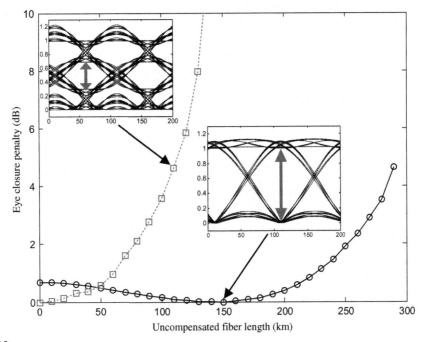

FIG. 10.3.10

Calculated eye closure penalties of binary NRZ modulated (open squares) and duo-binary modulated (open circles) optical signals as the function fiber length. Data rate is 10 Gb/s, and fiber type is standard single mode fiber with 16-ps/nm/km dispersion parameter.

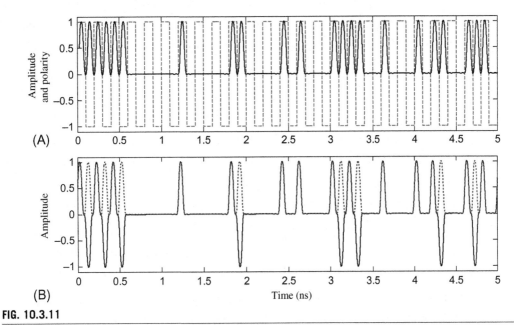

FIG. 10.3.11

(A) RZ waveform (*solid line*) and polarity of the optical field (*dashed line*). (B) RZ modulated waveform (*dotted line*) and the corresponding CSRZ waveform (*solid line*).

CSRZ modulation switches the polarity of each consecutive RZ symbol so that the average of the optical field waveform becomes zero. Fig. 10.3.11A shows the 10 Gb/s RZ modulated pulse train (black solid line) together with the optical phase which swings between $\pm\pi$, equivalent to a switch of polarity between ± 1 of every other RZ symbol. Fig. 10.3.11B is the comparison between the optical fields of the RZ modulated waveform (blue dotted line) and the converted CSRZ waveform (black solid line). With the statistically equal numbers of "0"s and "1"s, the average optical field of the CSRZ waveform is zero. An important feature of CSRZ waveform is that for continuous "1"s, because of the alternating polarities of the field between adjacent pulses, the fundamental modulation frequency of the optical field becomes $f_b/2$, where f_b is the symbol rate. This is in contrast to the fundamental frequency of f_b which is typical for a conventional RZ modulated waveform. Fig. 10.3.12 compares the spectra of RZ and CSRZ modulated optical waveforms both at 10 Gb/s data rate, and both with a 6.5-GHz (fifth-order Bessel) band-limiting filter applied. For the RZ spectrum shown in Fig. 10.3.12A, in addition to the strong DC component, discrete spectral lines exist at the clock frequency of 10 GHz. Whereas for the CSRZ spectrum shown in Fig. 10.3.12B, the DC component is eliminated, and the discrete spectral components are shifted to half the clock frequency at 5 GHz due to the alternating polarities of adjacent pulses. In fact, the continuous "1" pattern in a RZ waveform is most vulnerable in a system with limited optical and electrical bandwidths, and in comparison, CSRZ can tolerate additional bandwidth reduction due to electronic circuits or equivalently due to chromatic dispersion.

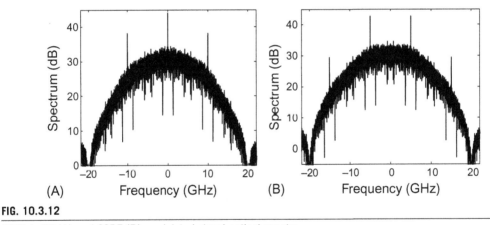

FIG. 10.3.12

10 Gb/s RZ (A) and CSRZ (B) modulated signal optical spectra.

To demonstrate the improved dispersion tolerance of CSRZ, Fig. 10.3.13 shows an example which compares 10 Gb/s RZ and SCRZ waveforms both band-limited by a fifth-order Bessel low-pass filter with 4.5 GHz bandwidth. This bandwidth is narrower than the optimum bandwidth of 6.5 GHz. These two waveforms have the same average value, but the CSRZ waveform has higher peak amplitude at points where decisions are performed, indicating a better receiver sensitivity.

A traditional way to generate a CSRZ optical waveform in an optical transmitter is to use two concatenated electro-optic modulators as schematically shown in Fig. 10.3.14. The first one is an electro-optic amplitude modulator which converts the input RZ voltage signal $V_1(t)$ into a modulated optical field $E_1(t)$ in the RZ format. The second modulator is an electro-optic phase modulator, which

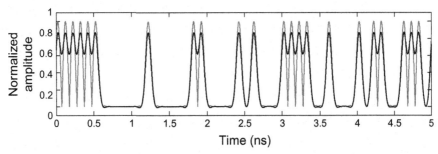

FIG. 10.3.13

Absolute values of 10Gb/s RZ *(black)* and CSRZ *(red; light gray in print versions)* waveforms band-limited by a 4.5-GHz bandwidth fifth-order Bessel filter.

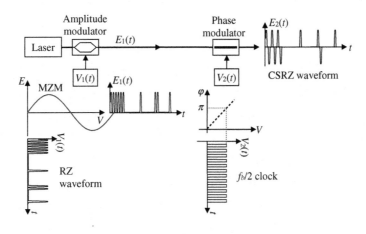

FIG. 10.3.14

Block diagram of a transmitter to create CSRZ optical waveform using an electro-optic amplitude modulator and an electro-optic phase modulator.

introduces an optical phase change linearly proportional to the applied voltage signal $V_2(t)$. In this setup, $V_2(t)$ is a square wave at half clock rate of $V_1(t)$, and with an amplitude to introduce π phase shift. So that every other pulses in the $E_1(t)$ waveform is phase modulated by π to create a CSRZ optical waveform.

Another way to generate CSRZ optical waveform is to use two electro-optic amplitude modulators as illustrated in Fig. 10.3.15. In this configuration, a NRZ waveform $V_1(t)$ is used to drive the first amplitude modulator to create a standard NRZ optical waveform $E_1(t)$. The second MZM is biased at the minimum power transmission point and driven by a sine wave at a frequency half the symbol rate of $V_1(t)$ and an amplitude of $2V_\pi$. Thus $E_1(t)$ is amplitude modulated so that the optical field of every other symbols switch the sign to create a CSRZ waveform.

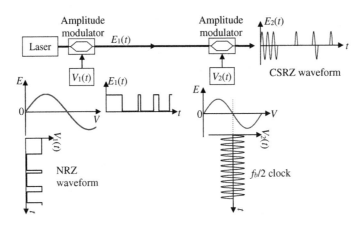

FIG. 10.3.15

Block diagram of a transmitter to create CSRZ optical waveform using two electro-optic amplitude modulators.

In the discussion of RZ and CSRZ modulation formats so far we have used 50% duty cycle, so that the first spectral null is at $2f_b$, where f_b is the clock frequency. In practical fiber-optic systems using RZ and CSRZ modulation, duty cycle may not be 50%, which usually needs to be optimized depending on the transmitter and receiver electronic circuits, as well as the property of the fiber system. With the availability of electro-optic I/Q modulator, generating CSRZ waveforms can be much simpler with a single modulator.

10.4 BPSK AND DPSK OPTICAL SYSTEMS

In the last section, we have discussed various optical modulation formats primarily for direct detection, so that only the information carried by the amplitude of the modulated optical waveform is used. The information capacity and the flexibility of an optical communication system can be greatly improved if the optical phase is also utilized to carry information. Because semiconductor laser sources commonly used in optical communication systems have significant phase noise, recovery of information carried on the optical phase can be quite challenging even with coherent detection. For example, the spectral linewidth of a single-longitudinal-mode laser diode, such as a DFB laser diode is typically tens of Megahertz, corresponding to a fast phase variation in the time scale on the orders of 10 to 100 ns. Thus, data carried by the optical phase may not be distinguishable from this phase noise background. In fact, a rule-of-thumb is that the symbol rate has to be at least 10^5 times higher than the spectral linewidth to guarantee a high quality coherent detection (Ip and Kahn, 2007). Narrow linewidth semiconductor lasers based on external cavities have been developed in 1980s, but they are typically bulky, high cost, and not reliable enough to meet telecommunications standard. In recent years, the rapid development in integrated optics and the maturation of micro-optic packaging technology enabled the creation of low cost and compact external cavity semiconductor lasers with the linewidth narrower than 100 kHz. Long cavity DBR lasers produced with integrated photonic circuits can also provide optical emission with a linewidth on the order of a few hundred kilohertz. For an optical system with symbol rate higher than

10 Gb/s, the symbol-rate/linewidth ratio can be higher than 10^5, so that the recovery of phase information is feasible. In this section, we discuss binary phase shift keying (BPSK) and differential phase shift keying (DPSK) optical modulation with coherent detection. These are two simplest optical phase modulation formats in fiber-optic systems (Winzer and Essiambre, 2006). Receiver sensitivity and the required OSNR will be calculated and compared with direct detection.

BPSK modulation uses a constant signal optical power, but the optical phase is modulated between 0 and π to represent digital "0"s and "1"s. BPSK is a binary modulation format, and the phase waveform can be either RZ or NRZ, but we only use NRZ as an example to discuss the receiver sensitivity and the comparison with intensity modulated optical systems.

For a BPSK modulated optical signal with a constant power P_s, the optical field $E_s(t)$ swings between $\pm\sqrt{P_s}$, as illustrated in Fig. 10.4.1. The balanced coherent detection combines $E_s(t)$ with the local oscillator (LO) optical field E_{LO} through a 2×2 optical coupler and two balanced photodiodes to produce a differential photocurrent signal $\Delta i(t)$. This photocurrent signal is then amplified by a transimpedance amplifier (TIA) to produce a voltage signal $v(t)$.

According to the discussion in Chapter 9, the differential photocurrent of coherent detection is

$$\Delta i(t) = i_1(t) - i_2(t) = 2\Re \sqrt{P_s P_{LO}} \sin\left[\Delta\varphi(t)\right] \tag{10.4.1}$$

where homodyne detection is assumed and the effect of polarization mismatch between signal and LO is neglected for simplicity. P_{LO} is the LO optical power, and $\Delta\varphi(t) = \varphi_s(t) - \varphi_{LO}(t)$ represents optical phase difference between the optical signal and the LO.

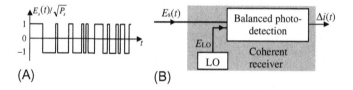

FIG. 10.4.1

(A) Normalized BPSK optical field waveform, and (B) Illustration of coherent detection.

For the BPSK modulated optical signal, digital "1"s and "0"s are represented by $\phi_s = \pi$ and $\phi_s = 0$, respectively. If we set the phase of the LO to be $\phi_{LO} = \pi/2$, differential photocurrents $\Delta i(t)$ corresponding to digital "1" and "0" are $i_+ = 2\Re \sqrt{P_s P_{LO}}$ and $i_- = -2\Re \sqrt{P_s P_{LO}}$, respectively. Since $P_{LO} \gg P_s$, we can assume that the shot noise caused by the LO with the variance of $\sigma_{sh}^2 = 2q\Re P_{LO}B_e$ is the dominant noise source at the receiver with q the electron charge, $\Re$ the photodiode responsivity, and B_e the receiver bandwidth. The electric SNR at the coherent receiver output is,

$$SNR_{BPSK} = \frac{2\Re P_s}{qB_e} \tag{10.4.2}$$

This *SNR* is 8 times higher than that predicted by Eq. (9.3.8) for intensity modulated optical signal. A factor of 2 is due to the well-controlled optical phase so that $\langle \sin^2\varphi(t)\rangle = 1$, and another factor of 4 is due to balanced detection which doubles the photocurrent corresponding to 4 times the signal electric power.

The Q-value of the receiver is

$$Q = \frac{i_+ - i_-}{2\sigma_{sh}} = \sqrt{\frac{2\Re P_s}{qB_e}} \tag{10.4.3}$$

where $i_+ - i_- = 4\Re\sqrt{P_s P_{LO}}$ is the signal, and $\sigma_{sh} = \sqrt{2q\Re P_{LO}B_e}$ is the shot noise standard deviation which is the same for digital "1" and "0." Since the photodiode responsivity is $\Re = \eta q/(hv)$, where η is the quantum efficiency and hv is the photo energy, Eq. (10.4.3) can also be expressed as $Q = \sqrt{2\eta P_s/(hvB_e)}$. For $Q=7$, equivalent to a BER of 10^{-12}, the receiver sensitivity is $P_{sen} = 24.5hvB_e/\eta$. For an ideal photodiode with 100% quantum efficiency, this receiver sensitivity corresponds to 24.5 photons per bit, which is 3-dB better than the quantum-limited receiver sensitivity for intensity modulated optical system discussed in Section 8.3.1. Because BPSK waveform carries optical power for both digital "0"s and "1"s, its receiver sensitivity is 3 dB better then on-off-keying (OOK) modulated waveform, in which only digital "1"s carry signal photons.

Waveform distortion can also exist in a phase modulated optical system so that $i_+ = 2A\Re\sqrt{P_s P_{LO}}$ and $i_- = -2B\Re\sqrt{P_s P_{LO}}$. Where, $0<A<1$ and $0<B<1$ represent eye closure of digital "1"s and "0"s at the decision event, which can be caused by phase distortion due to transmission impairments and the inaccuracy of LO phase reference. Because shot noise only depends on the LO power, thus,

$$Q = \frac{i_+ - i_-}{2\sigma_{sh}} = \sqrt{\frac{\Re P_s}{2qB_e}}(A+B) \tag{10.4.4}$$

In a fiber-optic system employing multiple in-line optical amplifiers, optical signal to noise ratio (OSNR) will become the limiting factor of transmission performance. Consider a BPSK modulated system with coherent homodyne detection, and assume that the signal optical power is P_s and the optical noise power spectral density is ρ_{ASE} at the receiver input. As the LO power is much higher than the signal power in a coherent receiver, the dominant noise source is the LO-ASE beat noise. In the Q-calculation, the electric signal is $i_+ - i_- = 4\Re\sqrt{P_s P_{LO}}$, and the LO-ASE beat noise after photodetection is $\sigma_1 = \sigma_0 = \sqrt{\Re^2 \rho_{ASE} P_{LO} B_e}$, so that

$$Q = 2\sqrt{\frac{P_s}{\rho_{ASE}B_e}} = 2\sqrt{\frac{OSNR}{B_e}} \tag{10.4.5}$$

where $OSNR = P_s/\rho_{ASE}$ has the unit of $[Hz]$, and optical noise only in the same polarization as LO is considered here. In fact, with the balanced coherent detection configuration using two photodiodes, the LO power and the ASE power spectral density received by each photodiode are $P_s/2$ and $\rho_{ASE}/2$, respectively, so that each photodiode produces LO-ASE beat noise power spectral density $2\Re^2(\rho_{ASE}/2)(P_{LO}/2)$. The differential output of the balanced detector adds up two equal contributions and thus the total electric noise power is still $\Re^2 \rho_{ASE} P_{LO} B_e$, which is the same as using a single photodiode.

To achieve the Q-value of 7, the required OSNR is simply, $ROSNR = 12.25B_e$, which is 6 dB better (lower) than that with binary intensity modulation shown in Eq. (8.3.11) without waveform distortion. As a numerical example, for a 10 Gb/s BPSK system using coherent homodyne detection with an electric bandwidth of 7.5 GHz, the required OSNR is 9.2×10^{10} Hz, which is approximately 8.7 dB when noise is measured with an optical spectrum analyzer (OSA) with 0.1-nm resolution bandwidth, or 12.5 GHz, at 1550-nm-wavelength window.

In order to achieve the maximum Q-values predicted by Eqs. (10.4.3) and (10.4.5), the optical phase of the LO has to be stabilized at $\phi_{LO} = \pi/2$ to obtain the maximum differential current for a BPSK modulated optical signal. Traditionally this can be done through active feedback control in a phase-locked loop. A more sophisticated solution is to use phase-diversity in the coherent receiver so that phase tracking and optimization can be accomplished through signal processing. As discussed in Section 9.4, in a coherent homodyne receiver with the combination of balanced detection and phase diversity, two photocurrent components are produced,

$$\Delta i_I(t) = \Re \sqrt{P_s P_{LO}} \sin[\Delta\varphi(t)] \tag{10.4.6}$$

$$\Delta i_Q(t) = \Re \sqrt{P_s P_{LO}} \cos[\Delta\varphi(t)] \tag{10.4.7}$$

Optical phase information can be obtained as

$$\varphi_s(t) = \tan^{-1}\left[\frac{\Delta i_I(t)}{\Delta i_Q(t)}\right] + \varphi_{LO}(t) \tag{10.4.8}$$

where, $\Delta\varphi(t) = \varphi_s(t) - \varphi_{LO}(t)$ has been used. As long as the variation of $\varphi_{LO}(t)$ is much slower than the signal phase modulation encoded in $\varphi_s(t)$, the impact of $\varphi_{LO}(t)$ can be minimized by signal processing and filtering which rejects the low-frequency components in Eq. (10.4.8). More details of carrier phase recovery in a coherent receiver will be provided in Chapter 11.

DPSK is another optical phase modulation format often used in long-distance fiber-optic systems, which can be detected by both coherent detection and direct detection. DPSK encoding is based on the optical phase change between consecutive bits. As illustrated in Fig. 10.4.2, a digital "1" is encoded as a π phase change, whereas a digital "0" is encoded as no phase change, between adjacent bits.

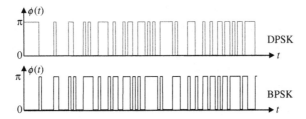

FIG. 10.4.2

Comparison between DPSK *(top)* and BPSK *(bottom)* waveforms.

Coherent homodyne detection of DPSK modulated optical signal is the same as that of BPSK, and the receiver sensitivity and the required OSNR can be calculated based on Eqs. (10.4.3) and (10.4.5). Unlike BPSK, DPSK can also be detected by a direct detection receiver through a one-bit-delay optical interferometer. In a direct detection receiver, since there is no optical LO to provide a phase reference, each bit of the optical signal itself has to be used as the phase reference for the next bit, so that a direct detection DPSK receiver can convert this differential phase of the received optical signal into an intensity waveform.

Direct-detection receiver of DPSK can be constructed with an optical interferometer as shown in Fig. 10.4.3A. A delay line is added in one of the two arms of the interferometer to introduce a delay of one bit T. Assume $3\,dB$ couplers are used at both the input and the output of the interferometer, the optical fields at the two output ports of the interferometer are

$$E_1(t) = [E_s(t+T) - E_s(t)]/2 \qquad (10.4.9)$$

$$E_2(t) = j[E_s(t+T) + E_s(t)]/2 \qquad (10.4.10)$$

Then the photocurrents are,

$$i_1(t) = |\Re E_1(t)|^2 = \Re\left[|E_s(t+T)|^2 + |E_s(t)|^2 - E_s(t+T)E_s^*(t) - cc\right]/4 \qquad (10.4.11)$$

$$i_2(t) = |\Re E_1(t)|^2 = \Re\left[|E_s(t+T)|^2 + |E_s(t)|^2 + E_s(t+T)E_s^*(t) + cc\right]/4 \qquad (10.4.12)$$

The differential photocurrent is

$$\Delta i(t) = i_2(t) - i_1(t) = 0.5\Re E_s(t+T)E_s^*(t) + cc \qquad (10.4.13)$$

where $E_s(t) = \sqrt{P_s}e^{j\varphi_s(t)}$ is the input signal optical field with P_s the optical power and $\phi_s(t)$ the optical phase. cc represents complex conjugate. For an ideally phase modulated optical signal, the signal optical power P_s is a constant, so that

$$\Delta i(t) = \Re P_s \cos[\varphi(t+T) - \varphi(t)] \qquad (10.4.14)$$

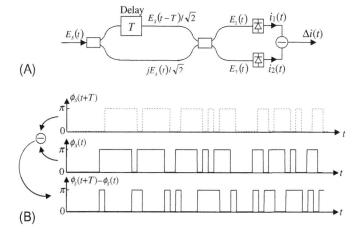

FIG. 10.4.3

(A) DPSK receiver architecture based on a Mach-Zehnder interferometer. (B) Illustration of DPSK waveform reconstruction at the receiver.

For a DPSK system with multiple in-line optical amplifiers, accumulated ASE noise that reaches the receiver is the major limit of the system performance, and the required OSNR is the most relevant measure to determine the quality of transmission. Assume that the optical signal average power and the ASE noise power spectral density are P_s and ρ_{ASE}, respectively, and the signal-ASE beat noise is the dominant noise source. Differential photocurrent from the balanced optical receiver is $2\Re P_s$ between digital "1" and "0" levels, and the noise standard deviation is identical during digital "1" and "0" levels which is $\sqrt{\Re^2 P_s B_e \rho_{ASE}}$. In this estimation, we have assumed that signal optical power and ASE noise power spectral density reaching to each photodiode are $P_s/2$ and $\rho_{ASE}/2$, respectively. In addition, the Mach-Zehnder interferometer (MZI) used before photo detectors has a differential delay of T between the two arms, and the optical power transfer function in the frequency domain is $\sin^2(\pi f T)$ with a free-spectral range (FSR) of $1/T$. A correlation is introduced by this interferometer for the noise between adjacent signal "1"s and "0"s. This equivalently reduces the noise bandwidth by approximately 3 dB so that the noise standard deviation becomes approximately $\sqrt{\Re^2 P_s B_e \rho_{ASE}/2}$. The Q-value can be found as,

$$Q = \frac{2\Re P_s}{2\sqrt{\Re^2 P_s B_e \rho_{ASE}/2}} = \sqrt{\frac{2P_s}{B_e \rho_{ASE}}} = \sqrt{2\frac{OSNR}{B_e}} \qquad (10.4.15)$$

Or equivalently, $OSNR/B_e = Q^2/2$. Fig. 10.4.4 shows the comparison of normalized OSNR requirements of DPSK modulated waveform using differential delay line based self-homodyne detection, and OOK modulated waveform using direct detection where $OSNR/B_e = Q^2$ (see Eq. 8.3.10) without considering waveform distortion. For the Q-value in the range from 3 to 12, the improvement of the required OSNR is 3 dB by using DPSK. This is because digital "0"s in OOK signal waveform carries no energy, while both "0"s and "1"s in DPSK carry the same optical power. However, in comparison to

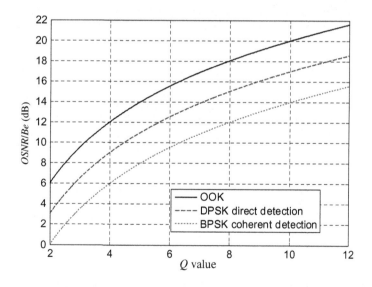

FIG. 10.4.4

Normalized OSNR as the function of Q-value for DPSK with differential delay line-based direct detection *(dashed line)*, OOK with direct detection, and BPSK with coherent homodyne detection *(dotted line)*.

BPSK with coherent homodyne detection where $OSNR/B_e = Q^2/4$, DPSK with differential delay line-based self-homodyne detection is 3 dB worse in terms of the required OSNR.

10.5 HIGH-LEVEL PSK AND QAM MODULATION

So far, we have discussed multilevel amplitude modulation such as M-ary and polybinary, as well as bipolar optical phase modulation such as BPSK and DPSK. In general, multilevel modulation can be applied on both the amplitude and the phase of a complex optical field. The number of constellation points M is usually even powers of 2 ($M = 2$, 4, 16, 64, ...), and positions of constellation points on the complex plane are typically distributed evenly on both real and imaginary axes to simplify implementation (Wilson, 1996). This allows the transmission of up to $b = \log_2(M)$ bits per symbol.

PSK modulation usually refers to optical phase modulation with a constant amplitude. The level of modulation M indicates the number of discrete phase levels within the 2π phase space. The constellation diagram of 16-PSK ($M = 16$) is shown in Fig. 10.5.1A, where the phase interval between adjacent constellation points is $\pi/8$.

Quadrature amplitude modulation (QAM) is the combination of amplitude modulation and phase modulation. In comparison to PSK, the introduction of amplitude modulation in QAM relaxes the angular space between bits for the same modulation level M. The constellation diagram of 16-QAM ($M = 16$) is shown in Fig. 10.5.1B, where there are three different amplitude and 12 different phase levels.

The complex optical field of both PSK and QAM can be expressed as $E(t) = \sqrt{P_s(t)} \exp[j\varphi(t)]$ with $P_s(t)$ the signal optical power. On the complex plane shown in Fig. 10.5.1, the optical field can be decomposed into the in-phase (I) and the quadrature (Q) components, so that

$$E(t) = I(t) + jQ(t) \tag{10.5.1}$$

For PSK, the amplitude is constant so that $I(t)^2 + Q(t)^2 = P_s$, whereas for the 16-QAM, both $I(t)$ and $Q(t)$ have discrete levels: ± 1 and ± 3. For equal probability of signal bits among constellation points, the power normalization factor is $(4 \times 18 + 8 \times 10 + 4)/16 = 9.75$, so that $E(t) = \sqrt{P_s}[I(t) + jQ(t)]/\sqrt{9.75}$ for $\{I, Q\} \in \{\pm 1 \pm 3, \pm 1 \pm 3\}$. This QAM constellation diagram ensures that the minimum separation between adjacent constellation points is 2, but the three optical power levels are $1.85P_s$, $1.03P_s$, and $0.103P_s$, respectively, with P_s the average signal optical power.

In a transmission channel with additive white Gaussian noise (AWGN), the maximum achievable spectral efficiency in terms of bit/s per Hz bandwidth is predicted by the well-known Shannon capacity formula,

$$C = \log_2(1 + \gamma_s) \tag{10.5.2}$$

where

$$\gamma_s = \frac{2 \cdot OSNR}{B_e} \tag{10.5.3}$$

represents optical SNR per symbol. Note that in the definition of $OSNR$ in Eq. (8.3.8), we have used ASE noise power spectral density ρ_{ASE} which includes energy in both polarizations. While in the definition of $OSNR$ per symbol in Eq. (10.5.2), only single polarized ASE noise is used, so that a factor 2 is introduced to maintain a self-consistency. For high-level modulation, each symbol carries more than one bit so that we can also define *OSNR* per bit as

$$\gamma_b = \frac{2 \cdot OSNR}{B_e \cdot \log_2(M)} \tag{10.5.4}$$

Although Eq. (10.5.2) provides an upper limit on the spectral efficiency, the actually achievable spectral efficiency is dependent on the modulation format, as well as transmitter and receiver configurations.

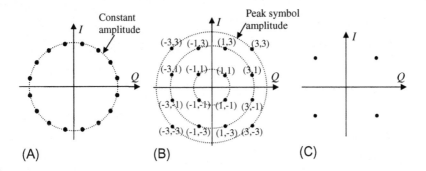

FIG. 10.5.1

Example constellation diagrams of (A) 16-PSK, (B) 16-QAM, and (C) QPSK.

It has been shown that for M-PSK modulation with large *M* and using coherent detection, the BER is approximately (Ip et al., 2008; Proakis, 2001),

$$BER_{M-PSK} \approx \frac{1}{\log_2(M)} erfc\left\{ \sqrt{\gamma_b \log_2(M)} \sin\left(\frac{\pi}{M}\right) \right\} \tag{10.5.5}$$

where, *erfc* is the complementary error function. For a special case of BPSK discussed in the last section with $M=2$, exact expression can be found as

$$BER_{BPSK} = \frac{1}{2} erfc\left(\sqrt{\gamma_b}\right) = \frac{1}{2} erfc\left\{ \sqrt{\frac{2 \cdot OSNR}{B_e}} \right\} \tag{10.5.6}$$

Using the *Q*-value definition in Eq. (8.1.14), this is equivalent to $Q_{BPSK} = 2\sqrt{OSNR/B_e}$, which is in agreement with Eq. (10.4.5). Note that the factor $1/\log_2(M)$ in Eq. (10.5.5) is only an approximation which is good for large *M* values. For $M=2$, this factor is 1/2 as shown in Eq. (10.5.6).

Increasing modulation level *M* will increase the required OSNR per bit γ_b to achieve the targeted BER. Fig. 10.5.2A shows the required OSNR per bit for *M*-PSK modulated signal using coherent detection for *M* varies from 2 to 128.

For *M*-QAM modulation with M≥4 and coherent detection, the BER is approximately (Ip et al., 2008; Proakis, 2001),

$$BER_{M-QAM} \approx \frac{2}{\log_2(M)} \left(\frac{\sqrt{M}-1}{\sqrt{M}}\right) erfc\left\{ \sqrt{\frac{3\gamma_b \log_2(M)}{2(M-1)}} \right\} \tag{10.5.7}$$

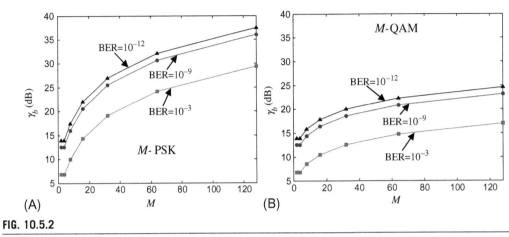

FIG. 10.5.2

Required OSNR per bit as the function of modulation level M for PSK (A) and QAM (B) modulated optical signals.

Fig. 10.5.2B shows the required OSNR per bit for M-QAM modulated optic system using coherent detection where M varies from 2 to 128. Thanks to the utilization of both amplitude and phase in coding, QAM has less degradation of required OSNR-per-bit with the increase of M in comparison to PSK which only uses phase modulation.

Note that for $M=4$, PSK and QAM have identical constellation diagram as shown in Fig. 10.5.1C, known as QPSK, and their BER expressions are also identical, which is

$$BER_{QPSK} = \frac{1}{2} erfc\left(\sqrt{\gamma_b}\right) = \frac{1}{2} erfc\left\{\sqrt{\frac{OSNR}{B_e}}\right\} \tag{10.5.8}$$

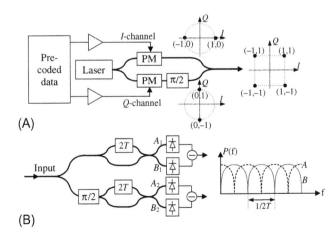

FIG. 10.5.3

(A) DQPSK transmitter and constellation diagrams, and (B) DQPSK receiver using direct detection with differential delay lines.

In fact, QPSK is one of the most often used modulation formats in optical systems with coherent detection. It has the same required OSNR per bit as BPSK, but has 2 bits per symbol so that the spectral efficiency is doubled in comparison to BPSK.

QPSK can also be pre-coded into a differential format known as DQPSK in the similar way as the conversion from BPSK to DPSK discussed in the last section. DQPSK can be considered as two parallel streams of DPSK each carrying half of the total bit rate.

Fig. 10.5.3A shows the configuration of a DQPSK optical transmitter, in which the laser output is equally split into two parts, and each part is modulated by an electro-optic phase modulator driven by a pre-coded binary data sequence. Each binary data sequence has a symbol duration $2T$ so that the combined data rate of the two channels is $1/T$. The two BPSK modulated optical signals are combined with a relative phase shift $\pi/2$ so that one channel is *in-phase* and the other channel is *quadrature*. The corresponding constellation diagrams are shown in Fig. 10.5.3A before and after the phase shifted beam combining. The DQPSK modulation can also be accomplished with a single I/Q electro-optic modulator.

This DQPSK optical signal has the same constellation diagram as QPSK, so that it can be coherently detected with the same performance as discussed in Eq. (10.5.8). Because of the differential coding, DQPSK can also be detected by direct-detection without the need of an optical local oscillator. In the direct detection receiver as shown in Fig. 10.5.3B, the received optical signal is equally split into two branches. Each branch has a MZI with a differential delay of $2T$ between the two arms, which is equal to the symbol time of each BPSK data stream. The optical transfer functions A and B before the two balanced photodiodes are also illustrated in Fig. 10.5.3B. A $\pi/2$ optical phase shifter is added between these two branches so that they detect the in-phase and the quadrature components of the phase modulated optical signal separately. If the optical system is limited by the OSNR due to signal-ASE beat noise, the required OSNR per bit for the DQPSK system using direct detection is (Proakis, 1968; Ip et al., 2008),

$$BER_{DQPSK} = Q_M(\alpha, \beta) - \frac{1}{2}I_0(\alpha\beta)e^{\frac{-1}{2}(\alpha^2 + \beta^2)} \tag{10.5.9}$$

where Q_M is the Marcum Q function, I_0 is the modified Bessel function of the *0th* order, and

$$\alpha = \sqrt{2\gamma_b\left(1 - \sqrt{\frac{1}{2}}\right)} \tag{10.5.10}$$

and

$$\beta = \sqrt{2\gamma_b\left(1 + \sqrt{\frac{1}{2}}\right)} \tag{10.5.11}$$

γ_b is the OSNR per bit as previously defined.

Fig. 10.5.4 compares the BER as the function of the normalized OSNR for QPSK with coherent detection and DQPSK with direct detection based on double differential delay lines. Direct detection receiver is simpler than coherent detection because no optical local oscillator is required, but the required OSNR is approximately 2.5 dB worse.

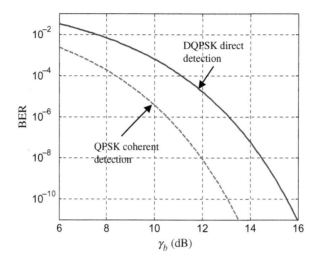

FIG. 10.5.4

BER as the function of OSNR per bit for QPSK with coherent detection *(dashed line)* and DQPSK with direct detection based on double differential delay lines *(solid line)*.

Error-vector-magnitude (EVM): For a system with intensity modulation, eye diagram is the most useful technique to visualize the signal quality. Both binary and multilevel AM signals can be represented by their eye diagrams as discussed in Section 10.3. Signal eye-opening with the consideration of both waveform distortion and random noise, as shown in Fig. 8.1.3, can also be used to quantify the signal quality and to find the BER. On the other hand, for digital signals encoded in both the amplitude and the phase of an optical carrier, constellation diagrams are often used to help estimate the signal quality. By acquiring a large number of signal symbols, a constellation diagram can be used to predict the SNR and BER. Error-vector-magnitude (EVM) is the most often used measure for the signal quality of complex modulated digital signals.

Fig. 10.5.5 shows constellation diagrams of QPSK and 16-QAM signals, where $E_{t,i}$ is the transmitted vector of the ith symbol, $E_{r,i}$ is the received vector of the ith symbol, and $E_{er,i}$ is the difference

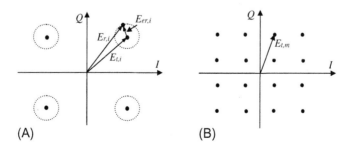

FIG. 10.5.5

Illustration of signal vector and vector errors in a QPSK (A) and a 16-QAM (B) constellation diagram.

between the transmitted vector and the received vector of the ith symbol, which represents the error vector. The magnitude square of this error vector is

$$|E_{er,i}|^2 = |E_{r,i} - E_{t,i}|^2 \tag{10.5.12}$$

The *EVM* is defined as the normalized root-mean-square of the error vector $E_{er,i}$ of N random data samples,

$$EVM_{RMS} = \frac{\sigma_{er}}{\sqrt{P_0}} = \sqrt{\frac{\sum_{i=1}^{N}|E_{r,i} - E_{t,i}|^2}{\sum_{i=1}^{N}|E_{t,i}|^2}} = \sqrt{\frac{\sum_{i=1}^{N}|E_{r,i} - E_{t,i}|^2}{P_0}} \tag{10.5.13}$$

where

$$\sigma_{er} = \sqrt{\frac{1}{N}\sum_{i=1}^{N}|E_{r,i} - E_{t,i}|^2} \tag{10.5.14}$$

is the root-mean-square of the error vectors for all N samples, and P_0 is the average power of all N symbols used for the calculation, that is

$$P_0 = \frac{1}{N}\sum_{i=1}^{N}|E_{t,i}|^2 \tag{10.5.15}$$

Under the Gaussian noise assumption, *EVM* is related to the electrical-domain *SNR* by,

$$EVM_{RMS} = \sqrt{\frac{1}{SNR}} \tag{10.5.16}$$

If we further use coherent detection as an example and assume that there is no waveform distortion, the relationship between Q-value, *SNR*, and *OSNR* for QPSK is $Q = 2\sqrt{SNR} = 2\sqrt{OSNR/B_e}$ as indicated by Eq. (10.5.8), and thus

$$EVM_{rms} = \sqrt{\frac{B_e}{OSNR}} = \frac{2}{Q} \tag{10.5.17}$$

The EVM defined by Eq. (10.5.13) is based on the assumption that the all the ideal (non-distorted) data vectors are known, so that this definition is referred to as *data-aided reception*. But in some cases, the position of each ideal constellation point is not known prior to the detection. In such cases the reference position of each constellation point has to be estimated from the recovered data, which may not be completely correct. This is referred to as the *non-data-aided reception*, and the EVM is related to the SNR in a slightly different way (Mahmoud and Arslan, 2009):

$$EVM_{rms} = \sqrt{\frac{1}{SNR} - \sqrt{\frac{96}{\pi(M-1)SNR}}\sum_{i=1}^{\sqrt{M}-1}\gamma_i\exp\left[-3\beta_i^2\frac{SNR}{2(M-1)}\right] + \frac{12}{M-1}\sum_{i=1}^{\sqrt{M}-1}\gamma_i\beta_i erfc\left[\sqrt{3\beta_i^2\frac{SNR}{2(M-1)}}\right]} \tag{10.5.18}$$

where M is the level of QAM (M-QAM), $\gamma_i = 1 - i/\sqrt{M}$, and $\beta_i = 2i - 1$.

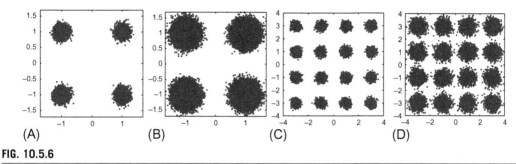

FIG. 10.5.6

Typical constellation diagrams of QPSK (A) and (B), and 16QAM (C) and (D) with EVM values of 11% (A), 17% (B), 7% (C), and 13% (D), corresponding to the SNR of 19, 15.4, 23, and 18 dB, respectively.

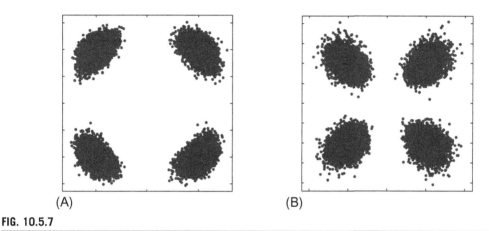

FIG. 10.5.7

Constellation diagrams of QPSK signals with some phase noise (A) and intensity noise (B).

Fig. 10.5.6 shows examples of constellation diagrams of QPSK and 16QAM with different EVM values. Horizontal and vertical axes represent the in-phase and quadrature components of the signal. A smaller EVM value corresponds to a better signal quality because the constellation points are more clearly separated. The shapes of constellation clusters can also help find sources of signal degradation. For example, the constellation diagram shown in Fig. 10.5.7A indicates impact of phase noise so that constellation points are stretched in the circular direction. Whereas for the constellation diagram shown in Fig. 10.5.7B, the impact of intensity noise is obvious, as constellation points are mostly stretched in the radial direction.

10.6 ANALOG OPTICAL SYSTEMS AND RADIO OVER FIBER

Analog modulation has been traditionally used in cable TV (CATV) systems in which a large number of TV channels are multiplexed together based on RF subcarrier multiplexing (Phillips and Darcie, 1997). In an analog lightwave system, the signals remain in the analog domain from the transmitter to the receiver which eliminates the need for analog to digital convertor (ADC) and digital to analog converter (DAC), and the associated digitizing errors. Analog systems have obvious advantages for signals with very high RF carrier frequencies but relatively narrow bandwidth. For example, to digitally create an analog signal around 50-GHz carrier frequency based on a 16-bit DAC, the clock rate has to be at least 100 GS/s and the bit rate has to be as high as 1.6 Tb/s.

In recent years, microwave photonics and radio-over-fiber (RoF) have been widely adopted. Thanks to the high frequency and wide bandwidth of optical signals and photonic techniques, microwave photonics is to use photonic techniques to generate, transmit, and process microwave and millimeter wave signals. Microwave photonics has the potential to enable various functionalities and breakthrough limitations that are not possible or too difficult to achieve with traditional RF and microwave techniques. Photonic devices are also much smaller in footprints and consume less power than microwave devices.

While BER is the ultimate measure of transmission performance in a digital system, analog system performance is measured by the carrier-to-noise ratio (CNR) and waveform distortion often quantified by the harmonic power ratio. In this section we discuss unique properties of analog lightwave systems including the definition and impact of CNR, inter-modulation distortion (IMD), and the impact of relative intensity noise (RIN).

10.6.1 ANALOG SUBCARRIER MULTIPLEXING AND OPTICAL SINGLE-SIDEBAND MODULATION

In a digital system, digitized signal with discrete levels is converted from electric domain to optical domain in the transmitter, and vice versa in the receiver. Linearity of the electrical-to-optic (E/O) and optical-to-electrical (O/E) conversions is not the most important concern especially for binary modulation with only two levels. On the other hand, analog modulation has much more stringent linearity requirement, because analog signal are much more susceptible to even very subtle waveform distortion. While digital modulation allows waveform reshaping and regeneration through decision circuits, analog signals cannot be easily reshaped.

An analog lightwave system usually uses frequency division multiplexing (FDM) in which multiple narrow bandwidth channels are multiplexed in the frequency domain each with a different subcarrier frequency as illustrated in Fig. 10.6.1, which is also often referred to as subcarrier-multiplexing (SCM).

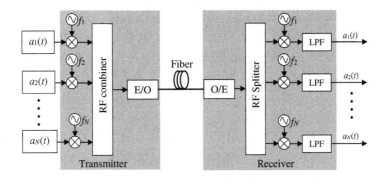

FIG. 10.6.1

Block-diagram of analog system based on subcarrier multiplexing.

In an analog SCM lightwave system, each analog signal channel $a_i(t)$ with $i = 1, 2, ...N$, is first up-converted to a subcarrier frequency f_i, and then combined with other channels by an RF $N \times 1$ power combiner. This composite RF signal is then converted into optical domain through an E/O converter, which can be accomplished either by direct modulation on a laser diode, or through external electro-optic modulation of a CW laser output. At the receiver side, the optical signal is first converted into the electric domain through an O/E converter that can either be direct detection by a photodiode, or co-herent detection in which the optical signal is mixed with a local oscillator. The recovered RF signal is then split into N copies, and the signal carried on the ith subcarrier channel $a_i(t)$ is down-converted to the baseband through mixing with an RF local oscillator of frequency f_i. A low pass filter (LPF) is used after the frequency down-conversion to remove the crosstalk from other channels.

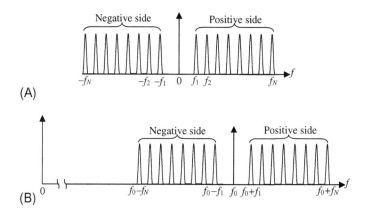

(A)

(B)

FIG. 10.6.2

Illustration of double-sideband SCM spectrum in the RF domain (A) and frequency up-converted version in the optical domain (B).

The analog signal of each channel can be AM, frequency modulated, or phase modulated in the RF domain (Way, 1999). Use AM as an example, the RF voltage signal can be expressed as,

$$V(t) = V_0 \left[1 + \sum_{i=1}^{N} m_i a_i(t) \cos(2\pi f_i t) \right] \tag{10.6.1}$$

where V_0 is the average signal voltage and m_i is the modulation index of the ith subcarrier channel at the central frequency of f_i, and N is the total number of subcarrier channels. This real voltage signal can be decomposed into two sidebands as

$$V(t) = V_0 \left\{ 1 + \frac{1}{2} \left[\sum_{i=1}^{N} m_i a_i(t) \cdot \exp(2\pi f_i t) + \sum_{i=1}^{N} m_i a_i(t) \cdot \exp(-2\pi f_i t) \right] \right\} \tag{10.6.2}$$

where the positive and the negative sidebands carry the same information $a_i(t)$. Through the E/O con-version process, this composite RF signal is up-converted into the optical domain with a central optical frequency f_0 as illustrated in Fig. 10.6.2B. The bandwidth of the optical signal is approximately $2f_N + B_N$ where B_N is the FWHM spectral width of the Nth subcarrier channel.

Considering the large number of subcarrier channels which can be used in a SCM lightwave system, the frequency separation between the positive and the negative sidebands of each subcarrier channel $2f_i$ can be much higher than the actual spectral bandwidth B_i of that channel. Thus, SCM lightwave system can be susceptive to dispersion of the optical fiber because the differential group delays caused by both chromatic dispersion and PMD are proportional to $2f_i$ which is much bigger than B_i. As the positive and the negative sidebands carry redundant information, removing one of the sidebands should help improve the spectral efficiency, and also increase the tolerance against chromatic dispersion and PMD.

Single-sideband modulation is equivalent to a Hilbert transform which can be accomplished by mixing the voltage signal with both $\cos(2\pi f_0 t)$ and its 90 degree shifted version $\sin(2\pi f_0 t)$, and then combine as shown in Fig. 10.6.3A, where f_0 is the carrier frequency.

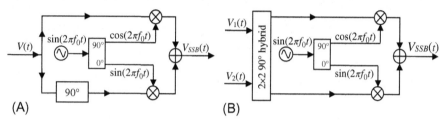

(A)   (B)

FIG. 10.6.3

Block diagram to create single-sideband modulation with one sideband utilized (A) and with both sidebands utilized to carry independent information (B).

Mathematically this results in a single sideband at the output of this mixing process,

$$V_{SSB}(t) = \frac{V_0}{2}\left\{\left[1 + \sum_{i=1}^{N} m_i a_i(t)\cos(2\pi f_i t)\right]\cos(2\pi f_0 t) + \left[1 - \sum_{i=1}^{N} m_i a_i(t)\sin(2\pi f_i t)\right]\sin(2\pi f_0 t)\right\}$$

$$= V_0\left\{\cos(2\pi f_0 t + \pi/4) + \sum_{i=1}^{N} m_i a_i(t)\cos[2\pi(f_i + f_0)t]\right\}$$

(10.6.3)

In this case, only the positive sideband in the spectrum remains while the negative sideband vanishes. Fig. 10.6.4 shows the optical SSB spectrum of 4 subcarrier channels loaded onto the lower sideband of an optical carrier with subcarrier frequencies of 3.6, 8.3, 13, and 18 GHz. Each subcarrier channel is modulated with a 2.5 Gb/s binary NRZ data sequence. This spectrum was measured with a scanning Fabry-Perot interferometer with 1 GHz resolution bandwidth (Hui et al., 2002).

It is also possible to upload two independent groups of subcarrier channels, $V_1(t)$ and $V_2(t)$, onto the positive sideband and the negative sideband of an optical carrier independently based on the configuration shown in Fig. 10.6.3B, where

$$V_1(t) = V_{10}\left[1 + \sum_{i=1}^{N} m_{1,i} a_{1,i}(t)\cos(2\pi f_i t)\right]$$

(10.6.4)

and

$$V_2(t) = V_{20}\left[1 + \sum_{i=1}^{N} m_{2,i} a_{2,i}(t)\cos(2\pi f_i t)\right]$$

(10.6.5)

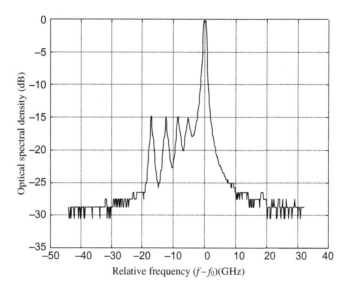

FIG. 10.6.4

Example of optical SSB spectrum with four subcarrier channels loaded in the lower sideband of an optical carrier. f_0 is the optical carrier frequency.

where $a_{1,i}(t)$ and $a_{2,i}(t)$ are signals of the ith subcarrier channels to be loaded onto the upper and lower sidebands of the optical carrier, respectively. A 2×2 90° RF hybrid coupler is used in this configuration, and the output optical signal has three parts: $V_{SSB}(t) = V_C + V_{UP}(t) + V_{LP}(t)$, where the upper sideband is

$$V_{UP}(t) = V_{1,0} \sum_{i=1}^{N} m_{1,i} a_{1,i}(t) \cos\left[2\pi(f_i + f_0)t\right] \qquad (10.6.6)$$

the lower sideband is

$$V_{LP}(t) = V_{2,0} \sum_{i=1}^{N} m_{2,i} a_{2,i}(t) \sin\left[2\pi(f_i - f_0)t\right] \qquad (10.6.7)$$

and the optical carrier component is

$$V_C = (V_{1,0} + V_{2,0}) \cos(2\pi f_0 t) \qquad (10.6.8)$$

Practically, optical single-sideband (SSB) modulation can be accomplished in the E/O conversion process by using a dual-drive electro-optic MZM as shown in Fig. 7.2.1 in Chapter 7, where the first optical beam splitter of the MZM creates 90 degree phase shift of the optical carrier at frequency f_0, and the mixing between RF signal and optical carrier happens at the two electrodes of the two MZM arms (Hui et al., 2002). In order to allow two independent groups of subcarriers to be uploaded onto the upper and the lower sidebands independently, Fig. 7.2.1 in Chapter 7 has to be modified by using a 2×2 90 degree RF hybrid coupler as shown in Fig. 10.6.5. This ensures a 90 degree relative phase shift between the RF signals $v_1(t)$ and $v_2(t)$ applied on the two MZM electrodes. Meanwhile there is also a 90 degree relative phase shift between the two groups of RF subcarrier channels $V_1(t)$ and $V_2(t)$.

Note that for optical SSB generated from a MZM biased at the quadrature point, there is a significant carrier component which is much bigger than the signal sidebands. This is mainly due to the small

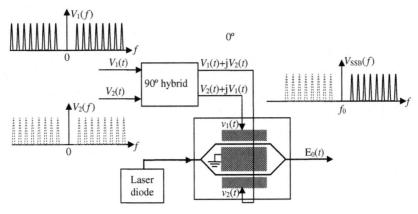

FIG. 10.6.5

Block diagram of optical single-sideband modulation with both sidebands utilized to carry independent information.

modulation index, m_i, of each subcarrier channel to avoid the nonlinearity of modulation. This strong carrier component can be reduced by narrowband optical filtering with a notch optical filter, or eliminated with a Sagnac loop as shown in Fig. 10.6.6A. A fiber Sagnac loop based on a 3-dB coupler has

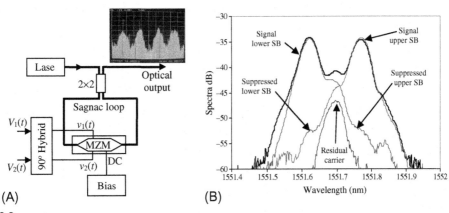

FIG. 10.6.6

(A) Block diagram of optical SSB modulation with two independent RF signals $V_1(t)$ and $V_2(t)$ to be loaded onto the lower and the upper bands of the optical carrier. An optical Sagnac loop is used to suppress the optical carrier component. (B) Measured optical SSB spectra. *Red*: spectrum of $V_2(t)$ indicated by "Suppressed lower SB", *brown*: spectrum of $V_1(t)$ indicated by "Suppressed upper SB", *blue*: residual optical carrier, and *bold black*: composite optical spectrum.

total reflection for the constant wave component as discussed in Chapter 6 (Example 6.2), which is the optical carrier, but high-frequency components can pass through. Fig. 10.6.6B shows an example of measured optical spectra with $V_1(t)$ loaded on the lower sideband and $V_2(t)$ loaded on the upper sideband of an optical carrier, where $V_1(t)$ and $V_2(t)$ are two independent RF signals. In Fig. 10.6.6B, the optical carrier component is suppressed and is more than 10 dB lower than the two signal sidebands. $V_1(t)$ and $V_2(t)$ can each carry a single channel or multiple subcarrier channels as shown in the inset of Fig. 10.6.6A, where both $V_1(t)$ and $V_2(t)$ are composed of two independent subcarrier channels.

For the double-sideband optical spectra shown in Figs 10.6.5 and 10.6.6, a narrowband optical filter has to be used in front of the receiver to separate the upper and the lower sidebands, so that they can be detected separately. Otherwise, coherent detection I/Q detection can also be used to recover complex optical field.

10.6.2 CARRIER-TO-SIGNAL RATIO, INTER-MODULATION DISTORTION AND CLIPPING

CNR vs. SNR: In high-speed digital modulation based on time division multiplexing, SNR at the receiver is commonly used which is defined as, $SNR = I_{ave}^2/(\rho_n B_e)$, where I_{ave}^2 is the average signal electric power which is the square of the average photocurrent, ρ_n is the noise power spectral density, and B_e is the electric bandwidth. Whereas for a SCM system with multiple subcarrier channels, SNR is not enough to determine the signal quality of each subcarrier channel. In a SCM receiver, the composite photocurrent signal, with N subcarrier channels, is

$$I(t) = I_0 \left[1 + \sum_{i=1}^{N} m_i a_i(t) \cos\left(2\pi f_i t\right) \right] \tag{10.6.9}$$

where m_i is the modulation index of the ith subcarrier with the normalized signal $a_i(t)$ and a RF subcarrier frequency f_i. Let us still assume that the Gaussian noise power spectral density is ρ_n, the carrier-to-noise ratio (CNR) of the ith subcarrier channel is defined as

$$CNR_i = \frac{m_i^2 I_0^2}{2\rho_n B_i} \tag{10.6.10}$$

where $m_i^2 I_0^2/2$ is the average signal power of the ith subcarrier channel, and $\rho_n B_i$ is the noise power within the electric bandwidth B_i of that particular subcarrier channel. If only a single subcarrier channel is used ($N = 1$), then $CNR = SNR$. But in general with $N > 1$, CNR is smaller than SNR because the reduction of modulation index m_i, which reduces the CNR. With the increase of the number of subcarrier channels N, modulation index of each channel m_i has to decrease accordingly to avoid clipping and modulation nonlinearity.

Clipping: Use direct modulation on a laser diode as an example in a SCM transmitter as shown in Fig. 10.6.7A, where I_{th} is the laser threshold current. For an ideally linear power-vs-current (PI) curve of a laser diode operating above threshold, the electric current signal is linearly converted into an optical signal. But the major concern is the clipping if the minimum signal electric current is lower than the threshold current of the laser. For an analog signal with N subcarrier channels, and with a DC bias current I_b on the laser diode, The driving current is

$$I(t) = I_{th} + (I_b - I_{th}) \left[1 + \sum_{i=1}^{N} m_i a_i(t) \cos\left(2\pi f_i t\right) \right] \tag{10.6.11}$$

and the normalized output optical power is

$$P(t)/P_b = 1 + \sum_{i=1}^{N} m_i a_i(t) \cos(2\pi f_i t) \tag{10.6.12}$$

where P_b is the average optical signal power determined by the biasing current I_b. In order to avoid clipping, $(\sum_{i=1}^{N} m_i) \leq 1$ has to be satisfied with a normalized signal $|a_i(t)| \leq 1$. Thus the modulation index is inversely proportional to the number of subcarrier channels. If all channels have the same modulation index m, this limit becomes $m \leq 1/N$. Then, based on Eq. (10.6.10), *CNR* is inversely proportional to the square of the number of channels ($CNR \propto N^{-2}$). This *CNR* reduction can be significant when there is a large number of subcarrier channels, in which the composite signal has a very large peak-to-average power ratio (PAPR). However, the probability for the instantaneous signal current in Eq. (10.6.9) to reach the minimum value, which is $-mN$, is very small for a large number of non-synchronized subcarrier channels. Therefore, practical system designs often use $m \leq \sqrt{N}$ as the limit of clipping, which is a tradeoff between signal degradation due to clipping and the improvement of CNR with a relaxed limit in the modulation index m.

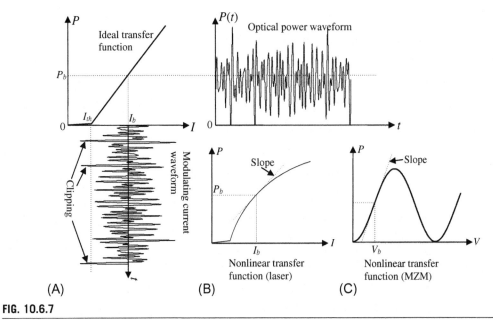

FIG. 10.6.7

(A) Illustration of direct modulation through an ideal laser diode with linear transfer function above threshold. (B) and (C) nonlinear transfer functions of laser diode and Mach-Zehnder modulator.

A more rigorous analysis on the impact of clipping has to consider the probability of clipping. Assume that all subcarrier channels have the same modulation index m, when the number of subcarrier channels is large enough ($N \gg 1$), the probability distribution function (PDF) of the composite current signal approaches Gaussian,

$$\mathrm{PDF} = \frac{1}{\sigma\sqrt{2\pi}}\exp\left[\frac{-(x-1)^2}{2\sigma^2}\right] \tag{10.6.13}$$

where $x = \sum_{i=1}^{N} m_i a_i(t)\cos(2\pi f_i t)$ is the normalized signal, and $\sigma^2 = Nm^2/2$ is the variance. The total power that is clipped-off from the signal can be found from integration (Saleh, 1989),

$$P_{clip} = \frac{P_b}{\sigma\sqrt{2\pi}}\int_{-\infty}^{0} x^2\exp\left[\frac{-(x-1)^2}{2\sigma^2}\right]dx \approx P_b\sqrt{\frac{2}{\pi}}\sigma^5\exp\left(\frac{-1}{2\sigma^2}\right) \tag{10.6.14}$$

Then an average carrier-to-clipping (CCL) ratio can be used to determine the clipping induced distortion, which is defined as,

$$\mathrm{CCL} = \frac{P_b m^2/2}{P_{clip}/N} \approx \sqrt{\frac{\pi}{2}}\sigma^{-3}\exp\left(\frac{1}{2\sigma^2}\right) \tag{10.6.15}$$

where $P_b m^2/2$ is the signal average power per carrier and P_{clip}/N is the clipped power per channel (Fig. 10.6.8).

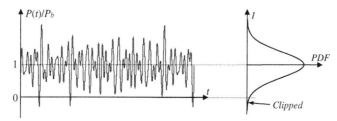

FIG. 10.6.8

Analog waveform consisting of many subcarrier channels *(left)* and the Gaussian probability distribution function (PDF).

In addition to clipping, nonlinear transfer function of EO conversion can also create waveform distortion with the increase of high-order harmonics. As illustrated in Fig. 10.6.7B and C, respectively, nonlinear transfer function can be caused by saturation of laser diode PI curve at high power levels, or by the nonlinear nature of a MZM-based external electro-optic modulator. A nonlinear transfer function can be expressed in a Taylor expansion around the bias point I_b, and the output optical power is related to the input electric current $I(t)$ (or voltage $V(t)$ for MZM external modulator) as

$$P(t) = P_b + \frac{dP}{dI}(I - I_b) + \frac{1}{2}\frac{d^2P}{dI^2}(I - I_b)^2 + \frac{1}{3!}\frac{d^3P}{dI^3}(I - I_b)^3 + \dots \tag{10.6.16}$$

where the normalized multi-carrier electric driving signal is

$$(I - I_b) = I_0\sum_{i=1}^{N} m_i a_i(t)\cos(2\pi f_i t) \tag{10.6.17}$$

with $I_0 = I_b - I_{th}$ the normalized amplitude of the modulating current. An ideal linear transfer function requires all the higher order derivatives such as d^2P/dI^2 and d^3P/dI^3 to be zero. Otherwise, waveform distortion may exist (Phillips et al., 1991). For each subcarrier channel, harmonic distortion can be created through the generation of high-order harmonics by the nonlinear transfer function. When multiple channels are considered, nonlinear mixing between different channels can also create inter-modulation distortion, known as IMD.

For example if $d^2P/dI^2 \neq 0$, $\left[\sum_{i=1}^{N} m_i a_i(t) \cos(2\pi f_i t)\right]^2$ will produce second-order mixing terms among different subcarriers at frequencies $(f_i \pm f_j)$ with $i, j = 1, 2, 3, ...N$. These new frequency components can create significant crosstalk among subcarrier channels. Composite-second-order (CSO) distortion of the ith channel is defined as

$$CSO_i = \frac{I_0^2 m_i^2 / 2}{\sum P_{2nd}(f_i)} \qquad (10.6.18)$$

where $(I_0^2 m_i^2)/2$ is the carrier power of the ith channel, and $\sum P_{2nd}(f_i)$ is the summation of all second order mixing products created at the frequency f_i.

Similarly, if $d^3P/dI^3 \neq 0$, $\left[\sum_{i=1}^{N} m_i a_i(t) \cos(2\pi f_i t)\right]^3$ can produce third-order mixing terms at frequencies $(f_i \pm f_j \pm f_k)$ with $i, j, k = 1, 2, 3, ...N$. Composite-triple-beat (CTB) is commonly used to specify the distortion of the ith channel due to the third-order mixing, which is defined as

$$CTB_i = \frac{I_0^2 m_i^2}{2 \sum P_{3rd}(f_i)} \qquad (10.6.19)$$

where $\sum P_{3rd}(f_i)$ is the summation of all third-order mixing products created at the frequency f_i which overlaps with the ith channel. Both CSO and CTB are important parameters in the specification of IMD in a multi-carrier analog system.

10.6.3 IMPACT OF RELATIVE INTENSITY NOISE

In an analog system with large number of subcarrier channels, the modulation index of each subcarrier has to be small enough to avoid clipping and other nonlinear effects, so that CNR is a major limiting factor of system performance. For the composite signal optical power shown in Eq. (10.6.11), the photocurrent produced at the receiver is

$$i(t) = I_p \left[1 + \sum_{i=1}^{N} m_i a_i(t) \cos(2\pi f_i t) \right] \qquad (10.6.20)$$

where $I_p = \Re G P_b$ is the average photocurrent with $\Re$ the responsivity of the photodiode, and G the optical gain if an optical preamplifier is used before the photodiode ($G = 1$ if without an optical preamplifier). So that the mean square of the signal photocurrent of subcarrier channel k is

$$\langle i_k^2 \rangle = \frac{1}{2} (m_k I_p)^2 \qquad (10.6.21)$$

The noise variance consists of a number of terms,

$$\langle i_n^2 \rangle = \sigma_{th}^2 + \sigma_{sh}^2 + \sigma_{S-ASE}^2 + \sigma_{RIN}^2 \qquad (10.6.22)$$

where σ_{th}^2, σ_{sh}^2, and σ_{S-ASE}^2 represent variances of thermal noise, shot noise, and signal-ASE beat noise, respectively defined in Eqs. (4.3.2) and (4.3.3), and Eq. (5.2.5). We have neglected dark current noise and ASE-ASE beat noise for simplicity, which are usually much smaller than other noise terms in a practical system. σ_{RIN}^2 is the variance of RIN defined in Eq. (3.3.42). The RIN contribution to the noise electric power spectral density σ_{RIN}^2 can be obtained from the RIN parameter of the laser, and the average signal electric power I_p^2 as

$$\sigma_{RIN}^2 = RIN \cdot I_p^2 B_e \tag{10.6.23}$$

where RIN is the laser RIN parameter defined by Eq. (3.3.42) and we assume it is frequency independent for simplicity, and B_e, is the signal electric bandwidth. Thus, if the RIN parameter of the source is known, the CNR of the kth subcarrier channel is

$$CNR_k = \frac{m_k^2 I_p^2 / 2}{\left[\dfrac{4kT}{R_L} + 2qI_p + 4\Re I_p n_{sp} h\nu (G-1) + RIN \cdot I_p^2\right] B_e} \tag{10.6.24}$$

where k is the Boltzmann's constant, T is the absolute temperature, R_L is the load resistance, q is the electron charge, $\Re$ is the responsivity of the photodiode, n_{sp} is the noise coefficient and G is the gain of optical preamplifier, $h\nu$ is the photon energy, B_e is the receiver electric bandwidth and B_o is the optical bandwidth.

Fig. 10.6.9 shows the calculated CNR for a subcarrier channel $B_e = 100$ MHz bandwidth. Other parameters are: $m_k = 1\%$, $n_{sp} = 1.58$, $G = 20$ dB, $R_L = 50\,\Omega$, $T = 300$ K, $\Re = 0.85$ A/W, $\lambda = 1550$ nm, and $RIN = -130$ dB/Hz.

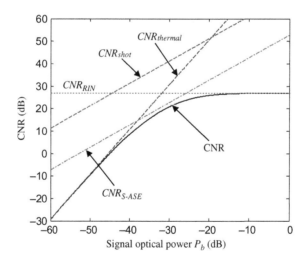

FIG. 10.6.9

CNR as the function of received signal optical power (*Black* solid line), and CNR limits due to different noise sources.

When signal optical power is very low, thermal noise is the limiting factor. As indicated in Eq. (10.6.24), under thermal noise limit (only considering thermal noise),

$$CNR_{thermal} = \frac{m_k^2 I_p^2 R_L}{8kTB_e} \propto m_k^2 I_p^2 \tag{10.6.25}$$

so that the $CNR_{thermal}$ vs P_b curve has a 20 dB/decade slope. With the increase of the signal optical power, signal-ASE noise becomes the limiting factor. Signal-ASE noise limited CNR is

$$CNR_{S-ASE} = \frac{m_k^2 I_p}{8\Re n_{sp}hv(G-1)B_e} \propto m_k^2 I_p \tag{10.6.26}$$

The CNR_{S-ASEI} vs. P_b curve also has a 20 dB/decade slope. Shot noise is not a limiting factor in this example because its value is always lower than the signal-ASE beat noise with the parameters we used. In a receiver without an optical preamplifier, shot noise may become a limiting factor in a certain signal power region.

Further increasing the signal power, CNR vs. P_b curve can be saturated by the RIN. In fact, RIN-limited CNR (only consider σ_{RIN}^2 in the noise) is independent of the signal optical power,

$$CNR_{RIN} = \frac{m_k^2}{2RIN \cdot B_e} \propto m_k^2 \tag{10.6.27}$$

The increase of signal optical power can improve the CNR caused by all other noise sources except the RIN. Therefore laser source with low RIN is critical for subcarrier multiplexed analog systems when the number of subcarrier channels is large and the modulation index m of each channel is small.

10.6.4 RADIO-OVER-FIBER TECHNOLOGY

In recent years rapid expansion of wireless communication systems and networks demand more and more bandwidth to satisfy the fast growth of data traffic and diversified services. However, spectrum resource in wireless domain is extremely precious because it has to be shared by many users for a large variety of applications ranging from civilian to military. On the other hand, optical fiber as a wire line is able to provide much broader spectral bandwidth, but less mobile compared to wireless communication. For high-speed and long-distance fiber communication networks, digital modulation formats are usually used for the superior transmission performance compared to analog modulation, and a digital

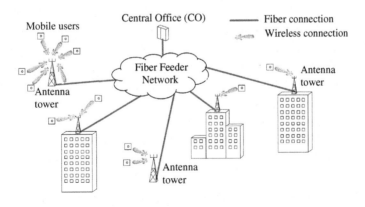

FIG. 10.6.10

Illustration of RoF network in which optical fibers are used to deliver high-frequency carriers from central office to antenna towers.

waveform has the ability of regenerating, reshaping, and retiming. But wireless communication uses microwave carriers to deliver signals through antennas. In order to make full use of the spectrum resources, high-frequency microwave carriers in multi-GHz range are more and more common, and expanding to the millimeter-wave frequency region for 5-G wireless networks. With the frequency increase of the microwave carriers, the coverage area of antenna will be reduced due to the increased propagation loss, and therefore more antennas have to be deployed to ensure the service coverage. Thus, antenna station has to be as simple as possible to reduce the network equipment cost and the maintenance effort. Instead of generating the microwave carrier and performing frequency upconversion of the baseband signal at each antenna tower, it makes more economic sense to consolidate these tasks in the central office and use optical fibers to deliver data-loaded microwave carriers to antennas. Analog system is appropriate for this situation because the bandwidths of data channels are usually much narrower in comparison to the frequency of the RF carrier. This type of systems using fiber to deliver radio frequency carriers is commonly referred to radio over fiber (RoF) (Novak et al., 2016), and a generic RoF network architecture is shown in Fig. 10.6.10.

Fig. 10.6.11 shows an example of bidirectional fiber link between a central office and an antenna station. The downstream data is uploaded onto a microwave carrier at frequency f_c through mixing with an oscillator, and is then converted into the optical domain through an external electro-optic modulator and a laser diode. The downstream signal may be composed of many low speed subcarrier channels using bandwidth efficient modulation formats commonly used for wireless communications. An optical circulator directs the downstream optical signal into the transmission fiber in the forward direction, and isolates it from the upstream optical signal propagating in the opposite direction. At the antenna, the downstream optical signal is detected by a photodiode, and the electric signal is bandpass amplified and directly fed to the antenna. Both the photodiode and the RF amplifier have to operate in the frequency window around the RF carrier which can be up to 100 GHz, but may only need less than 1 GHz bandwidth to accommodate the wireless data traffic. The cost of this type of narrowband RF amplifiers can be much lower than the broadband amplifiers commonly used in high-speed optical transceivers which normally start from 100 kHz and requiring flat frequency response.

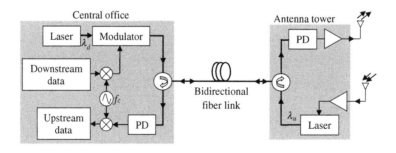

FIG. 10.6.11

Block diagram of a bidirectional fiber link between a central office and an antenna station.

The upstream traffic from wireless users back to the central office usually has lower data rate compared to the downstream traffic. The upstream microwave signal picked up by the receiving antenna is amplified and converted into optical domain through direct modulation on a laser diode.

The wavelength of the upstream optical signal, λ_u, can be different from that of the downstream optical signal λ_d, so that wavelength de-multiplexing can be used for the bidirectional fiber link and the optical circulators can be replaced by WDM couplers in Fig. 10.6.11. After arriving to the central office, the upstream optical signal is converted into an RF signal through direct detection by a photodiode. Then, it is frequency down-converted to the baseband through mixing with an RF local oscillator.

The ability of delivering signals on high-frequency carriers directly to the antenna tower with minimum loss is a unique advantage of RoF system, which cannot be achieved by coaxial cables. This is especially true when the carrier frequency is more than 50 GHz and into the millimeter wave region. Fig. 10.6.12 shows an example of generating ultrahigh-frequency RF carriers through heterodyne mixing of two optical carriers. In this configuration, two laser diodes are used, emitting at wavelengths λ_1 and λ_2, respectively, corresponding to optical frequencies of f_1 and f_2. The optical output of one laser is intensity modulated by the baseband electric signal through an external modulator before combining with the other laser output and sent to the transmission fiber. These two optical carriers mix at a high-speed photodiode, generating a heterodyne beating which creates a carrier at frequency $f_2 - f_1$, and the baseband electric signal is already loaded on this carrier through the external modulator in the central office. A band-pass filter (BPF) selects the RF carrier frequency and feed it to the antenna after an RF power amplifier. In the 1550-nm-wavelength window, 1-nm-wavelength separation between the two lasers is equivalent to approximately 125-GHz frequency difference which determines the heterodyne beating frequency $f_2 - f_1$. But the limitation of this technique is eventually set by the response speed of the photodiode which can approach 100 GHz for narrowband applications.

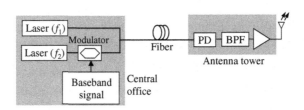

FIG. 10.6.12

Millimeter wave generation using optical heterodyne technique.

Because the frequency of the RF carrier f_c in an RoF system is much higher than the bandwidth that the signal occupies, chromatic dispersion tolerance is determined by the frequency separation between the positive and the negative modulation sidebands, which is $2f_c$ if double sideband modulation is used. The differential group delay between the two sidebands is $\delta t = 2f_c DL\lambda^2/c$, where D is the dispersion parameter, L is the fiber length, λ is the center wavelength and c is the speed of light. If this differential group delay reaches half-wavelength of the RF carrier, which is $c/(2f_c)$, carrier fading will happen. Thus optical SSB modulation described in Section 10.6.1, has to be used to avoid carrier fading if the accumulated chromatic dispersion of the fiber, DL, is large enough.

Another challenge of RoF system is the signal dynamic range. Due to the nonlinearity of the PI curve of laser diode, or the intrinsic nonlinear transfer function of an external electro-optic modulator described in Fig. 10.6.7, high modulation index may create large harmonic distortion and IMD. On the other hand, small modulation index will reduce CNR. Therefore the dynamic range of RoF signal is restricted by these two limits. Spurious-free dynamic range (SFDR) is a system parameter that is defined as the difference between the minimum signal that can be detected above the noise floor and the maximum signal that can be detected before reaching a specified distortion level. Various techniques have been used to increase SFDR on RoF systems, such as feed-forward linearization and pre-distortion (Yao, 2009).

10.7 OPTICAL SYSTEM LINK BUDGETING

Optical system design and specification require performance prediction through careful analysis based on the characteristics of all components used in the system, which is known as link budgeting. For signal optical power-limited optical systems, basic link budgeting of signal optical power limited optical systems considers the receiver sensitivity, the optical power emitted from the transmitter, and the total loss of the fiber system to find the system power margin. More sophisticated link budgeting has to consider the impacts of fiber chromatic dispersion, PMD, nonlinearity effects, and the accumulated amplified spontaneous emission (ASE) noise along the system if optical amplifiers are used.

10.7.1 POWER BUDGETING

In an optical system without inline optical amplifiers, receiver sensitivity is defined as the minimum signal optical power at the receiver to obtain the specified BER, as has been discussed in Section 8.3. In such a system, link budgeting is essentially a power budgeting to find the power margin between the receiver sensitivity and the signal optical power that actually reaches the receiver.

Fig. 10.7.1 shows an example of an optical system consisting of an optical transmitter, and a direct detection receiver. The fiber has five sections, linked together by four fiber connectors and three fiber splices. There is also a fiber tap in the system which taps out a portion of the optical power for monitoring purpose. Assume the signal optical power emitted by the transmitter is P_{tx}, and the loss of each fiber section is $\alpha_i L_i$ where α_i is the attenuation coefficient and L_i is the length of the ith fiber section, with $i = 1, 2, 3, 4, 5$. The loss of each fiber coupler is A_c, the loss of each fiber splice is A_s, and the loss of the fiber tap is A_{tap}. The signal optical power that reaches the receiver is then,

$$P_s = P_{tx} - \sum_{i=1}^{5} \alpha_i L_i - 4A_c - 3A_s - A_{tap} \tag{10.7.1}$$

where the optical power is in dBm, and the losses of different components are all in dB. If the receiver sensitivity is P_{sen}, then the system power margin is $P_{\text{margin}} = P_s - P_{sen}$ which has to be positive for the system to have a BER lower than the specified value. It is important to note that receiver sensitivity P_{sen}

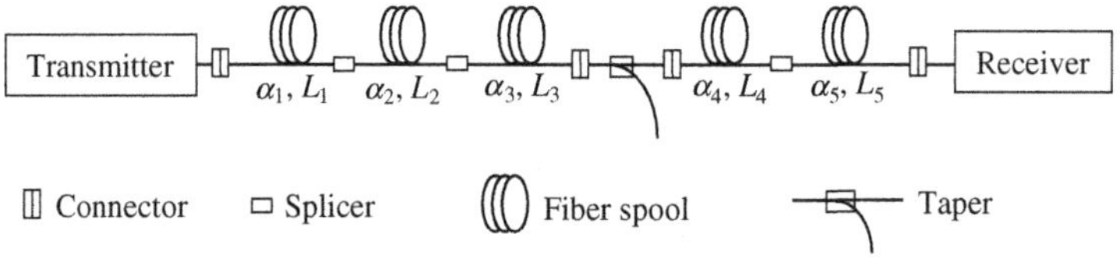

FIG. 10.7.1

Example of an optical system with five fiber sections, four couplers, three splices, and one fiber tap.

is determined by the target BER level of the receiver, which depends on the coding technique used in the system. Traditionally, BER levels of 10^{-12}–10^{-15} are required for most of the long-distance optical transmission systems. With the advance of forward error correction (FEC) coding, the required pre-FEC BER level can be much relaxed, to the level around 10^{-3} depending on the FEC coding strength. This helps reducing the power level of P_{sen}, and improving system power margin.

In the example shown in Fig. 10.7.1, assume attenuation coefficients of fiber sections are $\alpha_1=0.22$ dB/km, $\alpha_2=0.25$ dB/km, $\alpha_3=0.21$ dB/km, $\alpha_4=0.28$ dB/km, $\alpha_5=0.27$ dB/km, fiber lengths are $L_1=2.2$ km, $L_2=3.5$ km, $L_3=15$ km, $L_4=21$ km, and $L_5=12$ km, loss of each connector is $A_c=0.3$ dB, loss of each splice is $A_s=0.1$ dB, and the loss of the fiber tap is $A_{tap}=1.5$ dB. The total system loss is 16.63 dB. If the transmitter power is $P_{tx}=2$ mW, which is 3 dBm, and the receiver sensitivity is $P_{sen}=-20$ dBm, the system margin is $P_{margin}=6.37$dB. This means that the system can tolerate an additional 6.37 dB loss.

In practical system design, at least 3 dB margin is usually reserved to ensure the reliability of the system against unexpected events and system aging over the life time.

10.7.2 OSNR BUDGETING

In an optical system with multiple inline optical amplifiers, accumulated ASE noise along the link is most likely the limiting factor of the transmission quality, and thus receiver sensitivity is no longer a useful parameter to define the system performance as discussed in Section 8.3. For binary coded digital systems, BER is determined by both random noise and mostly deterministic waveform distortion.

In a transmission system employing multiple inline optical amplifiers, signal-ASE beat noise is often the dominant noise source, so that the BER of the received signal is determined by the OSNR at the receiver. Because of the statistical nature of the random noise, the relationship between the OSNR and the electric noise variance after photodetection can be calculated analytically for the evaluation of the Q-value and the BER.

On the other hand, waveform distortion can be caused by chromatic dispersion, PMD, self-phase modulation (SPM), and multi-path interference, as well as inter channel crosstalk such as cross-phase modulation (XPM) and four-wave mixing (FWM). All these effects are not completely random; instead they are dependent on the signal waveforms, so that they are deterministic. The evaluation of eye-closure penalty caused by waveform distortion is much more difficult, in comparison to the impact of random noise, because of the variety of sources and different nature of their impact in the signal waveforms. Numerical simulations, semi-analytic and analytic techniques described in Chapter 8 are often used to evaluate waveform distortions and the eye-closure penalty. Link budgeting has to consider all possible sources of waveform distortion and find the overall impact in the eye-closure. In the process of optical system design, it can also be beneficial to consider different sources of waveform distortion separately to identify major limiting factors, and to find most effective ways to improve the system performance. After finding the eye-closure penalty for a specific system, OSNR margin can be obtained, which is the difference between the actual OSNR and the required OSNR for a target BER.

Fig. 10.7.2 shows an example of Q-value (A) and BER (B) as the function of OSNR for a system with three different configurations and thus three different eye-closure penalties. The difference of eye-closure penalties can be caused by different types of fibers, different signal optical power levels, or different channel spacing in a WDM scenario. In order to achieve a BER of 10^{-15}, or equivalently $Q=8$, the three system configurations require OSNR levels of 12, 15, and 18 dB, respectively. Again,

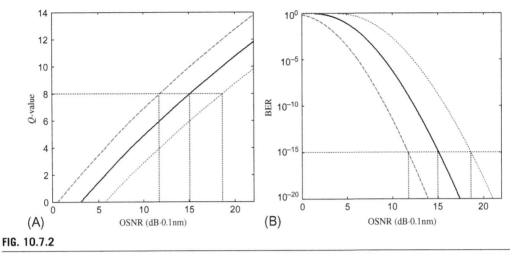

FIG. 10.7.2

An example of Q-value (A) and BER (B) as the function of OSNR for three different system configurations. OSNR is defined with 0.1-nm resolution bandwidth for the noise spectral density.

we emphasize that the unit of OSNR shown in Fig. 10.7.2 is [dB · 0.1nm] because the optical noise power spectral density is measured with 0.1-nm spectral resolution as usually used in industrial R&D practice. If one wants to use [dB · Hz] as the unit, a conversion factor of 12.5×10^9 (or approximately 101 dB) has to be used when operating in the 1550-nm-wavelength window because 0.1 nm is equal to 12.5 GHz.

With these BER vs. OSNR curves and the required OSNR values, it is straightforward to find the OSNR margins if he system configuration is defined. For example, for a system with N amplified fiber spans as shown in Fig. 10.7.3A assume that each fiber span has the same loss so that each EDFA has the same optical gain, G, which exactly compensates for the fiber loss, the OSNR at the receiver is

$$OSNR = \frac{P_{ave}}{2n_{sp}(hc/\lambda)\sum_{n=1}^{N}(G_i - 1)} \approx \frac{P_{ave}}{F(hc/\lambda)NG} \tag{10.7.2}$$

where P_{ave} is the average signal optical power, $n_{sp} \approx F/2$ is the noise parameter with F the noise figure of each EDFA, h is the Planck's constant, c is the speed of light and λ is the signal wavelength. Assume EDFA has a noise figure of 5 dB ($F = 3.16$) and the optical gain of each EDFA is $G = 20$ dB (compensates the loss of 80 km fiber with 0.25 dB/km attenuation parameter), average signal optical power is $P_{ave} = 0$ dBm, and the signal wavelength is $\lambda = 1550$ nm. OSNR is inversely proportional to the number of spans as shown in Fig. 10.7.3B. If the system has 20 fiber spans for a total reach of 1600 km, and if the required OSNR is 15 dB [the black-solid line shown in Fig. 10.7.2(b)], the OSNR margin will be 5 dB.

As indicated in Eq. (8.3.11) in Chapter 8, for signal-ASE beat noise limited optical systems, the required OSNR is inversely proportional to $(\sqrt{A} - \sqrt{B})^2$ where A and B are upper and lower inner levels of the normalized eye as defined by Fig. 8.2.2. Fig. 10.7.3C shows that the required OSNR

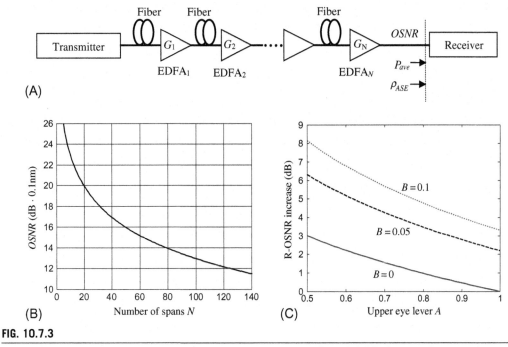

FIG. 10.7.3

(A) configuration of a fiber-optic system with N inline optical amplifiers, (B) OSNR as the function of the number of amplified fiber spans N, and (C) degradation of the required OSNR due to the decrease of A level and B level of the eye diagram.

increases as the decrease of A, with $B=0, 0.05$ and 0.1. Every 3 dB increase of the required OSNR will approximately halve the maximum number of fiber spans. Therefore, compensating the impact of waveform distortion caused by various sources is a major task in fiber optic system design.

10.8 SUMMARY

In this chapter we have discussed a number of modulation formats commonly used for optical communication systems. Traditionally binary NRZ has been the most popular choice for high-speed optical systems because it has relatively narrow spectral bandwidth and high tolerance to chromatic dispersion of the fiber. The relatively narrow electric bandwidth also makes it easier for the design of transmitters and the receivers. However, RZ coding has more concentrated signal peak power at the decision point of each bit, which helps improving the receiver sensitivity. As dispersion compensation become readily available, eye closure penalty due to chromatic dispersion can be largely removed, and RZ coding generally exhibits improved performance in dispersion compensated long-distance optical systems in comparison to NRZ.

We have discussed the generation of pseudo-random bit sequence (PRBS) in this chapter. The purpose of using PRBS is to mimic real data traffic as close as possible. The longest continuous bars and spaces specify the lowest frequency component of the data traffic and also provide stretch test of clock recovery circuit in the receiver.

In addition to binary modulation, we have also discussed phase correlated intensity modulation formats such as M-ary, polybinary, duobinary, and CSRZ with the purposes of reducing the spectral bandwidth and increasing the tolerance to chromatic dispersion in the fiber. In recent years, there is a significant interest in the development of four-level pulse amplitude modulation known as PAM4 for short reach optical interconnection. In fact, this is just a specific case of M-ary discussed in Section 10.3.1 with $M=4$.

Basic optical phase modulation formats such as BPSK and DPSK are introduced in Section 10.4. While BPSK requires coherent detection, DPSK can be received by both coherent detection and direct detection. Coherent detection helps improving the receiver sensitivity, but direct detection without the need of an optical local oscillator can be much simpler, and the receiver sensitivity can be improved by employing an optical pre-amplifier in the receiver. Multilevel phase shift keying (PSK) modulation and complex optical field modulation such as QAM, help further increase optical spectral efficiency, which are now quite commonly used for high capacity optical transmission with dense WDM. Similar to eye diagrams used for AM optical signals, constellation diagrams are used to evaluate the performance of complex modulated optical signals to display both amplitude and phase information of the recovered signal. A more quantitative measure of signal quality from a constellation diagram is the error victor magnitude (EVM) which is a good indication of BER.

Beside digital modulation, we also discussed analog modulated optical systems, which have been traditionally used for CATV, and are now more and more used to support wireless communication networks. RoF helps the delivery of high-frequency RF signals between central offices and wireless antenna towers. While BER is the ultimate measure of digital transmission system quality, CNR and linearity are most relevant quality measures of analog systems. Especially, the impacts of high-order harmonics and RIN on the performance of analog systems are discussed.

Finally, optical system link budgeting has been discussed in Section 10.7, which is one of the most important aspects of optical system design and engineering. In order to achieve the targeted system performance, in terms of BER for digital systems and CNR for analog systems, the required minimum signal optical power at the receiver is defined as the receiver sensitivity if the system is signal power limited. For optical systems employing multiple in-line optical amplifiers, required-OSNR will be a more appropriate criterion for link budgeting. Link budgeting is a comprehensive task which has to consider the level of optical signal as well as all possible mechanisms of performance degradation including noise and waveform distortion.

PROBLEMS

1. For a binary modulated NRZ signal, an isolated digital "1" is a square wave with a width $T=1/B$, with B the data rate.
   (a) What is the FWHM width of the power spectral density of this pulse? (use B as a variable)
   (b) If the coding is RZ with $T=0.5/B$, what is the FWHM width of the power spectral density of this pulse?

2. To obtain Eq. (8.2.10) in Chapter 8, NRZ was considered.

   (a) What will be the corresponding Q equation for RZ waveform assuming $A=1$ and $B=0$ (no waveform distortion)?

   (b) If the system only has signal-dependent noise, what will be the Q-value improvement (in dB) of RZ in comparison with NRZ?

3. For a 10 Gb/s binary PRBS, in order for the lowest frequency component to be lower than 50 kHz, what should be the lowest sequence length?

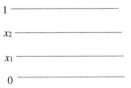

4. For an M-ary with $M=4$, if *signal-dependent* noise is the dominant noise source in the receiver, please find the eye levels x_1 and x_1 which would make all three eyes to have equal sensitivity.

5. Convert a binary data sequence $a_k=[0, 1, 1, 1, 0, 0, 1, 0, 1, 0, 1, 1, 1, 0, 0, 1, 0, 1, 1, 0]$ into a polybinary sequence for $M=4$. Write sequences of b_k and c_k, respectively, with $k=1, 2, ...20$. (note: use $b_1=b_2=0$ as the initial bits).

6. Convert a binary data sequence $a_k = [0, 1, 1, 1, 0, 0, 1, 0, 1, 0, 1, 1, 1, 0, 0, 1, 0, 1, 1, 0]$ into a dual-binary sequences which has 3 levels $[0, 1, 2]$. Realign these 3 levels to $[-1, 0, 1]$ and square the waveform, show that this is the same as the original binary waveform (may also be completely flipped: $0 \rightarrow 1$ and $1 \rightarrow 0$)

7. Consider a 40 Gb/s data sequence of $a_k = [0, 1, 0, 1, 0, 1, 0, 1, 0, 1, 0, 1, 0, 1, 0, 1, 0, 1, 0, 1]$ shown in the figure. If duo-binary coding is applied, please draw the waveform of c_k. What is the fundamental frequency of c_k?

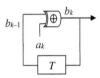

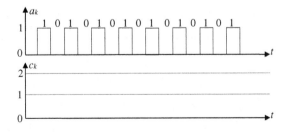

8. Use Matlab to create a binary bit sequence of 40 Gb/s, and convert it to duo-binary bit sequence. Plot the power spectra of both binary and duo-binary sequences at the same bit rate (40 Gb/s) similar to that in Fig. 10.3.7.

9. Consider an optically amplified fiber link operating in the C-band (1550 nm wavelength). At the receiver, OSNR=7.8 dB was measured with an optical spectrum analyzer with resolution bandwidth (RBW) of 0.1 nm. The optical signal carries 10 Gb/s BPSK data, and the receiver employs a balanced optical homodyne detection scheme with an electric bandwidth of 8 GHz. Assume that there is no waveform distortion, and LO-ASE beat noise is the only noise source of concern,

(a) What is the expected BER?

(b) If the receiver electric bandwidth is doubled to 16 GHz, what is the BER?

(c) Discuss why increasing receiver bandwidth in this case degraded BER performance.

10. BER test of a 40 Gb/s optical link has being running for 6 h without any error detected.

(a) What is the least expectation that can be made for the BER of this system?

(b) In this system the signal is DPSK modulated and the receiver is direct-detection with a Mach-Zehnder interferometer as shown in Fig. 10.4.3A, and the receiver electric bandwidth is $B_e=32$ GHz. What would be the OSNR (reference to 0.1 nm resolution bandwidth) of the received optical signal if the BER is 10^{-12}? (Only consider signal-ASE beat noise and neglect waveform distortion.)

11. Based on the Shannon capacity formula in Eq. (10.5.2), for a spectral efficiency of 1 bit/Hz, what are the BER values of M-QAM with $M=4$, 16, 64, and 128?

12. Consider a 1550-nm subcarrier multiplexed system using standard single mode fiber with 16 ps/nm/km chromatic dispersion, for a subcarrier frequency of 20 GHz which carries 1 Gb/s binary signal,

(a) if optical double-sideband modulation is used, what is the fiber length for carrier fading for direct detection?

(b) discuss the benefit of using optical SSB.

13. Based on configuration shown in Fig. 10.6.3A, create an OSSB subcarrier signal with two independent 1 Gb/s binary pseudorandom data sequences carried on subcarriers at 5 and 8 GHz. Please plot the OSSB spectrum.

14. In an analog AM subcarrier multiplexed optical system using a directly modulated laser diode emitting at 1550 nm. The number of SCM channels is 50. The laser threshold current is 10 mA and is biased at 30 mA.

(a) In order to avoid clipping, what is the maximum modulation index and the maximum AC modulation current (peak-to-peak) of each channel (assuming the same modulation index for all the channels)

(b) If the modulation current of each channel is 2 mA, what is the clipping probability?

15. For a simple SCM system with only 3 subcarrier channels, and the normalized modulating current is: $I(t) = \left[1 + \sum_{i=1}^{3} m\cos(2\pi f_i t)\right]$ with at $f_1=1$ GHz, $f_2=2$ GHz and $f_3=3$ GHz, and modulation index $m=0.3$. If the P-I curve of the modulator has $dP/dI=1$ and $d^2P/dI^2=0.1$, what is the composite-second-order (CSO) distortion of the middle channel at 2 GHz?

16. Consider an analog AM subcarrier multiplexed optical system with 20 subcarrier channels, and the laser relative intensity noise (RIN) is $-150\,\text{dB/Hz}$. In the receiver, the load resistance is $R_L = 50\,\Omega$, the temperature is $T = 300\,\text{K}$, the electric bandwidth of each channel is $6\,\text{MHz}$, and the photodiode responsivity is $1\,\text{A/W}$.

   (a) Assume the modulation index of each channel is $m = 1/20$ to avoid clipping, if we only consider thermal noise and RIN noise at the receiver, what is the minimum signal optical power required at the receiver to achieve a CNR of $50\,\text{dB}$?

   (b) Repeat (a) but with laser RIN of $-130\,\text{dB/Hz}$. In this case, what is the maximum achievable CNR?

17. Consider an analog AM subcarrier multiplexed optical system with 100 subcarrier channels each with $B_e = 5$ MHz electric bandwidth, and the laser relative intensity noise (RIN) is $-150\,\text{dB/Hz}$. Assume the receiver CNR is only determined by the RIN.

   (a) what is the optimum modulation index m at which the average carrier-to-clipping (CCL) ratio is equal to the CNR? (can be solved numerically).

   (b) at the optimum modulation index m what is the CNR value?

18. For a broadcasting optical system shown in the following figure, a 1×4 fiber coupler is used to split optical power into 4 output ports with intrinsic splitting loss of 6- and 0.5-dB excess insertion loss. Each connector has $0.2\,\text{dB}$ loss. Attenuation parameter of optical fiber is $0.22\,\text{dB/km}$. All receivers have the same sensitivity of $-30\,\text{dBm}$. If at least 3-dB power margin is needed,

   (a) what should be the signal optical power at the output of the transmitter?

   (b) at this transmitter power, what is the power margin of each receiver?

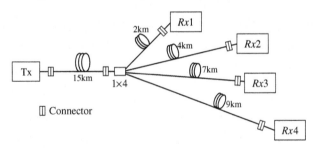

19. Consider a 40-Gb/s QPSK modulated system operating in 1550-nm-wavelength window over 2000 km fiber with 0.22-dB/km loss coefficient and $B_e = 15$-GHz receiver bandwidth. There are 20 EDFAs used in the system to compensate the loss of the fiber, and EDFA spacing is 100 km as shown in the following figure. Assume EDFA noise figure is 6 dB, there is no waveform distortion, signal optical power at each EDFA output is 0 dBm, and the required BER in the receiver is 10^{-4}. What is the OSNR margin of this system?

20. Repeat problem 19 for the same data rate of 40 Gb/s. Please find OSNR margins if 16-QAM and 64-QAM are used. (assume the receiver bandwidth B_e is equal to 0.75 times symbol rate).

REFERENCES

Bosco, G., Carena, A., Curri, V., Gaudino, R., Poggiolini, P., 2002. On the use of NRZ, RZ, and CSRZ modulation at 40 Gb/s with narrow DWDM channel spacing. J. Lightwave Technol. 20 (9), 1694–1704.

Caspar, C., et al., 1999. RZ vs NRZ modulation format for dispersion compensated SMF-based 10-Gb/s transmission with more than 100-km amplifier spacing. IEEE Photon. Technol. Lett. 11 (4), 481–483.

Hui, R., Zhu, B., Huang, R., Allen, C., Demarest, K., Richards, D., 2002. Subcarrier multiplexing for high-speed optical transmission. J. Lightwave Technol. 20, 417–427.

Ip, E., Kahn, J.M., 2007. Feedforward carrier recovery for coherent optical communications. J. Lightwave Technol. 25 (9), 2675–2692.

Ip, E., Lau, A.P.T., Barros, D.l.J.F., Kahn, J.M., 2008. Coherent detection in optical fiber systems. Opt. Express 16 (2), 753–791.

Lach, E., Idler, W., 2011. Modulation formats for 100G and beyong. Opt. Fiber Technol. 17 (5), 377–386.

Lender, A., 1964. Correlative digital communication techniques. IEEE Trans. Commun. Technol. 12 (4), 128–135.

Mahmoud, H.A., Arslan, H., 2009. Error vector magnitude to SNR conversion for nondata-aided receivers. IEEE Trans. Wirel. Commun. 8 (5), 2694–2704.

Miyamoto, Y., Yonenaga, K., Hirano, A., Toba, H., Murata, K., Miyazawa, H., 2001. Duobinary carrier-suppressed return-to-zero format and its application to 100 GHz-spaced 8/spl times/43-Gbit/s DWDM unrepeatered transmission over 163 km. In: Optical Fiber Communication Conference and Exhibit, OFC-2001, Anaheim, CA, USA, paper.

Novak, D., et al., 2016. Radio-over-fiber technologies for emerging wireless systems. IEEE J. Quantum Electron.. 52 (1) paper 0600311.

Olmos, J.J.V., Suhr, L.F., Li, B., Monroy, I.T., 2013. Five-level polybinary signaling for 10 Gbps data transmission systems. Opt. Express 21 (17), 20417–20422.

Penninckx, D., Chbat, M., Pierre, L., Thiery, J.P., 1997. The Phase-Shaped Binary Transmission (PSBT): a new technique to transmit far beyond the chromatic dispersion limit. IEEE Photon. Technol. Lett. 9, 259–261.

Phillips, M.R., Darcie, T.E., 1997. Lightwave analog video transmission. In: Kaminow, I.P., Koch, T.L. (Eds.), Optical Fiber Telecommunications IIIA. Academic Press.

Phillips, M.R., Darcie, T.E., Marcuse, D., Bodeep, G.E., Frigo, N.J., 1991. Nonlinear distortion generated by dispersive transmission of chirped intensity-modulated signals. IEEE Photon. Technol. Lett. 3 (5), 481–483.

Proakis, J.G., 1968. Probabilities of error for adaptive reception of M-phase signals. IEEE Trans. Commun. Technol. 16, 71–81.

Proakis, J.G., 2001. Digital Communications, fourth ed. McGraw-Hill, New York.

Saleh, A.A.A.M., 1989. Fundamental limit on number of channels in subcarrier multiplexed lightwave CATV system. Electron. Lett. 25 (12), 776–777.

Walklin, S., Conradi, J., 1999. Multilevel signaling for increasing the reach of 10 Gb/s lightwave systems. J. Lightwave Technol. 17 (11), 2235–2248.

Way, W.I., 1999. Broadband Hybrid Fiber/Coax Access System Technologies (Telecommunications). Academic Press.

Wilson, S.G., 1996. Digital Modulation and Coding. Prentice Hall.

Winzer, P.J., Essiambre, R.-J., 2006. Advanced optical modulation formats. Proc. IEEE 94 (5), 952–985.

Yao, J., 2009. Microwave photonics. J. Lightwave Technol. 27 (3), 314–335.

Yu, J., Dong, Z., Chien, H.-C., Jia, Z., Li, X., Huo, D., Gunkel, M., Wagner, P., Mayer, H., Schippel, A., 2013. Transmission of 200 G PDM-CSRZ-QPSK and PDM-16 QAM with a SE of 4 b/s/Hz. J. Lightwave Technol. 31, 515–522.

FURTHER READING

Ip, E., Kahn, J.M., 2006. Power spectra of return-to-zero optical signals. J. Lightwave Technol. 24 (3), 1610–1618.

APPLICATION OF HIGH-SPEED DSP IN OPTICAL COMMUNICATIONS

11

CHAPTER OUTLINE

Introduction 497
11.1 Enabling technologies for DSP-based optical transmission 498
11.2 Electronic-domain compensation of transmission impairments 499
    11.2.1 Dispersion compensation 500
    11.2.2 PMD compensation and polarization de-multiplexing 505
    11.2.3 Carrier phase recovery in a coherent detection receiver 507
11.3 Digital subcarrier multiplexing: OFDM and Nyquist frequency-division multiplexing 511
    11.3.1 Orthogonal frequency-division multiplexing 512
    11.3.2 Nyquist pulse modulation and frequency-division multiplexing 523
    11.3.3 Optical system considerations for DSCM 527
11.4 Summary 549
Problems 549
References 553

INTRODUCTION

The major advantage of fiber-optic communication is based on the utilization of the extremely wide spectral bandwidth and low loss of the optical fiber. Other advantages, in comparison to coaxial cables and twisted wires, also include lightweight and small size of the optical fiber, as well as high security of communication against electromagnetic interference and interception. With the introduction of in-line optical amplification, the wavelength-division multiplexing (WDM) technology for multiwavelength operation, and the bandwidth increase of each wavelength channel through high speed modulation, adversary effects such as chromatic dispersion, polarization mode dispersion (PMD), and fiber nonlinearity become more and more pronounced, which can severely limit the system performance.

The strategy of shifting signal processing tasks from the electric domain to the optical domain has been pursued for many years to improve the system performance. All-optical communication networking was once considered an ultimate goal of the optical communications industry. A number of optical domain processing techniques have been demonstrated, such as chromatic dispersion compensation module (DCM) based on the normal dispersion fiber or fiber Bragg gratings, PMD compensation techniques based on the adaptive optics techniques, all-optical wavelength conversion, and packet switching based on the nonlinear optics and semiconductor optical amplifiers. Some of these techniques have been successfully applied in commercial optical systems and networks.

On the other hand, the processing speed of digital electronics has been increased dramatically over the last three decades. As the Moore's law predicted that the number of transistors in CMOS-integrated circuits doubles approximately every 2 years, the processing speed of CMOS electronic circuits follows the same basic rule as well. Cost and power consumption per digital operation have also been reduced considerably. This provided an opportunity for the fiber-optic communications industry to reexamine its strategy by taking advantage of the available high-speed CMOS digital electronic technologies. In fact, some of the optical domain functionalities in fiber-optic communication networks have been shifted back to the electronic domain in the recent years, which provided much improved capability and flexibility. For example, electronic domain dispersion compensation and PMD compensation have been implemented in high-speed coherent optical transmission systems with unprecedented performance improvement (Roberts et al., 2009). In this chapter, we discuss the enabling technologies, configurations, and performance of high-speed optical systems utilizing electronic domain digital signal processing (DSP).

11.1 ENABLING TECHNOLOGIES FOR DSP-BASED OPTICAL TRANSMISSION

In optical communication systems, both WDM and time-division multiplexing (TDM) are commonly used to maximize the transmission capacity. The trade-off between WDM and TDM has enabled the full use of the wide low-loss spectral window of the optical fiber to achieve high transmission capacity while accommodating the use of electronic circuits operating at relatively low speed. Use commercial synchronous optical networking (SONET) optical network as an example, data rate per wavelength channel has been increased from 2.5 Gb/s (OC48) to 10 Gb/s (OC192), and 40 Gb/s (OC768) over the past two decades. In all, 100 and 400 Gb/s per wavelength optical transmission and switching equipments are on their way to becoming new industrial standards. With the per-wavelength capacity increase, tolerance to degradation effects, such as chromatic dispersion and PMD, reduces significantly, and compensation of these effects has to be made more and more precise. Use chromatic dispersion as an example: with the increase of per-channel data rate, the differential group delay within a wavelength channel increases because of the increased spectral bandwidth. At the same time, the tolerance to the differential group delay decreases because the temporal pulse width of each data symbol decreases. Thus, the system tolerance to uncompensated chromatic dispersion is inversely proportional to the square of the symbol rate. From this point of view, by increasing the symbol rate from 10 to 40 Gbaud/s, the system would become approximately 16 times more vulnerable to uncompensated chromatic dispersion. Furthermore, for a dynamically switchable optical network, distances and routing paths between the transmitter and the receiver can be switched according to the network status and traffic dynamics, so that precise compensation of accumulated chromatic dispersion, PMD, and other effects can become extremely challenging.

The electronic domain compensation of various degradation effects for a high-speed optical system is enabled by two key technologies, namely coherent detection, and advanced DSP algorithms based on the high-speed CMOS digital electronic circuits. As the major impact of chromatic dispersion and PMD is the introduction of differential group delays between frequency components and polarization modes of the optical field, compensation has to be performed on the signal optical field instead of the signal optical power. In a traditional optical system based on the direct detection, signal optical phase information is not preserved in the square-law detection process of the photodiode (PD) in the receiver. On the other hand, for a coherent optical receiver, the signal optical field is linearly downshifted to the electric domain through mixing with an optical local oscillator (LO), in which the LO acts as a frequency or phase reference. In the electronic circuit following the coherent detection, a frequency-dependent differential group delay can be introduced either through a radio frequency (RF) circuit or through a digital filter. While it is not easy to dynamically adjust the dispersion introduced by an RF circuit, the complex transfer functions of digital filters can be realized and dynamically controlled through relatively simple DSP algorithms. In order for DSP algorithms to be applied, analog-to-digital convertor (ADC) and digital-to-analog convertor (DAC) are required to make conversions between analog and digital domains. With the rapid advance of silicon CMOS technology, ADC and DAC operating at >50 GS/s sampling rate are now commercially available.

Practical application of coherent detection, as one of the enabling technologies, is made possible by the availability of low-cost and low-phase-noise tunable diode lasers, and 90 degree optical hybrid couplers based on the integrated photonic circuits. High-speed ADC/DAC and DSP capability is another major enabling technology for electronic domain processing of optical systems and networks which was not conceivable even a decade ago.

Optical transmission systems based on the electronic domain DSP allows the use of bandwidth efficient modulation formats such as orthogonal frequency-division multiplexing (OFDM) normally used for wireless communications (Shieh and Djordjevic, 2010; Goodsmith, 2005). Based on the orthogonal data encoding of adjacent channels, controlled spectral overlapping between them is allowed and no spectral guard band is necessary. This can significantly increase the spectral efficiency compared to a conventional WDM system, or an analog subcarrier multiplexed system.

11.2 ELECTRONIC-DOMAIN COMPENSATION OF TRANSMISSION IMPAIRMENTS

In traditional optical transmission systems based on the direct detection, dispersion compensation has to be accomplished in the optical domain through DCMs implemented in the system. PMD compensation has also been demonstrated through adaptive polarization tracking and compensation techniques in the optical domain. The accuracy requirements of both chromatic dispersion compensation and PMD compensation become extremely stringent as the data rate per wavelength channel increases beyond 40 Gb/s. Coherent detection and high-speed digital electronic circuits enabled chromatic dispersion compensation and PMD compensation in the electrical domain. As both of these two effects are linear processes, digital filtering can be effective for their compensation.

Fig. 11.2.1 shows the block diagram of a *polarization-division multiplexed* (PDM) coherent detection receiver with electronic domain compensation of chromatic dispersion and PMD (Roberts et al., 2009). In this configuration, a coherent I/Q receiver is used with the capability of both optical phase diversity and polarization de-multiplexing as discussed in Chapter 9. In a PDM coherent optical transmission system, the two optical channels are orthogonally polarized at the output of the transmitter with A_{xI}, A_{xQ}, A_{yI}, and A_{yQ}, representing the in-phase (I) and the quadrature (Q) components of the complex optical field of the x- and the y-polarized channels, respectively. After transmitting over a long optical fiber, both the originally x- and y-polarized channels experience random state-of polarization (SOP)

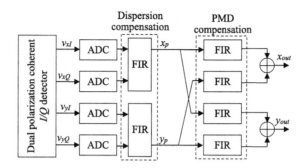

FIG. 11.2.1

Block diagram of a coherent receiver with electronic domain compensation of chromatic dispersion and PMD.

rotation and mode coupling between them through the random birefringence of the fiber. Thus, the complex optical field vector can be randomly oriented at the receiver. In Fig. 11.2.1, v_{xI}, v_{xQ}, v_{yI}, and v_{yQ} are the photo-detected I- and Q-components of the x- and y-polarized signal optical fields, respectively, at the input of the coherent I/Q receiver, which in principle are not the same as A_{xI}, A_{xQ}, A_{yI}, and A_{yQ} because of the impact of both chromatic dispersion and PMD. The job of DSP in the receiver is to undo the birefringence and chromatic dispersion introduced by the transmission fiber. As both chromatic dispersion and PMD are linear processes, they can be compensated independently. In the following, we separately discuss chromatic dispersion compensation and PMD compensation.

11.2.1 DISPERSION COMPENSATION

First, let us neglect the birefringence of the fiber and only consider the x-polarized optical field component, and also assume that the optical LO is ideal without any phase noise. Here, we have assumed that the impact of chromatic dispersion is the same also for the y-polarized component, which is valid because the differential delays between the x- and the y-polarized components caused by fiber birefringence is much less than their common propagation delay of the fiber. Based on Eqs. (9.4.19) and (9.4.20), $v_{xI} = \xi A_s(t)\cos(\omega_{IF}t + \phi_s)$, and $v_{xQ} = \xi A_s(t)\sin(\omega_{IF}t + \phi_s)$, where $A_s(t)$ is the amplitude and ϕ_s is the phase of the optical signal, and ω_{IF} is the intermediate frequency (IF). ξ is a proportionality constant determined by PD responsivity, the trans-impedance amplifier (TIA) gain, and the power of the optical LO. If we further assume a coherent homodyne detection so that $\omega_{IF}=0$, we can have

$$v_x(t) = v_{xI} + jv_{xQ} = \xi A_s(t)e^{j\varphi_s(t)} \tag{11.2.1}$$

which recreates the complex field of the optical signal in the electric domain.

Both v_{xI} and v_{xQ} have to be first digitized before been sent to the digital circuits to perform DSP on $v_x(t)$ and to obtain the digital output x_p as shown in Fig. 11.2.1. As has been discussed in Chapter 2, Eq. (2.6.6), the complex field transfer function due to fiber chromatic dispersion is an all-pass filter is

$$H(\omega, L) = \frac{\tilde{A}(\omega, L)}{\tilde{A}(\omega, 0)} = \exp\left(-j\frac{\omega^2}{2}\beta_2 L\right) \tag{11.2.2}$$

where $\widetilde{A}(\omega,0)$ and $\widetilde{A}(\omega,L)$ are complex optical fields at the input and the output of the fiber. β_2 is the chromatic dispersion coefficient of the fiber and L is the fiber length. Eq. (11.2.2) can also be expressed as

$$H(\omega,L) = \exp\left(j\frac{\lambda^2\omega^2 D}{4\pi c}L\right)$$
(11.2.3)

where $D = -2\pi c\beta_2/\lambda^2$ is the dispersion parameter of the fiber and λ is the signal center wavelength.

In order to remove the impact of chromatic dispersion, the DSP circuit has to provide an all-pass filter transfer function $G(\omega,L)=[H(\omega,L)]^{-1}$, which is

$$G(\omega,L) = \exp\left(-j\frac{\lambda^2\omega^2 D}{4\pi c}L\right)$$
(11.2.4)

so that $G(\omega,L)H(\omega,L)=1$. Instead of performing this operation in the frequency domain, which would require Fourier transforming $v_x(t)$ into frequency domain and multiplying it with the frequency-domain transfer function (11.2.4), a finite impulse response (*FIR*) digital filter can be used to perform this operation directly in the time domain. To do this, the desired all-pass filter transfer function of Eq. (11.2.4) can be converted into the time domain through an inverse Fourier transform, which is

$$g(t,L) = F^{-1}[G(\omega,L)] = \sqrt{\frac{c}{j\lambda^2 DL}}\exp\left(j\frac{\pi c}{\lambda^2 DL}t^2\right)$$
(11.2.5)

where F^{-1} represents inverse Fourier transform.

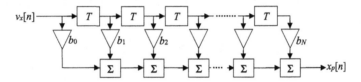

FIG. 11.2.2

Block diagram of a FIR filter based on the multiple tapped delay lines. T is a unit delay which is the sampling period and b_k ($k=1, 2, ...N$) is the weight factor after the kth delay line.

This time-domain transfer function can be realized by a digital *FIR* filter based on the tapped delay lines as shown in Fig. 11.2.2, where $v_x[n]$ is the nth discrete sample of the input signal, and the filter output $x_p[n]$ is computed as a weighted finite-term sum of past, present, and future values of the filter input. Thus, the filter output is related to the input by $x_p[n] = \sum_{k=-N/2}^{N/2} b_k \cdot v_x[n-k]$, with b_k the weight of each tap. Without getting into details of FIR filter design, which is outside the scope of this book, the following equation gives the weight of each tap (Roberts et al., 2009):

$$b_k = \sqrt{\frac{cT^2}{j\lambda^2 DL}}\exp\left(j\frac{\pi c}{\lambda^2 DL}T^2 k^2\right) \text{ with } -\left\lfloor\frac{N}{2}\right\rfloor \le k \le \left\lfloor\frac{N}{2}\right\rfloor$$
(11.2.6)

where $\lfloor x \rfloor$ represents the integer part of x rounded toward $-\infty$.

When the input to the FIR filter is an impulse, that is, $v_x(0)=1$ and $v_x(n-k)=0$ for $n\neq k$, the impulse response of the FIR filter is equal to the coefficient b_k. Fig. 11.2.3A shows the real (black open circles) and the imaginary (red squares) parts of the impulse response of an FIR filter as the function of the normalized time. This filter is designed to compensate the accumulated chromatic dispersion of $D \cdot L = 27,200$ps/nm. Consider a standard single-mode fiber (SMF) with dispersion parameter $D = 17$ps/nm/km, this accumulated dispersion corresponds to a fiber length of $L = 1600$km. Fig. 11.2.3B shows the phase of the impulse response as the function of the normalized time. Solid lines in Fig. 11.2.3 shows $T \cdot g(t, L)$ with $g(t, L)$ obtained from Eq. (11.2.5) analytically, while open dots were impulse responses obtained with an FIR filter using weight coefficients of Eq. (11.2.6) with the filter order of $N = 174$.

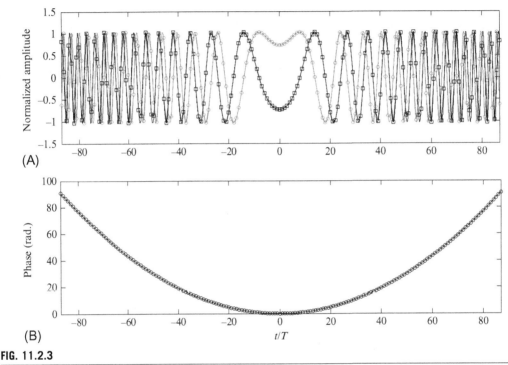

FIG. 11.2.3

Real *(black)* and imaginary *(red)* parts (A) and the phase (B) of the impulse response as the function of the normalized time. Solid lines were obtained by Eq. (11.2.5) and open dots were obtained by an FIR filter using weight coefficients of Eq. (11.2.6) with the filter order of $N = 174$.

It is also important to point out that the unit delay time T of the FIR filter represents the time sampling interval so that the Nyquist frequency is $f_n = 1/(2T)$. Since the impulse response of the FIR filter shown in Eq. (11.2.5) has a time-dependent phase $\phi(t) = \pi c t^2/(\lambda^2 DL)$, the corresponding frequency shift is

$$f = \frac{1}{2\pi} \frac{d\phi(t)}{dt} = \frac{ct}{\lambda^2 DL} \tag{11.2.7}$$

In order to avoid spectral aliasing, this frequency shift cannot exceed the Nyquist frequency f_n, that is,

$$\left| \frac{ct}{\lambda^2 DL} \right| \leq \frac{1}{2T} \tag{11.2.8}$$

This sets the maximum truncation time window for the signal,

$$-\frac{\left|\lambda^2 DL\right|}{2cT}\leq t\leq\frac{\left|\lambda^2 DL\right|}{2cT}$$

(11.2.9)

As the sampling interval is T, the maximum allowed truncation window determined by Eq. (11.2.9) limits the maximum number of taps of the FIR filter as

$$N_{\max}=2\left\lfloor\frac{\lambda^2 DL}{2cT^2}\right\rfloor+1$$

(11.2.10)

For the example of impulse response shown in Fig. 11.2.3 with the accumulated dispersion of $D\cdot L=27,200\text{ps/nm}$ in the 1550-nm wavelength window, and assume $T=50\text{ps}$ (for 10 Gbaud/s symbol rate with two samples per symbol), the filter order should be no more than 174. This provides the desired dispersion compensation within a frequency range of $-1/(2T)\leq f\leq 1/(2T)$. In practice, the number of taps used for the FIR filter is often less than the maximum value to reduce the complexity of the digital circuit. As the result, the dispersion value created by the digital circuit will be in a reduced spectral window of less than $1/T$.

It is also important to note that the FIR filter for dispersion compensation is a phase-only all-pass filter, and a low-pass filter (LPF) has to be superimposed to suppress the high-frequency components, which limits the signal spectrum within $-1/(2T)\leq f\leq 1/(2T)$. Because chromatic dispersion in a fiber system is relatively stable after the fiber length is determined, FIR filter in the DSP circuit does not have to change dynamically unless the fiber length is dynamically switched.

In addition to dispersion compensation performed in the receiver as described in Fig. 11.2.1 which is known as post-compensation, digital compensation can also be performed in the transmitter, known as pre-compensation as schematically shown in Fig. 11.2.4. In this transmitter, digital signal to be transmitted is pre-distorted through DSP and converted to the analog domain by a pair of DAC. These electric domain analog signals are amplified and converted into an optical signal through an electro-optic modulator (EOM). Based on an I/Q modulator with the functionality discussed in Section 7.3, the voltage signal waveforms $v_1(t)$ and $v_2(t)$ can be linearly translated into the in-phase (I) and the quadrature (Q) components of the modulated complex optical field.

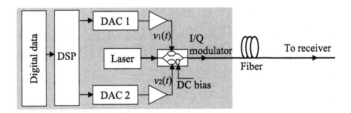

FIG. 11.2.4

Block diagram of an optical transmitter with the capability of digital electronic domain pre-compensation.

When the accumulated dispersion along the fiber link is known, the transfer function of the fiber can be predicted by Eq. (11.2.3) which is deterministic. Electronic domain pre-compensation (McNicol et al., 2005) is performed by generating two arbitrary waveforms which are pre-distorted versions of the real and the imaginary parts of the optical field by taking into account the dispersion effect of the fiber based on Eq. (11.2.4). Fig. 11.2.5 shows an example of transmitted and received waveforms of a 10 Gb/s binary nonreturn-to-zero (NRZ)-modulated optical system over 1600 km standard SMF with an accumulated chromatic dispersion of 27,200 ps/nm. Fig. 11.2.5A shows the real (solid line) and the imaginary (dotted line) parts of the signal optical field generated by the optical transmitter. Fig. 11.2.5B shows the normalized signal optical power emitted by the transmitter, which is severely distorted. Fig. 11.2.5C is the waveform obtained at the direct-detection optical receiver, which turns out to be an ideal waveform which has a completely opened eye diagram. There is no need of optical domain dispersion compensation.

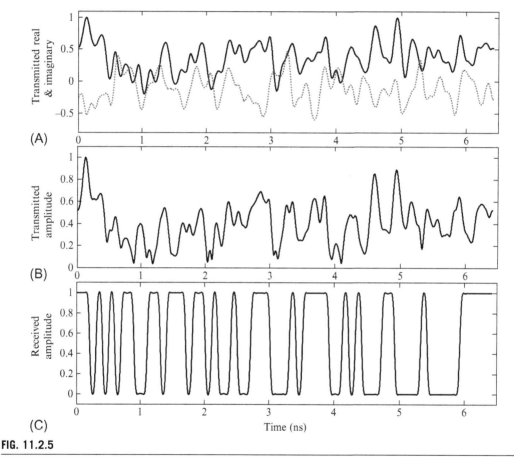

FIG. 11.2.5

(A) Real *(solid line)* and imaginary *(dotted line)* of the optical field created by the transmitter, (B) normalized optical power generated by the transmitter, and (C) normalized amplitude waveform obtained at the receiver.

In fact, if all parameters of an optical system are known, electronic domain digital pre-compensation can compensate the impacts of both chromatic dispersion and self-phase modulation (SPM) (Roberts et al., 2006). This can be done by obtaining the required pre-distortion transfer function through a numerical simulation based on the split-step Fourier method (SSFM) as discussed in

Section 8.5.2. The only difference is that this SSFM has to be performed in the reverse direction (negative z) along the fiber and with the sign of chromatic dispersion reversed. In this way, with the desired *ideal* waveform at the receiver, a distorted complex waveform can be calculated for the transmitter. This "distorted" complex arbitrary waveform can be generated by the optical transmitter which is equipped with high-speed DSP, DAC, and complex optical field modulation capabilities. Propagating this distorted optical waveform along the fiber to the receiver, the impacts of both chromatic dispersion and SPM are reversed so that an ideal waveform is recreated at the receiver.

The disadvantage of transmitter-side digital pre-compensation, in comparison to the receiver-side post-compensation, is the difficulty, if not impossible, to perform PMD compensation which will be discussed next. As fiber birefringence is random, it is much easier to run adaptive optimization and compensation algorithms at the receiver side with the signal impacted by time-varying and polarization-dependent transmission properties.

11.2.2 PMD COMPENSATION AND POLARIZATION DE-MULTIPLEXING

In comparison to chromatic dispersion, PMD and signal SOP variation are more dynamic due to the random variation of birefringence along the fiber. PMD compensation has to follow the variation of fiber birefringence overtime (Savory, 2008). In general, the polarization effect of a fiber channel can be described by a frequency-dependent Jones Matrix,

$$\begin{bmatrix} x_p \\ y_p \end{bmatrix} = \begin{bmatrix} H_{xx}(\omega) & H_{xy}(\omega) \\ H_{yx}(\omega) & H_{yy}(\omega) \end{bmatrix} \begin{bmatrix} A_x \\ A_y \end{bmatrix} \qquad (11.2.11)$$

where x_p, y_p, A_x, and A_y are x- and y-polarized optical field components at the output and input, respectively, of an optical fiber, and ω is the optical frequency. The four Jones matrix elements represent the frequency-dependent relations between the input and the output complex optical field components, and coupling between x- and y-polarized fields. PMD compensation is to create an inverse Jones matrix to undo the transfer function of Eq. (11.2.11).

Fig. 11.2.6 shows a DSP structure for PMD compensation and polarization de-multiplexing. In this configuration, the input signals are x_p and y_p which represent the digitized sequences of x- and y-polarized components detected by the receiver. Theoretically, the x- and y-polarized components of the original optical signals can be derived from the measured x_p and y_p through an inverse operation of Eq. (11.2.11), that is,

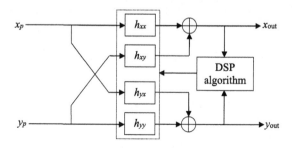

FIG. 11.2.6

DSP structure for PMD compensation and polarization de-multiplexing.

$$\begin{bmatrix} A_x \\ A_y \end{bmatrix} = \frac{1}{|Det|} \begin{bmatrix} H_{yy}(\omega) & H_{xy}(\omega) \\ H_{xx}(\omega) & H_{yx}(\omega) \end{bmatrix} \begin{bmatrix} x_p \\ y_p \end{bmatrix} \tag{11.2.12}$$

This is an expression in frequency domain, where $|Det|$ is the determinant of the Jones matrix. Because the fiber birefringence is time varying, it is convenient to perform DSP in the time domain based on the digitized signal sequences $x_p[n]$ and $y_p[n]$. As a frequency domain multiplication is equivalent to a time-domain convolution, the operation shown in Eq. (11.2.12) can be accomplished by digital filters performing the following operation:

$$x_{out}(k) = \sum_{m=0}^{M-1} \left[h_{xx}(m)x_p(k-m) + h_{xy}(m)y_p(k-m) \right] \tag{11.2.13a}$$

$$y_{out}(k) = \sum_{m=0}^{M-1} \left[h_{yx}(m)x_p(k-m) + h_{yy}(m)y_p(k-m) \right] \tag{11.2.13b}$$

where h_{xx}, h_{xy}, h_{yx}, and h_{yy} are weight factors of the four digital filters after the mth delay line (with the FIR filter structure shown in Fig. 11.2.2). These filter transfer functions can be considered discrete Fourier transforms (DFTs) of $H_{yy}/|Det|$, $H_{yx}/|Det|$, $H_{xx}/|Det|$, and $H_{yx}/|Det|$, respectively, with respect to Eq. (11.2.12), and M is the number of taps of each digital filter. Ideally, x_{out} and y_{out} should be as close as possible to the original optical fields A_x and A_y. However, since the Jones matrix of the fiber is unknown and is constantly changing, the determination and optimization of filter transfer functions are most important tasks in the receiver DSP. There are a number of DSP algorithms for adaptive equalization based on the specific signal and modulation formats. For example, for a PDM-quadrature phase shift keying (QPSK) system, the signal of each orthogonally polarized channel should have constant amplitude. This constant modulus condition can be used to adaptively optimize digital filter transfer functions, which is known as the *constant modulus algorithm* (CMA). For a normalized optical signal, the power of both x_{out} and y_{out} should be equal to 1, so that $\langle \varepsilon_x^2 \rangle = \langle 1 - |x_{out}|^2 \rangle$ and $\langle \varepsilon_y^2 \rangle = \langle 1 - |y_{out}|^2 \rangle$ represent the mean-square error magnitude. Based on CMA, the DSP algorithm will attempt to minimize both $\langle \varepsilon_x^2 \rangle$ and $\langle \varepsilon_y^2 \rangle$ by adjusting the transfer functions of the digital filters. The CMA criteria can be expressed by minimizing the gradients of the mean-square errors against the tap weights of digital filters,

$$\frac{d\langle \varepsilon_x^2 \rangle}{dh_{xx}} = 0; \frac{d\langle \varepsilon_x^2 \rangle}{dh_{xy}} = 0; \frac{d\langle \varepsilon_y^2 \rangle}{dh_{yx}} = 0; \frac{d\langle \varepsilon_y^2 \rangle}{dh_{yy}} = 0 \tag{11.2.14}$$

where h_{xx}, h_{xy}, h_{yx}, and h_{yy} represent tap weights of digital filters, and each of them is a victor with a length M. The digital process of optimizing tap weights can be based on the Wiener-Hopf equations for linear transversal digital filters (Haykin, 2002). Through each recursive adaption process, the updated tap weights can be obtained,

$$h_{xx,k+1} = h_{xx,k} + \mu \varepsilon_x x_{out} \cdot x_p^* \tag{11.2.15a}$$

$$h_{xy,k+1} = h_{xy,k} + \mu \varepsilon_x x_{out} \cdot y_p^* \tag{11.2.15b}$$

$$h_{yx,k+1} = h_{yx,k} + \mu \varepsilon_y y_{out} \cdot x_p^* \tag{11.2.15c}$$

$$h_{yy,k+1} = h_{yy,k} + \mu \varepsilon_y y_{out} \cdot y_p^* \tag{11.2.15d}$$

where x_p^* and y_p^* are complex conjugates of the received data sequences of x_p and y_p with length M, and μ is a convergence parameter for the numerical process. Detailed numerical process and examples can be found in Savory (2008).

Although the CMA algorithm requires constant modulus of optical signal such as QPSK which has no intensity modulation, it has been shown effective also for other modulation formats such as higher-order QAM whose modulus are not necessarily constant (Johnson et al., 1998). The reason is that the variation of birefringence, and thus the signal SOP, along an optical fiber is a relatively slow process with a time constant typically longer than a microsecond. For high-speed modulated optical signal, the modulus averaged within the microsecond time window can be considered constant.

11.2.3 CARRIER PHASE RECOVERY IN A COHERENT DETECTION RECEIVER

In a coherent receiver based on I/Q detection, the in-phase (I) and the quadrature (Q) components of the complex optical field can be converted into electric domain. In the coherent detection process discussed in Section 9.4, mixing between the received optical signal and the optical LO in balanced PDs after a 90° optical hybrid coupler produces the I- and Q-components of the photocurrent,

$$i_I(t) = \mathfrak{R}A_s(t)A_{LO} \sin[\omega_{IF}t + \varphi_s(t) + \delta\varphi_n(t)] \tag{11.2.16a}$$

$$i_Q(t) = \mathfrak{R}A_s(t)A_{LO} \cos[\omega_{IF}t + \varphi_s(t) + \delta\varphi_n(t)] \tag{11.2.16b}$$

where $\mathfrak{R}$ is the PD responsivity, ω_{IF} is the IF of heterodyne detection, and $\omega_{IF}=0$ when homodyne detection is used. $A_s(t)$ and A_{LO} are optical field amplitudes of the optical signal and the LO. $\varphi_s(t)$ represents phase information carried by the modulated optical field, and $\delta\varphi_n(t)=\varphi_{n,s}(t)-\varphi_{n,LO}(t)$ is the relative phase noise between the signal laser phase noise $\varphi_{n,\,s}(t)$ and the LO phase noise $\varphi_{n,LO}(t)$. If ideal lasers are used both in the transmitter and the receiver without phase noise, we should have $\delta\varphi_n(t)=0$ so that both $A_s(t)$ and $\varphi_s(t)$ can be recovered from $i_I(t)$ and $i_Q(t)$. However, in practical optical systems, both lasers used in the transmitter and the LO in the receiver have phase noise. For example, the spectral linewidths of typical DFB lasers and external cavity semiconductor lasers are on the orders of a few MHz, and $\sim$100 kHz, respectively. This phase noise has to be removed in order to recover the phase information $\varphi_s(t)$ carried on the optical signal.

There are various different algorithms for the removal of phase noise in coherent detection optical receivers. Let us first use QPSK-modulated optical signal as an example to explain how DSP can help in carrier phase recovery. Based on the photocurrent signals $i_I(t)$ and $i_Q(t)$ in Eqs. (11.2.16a) and (11.2.16b), and assume homodyne detection is used so that $\omega_{IF}=0$ (in practice the intermediate frequency ω_{IF} can be removed by a simple digital frequency offset algorithm in digital processing), the complex optical field can be reconstructed as

$$E[n] = i_I[n] + ji_Q[n] = A[n]e^{j\theta[n]} \tag{11.2.17}$$

where $i_I[n]$ and $i_Q[n]$ are the digitized data sequences of $i_I(t)$ and $i_Q(t)$, $A[n]=2\mathfrak{R}A_s[n]A_{LO}$ and $\theta[n]=\varphi_s[n]+\delta\varphi_n[n]$ are the nth sample of the amplitude and phase of the complex signal, respectively. $\delta\varphi_n[n]$ is the phase noise component in the nth phase sample which can be modeled as a Wiener process. Under the Gaussian noise approximation, the variance of $\delta\varphi_n[n]$ is related to the combined linewidth, $\Delta\nu$, of the transmitter laser and the LO by: $\sigma_n^2=2\pi\Delta\nu\cdot T_s$, where T_s is the data sampling period.

For a QPSK-modulated signal, $A[n]$ should be a constant so that the amplitude can be normalized as $A[n] = 1$, while the correct signal phase $\varphi_s[n]$ should only have four possible values: 0, $\pi/2$, π, and $3\pi/2$. The fourth power of $E[n]$ will be, $E^4[n] = e^{j4\theta[n]}$, so that the phase of $E^4[n]$ is

$$4\theta[n] = 4\varphi_s[n] + 4\delta\varphi_n[n] \tag{11.2.18}$$

Interestingly, for a QPSK-modulated signal, $4\varphi_s[n]$ of the four correct constellation points should be 0, 2π, 4π, and 6π, respectively, and they all locate at the same point $(1,0)$ on the complex plane. Any deviation of $4\theta[n]$ from this point must be caused by the phase noise $4\delta\varphi_n[n]$. Then, the actual phase error $\delta\varphi_n[n]$ can be calculated and its impact can be compensated by multiplying the originally received symbols with $\exp\{-j\delta\varphi_n[n]\}$. This is commonly referred to as the Viterbi-and-Viterbi phase estimation algorithm (Viterbi and Viterbi, 1983).

Consider that the phase noise $\delta\varphi_n(t)$ produces a spectral linewidth of 500 kHz, for example, which results from the mixing between the signal laser and the LO. This means roughly $\delta\varphi_n(t)$ can vary for $\pm\pi$ radians over a time of 2 µs which is the inverse of the spectral linewidth. If the optical signal has a symbol rate of 10 Gbaud/s, this time window of 2 µs corresponds to about 20,000 data symbols. Now if we take 500 data symbols $E[n]$ ($n = 1, 2, \ldots 500$) within a time window $t_w = 0.05$ µs, the phase variation of $E[n]$ due to the phase noise should be small enough with $|\delta\varphi_n[n]| < <\pi/4$. The phase noise of each sample can be estimated by

$$\delta\varphi_n[n] = \frac{\text{Arg}\{E^4[n]\}}{4} \tag{11.2.19}$$

where $\text{Arg}\{E^4[n]\}$ represents the phase of $E^4[n]$, which is $4\theta[n]$. If the window is small enough, all $\delta\varphi_n[n]$ should be located well within $\pm\pi/4$. An average phase deviation can be found within the window as

$$\Delta\varphi_n = \frac{1}{4N}\text{Arg}\left\{\sum_{n=1}^{N} E^4[n]\right\} \tag{11.2.20}$$

where N is the number of samples within the window. Then, the correct carrier phase within this window can be recovered through

$$\varphi_s[n] = \text{Arg}\{E[n]\} - \Delta\varphi_n \tag{11.2.21}$$

This $\varphi_s[n]$ is then fed into a decision circuit which converts $\phi_s[n]$ to the closest constellation point, similar to the decision circuit used for binary-modulated signals illustrated in Fig. 8.1.2.

In this carrier phase recovery process, the window width t_w can be optimized based on the symbol rate, signal-to-noise ratio (SNR), and the level of phase noise. t_w has to be small enough so that phase variation due to phase noise is negligible within the window. If the phase variation within a window spreads beyond $\pm\pi/4$, phase jump can occur, shown as $\pm\pi/2$ cycle slip on the constellation diagram, which can result in bursts of bit errors. On the other hand, if t_w is too small, $\Delta\varphi_n$ calculated from the average in Eq. (11.2.20) can be affected by the uncertainty of the $\text{Arg}\{E^4[n]\}$ due to the limited SNR.

In a real-time DSP, the continuously sampled data sequence is sequentially divided into parallel data blocks of length N for carrier phase recovery as illustrated in Fig. 11.2.7. After carrier phase recovery, all parallel sections of samples are resynchronized into a serial data sequence through a digital parallel-to-serial converter.

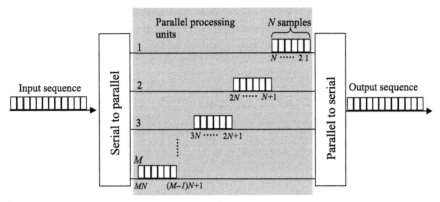

FIG. 11.2.7

Carrier phase estimation with *M* parallel processing channels.

The functionality of carrier phase estimation processing unit in each data block is described by Equations. (11.2.17) through (11.2.21), and is schematically shown in Fig. 11.2.8. As there is an averaged phase noise obtained within each data block, the variation of phase noise as the function of time $\varphi_n(t)$ can be derived from this process with an equivalent phase noise sampling time interval of $t_w = T_r N$, which is the length of the window, where T_r is the data sampling interval. While the carrier phase estimation technique discussed above is straightforward, it was based on QPSK-modulated

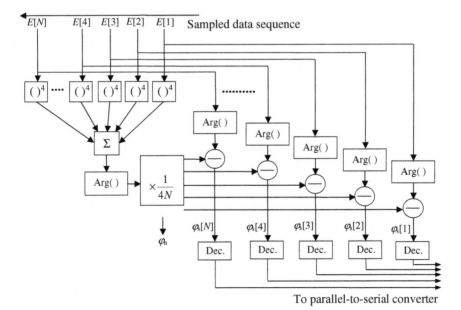

FIG. 11.2.8

Functionality of a carrier phase estimation processing unit.

optical signal, and it also introduces processing delay proportional to $M \cdot t_w$, where M is the number of parallel processing channels. A modified version of Viterbi-Viterbi has been demonstrated for carrier phase recovery of high-order QAM by partitioning the constellation diagram into groups of equal amplitude PSK subsets (Fatadin et al., 2010) as shown in Fig. 10.5.1B.

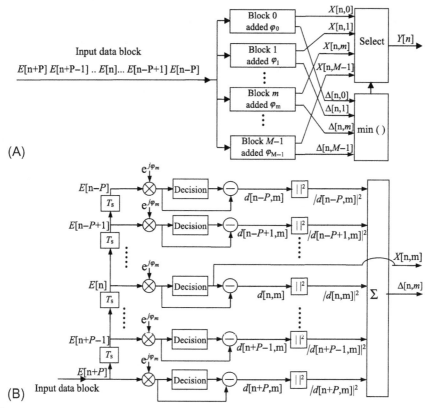

FIG. 11.2.9

(A) Block diagrams of blind phase search (BPS) algorithm, and (B) configuration of block m in (A) in which each sample is phase rotated by φ_m.

There is another example of carrier phase recovery, also based on the feed-forward algorithm, known as blind phase search (BPS) (Pfau et al., 2009), and the block diagram of BPS is shown in Fig. 11.2.9. A block of $(2P+1)$ samples is used at the input of the DSP. Each data sample in the block is rotated by a group of M progressive phase shifters with $\varphi_m = m\pi/(2M)$, where $m = 0, 1, 2, \ldots M-1$. As shown in Fig. 11.2.9B, which represents a block m in Fig. 11.2.9A, each sample phase rotated by ϕ_m goes to a decision circuit. At the output of each decision circuit, the signal is forced to the closest constellation point. Then, the distance between the phase-shifted sample and the closest constellation point can be obtained by a subtraction, and this distance is squared to obtain the error vector magnitude (EVM),

$$|d(n+x,m)|^2 = \left| E[n+x]e^{j\varphi_m} - \lfloor E[n+x]e^{j\varphi_m} \rfloor_D \right|^2 \tag{11.2.22}$$

where x is an integer ($-P \le x \le P$) and $\lfloor \ \rfloor_D$ represents the decision operation. In order to remove the random noise contributions, the EVMs of the $2P+1$ consecutive symbols rotated by the same carrier phase angle φ_m are summed up to obtain an estimation of the error,

$$\Delta[n,m] = \sum_{x=-P}^{P} |d(n+x,m)|^2 \tag{11.2.23}$$

Meanwhile, a constellation point at the middle of the data block is recorded as

$$X[n,m] = \lfloor E[n]e^{j\varphi_m} \rfloor_D \tag{11.2.24}$$

The length, $2P+1$, of the data block is similar to the data block length N used in the carrier phase estimation technique previously discussed and shown in Fig. 11.2.7. Likewise the selection of the P-value depends on the spectral linewidth of the laser and the symbol rate of the data.

Then, all the error values $\Delta[n,m]$ with $m=0, 1, 2, \dots M$, are compared to find a minimum value, which corresponds to, say, $m=\zeta$. Then, we know φ_ζ is the correct carrier phase, so that a selection circuit should select

$$Y[n] = X[n, \zeta] \tag{11.2.25}$$

as the correct signal constellation point.

In this feed-forward carrier phase recovery operation, the data sequence is sliding sample-by-sample and no feedback is required.

11.3 DIGITAL SUBCARRIER MULTIPLEXING: OFDM AND NYQUIST FREQUENCY-DIVISION MULTIPLEXING

High spectral efficiency multiplexing formats such as OFDM and Nyquist pulse modulation based on the sharp edge digital filters have been used for wireless communication for many years, primarily due to the scarcity of spectrum resource. The speed of digital electronic components such as ADC, DCA, and DSP is also fast enough to support most of the wireless applications with the bandwidth on the order of a few hundred megahertz. For fiber-optic systems, bandwidth efficiency has traditionally not been considered an important issue. WDM has been very successful allowing a large number of wavelength channels to be carried by an optical fiber, which makes much more efficient use of the available optical bandwidth compared to a single-wavelength system. Channel spacing of commercial WDM systems went down from 200 to 50 GHz and even 25 GHz so that more wavelength channels can be accommodated within the optical bandwidth of in-line optical amplifiers. With the rapid growth of capacity demand, increasing bandwidth efficiency in optical communication systems has also become an important issue. In general, bandwidth efficiency can be increased by (1) reducing spectral guard band between adjacent wavelength/frequency channels and (2) increase bit-per-symbol through high-level modulation. High-speed digital electronics is an enabling technology that allows advanced multiplexing techniques such as OFDM and Nyquist pulse modulation, to be extended to applications in optical systems which usually require much higher speed than wireless applications. Both OFDM and Nyquist pulse modulation are based on the subcarrier multiplexing. While OFDM relies on the orthogonality between spectrally overlapped adjacent subcarrier channels, Nyquist pulse modulation relies on sharp-edged digital filters to reduce the required spectral guard band. As both of these multiplexing techniques create subcarriers using DSP, they can both be categorized as digital subcarrier multiplexing (DSCM).

11.3.1 ORTHOGONAL FREQUENCY-DIVISION MULTIPLEXING

In analog subcarrier multiplexing (SCM), a spectral guard band has to be reserved between adjacent subcarrier channels and no spectral overlap between them is allowed so that they can be separated by RF filters in the de-multiplexing process. Thus, channel spacing Δf, which is the frequency difference between adjacent subcarriers, has to be much larger than symbol rate carried by each subcarrier channel, that is, $\Delta f >> 1/T_N$, where T_N is the symbol length of data on each subcarrier channel as illustrated in Fig. 11.3.1A. The obvious disadvantage is that a significant portion of the spectrum, reserved as the guard band, is not efficiently utilized, and thus the spectral efficiency is low.

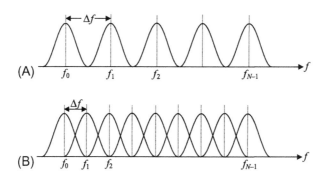

FIG. 11.3.1

Spectra of analog SCM (A) where spectral guard band is required and OFDM (B) with significant spectral overlap between adjacent subcarrier channels.

For OFDM, the channel spacing is exactly $\Delta f = 1/T_N$ which is much tighter than analog SCM. In fact in OFDM, there is significant spectral overlap between adjacent subcarrier channels as illustrated in Fig. 11.3.1B. Then, the question is that how these subcarrier channels can be separated in the receiver. To answer this question, let us first define the orthogonality between a set of subcarriers. Consider the mth and the nth subcarriers $\cos(2\pi f_m t + \varphi_m)$ and $\cos(2\pi f_n t + \varphi_n)$ with their frequencies at $f_m = f_0 + m/T_N$ and $f_n = f_0 + n/T_N$, where m, $n, = 0, 1, 2, 3, \ldots N-1$, f_0 is the zeroth subcarrier frequency as shown in Fig. 11.3.1, and φ_m and φ_n are the phases of these two subcarriers. Performing integration over a time window $T_N = 1/\Delta f$,

$$\frac{1}{T_N} \int_0^{T_N} \cos(2\pi f_m t + \varphi_m) \cos(2\pi f_n t + \varphi_n) dt = \frac{1}{2}\delta(m-n) \tag{11.3.1}$$

which is nonzero only when $m=n$. This defines a set of subcarriers which are mutually orthogonal.

Fig. 11.3.2 shows the generic block diagram of a communication system with subcarrier multiplexing and de-multiplexing. At the transmitter side, multiple data streams $s_0, s_1, \ldots s_{N-1}$ are frequency

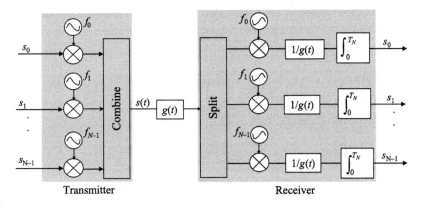

FIG. 11.3.2

Block diagram of an OFDM system with N subcarrier channels.

up-converted by mixing with a set of mutually orthogonal subcarriers $f_0, f_1, \ldots f_{N-1}$, and they are combined to form a composite signal $s(t)$. This composite signal is transmitted through the communication channel with a time-domain response $g(t)$. At the receiver, the signal is equally split into N copies and mixing with the same set of orthogonal subcarriers for frequency down-conversion of data streams to the baseband. The impact of the transmission impairments can be removed by passing through an inverse transfer function $1/g(t)$. Then, the baseband signal, s_0, $s_1, \ldots s_{N-1}$, on each subcarrier can be recovered by an integration over a time window T_N, which is equivalent to low-pass filtering. This integration also effectively removes the cross talk between adjacent channels.

Mathematically, the composite signal consisting of N subcarrier channels arriving at the receiver can be expressed as

$$s(t) = \sum_{k=0}^{N-1} s_k g(t) \cos\left(2\pi f_k t + \varphi_k\right) \tag{11.3.2}$$

where $g(t)$ is the channel transfer function, which may include the impact of chromatic dispersion and PMD. In order to recover the mth channel, this received composite signal mixes with the corresponding subcarrier $\cos(2\pi f_m + \varphi_m)$ and the inverse transfer function of the communication medium $1/g(t)$. After integration over the time window T_N, the output is

$$s_i^{(out)} = \int_0^{T_N} \left[\sum_{n=0}^{N-1} s_n g(t) \cos\left(2\pi f_n t + \varphi_n\right) \right] \frac{1}{g(t)} \cos\left(2\pi f_m t + \varphi_m\right) dt$$

$$= \int_0^{T_N} \left\{ \sum_{n=0}^{N-1} s_n \cos\left(2\pi\left(f_0 + \frac{n}{T_N}\right) t + \varphi_n\right) \cos\left(2\pi\left(f_0 + \frac{m}{T_N}\right) t + \varphi_m\right) \right\} dt = \sum_{j=0}^{N-1} s_n \delta(m-n) = s_n^{(in)} \tag{11.3.3}$$

Note that in the frequency down-conversion process, the LO in the receiver has to have the same frequency f_m and phase φ_m as those used in the transmitter. In comparison to the analog SCM system shown in Fig. 10.6.1, the major difference of OFDM is the use of a set of mutually orthogonal subcarriers, and the ability of performing integration precisely over a time window T_N, which can be done in the digital domain but quite challenging with analog electronics.

It is well known that the definition of DFT is

$$DFT[x[n]] = X[m] \equiv \frac{1}{\sqrt{N}} \sum_{n=0}^{N-1} x[n] \exp\left(-j\frac{2\pi nm}{N}\right) \quad \text{for } 0 \leq m \leq N-1 \qquad (11.3.4)$$

and the definition of inverse discrete Fourier transform (IDFT) is

$$IDFT[X[m]] = x[n] \equiv \frac{1}{\sqrt{N}} \sum_{m=0}^{N-1} X[m] \exp\left(j\frac{2\pi nm}{N}\right) \quad \text{for } 0 \leq n \leq N-1 \qquad (11.3.5)$$

where $X[m]$ and $x[n]$ are a DFT/IDFT pair and N is the DFT/IDFT size. If we consider $f_n = n/N$ and $f_m = m/N$ for two subcarrier frequencies, Eqs. (11.3.4) and (11.3.5) would represent the same functions of frequency up- and down-conversion for subcarrier multiplexing and de-multiplexing shown in Fig. 11.3.2, where N is the total number of subcarriers. If we use IDFT for frequency up-conversion in the transmitter, $X[m]$ will represent data in the frequency domain, and $x[n]$ will represent data in the time domain.

Fig. 11.3.3 illustrates the operation of an OFDM transmitter, where a high-speed data stream is first partitioned into multiple parallel channels through a serial-to-parallel converter. Then, IDFT is performed on each channel converting $X[m]$ from frequency domain to $x[n]$ in the time domain. The time-domain parallel sequences are then changed back to a serial data stream through a parallel-to-serial conversion before been translated into analog domain using a high-speed DAC. For many applications, an external RF carrier can be used for another frequency up-conversion which moves the entire composite signal spectrum to f_c as the central frequency. This helps reducing the bandwidth requirement for DSP and DAC.

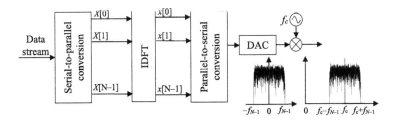

FIG. 11.3.3

Block diagram of an OFDM transmitter using IDFT for frequency up-conversion.

For an input data sequence $\{X[0]\ X[1]\ X[2]\ X[3]\dots X[MN]\}$, serial-to-parallel conversion put $\{X[0]\ X[1] \dots X[N-1]\}$ into the first row, $\{X[N]\ X[N+1] \dots X[2N-1]\}$ into the second row, and so on, as shown in Fig. 11.3.4. This forms a $M \times N$ matrix with $\{X[iN]\ X[iN+1] \dots X[iN-1]\}$ in the ith row, where $i = 1, 2, \dots M$. Performing IDFT on this matrix up-converts the nth column to a subcarrier frequency f_n, which carries data $\{X[n]\ X[2N+n] \dots X[MN+n]\}^T$. After IDFT, the $M \times N$ matrix is converted back to a serial sequence $\{x[0]\ x[1]\ x[2]\ x[3]\dots x[MN-1]\}$. This serial data stream is finally converted into analog domain as a real voltage waveform by an ADC. The double-sideband (DSB) spectrum of OFDM is flat top with the maximum subcarrier frequency of $\pm f_{N-1}$ as shown in the insets

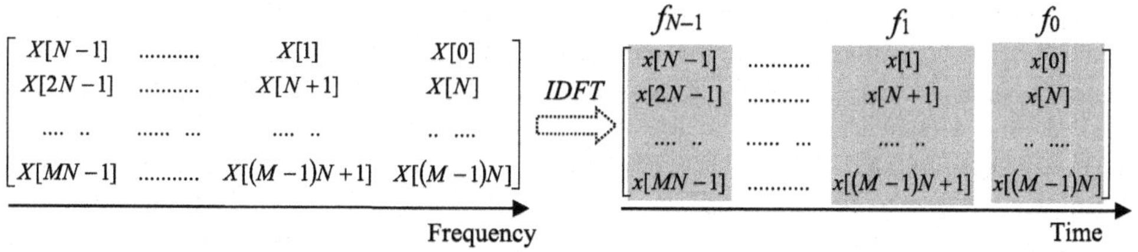

$$\begin{bmatrix} X[N-1] & \cdots\cdots\cdots & X[1] & X[0] \\ X[2N-1] & \cdots\cdots\cdots & X[N+1] & X[N] \\ \cdots\cdots & \cdots\cdots & \cdots\cdots & \cdots\cdots \\ X[MN-1] & \cdots\cdots\cdots & X[(M-1)N+1] & X[(M-1)N] \end{bmatrix}$$
$$\xrightarrow{IDFT}$$

$$\begin{array}{ccc} f_{N-1} & f_1 & f_0 \\ \begin{bmatrix} x[N-1] & \cdots\cdots\cdots & x[1] & x[0] \\ x[2N-1] & \cdots\cdots\cdots & x[N+1] & x[N] \\ \cdots\cdots & \cdots\cdots & \cdots\cdots & \cdots\cdots \\ x[MN-1] & \cdots\cdots\cdots & x[(M-1)N+1] & x[(M-1)N] \end{bmatrix} \end{array}$$

Frequency   Time

FIG. 11.3.4

Matrices of data mapping in an OFDM transmitter. The matrix on the left side is in frequency domain, and the matrix on the right side is in time domain.

of Fig. 11.3.3. This spectrum can be further up-converted to an RF central frequency f_c through mixing with another oscillator.

Block diagram of an OFDM receiver using DFT for frequency down-conversion is shown in Fig. 11.3.5. The signal spectrum is first frequency down-converted to the baseband by mixing with an RF oscillator of frequency f_c. It is then filtered by a LPF to eliminate high-frequency components. The baseband analog signal is converted into digital domain by a high-speed ADC to obtain a sampled serial digital sequence $\{y[0]\ y[1]\ y[2]\ y[3]\ldots y[MN-1]\}$. Then, this serial digital sequence is partitioned into N parallel channels with the time-domain data matrix shown on the left side of Fig. 11.3.6.

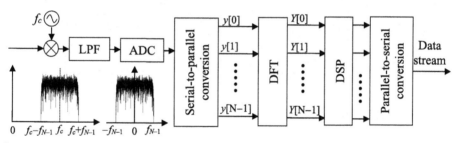

FIG. 11.3.5

Block diagram of an OFDM receiver using DFT for frequency down-conversion.

$$\begin{array}{ccc} f_{N-1} & f_1 & f_0 \\ \begin{bmatrix} y[N-1] & \cdots\cdots\cdots & y[1] & y[0] \\ y[2N-1] & \cdots\cdots\cdots & y[N+1] & y[N] \\ \cdots\cdots & \cdots\cdots & \cdots\cdots & \cdots\cdots \\ y[MN-1] & \cdots\cdots\cdots & y[(M-1)N+1] & y[(M-1)N] \end{bmatrix} \end{array}$$
$$\xrightarrow{DFT}$$
$$\begin{bmatrix} Y[N-1] & \cdots\cdots\cdots & X[1] & Y[0] \\ Y[2N-1] & \cdots\cdots\cdots & X[N+1] & Y[N] \\ \cdots\cdots & \cdots\cdots & \cdots\cdots & \cdots\cdots \\ Y[MN-1] & \cdots\cdots\cdots & Y[(M-1)N+1] & Y[(M-1)N] \end{bmatrix}$$

Time   Frequency

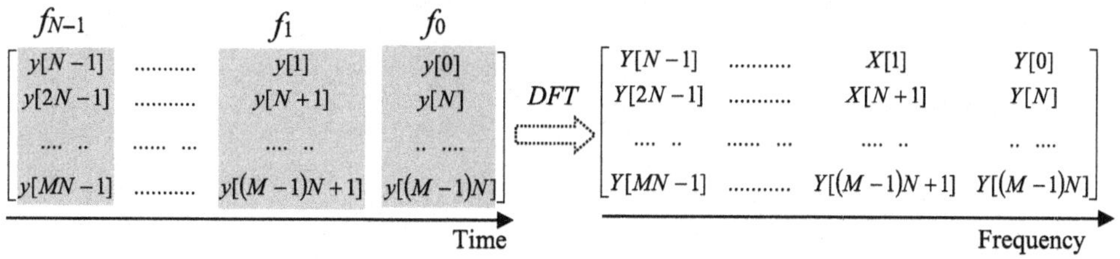

FIG. 11.3.6

Matrices of data mapping in an OFDM receiver. The matrix on the left side is in frequency domain, and the matrix on the right side is in time domain.

The data matrix after serial-to-parallel conversion in Fig. 11.3.5 is treated as in time domain and arranged column-wise, where each column represents a subcarrier channel. The DFT operation equivalently mixes each subcarrier channel with a LO at frequency f_n, with $n=0, 1, 2, N-1$, and converts data carried on that subcarrier back to the frequency domain, which is $\{Y[0]\ Y[1]\ Y[2]\ Y[3]...Y[MN-1]\}$ as shown in Fig. 11.3.6. This recovered data stream may not be the same as that sent by the transmitter because of the transmission impairments of the fiber system represented by $g(t)$ in Fig. 11.3.2.

In fact, after serial-to-parallel conversion in the receiver, the actually received signal should be $y[n]=x[n]g[n]$, where $g[n]$ is the transfer function of the transmission medium in time domain.

Converting back into the frequency domain through DFT, one can define a frequency domain system transfer function $G[m]=DFT[g[n]]$, so that $Y[m]=X[m]\otimes G[m]$, where $\otimes$ represents convolution. With this linear system transfer function in mind, the effect of transmission-induced impairments can be removed in by a DSP unit through digital processing (Molisch, 2011; Shieh and Djordjevic, 2010).

One of the key requirements of orthogonality between OFDM subcarriers and the ability of demultiplexing without cross talk is the integration over exactly a symbol period T_N, which is the inverse of the channel spacing Δf. However, chromatic dispersion in the transmission fiber can introduce significant differential group delay between subcarrier channels of an OFDM-modulated signal when the total bandwidth is large enough. Fig. 11.3.7 illustrates the impact of differential group delay between subcarrier channels, where T_N is the symbol length and ΔG is the maximum differential delay between all subcarrier channels. Temporal misalignment between subcarriers creates relative phase delay so that integration over a fixed time window T_N may not cover a complete 2π cycle, and thus the orthogonality condition may not be maintained. *Cyclic prefix* is a method that copies a section of the waveform at the front of the period and appends it to the end of the period as illustrated in Fig. 11.3.7. The length of the duplicated section is determined by the maximum differential group delay ΔG of the system, and thus

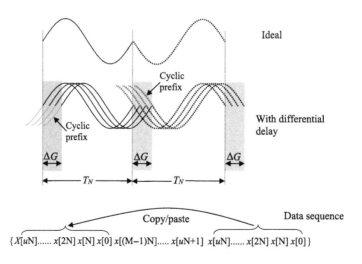

FIG. 11.3.7

Illustration of cyclic prefix to mitigate the impact of differential delay ΔG between subcarrier channels. The bottom part shows the data sequence of one subcarrier with cyclic prefix of length u.

the bandwidth efficiency is reduced to $(T_N - \Delta G)/T_N$ due to the redundancy of the appended section. The data sequence of a subcarrier is shown at the bottom of Fig. 11.3.7 with cyclic prefix of length u.

In an OFDM-based fiber-optic system, the OFDM signal created in the electric domain has to be converted into the optical domain in the optical transmitter, and converted back to the electric domain in the optical receiver. Both intensity modulation and direct detection (IMDD), and optical field modulation and coherent detection can be used in the optical system.

The block diagram of an OFDM optical system using IMDD is shown in Fig. 11.3.8, where the electric signal generated by the digital circuit is first converted into an analog waveform through a DAC, and then it is used to modulate the intensity of an optical carrier produced by a laser diode through an EOM. Direct modulation on the injection current is also possible if modulation chirp of the laser diode is not a big concern. At the receiver side, the optical signal is detected by a PD, amplified by a TIA, and digitized by an ADC. This digital signal is then processed by the DSP circuit in the receiver, as discussed above, to recover the data.

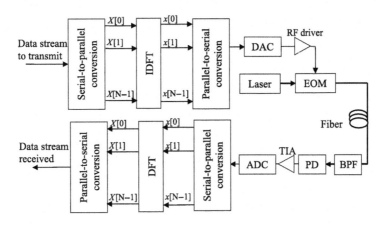

FIG. 11.3.8

Block diagram of an OFDM optical system with intensity modulation and direct detection (IMDD).

One important note is that since the electric voltage signal sent to the DAC in the transmitter has to be real, the data matrix $[X]$ has to maintain a Hermitian symmetry. This is usually accomplished by doubling the size of the $M \times N$ data matrix into the size of $M \times 2N$ with one side to be the complex conjugate of the other side, such as

$$
\begin{bmatrix}
X[N-1] & \cdots\cdots & X[1] & X[0] & jX[0] & jX[1] & \cdots\cdots & jX[N-1] \\
X[2N-1] & \cdots\cdots & X[N+1] & X[N] & jX[N] & jX[N+1] & \cdots\cdots & jX[2N-1] \\
\cdots\cdots & \cdots\cdots & \cdots\cdots & \cdots\cdots & \cdots\cdots & \cdots\cdots & \cdots\cdots & \cdots\cdots \\
X[MN-1] & \cdots\cdots & X[(M-1)N+1] & X[(M-1)N] & jX[(M-1)N] & jX[(M-1)N+1] & \cdots\cdots & jX[MN-1]
\end{bmatrix}
$$

This ensures that after IDFT, the waveform $[x]$ sending to the DAC is real, which creates a real voltage waveform $V(t)$. There are also a few other considerations in data matrix creation. First of all, there is a need of over sampling. By the Nyquist sampling theorem, at least two samples are needed for each data symbol, and the analog bandwidth is half the sampling rate. In a practical OFDM system, the number of samples per symbol has to be more than two to represent the arbitrary waveform of the OFDM signal.

This oversampling can be done by zero-padding on each sides of the data matrix. For example, if $[\hat{X}]$ is the $M \times N$ OFDM data matrix, the zero-padded matrix has to be $\left[[O] \; [\hat{X}] \; j[\hat{X}] \; [O] \right]$, where $[O]$ on each side is a $M \times N$ matrix filled with zeros for a two-time oversampling.

Fig. 11.3.9A shows an example of OFDM modulated voltage signal $V(t)$ which is obtained after IDFT, parallel-to-serial conversion, and DAC. This signal in the analog domain looks like an arbitrary waveform. Converting this arbitrary waveform into the frequency domain through Fourier transform reveals a typical OFDM spectrum as shown in Fig. 11.3.9B. In this example, the signal bandwidth is 2.5 GHz, and the sampling rate is 10 G Sample/s, representing a twofold oversampling beyond the Nyquist sampling rate.

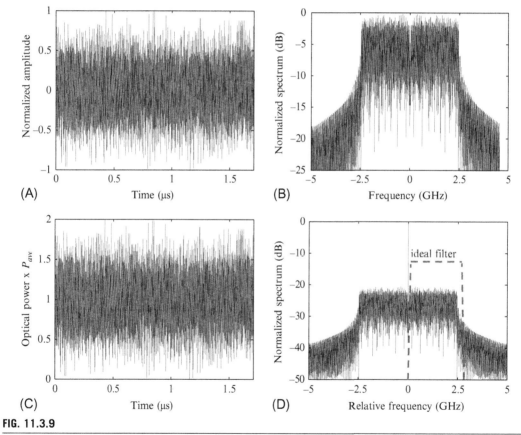

FIG. 11.3.9

(A) Example of an OFDM electric waveform generated at the DAC output, (B) the Fourier transform of this OFDM electric waveform, (C) OFDM waveform of the intensity-modulated optical signal (P_{ave} is the average optical power), and (D) Normalized optical power spectral density with a strong carrier component. Dashed line in (D) shows the transfer function of an ideal optical band-pass filter (BPF) to select a single sideband.

With an intensity modulation on the optical signal, the waveform of signal optical power shown in Fig. 11.3.9C should be the same as the electric waveform shown in Fig. 11.3.9B except for the all-positive values as the power cannot be negative. As the consequence, the spectrum of the optical intensity waveform shown in Fig. 11.3.9D has a strong carrier component and the center frequency is up-converted to the frequency of the optical carrier. This modulation scheme is simple, but with an obvious disadvantage which is the DSB structure of the optical spectrum, typically susceptible to the impact of chromatic dispersion. In fact, the spectral width of the optical signal is twice the width of the baseband determined by the symbol rate. In the direct-detection process, the negative side of the spectrum will fold to the positive side creating signal fading at frequency components f_n with

$$f_n = \frac{1}{\lambda}\sqrt{\frac{2c}{DL}} \tag{11.3.6}$$

where D is the fiber chromatic dispersion parameter, L is the fiber length, c is the speed of light, and λ is the signal wavelength.

This problem can be solved by using an optical band-pass filter (BPF) at the receiver in front of the PD to select only one single sideband (SSB) of the optical spectrum. However, because the edges of an optical filter transfer function are usually not sharp enough, a spectral guard band has to be reserved between the two optical sidebands. In the waveform generation, this can be done by padding extra zeros as $\left[[O] \quad [\hat{X}] \quad [O] \quad [O] \quad j[\hat{X}] \quad [O] \right]$, which would further reduce the spectral efficiency.

As an example, Fig. 11.3.10 shows an OFDM spectrum which has 5 GHz guard band ($-2.5\,G < f < 2.5\,G$) between the two signal-carrying sidebands. Thus, the positive and the negative sidebands can be easily separated by a bandpass optical filter. Two-times oversampling beyond the Nyquist sampling rate is used also in this case, so that the bandwidth efficiency is 25%.

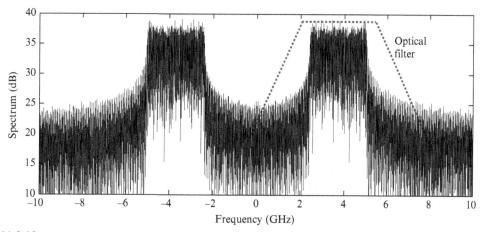

FIG. 11.3.10

Spectrum of an OFDM signal with guard band in the low frequency, and two times oversampling.

Note that in the serial-to-parallel conversion process shown in Fig. 11.3.4, data are mapped into rows of the matrix, but each column is carried by a subcarrier after IDFT. As a consequence, consecutive data points are carried by different subcarriers. The advantage of this data mapping is to have stable bit error rate (BER) even though different subcarrier channels may not have the same transmission impairments. This is useful for wireless communication systems because of the strong multipath interference of antennas which can cause frequency-dependent losses. However, any section of the original data sequence cannot be selectively recovered until the entire matrix is recovered in the receiver.

Alternatively, data sequence can be mapped into columns of an OFDM frame, as shown in Fig. 11.3.11, so that a continuous section of signal data sequence can be carried by the same subcarrier, and can be recovered independently without recovering other subcarrier channels. This may be suitable for optical systems with multiple low data-rate channels, and these mutually independent parallel data channels can be selectively detected at the receiver. As an example, Fig. 11.3.12 shows an electric power spectral density measured at an OFDM receiver. In this system, nine subcarrier channels are used, each carrying 1 Gbaud/s symbol rate, and thus the frequency spacing between adjacent subcarrier channels is 1 GHz. The sampling rate of the DAC is 20 GS/s, so that there are four sampling points per period for the highest signal frequency components (two-times oversampling beyond the Nyquist sampling rate). After photodetection, each subcarrier channel can be frequency downshifted to the baseband by mixing with a LO, and then integrated over one-bit period to remove the cross talk from adjacent subcarrier channels. In this particular example, 75 km standard SMF was used between the

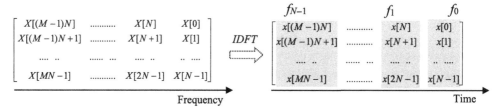

FIG. 11.3.11

Data are mapped into columns of OFDM matrix frame.

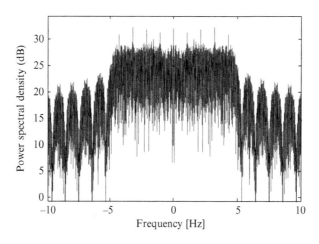

FIG. 11.3.12

Power spectral density of an OFDM signal carrying nine subcarrier channels with 1 Gb/s NRZ data on each subcarrier channel.

transmitter and the receiver, and an adjustable ASE noise is added to the optical signal before the receiver which sets the optical carrier-to-noise ratio of OCNR $=-4.5$ dB to mimic a practical optical system condition. The definition of OCNR is based on 0.1 nm optical bandwidth for the ASE noise and 1 GHz bandwidth for each subcarrier.

Ideally, after frequency down-conversion which shifts a specific subcarrier to the baseband, the receiver bandwidth only needs to be the same as the spectral width of that particular subcarrier channel. However, a much wider receiver bandwidth is usually required for the receiver to minimize the BER. Fig. 11.3.13 shows the BER as the function of the receiver bandwidth set by a raised-cosine filter. It is noticeable that residual cross talk exists when the receiver bandwidth is exactly equal to the channel spacing (normalized receiver bandwidth $=1$). Local minima of BER can be found when the normalized receiver bandwidth is an integer. Residual cross talk still exist at these points which is attributed to the existence of noise and the limited sampling points per period which may have affected the accuracy of integration to minimize the cross talk. In comparison, better BER performance can be obtained when the receiver bandwidth is much wider than the bandwidth of each subcarrier. In a practical system, if the receiver has a limited RF bandwidth, orthogonality between subcarrier channels can be affected, so that cross talk between them can be increased.

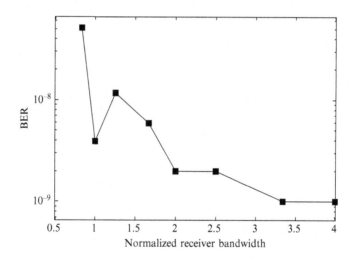

FIG. 11.3.13

Measured BER of the center subcarrier channel (referring to the spectrum shown in Fig. 11.3.12) as the function of the receiver bandwidth.

To explain this residual cross talk, Fig. 11.3.14A illustrates the spectra of the signal channel to be selected (red solid line) and the nearest cross talk channel (blue dashed line). If the receiver bandwidth is infinitely wide, the waveform of the cross talk channel is shown in Fig. 11.3.14B, in which the carrier of the cross talk channel at a digital '1' has a complete 2π cycle of a sine wave which can be averaged to zero after integration over a bit period, and there is no energy in time slot of "0" bit. However, if the receiver is band limited to the width of a single subcarrier channel, for example, with a raised-cosine transfer function shown as the black dotted line in Fig. 11.3.14A, the time-domain waveform of the cross talk channel will be distorted, shown as the spreading of the nearest cross talk subcarrier outside

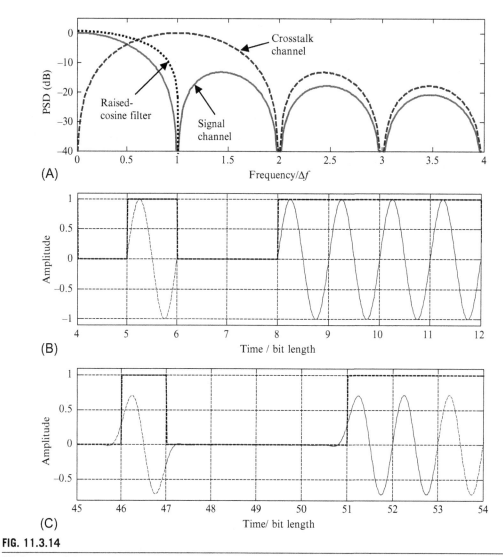

FIG. 11.3.14

(A) Spectra of the signal channel to be selected (*red* solid line) and the nearest cross talk channel (*blue* dashed line) as the function of the normalized frequency (Δf is the data rate on each subcarrier channel), (B) time-domain waveform of the cross talk channel when the receiver bandwidth is infinite, and (C) time-domain waveform of the cross talk channel when the receiver bandwidth is equal to Δf [receiver transfer function is shown as the dotted line in (A)].

the '1' bit slot as indicated in Fig. 11.3.14C. The energy leakage into the previous and the following time slots will create residual cross talk between the signal and adjacent subcarrier channels. This is because the average over a bit period may not produce a zero for every "0" bit of the cross talk channel. Therefore, in practical implementation of an OFDM system, the receiver bandwidth has to be at least three times the bandwidth of each subcarrier channel to effectively remove the inter-subcarrier cross talk as shown in Fig. 11.3.13.

11.3.2 NYQUIST PULSE MODULATION AND FREQUENCY-DIVISION MULTIPLEXING

OFDM discussed above is based on the orthogonality between adjacent subcarrier channels, and demultiplexing of adjacent subcarrier channels is accomplished with integration over one bit period. Another, yet equivalent, approach is to use Nyquist pulses for multiplexing. Fig. 11.3.15 summarizes the similarity and difference between OFDM and Nyquist pulse modulation approaches. Fig. 11.3.15A shows an ideal square pulse representing an isolated digital "1" in the time domain with a temporal pulse width T_N. Its spectrum obtained by a Fourier transform is shown in Fig. 11.3.15B which is a *sinc* function centered at zero and with the first null at frequency $\Delta f = \pm 1/T_N$. When this time-domain pulse is loaded onto a subcarrier of frequency f_n, the center frequency of the *sinc* function becomes f_n. Then, OFDM can be composed of a number of subcarriers with a frequency spacing Δf as shown in Fig. 11.3.15C.

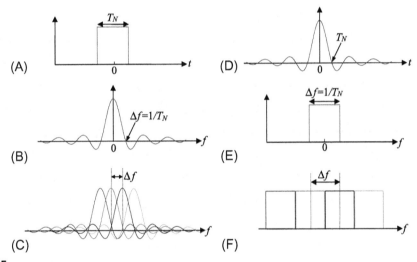

FIG. 11.3.15

Comparison between time-domain waveforms and frequency spectra of OFDM (*left* column) and Nyquist frequency-division multiplexing (*right* column).

In a similar way, Fig. 11.3.15D shows a time-domain pulse with a shape of *sinc* function, known as the Nyquist pulse, and the first null is at $\pm T_N$ [the full-width at half-maximum (FWHM) pulse width in the time domain is approximately $1.2T_N$]. In the frequency domain, which can be obtained by a Fourier transform, this Nyquist pulse has an ideal rectangular spectrum with a spectral with $\Delta f = 1/T_N$, and sharp edges on both sides as shown in Fig. 11.3.15E. Similarly, when this time-domain *sinc* pulse is loaded onto a subcarrier of frequency f_n, the center frequency of the square-shaped spectrum will be shifted to f_n. Then, Nyquist frequency-division multiplexing (Nyquist-FDM) can be realized with a number of subcarriers frequency separated by Δf as shown in Fig. 11.3.15F. These subcarrier channels can be de-multiplexed by using Nyquist filters with sharp cutting edges.

A generalized Nyquist filter has the transfer function of

$$H_{TX}(f) = \begin{cases} \dfrac{\pi f T}{\sin(\pi f T)} & \text{for} \quad |f| \le \dfrac{1-\beta}{2T} \\[3mm] \dfrac{\pi f T}{\sin(\pi f T)} \cos\left(\dfrac{\pi T}{2\beta}|f| - \dfrac{1-\beta}{2T}\right) & \text{for} \quad \dfrac{1-\beta}{2T} < |f| < \dfrac{1+\beta}{2T} \\[3mm] 0 & \text{for} \quad |f| \ge \dfrac{1+\beta}{2T} \end{cases} \qquad (11.3.7)$$

where $0 \le \beta \le 1$ is the roll-off factor which determines the sharpness of the transfer function at the edges of the filter, T is the length of a data bit so that the FWHM bandwidth of the filter is approximately $1/T$, but increases with the increase of the β factor. Within the pass band, the filter has an inverse *sinc* transfer function, and the purpose is to produce flat-top spectrum after multiplying with the signal spectrum which is typically a *sinc* function. Fig. 11.3.16A shows the transfer functions of Nyquist filters with $T=1$ ns and $\beta=0$, 0.2, 0.4, and 0.6, respectively. This type of Nyquist filters is typically used in the transmitter side for signal bandwidth limiting.

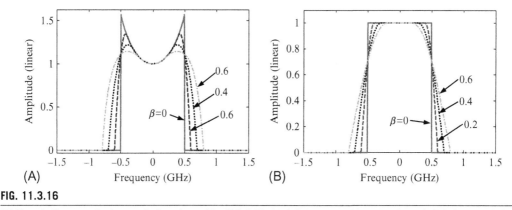

FIG. 11.3.16

(A) Nyquist filter amplitude transfer functions of Eq. (11.3.7) with 1 GHz bandwidth and with $\beta=0, 0.2, 0.4,$ and 0.6 and (B) raised-cosine filter transfer functions of Eq. (11.3.8) with 1 GHz bandwidth and with $\beta=0, 0.2, 0.4,$ and 0.6.

For OFDM, a sharp-edged signal spectrum can be obtained from a large number of subcarrier channels with relatively low data rate of each subcarrier as shown in Fig. 11.3.15C. Sharp-edged spectrum of Nyquist pulse modulation does not need to split the signal into a large number of subcarrier channels. The roll-off rate at the edges of the Nyquist pulse-modulated spectrum only depends on the β-factor of the Nyquist spectral shaping filter.

Multiplexing a large number of Nyquist pulse-modulated channels into a subcarrier system can be accomplished either in the electric domain using DSCM, or in the optical domain using WDM. The sharp-edged signal spectrum minimizes the required channel spacing and avoids spectral overlapping between adjacent subcarrier channels. For electric domain multiplexing, each Nyquist pulse-modulated channel is loaded onto a subcarrier of frequency f_n, so that the center frequency of the spectrum is up-converted to f_n. Then, Nyquist-FDM can be created by adding a number of subcarriers with frequency spacing Δf as shown in Fig. 11.3.15F. Because of the rectangular spectral shape of each subcarrier channel, ideally there is no guard band required between adjacent subcarrier channels, so that the bandwidth efficiency can approach 100%. While OFDM allows spectral overlapping between adjacent subcarrier channels and de-multiplexing based on their orthogonality, Nyquist-FDM eliminates the spectral overlap, and subcarrier channels can be de-multiplexed by spectral filtering by high-order BPFs with sharp edges.

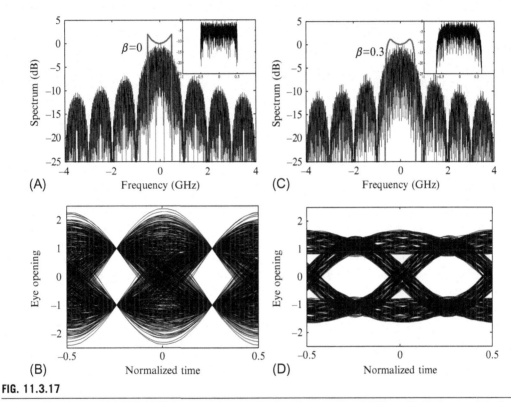

FIG. 11.3.17

(A) and (C): Spectra of 1 Gb/s NRZ data and the transfer function of a Nyquist filter with $T=1$ ns and $\beta=0$ and $\beta=0.3$, respectively. Insets show spectra after filtering. (B) and (D): Eye diagrams of the NRZ-modulated signal filtered by Nyquist filters with $\beta=0$ and $\beta=0.3$ as shown in (A) and (C).

Another Nyquist filter known as the raised-cosine filter typically used in the receiver for channel selection has the transfer function defined as

$$
H_{RX}(f) = \begin{cases} 1 & \text{for} & |f| \le \dfrac{1-\beta}{2T} \\ \cos\left(\dfrac{\pi T}{2\beta}|f| - \dfrac{1-\beta}{2T}\right) & \text{for} & \dfrac{1-\beta}{2T} < |f| < \dfrac{1+\beta}{2T} \\ 0 & \text{for} & |f| \ge \dfrac{1+\beta}{2T} \end{cases} \tag{11.3.8}
$$

which has a flat-top transfer function within the pass band as shown in Fig. 11.3.16B.

Fig. 11.3.17 illustrates the impact of Nyquist filtering on the signal spectral shapes and eye diagrams. Fig. 11.3.17A and C shows bipolar NRZ-modulated spectra (signal waveform swings between

± 1 so that there is no DC component) at 1 Gb/s data rate ($T=1$ ns), and transfer functions of Nyquist filters with $\beta=0$ and $\beta=0.3$, respectively. Insets in Fig. 11.3.17A and C show the spectra after filtering. For an ideal Nyquist filter with a roll-off factor $\beta=0$ shown in Fig. 11.3.17A, the edges are infinitely sharp so that no spectral component exists outside the frequency band of ± 0.5 GHz. The corresponding signal eye diagram after the filter is shown in Fig. 11.3.17C, which is completely open at decision time. Thus, $\beta=0$ provides the optimum filter transfer function with the strongest spectral confinement without reducing signal eye opening. However, considering digital filter implementation, a Nyquist filter with $\beta=0$ is not practically realizable. In fact, the common technique to realize a Nyquist filter is to use multiple tapped delay line structure shown in Fig. 11.2.2 to realize the required transfer function. A Nyquist filter with $\beta=0$ would require infinite number of tapped delay lines. A practical design has to make trade-offs between the spectral confinement and the complexity of filter implementation, such as the number of taps of a FIR filter. Fig. 11.3.17D shows the eye diagram for the spectrum shaped by a Nyquist filter with $\beta=0.3$. Because of the increased bandwidth in comparison with $\beta=0$, the eye opening has an increased timing jitter window. This increased bandwidth and the degraded spectral confinement may also result in spectral overlap between adjacent subcarrier channels if the channel spacing is still set to be $1/T$. In order to avoid spectral overlap, frequency spacing between adjacent subcarrier channels has to be increased to $(1+\beta)/T$, which reduces the spectral efficiency by a factor of $1/(1+\beta)$.

Fig. 11.3.18 shows the block diagram of a Nyquist-FDM optical system with N-independent subcarrier data channels, $a_1(t)$ through $a_N(t)$. A multiple-delay-line FIR filter at each data channel imposes Nyquist transfer function as shown in Eq. (11.3.7) applied on the signal reshapes the signal spectrum and converts signal pulses into a Nyquist pulse train. Then, each subcarrier channel is up-converted to a subcarrier frequency f_n (with $n=1, 2, ...N$) before they are combined. This process can be accomplished in the digital domain based on DSP. The composite digital signal including all subcarrier channels is then converted into an analog waveform through a high-speed DAC, and modulates onto an optical carrier through an electro-to-optical converter (E/O). At the receiver side, an optical-to-electric (O/E) converter, such as a PD for direct-detection receiver, converts the optical signal back into an

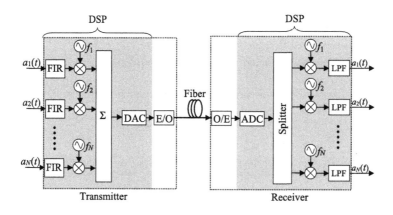

FIG. 11.3.18

Block diagram of a Nyquist-FDM optical system based on the high-speed DAC, ADC, and DSP. $a_1(t)$, $a_2(t)$, $...a_N(t)$ are digital data channels.

electric domain waveform and then it is translated into a digital sequence by a high-speed ADC. This digital signal is split into N copies for frequency down-conversion, which translates each data channel into the baseband. A digital LPF is used after the frequency down-conversion of each data channel to remove high-frequency components.

In the configuration shown in Fig. 11.3.18, most of the signal processing is done in the digital domain including FIR filtering, frequency up-conversion and down-conversion, and LPF. This requires DAC and ADC, as well as digital electronic circuits to perform high-speed operations to include all subcarrier frequencies.

An alternative way is to use the combination of digital electronics and analog RF electronics to extend the overall bandwidth of Nyquist-FDM, and the block diagram is shown in Fig. 11.3.19. In this configuration, each digital data channel is converted into a Nyquist pulse train through an FIR filter, and converted into analog domain using a DAC. Then, each analog waveform is up-converted into a subcarrier frequency f_n (with $n = 1, 2, ...N$) by analog mixing with an RF oscillator before combining into a FDM composite signal and modulating onto an optical carrier. Similarly in the receiver, frequency down-conversion and LPF are performed in the analog domain through RF mixing, and an ADC is used only to convert the recovered baseband waveform on each subcarrier channel into digital data. In this digital-analog hybrid approach, the speed requirement of DAC, ADC, and DSP can be greatly reduced, which is determined only by the signal bandwidth on each subcarrier instead of the aggregated bandwidth of the composite Nyquist-FDM.

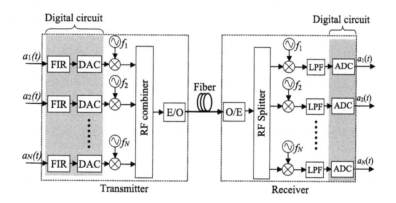

FIG. 11.3.19

Block diagram of a digital-analog hybrid approach of Nyquist-FDM optical system based on the parallel processing of relatively low-speed subcarrier channels in the digital domain, and multiplexing/de-multiplexing them by analog FR circuits.

11.3.3 OPTICAL SYSTEM CONSIDERATIONS FOR DSCM

In the last two sections, we have discussed high spectral efficiency DSCM techniques such as OFDM and Nyquist-FDM. Despite their difference in the implementation details, they all utilize subcarriers created through DSP, so that they belong to the same category of DSCM.

Optical systems based on the DSCM can use either IMDD, or complex optical field modulation and coherent detection. Complex optical field modulation and direct detection can also be used depending on specific circumstance, system performance requirement, and complexity constrain. Various algorithms have been developed for analog and DSP of DSCM optical systems. Spectral efficiency, receiver sensitivity [or required–optical signal-to-noise ratio (OSNR) for optically amplified systems], and frequency/phase recovery are among the important issues.

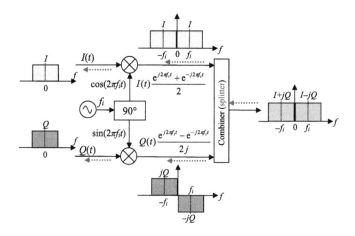

FIG. 11.3.20

Block diagrams of I/Q frequency up-conversion of two data channels. Frequency down-conversion uses the same block diagram but follows reverse direction (dashed red arrows).

FDM has been used for communication networks for many years for network resources sharing among many users. In a FDM-based communication network, each pair of users is assigned a dedicated frequency channel for communication. In an all-analog FDM system such as subcarrier multiplexing, a substantial spectral guard band has to be reserved between adjacent frequency channels to avoid cross talk in the de-multiplexing process, so that the bandwidth efficiency is low. That is one of the reasons that packed-based switching has become the dominate technique for resource sharing in communication networks. However, with the use of DSCM which eliminates the need for a spectral guard band between subcarrier channels, the spectral efficiency can be increased significantly. Network resources sharing based on the FDM may become both efficient and flexible if the data rate granularity on each digital subcarrier channel can be fine enough. For a DSCM optical system with both IMDD and coherent detection, SSB modulation is important on both electric level and optical level to achieve high spectral efficiency, and to avoid signal fading caused by fiber chromatic dispersion manifested as destructive interference between the positive and the negative signal sidebands.

11.3.3.1 I/Q mixing for subcarrier up-conversion and down-conversion

Frequency up-conversion can be achieved by I/Q mixing as illustrated in Fig. 11.3.20. Assume that two independent signal waveforms $I(t)$ and $Q(t)$ are created through Nyquist pulse modulation so that they both have rectangle spectral shapes. In the I/Q mixing configuration, $I(t)$ and $Q(t)$ channels mix with the in-phase and the quadrature components of an RF carrier at frequency f_i independently, and then combine. This complex frequency up-conversion moves the central frequency of $(I - jQ)$ into f_i, and $(I + jQ)$ into $-f_i$. By setting f_i to be half the channel full bandwidth, there will be no spectral gap between the newly created $(I - jQ)$ and $(I + jQ)$ channels at the output of this I/Q frequency up-converter. In this case, the upper and the lower sidebands of the composite signal spectrum are used to carry different data information. Frequency down-conversion in the receiver uses exactly the same block diagram but in a reversed process, following the red arrows in Fig. 11.3.20, so that the baseband signal can be recovered.

FDM can be performed based on the same block diagrams shown in Figs. 11.3.18 and 11.3.19, but with the complex (I/Q) frequency up-conversion and down-conversion to create a FDM spectrum. As illustrated in Fig. 11.3.21A, the spectra of N pairs of data channels are closely packed in the frequency domain without any spectral gap. Then, this electric domain spectrum can be converted into optical domain through optical single-sideband (OSSB) modulation with the OSSB spectrum illustrated in Fig. 11.3.21B.

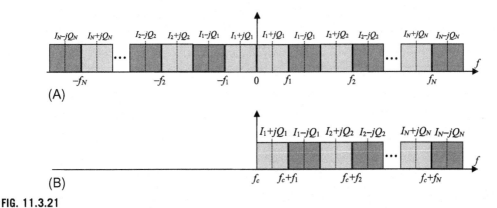

FIG. 11.3.21

(A) N pairs of channels multiplexed through FDM in the electric domain and (B) electric spectrum is converted into optical domain through optical single-sideband modulation.

In the I/Q frequency up-conversion process shown in Fig. 11.3.20, and the subsequent FDM and OSSB modulation, the data streams of the I and the Q channels are superimposed in the optical spectrum. The quality of I/Q mixing such as imaging band rejection ratio cannot be evaluated directly from the spectrum. In addition, if there is a need of channel add/drop in an intermediate node, the I and the Q channels cannot be dropped separately. Fig. 11.3.22 shows a modified technique for I/Q up-conversion which utilizes a 90 degree hybrid coupler at the input, which is able to up-convert the I and the Q channel onto the upper and the lower sidebands separately. The spectral separation between the I and the Q channels makes de-multiplexing much simpler, which can be accomplished simply by spectral filtering.

As an example, Fig. 11.3.23 shows the experiment setup used to test a hybrid-SCM system (Hui et al., 2016). In this example, four RF subcarriers are used at frequencies of $f_1 = 1\,\text{GHz}$, $f_2 = 3\,\text{GHz}$, $f_3 = 5\,\text{GHz}$, and $f_4 = 7\,\text{GHz}$, respectively. Eight independent data channels are created by an arbitrary waveform generator (AWG) which produces Nyquist pulse-modulated data streams through FIR filtering, and each channel has 1 GHz full bandwidth. Two data channels are frequency up-converted by each RF subcarrier through I/Q mixing be translated to the upper and lower sidebands of the subcarrier. The four data-loaded subcarriers are multiplexed through a RF combiner and converted into an optical signal through an OSSB modulation which can be accomplished with a dual-drive EOM.

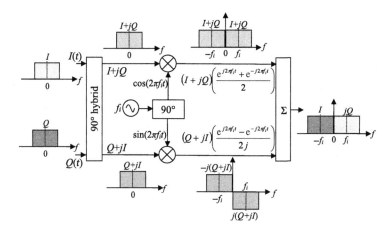

FIG. 11.3.22

Block diagrams of I/Q frequency up-conversion, shifting the *I* and the *Q* channels to the upper and lower sidebands independently.

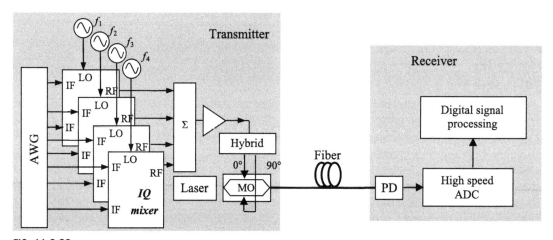

FIG. 11.3.23

Block diagram of a hybrid digital-RF subcarrier transmitter generating four4 subcarrier channels. Broadband receiver with high- speed ADC is used in the receiver to recover all subcarrier channels.

Fig. 11.3.24A shows the RF spectrum with all the four subcarriers fully loaded with data. In this figure, each subcarrier channel is shown with a different color. Thanks to the digital filtering which created tightly confined spectrum of each data channel, there is no spectral overlap between adjacent subcarrier channels even though no spectral guard band between then is reserved. This maximizes the spectral efficiency. Fig. 11.3.24B shows a subset of the spectrum in which the upper sidebands of the 1, 3, and 5 GHz subcarriers, and the lower sideband of the 7 GHz subcarrier are loaded with data, but the opposite sidebands of these subcarriers are empty. The independence of the upper and lower sidebands

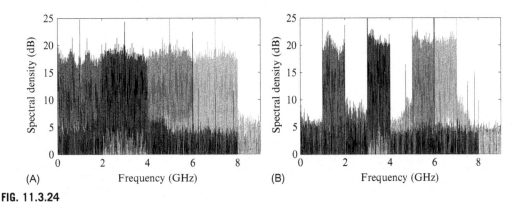

FIG. 11.3.24

RF spectra of hybrid digital-RF multiplexing with four subcarriers. (A) All four subcarrier channels are fully loaded with data, and (B) the upper sideband if 1, 3, and 5 GHz subcarriers and the lower sideband of the 7 GHz subcarrier are loaded with data while the opposite sidebands of these subcarriers are empty.

of each subcarrier provides the full flexibility for channel multiplexing, which also allows the evaluation of image sideband rejection ratio through spectral measurement. The digital-RF hybrid approach effectively reduces the speed requirement on the DAC by parallel processing. In fact in this example, each DAC only needs to handle 1 GHz data bandwidth to support the overall 8 GHz data bandwidth of the system.

The composite RF spectra shown in Fig. 11.3.24 is then converted into optical domain through an OSSB modulation based on a dual-electrode Mach-Zehnder EOM and a laser diode operating in constant wave (CW). As the spectral resolution of a typical optical spectrum analyzer (OSA) is not better than 0.01 nm (equivalent to 1.25 GHz in the 1550-nm wavelength window), the OSSB spectrum measured by an OSA shown in Fig. 11.3.25A does not clearly display the spectral shape. A coherent

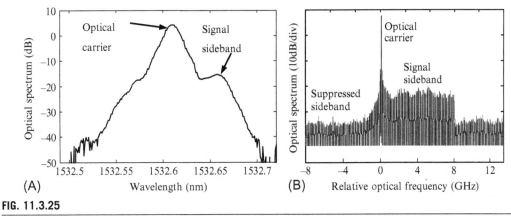

FIG. 11.3.25

OSSB optical spectrum measured with an optical spectrum analyzer with 0.01 nm resolution bandwidth (A), and coherent heterodyne detection (B).

heterodyne detection can be used to convert the optical spectrum to the RF domain for the spectrum measurement with a much better spectral resolution. Fig. 11.3.25B shows the heterodyne spectrum of the OSSB-modulated optical signal with 8 GHz bandwidth on the upper side of the optical carrier. In this figure, the horizontal axis is the relative optical frequency with the frequency of the optical carrier set at zero.

For the OSSB modulation using a dual-drive Mach-Zehnder modulator as discussed in Chapter 7, the modulator is biased at the quadrature point. Small modulation index has to be used to guarantee the linearity of modulation and to achieve high image-side rejection ratio. On the other hand, the disadvantage of small modulation index is the low energy in the signal sideband in comparison to that in the optical carrier, which is clearly shown in Fig. 11.3.25.

If a high-speed ADC is available as the setup shown in Fig. 11.3.23, all RF subcarrier channels can be digitized and processed in the digital domain to recover the data carried on all subcarriers. This includes frequency down-conversion of each subcarrier channel by mixing with a digitally generated subcarrier in the DSP unit with phase synchronization. Fig. 11.3.26A shows an example of measured RF spectrum at the receiver after photodetection and a high-speed ADC. In this example, only two RF subcarrier channels are used at 5 and 7.2 GHz, each carrying an I-channel on the lower sideband and a Q-channel on the upper sideband. Both the I and the Q channels are digitally created, and each of them has 1 GHz bandwidth consisting 10 QPSK-modulated subcarrier channels of 100 Mb/s symbol rate. Each Nyquist filter used to generate these tributary digital subcarrier channels has 100 MHz bandwidth and $\beta = 10\%$ roll-off rate. The spacing between centers of adjacent digital subcarrier channels is 110 MHz, resulting in a bandwidth efficiency of 90%. The inset of Fig. 11.3.26A shows spectrum details around the 5 GHz subcarrier, revealing narrow spectral gaps between low rate tributary digital subcarriers.

In this particular system, the overall transfer function is not ideally flat across the pass band, which is largely attributed to residual reflections from RF components and connectors commonly exist in practical systems. In fact, in this example as much as 3 dB transfer function ripple is observed in the spectrum across the 4 GHz bandwidth as shown in Fig. 11.3.26A. The nonuniformity of the RF transfer function can be minimized with careful engineering of RF circuits. Fig. 11.3.26B shows the EVM of all 40 recovered digital subcarrier channels. The average EVM is approximately 11%, with about 5% variation from channel to channel. In addition to the impact of chromatic dispersion and SNR, the EVM and its variation across channels can be caused by the spurious reflections in the RF circuit, the cross talk between RF subcarrier channels, and between I and Q channels of the same RF subcarrier due to the nonideal IQ mixing.

In order to identify the impact of opposite sideband rejection ratio in the RF I/Q mixing process, one of the two sidebands on each subcarrier channel can be set to zero for comparison. Fig. 11.3.27A shows the RF spectrum in the receiver in which the I-channel on the 5 GHz carrier and the Q-channel on the 7.2 GHz carrier are turned-off, while in Fig. 11.3.27B, the Q-channel on the 5 GHz carrier and the I-channel on the 7.2 GHz carrier are turned-off. The opposite sideband rejection ratio is on the order of 10 dB in this experiment. This nonideal sideband rejection is primarily caused by the inaccuracy of the RF 90° hybrid, the efficiency mismatch of the two mixers, as well as the spurious reflections and interferences caused by RF connectors used in constructing the IQ mixer. Fig. 11.3.27C shows the values

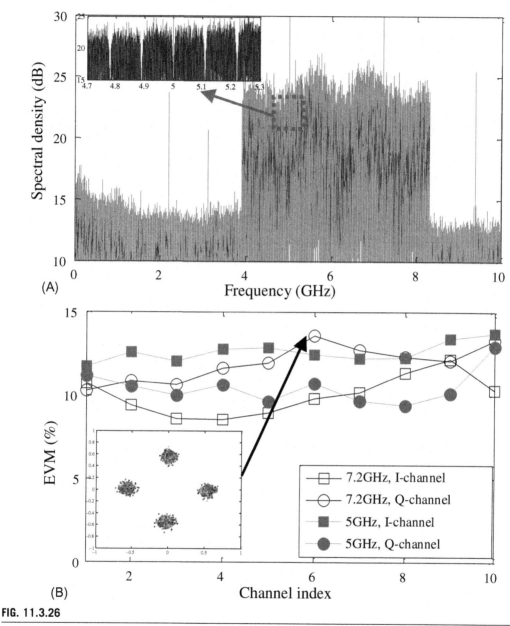

FIG. 11.3.26

Two double-sideband RF subcarriers at 5 and 7.2 GHz, each carrying *I* and *Q* channels. (A) Spectrum measured with a high-speed receiver. (B) EVM of recovered digital subcarrier channels.

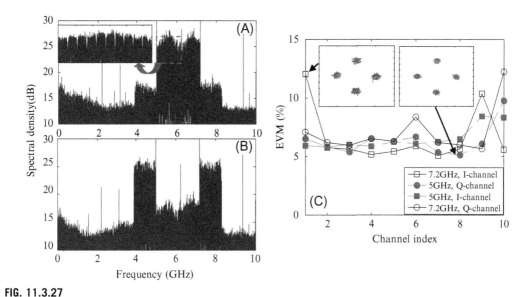

FIG. 11.3.27

Two single-sideband RF subcarriers at 5 GHz and 7.2 GHz, each carrying an *I* or *Q* channel. (A) Spectra measured with a high-speed receiver. (B) EVM of recovered digital subcarrier channels.

of EVM of all 20 tributary subcarrier channels in each spectral setting shown in Fig. 11.3.27A and B. The overall average EVM in the SSB-modulated system is 6.6%, which is lower than the 11% EVM shown in Fig. 11.3.26B, mainly because there is no data loaded in the opposite sideband of each subcarrier. The cross talk between the *I* and the *Q* channels of the same RF carrier due to the nonideal sideband rejection of the IQ mixer can have significant contribution to EVM degradation. Although Fig. 11.3.27B has widely separated spectral components between the two subcarriers, the EVM values are very similar to those with spectrum shown in Fig. 11.3.27A where spectral components of the two RF subcarriers are separated only by 10 MHz. This indicates that the cross talk between different RF subcarrier channels is negligible thanks to the sharp cutting edge of digital filters.

Although a broadband optical receiver with a high-speed ADC can be used to convert the entire signal optical spectrum into the digital domain, the digital/RF hybrid system is most suitable for narrowband receivers based on the RF frequency down-conversion of selected subcarrier channels. Fig. 11.3.28A shows the block diagram of RF frequency down-conversion using an IQ mixer which can replace the receiver used in Fig. 11.3.23. This receiver can selectively detect any subcarrier channel by tuning the frequency f_i (with $i = 1, 2, 3, 4$) of the RF LO for frequency down-conversion. A LPF (fifth-order Bessel with 1.75 GHz 3 dB-bandwidth) is placed before each ADC to prevent aliasing of high-frequency components. In this experiment, only 1 GHz analog bandwidth is required for each ADC to accommodate the symbol rate carried on each subcarrier channel. Fig. 11.3.28B shows the spectra of the down-converted *I* and *Q* channels from the 7.2 GHz subcarrier. In this example, both the upper and the lower sidebands are loaded with data for both the 5 GHz and 7.2 GHz subcarrier channels, and the spectrum is the same as that shown in Fig. 11.3.26A. The recovered *Q*-channel carried on the upper sideband of the 7.2 GHz subcarrier only has approximately 1.1 GHz bandwidth as

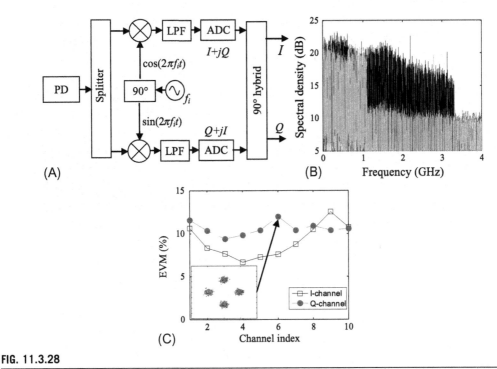

FIG. 11.3.28

(A) Block diagram of frequency down-conversion based on the IQ mixing in the receiver, (B) spectra of recovered *I* *(black)* and *Q* channel *(yellow)* on the 7.2 GHz RF carrier, and (C) measured EVM of I and Q channels.

shown in Fig. 11.3.28B. But the recovered *I*-channel carried on the lower sideband of the 7.2 GHz has much wider bandwidth because the spectral components from 1.1 to 3.3 GHz (as there is a 0.1 GHz guard band between subcarrier channels) belong to channels carried by the 5 GHz subcarrier which can be removed by low-pass filtering. The EVM of the recovered constellation diagrams is shown in Fig. 11.3.28C with an average value of approximately 10%, similar to that obtained with the wideband receiver.

In order to demonstrate the impact of OSNR, a noise-loading experiment is carried out, in which a controllable optical noise from an erbium-doped fiber amplifier (EDFA) is combined with the optical signal through a variable optical attenuator (VOA) and a fiber directional coupler. This is a common technique for system performance testing by varying OSNR. Fig. 11.3.29A shows the measured EVM as the function of OSNR. Again, two RF subcarriers at 5 and 7.2 GHz were used in this measurement with spectrum shown in Fig. 11.3.26A. For the SSB-modulated system, in which the *Q*-channel on the 5 GHz subcarrier and the *I*-channel on the 7.2 GHz subcarrier are turned off, so that cross talk due to the nonideal opposite-sideband rejection does not exist. For the case of DSB modulation, both the 5 and 7.2 GHz subcarriers are fully loaded with the *I*- and the *Q*-channels, and thus cross talk due to nonideal opposite-sideband rejection contributes to additional EVM degradation. With QPSK coding, a BER of 10^{-3} corresponds to an EVM of approximately 33%. Overall, the relatively high value of the required OSNR in this experiment is primarily due to the low optical modulation index with an optical signal to

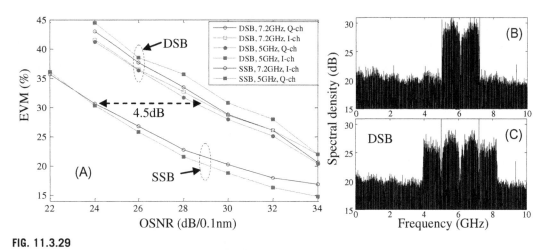

FIG. 11.3.29

(A) EVM as the function of OSNR for DSB and SSB-modulated subcarriers, (B) SSB-modulated RF spectrum at receiver, and (C) DSB-modulated RF spectrum at receiver.

carrier power ratio of approximately −20 dB as indicated in Fig. 11.3.25. Assume each subcarrier has 1 GHz bandwidth, the OCNR should be 6 and 3 dB lower for the DSB and SSB-modulated signals, respectively, compared to the OSNR shown in the horizontal axis of Fig. 11.3.29A. Notice that there is an approximately 4.5 dB difference in the required OSNR between SSB and DSB-modulated system as shown in Fig. 11.3.29A. Within which, 3 dB is due to the increased OCNR in the SSB system, and the rest 1.5 dB is the penalty attributed to the nonideal IQ mixing and the cross talk introduced between the opposite sidebands of the same RF carrier.

11.3.3.2 Signal-signal beat interference in DSCM with direct detection

A coherent optical receiver based on the I/Q detection is able to linearly translate the optical field into electronic domain, and is able to recover both the amplitude and the phase information carried by the optical signal. However, an optical LO has to be used, which has to be phase synchronized to the transmitter laser, making a coherent receiver expensive. On the other hand, direct detection using a single PD in the receiver can be much simpler. Nonetheless, for systems based on the IMDD, optical signal is DSB modulated, and the transmission performance is susceptible to chromatic dispersion. The impact of chromatic dispersion cannot be compensated in the electronic domain due to the spectral folding of direct detection.

OSSB modulation can be used to improve the spectral efficiency as discussed above. In a direct-detection receiver, the SSB optical signal mixes with the unsuppressed optical carrier at the PD, so that signal optical spectrum can be linearly translated to the electronic domain, which avoids spectral folding. This is similar to coherent detection, except that the reference optical signal is not from a LO at the receiver; rather it is received with the optical signal from the transmitter. This "self-coherent" detection allows dispersion compensation and channel equalization to be performed in the electronic domain.

In a "real" coherent detection receiver, the power of LO is much higher than the received optical signal so that the useful photocurrent signal is the mixing between the LO and the optical signal, and the direct-detection contribution of the optical signal is negligible. For self-coherent detection, the carrier component is not much stronger than the signal sideband, so that intermixing among different spectral components of the signal can be significant, which is known as signal-signal beat interference (SSBI). To generalize, for a multi-carrier system with optical SSB modulation, the optical field can be expressed as

$$E(t) = \sqrt{P_{ave}}\left[1 + \sum_{k=1}^{N} m_k a_k(t) e^{-j2\pi f_k t}\right] \tag{11.3.9}$$

where P_{ave} is the average optical power and N is the total number of subcarriers. f_k is the subcarrier frequency, m_k is the modulation index, and $a_k(t)$ is the normalized data bits carried by the kth subcarrier channel. Here, $a_k(t)$ is assumed to be real for simplicity. In Eq. (11.3.9), the negative sideband is rejected, which can be accomplished by using a dual-drive, or I/Q, Mach-Zehnder modulation as discussed in Chapter 7, or by using optical filtering. At a direct-detection receiver, the photocurrent is proportional to the square of the optical field, that is,

$$I(t) = \Re P_{ave}\left|1 + \sum_{k=1}^{N} m_k a_k(t) e^{-j2\pi f_k t}\right|^2 \tag{11.3.10}$$

where $\Re$ is the responsivity of the PD. The useful part of the photocurrent is the mixing between the carrier and the signal sidebands,

$$i_{linear}(t) = 2\Re P_{ave}\sum_{k=1}^{N} m_k a_k(t)\cos(2\pi f_k t) = \Re P_{ave}\sum_{k=1}^{N} m_k a_k(t)\left(e^{2\pi f_k t} + e^{-2\pi f_k t}\right) \tag{11.3.11}$$

which is a *real* photocurrent, and the amplitude is linearly proportional to the data signal on the subcarriers.

Meanwhile, signal subcarriers also mix among themselves to create SSBI,

$$i_{SSBI}(t) = \Re P_{ave}\sum_{i=1}^{N}\sum_{k=1}^{N} m_i m_k a_i(t) a_k(t) e^{j2\pi(f_i - f_k)t} \quad (i \neq k) \tag{11.3.12}$$

This indicates that the frequency components of SSBI will range from DC to $\pm f_N$, with f_N the highest subcarrier frequency.

As an example, Fig. 11.3.30A shows an OFDM spectrum with 2.5 GHz bandwidth of subcarriers on both positive and negative frequency sides, and $-2.5\,\text{GHz} < f < 2.5\,\text{GHz}$ is reserved as a guard band to observe the impact of SSBI. This electric domain OFDM signal can be uploaded onto an optical carrier through OSSB modulation, which creates an optical field shown in Eq. (11.3.9) with a relatively strong optical carrier component and a single optical sideband. In the direct-detection receiver, this optical field is square law detected by a PD to recover the signal carried on subcarriers, but this process also produces unwanted SSBI. The spectrum of the photocurrent is shown in Fig. 11.3.30B, where the SSBI noise caused by intermixing among subcarriers is quite pronounced within the low-frequency guard band between -2.5 and $2.5\,\text{GHz}$. If subcarrier channels are assigned within this low-frequency band, cross talk with SSBI can be significant, degrading signal quality. In a DSCM system employing direct detection, a spectral guard band has to be reserved to avoid the signal subcarriers to interact with the SSBI components, and thus the spectral efficiency will be reduced by 50%.

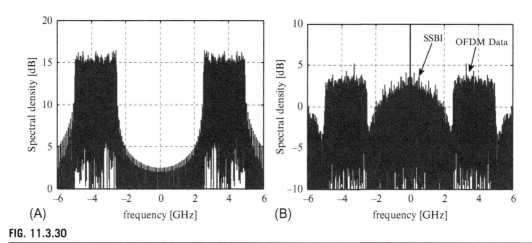

FIG. 11.3.30

(A) OFDM spectrum with 2.5 GHz signal bandwidth, and a 2.5-GHz guard band is reserved at low frequency. (B) Spectrum of photocurrent in which noise level is increased significantly within the guard band due to SSBI.

As the signal components in the photocurrent is produced by the mixing between the optical carrier and the signal subcarriers, while SSBI is created by intermixing among subcarriers, the ratio between signal photocurrent and SSBI is dependent on the modulation index m_k in Eq. (11.3.10). High modulation index will increase SSBI as indicated by Eq. (11.3.12), but low modulation index will reduce the modulation efficiency as most of the signal power will remain in the optical carrier, which degrades the carrier-to-noise ratio (CNR).

Various techniques have been proposed to reduce the impact of SSBI in multi-carrier optical systems with direct detection. Compatible-OSSB modulation is one of such techniques in which the multi-carrier signal is loaded onto the exponent of the optical field in the optical transmitter (Zhang et al., 2010),

$$E(t) = \sqrt{P_{ave}} \exp\left\{ \sum_{k=1}^{N} m_k a_k(t) e^{-j2\pi f_k t} \right\} \tag{11.3.13}$$

With square-law detection of a PD at the receiver, the photocurrent is

$$I(t) = \Re P_{ave} \exp\left\{ 2\sum_{k=1}^{N} m_k a_k(t) e^{-j2\pi f_k t} \right\} \tag{11.3.14}$$

The multi-carrier signal can be recovered from this photocurrent by a nature logarithm operation in the digital domain,

$$i(t) = \ln[I(t)] = \ln[\Re P_{ave}] + 2\sum_{k=1}^{N} m_k a_k(t) e^{-j2\pi f_k t} \tag{11.3.15}$$

There is no intermixing among subcarrier channels, so that no SSBI is introduced. However, putting the multi-carrier signal into the exponent tends to increase the peak-to-average power ratio (PAPR) of the modulated signal, and the signal waveform in the time domain becomes highly asymmetric which often affects the receiver sensitivity and increases the required OSNR.

Another technique is to digitally compensate the impact caused by SSBI (Peng et al., 2009), which can be done both in the transmitter for pre-compensation and in the receiver for post-compensation. Fig. 11.3.31A illustrates the iterative procedure of SSBI compensation. In this process after direct detection, the data on the OFDM sideband are first recovered through the normal procedure of fast Fourier transform (FFT), channel equalization, and decision with the impact of SSBI. Then, the multi-subcarrier OFDM waveform is recreated and squared to mimic the impact of a PD. This allows the creation of the SSBI noise waveform in the digital domain. This SSBI waveform is then subtracted from the originally received signal waveform. In principle, this process can be repeated multiple times to minimize the effect of SSBI, but the effectiveness diminishes as the number of iteration increases.

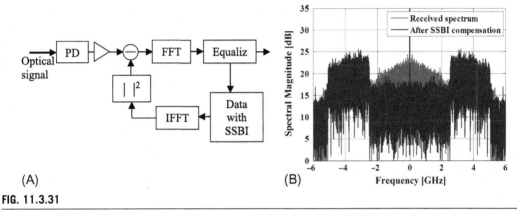

(A)   (B)

FIG. 11.3.31

(A) Block diagram of iterative digital SSBI compensation in a direct-detection OFDM receiver and (B) received signal spectra before *(red)* and after *(blue)* SSBI compensation. (Note: the signal sidebands of the red spectrum without compensation are blocked by the blue spectrum.)

Fig. 11.3.31B shows a gapped OFDM spectrum with 2.5 GHz spectral width of the signal sideband on each side, and a gap of $-2.5\,\text{GHz} < f < 2.5\,\text{GHz}$ in the middle for demonstration purpose. Upon direct detection, SSBI creates significant cross talk components inside the guard band $(-2.5\,\text{GHz} < f < 2.5\,\text{GHz})$ shown as the red spectrum in Fig. 11.3.31B. The SSBI has a signature triangular spectral shape originated from the convolution process of signal-signal interference predicted by Eq. (11.3.12). The electric spectrum after a single iteration of SSBI correction is also shown in Fig. 11.3.31B as the blue line spectrum, which clearly demonstrates significant reduction of SSBI noise within the guard band. Although iterative SSBI compensation in the digital domain can be effective, it requires significant digital processing which increases the complexity of the system. This to some extent diminishes the major advantage of direct detection which is the simplicity.

11.3.3.3 The Kramers-Kronig algorithm

For optical systems based on the OSSB modulation and direct detection, SSBI is a major problem limiting the system performance. In addition to the iterative digital compensation technique as discussed in Section 11.3.3.2, a *Kramers-Kronig* (KK) technique (Mecozzi et al., 2016) is more promising, which is able to reconstruct the complex optical field from the direct-detection photocurrent if the optical signal satisfies a *minimum-phase* condition.

Let us start from the complex envelope of the OSSB-modulated optical field defined in Eq. (11.3.9)

$$E(t) = E_c[1 + s(t)] \tag{11.3.16}$$

where $E_c = \sqrt{P_{ave}}$ is a constant optical carrier with P_{ave} the average signal optical power, and $s(t) = \sum_{k=1}^{N} a_k(t) e^{-j2\pi f_k t}$ represents the normalized composite signal complex field carried in the OSSB. Since $[1 + s(t)]$ is complex, it can be defined by its magnitude and phase, so that

$$[1 + s(t)] = |1 + s(t)| e^{-j\varphi(t)} \tag{11.3.17}$$

Applying a nature logarithm operation on both sides of Eq. (11.3.17), we have

$$\ln[1 + s(t)] = \ln|1 + s(t)| - j\phi(t) = u_r(t) + ju_i(t) \tag{11.3.18}$$

where $u_r(t) = \ln|1 + s(t)|$ and $u_i(t) = -\phi(t)$ are real and imaginary parts of $\ln[1 + s(t)]$.

Under the condition of $|s(t)| < 1$, $\ln[1+s(t)]$ can be expended into a series as $\ln[1 + s(t)] = \sum_{n=1}^{\infty} n^{-1}(-1)^{n+1} s^n(t)$. We have assumed that the spectrum $\tilde{s}(\omega)$ of the time-domain signal $s(t)$ has only a SSB, that is, $\tilde{s}(\omega) = 0$ for $\omega < 0$. The spectrum of $s^n(t)$, which is the nth-order autocorrelation of $\tilde{s}(\omega)$, should also be SSB. The classic KK relation (Kronig, 1926; Kramers, 1927) implies that if the spectrum of $\ln[1 + s(t)]$ is SSB, its real and imaginary parts, $u_r(t)$ and $u_i(t)$, are not independent, instead they are related by

$$u_r(t) = -\int_{-\infty}^{\infty} \frac{u_i(\tau)d\tau}{\pi(t - \tau)} \tag{11.3.19a}$$

and

$$u_i(t) = \int_{-\infty}^{\infty} \frac{u_r(\tau)d\tau}{\pi(t - \tau)} \tag{11.3.19h}$$

Based on this KK relation and Eq. (11.3.18), we have

$$\phi(t) = -\int_{-\infty}^{\infty} \frac{\ln|1 + s(\tau)|}{\pi(t - \tau)} d\tau \tag{11.3.20}$$

In a direct-detection receiver, the photocurrent is proportional to the signal optical power,

$$i(t) = \Re P_{ave}|1 + s(t)|^2$$

where $\Re$ is the PD responsivity. Since

$$\ln\sqrt{i(t)} = \ln\left\{\sqrt{\Re P_{ave}}|1 + s(t)|\right\} = \frac{1}{2}\ln(\Re P_{ave}) + \ln|1 + s(t)|$$

and

$$\int_{-\infty}^{\infty} \frac{\ln(\Re P_{ave})}{2\pi(t - \tau)} d\tau = 0$$

Eq. (11.3.20) can be expressed as the function of the photocurrent as

$$\phi(t) = -\frac{1}{\pi} \int_{-\infty}^{\infty} \frac{\ln \sqrt{i(\tau)}}{t - \tau} d\tau = -H\left[\ln\left(\sqrt{i(t)} \right) \right] \tag{11.3.21}$$

where $H[f(t)] = \frac{1}{\pi} \int_{-\infty}^{\infty} \frac{f(\tau)}{t-\tau} d\tau$ represents a Hilbert transform of $f(t)$.

This Hilbert transform shown in Eq. (11.3.21) can also be conveniently implemented in the frequency domain as

$$\Phi(\omega) = j \cdot \text{sign}(\omega) \cdot F\left\{ \ln\left(\sqrt{i(t)} \right) \right\} \tag{11.3.22}$$

where $F\{f(t)\}$ represents the Fourier transform of $f(t)$, and $\text{sign}(\omega)$ is the sign of ω, that is, $\text{sign}(\omega) = -1$ for $\omega < 0$, and $\text{sign}(\omega) = 1$ for $\omega \geq 0$. Then, the time-domain phase $\phi(t)$ can be obtained from $\Phi(\omega)$ through an inverse Fourier transform. This allows the reconstruction of the complex optical field by $E(t) = E_c[1 + s(t)]$

$$E(t) = \sqrt{\frac{i(t)}{\Re}} e^{-j\phi(t)} \tag{11.3.23}$$

The baseband complex data signal $s(t)$ can be recovered through $s(t) = E(t)/E_c - 1$.

Throughout this analysis, it is apparent that in order for the KK technique to be effective, there are two major constrains. The first necessary condition is that the optical spectrum has to be SSB, so that the real and the imaginary parts of the optical field are related through the KK relation. The second necessary condition is that the carrier component has to be stronger than the signal sideband so that $1 + s(t) > 0$ is always true for the nature logarithm operation. This second condition is also known as the minimum-phase condition (Mecozzi et al., 2016).

While OSSB has been discussed previously, it might be helpful to have more detailed discussion on the implication of the minimum-phase condition: the optical field $E(t)$ is considered *minimum phase* only when its time trajectory does not wind around the origin of the complex plane. To satisfy this condition, the modulation index has to be small enough so that $|s(t)| < 1$ at any time,

$$\max\left\{ |s^2(t)| \right\} < 1 \tag{11.3.24}$$

In practice, since the normalized signal in the sideband $s(t)$ is time varying, statistic values of the signal such as the carrier-to-signal power ratio (CSPR) and the PAPR are often used to specify the optical field and the modulated optical signal, respectively, which are defined as

$$CSPR\{E(t)\} = \frac{1}{\langle s^2(t) \rangle} \tag{11.3.25}$$

$$PAPR\{S(t)\} = \frac{\max\left\{ |s(t)|^2 \right\}}{\langle s^2(t) \rangle} \tag{11.3.26}$$

so that

$$CSPR\{E(t)\} = \frac{PAPR\{S(t)\}}{\max\left\{ |S(t)|^2 \right\}} \tag{11.3.27}$$

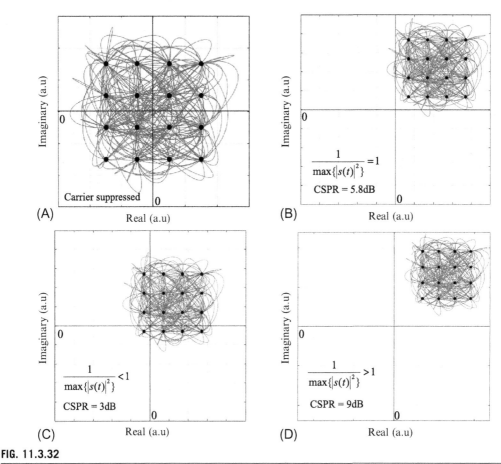

FIG. 11.3.32

Examples of time trajectories of optical field on the complex plane with four different values of $1/\max\{|s(t)|^2\}$.

where $\langle s^2(t) \rangle$ is the ensemble average of the signal carried by the sideband. For the same value of $1/\max$ $\{|s(t)|^2\}$, CSPR may not be the same which depends on the PAPR of the signal. For DSCM, PAPR increases with the increase of the number of subcarrier channels N, and thus higher CSPR is required to satisfy the minimum-phase condition.

Fig. 11.3.32 shows four examples of time trajectories of optical field with different values of $1/\max$ $\{|s(t)|^2\}$. To obtain this figure, only a single subcarrier channel is used on one side of the carrier, and with the bandwidth equal to the subcarrier frequency. $s(t)$ is a 16-QAM signal shaped by a raised-cosine filter with the transfer function of Eq. (11.3.8), and with a roll-off factor of $\beta = 0.1$. This results in a PAPR of 3.8, which is equivalent to 5.8 dB.

For Fig. 11.3.32A, the optical carrier component is completely suppressed with $1/\max\{|s(t)|^2\} = 0$, so that the trajectory of the optical field encircles the origin. Fig. 11.3.32B was obtained with $1/\max$ $\{|s(t)|^2\} = 1$ which narrowly satisfies the minimum-phase condition, and the trajectory of the optical field does not encircle the origin. At that condition, CSPR = PAPR which is 5.8 dB. Fig. 11.3.32C corresponds to a CSPR $=3$ dB, so that $1/\max\{|s(t)|^2\}\approx0.53$ based on Eq. (11.3.27), which is smaller than 1, and therefore, the trajectory of the optical field encircles the origin. Fig. 11.3.32D shows a situation where the optical carrier component is much stronger than the signal sideband with CSPR $=9$ dB, corresponding to $1/\max\{|s(t)|^2\}\approx2.1$. In this case, the trajectory of the optical field is kept far away from the origin on the complex plane.

Fig. 11.3.33 shows the block diagram of a KK receiver. After direct detection through a PD and amplified by a TIA, the photocurrent $i(t)$ is digitized by an ADC and sent to the KK processing unit. In the KK process, the positive and real-valued digital sequence of the photocurrent from the ADC is first normalized and resampled to a higher sampling rate. Usually >4 samples/symbol is required because the logarithm operation may expand the bandwidth considerably. Next, the phase of the optical signal $E(t)$ is calculated by applying the Hilbert transform to the natural log of the square root of the sampled photocurrent. Then, the phase of the optical field calculated from this process is applied to the square root amplitude of photocurrent so that the complex optical field $E(t)$ can be reconstructed. A DSP unit is able to compensate for the impact of chromatic dispersion following the standard algorithms developed for coherent receivers. In the phase recovery process, the bandwidth of the recovered optical field is reduced, and thus a data sequence can be down-converted to two samples per symbol to save the DSP resources.

Fig. 11.3.34 shows an example of signal spectra and constellation in a system based on the KK receiver. The signal has a baud rate of 25 Gbaud/s modulated with 16-QAM, so that the bit rate is 100 Gb/s. Fig. 11.3.34A shows the SSB optical spectrum with 25 GHz bandwidth which is shaped by a Nyquist filter with $\beta=0.1$. A CSPR of 9 dB is used to guarantee the minimum-phase condition. In this example, OSNR $=28$ dB is assumed at the receiver (optical noise is measured with 0.1 nm resolution bandwidth). After direct detection with a PD, the photocurrent is a real waveform so that the spectrum is DSB as shown in Fig. 11.3.34B. This photocurrent is digitized and sent to the KK process. Fig. 11.3.34C shows the recovered complex optical field spectrum which has the total bandwidth of 25 GHz. Although the center of the spectrum shown in Fig. 11.3.34C is at zero frequency, its positive and negative sidebands are not redundant. The constellation diagram is then reconstructed from this recovered complex optical field which is shown in Fig. 11.3.34D.

The KK algorithm provides better performance in suppressing the effect of SSBI compared to the digital iteration schemes discussed above. However, there also have been some concerns in terms of implementation complexity in the DSP. The complexity mainly arises from the requirement of upsampling to deal with the expanded bandwidth caused by nonlinear transfer functions of the natural logarithm operation. Digital implementation of logarithm operation based on the lookup tables can also

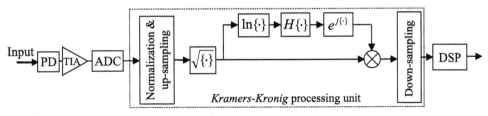

FIG. 11.3.33

Block diagram of a KK receiver.

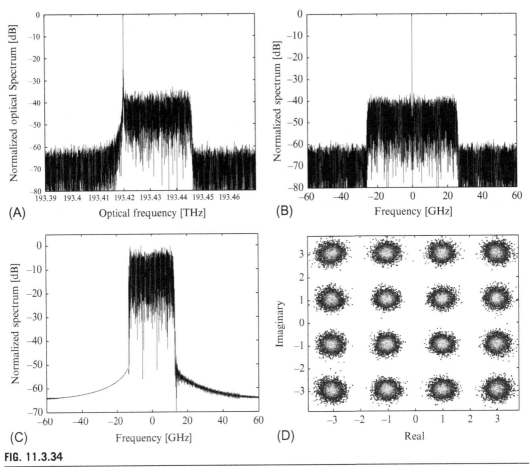

FIG. 11.3.34

(A) OSSB spectrum, (B) electric spectrum after direct detection with a photodiode, (C) spectrum of recovered signal optical field amplitude, and (D) recovered constellation diagram.

be expensive in terms DSP resources. Nevertheless, the KK receiver can still be a viable solution with relatively low cost for the recovery of complex optical field, and to avoid SSBI in a direct-detection receiver.

11.3.3.4 DSCM with coherent detection

Coherent detection has been shown to have many advantages compared to direct detection. Coherent detection can also be used in complex optical field modulated DSCM systems with the block diagram shown in Fig. 11.3.34. In the transmitter, an I/Q EOM is used which converts the complex electric signal into the complex optical field. The real and the imaginary parts of the driving electric signal, $V_I(t)$ and $V_Q(t)$ are obtained after digital processing of the input data sequence and converted into

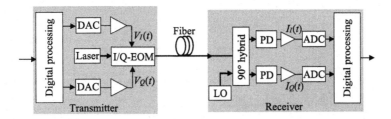

FIG. 11.3.35

Block diagram of a DSCM optical system with complex optical field modulation and coherent detection.

analog domain through two DACs. Both OFDM and Nyquist-FDM can be applied in the digital processing with IFFT to create OFDM channels, or with FIR filters to generate Nyquist pulse trains. At the receiver, a 90 degree optical hybrid and two PDs are used which performs I/Q mixing between an optical LO and the received signal optical field. The in-phase and the quadrature components $I_I(t)$ and $I_Q(t)$ of the photocurrent are digitized by two ADCs, and the digital signals are processed to recover the data. For simplicity with the purpose of demonstrating the concept, polarization diversity and balanced detection, commonly used in coherent receivers, are not shown in Fig. 11.3.35.

As an example, Fig. 11.3.36 shows the spectra of an OFDM system with different number of subcarrier channels (Zhang et al., 2011). In this example, 10 independent QPSK data streams are used, each digitally uploaded onto a subcarrier in the DSP module of the transmitter to form a composite OFDM signal. The real and the imaginary parts of the OFDM signal are converted into analog domain through two DACs with 22 GS/s sampling rate. The analog voltage waveforms $V_I(t)$ and $V_Q(t)$ are amplified to drive an electro-optical IQ modulator. The overall analog bandwidth of the transmitter is approximately 10 GHz. At the output of the modulator, an EDFA boosts the optical signal power to approximately 1 dBm before it is launched into 75 km standard SMF for transmission. The lasers used in both the transmitter and the coherent receiver have narrow spectral linewidth of less than 100 kHz to minimize phase errors in the optical field detection. Based on the same principle if I/Q mixing for frequency up-conversion discussed above, two parallel voltage waveforms used to drive the two arms of the IQ EOM are

$$V_I(t) = \sum_{k=1}^{m} \{Q_{kL}(t) - Q_{kU}(t)\} \cos\left(2\pi \frac{\Delta f \cdot k}{2}\right) + \sum_{k=1}^{m} \{I_{kU}(t) - I_{kL}(t)\} \sin\left(2\pi \frac{\Delta f \cdot k}{2}\right) \qquad (11.3.28)$$

$$V_Q(t) = \sum_{k=1}^{m} \{I_{kL}(t) + I_{kU}(t)\} \cos\left(2\pi \frac{\Delta f \cdot k}{2}\right) + \sum_{k=1}^{m} \{Q_{kU}(t) + Q_{kL}(t)\} \sin\left(2\pi \frac{\Delta f \cdot k}{2}\right) \qquad (11.3.29)$$

where I_{KU} and Q_{KU} are in-phase and quadrature components of the k_{th} subcarrier in the upper sideband of the optical carrier; similarly, I_{KL} and Q_{KL} are in-phase and quadrature components of the k_{th} subcarrier in the lower sideband of the optical carrier; and Δf represents the bandwidth of each subcarrier. QPSK modulation with 2.22 Gb/s data rate is applied on each subcarrier channel. A total of 10 subcarrier channels were used, with 5 on each side of the optical carrier and the subcarrier channel spacing was 1.11 GHz. The optical bandwidth efficiency is approximately 2 bits/s/Hz for QPSK modulation.

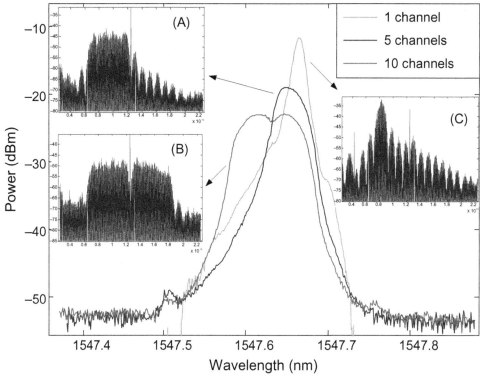

FIG. 11.3.36

Optical spectra of an OFDM-modulated signal with 5 *(blue)*, 10 *(red)*, and 1 *(green)* subcarrier channels measured by an OSA. Insets are the corresponding optical spectra with 5 channels (A), 10 channels (B), and one channel (C) measured with coherent heterodyne detection for fine spectral resolution.

Because of the complex optical field modulation, the upper and the lower sidebands of the optical signal carry different data channels which can be separated in the coherent detection receiver which linearly down-converts the complex optical field into the electric domain.

Inset (A) in Fig. 11.3.36 shows the optical spectrum in which five data channels are loaded onto the lower sideband of the optical carrier, while the upper optical sideband is empty. The spectrum shown in the inset (B) of Fig. 11.3.36 has all 10 subcarrier channels (five on the upper sideband and five on the lower sideband) fully loaded with data so that the total data rate is 22.2 Gb/s. Inset (C) in Fig. 11.3.36 shows an example of the optical spectrum with only one subcarrier channel loaded with data. Spectra shown in the insets of Fig. 11.3.36 were obtained with coherent detection so that the spectral resolution is high enough to resolve detailed features. Blue, red, and green lines in Fig. 11.3.36 show optical spectra corresponding to those shown in insets (A), (B), and (C) but measured with an OSA with 0.01 nm resolution bandwidth.

In the coherent detection receiver, the bandwidths of the PD and the RF preamplifiers are both wider than 30 GHz, and the bandwidth limitation is usually set by the speed of the ADCs. In this experiment,

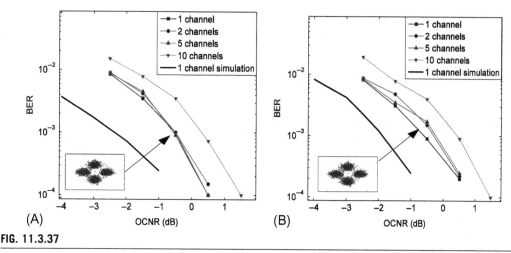

FIG. 11.3.37

Examples of measured BER vs. OCNR with different number of subcarrier channels. (A) Back-to-back and (B) over 75-km standard single-mode fiber.

the sampling speed of the ADC is 20 GS/s with an approximately 6 GHz analog bandwidth. To measure the system transmission performance, an adjustable amount of ASE noise is added to the optical signal before the coherent receiver, and to evaluate the required optical signal to noise ratio (R-OSNR) for the target BER.

In an optical system with multiple subcarriers, the signal optical power is divided into multiple subcarrier channels, and therefore, the performance is determined primarily by the required OCNR (R-OCNR), which is defined as the power ratio between signal power of each subcarrier and the optical noise within 0.1 nm resolution bandwidth.

Fig. 11.3.37 shows the measured BER vs. OCNR with different number of channels. There is negligible increase in OCNR penalty when the number of subcarrier channels increases from 1 to 5. Note that in these measurements, all the five subcarrier channels are located on the upper sideband with respect to the center optical carrier. When the other five subcarrier channels on the lower sideband of the spectrum are added to make the total channel count to 10, an approximately 1 dB OCNR penalty is introduced. This can be partly attributed to the imperfect sideband suppression in the OSSB modulation process as described above. Compare Fig. 11.3.37B with Fig. 11.3.37A, there is negligible OCNR degradation introduced after 75 km fiber transmission. Results of numerical simulation are also shown in Fig. 11.3.37. In the simulation, the linewidth of the laser was 100 kHz, the fiber dispersion was $D = 16 \times 10^{-6}$ s/m^2, nonlinear index of fiber was $n_2 = 2.6 \times 10^{-20}$ m^2/W, fiber attenuation parameter was $\alpha = 0.2$ dB/km, and the effective core area was $A_{eff} = 80$ μm^2. The sampling rate of DAC and ADC was both 22.2 GS/s, and other components were considered ideal with infinite bandwidths. For a single-channel system, simulation predicted an approximately 1.5 dB lower R-OCNR compared to the experiment for the BER of 10^{-3}. This discrepancy can be attributed to the distortion introduced by the passband ripple of RF amplifiers, multiple reflections in the RF and optical paths, as well as time jitter in the ADC process. For an ideal coherent detection receiver only considering LO-ASE beat noise in the photodetection process, an oversimplified analytical solution can be obtained as

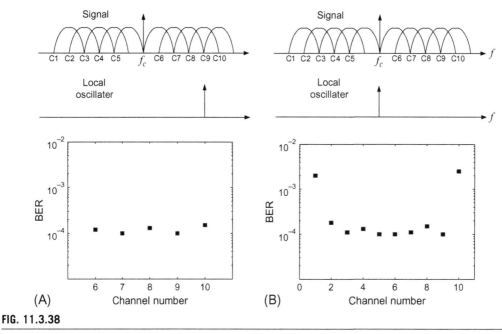

FIG. 11.3.38

BER measured as a function of the channel index for (A) when the local oscillator was tuned to channel 9 and (B) when the local oscillator was tuned to the center optical carrier.

$OCNR = Q^2 B_e/(2B_0)$ which provides the ultimate OCNR requirement for the QPSK-modulated subcarrier channels. For the optical bandwidth $B_0 = 12.5\,GHz$ (0.1 nm) used to measure OCNR and electrical bandwidth $B_e = 1.11\,GHz$ (for each 2.22 Gb/s QPSK subcarrier) for the signal, the required OCNR for $Q = 3.08$ (BER $\approx 10^{-3}$) should be approximately $-3.75\,dB$.

Similar to the previous discussion of DSCM with direct detection, DSCM can be flexible with fine data rate granularity of subcarrier channels also for coherent detection. When a certain data rate is requested by an end user, multiple subcarriers can be delivered to the same user without changing the network architecture. The end user only needs to tune the LO to the targeted subcarrier frequency in the OFDM spectrum, and coherent IQ detection can translate the optical spectrum to the electric domain with the selected subcarrier channel in the center. A selected group of subcarrier channels within the receiver bandwidth can all be detected and recovered by the receiver DSP. In Fig. 11.3.38A, the wavelength of the LO is set at subcarrier channel C9, and the BER of five adjacent channels are measured, indicating no significant performance variation across these channels. One can also set the LO at the center wavelength of the optical carrier as shown in Fig. 11.3.38B in an attempt to detect all the 10 subcarrier channels. The results show reasonably uniform BER performance except for the two outmost channels C_1 and C_{10}. The increased BER in these two channels is due to the bandwidth limit of the ADC which is 6 GHz, and potions of the spectra of these two channels are already outside this receiver bandwidth.

11.4 SUMMARY

In this chapter, we have discussed some applications of DSP in fiber-optic communication systems. With the rapid technological advance in CMOS digital electronics, high-speed ADC, DAC, and DSP become available and affordable, enabled many new functionalities in the recent years which were not even conceivable two decades ago. The introduction of DSP has changed many design perspectives as well as architectures of optical systems and networks. Some important optical domain functionalities are now pushed to the electronic domain with much improved flexibility and accuracy. The combination of coherent detection and DSP has enabled the compensation of chromatic dispersion and PMD adaptively in the electronic domain. This elimites the need for optical domain compensation of these impairments. Adaptive DSP allows polarization multiplexing to be used to further increase the spectral efficiency by creating an inverse Jones matrix in the receiver DSP which effectively removes the random mode coupling effect in the fiber. DSP is also indispensable in optical-phase tracking for carrier phase recovery in coherent detection. This relaxes the otherwise very stringent requirement on the laser phase noise.

Most of the DSP functionalities in an optical receiver are based on the digital filters with desired complex transfer functions. Many signal processing techniques previously developed for RF and wireless communication are now utilized for optical systems. High-order digital filters, such as Nyquist filters, with sharp cutoff edges in the transfer functions allow the tight confinement of optical signal spectra, which enabled FDM with minimum or nonspectral guard band between adjacent channels. Digital Fourier transform (DFT) and inverse DFT (IDFT) can also be used to for OFDM with high spectral efficiency, and high tolerance to various transmission impairments such as chromatic dispersion and PMD. Dividing a high-speed and broadband optical channel into multiple subcarrier channels through DSP without decreasing spectral efficiency is known as DSCM. Complementary to TDM, this FDM mechanism can also be used for optical cross-connection and routing, which will be discussed in Chapter 12.

PROBLEMS

1. Consider a 100 Gb/s dual-polarization QPSK-modulated optical system operating in 1550 nm wavelength with a baud rate of 25 Gbaud/s, and a clock period of $T = 20$ ps. The accumulated chromatic dispersion of the system is 10,000 ps/nm. A FIR filter is used in the transmitter to perform pre-compensation of chromatic dispersion.
   (a) please find the maximum number of taps of the FIR filter (without spectral aliasing),
   (b) plot filter coefficient (real and imaginary parts) as the function of k, and
   (c) plot the phase of the FIR filter as the function of relative frequency (set center frequency to be zero).

2. Consider a 1600-km long fiber optic link operating in the 1550-nm wavelength window having a dispersion of 17 ps/km/nm. If information is signaled at a rate of 10 Gbaud using an ideal Nyquist filter with roll-off factor of 0,

(a) Determine the maximum amount pulse broadening (in ps) (assume only two frequency components separated by the baud rate). What will be the width of the broadened pulse in terms of number of symbols?

(b) If we use a time-domain FIR filter for compensating the dispersion effect, determine approximately, the minimum required number of symbol-spaced taps. Compare this result with the solution from (a.)

3. A fiber-optic system with 10 Gbaud symbol rate. In the process of PMD equalization using the butterfly structure using "five" tap symbol-spaced adaptive FIR filter,

   (a) What is the amount of peak DGD that can be compensated (*in ps*)?

   (b) If the fiber PMD parameter is $0.5 ps/\sqrt{km}$ which can be used to determine the mean-DGD, and assume the system has to tolerate four times the mean DGD, what is the maximum PMD-limited fiber length? (Read Section 8.5.1.2 for the property of PMD.)

   (c) Compare problems 2 and 3, please discuss which compensating circuit (dispersion or PMD) is more complex and which one usually needs to be dynamically adjusted?

4. Consider a 40-Gb/s PDM-QPSK coherent transmission system equipped with DSP, and the laser in the transmitter has a spectral linewidth of 500 kHz. Assume the received signal has very high SNR so that the phase error is mainly caused by the phase noise of laser sources. Viterbi-Viterbi algorithm is used for digital phase recovery at 1 Sample/Symbol and with a window width of 1000 symbols. Also assume the worst case that the noise phase linearly grows with time, and the accumulated noise phase within the window of 1000 symbols [i.e., noise phase is $\varphi(T) = 2\pi\Delta v \cdot T$, where Δv is the combined linewidth of the transmitter laser and the LO and T is the window width] has to be less than $\pi/2$ to avoid cycle slip.

   (a) What is the symbol rate of this system?

   (b) What is the maximum allowable linewidth for the LO at the receiver side to avoid cycle slip?

5. In an optical system with coherent homodyne receiver, assume the combined spectral linewidth of the transmitter laser and the LO in the receiver produce a Gaussian phase noise with the probability distribution function (PDF) of phase $PDF(\varphi) = \frac{1}{\sigma\sqrt{2\pi}} \exp\left(-\frac{\varphi^2}{2\sigma^2}\right)$, where the variance is $\sigma^2 = 2\pi\Delta v \cdot T$ with T the time interval in which phase noise is measured.

   (a) For a linewidth-symbol rate product $\Delta v T_s = 10^{-5}$, what is the probability that the differential phase (caused by laser linewidth) between adjacent data symbols is higher than 0.01π *rad*?

   [Hint: use error function $\text{erfc}(x) = (2/\sqrt{\pi}) \int\limits_x^\infty \exp(-y^2) dy$]

   (b) If data are QPSK modulated, and use Viterbi-Viterbi for phase recovery with a window width consisting of 1000 symbols, what is the probability for the phase variation (caused by laser linewidth) to be larger than $\pi/2$ within this time window? (This is the probability of cycle slip.)

6. Create a QPSK data sequence $X[n] = A_0 \exp\{j[\varphi_s(n) + \varphi_n(n)]\}$ for $n = 0, 1, ..., 5000$, where A_0 is the amplitude, $\varphi_s(n)$ represents data phase with four possible values $\{\pm 1, \pm j\}$, and $\varphi_n(n)$ is the phase noise associated with the nth data sample. Assume the phase noise is created by the mixing between the transmitter laser and the LO in the receiver with $\varphi_n(n) = 5 \times 10^{-6} \pi \sum_0^n \text{randn}(m)$,

where randn(m) with ($0 \leq m \leq 5000$) is the normalized random Gaussian noise [can be generated in Matlab using randn(1,5000)]. The impact of SNR can be neglected.

(a) Use Viterbi-Viterbi phase recovery algorithm with window length of 50 symbols, please compare the original phase noise waveform and the recovered noise waveform.

(b) Repeat (a) but with window length of 500 symbols.

7. Repeat problem 3, but using BPS algorithm described in Fig. 11.2.9.

8. Consider a DSP-based fiber optic transmission system designed for transporting data at 10 Gbaud symbol rate. Using 16-QAM as the modulation format and OFDM as a subcarrier multiplexing scheme, compute the following.

(a) What is the achieved data rate in bits/s using the above transmission system?

(b) Assuming a DAC sampling frequency of 20 Gsps, and the number of subcarriers is 256, what is the symbol rate carried by each subcarrier and what is the frequency spacing between adjacent subcarriers?

(c) Determine the OFDM symbol time in ps.

9. Consider a 10 Gbaud OFDM system using 256 subcarriers, and the DAC sampling rate is 20 Gs/s. The accumulated fiber dispersion in the fiber system is 10,000 ps/nm in the 1550 nm operating wavelength window. When OSSB modulation is used, the lowest and the highest-frequency subcarriers are separated by the total symbol rate (10 GHz in this case). Using cyclic prefix to combat the dispersion problem,

(a) What is the required length of cyclic prefix duration in (ps) that is just enough to compensate the dispersion induced differential delay? What is the percentage of this cyclic prefix in terms of the symbol time of each subcarrier?

(b) In this system, if the number of subcarriers, N, can be changed, express the percentage of cyclic prefix duration in terms of the symbol time of each subcarrier.

10. *Simulation:* This exercise will demonstrate the impact of fiber dispersion on Generate a 10 Gbaud NRZ signal and transmit it through a dispersive fiber with transfer function shown in Eq. (11.2.4). Assume the following: NRZ signal has a length of 100,000, sampling frequency of 50 Gsps, fiber length of 80 km and $D = 17$ ps/km/nm. Plot the eye diagram of NRZ signal after fiber transmission. Repeat the above simulation for baud rates of 1, 2, and 5 Gbaud. What do you observe from the eye diagrams? Now, determine the number of symbol-spaced FIR taps needed to compensate the dispersion for all the four baud rates.

11. Numerically recreate Fig. 11.3.17 with the following steps: (1) create 1 Gb/s NRZ 2^9 PRBS with 512 bits, and up-sample to 16 data samples per bit, (2) apply FFT to convert this time-domain waveform into frequency, (3) multiply with the Nyquist filter transfer function shown in Eq. (11.3.7) with 1 GHz FWHM bandwidth and roll-off factor $\beta = 0.1$, and (4) apply IFFT to convert the Nyquist filtered spectrum back into time domain and plot the eye diagram.

12. Numerically create a *sinc* function $A(t) = \frac{\sin(\pi t/T)}{\pi t/T}$ to represent a signal pulse of digital "1" with a pulse width T. Use 20 samples per pulse width (i.e., $\delta t = T/20$) for a time window $-50T \leq t \leq 50T$. Apply FFT, plot find the power spectral shape, and find the spectral bandwidth.

(a) Construct a PRBS of 30 bits with each bit represented by $U(t) = \sum_{n=1}^{30} a_n \frac{\sin(\pi(t-nT)/T)}{\pi(t-nT)/T}$

(b) Plot eye diagrams.

13. Consider frequency up-conversion based on the I/Q mixing with the block diagram shown in Fig. 11.3.20. If the two complex baseband signals are $I(t) = e^{j2\pi F_1 t}$ and $Q(t) = e^{-j2\pi F_2 t}$, and they need to be carried by a subcarrier at frequency f_i.
    (a) Write the expression of the output complex waveform after frequency up-conversion.
    (b) What is the minimum value of f_i to avoid spectral aliasing?
    (c) If $F_1 = 10\,\text{MHz}$, $F_2 = 15\,\text{MHz}$, and $f_i = 50\,\text{MHz}$, please sketch the amplitude of the up-converted spectrum.

14. Similar to problem 13, but now the frequency up-conversion is based on the I/Q mixing with the block diagram shown in Fig. 11.3.22 (with an extra 90 degree hybrid compared to Fig. 11.3.20). Again, assume the two complex baseband signals are $I(t) = e^{j2\pi F_1 t}$ and $Q(t) = e^{-j2\pi F_2 t}$, and they need to be carried by a subcarrier at frequency f_i.
    (a) Write the expression of the output complex waveform after frequency up-conversion.
    (b) If $F_1 = 10\,\text{MHz}$, $F_2 = 15\,\text{MHz}$, and $f_i = 50\,\text{MHz}$, please sketch the amplitude of the up-converted spectrum.

15. *Simulation*: Create four independent waveforms following Problem 11 (Nyquist channel with 1 GHz bandwidth). Then, frequency up-convert them onto subcarrier frequencies of 1, 2.2, 3.3, and 4.4 GHz based on the complex I/Q mixing block diagram shown in Fig. 11.3.22.

16. Consider a SSB-modulated optical signal with the optical field expressed by Eq. (11.3.9). The signal has five equal-amplitude $[a_k(t) = 1]$ frequency components at $f_1 = 1\,\text{GHz}$, $f_2 = 1.1\,\text{GHz}$, $f_3 = 1.2\,\text{GHz}$, and $f_4 = 1.3\,\text{GHz}$. Assume $a_k(t) = 1$, and $m_k = 0.2$ with $k = 1$, 2, ...5. After direct detection at the receiver, what are the relative amplitudes of SSBI components at 0.1, 0.2, and 0.3 GHz in comparison to the signal photocurrent at 1 GHz? (Assume all the SSBI components at the same frequency coherently add up for the worst case.)

17. In a multi-subcarrier optical transmitter, if the information is carried by the power of the optical signal (without using OSSB modulation), is there SSBI problem at the direct-detection receiver? Please explain the reason.

18. *Simulation*: For the five-channel subcarrier system described in problem 16, but now the modulation is compatible-OSSB. Please create the five discrete frequency tones numerically, multiplex them together into an optical field based on Eq. (11.3.13). Plot the electric power spectral density after direct-detection and nature logarithm operation.

19. Consider a DSCM system with KK receiver. The system has five equal amplitude subcarrier channels so that the normalized signal is $s(t) = \sum_{k=1}^{5} a \cdot e^{-j(2\pi f_k t + \phi_k)}$ with $f_1 = 1\,\text{GHz}$, $f_2 = 2\,\text{GHz}$, $f_3 = 3\,\text{GHz}$, $f_4 = 4\,\text{GHz}$, and $f_5 = 5\,\text{GHz}$. The amplitude, a, of each subcarrier channel is constant and the phase, φ_k, is random.
    (a) What is the maximum value of a to satisfy the minimum phase condition?
    (b) With the maximum allowable amplitude, a, of each channel, what is the carrier to signal power ratio (CSPR) of the optical signal?

20. Consider a DSCM system with on 10 subcarrier channels, and each channel is QPSK modulated at 4 GS/s baud rate. With coherent homodyne detection, the required OCNR is related to the targeted Q value as $OCNR = Q^2 B_e/(2B_0)$, where $B_e = 4\,\text{GHz}$ is the electric bandwidth of each DSCM channel and $B_0 = 12.5\,\text{GHz}$ (0.1 nm) is the optical bandwidth to specify OCNR. In order to achieve $Q = 3$ for all DSCM channels, what is the required OSNR?

REFERENCES

Fatadin, I., Ives, D., Savory, S.J., 2010. Laser linewidth tolerance for 16-QAM coherent optical systems using QPSK partitioning. IEEE Photon. Technol. Lett. 22 (9), 631–633.

Goodsmith, A., 2005. Wireless Communications. Cambridge University Press.

Haykin, S.O., 2002. Adaptive Filter Theory, fourth ed. Prentice Hall Information and System Science Series.

Hui, R., Kaje, K., Fumagalli, A., July 2016. Digital-analog hybrid SCM for fine-granularity circuit-switched optical networks. In: 18th International Conference on Transparent Optical Networks (ICTON). Trento, Italy. 10–14 July 2016.

Johnson, C.R., Schniter, P., Endres, T.J., Behm, J.D., Brown, D.R., Casas, R.A., 1998. Blind equalization using the constant modulus criterion: a review. Proc. IEEE 86, 1927–1950.

Kramers, H.A., 1927. La diffusion de la lumiere par les atomes. Atti Cong. Intern. Fis. 2, 545–557.

Kronig, R.L., 1926. On the theory of the dispersion of x-rays. J. Opt. Soc. Am. 12, 547–557.

McNicol, J., O'Sullivan, M., Roberts, K., Comeau, A., McGhan, D., Strawczynski, L., 2005. Electrical domain compensation of optical dispersion. In: Proceedings of Optical Fiber Communication Conference, OFC'2005, Anaheim, CA. paper OThJ3.

Mecozzi, A., Antonelli, C., Shtaif, M., 2016. Kramers-Kronig coherent receiver. Optica 3 (11), 1220–1227.

Molisch, A.F., 2011. Wireless Communications. Wiley.

Peng, W.-R., et al., 2009. Theoretical and experimental investigations of direct-detected RF-tone-assisted optical OFDM systems. J. Lightwave Technol. 27 (10), 1332–1339.

Pfau, T., Hoffmann, S., Noé, R., 2009. Hardware-efficient coherent digital receiver concept with feedforward carrier recovery for M-QAM constellations. J. Lightwave Technol. 27 (8), 989–999.

Roberts, K., Li, C., Strawczynski, L., O'Sullivan, M., Hardcastle, I., 2006. Electronic pre-compensation of optical nonlinearity. IEEE Photon. Technol. Lett. 18 (2), 403–405.

Roberts, K., O'Sullivan, M., Wu, K.-T., Sun, H., Awadalla, A., Krause, D.J., Laperle, C., 2009. Performance of dual-polarization QPSK for optical transport systems. J. Lightwave Technol. 27 (16), 3546–3559.

Savory, S.J., 2008. Digital filters for coherent optical receivers. Opt. Express 16 (2), 804–817.

Shieh, W., Djordjevic, I., 2010. OFDM for Optical Communications. Academic Press.

Viterbi, A.J., Viterbi, A.M., 1983. Non-linear estimation of PSK-modulated carrier phase with application to burst digital transmission. IEEE Trans. Inf. Theory 29, 543–551.

Zhang, Y., O'Sullivan, M., Hui, R., 2010. Theoretical and experimental investigation of compatible SSB modulation for single channel long-distance optical OFDM transmission. Opt. Express 18 (16), 16751–16764.

Zhang, Y., O'Sullivan, M., Hui, R., 2011. Digital subcarrier multiplexing for flexible spectral allocation in optical transport network. Opt. Express 19 (22), 21880–21889.

OPTICAL NETWORKING

CHAPTER OUTLINE

Introduction 555
12.1 Layers of communication networks and optical network topologies 556
    12.1.1 Layers of an optical network 557
    12.1.2 Optical network topologies 559
    12.1.3 SONET, SDH, and IP 567
12.2 Optical network architectures and survivability 573
    12.2.1 Categories of optical networks 573
    12.2.2 Optical network protection and survivability 574
12.3 Passive optic networks and fiber to the home 579
    12.3.1 TDM-PON 580
    12.3.2 WDM-PON 584
12.4 Optical interconnects and datacenter optical networks 586
    12.4.1 Optical interconnection 586
    12.4.2 Datacenter optical networks 588
12.5 Optical switching and cross connection 596
    12.5.1 Reconfigurable optical add/drop multiplexing 596
    12.5.2 Optical cross-connect circuit switching 599
    12.5.3 Elastic optical networks 602
    12.5.4 Optical packet switching 604
12.6 Electronic circuit switching based on digital subcarrier multiplexing 606
12.7 Summary 612
Problems 613
References 618
Further reading 619

INTRODUCTION

Optical networks provide the communication backbone of the internet. As the internet traffic has been growing exponentially, with more and more diversified services and applications, the capacity of optical networks has to expand accordingly. In the previous chapters, we have discussed properties of

Introduction to Fiber-Optic Communications. https://doi.org/10.1016/B978-0-12-805345-4.00012-3

physical layer optical components and performance evaluation criteria of optical transmission systems. While an optical transmission system usually refers to a point-to-point optical link between a transmitter and a receiver, a communication network is much more general, including communication among a large number of users at many different locations, and with various different types of services and applications. In addition to ensuring the communication quality between users, optical network has to take into account the efficiency and flexibility of resources sharing when many users have access to the same network at the same time. Low probability of traffic jam and blocking, large available communication capacity to users, and high reliability are among desired properties of optical networking. In order for an optical network to be efficient and to provide various types of services to a large number of users, a set of rules and regulations has to be agreed upon by network developers, operators, and users. Depending on application requirements, various different optical network architectures have been developed and standardized. This chapter introduces basic optical network concepts which include layers of optical network, data encapsulation and connection mechanisms and standards, as well as optical network topologies, architectures, protection mechanisms, and survivability. We also discuss a number of special optical network architectures including passive optical networks (PONs), datacenter optical networks, and short distance optical interconnection. Data traffic switching and routing are highly dynamic in optical networks just like traffic in complex road networks with conjunctions, interchanges and bridges. Application of circuit switching and packet switching in optical networks, and their advantages and limitations are discussed. Optical circuit switching can handle wide spectral bandwidth, and transparent to modulation format of the optical signal, it has clear advantages in switching high-speed optical channels. However, optical switching is relatively slow and bandwidth granularity is usually coarse, and thus packet switching has to be used to further improve dynamic bandwidth sharing efficiency and flexibility among many users. At the end of this chapter, digital subcarrier cross-connect (DSXC) switching is introduced as an alternative to packet switching. DSXC is based on circuit switching but is able to provide fine enough data rate granularity and high spectral efficiency.

12.1 LAYERS OF COMMUNICATION NETWORKS AND OPTICAL NETWORK TOPOLOGIES

A communication network is shared by many end users of various applications as illustrated by Fig. 12.1.1, where lines connecting different nodes can be fibers, wavelengths channels, or even virtual paths from a network's point of view. Information to be communicated between end-users has to be first translated into formats that can be recognized by the network. The network then selects the best path for the dialog between each pair of users (sender and receiver) depending on the availability of routes in the network. For example, communication between users a and c can go through nodes $A \rightarrow B \rightarrow D \rightarrow F$, or through $A \rightarrow C \rightarrow F$, which is decided by network management and routing algorithms. In a circuit-switched network, the route, once established, is dedicated for the communication between the user pair and for a relatively long time without interruption. In a packet-switched network, on the other hand, data stream coming out from each user is chopped into short pieces in time known as packets. Consecutive data packets coming from a user may travel to the destination via different routes, allowing more efficient algorithms to be used to maximize the efficiency of the entire network resources. Each packet has an overhead section containing information such as source and destination identification, priority level in the network, routing/switching algorithm, and so on. At the receiver side, these packets are re-aligned into the original data sequence.

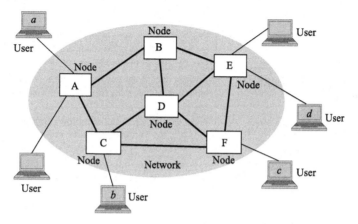

FIG. 12.1.1

Illustration of a data communication network. *A, B, C, D* ... are network nodes and *a, b, c, d* ... are users.

12.1.1 LAYERS OF AN OPTICAL NETWORK

With the rapid increase of the number of users and the types of applications, the complexity of networks increases exponentially to accommodate different applications and their bandwidth requirements, to meet the minimum acceptable quality of service (QoS), and to minimize the cost. The introduction of optical fibers into communication networks has provided drastically increased capacity compared to the use of copper wires. Wavelength division multiplexing (WDM) further increased network capacity and flexibility, but at the same time increased the complexity of networking. Both network architectures and network management algorithms have to be extensively modified to make the optimum use of optical network's capabilities.

In order to make a highly complex communication network manageable, a communication network is usually divided into different layers based on their functionalities, and in principle, the resources of lower layers are used to support applications of upper layers. Table 12.1.1 shows the classic layer hierarchy of a network (Ramaswami and Sivarajan, 2002) defined by the International Standard Organization (ISO).

Table 12.1.1 Layers of an optical network			
Layer number	Name of layer	Sublayers	Functionalities
4	Higher layers		Including session, presentation, and application layers to support user applications and data format interpretation
	Transport layer		Provides end-to-end communication links between users with correct data sequences. Responsible also for the management of error correction, providing reliable and error-free service to end users

Continued

Table 12.1.1 Layers of an optical network—cont'd

Layer number	Name of layer	Sublayers	Functionalities
3	Network layer		Provides end-to-end circuits between users across multiple network nodes and links, and in some cases defines quality of service such as bandwidth and BER
2	Data link layer	Logical link control layer	Defines data structures for transmission in the physical layer, including data framing, flow control, and multiplexing for logical links
		Media access control layer	Provides flow control and multiplexing for the transmission medium and coordinates the operations of nodes when they share the same transmission resources such as spectrum or time slots
1	Physical layer	Optical layer (Optical transport)	Provides wavelength based end-to-end optical connectivity known as lightpath, optical transceivers, bit-error corrections
		Fiber infrastructure	Optical fibers, optical switches, inline optical amplifiers, multiple fibers in cable

From the bottom-up, the physical layer is usually divided into two sublayers: the *fiber infrastructure layer* provides installed optical transmission media including optical fiber cables, inline optical amplifiers such as EDFAs, optical switches, wavelength-selective routers, and wavelength division multiplexers and demultiplexers. The *optical layer* runs on the fiber infrastructure providing end-to-end optical connectivity, known as lightpath, between optical nodes. Optical transceiver can be part of the optical layer which serves as an interface between optical and electric domains. The optical layer also needs to have mechanisms to guarantee the quality of optical transmission such as transmission impairment mitigation and bit-error correction. Fig. 12.1.2 illustrates a more detailed physical layer configuration and its relation to higher network layers.

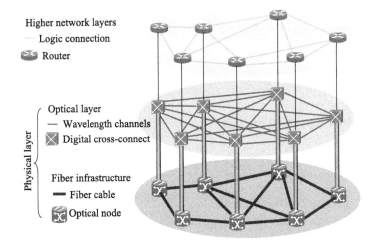

FIG. 12.1.2

Illustration of physical layer configuration and its relation to higher network layers.

In most data communication networks, a data stream is usually divided into short frames for transmission over the physical layer. The *data link layer* (Ramaswami and Sivarajan, 2002) performs formation, multiplexing, and de-multiplexing of data frames. Framing protocol defines how data is encapsulated into frames and how these frames are transmitted in the physical layer. Each frame typically includes overhead information identifying the link and also allowing the detection of link errors which may occur. Examples of data link protocols include *point-to-point protocol* (PPP) and *high-level data link protocol* (HDLC). In IEEE 802 LAN/MAN standards, the data link layer is further divided into a *logical link control* (LLC) sublayer and a *media access control* (MAC) sublayer. While LLC provides flow control and multiplexing/de-multiplexing for the logical link, MAC provides a control abstraction of the physical layer such that the complexities of physical link control are invisible to the LLC layer and other upper layers of the network. MAC encapsulates higher-level frames into frames appropriate for the transmission medium.

The *network layer* responds to service requests from the transport layer and translates the service requests to the data link layer. It performs the end-to-end routing function which links each data packet from its source to its destination via one or more network nodes. The most relevant example to explain network layer functionalities is the *Internet Protocol* (IP). It is commonly used as TCP/IP because the foundational protocol is the transmission control protocol (TCP) in the transport layer which is above the network layer. IP in the network layer supports end-to-end data communication between users, and specifies how data is split into packets, how address information is added into each packet, how packets are transmitted, routed, and received by users. These packets are wrapped into frames in the data link layer suitable for transmission.

The *transport layer* is connection-oriented, which defines the logical connections, error control mechanisms, flow control, congestion avoidance, and many more services which are highly critical for the network. In the TCP protocol, the entire path is predetermined before sending out a packet. While IP is connectionless and each packet may go through a different route, TCP coordinates and rearrange the sequence of packets. Flow control function provided by the transport layer is used to control the rate at which a user sends out data packets. Depending on the probability of network congestion, the status of data buffering in the receiver and other conditions of the network, the packet delivering rate may change with time. Transport layer is also used to ensure the reliability of data delivery. For example, if some of the data packets do not arrive at the receiver due to network congestion, the transport layer will have to schedule a retransmit of the lost or corrupted packets.

While the physical layer of an optical network provides wideband optical transmission medium, upper layers of the network above the transport layer are more application oriented, which include session layer, presentation layer and application layer. These higher layers support various user applications and data format interpretation, which are outside the scope of this book. There is a variety of different optical network architectures and topologies suitable for different network scales and applications, which are discussed in the following sections.

12.1.2 OPTICAL NETWORK TOPOLOGIES

Optical network is the backbone of the current internet which supports a large variety of applications worldwide. Architectural design of an optical network has to consider geographic coverage of the network, bandwidth requirement of each section, efficiency of resources sharing with minimum blocking probability, and the quality and flexibility of services provided to the users at lowest cost.

From the point of view of physical layer optical network architecture, the most basic network topologies include point-to-point, bus, star, ring, mesh, and tree, as illustrated in Fig. 12.1.3. Each topology has its own advantages and disadvantages, and the selection of network topology should be based on specific applications and geometric distributions of users. Different topologies can also coexist to construct a larger network.

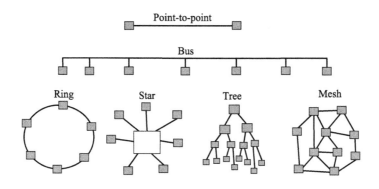

FIG. 12.1.3

Illustration of point-to-point, bus, ring, star, tree, and mesh topologies of networking.

First, let us look at the folded bus shown in Fig. 12.1.4. In this configuration, each transceiver is connected to the bidirectional bus through two optical couplers, and thus any transceiver can communicate with any other transceiver through the bus. Consider that each fiber coupler has a power coupling coefficient ε, the transmission of each coupler defined by Eq. (6.2.17) is $1 - \varepsilon$. Assume the fiber length in each direction is L, and fiber loss is α, the number of users is N, and assume all the users are equally spaced along the bus, the transmission loss between the mth transmitter Tx_m and the nth receiver Rx_n is,

$$\eta_{n,m} = -10\log_{10}\left[\varepsilon^2(1-\varepsilon)^{m+n-2}e^{-\alpha(m+n-2)L/N}\right] \tag{12.1.1}$$

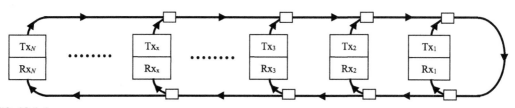

FIG. 12.1.4

Configuration of a folded bus connecting N users.

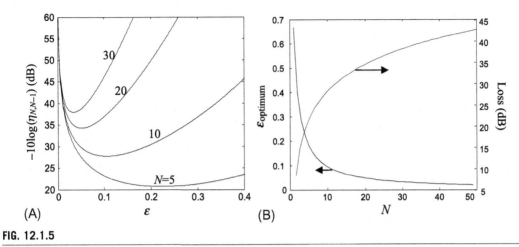

FIG. 12.1.5

(A) Transmission loss between Tx_N and Rx_{N-1} as the function of ε calculated with the user number $N=5$, 10, 20, and 30. (B) Optimum coupling coefficient of couplers (left y-axis) and minimum transmission loss between Tx_N and Rx_{N-1} (right y-axis) as the function of N.

The highest transmission loss is between the last transmitter Tx_N and its adjacent receiver Rx_{N-1}, which is

$$\eta_{N,N-1} = -10\log_{10}\left[\varepsilon^2(1-\varepsilon)^{2N-3}e^{-\alpha L(2N-3)/N}\right] \tag{12.1.2}$$

Fig. 12.1.5A shows the maximum loss as the function of ε for 5, 10, 20, and 30 users. As the fiber attenuation is independent of ε, and almost independent of N for $N \gg 1$, we have only considered the loss due to optical couplers in Fig. 12.1.5A, or in other words, we have assumed $\alpha = 0$ in this case for simplicity. When ε is high, the efficiency of sending and receiving power into and from the bus is high, but the transmission loss through each coupler along the bus is also high. Thus there is an optimum coupling coefficient to minimize the combined loss, which is dependent on the number of users. Based on Eq. (12.1.2), the optimum coupling coefficient of coupler can be found as, $\varepsilon_{\text{optimum}} = 2/(2N-1)$ which is shown in Fig. 12.1.5B. With the optimum coupling coefficient, the transmission loss between Tx_N and Rx_{N-1}, as the function of N, is

$$\eta_{N,N-1} = -10\log_{10}\left[\left(\frac{2}{2N-1}\right)^2\left(1-\frac{2}{2N-1}\right)^{2N-3}e^{-\alpha L(2N-3)/N}\right] \tag{12.1.3}$$

.

This transmission loss is shown in Fig. 12.1.5B (right y-axis), also without considering the fiber loss.

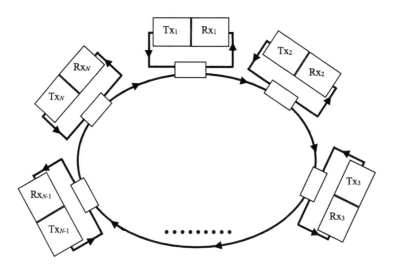

FIG. 12.1.6

Configuration of a unidirectional ring connecting N users.

Second, let us look at an example of unidirectional ring configuration connecting N users shown in Fig. 12.1.6. In this configuration, each user is connected to the fiber ring through a 2×2 fiber directional coupler. Again consider that each fiber coupler has a power coupling coefficient ε, and thus the power transmission of each coupler is $1 - \varepsilon$. Because the ring is unidirectional, transmission loss between user m and user n, with $n > m$, is

$$\eta_{m,n} = -10\log_{10}\left[\varepsilon^2(1-\varepsilon)^{n-m-2}e^{-\alpha L(n-m)/N}\right] \tag{12.1.4}$$

Here we have assumed that the users are equally spaced along the ring for simplicity. The maximum transmission loss of this unidirectional ring is between the transmitter in node 1 and the receiver in node N, which is

$$\eta_{1,N} = -10\log_{10}\left[\varepsilon^2(1-\varepsilon)^{N-3}e^{-\alpha L(N-1)/N}\right] \tag{12.1.5}$$

Fig. 12.1.7A shows the maximum loss, based on Eq. (12.1.5), as the function of ε for $N=5$, 10, 20, and 30 users without considering fiber loss ($\alpha = 0$). The optimum coupling coefficient of fiber directional coupler to minimize this maximum transmission loss can be found as $\varepsilon_{\text{optimum}} = 2/(N-1)$ which is shown in Fig. 12.1.7B (left y-axis). With this optimum coupling coefficient, the transmission loss between Tx_1 and Rx_N, which is the highest possible loss in the network, as the function of N, is

$$\eta_{N,N-1} = -10\log_{10}\left[\left(\frac{2}{N-1}\right)^2\left(1-\frac{2}{N-1}\right)^{N-3}e^{-\alpha L(N-1)/N}\right] \tag{12.1.6}$$

This transmission loss is shown in Fig. 12.1.6B (right y-axis).

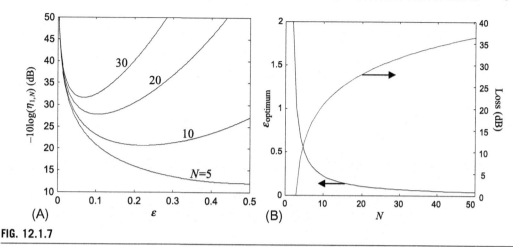

FIG. 12.1.7

(A) Transmission loss between Tx_1 and Rx_N as the function of ε calculated with the number of users $N=5$, 10, 20, and 30. (B) Optimum coupling coefficient of couplers (left y-axis) and minimum transmission loss between Tx_N and Rx_{N-1} (right y-axis) as the function of N.

In comparison, a unidirectional ring has lower transmission loss than a folded bus of the same number of nodes. This is because unidirectional ring configuration uses a 2×2 fiber coupler for each transceiver, whereas folded bus uses two 1×2 fiber couplers for each transceiver.

The third example that we discuss is the star topology. Fig. 12.1.8A shows the configuration of a star network based on an $N \times N$ star coupler. Assume an equal splitting ratio of all the input and output ports, the intrinsic splitting loss of the $N \times N$ star coupler is

$$\eta_{NxN} = -10\log_{10}[1/N] \tag{12.1.7}$$

In this case the transmission loss between any transmitter-receiver pair due to fiber coupler is identical, which is the function of user number N, as shown in Fig. 12.1.8B. In comparison, the network based on star coupler has much smaller splitting loss than both folded bus and unidirectional ring. But the selection of network architecture depends on specific network requirements, user distribution, as well as network reliability and protection concerns.

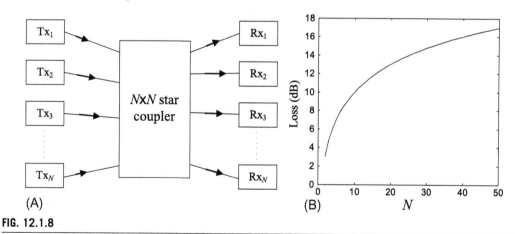

FIG. 12.1.8

(A) Configuration of a star network based on an $N \times N$ star coupler, and (B) splitting loss of the star coupler as the function of N.

Although a star network can be built based on an $N \times N$ star coupler, it would be more flexible to use 2×2 optical directional couplers as building blocks. Fig. 12.1.9 shows two examples of 16×16 shuffle-net (Hluchyj and Karol, 1991) built with 2×2 optical couplers with different coupler arrangements.

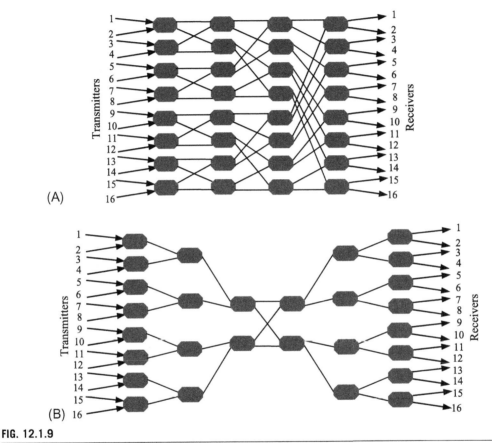

FIG. 12.1.9

Two examples of 16×16 star network based on the combination of 3 dB directional couplers in four layers (A) and six layers (B).

The 16×16 shuffle-net shown in Fig. 12.1.9A uses 32 2×2 directional couplers arranged into four layers, and if all couplers are 3 dB, the total splitting loss is 2^{-4} or equivalently -12 dB (the same as using a single 16×16 fiber coupler). While also for a 16×16 star network, the number of 3 dB couplers can be reduced by rearranging the configuration as shown in Fig. 12.1.9B. In this modified configuration, 28 couplers are used with 12 2×1 couplers and 4 2×2 couplers. Since six layers of couplers are used, the total splitting loss is increased to 2^{-6}, which is -18 dB.

Note that for the network topologies shown in Figs. 12.1.4 and 12.1.6, and Fig. 12.1.8, any receiver can receive signals from all transmitters simultaneously, and thus time sharing has to be scheduled carefully to avoid signal contention. WDM can also be used for the signal selection and to improve the network efficiency, which is discussed later.

In addition, for the three network topologies discussed above, the traffic is assumed unidirectional. Although unidirectional network is simple, they are quite vulnerable to accidental fiber cut in the network. Especially for the bus and ring topologies, if the fiber is accidentally cut somewhere in the bus (or ring), the entire network will fail. To provide better protection, bidirectional network architectures are often used in practice.

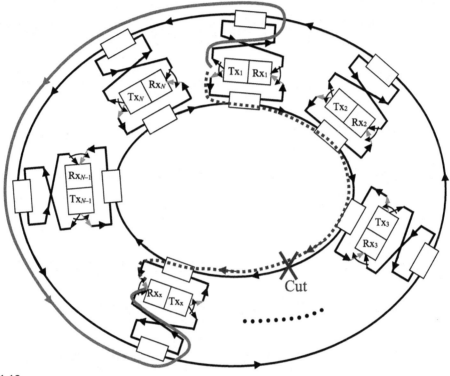

FIG. 12.1.10

Bidirectional double ring configuration. The dotted line is the normal path from node 1 to node x. If there is a fiber cut in the clockwise ring, the path is switched to the solid line through the counter-clockwise ring.

Fig. 12.1.10 shows an example of bidirectional fiber ring, in which the output from each transmitter can selectively send its output to one of the two parallel rings propagating in the opposite directions. In a normal operation condition with no fiber cut anywhere, data traffic between two users can go through either one of the two rings, but usually selects the one with a shorter path length and thus a lower loss. This is illustrated as the dotted line (blue color) in Fig. 12.1.10 between node 1 and node x. However, if the fiber along the primary path is accidentally cut, the traffic can be rerouted through the other ring along the opposite direction as illustrated by the solid line (red color). Similar bidirectional configuration can also be used for the bus. There are more detailed discussions in ring network protection in the next section.

One of the key requirements of communication networking is the efficient sharing of network resources by a large number of users. This resources sharing can either be in time domain or in frequency domain. For the three examples discussed earlier, for time domain sharing, communication between each user pair can be made within specific time slots scheduled by the network management so that there is no interference or contention with other users. For wavelength domain sharing, each transmitter can have a unique but fixed wavelength, and each receiver has a tunable optical filter which selects the specific wavelength from the desired transmitter. Or alternatively, wavelength domain sharing can also be based on wavelength-tunable transmitters and fixed-wavelength receivers.

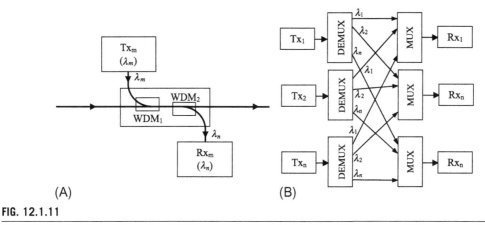

(A)   (B)

FIG. 12.1.11

(A) Optical network node based on WDM couplers to avoid intrinsic splitting loss, and (B) an $n \times n$ star network with wavelength tunable transmitters.

In WDM optical networks based on bus or ring topologies, switchable WDM multiplexers (MUX) and de-multiplexers (DEMUX) can be used to reduce insertion losses in each node. As shown in Fig. 12.1.11A, each node can be equipped with two WDM couplers, WDM_1 and WDM_2. For the network with fixed wavelength transmitters and tunable wavelength receivers, at node m, WDM_1 can be a fixed wavelength multiplexer (high coupling ratio for λ_m), while WDM_2 has to be switchable with high coupling ratio at λ_n where λ_n can vary. This allows the transmitter at λ_m to couple light into the network, and the signal at wavelength λ_n originated from node n can be received. Or alternatively, a wavelength switchable WDM multiplexer (for WDM_1) and a fixed-wavelength DEMUX (for WDM_2) can be used in each node for a network based on wavelength-tunable transmitters and fixed wavelength receivers. Because WDM MUX and DEMUX do not have intrinsic splitting loss as in optical directional couplers, loss budget can be significantly relaxed; especially when the number of users N is high.

With wavelength switchable optical transmitters, an $n \times n$ star network can also be built based on WDM MUX and DEMUX as shown in Fig. 12.1.11B. In this configuration, by switching the wavelength of an optical transmitter to λ_i ($i = 1, 2, \ldots, n$), signal of this transmitter will be routed to the ith receiver Rxi. This WDM-based star coupler does not suffer from intrinsic splitting loss of a conventional star coupler.

12.1.3 **SONET, SDH, AND IP**

Although optical fibers in the physical layer can provide wideband transmission mechanisms, data to be delivered from the source to the destination has to be packaged into a certain format and multiplexed with data streams of other users for efficient sharing of network resources. This is considered as a client layer within the optical layer, which processes data in electrical domain and aggregating a variety of low-speed data streams of various applications and services into the network.

SONET (synchronous optical network) is one of the most important multiplexing standards for high-speed transmission in North America. An equivalent standard widely used outside North America is *SDH* (synchronous digital hierarchy). Both SONET and SDH are *synchronous* multiplexing structures in which all clocks in the network are perfectly synchronized. This allows the rates of SONET/ SDH to be defined as integer multiples of a base rate. Multiplexing multiple low-speed data streams into a higher speed data stream for transmission, and de-multiplexing it back to low-speed streams at the receiver are straightforward by interleaving. Note that for *asynchronous* multiplexing such as plesiochronous digital hierarchy (PDH) prior to SONET, bit stuffing is often needed to compensate the frequency mismatch between tributary data streams.

Table 12.1.2 Transmission data rate of SONET and SDH		
SONET	**SDH**	**Bit rate (Mb/s)**
STS-1 (OC1)		51.84
STS-3 (OC3)	STM-1	155.52
STS-12 (OC12)	STM-4	622.08
STS-48 (OC48)	STM-16	2488.32
STS-192 (OC192)	STM-64	9953.28
STS-768 (OC768)	STM-256	39,814.32
STS-1920 (OC1920)	STM-640	99,535.80
STS-3840 (OC3840)	STM-1280	199,071.60

Table 12.1.2 shows the data rates of SONET and SDH which are often used in optical networks. The basic rate of SONET is 51.84 Mb/s which is also known as synchronous transport signal level-1 (STS-1), or optical carrier 1 (OC1) to specify the optical interface. Higher data signals STS-N ($N = 3, 12, 24, 48, 192, 768, 1920, 3840, ...$) can be obtained by N-interleaving of N STS-1 data streams. Since SONET is designed based on STS-1 frames, if one wants to multiplex M STS-N into STS-$N \times M$, the common practice is to de-multiplex each STS-N into N STS-1 frames and then multiplex $M \times N$ STS-1 frames into a high rate STS-$N \times M$. For SDH, the basic rate is 155.52 Mb/s, which is higher than the basic rate of SONET.

Before discussing the data frame structures of SONET/SDH, it is necessary to specify the four layers of SONET/SDH, namely optical layer, path layer, line layer, and section layer as illustrated in Fig. 12.1.12.

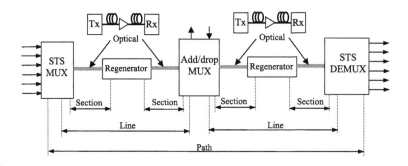

FIG. 12.1.12

Definitions of SONET/SDH layers.

In Fig. 12.1.12, STS MUX multiplexes multiple low-rate incoming signal streams into a high data rate STS-N signal, such as STS-192, and then converts the electrical signal into optical signal through an optical transmitter for optical transmission. Regenerator is known as repeater, which detects the incoming optical signal through an optical receiver, resamples and regenerates the electrical signal, and then convert the regenerated electrical signal back to the optical domain through an optical transmitter. Thus a regenerator performs O-E-O (optical-electrical-optical) transformations. In a SONET/SDH network, low-rate tributary data streams can be dropped from or added into a high-rate data stream without de-multiplexing the entire high-rate data stream. For example, an STS-12 data stream can be dropped from an STS-192 channel through bit interleaving. In this case, header information is used to identify individual low-rate data stream. Note that the add/drop functionality in Fig. 12.1.12 is in the electrical domain. Layers in Fig. 12.1.12 are defined as follows:

Optical layer: it provides physical transmission medium which includes optical transmitter, receiver, optical fiber, and multiple optical amplifiers if the distance is long enough.
Section layer: represents electrical domain data link between regenerators through optical layer transmission medium.
Line: is the portion of a network between neighboring MUX/DEMUX including add/drop multiplexer (ADM). In other words, the line layer may consist of multiple sections and is terminated at each intermediate line terminal which can be MUX, DEMUX, or ADM.
Path: is the end-to-end connection from the point where a STS-1 frame is created to where the STS-1 frame ends. In a simple point-to-point link, section, line, and path are the same.

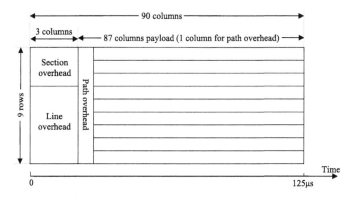

FIG. 12.1.13

STS-1 frame structure.

In the layered structure of SONET/SDH, each layer has a set of associated overhead information within the data frame structure, which is added when the layer starts and removed when the layer is terminated. The basic rate of SONET is STS-1. As shown in Fig. 12.1.13, each STS-1 frame has 810 bytes consisting of 90 columns and 9 rows. With each byte carrying 8 bits, an STS-1 frame carries $8 \times 810 = 6480$ bits. The first three columns carry transport overhead bytes, within which three rows carry section overhead and six rows carry line overhead. The 4th column carries path overhead, which left the rest 86 columns carry the payload. Each STS-1 frame is 125 µs long in time, corresponding to 8000 frames per second, and thus the data rate is $6480 \times 8000 = 51.84$ Mb/s.

A higher speed STS-N frames can be multiplexed from N STS-1 frames through byte-interleaving. The frame structure of STS-N is shown in Fig. 12.1.14, which has 9 rows and $90 \times N$ columns. Section overhead and line overhead together occupy $3 \times N$ columns, $87 \times N$ columns are used for payload, and one column within the payload is dedicated for path overhead. The frame length in time is still 125 µs, so that bit rate of STS-N is $N \times 51.84$ Mb/s.

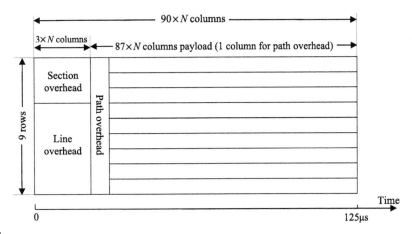

FIG. 12.1.14

STS-N frame structure.

Information carried in the section overhead and line overhead can be monitored for various purposes, and each overhead byte has its own special function, which makes SONET/SDH attractive for network management. More detailed description of overhead bytes and their specific functionalities can be found in (Ramaswami and Sivarajan, 2002).

While SONET and SDH have highly regulated frame and network structures, IP is more flexible, which becomes popular in modern dynamic optical networks. IP is a packet-switching protocol in which information is delivered in packets, and the length of each packet can be variable. In a classic layered network structure shown in Table 12.1.1, the position of IP is situated on top of the data link layer, which is able to work with a variety of data link types and protocols such as Ethernet and token ring as illustrated in Fig. 12.1.15, where *PPP* stands for point-to-point protocol, *HDLC* stands for high-level data link control, TCP stands for transmission control protocol, and UDP stands for user datagram protocol. Note that IP by itself does not provide an in-sequence delivery of data stream from the sender to the destination, and it has to be combined either with TCP or UDP which sits on top of the IP layer as shown in Fig. 12.1.15 to support the logical connections of data delivery. This is the reason why terms such as TCP/IP or UDP/IP are often used in IP networks.

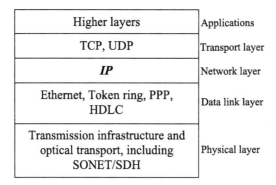

Higher layers	Applications
TCP, UDP	Transport layer
IP	Network layer
Ethernet, Token ring, PPP, HDLC	Data link layer
Transmission infrastructure and optical transport, including SONET/SDH	Physical layer

FIG. 12.1.15

IP in the layered hierarchy.

In an IP network, each packet originated from a sender can be routed into its destination via a number of intermediate routers. When a packet arrives in a router, the router decides the next adjacent router that the packet needs to be forwarded to. The routing decision is based on the information carried in the packet header and a routing table provided by the network management. The routing decision may change dynamically depending on the traffic situation and network resources availability at each moment. For example, for the IP network shown in Fig. 12.1.16, a packet at router 1 with the intended destination of router 8 can be forwarded to routers 2, 4, or 5 depending on the availability of network resources and routing algorithm. If router 1 decided to forward the packet to router 5, then the next routing decision that router 5 makes should be forwarding the packet to is final destination at router 8. However, if the link between router 5 and router 8 is busy at the moment, the packet can also be forwarded to 3 or 7 before being forwarded to router 8.

In order to make valid forwarding decision, each router in an IP network has to be aware of all other routers in the network and their availability of connectivity at each moment. As network configuration and traffic situation can be dynamic, network maintenance process has to be automatic. This maintenance is done through a distributed routing protocol: each router broadcasts link-state packets as soon as it detects any change of its link with adjacent routers. This allows all routers in the network to be informed about any change of other routers, and to update look-up routing tables dynamically.

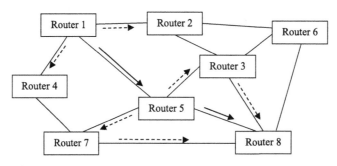

FIG. 12.1.16

Example of routing a packet from router 1 to router 8. Solid arrows: actual routing paths, and dashed arrows: possible routing paths.

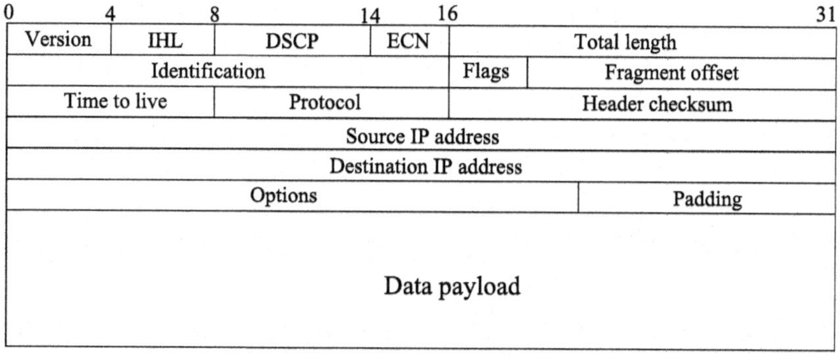

0 4 8 14 16 31					
Version	IHL	DSCP	ECN	Total length	
Identification				Flags	Fragment offset
Time to live		Protocol		Header checksum	
Source IP address					
Destination IP address					
Options				Padding	
Data payload					

FIG. 12.1.17

An IP packet structure.

Fig. 12.1.17 shows the IP packet structure (Huitema, 1999) known as IPV4. IP packet encapsulates data received from the higher layer and adds header which contains all necessary information for the delivering of the packet to the end users and for the intermediate node to make optimum forwarding decisions. More specifically the header should include the following information:

Version: specifies the version number of IP protocol used in the network, such as IPV4 (the fourth version of the internet protocol).

IHL: (Internet header length) specifies header length of the IP header.

DCSP: (Differentiated service code point) defines the type of service.

ECN: (Explicit congestion notification) contains information about the estimated congestion in the expected route of packet.

Total Length: specifies the total length of the Packet, which includes both header and payload.

Identification: is an identification of the original IP packet. In case if the packet is fragmented during the transmission, all the fragments should carry the same identification number.

Flags: tells if the packet is too large for the network resources to handle, and if the packet needs to be fragmented or not. If the packet is not too large, this 3-bit flag is set to "0."

Fragment Offset: tells the position where the original packet was fragmented.

Time to Live: specifies the maximum number of routers this packet is expected to travel across. After going through each router in the network, its value is reduced by one, and thus when the value of "time to live" reaches zero, the packet should be discarded. This is designed to avoid packet looping in the network which may cause problems.

Protocol: identifies specific protocol that should be used for the next layer, such as ICMP (Internet Control Message Protocol), TCP (transmission control protocol), or UDP (user datagram protocol).

Header Checksum: is the digital value of the sum of the entire header, which is used to verify if the packet is received error-free by simply checking bit parity.

Source Address: is a 32-bit IP address of the packet sender.

Destination Address: is a 32-bit IP address of the packet receiver. Both source and destination addresses are keys to identify the IP packet origin and destination, which allow each router to make correct forwarding decision.

Options: although not often used, it can be used if the value of header length is too high (say, greater than 5). These options may also be used to specify options such as security and time stamp.

IPV6 packet structure has also been standardized which is two time larger than IPV4. In IPV6, fragmenting packets are not used, and a QoS mechanism is built-in that distinguishes delay-sensitive packets (Stallings, 1996).

In an IP network different packets can be routed through different routes to maximize network efficiency. Even consecutive packets from the same sender to the same destination can be routed differently through the network, and sequenced at the receiver to reconstruct the original data stream. One important concern in an IP network is the lack of QoS mechanism. As different packets may take different routes through the network with different degradation characteristics and different delays, some packet may be dropped due to transmission errors or due to traffic congestions in the intermediate nodes, QoS cannot be universally guaranteed. Among various efforts to improve QoS, *Differentiated Services* has been used extensively in IP networks. Because different applications may have very different requirements on the signal quality, packets can be grouped into different classes and treated with different levels of priorities. The class-type indicated in each packet header specifies how this packet needs to be treated in each router. Although differentiated services helped different customers get what they have paid for, the overall average quality of the network largely remains the same. In order to further improve the QoS of IP network MPLS (multiprotocol label switching) is introduced.

MPLS is a layer sandwiched between the IP layer and the data link layer, or simply considered as part of the IP layer. As a connection-oriented network technique, MPLS provides a label-switched path (LSP) between nodes in the network. In an MPLS network, each data packet is assigned with a label, and packet-forwarding decision in a participating router is made solely based on the contents of this label without the need of reading into the packet itself. Thus, end-to-end circuits can be established across the transport medium, independent of particular model used in the data link layer, such as SONET or Ethernet. A router implemented with MPLS functionality is commonly referred to as a label-switched router (LSR). MPLS network has completely separate control plane and data plane so that the process of label switching and forwarding is independent of LSP creation and termination. The process of setting end-to-end LSP is a control plane function, and each LSR maintains a label forwarding table. An LSR extracts the label from each incoming packet, replace the incoming label by the outgoing label, and forward the packet to the next route based on the forwarding table.

In a conventional IP network, packet forwarding in a router is largely autonomous without a notation of end-to-end path, and thus different packets between the same source-destination pair can take different routes based on the current status of the routing table of the routers. MPLS, on the other hand, has the ability to specify end-to-end packet paths in the network. This allows network planner to optimize routing paths based on various criteria such as balancing network loads, minimizing congestion, and minimizing packet delays. The ability of having explicitly selected routing paths also allows the guarantee of QoS for high priority traffic in the MPLS network. Traffic of different priority levels can be assigned with different routing paths. More specifically, highest priority traffic can be assigned routing paths with maximum available bandwidth, minimum packet traveling latency, and minimum rate of packet loss. Thus, the QoS of MPLS is much more deterministic and less random compared to the conventional IP.

12.2 OPTICAL NETWORK ARCHITECTURES AND SURVIVABILITY
12.2.1 CATEGORIES OF OPTICAL NETWORKS

The architecture of an optical network can be designed and optimized based on many parameters such as geometric distribution of users, communication capacity requirements, availability of fiber plants, locations of transmission and switching equipments, and so on. In general, optical networks can be divided into a number of categories based on the geometric locations and the nature of services as illustrated in Fig. 12.2.1.

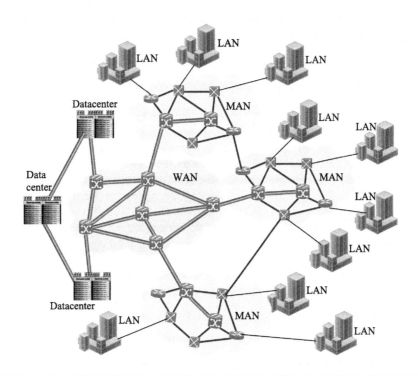

FIG. 12.2.1

Categories of optical networks: WAN: wide-area network, MAN: Metro-area network, LAN: local-area network.

Wide-area network (WAN) usually refers to a large scale network with high-speed backbone fiber-optic links connecting between cities, regions, and even countries. The lengths of these fiber links are normally more than 100 km, which can be terrestrial or transoceanic. Multiple inline optical amplifiers may be used to compensate for transmission losses in the optical fibers. Based on DWDM technology, transmission capacities in backbone fiber links can be multi-Terabits per second. Because of the high capacity and long distance, equipment used for backbone fiber links are most expensive, requiring high-quality lasers, external modulators, low-noise amplifiers, as well as the requirement of compensation of various linear and nonlinear transmission impairments.

Metropolitan area network (MAN), or simply metro network, connects business enterprises and buildings within a city. Depending on the size of the city, fiber links in a MAN can range from a few kilometers to tens of kilometers, but usually less than 100 km. In comparison with wide-area networks (WANs), a metro network needs more flexibility in terms of bandwidth and fiber link distance to support a large variety of different demands and applications. Surprisingly, the demand for transmission capacity in metro networks has been growing more than that in the backbone because of the increased percentage of local traffic in the internet. In many circumstances, installing new fibers in a city, especially in the metropolitan area of a big city, is cost prohibitive. Thus increasing transmission capacity and improve bandwidth efficiency in existing fibers are among important issues in metro networks.

Local-area network (LAN) is a network that connects users in a local area which can be as small as a house and as large as an area with a small number of buildings. Campus network, which provides interconnection among several LANs within a single organization of business, can be considered as an extension of LAN, but the area of coverage may be larger. In addition to the connectivity and available bandwidth to the end user, cost effectiveness is an important concern for a LAN. Ethernet is the most popular format that is currently used in local-area networks which allows users to share the cost and the resources of the network. Because of the short distance, normally not more than a few hundred meters, LED-based transmitters are often used in LAN to reduce the cost. Large core multi-mode or plastic fibers can also be used to relax optical connection accuracy requirements and reduce system maintenance cost.

Access network is a network that connects local-area networks to the nearest metro network. As both WAM and MAN are generally owned by one or more telecom or network service providers, while a LANs is usually owned by the user or an enterprise that runs the LAN, access network is considered a gateway and an interface that connects subscribers to their immediate service provider. Internet traffic that originates from users in the LAN is aggregated and sent upstream into the MAN, and network traffic from outside world travels downstream from MAN to the LAN all through the access network. In recent years, bandwidth-intensive applications, such as video on demand, cloud computing and smart phones, made internet traffic increasingly heterogeneous and unpredictable. Although traffic dynamics can be averaged out in large networks because of the large number of users, the "bursty" nature of data traffic can overwhelm the access network if the bandwidth is not wide enough to accommodate instantaneous traffic surge. Access network, commonly known as the last-mile, is often recognized as the bottleneck of the optical network.

Data Center network (DCN) provides interconnection between computational and storage resources within a data center and communicate between different data centers. In order to meet the growing demands of cloud computing, both the number and the size of data centers around the world grow rapidly. Although the geometric area of a data center can be relatively small, data traffic within a data center connecting hundreds of thousands of servers can be very heavy and highly dynamic. In addition to the ability of providing very wide interconnection bandwidth, DCN needs to be highly scalable, energy efficient, and highly reliable.

12.2.2 OPTICAL NETWORK PROTECTION AND SURVIVABILITY

Reliability is one of the most important concerns of communication systems and networks. Telecommunication service providers usually guarantee 99.999% (five 9's) availability of the connection for the users. That corresponds to less than 2 h interruption of communication within 20 years. Practically network failure may happen due to a number of reasons such as device/component and hardware failure, software malfunction, fiber cut due to construction, and human errors in operation, to name a few.

In practical network design, the most common way of protecting network from failing is to provide a redundant protect path (or even multiple redundant paths) on top of each working path. When the working path fails, the network immediately switches to the protect path. However, the requirement of redundant paths not only makes the network more complex but also increases the overall cost. The reliability of the network obviously depends on the architecture of the network and degree of protection redundancy. Tradeoff has to be made between the reliability and the cost of the network. In the following, we discuss a few protection techniques in different network architectures.

Fig. 12.2.2A shows a bidirectional point-to-point link between transceivers A and B, over a working fiber pair 1 and 2. Fiber 3 and 4 are reserved as protection channels. If only one fiber, say fiber 1, is accidentally cut, the traffic from TX_A to RX_B can be switched to fiber 3. At this point, the return traffic from TX_B to RX_B can either stay with fiber 2 or switched to fiber 4 depending on the locations of fiber pairs. In general, communication paths 1, 2, 3, and 4 shown in Fig. 12.2.2A do not have to be actual fibers; they can be wavelength channels in a WDM scenario. In that case, failure of a particular channel can be caused by crosstalk from neighboring channels, wavelength dependent gain/loss of EDFA, or problems related to WDM MUX and DEMUX.

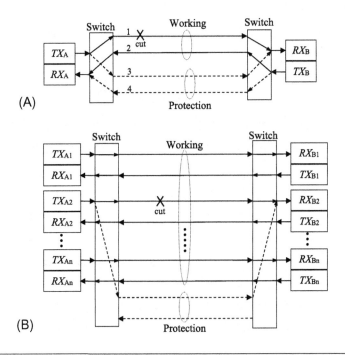

FIG. 12.2.2

Dedicated (A) and shared (B) protection schemes. Solid lines: working channels, dashed lines: protection channels.

For *dedicated protection*, each working channel is associated with a redundant protection channel. Another option is to use *shared protection* to reduce the cost, as shown in Fig. 12.2.2B. For shared protection, the number of protection channels is less than that of working channels. In the example shown in Fig. 12.2.2B, only one pair of protection channels is reserved in the link while there are *n* working channels. If one of the working channels fails the traffic can be switched to the protection channel. Shared protection fails if multiple working channels fail simultaneously and there are not enough protection channels to back them up. In that case, the network has to protect channels based on their priority levels.

In addition to the protection again fiber channel failures, optical transmitter and receiver may also fail which need to be protected as well. Fig. 12.2.3 shows a unidirectional WDM transmission system with *n* wavelength channels. In the dedicated protection scheme shown in Fig. 12.2.3A, each transmitter and receiver has its dedicated backup at the same wavelength. As soon as the working transmitter or receiver fails, the backup device switches on to replace the failed device. Fig. 12.2.3B shows the shared protection scheme in which only one backup transmitter and one receiver are used to protect the entire WDM system. While optical receivers are usually wavelength insensitive, optical transmitters in a WDM system are wavelength selective. Thus, the backup transmitter has to be wavelength tunable, which can be switched to any wavelength when a transmitter at that wavelength fails.

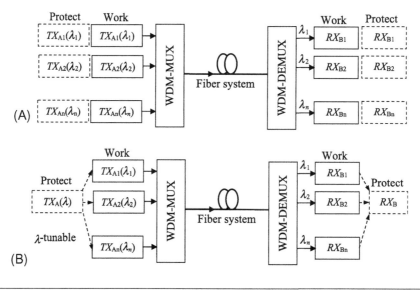

FIG. 12.2.3

Dedicated (A) and shared (B) protection schemes for transmitters (TX) and receivers (RX).

Although a wavelength tunable transmitter is more costly than a fixed wavelength transmitter, using one wavelength tunable transmitter to backup *n* fixed wavelength transmitter still makes sense in terms of cost reduction and the reduced complexity.

Ring configuration has become very popular in optical networks mainly because of its simplicity and the effectiveness of protection schemes known as *self-healing*. Both two fiber ring and four fiber ring are discussed which are classic examples of self-healing feature of network protection.

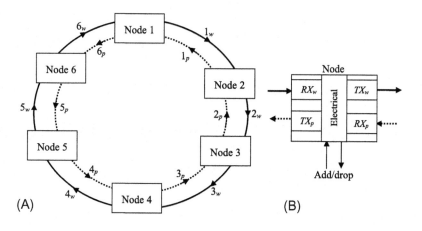

FIG. 12.2.4

(A) Configuration of unidirectional path-switched rings (UPSR) with self-healing protection functionality and (B) configuration of each node. Solid line: working links and dashed line: protection links.

Fig. 12.2.4A shows a SONET ring with two fibers connecting six nodes. The fiber ring shown in solid line provides working links in the clockwise direction, and that shown in dashed line is the protection ring with the transmission direction counterclockwise. Each node with the configuration shown in Fig. 12.2.4B is a SONET ADM with the ability of electronic add/drop in the line level. As both working and projection rings are unidirectional, this configuration is known as unidirectional path-switched rings (UPSRs). The transmitter sends the optical signal directly to both working and protection rings in the opposite directions, and the receiver also detects optical signals from both rings. Now let us use communication between nodes 1 and 5 as an example to explain how this protection scheme works. Traffic from node 5 to node 1 goes through working links 5_w, and 6_w via node 6, and the return traffic from node 1 to node 5 needs to go through links 1_w, 2_w, 3_w, and 4_w, via nodes 2, 3, and 4, all in the clockwise direction. Meanwhile, traffic between nodes 5 and 1 also travels counterclockwise through the protection ring. In normal operation condition, the receiver in each node compares signals received from both directions and chooses the one with the better transmission quality such as higher signal-to-noise ratio. The simultaneous use of the bidirectional traffic in the 2-fiber-ring configuration also allows the balance of propagation delay in the two directions. For example, there is no intermediate node when sending signal from node 2 to node 3 through link 2_w. However, the return signal from node 3 to node 2 would have to go through nodes 4, 5, 6, and 1 if only the working ring is used. In this situation, the protection link 2_p will be a better choice for sending signal from node 3 to node 2.

Now, if there is a fiber cut, for example at links 6_w and 6_p as shown in Fig. 12.2.5, or equivalently if node 6 fails, traffic from node 5 to node 1 will be rerouted through links 4_p, 3_p, 2_p, and 1_p, while the return traffic from node 1 to node 5 still goes through links 1_w, 2_w, 3_w, and 4_w. Switching and selection of the better transmission path is automatically accomplished with feedback electronic circuits and software control.

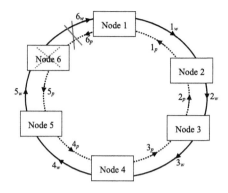

FIG. 12.2.5

UPSR fault protection against fiber cut (at 6_w and 6_p), or node failure (node 6).

Another example of is the four-fiber ring configuration shown in Fig. 12.2.6, where both working paths and protection paths use two-fiber bidirectional fiber rings. This is commonly known as bidirectional line-switched rings (BLSRs). In this configuration, each node has four pairs of transceivers, so that the total capacity is doubled in comparison to the two-fiber ring shown in Fig. 12.2.4. In a normal operation condition, bidirectional connection allows the selection of shortest distance for communication between two nodes. For example, bidirectional data traffic between node 5 and node 1 only use 5_w and 6_w instead of going through the longer route.

There are two types of fault protection mechanisms that are usually used for BLSRs, namely span switching and ring switching. If a transmitter or a receiver in a node that is associated with the working ring fails, data traffic for that particular span will be switched from working ring to the protection ring. This is known as span switch. For example, if the transmitter in node 1 going clockwise to the working ring fails, the traffic from node 1 to node 2 will be switched to the protection transmitter that sends data traffic from node 1 to node 2 through the protection fiber 1_p. In this case, only one span is impacted without affecting other spans. Span switch can also be used to protect working fiber cut. However, if the entire fiber cable is cut which carries both working fiber and protection fiber in a certain span, ring protection has to be used. For example, if both 6_w and 6_p are cut, traffic between node 6 and node 1 will be rerouted through protection fibers 1_p, 2_p, 3_p, 4_p, and 5_p via nodes 2, 3, 4, and 5.

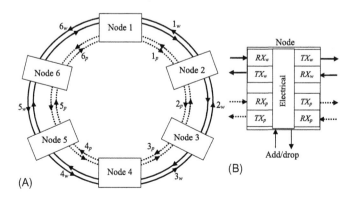

FIG. 12.2.6

(A) Configuration of bidirectional line-switched rings (BLSR) with self-healing protection functionality, and (B) configuration of each node. Solid line: working links and dashed line: protection links.

Because the protection capacity in BLSR is shared among all the connections, it is generally more efficient than UPSR in protecting distributed traffic patents (Ramaswami and Sivarajan, 2002).

12.3 PASSIVE OPTIC NETWORKS AND FIBER TO THE HOME

Access networks, connecting end users to telecommunication service providers, are commonly known as the last-mile networks. In order to meet the needs of a large variety of user applications, access networks have to be flexible in terms of capacity and signal format. At the same time, access networks also have to be low cost and requiring low maintenance efforts so that they can be owned or at least partially owned and maintained by the consumer. Historically, voice with 4 kHz analog bandwidth has been the major service for many years through twisted copper telephone wires. Digital subscriber line (DSL) is then introduced to support digitized voice channel at 64 kb/s data rate, known as DS0. With the introduction and rapid growth of internet, the demand of data rate increases dramatically, and broadband DSL based on frequency division multiplexing (FDM) with discrete multi-tone (DMT) modulation has been used to increase the transmission capacity and reduce the impairments of twisted copper wires. At the same time, cable TV (CATV) is another driver for the demand of broadband access. Broadcast CATV with large number of FDM channels may need up to 1 GHz bandwidth. Nowadays, internet traffic has been extremely diversified, including digital videos, teleconferencing, gaming, and high-speed data download and upload. The bandwidth demand on access networks can no longer be handled by twisted copper wires in many circumstances, and thus fiber-optic access networking becomes the best choice.

Fiber to the home (FTTH) refers to fiber connection between a service-provider and many household users to provide virtually unlimited optical layer bandwidth dedicated to each user. Whereas, fiber to the curb (FTTC) refers to fiber connection between a service provider and the neighborhood (curb) of a community which may include a group of users. In FTTC, although optical signals at the curb has to be converted to electrical signal and distributed to users through twisted copper wires or even broadband wireless systems, the curb is much closer to the users compared to the service provider. A general terminology FTTx is usually used, where "x" can be *home* for FTTH, and *curb* for FTTC.

As an access network can be point-to-point (PTP) or point-to-multipoint (P2MP), PONs with different power-splitting schemes, or even employing WDM, are most suitable for these applications.

By definition, PON is a fiber access network in which passive optical splitters are used to enable the communication between the optical line terminal (OLT) in the service provider's central office and multiple network users, known as optical network units (ONU). The most important advantage of using PON as optical distribution network (ODN) is the elimination of outdoor active devices and equipment. Although the area of coverage is usually smaller than access networks that involve active components, PONs usually have lower cost, without the need to sit inside a room, and thus are easy to maintain. The transmission distance of a PON is usually less than 20 km.

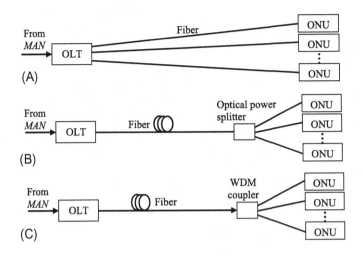

FIG. 12.3.1

Basic optical configurations of PON including (A) direct fiber connection between OLT and each user, (B) TDM-PON, and (C) WDM-PON.

Fig. 12.3.1 shows three basic optical configurations of PON. In Fig. 12.3.1A, a dedicated fiber is used to connect between OLT and each ONU. This configuration can provide the widest transmission capacity, but would require a large number of fibers between the central office (CO) and users. Alternatively, for the configuration shown in Fig. 12.3.1B, only one fiber delivers optical signal from OLT to the vicinity of ONUs, and a passive optical star coupler is used to distribute signal optical power to multiple users. In that case, time sharing can be used between users based on time division multiplexing (TDM), known as TDM-PON. Each user recognizes its own time slot of the data traffic through address labels embedded in the TDM data packets. The configuration shown in Fig. 12.3.1C is known as WDM-PON, in which multiple wavelength channels are carried in the fiber between OLT and a WDM coupler (MUX for upstream traffic and DEMUX for downstream traffic). Each user can have a dedicated wavelength channel so that the transmission capacity can be much higher and the power budget can be much relaxed in comparison to a TDM-PON.

12.3.1 TDM-PON

TDM-PON shown in Fig. 12.3.2 is the most popular PON configuration. For the downstream traffic, OLT broadcasts time-interleaved data frames as a continuous traffic stream to all ONUs, and each ONU selects the desired packets within each frame with a repetition time t_f which is typically 125 μs. This selection is based on the header address bytes associated with data packets imbedded in the frame. Statistically, the downstream data packet rate addressed to each ONU is inversely proportional to the frame repetition time t_f, and also inversely proportional to the number of ONUs because of the time sharing. However, the actual packet rate of each ONU may change dynamically because the number of packets addressed to different ONUs within a particular frame may differ depending on the dynamic demand of each user. In addition, the OLT can also multicast same packets to multiple ONUs simultaneously. In that case, the packet rate for each ONU can be significantly higher than $1/t_f$.

On the other hand, as all ONUs are operated independently, upstream traffic from ONU to OLT needs to be carefully coordinated to avoid contention at the receiver of OLT. In a burst-mode upstream transmission format, every time when an ONU needs to send a data packet to OLT, it has to send a preamble sequence in advance, so that the OLT can synchronize upstream packets from all ONUs into frames and to streamline data reception process in the receiver. Generally, a time gap has to be reserved between burst data packets from different ONUs in the frame structure to avoid interference between them in the OLT.

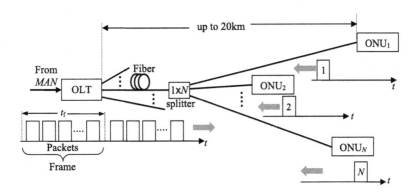

FIG. 12.3.2

Illustration of TDM-PON and frame scheduling.

Optoelectronic structures of OLT and ONU are shown in Fig. 12.3.3. In TDM-PON, 1490 and 1310nm wavelengths are usually used to carry downstream and upstream data traffics, respectively. The 1550nm wavelength is usually reserved for broadband video transmission. The physical layer components include a transmitter (Tx), a receiver (Rx), and a simple WDM coupler which separates 1490 and 1310nm optical signal components traveling in opposite directions. A medium access control (MAC) layer is used in both OLT and ONU, which is responsible for time-slot scheduling of data frames among different ONUs to avoid contention. The MAC layer in the OLT, serving as the master, specifies the starting and ending time that a particular ONU is scheduled to transmit, whereas, the MAC layer in each ONU is synchronized with that in the OLT for scheduling. An OLT may have a number of transceiver pairs to support multiple PON distribution systems, which is m in Fig. 12.3.3, and each transceiver pair needs to have its own MAC. A cross-connect layer in an OLT provides interconnection and switching functionalities between these PON distribution fiber systems and the backbone network. The service adaption layers in OLT and ONU are used to translate between PON system frame formats and signal formats used in backbone and those used by the clients. Service MUX/DEMUX in ONU provides the ability for each ONU to support several different clients with different types of applications.

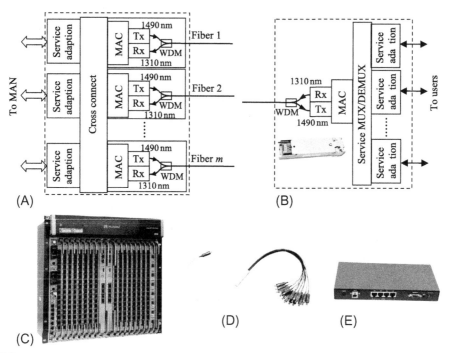

FIG. 12.3.3

Block diagrams of (A) OLT, and (B) ONU in a TDM-PON, (C) an image of OLT, (D) image of a fiber 1×16 star coupler, (E) an image of ONU. (C) Used with permission from Huawei SmartAX MA5800.

Depending on the geometric locations, the distance between an OLT and different ONUs can be very different. Use a maximum range difference of 20 km as an example, it corresponds to a time delay difference of approximately 100 μs assuming that the refractive index of fiber is $n \approx 1.5$. To be able to coordinate upstream traffic originated from different ONUs in a TDM-PON, it is essential to know the distance and propagation delay between OLT and each ONU. This is commonly accomplished by a ranging processing, in which the OLT sends out a ranging request to all associated ONUs, and each ONU replies with a ranging response back to the OLT. This allows roundtrip time delay between OLT and each ONU to be determined and stored in the OLT. This roundtrip time is then sent to each ONU and used to schedule and synchronize burst upstream data packets from all ONU into fixed-length frames so that contention can be avoided in the OLT. This ranging process has to be repeated every time when a new ONU is added to the PON.

In the PON system discussed so far, 1490 and 1310 nm wavelengths are used for downstream and upstream data delivery, respectively. The 1550 nm wavelength has not been discussed so far in PON systems. The major reason is that historically 1550 nm wavelength was reserved for analog TV delivery which has much stringent link budget as well as channel linearity requirement. A straightforward way to broadcast broadband TV channels to multiple end users through a PON system is shown in Fig. 12.3.4, where the TV channels carried by 1550 nm optical wavelength are added to the transmission fiber through a WDM coupler. At each ONU, signal carried by the 1550 nm wavelength is demultiplexed before the receiver.

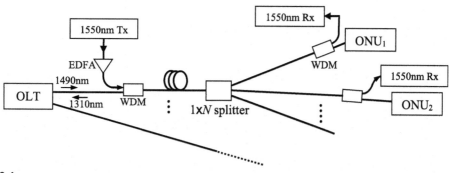

FIG. 12.3.4

Overlay of 1550 nm analog TV broadcast on a TDM-PON.

With the rapid growth of on-demand TV, internet gaming and internet video, bandwidth requirement on access networks and PON has been increasing exponentially. The expansion of PON into 1550 nm wavelength, window is highly anticipated to carry internet data traffic, as well as digital TV and other broadband applications.

Detailed data formats of TDM-PON can be found in various IEEE standards such as ITU-T G.983.1 and G.983.2 [ITU Standard] and are not discussed in this book. Here we only provide brief discussion of a few often used terminologies in the PON.

APON/BPON: APON stands for ATM-PON which is the first commercially deployed PON with an electrical layer built on asynchronous transfer mode (ATM). BPON stands for Broadband-PON, which can be considered as an extended version of APON, with higher transmission speed and the addition of dynamic bandwidth distribution, protection, and other functions. APON/BPON systems typically have downstream and upstream capacities of 1244 Mb/s and 622 Mb/s, respectively.

GPON/XGPON: GPON standards for Gigabit-PON, which is based on the ITU-T G.984 standard defining gigabit PONs. In comparison to the APON standards, GPON specifies much higher bandwidths of up to 2.488 Gb/s and 1.244 Gb/s for downstream and upstream, respectively. A connection-oriented GPON-encapsulation-method (GEM) has been introduced as a data frame transport scheme for GPON. GEM supports fragmentation of the network administrator data frames into variable sized transmission fragments. GPON also provides improved QoS and protection mechanisms than APON. XGPON, or 10-GPON, is an extended version of GPON with the downstream and upstream data rates of 10 Gb/s and 2.5 Gb/s, respectively, as specified by ITU-T G.987 standard.

EPON stands for Ethernet PON, which differs from APON by using Ethernet packets instead of ATM cells in PON systems. Standard GEPON supporting a raw data rate of 1 Gb/s both upstream and downstream in a PON with a range of up to 20 km, as specified by IEEE EPON 802.3ah standard. It can be considered a fast Ethernet over PONs, and thus often been referred to as Gigabit Ethernet PON or GEPON. EPON data format is fully compatible with other Ethernet standards, so that no conversion or encapsulation is needed for service adaptation when connecting to Ethernet-based MAN at the OLT side or Ethernet users at the ONU side. There is also a 10-Gbit/s Ethernet defined by IEEE 802.3 standards. The standard 10/10G-EPON is symmetric with 10 Gb/s data rate both upstream and downstream. The asymmetric 10/1G-EPON is an alternative format with 10 Gb/s downstream data rate and 1 Gb/s upstream data rate.

12.3.2 **WDM-PON**

In TDM-PON based on time sharing, the packet rate of each user has to be reduced with the increase of OTUs. Similar to the concept of wavelength-division multiplexing (WDM) in long-haul and metro optical networks, the application of WDM in PON allows the significant increase of the network capacity. Utilizing the existing TDM-PON infrastructure based on wavelength independent power splitting, multiple wavelengths can be added to create WDM-PON. In a WDM-PON as shown in Fig. 12.3.5, the OLT has N WDM transceivers, and each ONU receives all the N wavelength channels carrying the downstream traffic. An ONU in a WDM-PON usually has a wavelength selective receiver to detect the desired wavelength channel, and a wavelength tunable transmitter to send the upstream traffic. An ONU can also be equipped with multiple receivers to detect several downstream wavelength channels depending on the need. However an ONU can only transmit a single wavelength channel to avoid wavelength contention of the upstream traffic in the transmission fiber.

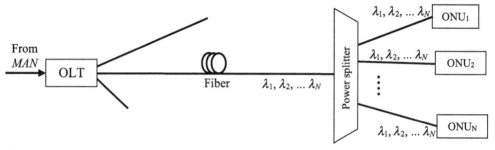

FIG. 12.3.5

Configuration of a WDM-PON based on the TDM-PON infrastructure, (that is wavelength independent power splitter).

In both the TDM-PON shown in Fig. 12.3.2, and WDM-PON shown in Fig. 12.3.5, optical signal power budget is often the most important limiting factor because of power splitting to a large number of ONUs. Intrinsic splitting loss of a star coupler with N ports is $10 \log_{10}(1/N)$ in dB, which amounts to an approximately 21-dB splitting loss for a 1:128 power splitter without considering other losses such as extrinsic loss of the coupler, fiber attenuation, connector loss, and so on. For the purpose of improving the system power budget, a WDM coupler can be used to replace the power splitter so that the intrinsic splitting loss of the optical splitter can be eliminated. In this type of WDM-PON with the configuration shown in Fig. 12.3.6, each ONU can have a dedicated wavelength for both downstream and upstream traffic. For the downstream traffic, the WDM coupler is used as a wavelength DEMUX which separates WDM channels going into different ONUs. While for the upstream traffic, this WDM coupler acts as a wavelength MUX, aggregating optical signals from all ONUs into the same fiber for transmission to the OLT.

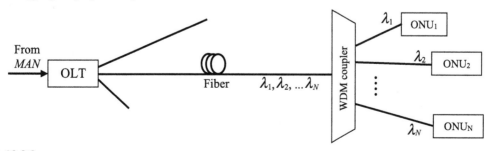

FIG. 12.3.6

Configuration of a WDM-PON based on WDM coupler.

For bidirectional link in a single fiber, directing the upstream and downstream traffics into and from the same fiber can be accomplished simply by optical directional couplers, or by optical circulators to avoid combining/splitting loss. Optoelectronic circuit block diagrams of OLT and ONU are shown in Fig. 12.3.7. For an OLT supporting N WDM channels, N WDM-grade optical transmitters are used each having its own wavelength, λ_i, with $i=1, 2, ..., N$. These downstream WDM channels are aggregated and sent into transmission fiber through a WDM MUX. In the opposite direction, different WDM channels of the upstream traffic received by the OLT are separated by a WDM DEMUX before been detected by N optical receivers. An optical circulator is used in this setup to separate the bidirectional optical signals. For an ONU designated to operate at wavelength of λ_i, as shown in Fig. 12.3.7B, a fixed-wavelength transmitter sends upstream optical signal to the OLT through an optical circulator, and an optical receiver detects the downstream optical signal of the λ_i channel.

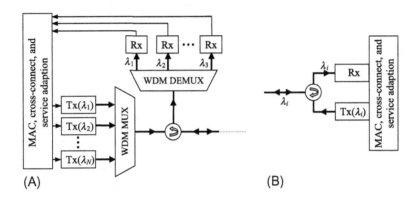

FIG. 12.3.7

Block diagrams of (A) OLT, and (B) ONU in a WDM-PON.

It is also possible to combine WDN-PON with TDM-PON as shown in Fig. 12.3.8. In this hybrid PON configuration, a wavelength channel can be further shared by several ONUs through power splitting and time interleaving.

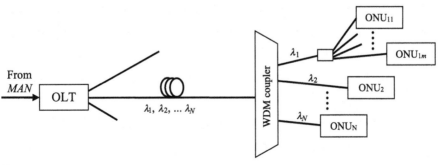

FIG. 12.3.8

Block diagrams of a hybrid TDM-WDM-PON.

12.4 OPTICAL INTERCONNECTS AND DATACENTER OPTICAL NETWORKS

In recent years, various new applications based on cloud-computing, big-data, and internet-of-things emerged rapidly, and huge mesh data networks with extremely high capacity will need to be implemented to support all these applications. Advanced optical communication technologies will play an even more important role in scaling up the performance of networks of all scales range from long-distance, inter- and intra-datacenter connections to short-reach rack-to-rack communications including on-board and chip-to-chip connections.

12.4.1 OPTICAL INTERCONNECTION

Although long distance fiber-optic systems can be considered part of optical interconnection between terminals geographically located far apart, *optical interconnects* usually refer to optical connections of relatively short distance, high speed, low cost, and especially with large number of parallel channels. In addition to enabling components, such as low-cost vertical cavity surface emitting lasers (VCSEL), VCSEL arrays, and high density integrated silicon photonic circuits, new technologies in systems, sub-systems and networks have been developed for optical interconnects.

The simplest example of low-cost optical interconnects is the active optical cable (AOC), which accepts the same electrical inputs and delivers the same electrical outputs as a traditional copper coaxial cable, but uses optical fiber as the transmission medium. Because optical fiber has low loss, wide bandwidth, smaller size, and less weight compared to a typical RF cable, AOC is a good candidate to replace coaxial cable for many applications including storage, data transfer, and high-performance computing interconnectivity.

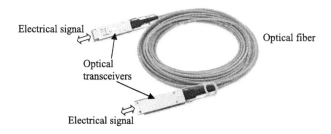

FIG. 12.4.1

High-speed active optical cable (AOC). *Used with permission from Finisar.*

Fig. 12.4.1 shows an example of a FINISAR 100 Gb/s AOC which is able to transmit 4 parallel channels of 25 Gb/s data over a multimode fiber up to 100 m. The optical transceiver on each end of the fiber, based on integrated VCSEL array technology, performs E-O-E (electrical-optical-electrical) conversions, with a power consumption of less than 3 W.

Optical communication can also be used to interconnect between circuit boards, as well as between chips within a circuit board, through free-space or via planar optical waveguides (Wu et al., 2009; Immonen et al., 2007). Fig. 12.4.2A illustrates optical interconnection between printed circuit boards

(PCB). In this case, each PCB is equipped with an optical transceiver, which aggregates all signals that needs to be communicated with other boards and delivers the signal optically through free-space interconnection. Free-space optical links can also be accomplished with two-dimensional VCSEL and detector arrays to form parallel optical channels between computer racks which may require higher capacity data exchange as shown in Fig. 12.4.2B. Free-space optical interconnection has the advantage of flexibility without the requirement of wiring. But alignment of optical beams between the transmitter and receiver has to be adjusted and stabilized against motion and mechanical vibration. Free-space optical beams can also be blocked by objects which may disrupt the interconnection, which needs to be avoided.

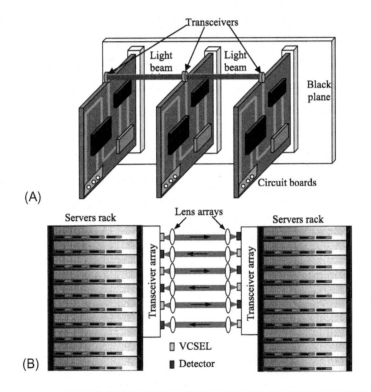

FIG. 12.4.2

Free-space optical interconnects between printed circuit boards (A) and between computer servers (B).

According to Moore's law, the number of transistors per unit area on integrated circuits doubles every year, and processor speeds of CMOS chips doubles every two years. The bandwidth for data interconnection between integrated CMOS chips will also need to increase accordingly. Bandwidth limit of copper strips on PCBs may become bottlenecks for interconnecting these high-speed chips. Optical interconnection may provide a better solution to this problem with high bandwidth and low power dissipation. Fig. 12.4.3A shows an example of inter-chip interconnection on a PCB, in which each chip is connected to a group of VCSELs and photodetectors at its edge (Kim et al., 2008). An array of parallel planar waveguides is used for optically interconnecting VCSELs and detectors associated with a pair of chips. These optical waveguides can be made by polymer materials through molding and casting (Ma et al., 2002), and they are flexible enough to be made on the surface or embedded inside different substrates such as PCB, and made into various different geometric configurations. Polymer waveguides can be made by materials such as PDMS (Polydimethylsiloxane) and PMMA (Polymethyl-methacrylate), and the propagation loss can be less than 0.1 dB/cm (Cai et al., 2008) at 850 nm wavelength. Although this loss is still much higher than that in an optical fiber, it is low enough for inter-chip optical interconnections as where the distance is usually shorter than a few centimeters.

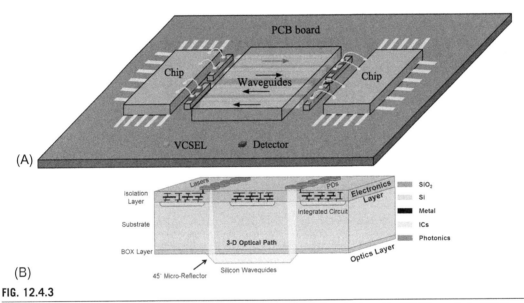

FIG. 12.4.3

Examples of optical interconnections inter-chips on a PCB board (A) and intra-chip (B) (Shen et al., 2014).

Optical interconnection can also be used for intra-chip data links between different areas of the chip on the same silicon substrate as shown in Fig. 12.4.3B. In this particular example, light beams are guided through silicon waveguides on the backside of the substrate, and these waveguides are optically connected with micro structured lasers and photodiodes through holes, known as through-silicon via (TSV) on the silicon substrate.

With the technological advances in power efficient microstructure lasers and VCSELs, as well as CMOS-compatible photonic integration, inter-chip and intra-chip optical interconnects with large number of parallel channels will be more and more popular in high-speed CMOS digital electronic circuits.

12.4.2 DATACENTER OPTICAL NETWORKS

Datacenter networks (DCNs) is a special category of optical network to address the need of highly dynamic data traffic between large numbers of computer servers. With cloud-based services become more and more adopted by the community, most of the computational-intensive tasks are moved from terminal devices, such as personal computers, cell phones, robots, and other emerging handheld devices, to centralized warehouse-scale computer clusters, which are commonly known as mega datacenters. The unprecedented amount of data that needs to be communicated among computers and servers within each datacenter and between different datacenters imposes unique requirements and unprecedented challenges to data communication networks of all scales.

Based on Cisco Global Cloud Index 2015–2020, in 2020 the global datacenter traffic will reach approximately 15 Zetta-bytes per year (Zetta: 10^{21}) as shown in Fig. 12.4.4A. In which, the major portion (77%) is between computers and servers within each datacenter, while 9% is between different datacenters and 14% is between users and datacenters as shown in Fig. 12.4.4A. Intra datacenter interconnection can be within one building or between datacenter buildings on a campus, and the distance of intra datacenter interconnection can be a few meters up to 10 km. Whereas, communication between different datacenters, known as inter datacenter interconnection, can have distances be from 10 up to 80 km. Longer distance connections can then be categorized as metro or long-haul as discussed earlier.

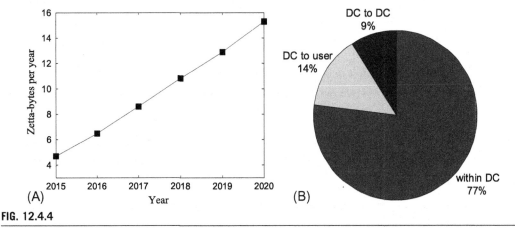

FIG. 12.4.4

(A) Trend of global datacenter traffic increase between 2015 and 2020, and (B) prediction of percentages of data traffic within datacenters, between datacenters and between datacenter and users for year 2020 (Barnett et al., 2016).

Fig. 12.4.5A shows the breakdown of applications of datacenter traffic. Major applications include internet search (22%), social networking (22%), web pages (18%), big data (17%), video streaming (10%), and other applications (12%). Among these applications, big data has the highest growth rate in recent years. Different applications may have different requirements in terms of bit error rate and latency tolerance, which have to be considered in the design of DCNs.

In order to simplify terminal devices, datacenters not only provide computational serviceplatforms, they also provide storage of vast amount of data which can be accessed by users. Fig. 12.4.5B shows the global data storage by datacenters in Exa-bytes (10^{18}). For the 5-year period from 2015 to 2020, data stored in datacenters is expected to increase by more than 5-fold.

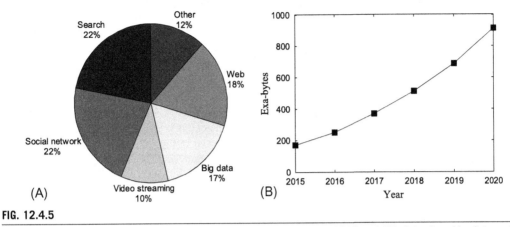

FIG. 12.4.5

(A) Breakdown of applications of datacenter traffic predicted for year 2020 and (B) data stored in datacenters (Barnett et al., 2016).

In order to store the vast amount of data and support various applications and services in the cloud, a datacenter has to host a large number of servers. In general, a facility hosting hundreds or thousands of servers is considered small- to medium-scale datacenter. A large-scale datacenter may support servers in hundreds of thousands.

In order to interconnect such large amount of computer servers, DCNs have to be properly designed and maintained. One of the most important issues in DCN design and operation is the energy efficiency. In fact, to handle the vast amount of data flow in DCN, electric power budget is often the most severe limiting factor. For this reason, datacenters are commonly built near electrical power plants. Real-time power monitoring, dynamic network load management and scheduling are necessary in DCN design and operation. A DCN also has to have efficient and cost-effective routing mechanisms, provisioning and control algorithms in order to move high-speed data flows among a large number of servers with minimum congestion.

Fig. 12.4.6 shows a typical DCN architecture based on multilayer tree topology. In a datacenter, computer servers are mounted inside many racks with each rack holding a few tens servers (typically up to 48). All servers within a rack are interconnected through a Top-of-Rack (ToR) switch, which also serves as an access node to other racks and outside networks. A large number of ToRs are then interconnected through aggregation switches, which can be arranged into one or more layers. The aggregation switches are further interconnected through core switches, which also provide gateways into WANs. In this tree topology, connection data rates go up at higher layers. Data rates for interconnection within a rack between servers typically range from 1 to 10 Gb/s using Gigabit Ethernet, while interconnection data rates in the aggregation layer and core layer can be from 10 to 100 Gb/s and from 100 Gb/s to 1 Tb/s, respectively, which increase steadily over years with the availability of new technologies.

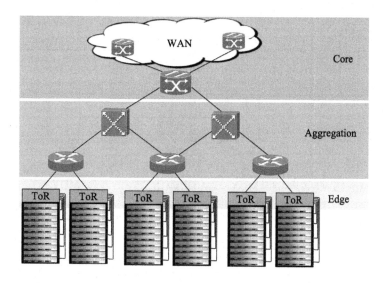

FIG. 12.4.6

Datacenter network architecture based on three-layer tree topology. ToR: Top-of-Rack.

In early days of datacenters with relatively small sizes and low data rates, all electronic interconnections and switches were used. Over the years, especially after 2000, electronic interconnection based on copper wires cannot keep up with the rapid grow of data rate demand in datacenters, and energy consumption of electronic switches become too high with the rapid expansion of datacenter sizes. Then the benefit of introducing optics into DCNs became obvious in terms of both broadband interconnection and switching with reduced energy consumption. Miniaturized pluggable modules of optical transceivers can be easily plugged into switch racks and network interface controllers in a datacenter. Cross-connect optical switches can provide ultra wide bandwidth yet with very low power consumption.

Fig. 12.4.7 shows an example of DCN based on optical switching architecture (Chen et al., 2014). This DCN has N ToRs to be interconnected. Each ToR has W optical transceivers operating at wavelengths λ_1, λ_2, ..., λ_W. Optical output from all W transmitters are multiplexed together into one fiber through a WDM multiplexer, and sent to a wavelength selective switch (WSS) with k output ports. This $1 \times k$ WSS selectively partitions the set of W wavelengths coming in through the common port into k groups at the output. The k output fibers from each WSS are sent to the input of an optical cross-bar switch which can be made by micro-electro-mechanical systems (MEMS). The MEMS cross-bar switch has the advantage of low energy consumption and colorless broad bandwidth which is capable of connecting any input port to and output port by mechanical actuation. Each ToR can also receive W wavelength channels carried by k fibers from the output of the MEMS switch.

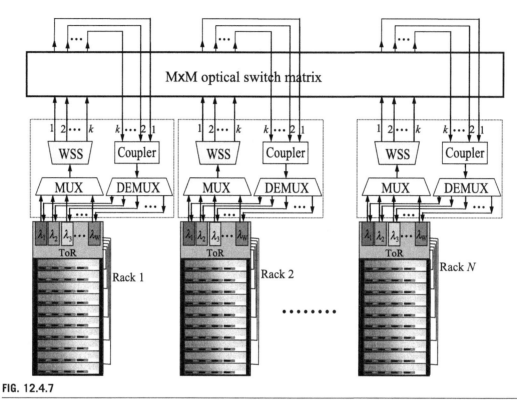

FIG. 12.4.7

Example of datacenter network based on optical interconnection.

In the optically switched datacenter architecture shown in Fig. 12.4.7, if the number of ToRs is N, and each ToR has W wavelength channels grouped into k fiber ports through WSS, the size of the MEMS switch will have to be $M = k \times N$. Assume each wavelength channel supports a server in the rack, the total servers this DCN can interconnect is $W \times N$. The selection of k depends on the number of ToRs and the available port count of MEMS switch. A larger k value can enable each ToR to connect to more other ToRs simultaneously. However, for a limited port count, M, of the MEMS switch, the total number of ToRs, M/k, that the network can support will reduce with the increase of k. Thus, tradeoffs have to be made between k, M, W, and N, as well as the data rate of each wavelength channel.

While the optical switching architecture shown in Fig. 12.4.7 is particularly effective for optical level interconnection between racks in a datacenter, interconnection among servers within a rack still relies on electronics. In practical DCNs, a significant percentage of traffic is among servers in the same rack, and the interconnection between them often consumed the majority of the electrical power.

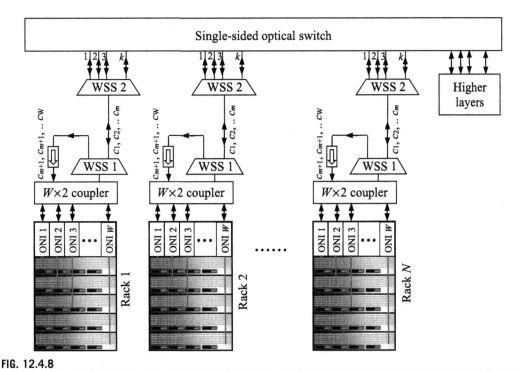

FIG. 12.4.8

Datacenter network based both inter-rack and intra-rack optical interconnection with dynamic traffic adjustment.

The DCN architecture shown in Fig. 12.4.8 addresses optical interconnections for both inter-racks and intra-rack among different servers (Fiorani et al., 2014). In this configuration, each rack holds N servers, and each server is connected to an optical network interface (ONI) card. The ONI is a bandwidth variable optical transceiver whose wavelength, bandwidth, and operation state are determined by a centralized control plane. The W optical outputs ($c_1, c_1, ..., c_W$) from a ToR are combined by an $W \times 2$ optical coupler. Assume that m wavelength channels ($c_1, c_1, ..., c_m$) are needed to communicate with other racks and higher layers of the network, the rest of the channels ($c_m, c_{m+1}, ..., c_W$) can be used for interconnection among servers within the same rack. In order to separate inter-rack traffic from intra-rack traffic, a flexible-grid WSS (WSS 1), which supports elastic channel spacing, is connected to one of the two outputs of the $W \times 2$ optical coupler. Channels $c_m, c_{m+1}, ..., c_W$ selected by WSS 1 are broadcast back to the ToR through the $W \times 2$ coupler via an optical isolator, while channels $c_1, c_1, ..., c_m$ are sent to the core switch via another flexible-grid WSS (WSS 2). Note that for a typical MEMS optical switch with the size of $M \times M$, there are usually M input and M output ports, and optical signals can be routed from any input port to any output port. However, the optical switch in the architecture shown in Fig. 12.4.8 is single-sided, and any pair of ports can be optically connected bi-directionally through switching control. This can be obtained by folding output ports of a conventional double-sided optical switch back to the input side. By using flexible-grid WSS, the percentage of traffic, the number of channels, and the bandwidth used for intra-rack and inter-rack interconnections can be dynamically adjusted to optimize the network performance.

It is important to note that data traffic in DCN is highly bursty and dynamic. If the highest data rate of each computer server can reach 10 Gb/s, truly non-blocking interconnection among all servers across a datacenter with hundreds of thousands of servers can be cost prohibitive and would require tremendous bandwidth which is hard to satisfy even with optical switching architectures. Oversubscription is usually used in DCN which takes advantage of the statistic nature of datacenter traffic. For example, if a DCN is oversubscribed by a factor 10, only 10% of the maximum data rate of each server is guaranteed when the DCN is fully loaded, or not more than 10% of servers in one rack can simultaneously communicate with servers in other racks. DCN built with oversubscription would require dynamic provisioning of interconnection bandwidth according to network traffic status and requirements. Although MEMS-based optical circuit switches can provide wide optical bandwidth with wavelength transparency, their switching speed is usually slow, on the millisecond level. On the other hand, electronic packet switching can be much more flexible for dynamic provisioning and reconfiguration with much faster switching speed. The combination of fast configurable electronic packet switching and wideband optical circuit switching led to a hybrid electrical/optical switch architecture as illustrated in Fig. 12.4.9. In this hybrid architecture, servers in each rack

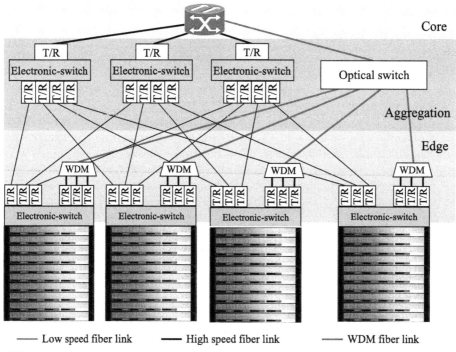

FIG. 12.4.9

Hybrid architecture with the combination of electronic packet switch and optical circuit switch. T/R: optical transceiver, WDM: WDM MUX and DEMUX depending on the directions. All fiber links are bidirectional.

are connected to a group of optical transceivers through an electronic packet switch. Traffic from each rack is divided into two different categories: low speed but highly dynamic traffic and high speed but less dynamic traffic. The low-speed traffic is interconnected by electronic packet switch in the aggregation layer, whereas it more advantageous to interconnect high-speed traffic in the optical domain through WDM multiplexers and de-multiplexers and interconnection-switched by MEMS-based optical circuit switch.

With more and more new applications emerging which require cloud computing and storage, modular datacenter has become a viable approach. A modularized datacenter, commonly referred to as a POD, is a self-contained shipping container with a group of computer servers, a self-consistent network interconnecting these servers, and a cooling/fan system to support its operation. A mega data-center can be constructed by interconnecting hundreds to thousands of such PODs depending on the need, so that the mega datacenter can be modular and easy to expand.

Typically a POD can hold between 250 and 1000 servers, so that a non-blocking switch fabric to interconnect these servers can be reasonably manageable. However, even with the rapid technological advancement in both CMOS electronic circuits and optical communication, interconnecting a large number of PODs to form a mega datacenter still remains a significant challenge. Fig. 12.4.10 shows the hybrid electrical/optical switching (HELIOS) architecture for a modular DCN which is able to interconnect multiple PODs (Farrington et al., 2010). In comparison to the hybrid architecture inter-connecting multiple ToRs shown in Fig. 12.4.9, data rates for both electronic packet switch and optical circuit switch will have to be much higher for HELIOS interconnecting multiple PODs to form a mega datacenter.

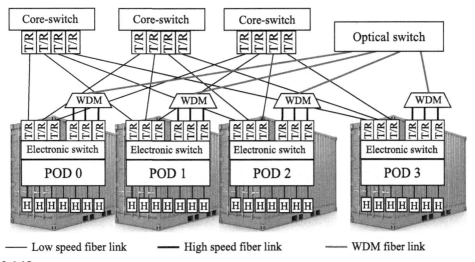

———— Low speed fiber link   ▬▬▬ High speed fiber link   ———— WDM fiber link

FIG. 12.4.10

Modular DCN based on HELIOS architecture interconnecting multiple PODs. T/R: optical transceiver, WDM: WDM MUX and DEMUX depending on the directions. All fiber links are bidirectional.

12.5 OPTICAL SWITCHING AND CROSS CONNECTION

In dynamic optic networks, optical domain switching is one of the most important functionalities that enabled many network architectures. Optical switches can be categorized as optical *circuit switching* and optical *packet switching*. Optical packet switching requires high switching speed on the order of nanoseconds or less. Although it has been an important research topic in the fiber-optic networks community for many years, optical domain packet processing, buffering, sorting, and switching are still challenging, and optical packet switching has not been widely used in commercial optical networks. On the other hand, optical circuit switching can tolerate relatively slow speed. Microseconds to milliseconds switching speed can be sufficient for many circuit-switched network applications. Optical circuit switching is transparent to both data rate and modulation format, and the electrical energy required to switch each data bit can be several orders of magnitude smaller than that of both electrical and optical packet switches. Thus, optical circuit switching is becoming more and more attractive for applications in broadband WDM optical networks. In the following, we discuss a few examples of optical network applications based on optical circuit switching.

12.5.1 RECONFIGURABLE OPTICAL ADD/DROP MULTIPLEXING

Optical add/drop multiplexing (OADM) is an important optical network functionality which allows some wavelength channels to be dropped from an optical node, and the spectral slots of these wavelengths can be reused by added channels back to the system with different data traffic. As shown in Fig. 12.5.1A, an optical fiber carrying WDM signal at wavelength channels, $\lambda_1, \lambda_2, \lambda_3, \ldots, \lambda_n$, enters an add/drop optical node. These WDM channels are first spatially separated into n fibers through a WDM DEMUX, and then each wavelength channel passes a 2×2 optical switch. The optical switch has two possible states, *pass* or *cross*, as shown in Fig. 12.5.1B, determined by an electric control signal. If this particular channel needs to be dropped at the node, the switch is set to the *cross* state, so the optical signal is routed to an optical receiver, and the information carried by the channel is downloaded to the local electrical circuits for processing. The node can also add new data traffic back to the system at the same wavelength through the switch. If the network controller decides not to drop a particular channel, the switch for that channel will be set to the *pass* state so that the optical signal passes to the output straightforwardly without been affected. Outputs from all switches are then combined through a WDM MUX and sent to the output fiber with all the wavelength channels.

This add/drop operation is reconfigurable because the decision of add and drop of each wavelength channel is determined by a network controller. Therefore this node is commonly referred to as a *reconfigurable optical add/drop multiplexer* (ROADM), which provides a high-level flexibility for network operation and management. On the other hand, optical add/drop multiplexing can also be fixed for simplicity, which is thus known as OADM. In that case, no control is required and each optical switch can be replaced by two optical circulators as schematically shown in Fig. 12.5.1C. If a channel does not need to be dropped at the node, one has to manually remove the corresponding circulators so that the optical signal can bypass.

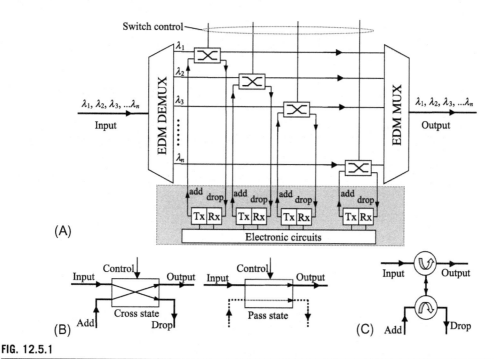

FIG. 12.5.1

(A) Configuration of an optical add/drop multiplexer, which can be reconfigurable (ROADM) by using a 2×2 optical switch shown in (B) for each wavelength channel, or non-reconfigurable (OADM) by using optical circulators as shown in (C).

OADM can be made more efficient by grouping those wavelength channels which are not expected to be dropped at the node, and let them pass through as express channels. Fig. 12.5.2 illustrates this option, where WDM channels, λ_1, λ_2, λ_3, ..., λ_n, in the input fiber are prearranged into several groups based on their add/drop requirements. Channels that need to be dropped in the current node are assigned wavelengths λ_{m2+1}, λ_{m2+2}, ..., λ_n, which pass through switches in the ROADM. While other channels are divided into two groups λ_1, λ_2, ... λ_{m1}, and λ_{m1+1}, λ_{m1+2}, ... λ_{m2}, which go through the ROADM along the express paths without switches. These two groups of channels can be dropped later, possibly each at a different node. This avoids unnecessary switching operations and the insertion losses associated. However, the network may become less flexibility because wavelength assignment has to be prearranged, and the maximum number of add/drop channels at each node is fixed by design.

Fig. 12.5.3 shows OADM based on wavelength selective optical components connected in series, which can be fiber Bragg gratings (FBGs) or band-reflecting filters based on multilayer optical films. In Fig. 12.5.3A, each FBG has high reflectivity at the wavelength of a particular channel, and the optical signal of that channel is selected by the FBG. The dropped channel is redirected to a local receiver through an optical circulator at the input side of the FBG. All other channels pass through the FBG with minimum loss. As the spectral content of the dropped channel is blocked

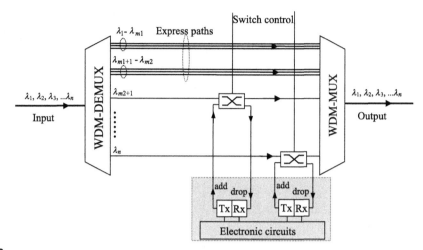

FIG. 12.5.2

ROADM with pre-grouped channels and express paths. WSS can be used to replace DEMUX and MUX for flexible channel selection and grouping.

by the FBG, this wavelength can be reused by adding a new channel at the same wavelength through the optical circulator after the FBG. Similar optical add/drop functionality can also be accomplished by a thin-film based 2×2 WDM coupler as shown in Fig. 12.5.3B. The thin-film is designed to have high reflectivity within a certain wavelength window. The channel that needs to be dropped is reflected by the WDM coupler, and all other channels pass through. Signal of the dropped wavelength can be added back to the system from the opposite side of the coupler. This configuration does not need optical circulators for redirecting add and drop of each optical channel. In the example shown in Fig. 12.5.3, n WDM channels reach the OADM, but only three channels at λ_j, λ_k, and λ_l are dropped. Although the series configuration shown in Fig. 12.5.3 is relatively simple, it is only suitable for OADM with relatively small number of add/drop channels because total optical loss linearly increases with the number of concatenated add/drop units. Wideband FBGs and WDM couplers can also be used to perform group add/drop of adjacent WDM channels, so that the total add/drop units can be reduced.

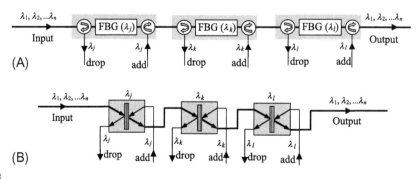

FIG. 12.5.3

OADM in series configurations based on FBGs (A) and 2×2 WDM couplers (B).

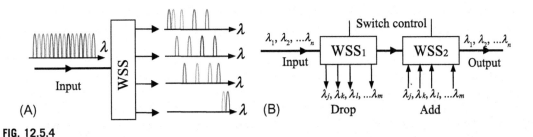

FIG. 12.5.4

(A) Illustration of WSS functionality and (B) ROADM based on WSS.

With the availability of WSS of large channel counts, the configuration of ROADM can be significantly simplified. As discussed in Chapter 6, a WSS has flexible bandwidth selection at each output port, and any wavelength channel from the input can be routed to any output port as shown in Fig. 12.5.4A. Which wavelength channel is routed to which output port is determined by an electric control. This flexibility enables the simplification of ROADM design as shown in Fig. 12.5.4B, where two WSS devices are arranged in series. In this configuration, channels that need to be dropped can be selected by WSS_1, and new signals can be inserted back into the fiber at these wavelengths through WSS_2. This architecture allows dynamic reconfiguration of add/drop channels with flexibility of channel grouping.

12.5.2 OPTICAL CROSS-CONNECT CIRCUIT SWITCHING

In addition to ROADM, optical cross connect (OXC) switching is another important optical domain functionality which is usually used in optical nodes with multiple input and output fibers.

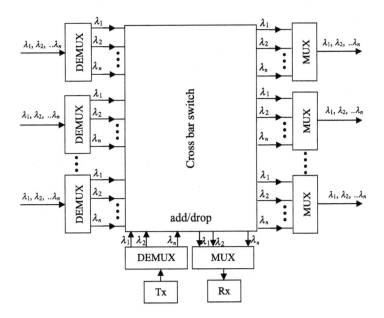

FIG. 12.5.5

A wavelength-space hybrid optical switching node which has m optical input ports each carrying n WDM channels.

Fig. 12.5.5 shows a generic OXC node architecture which has m optical input ports each carrying n WDM channels, and thus the total number of wavelength channels is $m \times n$. This can be used as a node connecting to multiple optical fibers in mesh optical networks. In addition to OXC switch between any input fiber to any output fiber, some wavelength channels can also be dropped, processed, and re-inserted back into the switch known as channel add/drop. The crossbar switch size should be $(m \times n) \times (m \times n)$.

For this wavelength-space hybrid optical switching without the capability of wavelength conversion, the n channels grouped together by each WDM MUX at the output side must have different wavelengths to avoid contention. This wavelength contention issue reduces the degree of freedom of this crossbar switch from $(m \times n) \times (m \times n)$ to $n \times (m \times m)$. In fact the OXC switching network shown in Fig. 12.5.5 can be simplified by n cross-bar switches each with a size of $m \times m$, and each is responsible for a specific wavelength as shown in Fig. 12.5.6, and in this case the total number of input and output ports of the switching fabric is $n \times (m \times m)$.

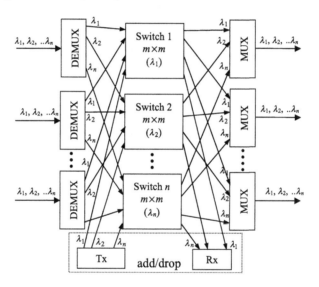

FIG. 12.5.6

A $(m \times n) \times (m \times n)$ wavelength-space hybrid optical switching node using $n \times (m \times m)$ crossbar switches, and each switch is dedicated to a specific wavelength.

Wavelength conversion: In an optical network with a large number of switching nodes, a *lightpath*, which carries a communication channel from the source to the destination, has to travel through many nodes. Suppose each wavelength only carry one independent signal channel in an optical fiber, the total number of wavelength channels that can be supported by each fiber is determined by the WDM equipment. The total number of users that the network can support can be severely limited by wavelength contention. With the capability of wavelength conversion, if two or more channels of the same wavelength from the input side of a node need to be switched to the same output fiber, they can to be converted into different wavelengths.

Assume that a *lightpath* has to go through $k+1$ OXC nodes over k fiber hops, as shown in Fig. 12.5.7A, and each fiber can carry a maximum of F WDM channels. Without wavelength conversion, a lightpath must keep the same wavelength on all fiber sections along the route, which excludes this wavelength to be used by other lightpaths. Define η as the probability that a wavelength is used in a fiber hop, which represents the efficiency of wavelength utilization. ηF is the expected number of busy wavelengths in that hop, $1-\eta$ is the probability that a particular wavelength is not occupied in that hop, and thus $(1-\eta)^k$ is the probability that the wavelength slot is not blocked for all k fiber hops from the source to the destination. Since there are F wavelength slots, in each fiber hop, the probability that all wavelengths are used in at least one fiber hop, known as the blocking probability is (Barry and Humblet, 1996).

$$P_{block} = \left[1 - (1 - \eta)^k\right]^F \tag{12.5.1}$$

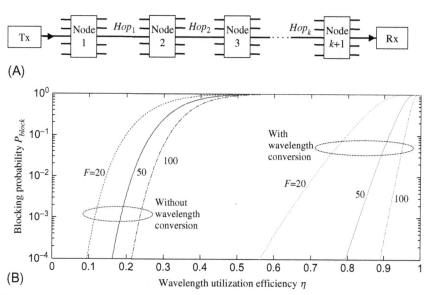

FIG. 12.5.7

(A) A lightpath across $k+1$ OXC switching nodes and k fiber hops, and (B) blocking probabilities as the function of wavelength utilization efficiency η with and without the capability of wavelength conversion for $k=10$ hops. Three different number of wavelengths are used with $F=20$ (dotted lines), $F=50$ (solid lines), and $F=100$ (dash-dotted lines).

Now let us find out how wavelength conversion can help. If each node has the capability of wavelength conversion, the probability that all the F wavelengths in a fiber hop are used is η^F, and the probability that at least 1 wavelength slot is available in each fiber hop along the lightpath is $(1 - \eta^F)^k$. Therefore, the blocking probability for a network lightpath with the capability of wavelength conversion at each node is

$$P_{block} = 1 - \left(1 - \eta^F\right)^k \tag{12.5.2}$$

Fig. 12.5.7B shows the comparison of blocking probabilities as the function of wavelength utilization efficiency η for a lightpath over $k=10$ hops without and with wavelength conversion. Three different number of wavelength channels are used with $F=20$, 50, and 100. In both cases, increasing the number of maximum wavelength channels that each fiber hop can support helps reducing the blocking probability. Without the capability of wavelength conversion, the blocking probability increases dramatically when wavelength utilization of each hop reaches to about 20%, whereas with the used of wavelength conversion, wavelength utilization can be increased to almost 90% with $F=100$ for a blocking probability of 10^{-4}. This significantly improves the network efficiency.

Unfortunately, wavelength conversion is an expensive operation, which can be done either through O/E/O (optical transceiver) or all-optically based on nonlinear optics. In addition to applying wavelength conversion for all WDM channels and all switching nodes, wavelength conversion can also be used for selected network nodes and small group of WDM channels. Practical network design has to make tradeoffs between performance and cost. In a WDM optical network, setting up lightpaths between source/receiver pairs is accomplished through network management with optimum routing algorithms based on the existing physical infrastructure. Wavelength assignment of each lightpath and wavelength conversion (if available) at each node have to be dynamically optimized to maximize the efficiency of network resources sharing and minimize blocking probability. Specific wavelength routing algorithms and network optimization are outside the scope of this book. More details can be found at Ramaswami and Sivarajan (2002) for interested readers.

12.5.3 ELASTIC OPTICAL NETWORKS

Traditionally, WDM optical networks have been primarily based on fixed spectral grid, which is specified by the International Telecommunications Union (ITU), known as the ITU grid. Most commonly used ITU grids in commercial WDM networks are 200, 100, 50, and 25 GHz, corresponding to 1.6, 0.8, 0.4, and 0.2 nm in the 1550 nm wavelength window. The evolution of ITU grids used in commercial WDM networks which went down from 200 GHz all the way to 50 GHz and even 25 GHz has been driven by the increased demand of transmission capacity in optical fiber, which requires the increase of optical spectral efficiency. The reduction of channel spacing over the years was also enabled by the improved spectral selectivity of optical filters and performs of WDM MUX and EDMUX. Various high-speed optical transceivers with data rates of 10 Gb/s, 40 Gb/s, 100 Gb/s, 400 Gb/s, and up to 1 Tb/s have been developed to meet more and more diversified demand of different applications. Different modulation formats and detection methods are also readily available to provide different spectral efficiency and different levels of tolerance to transmission impairments. A WDM optical network has to support various different data rates and modulation formats for different applications and to maximize the efficiency. For example, if a user needs 400 Gb/s communication capacity, it is more efficient to provide a single 400 Gb/s channel than ten 40 Gb/s channels. The use of a small number of high-speed channels not only helps reducing the number of transceivers and MUX/DEMUX ports, but also reduces the equipment footprint and maintenance efforts.

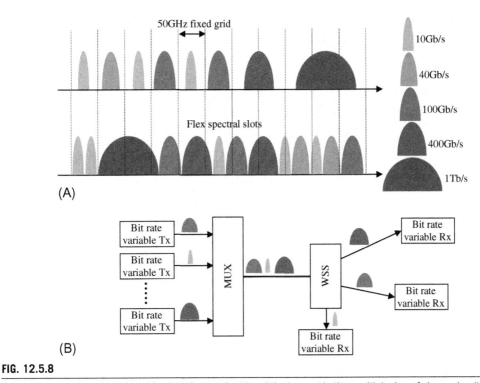

FIG. 12.5.8

(A) Illustration of fixed *(top)* and flexible *(bottom)* grid and the impact in the multiplexing of channels with different spectral bandwidths and (B) elastic network based on bandwidth variable transceivers (BVT) and WSSs.

Optical channels with different data rates and spectral widths cannot efficiently fit into fixed spectral grid. As an example, Fig. 12.5.8A illustrates wavelength channels with 10 Gb/s, 40 Gb/s, 100 GHz/s, 400 Gb/s, and 1 Tb/s, and their spectral widths depend on the specific modulation format. For a fixed grid, as shown in Fig. 12.5.8A (top row), 10, 40, and 100 Gb/s (based on dual-polarization QPSK modulation) each can fit into a single 50 GHz grid. A 400 Gb/s channel has to occupy two 50 GHz slots and a 1 Tb/s channel has to occupy three 50 GHz slots. Different spectral gaps between channels are created due to the mismatch between the fixed grid and the discrete channel spectral widths. These spectral gaps can be filled by using flexible channel spacing as shown in Fig. 12.5.8A (bottom row), known as flexible grid. Using flexible grid, one can choose to have fixed spectral gap, say 10 GHz, between adjacent channels, while both the central wavelength and the spectral width of each channel can vary.

Elastic optical networks based on flexible grid require bandwidth variable transceivers (BVTs) to create optical channels and WSS to dynamically select and route wavelength channels with different spectral bandwidths, as shown in Fig. 12.5.8B. In addition to the tighter channel spacing, elastic optical network also allows the tradeoff between the spectral efficiency and the transmission distance by dynamically selecting the modulation format in the BVT commensurate with the reach of that particular channel.

12.5.4 OPTICAL PACKET SWITCHING

Optical circuit switching including ROADM discussed above have the advantage of handling channels of high data rates in the optical domain with low energy consumption. However, the bandwidth granularity is limited by the spectral selectivity of optical filters, which is usually tens of GHz, and the switching speed is also relatively slow. On the other hand, electronic packet switching is much more flexible and faster in terms of network resources sharing. But this requires transceivers for optical to electric (O/E) and electric to optical (E/O) conversions, as well as electronic packet sorting and processing. Optical packet switching is a mechanism to switch and route data packets in the optical domain, to improve the network flexibility and minimize traffic congestion, but without the need of O/E and E/O.

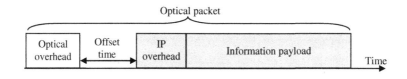

FIG. 12.5.9

Structure of an optical packet.

Fig. 12.5.9 shows the basic structure of an optical packet, which includes an optical overhead, a standard IP overhead, and an information payload. The optical overhead provides the control information that is used in the intermediate optical nodes for routing decision, and the standard IP overhead contains the identity information of the payload in an IP network as previously discussed.

In an optical network using optical packet switching, long and continuous message from each host has to be divided into short sections and put into packets. Optical path is not predetermined, and thus it is usually referred to as *connectionless*. Routing decisions are made by intermediate nodes, known as optical packet routers, based on a routing table which is frequently updated based on network traffic situation. A set of routing algorithms has to be agreed upon by all the nodes for their coordination to form a network. As shown in Fig. 12.5.10, if a packet needs to be transmitted from host *A* to host *D*, node 1 first reads the destination information in the optical overhead and routes the packet to node 4 according to the routing table of node 1. Then according to its own routing table, node 4 routes the packet to node 3, and finally node 3 routes the packet to node 7, and then to its destination of host *D*.

In an optical packet switched network, routers are shared by a large number of users. In order to avoid congestions packets can go through alternative routes through the optimization of routing tables. Optical packet buffers also help reducing congestion.

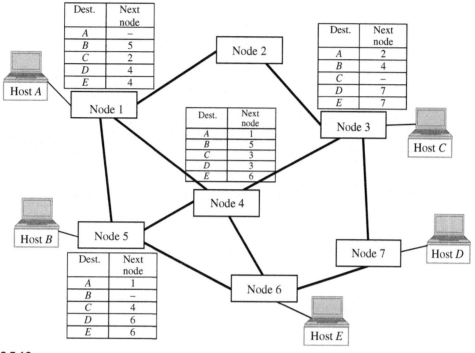

FIG. 12.5.10

Illustration of optical packet routing through intermediate nodes and routing tables (only show routing tables for nodes 1, 3, 4, 5, and 7).

In an optical packet, although data rate carried by the payload can be very high, optical overhead often uses relatively low data rate which can be easily converted into electric domain in each switching node for routing decision. Information carried by the optical overhead also needs to be replaced or updated at each node, known as label swapping. Many different techniques for optical overhead extracting and replacing have been demonstrated. For illustration purpose, Fig. 12.5.11 shows an example in which optical overhead is combined with the payload as a subcarrier tone which "labels" the optical packet. When arriving at a switching node, a narrow band FBG extracts the subcarrier tone and redirects it to an optical receiver through an optical circulator. A low-speed direct detection receiver is usually sufficient for overhead detection. Meanwhile the optical signal carried by the payload passes through the FBG but with the overhead subcarrier removed. The overhead information is processed by an electronic circuit and used for routing decision. The electronic circuit also creates a new overhead to be added back to the packet which will be read by the next node. Because the electronic circuit has processing delay for the header, the payload also has to be delayed by an optical fiber delay line before combining with the new overhead.

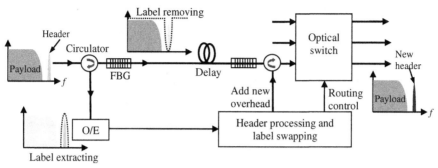

FIG. 12.5.11

Configuration of an optical packet switching node with optical overhead extracting and label swapping.

In practice, the optical overhead does not have to be at the same wavelength as the payload. In fact, if a different wavelength is used for the overhead, extracting, and reinserting of overhead in a switching node can be easier.

Note that optical packet switching requires much higher switching speed compared to optical circuit switching. As an example, for an optical packet carrying 1500 data bytes, the payload length is approximately 150 ns for 10-Gb/s data rate and 15 ns for 100 Gb/s data rate. Thus, fast switching time on the order of nanosecond or less is required for optical packet switching. Packet header processing in optical domain has also been demonstrated to further reduce the processing delay (Calabretta et al., 2010).

12.6 ELECTRONIC CIRCUIT SWITCHING BASED ON DIGITAL SUBCARRIER MULTIPLEXING

Transparent optical networks based on optical domain cross-connection (OXC) and switching have the advantage of reduced complexity, signal transparency, and low energy consumption in comparison to packet-based switching techniques. However, data rate granularity of OXC and routing, including elastic optic networks based on flexible grid, tends to be coarse due to the limited spectral selectivity of optical filtering, which may not fit the requirements of certain applications. Thus, electronic multiplexing of the optical signal has to be added to achieve finer bandwidth granularities and high resources sharing efficiency among users of diverse applications. Electronic domain digital cross-connection (DXC) typically uses fixed bandwidth TDM. For example an optical transport network (OTN) standard such as SONET and SDH can groom together multiple sub-wavelength tributaries, and provides edge routers with links of desired rates to support various client port types. DXC routing based on byte interleaved TDM requires bit synchronization, buffering, grooming, and forwarding engine. The complexity of these operations usually leads to high levels of electrical power consumption.

FDM, refers to sub-wavelength channels aggregated in frequency domain, can be used in a circuit-switched network to provide a dedicated spectral bandwidth between each pair of end users (clients ports). In order to maximize network efficiency, the specific bandwidth allocated to each pair of users has to be commensurate with the actual user demand which may vary over time. Thus fine enough bandwidth granularity and dynamic bandwidth allocation are especially important. In the pre-SONET/SDH/OTN era, FDM technology was widely used to aggregate multiple transport channels into the same transmission medium. End-to-end circuits were routed and switched across the network by reserving the required RF bandwidth of each channel. These FDM transport solutions were mostly implemented using analog technologies based on RF filters, oscillators, amplifiers, and mixers for channel selection, up- and down-conversions. Traditional FDM based on analog RF techniques such CATV has very low tolerance to the accumulated signal distortion and crosstalk, and has stringent requirement on the carrier-to-noise ratio (CNR). Significant spectral guard band is usually required between adjacent frequency channels of an analog FDM system so that these channels can be separated by RF filters at the receiver. This limits the bandwidth efficiency of analog FDM. In addition, analog RF devices are not particularly suitable for dynamic provisioning of parameters such as subcarrier channel frequency allocation and bandwidth.

Digital subcarrier multiplexing (DSCM) (Zhang et al., 2011) is a FDM technique which produces multiple digitally created subcarriers on each wavelength. DSCM, which includes OFDM and Nyquist FDM, has been widely used thanks to the rapid technological advances in high-speed CMOS digital electronics and efficient DSP algorithms as have been discussed in Chapter 11. This circuit-based approach combined with elastic optic networking has the potential to provide high bandwidth efficiency, sub-wavelength level flexible data-rate granularity, and electronic compensation of transmission impairments. DSCM can also be used to build a circuit-switched cross-connect (DSXC) in optical networks to interconnect users with dedicated sub-wavelength circuits, end-to-end across the transport network, with dynamic channel bandwidth allocation and signal power provisioning. A particularly promising application of DSCM can be found in the radio-over-fiber (RoF) wireless fronthaul. In this scenario wireless baseband-unit (BBU) and remote-radio-heads (RRH) are the end users of the DSCM fronthaul network. Each baseband signal can be transmitted using a dedicated subcarrier channel, whose bandwidth and power level can be customized to best fit the radio channel requirements.

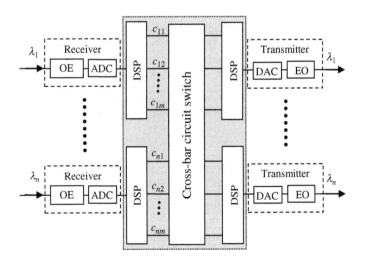

FIG. 12.6.1

Block diagram of a digital cross-connect switching node.

The basic block diagram of a DSXC is shown in Fig. 12.6.1. In this example, m subcarrier channels are carried by each wavelength in a system of n wavelength channels for WDM. Each digital subcarrier channel c_{ij}, ($i = 1, 2, ..., n$, and $j = 1, 2, ..., m$) can be routed to any output wavelength and subcarrier through a cross-bar circuit-switching architecture. Each receiver on the left side of Fig. 12.6.1 performs optical-to-electrical (OE) conversion, and each receiver on the right side of Fig. 12.6.1 performs electrical-to-optical (EO) conversion. These OE and EO conversions can be either intensity modulation and direct detection (IMDD), or complex optical field modulation and coherent detection. An analog-to-digital converter (ADC) is used at the receiver to convert the received signal waveform into digital signal, which is then processed by a DSP module. Similarly, a digital-to-analog converter (DAC) is used at the transmitter which converts the processed digital signal into an analog waveform.

Through DSP it is possible to achieve precise frequency and phase control of the digitally generated subcarriers. For example, DSP algorithms can enable the application of orthogonal frequency division multiplexing (OFDM), and Nyquist-frequency division multiplexing (Nyquist-FDM) as have been discussed in Chapter 11. Both of these digital multiplexing formats are known to provide high spectral efficiency, as well as the ability to digitally compensate a number of transmission impairments in fiber-optic systems such as chromatic dispersion and polarization mode dispersion (PMD), thus assuring robust signal quality. OFDM relies on the phase orthogonality between adjacent subcarrier channels so that crosstalk is minimized even though there is no spectral guard band between them. OFDM multiplexing and de-multiplexing can be implemented by utilizing highly efficient FFT algorithm in DSP. Nyquist-FDM, on the other hand, uses high order digital filters to tightly confine the spectrum of each subcarrier channel. Ideally, as the spectrum of each subcarrier channel approaches infinitely sharp edges, crosstalk between adjacent subcarrier channels is circumvented even in the absence of spectral guard bands. This is equivalent to reshaping signal bits into Nyquist pulses in the time domain so that the spectral shape tends to a square. While OFDM needs to maintain a precise phase relation between adjacent subcarrier channels to minimize the crosstalk, subcarrier channels in Nyquist-FDM are spectrally isolated so that there is no need for phase synchronization between them.

Using the DSXC architecture shown in Fig. 12.6.1, a network operator can stitch a dedicated subcarrier at the input to a dedicated subcarrier at the output, forming a pass-through circuit. At the DSXC there is no need to provide any form of forwarding engine once the subcarriers are stitched together (the subcarrier circuit switch is set up across the DSXC). As a circuit-based switching mechanism, DSXC can work with optical connect (OXC) as shown in Fig. 12.6.2 to provide an additional level of data rate granularity for selected wavelength channels, and to enable end-to-end circuits with guaranteed bandwidths between the user pairs it supports.

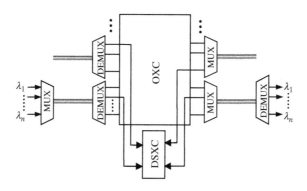

FIG. 12.6.2

Combination of optical cross connect (OXC) and digital subcarrier cross-connect (DSXC).

The circuit-based DSXC architecture is particularly suitable for high-speed and latency-sensitive applications such as video streaming, cloud robotics, and 5G wireless fronthaul, to name a few. The flexibility of DSP in the DSXC is expected to offer any-to-any dynamic routing, as well as channel combining and splitting operation with minimum processing latency.

Use 5G mobile wireless system as an example which offers unprecedented services and connectivity among a diverse pool of devices and applications. The broadband nature of 5G networks requires significant technological advances, both in the wireless infrastructure and the transport network that supports it. A promising radio access network (RAN) architecture leverages the Cloud resources (C-RAN) in order to achieve a simplified design of the remote radio head (RRH) in the cell tower (Pizzinat et al., 2014; Larsson et al., 2014) as shown in Fig. 12.6.3A. The term "fronthaul" is used to describe the network which provides connectivity between the RRH and the BBU, the latter being implemented in the Cloud. The up-link of the fronthaul collects the waveforms received from the antenna arrays in the RRH and sends them transparently (i.e., without decoding) to a centralized BBU, and the downlink, follows the reverse path.

As radio signals received and transmitted through antenna arrays are analog in nature, analog RoF is the simplest format of transmission between RRH and BBU, which keeps the analog waveforms largely untackled. However, analog RoF is known to be very sensitive to waveform distortion which may be introduced from linear and nonlinear impairments and inter-channel crosstalk in the transmission. In addition to the requirement of high carrier-to-noise ratio (CNR), the performance of analog RoF is also limited by the relative intensity noise (RIN) when the signal power is high. For broadband applications such as in 5G wireless networks, the required quality of transmission of analog RoF over fronthaul is quite challenging to guarantee.

Digital transmission over fronthaul, on the other hand, digitizes the received analog wireless waveforms and encodes them into digital bits for transmission. In this analog to digital conversion process, the data rate and thus the bandwidth required for transmission over the fronthaul is scaled roughly by b times higher than the original analog signal bandwidth where b is the bit resolution of the ADC. For example, for eight channels of 20 MHz LTE signals using 40 MS/s ADC sampling rate at 15-bit resolution for each I- and Q-component of the complex RF waveform, the digital data rate will be approximately $8 \times 40 \times 15 \times 2 = 12,000$ Mb/s (Liu et al., 2016). More detailed data rate estimation also needs to take into account control words [(CWs) and line coding]. Further considering multiple antenna elements for MIMO beam forming, the required data rate can easily reach to 100 Gb/s.

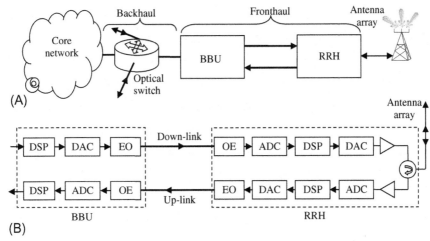

FIG. 12.6.3

(A) wireless fronthaul and (B) block diagram of DSCM-based fronthaul. BBU: Baseband Unit, and RRH: Remote Radio Head.

DSCM combines the advantages of analog RoF and digital processing by converting each analog wireless channel into a digital subcarrier occupying the same bandwidth in the optical spectrum. Fig. 12.6.3B shows the configuration of DSCM-based fronthaul. For the upstream-link, the analog waveform received from the antenna is first digitized and processed by the DSP and then converted back to the analog domain. The DSP is used for subcarrier channel aggregation, compensation of waveform distortions, and interference mitigation after radio wave propagation and detection by the antenna array. Then this analog signal is converted into optical domain through an optical transmitter (EO). At BBU, the received optical signal is converted into electrical domain by an optical receiver (OE), the waveform is digitized by an ADC and fed into the DSP for processing. This DSP can be used to compensate for transmission impairments of the fronthaul fiber system. Similar process can be described also for the downstream link but following the reverse path. In this system, although high-speed data is digitally processed inside the BBU, the signal transmitted through the fronthaul remains in the analog format with the same spectral bandwidth as the original RF signal from the antenna. However, the major advantage of DSCM over analog RoF in fronthaul application is that advanced DSP algorithms can help correct waveform distortion, inter channel crosstalk, and other transmission impairments, and thus signal integrity can be assured.

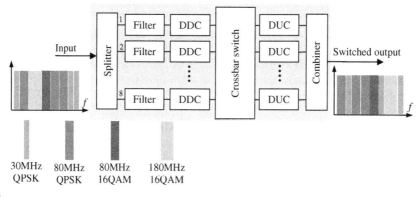

FIG. 12.6.4

DSP process of an 8 × 8 DSXC to cross-connect switch subcarrier channel.

Fig. 12.6.4 shows an example of an 8 × 8 DSXC implemented on a FPGA platform (Xu et al., 2018). Three different channel bandwidths (30, 80, and 180 MHz) are used with two different modulation formats (QPSK and 16QAM) as shown in the figure. Nyquist FDM channels are digitally created with sharp transition edges on both sides of the spectrum so that minimum spectral guard band is allowed between subcarrier channels. The input signal is first split into eight copies, and each subcarrier channel is selected by a band-pass filter and digitally down converted (DDC) to the baseband. After cross-bar switching, a group of digital up converters (DUC) up-shifts these channels to their target subcarrier frequencies. Through this process, any input subcarrier channel can be switched to any output subcarrier frequency.

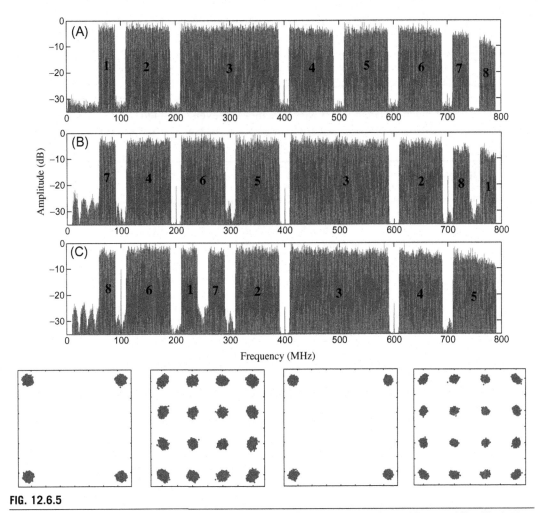

FIG. 12.6.5

Input (A) and output [(B) and (C)] spectra of an 8 × 8 DSXC indicating any-to-any switching capability. The bottom row shows the typical constellation diagrams of QPSK and 16QAM after cross-connect switching.

Fig. 12.6.5 shows the measured input and output spectra of the 8 × 8 DSXC with two different switching settings. The original subcarrier channels [1 2 3 4 5 6 7 8] shown in Fig. 12.6.5A have been switched to [7 4 6 5 3 2 8 1] and [8 6 1 7 2 3 4 5], respectively, as shown in Fig. 12.6.5B and C. Equiripple 108th order FIR filters are used in this example for DDC and DUC. The ripple in the filter passband is 0.5 dB and the stopband attenuation is 30 dB. Typical constellation diagrams of QPSK and 16QAM modulated channels after cross-connect switching are shown in the bottom row of Fig. 12.6.5 indicating no visible degradation of signal quality through the DSXC process. The spectral guard band reserved between adjacent subcarrier channels is set to 20 MHz primarily for the demonstration purpose. Narrower guard band of less than 5 MHz would not significantly increase the crosstalk as the spectral confinement of each channel is tight enough, which is determined by the order of digital filters.

As this DSXC is operated in the electric domain, the overall bandwidth is limited by the speed of ADC and DAC. It is also important to note that processing delay is primarily determined by the digital filters implemented in the DSP. Higher order finite-impulse-response (FIR) filters with large number of taps provide better spectral selectivity, but would introduce longer processing delays. Thus, a tradeoff has to be made between spectral selectivity and processing delay of FIR filters. Nevertheless, as the delay of each tap is determined by the clock period which is on the order of several hundred Megahertz, processing delay of much less than 1 µs is easy to achievable even with high order (>100) FIR filters.

12.7 SUMMARY

In this chapter, we have discussed and explained basic concepts of optical networking. An optical network can be divided into several different layers based on their functionalities. While lower layers provide physical transmission medium such as optical fibers and optical transceivers, higher layers are more application oriented providing connection and routing control, scheduling and dada format interpretation, quality-of-service control, and network management. Note that the definition of each network layer may change based on the specific network architecture. Boundaries between layers may also become less clear as cross-layer design and optimization become more common in the network design (Subramaniam et al., 2013), which helps simply the overall network architecture. A number of physical layer network topologies for connections between optical nodes have been discussed, which include point-to-point, bus, star, ring, mesh, and tree. The selection of physical layer network topologies in the network design depends on many factors, such as the number of users and the geological locations of these users.

SONET and SDH are the two most important data multiplexing standards for high-speed transmission which have been used for several decades. Both SONET and SDH are *synchronous* multiplexing structures which allows the rates of SONET/SDH to be defined as integer multiples of a base rate. It allows multiple low-speed data streams to be multiplexed into a higher speed data stream in a straightforward way. The overhead bits embedded SONET/SDH frames provide necessary information for connection control, frame interpretation, and network management. In comparison to SONET/SDH, IP is a packet-switching protocol in which information is delivered in packets with variable packet lengths. Because of its flexibility, IP has become the more preferable choice for dynamic optical networks.

Depending on the geographic coverage, the number of users, the capacity requirements, and functionalities, optical networks can be divided into several categories, such as WAN, MAN, LAN, access network, and DCN. Each network category may have its priority in terms of requirements and concerns. For example, a WAN has to provide high transmission capacity over long distances, whereas a local-area network may have more emphasis in the flexibility and the cost. Although multi-terabit transmission capacity can be realized in long haul optical systems, aggregation and access can often be bottlenecks for network users. PON discussed in Section 12.3 is one of the most popular architectures for FTTH or FTTC.

Optical interconnection usually refers to short distance optical connection and often with multiple parallel channels. AOC is a good example of optical interconnection which can provide 100 Gb/s connection capacity for short distance but at very low cost. Optical interconnections between printed electronic circuit boards and between IC chips on the same board are very promising for high-speed electronic circuits to increase the connection bandwidth and reduce the energy consumption. Optical interconnection can be benefited from photonic integration with VCSEL and photodiode arrays.

DCNs are a special category of optical network to address the need of highly dynamic data traffic between large numbers of computer servers. The unprecedented amount of data that needs to be communicated among computers and servers within each datacenter and between different datacenters imposes unique requirements and unprecedented challenges to data communication networks of all scales. Several DCN architectures have been discussed in Section 12.4 including traditional three layer tree architecture, optical switching architecture, and an optical architecture including both inter-rack and intra-rack interconnections. As optical switches are usually slow, a hybrid architecture is introduced which combines electronic packet switching and optical circuit switching. A mega datacenter architecture is also discussed, which interconnects a large number of identical self-contained shipping containers known as PODs. Each POD encloses a group of computer servers, a self-consistent network interconnecting these servers, and a cooling/fan system to support its operation. This mega datacenter architecture is modular and easy to expand.

Switching and cross connection are the most important functionalities in dynamic optical networks. Optical circuit switching is transparent to signal modulation format and insensitive to spectral bandwidth, but the switching speed is relatively slow. Thus, optical circuit switching is most efficient for the routing and cross-connection of broadband optical signals. Both ROADM and cross-bar optical switching can be space-based switching or wavelength-based switching, or their combination. Traffic congestion and blocking is the most important issue in optical network design, especially with circuit switched optical networks. Wavelength conversion, although costly, is shown to be effective in reducing blocking probability. Elastic optical networks with flexible wavelengths and spectral bandwidths is also discussed which can help improve the spectral efficiency. Optical packet switching, on the other hand, can be much more flexible and dynamic for traffic routing. But an optical packet switching has to be much faster, and a switching node with optical overhead extracting and label swapping is usually more complex than that of circuit switching.

A digital subcarrier cross connect (DSXC) switch architecture is discussed in Section 12.6. As an alternative to electronic packet switching, DSXC is based on circuit switching, it is able to provide fine data rate granularity for efficient network resource sharing among users while maintaining high spectral efficiency.

PROBLEMS

1. Consider a lightwave broadcasting system operating in 1320 nm wavelength using standard single-mode fibers with negligible chromatic dispersion. There are 10 fiber spans as shown in the following figure, and each span has 4 km fiber with attenuation coefficient $\alpha_{dB}=0.5$ dB/km. There is also an optical splitter at the end of each fiber span (except the last span) with 3-dB power-splitting ratio ($\alpha=3$ dB). The receiver electric bandwidth is $B_e=1.5$ GHz. The optical power launched from the transmitter is 10 dBm. The PIN photo diode used in the receiver has a responsivity of 0.75 A/W and negligible dark current. The load resistor is 50 Ω.

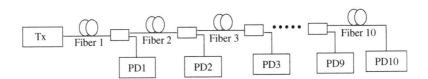

(a) Find the signal electrical currents generated at the output of the 1st photodiode (PD1) and the last photodiode (PD10).

(b) Consider only thermal noise and shot noise, calculate the electrical SNR at the output of PD1 and PD10, respectively. Discuss which photodiode is limited by thermal noise and which photodiode is limited by shot noise

(c) Obviously in this system, the 3-dB splitting ratio of each fiber coupler is not optimum and the last receiver always receives the lowest optical power. Find the optimum splitting ratio α (assume all the fiber couplers have the same coupling ratio) such that the 1st and the 10th photodiodes receive the same signal optical power. (at lease write an equation which leads to the finding of the optimum α value)

2. The following unidirectional bus network has N nodes. All 2×2 directional couplers are identical and each has an excess loss $\beta = 0.5$ dB. The total fiber length in the bus is $l_{fiber} = 40$ km with the attenuation coefficient $\alpha_{dB} = 0.3$ dB/km (neglecting connector and splice losses).

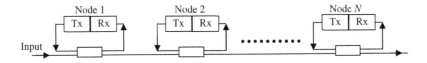

(a) Suppose the power-splitting ratio of each coupler is α, what is the total loss between the transmitter of node 1 and the receiver of node N? (Write the expression in terms of α and N).

(b) Find the *optimum* power-splitting ratio α of the fiber coupler (suppose all couplers are identical).

(c) Suppose that each receiver has a sensitivity of 100 nW. What is the minimum optical power of each transmitter to support the network with 12 nodes with zero margin.

3. There are six users located in a square grid separated by 2 km between the nearest neighbors, as shown in the following figure. Each user is equipped with an optical transceiver, and assume the fiber loss is 2.5 dB/km (multi-mode fiber at 850 nm). To design a network which provides bidirectional interconnection (any-to-any) among these users, three configurations are considered, including folded bus (shown in Fig. 12.1.4), unidirectional fiber ring (shown in Fig. 12.1.6), and star (shown in Fig. 12.1.8A), and with a 6×6 star coupler with equal splitting ratio 1/6 and co-located with user 3).

(a) What is the total fiber length required for each network configuration

(b) With the optimum coupling coefficient of optical couplers, which configuration has the lowest loss (worst case)?

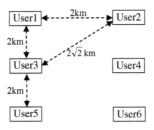

4. Repeat problem 3, but now the 6 users are located along a straight line as shown in the following figure (assume the 6×6 star coupler is co-located with user 3 equal splitting ratio of 1/6)

5. For the following 16×16 shuffle-net with five layers of optical couplers, what is the power-splitting loss?

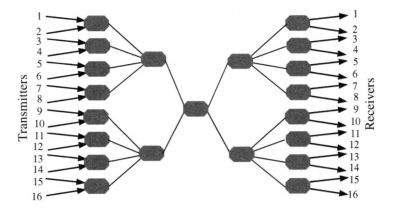

6. Consider an 8×8 optical star network using configuration similar to Fig. 12.1.11B. However, each wavelength-tunable transmitter only has four wavelengths to choose from. If you are given four DEMUX, four MUX, and many 1×2 (and 2×1) optical couplers
   (a) Draw the star network configuration
   (b) Discuss possible disadvantages compared to having eight wavelengths to choose.

7. In SONET frame structure (a) what is the percentage of signal data in each STS-1 frame? and what is the percentage of signal data in each STS-N frame ($N=3$, 12, 48, and 192)?

8. Consider shared protection scheme shown in Figs. 12.2.2B, not only fiber protection is shared, protection of transmitters and receivers is also shared. Assume there are N pairs of working fibers and one pair of protection fiber, N transmitters and N receivers. If the failing probabilities for each fiber, transmitter, and receiver are η, ξ, and ε, what is the failing probability of the network?

9. Two unidirectional path-switched rings (UPSRs) are connected through a shared node. Each node has two transceivers, except for node 3 which has four transceivers. Solid lines are working paths and dashed lines are protection paths.

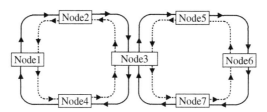

(a) If both fibers are cut between node 2 and node 3, draw signal path from node 2 to node 5.

(b) If both fibers are cut between node 2 and node 3, and between node 3 and node 5, draw signal path from node 2 to node 5.

10. Consider TDM PON shown in Fig. 12.3.2 in which 1490 nm wavelength is used for downstream (from OLT to ONU) and 1310 nm wavelength is used for upstream (from ONU to OLT). Fiber loss is 0.3 dB/km for 1490 nm and 0.5 dB/km for 1310 nm. Other parameters are: fiber length between OLU to star coupler is 10 km, maximum length from star coupler to ONU is 5 km. Only consider intrinsic splitting loss of $1 \times N$ star coupler.

(1) What are the maximum optical losses for $N=16$, 32, and 64, respectively at these two wavelengths?

(2) For a direct detection receiver limited by thermal noise, the receiver sensitivity is related to the electric bandwidth. Assume transmitters in the OLT and ONU have the same optical power, what is the difference of maximum allowable transmission bandwidth between the downstream and upstream channels.

11. Consider a gigabit PON (GPON) with a $1 \times N$ star coupler connecting OLT with N ONUs. Data rates are 2.5 Gb/s for downstream and 1.25 Gb/s for upstream.

(a) Based on the TDM-PON configuration shown in Fig. 12.3.2, what are the maximum and the average data rates each ONU can use for downloading and uploading?

(b) If WDM-PON is used with wavelength-independent $1 \times N$ star coupler (Fig. 12.3.5), and five wavelengths are used in each ONU, what are the maximum and the average data rates each ONU can use for downloading and uploading?

(c) If WDM-PON is used with a *N*-channel WDM coupler as shown in Figs. 12.3.6 and 12.3.7, and each ONU has a dedicated wavelength, what are the maximum and the average data rates each ONU can use for downloading and uploading?

12. Consider a datacenter network (DCN) interconnecting 20 ToRs based on the architecture shown in Fig. 12.4.7, in which an $M \times M$ cross-bar optical switch is used, and at any moment of time each input port can be connected to any one of the output ports. Assume 10 wavelength channels are used for each ToR, but each WSS has 4 output ports.
    (a) What is the minimum value of M for the $M \times M$ cross-bar optical switch?
    (b) Can this DCN guarantee any-to-any interconnection between ToRs, and why?
    (c) For a given ToR, what is the maximum number of ToRs it can simultaneously connect to?

13. Design a RODAM similar to the configuration shown in Fig. 12.5.2, but the WDM MUX and DEMUX are each replaced by a WSS with the functionality shown in Fig. 12.5.4A, and a WSS can also be used in the reverse direction. Assume there are 20 wavelength channels and the ROADM allows a maximum of 4 channels to be add/dropped.
    (a) What is the minimum number of optical ports each WSS must have in this case?
    (b) Sketch the ROADM configuration, and explain how it works.

14. For the crossbar optical switch architecture shown in Fig. 12.5.5 with m input fiber, m output fiber and, each fiber carries n wavelength channels, explain, without the wavelength conversion capability, why the degree of freedom is only $n \times (m \times m)$, instead of $(m \times n) \times (m \times n)$.

15. Consider a WDM optical network as shown in Fig. 12.5.7A with the capability of carrying a maximum of 40 wavelength channels in each hop, and there are 10 hops. In order for the blocking probability to be less than 10%, what is the maximum number of wavelength that can be actually used in the following two cases?
    (a) Without wavelength conversion
    (b) With wavelength conversion

16. Consider a WDM optical network again as shown in Fig. 12.5.7A without the capability of wavelength conversion. The network has 10 hops, 9 of them can carry a maximum of 40 wavelength channels, but one of them can only carry 30 wavelength channels. If this network utilizes 15 wavelength channels, what will be the blocking probability?

17. Same as problem 16, but now 27 wavelengths are used and all switch nodes have wavelength conversion capability. Find the blocking probability?

18. Consider a multi-wavelength optical network in a ring configuration as shown in the following figure with five nodes, and five fiber links connecting them. Assume each fiber link is bidirectional and three wavelengths λ_1, λ_2, and λ_3, can be used in each direction. At the time a request arrives to setup a lightpath, the network is in the following status:

λ used by others	In links
λ_1	F_0, F_2, F_4
λ_2	F_1, F_3, F_2
λ_3	F_0, F_2, F_4

(a) What is the set of available wavelengths on each link?

(b) For a request of setting up a lightpath from node 1 to node 4 at this moment, which path and wavelength you need to use (there is no wavelength conversion)?

19. Consider an optical network considering of multiple channels with 10, 12.5, 32, 130, and 330 GHz spectral bandwidths for 10, 40 Gb/s (QPSK+dual polarization), 100, 400 Gb/s, and 1 Tb/s, respectively. Assume the network has a total of 50 channels: 20×10 Gb/s, 10×40 Gb/s, 10×100 Gb/s, 8×400 Gb/s, and 2×1 Tb/s.

(a) Consider these channels are arranged on a regular 50 GHz WDM grid, what is the efficiency of spectral utilization?

(b) If flexible grid is used and reserving 2-GHz guard band between adjacent channels, what is the efficiency of spectral utilization?

20. For the optical packet switched network shown in Fig. 12.5.10, complete routing tables for nodes 2 and 6 (which may not be unique).

21. In comparison between optical packet switching and optical circuit switching list their advantages and weaknesses.

REFERENCES

T. Barnett, S. Jain, U. Andra, and T. Khurana, 2016. Cisco Global Cloud Index 2015–2020, Online Presentation: https://www.cisco.com/c/dam/m/en_us/service-provider/ciscoknowledgenetwork/files/622_11_15-16-Cisco_GCI_CKN_2015-2020_AMER_EMEAR_NOV2016.pdf.

Barry, R.A., Humblet, P.A., 1996. Models of blocking probability in all-optical networks with and without wavelength changers. IEEE J. Sel. Areas Commun. 14 (5), 8568–8867.

Cai, D.K., Neyer, A., Kuckuk, R., Heise, H.M., 2008. Optical absorption in transparent PDMS materials applied for multimode waveguides fabrication. Opt. Mater. 30, 1157–1161.

Calabretta, N., Jung, H.-D., Tangdiongga, E., Dorren, H., 2010. All-optical packet switching and label rewriting for data packets beyond 160 Gb/s. IEEE Photon. J. 2 (2), 113–129.

Chen, K., Singla, A., Singh, A., Ramachandran, K., Xu, L., Zhang, Y., Wen, X., Chen, Y., 2014. OSA: an optical switching architecture for data center networks with unprecedented flexibility. IEEE/ACM Trans. Networking 22 (2), 498–511.

Farrington, N., et al., 2010. Helios: A Hybrid Electrical/Optical Switch Architecture for Modular Data Centers. ACM SIGCOMM.

M. Fiorani, S. Aleksic, M. Casoni, L. Wosinska, and J. Chen, "Energy-efficient elastic optical interconnect architecture for data centers," IEEE Commun. Lett., VOL. 18, NO. 9, 2014, pp 1531-1534.

Hluchyj, M.G., Karol, M.J., 1991. ShuffleNet: an application of generalized perfect shuffles to multihop lightwave networks. J. Lightwave Technol. 9 (10), 1386–1397.

Huitema, C., 1999. Routing in the internet, second ed. Prentice Hall, Upper Saddle River, NJ.

Immonen, M.P., et al., 2007. Investigation of environmental reliability of optical polymer waveguides embedded on printed circuit boards. Microelectron. Reliab. 47, 363–371.

Kim, J.T., Ju, J.J., Park, S., Kim, M.-s., Park, S.K., Lee, M.-H., 2008. Chip-to-chip optical interconnect using gold long-range surface plasmon polariton waveguides. Opt. Express 16 (17), 13133–13138.

Larsson, E., Edfors, O., Tufvesson, F., Marzetta, T., 2014. Massive MIMO for next generation wireless systems. IEEE Commun. Mag. 52, 186–195.

Liu, X., Zeng, H., Chand, N., Effenberger, F., 2016. Efficient mobile fronthaul via DSP-based channel aggregation. J. Lightwave Technol. 34, 1556–1564.

Ma, H., et al., 2002. Polymer-based optical waveguides: materials, processing, and devices. Adv. Mater. 14, 1339–1365.

Pizzinat, A., Chanclou, P., Diallo, T., Saliou, F., 2014. Things you should know about fronthaul. In: Proceedings of the European Conference on Optical Communications, Cannes, France. Paper Tu.4.2.1.

Ramaswami, R., Sivarajan, K.N., 2002. Optical Networks: A practical Perspective, second ed. Morgan Kaufmann Publishers.

Shen, P.-K., Chen, C.-T., Chang, C.-H., Chiu, C.-Y., Li, S.-L., Chang, C.-C., Wu, M.-L., 2014. Implementation of chip-level optical interconnect with laser and photodetector using SOI-based 3-D guided-wave path. IEEE Photon. J. 6 (6). paper 2500310.

Stallings, W., 1996. IPv6: the new Internet protocol. IEEE Commun. Mag. 34 (7), 96–108.

Subramaniam, S., Brandt-Pearce, M., Demeester, P., Vijaya Saradhi, C. (Eds.), 2013. Cross-Layer Design in Optical Networks. Springer.

Wu, F., Logeeswaran, V.J., Islam, M.S., Horsley, D.A., Walmsley, R.G., Mathai, S., Houng, D., Michael, R.T.T., Wang, S.-Y., 2009. Integrated receiver architectures for board-to-board free-space optical interconnects. Appl. Phys. A Mater. Sci. Process. 95, 1079–1088.

Xu, T., Fumagalli, A., Hui, R., 2018. Real-time DSP-enabled digital subcarrier cross-connect based on resampling filters. J. Opt. Commun. Netw. 10 (12), 937–946.

Zhang, Y., O'Sullivan, M., Hui, R., 2011. Digital subcarrier multiplexing for flexible spectral allocation in optical transport network. Opt. Express 19 (22), 21880–21889.

FURTHER READING

Finisar optical cable: https://www.finisar.com/active-optical-cables.

ITU: https://www.itu.int/rec/T-REC-G.983.2-200507-I/en.

Index

Note: Page numbers followed by *f* indicate figures, *t* indicate tables, and *b* indicate boxes.

A

Access networks, 574, 579
Acoustic phonons, 56
Acousto-optic filters (AOFs), 246, 246*f*
    application, 248
    Bragg angle, 247
    colinear configuration, 248–249, 248*f*
    frequency modulation, 247–248
    frequency shift, 249
    spectral resolution, 247
Acousto-optic (AO) switch, 278, 278*f*
Active optical cable (AOC), 586, 586*f*, 612–613
Active quenching, 144, 145*f*
Active region, 83
Add/drop multiplexer (ADM), 568
Adiabatic chirp, 98
All-optical signal processing, 172–176
Ampere's law, 261
Amplified spontaneous emission (ASE), 160, 486
Amplified spontaneous emission (ASE) noise, 9–11
Analog modulation
    analog subcarrier multiplexing and optical single-sideband modulation
        analog lightwave system, 473
        double-sideband optical spectra, 478
        E/O conversion process, 474
    carrier-to-signal ratio, inter-modulation distortion and clipping
        clipping, 478–480
        CNR *vs.* SNR, 478
    radio-over-fiber technology, 483–486
    relative intensity noise, 481–483
Analog-to-digital converter (ADC), 498–499, 532, 608
Angled physical contact (APC) connectors, 42*b*, 43
Angle polished fiber surface, 42–43, 42*f*
Antireflection coating, 109–110
Arrayed waveguide grating (AWG), 210, 239, 294
    free-spectral range, 244–245
    star coupler configuration, 244–245, 244*f*
    structure, 243, 243*f*
    transfer function, 245, 245*f*
    uses, 245
ASE-ASE beat noise, 161, 201
    in electrical domain, 163
    and power spectral density, 162*f*, 163
    spontaneous-spontaneous generation, 163, 163*f*

unit of, 163
Asynchronous multiplexing, 567
Asynchronous transfer mode (ATM)-PON, 583
Attenuation, 44–48
Attenuation coefficient, 46
Automatic gain control (AGC), 191–192
Automatic power control (APC), 191–193
Avalanche effect, 142
Avalanche photodiode (APD), 125–126, 150, 153
    applications
        as linear detector, 142–143
        single-photon detector, 144
    carrier multiplication process, 142, 142*f*
    junction structure, 141–142, 141*f*
Avalanche region, 141–142

B

Backward pumping, 190–191, 190*f*
Band-pass filter (BPF), 178, 485
Bandwidth variable transceivers (BVTs), 603, 603*f*
Bench-top external cavity tunable laser, 118*f*, 119
Benzocyclobuthene (BCB), 111*f*, 112
Bessel equation, 35
Bessel functions, 35, 36*f*
Biasing circuit
    laser diodes (LDs), 117, 117*f*
    quenching, 144, 145*f*
Bidirectional double ring, 565–566, 565*f*
Bidirectional lineswitched rings (BLSRs), 578, 578*f*
Bidirectional pumping, 191, 191*f*
Birefringence, 54
Birefringence beat length, 54–55
Bit-error rate (BER), 8
    definiton, 338
    vs. Q-values
        intensity modulation and direct detection (IMDD) mode, 339–342
        receiver BER, 342–347
BLSRs. *See* Bidirectional lineswitched rings (BLSRs)
Bowtie fiber, 66–67, 67*f*
Bragg grating, 247
Bragg wavelength, 105–106, 234–235, 294
Brewster angle, 29–30
Broadband-PON, 583
Buried-tunnel-junction (BTJ) VCSEL, 112

C

Cable TV (CATV), 579
Campus network, 574
Carrier confinement, 83–84
Carrier drift, 79–80
Carrier fading, 74–75
Carrier lifetime, 86–87
Carrier-suppressed return-to-zero (CSRZ), 456–460
Carrier-to-noise ratio (CNR), 473, 478, 606–607
Carrier transient, 134
Causs's Law, 127
C-band, 14
Characteristic equations of modes, 262–263
Charge-coupled device (CCD), 148, 151
    parameters, 149
    row and pixel shift, 148–149, 149f
    structure, 148–149, 148f
    uses, 149
Chirp parameter, 329–330
    electro-absorption modulator, 328, 329f
    electro-optic (EO) modulators, 309–310
Chromatic dispersion, 20–21, 73
    definition, 48
    digital signal processing (DSP), 498
    eye closure penalty, 361–365
    of non-zero dispersion-shifted fibers, 64–65, 65f
    orthogonal frequency-division multiplexing, 516–517
    sources, 51–53
Circuit-based DSXC architecture, 609
Circuit-switched network, 556–557
Circular polarization, 23
Circulators. *See* Optical circulators
Cisco Global Cloud Index, 589
Cloud resources (C-RAN), 609
CMOS electronic circuits, 11
Coaxial cables, 4, 4f
Coherence time, 104
Coherent detection
    balanced coherent detection
        direct-detection components, 428
        direct-detection terms, 428
        and electric circuit, 428, 428f
        head-to-toe electric configuration, 429
        headto-toe photodetectors, 428
        polarization diversity, 429, 430f
    block diagram, in radio communications, 418–419, 419f
    carrier phase recovery, 507–511
    composite optical signal, 420
    erbium-doped fiber amplifier, 418
    heterodyne configuration, 421–422
    homodyne configuration, 422
    in lightwave systems, 419–420, 419f

    low-pass filter, 419
    optical frequency and phase, 418
    phase diversity and I/Q detection
        balanced detection coherent receiver, 431
        hybrid coupler, 434
        integrated-optic approach, 433
        optical phase noise, 430
        phase-locked feedback loop, 430, 431f
        planar lightwave circuits (PLCs), 433
        polarization-division multiplexing (PDM), 435
        x- and *y*-polarized optical signal components, 435
    photocurrent, 420–421
    signal-to-noise-ratio, optical receiver with
        and direct detection, 423, 424f
        EDFA preamplifier, 424–425, 424f
        Gaussian statistics, 424
        heterodyne detection scheme, 422–423
        homodyne detection, 422–423, 425
        long-distance optical transmission system, 426, 426f
        phase-encoded optical systems, 427
        phase-locked homodyne detection, 424–425
        pre-amplified direct detection, 425
        thermal noise and shot noise, 422–423
    square-law detection device, 419–420
    transmission impairments, 418
Coherent optical communication systems
    direct injection current modulation, 418
    erbium-doped fiber amplifier, 418
    in-phase/quadrature (I/Q) electro-optic modulator, 418
    local oscillator (LO), 418
    phase modulation and intensity modulation, 418
    wavelength-division multiplexing, 417–418
Coherent optical detection, 8, 11–12
Collinear configuration, 226
Communication, 556–557
Communication network, 556–557, 557f
Composite-second-order (CSO) distortion, 481
Composite-triple-beat (CTB), 481
Contrast, 228–230
Counter propagation, 249
Critical angle, 27
Crossbar optical switch, 289, 290f
Cross-phase modulation (XPM), 61, 70–71, 174
Cross talk, 274
Cutoff wavelength, 41, 69, 131–132
Cylindrical packaged diode laser, 117–118, 117f

D

Dark charge, 149
Dark current, 135–138
Dark current noise, 138
    in avalanche photodiode, 143

signal-to-noise ratio, 140
Datacenter networks (DCNs), 574, 588–595, 613, 617
    architecture, 590, 591*f*, 593–594
    based on optical interconnection, 591, 592*f*
    dynamic network load management, 590
    electric power budget, 590
    inter-rack and intra-rack interconnections, 593–594, 593*f*
    oversubscription, 594–595
    real-time power monitoring, 590
Datacenter traffic, 589, 589–590*f*
Data link layer, 559
Data link protocols, 559
Data stream, 559
DCNs. *See* Datacenter networks (DCNs)
Decibel units, 15
Dedicated protection scheme, 575–576, 576*f*
Degenerate FWM, 61, 61*f*, 176–177
Degenerate modes, 53, 54*f*
Degree of polarization (DOP), 23
De-multiplexers (DE-MUXS), 10, 239, 242, 242*f*, 566, 584
Depletion region, 79–80, 127–128
DFB laser
    grating, 106, 106*f*
    single-mode operation, 107, 107*f*
    structure, 105, 105*f*
Differential group delay (DGD), 7, 9–10, 13, 54–55
Differentiated service code point (DCSP), 571
Digital cross-connection (DXC), 606, 607*f*
Digital Light Processing (DLP), 286
Digitally down converted (DDC), 610–611
Digital Micromirror Device (DMD), 286, 286*f*
    design, 287, 288*f*
    diffraction angles, 286–287, 287*f*
    uses, 286
Digital modulation, 338, 439–440, 478
Digital signal processing (DSP), 11–13
    analog-to-digital convertor (ADC), 498–499
    chromatic dispersion, 498
    coherent detection, 498–499
    coherent optical receiver, 498–499
    digital subcarrier multiplexing (DSCM)
        with coherent detection, 544–548
        I/Q mixing, for subcarrier up-conversion and down-
            conversion, 528–536
        Kramers-Kronig algorithm, 539–544
        signal-signal beat interference, 536–539
    digital-to-analog convertor (DCA), 498–499
    electronic-domain compensation, transmission impairments
        carrier phase recovery, coherent detection receiver,
            507–511
        dispersion compensation, 500–505
        PDM coherent detection receiver, 499–500

PMD compensation and polarization de-multiplexing,
    505–507
random state-of polarization rotation and mode coupling,
    499–500
frequency-dependent differential group delay, 498–499
integrated photonic circuits, 499
Nyquist pulse modulation and frequency-division
    multiplexing
    digital-analog hybrid approach, 527, 527*f*
    Nyquist-FDM optical system, 526–527, 526*f*
    Nyquist filter amplitude transfer functions, 523, 524*f*
    sharp-edged spectrum, 524
    time-domain waveforms and frequency spectra of OFDM,
        523, 523*f*
orthogonal frequency-division multiplexing, 499
    band-pass filter (BPF), 519
    BER, receiver bandwidth functions, 521, 521*f*
    bit error rate (BER), 519–520
    chromatic dispersion, 516–517
    cyclic prefix, 516–517, 516*f*
    data sequence mapping, 520–521, 520*f*
    double-sideband (DSB) spectrum, 514–515
    electric power spectral density, 520–521
    Fourier transform, 518
    frequency down-conversion, DFT for, 515, 515*f*
    guard band in low frequency, 519, 519*f*
    IDFT for frequency up-conversion, 514, 514*f*
    intensity-modulated optical signal, 518, 518*f*
    intensity modulation and direct detection, 516–517*f*, 517
    with *N* subcarrier channels, 512–513, 513*f*
    Nyquist sampling theorem, 517–518
    optical intensity waveform, 518–519
    serial-to-parallel conversion process, 519–520
    spectral guard band, 512, 512*f*
    spectral overlap, subcarrier channels, 512, 512*f*
    spectra of the signal channel, 521–523, 522*f*
    time-domain data matrix, 515, 515*f*
    time-domain parallel sequences, 514, 514*f*
    time-domain waveform, 521–523, 522*f*
    transmission impairments, 512–513
radio frequency (RF) circuit, 498–499
system tolerance, 498
WDM and TDM, 498
Digital subcarrier cross-connect (DSXC), 608, 612
    architecture, 608, 613
Digital subcarrier multiplexing (DSCM), 607
    Nyquist pulse modulation, 523–527
    orthogonal frequency-division multiplexing, 512–523
Digital subscriber line (DSL), 579
Digital transmission, 446, 609
Digital up converters (DUC), 610
Direct injection-current modulation, 6–7

Directional coupler, 267–270, 269f
Direct modulation, 77–78, 299
Direct semiconductors, 81
Dispersion, 70. *See also* Chromatic dispersion
Dispersion compensating fibers (DCFs), 9, 63, 66, 341
Dispersion compensating modules (DCMs), 9, 66
Dispersion compensation (DC), 341, 497–498
    electronic domain pre-compensation, 503–504, 503f
    frequency domain, 501
    impulse response, normalized time functions, 501–502, 502f
    receiver-side post-compensation, 505
    split-step Fourier method (SSFM), 504–505
    time-domain transfer function, 501
    x-polarized optical field component, 500
    y-polarized component, 500
Dispersion relation, 259–260, 259f
Dispersion-shifted fibers (DSFs), 7–9, 62–63, 71, 74, 341
    core diameter, 64
    vs. fiber gratings, 66
Distributed Bragg reflector (DBR), 7, 78, 107, 111, 120
Distributed feedback (DFB), 7, 78, 104–105, 108f, 120
Distributed Raman amplification, 157, 199–200
Doppler frequency shift, 249
Double-sideband modulation
    definition, 312–313
    power spectral density, 314, 315f
Drop state, 288
Dynamic gain tilt, 189, 193
Dynamic range, 149

E
EDFAs. *See* Erbium (Er)-doped fiber amplifiers (EDFAs)
Edge-emitting diodes, 84f, 85
    vs. vertical-cavity surface-emitting laser, 112
Elastic optical networks, 602–603, 613
Electrical-to-optical (EO) conve, 608
Electro-absorption modulator (EAM), 330
    applications, 325
    attenuation and phase parameter, 326, 327f
    chirp parameter, 328, 329f
    with DFB laser, 324, 324f
    laser diode integrated, 300–301, 300f
    Mach-Zehnder configuration, 325, 326f
    phase and absorption coefficients, 324–325, 325f
    power and phase transfer function, 326, 327–328f
Electronic-domain compensation
    carrier phase recovery, coherent detection receiver, 507–511
    dispersion compensation, 500–505
    PDM coherent detection receiver, 499–500
    PMD compensation and polarization de-multiplexing, 505–507

    random state-of-polarization rotation and mode coupling, 499–500
Electronic packet switching, 594–595
Electro-optic modulation, 95
Electro-optic (EO) modulators, 10, 12–13
    field transfer function, 310, 311f
    frequency doubling, 311–312
    high-order harmonic generation, 311–312, 312f
    intensity modulation, 307
        chirp parameter, 309–310
        vs. field modulation, 310–311
        frequency chirp, 309
        input-output power relationship, 307
        MZI configuration, 307, 307f, 310, 310f
        optical power relation, 308
        power transfer function, 308–309
    I-Q modulation, 300, 316
        field transfer function, 316–317
        MZI configuration, 316, 316f
        output optical field, 317, 318f
        voltage waveforms, 317
    $LiNbO_3$, 300, 300f
        EO coefficients, 304–305
        index ellipsoid equation, 302
        longitudinal modulation, 212f, 301–302
        modulation efficiency, 306
        phase modulation, 305–306, 306f
        refractive indices, 305, 306f
        tensor matrix, 301–302
        transverse modulation, 213f, 303
        waveguides, 215f, 304
    ring-resonators
        configuration, 318, 318f
        free-spectral range, 320
        light source modulation, 321, 321f
        normalized carrier density, 323, 323f
        optical signal transmission loss, 322–323, 322f
        photon lifetime limited modulation bandwidth, 321–322
        power transfer function, 319–320, 319f
        Q-value, 318–319
        transfer function, 322
        wavelength, 319–320
    single-sideband modulation, 313, 313f
        vs. double-sideband modulation, 315, 315f
        Jacobi-Anger identity, 314
        output optical field, 313
        power spectral density, 314, 315f
        power transfer function, 314
Electro-optic phase modulator, 305–306, 306f
Elliptical polarization, 23
Emission cross section, 181–182, 182f
Equivalent electrical circuit
    LDs, 115, 116f

photovoltaic (PV), 147, 148*f*
Erbium (Er)-doped fiber amplifiers (EDFAs), 8, 10, 14, 156, 180, 180*f*, 202, 418
  advantages, 179
  carrier density, 187–188
  configuration, 179, 180*f*
    automatic gain control, 192, 192*f*
    automatic power control, 192, 193*f*
    bidirectional pumping, 191, 191*f*
    with forward and backward pump, 190, 190*f*
  emission cross section, 181–182, 182*f*
  energy systems, 180–181, 180*f*
  gain flattening
    optical filter, 193, 194*f*
    passive filter and spatial light modulator, 193, 194*f*
  net emission rate, 182
  optical gain *vs.* wavelength, 189, 190*f*
  performance analysis, 189
  pump absorption efficiency, 181, 182*f*
  rate equations, 183–186
  signal and ASE noise, 187, 187*f*
Ethernet, 574
Ethernet PON, 583
Evanescent field, 25*f*
  attenuation parameter, 27–28
  penetration depth, 28–29, 28*f*
Excess APD noise, 150–151
Explicit congestion notification (ECN), 571
Extended-cavity, 110
External cavity, 104–105
External cavity laser
  configuration, 108–109, 108*f*
  grating-based, 110
  linewidth, 109–110
  resonance losses, 109, 109*f*
External electro-optic modulators, 418
External modulation, 299
External quantum efficiency, of LED, 85, 86*b*
Extinction ratio, 228–230, 274
  EA modulator, 325
  Fabry-Perot interferometers, 228–230
  Michelson interferometer, 223
Extrinsic perturbation, 54
Eye closure penalty
  chromatic dispersion, 361–365
  polarization mode dispersion, 365–368

F

Fabry-Perot interferometers (FPIs), 210, 293
  application, 228–230*b*
  configuration, 223, 223*f*
    finesse, 227
    free-spectral range, 227
    fringes pattern, 225, 225*f*
    half-power bandwidth, 227
    power transfer function, 224, 226, 226*f*
    transmission *vs.* beam incident angle, 225, 225*f*
Fabry-Perot (FP) lasers, 104
Fabry-Perot (FP) resonator, 104
Faraday rotator, 250–252
Fast Fourier transform (FFT), 11
Fermi-Dirac distribution, 82, 90
Fiber-Bragg gratings (FBGs), 9, 66, 210, 234, 234*f*, 597–598, 598*f*
  applications, 238–239
  coupled-wave equations, 234–235
  coupling coefficient, 237, 238*f*
  reflectivity, 236, 237*f*
  transfer matrices, 238
Fiber infrastructure layer, 558
Fiber-optic communication, 3
  block diagram, 6, 6*f*
  first-generation, 6–7
  fourth generation, 8, 10
  performance evaluation, 10–11
  second generation, 7
  third-generation, 7
Fiber to the curb (FTTC), 579
Fiber to the home (FTTH), 579
Finesse
  Fabry-Perot interferometers, 227–228
  ring resonator, 231–232, 232*f*
Finite-difference time-domain method (FDTD), 268–269
Finite element method (FEM), 268–269
Finite-impulse-response (FIR) filters, 612
First-generation fiber-optic communication, 6–7
5G mobile wireless system, 609
Fixed grid, 603, 603*f*
Flexible grid, 603, 603*f*
Folded bus, 560, 560*f*
Forward error correction (FEC) coding, 486–487
Forward pumping, 190, 190*f*, 200–201
Fourth generation fiber-optic systems, 8, 10
Four-wave mixing (FWM)
  in optical fiber, 61–63, 63*f*, 70–71
  in semiconductor optical amplifiers, 176–179
Free-spectral range (FSR), 112, 169, 295
  arrayed waveguide grating, 244–245
  Fabry-Perot interferometers, 227
  ring resonator, 231, 320
Frequency, 14
Frequency chirp, 78, 96, 299
Frequency conversion, 178, 179*f*

Frequency division multiplexing (FDM), 579, 606–607
Frequency-shift key (FSK), 99
Fresnel reflection coefficients, 25–26
"Fronthaul", 609
Fused fiber directional couplers, 214
FWM. *See* Four-wave mixing (FWM)

G

Gain medium, 157, 157*f*
Gain saturation, 201
   in erbium-doped fiber amplifiers, 202
   of optical amplifiers, 160
   semiconductor optical amplifiers, 169, 172, 172*f*
Gaussian approximation, 346
Geiger mode, 144, 145*f*
Geometric birefringence, 54
Geometric optics analysis, 31–34
Gigabit PON (GPON), 583, 616
GPON-encapsulation-method (GEM), 583
Graded-index fiber, 31, 31*f*
Graded index profile, 20
Group velocity, 49
Group velocity dispersion, 50–51
Guided mode, 39

H

Half-power bandwidth (HPBW), 227
Helmholtz equation, 34–35
Higher-order carriers, 142
High-level data link protocol (HDLC), 559
Highly nonlinear PCF, 68*f*, 69
Holey fiber, 68
Hollow-core PCF, 68, 68*f*
Homeotropic liquid crystal, 279, 279*f*
Homogeneous broadening, 82–83
Homogenous liquid crystal, 279, 279–280*f*
Homojunction, 83
Hybrid electrical/optical switching (HELIOS), 595, 595*f*
Hybrid mode (HE mode), 37, 39

I

Impact ionization, 142
Indirect semiconductors, 81
InGaAsP, 7, 45
   carrier drift, 134
   refractive index, 105–106
In-phase/quadrature (I/Q) modulator, 300, 316, 329, 332
   field transfer function, 316–317
   MZI configuration, 316, 316*f*
   output optical field, 317, 318*f*
   voltage waveforms, 317

Input-referred rms noise, 138–139
Insertion loss
   fiber coupler, 214
   optical circulators, 255
   optical isolator, 252
   switch, 274
Integrated Tunable Laser Assembly (ITLA), 118*f*, 119
Intensity modulation and direct detection (IMDD) mode
   amplified multispan WDM optical system, 341
   dispersion compensation, 341
   dispersion shifted fibers, 341
   frequency shift key (FSK), 339
   FWM-induced crosstalk, 391–396
   intensity-modulated optical signal, 339–340
   intersymbol interference (ISI), 341
   linear and nonlinear impairments, 341–342
   modulation transmission and photodetection, 339–340
   O/E converter, 339–340
   and phase shift key (PSK), 339
   receiver sensitivity, 341
   wavelength information, 339
   XPM-induced intensity modulation, 376–389
Intensity modulators, 307, 331
   chirp parameter, 309–310
   vs. field modulation, 310–311
   frequency chirp, 309
   input-output power relationship, 307
   MZI configuration, 307, 307*f*, 310, 310*f*
   optical power relation, 308
   power transfer function, 308–309
Inter datacenter interconnection, 589
Internal quantum efficiency, of LED, 85–86
International Standard Organization (ISO), 557
International Telecommunications Union (ITU), 63, 602
Internet Control Message Protocol (ICMP), 571
Internet header length (IHL), 571
Internet Protocol (IP), 559, 569–570, 570*f*, 612
   packet structure, 571–572, 571*f*
Internet traffic, 555–556, 574
Intrinsic perturbation, 54
IP. *See* Internet Protocol (IP)
IPV4, 571–572
Isolation, 252
Isolators. *See* Optical isolators
ITU-T G.652 fiber, 64
ITU-T G.651 MMF, 63–64

J

Jacobi-Anger identity, 314–315
Johnson noise. *See* Thermal noise
Junction capacitance, 134
Junction temperature, LDs, 116–118

K

Kerr-effect nonlinearity, 20–21, 58–63, 70–71

L

Label swapping, 605
Label-switched path (LSP), 572
Label-switched router (LSR), 572
Lambertian, 84
Large core area PCF, 68f, 69
Large effective area fiber (LEAF), 9, 64, 65t, 71
Laser diodes (LDs), 77–78
   amplitude and phase conditions, 88–90
   biasing circuit, 117, 117f
   configuration, 118–119, 118f
   cylindrical packaged, 117–118, 117f
   equivalent electrical circuit, 115, 116f
   external cavity laser, 108–111
   heat sink mounted, 117–118, 117f
   mode partition noise, 104
   modulation
     adiabatic chirp, 98
     direct intensity, 95, 96f
     small-signal response, 97
     transient chirp, 98
     turn-on delay, 96, 97f
   optical feedback, 88
   packaging, 119
   phase noise, 102, 102f
   $P{\sim}I$ curves, 116–117, 116f
   rate equations, 91
   relative intensity noise, 99, 100f
   side-mode suppression ratio, 94–95
   steady state solutions of rate equations, 92–94
   temperature stability, 118
Last-mile networks. *See* Access networks
L-band, 14
LDs. *See* Laser diodes (LDs)
Light-emitting diodes (LEDs), 77–78
   modulation dynamics, 86–88
   $P{\sim}I$ curve, 85–86
   surface- and edge-emitting, 84
Lightpath, 558, 600–601, 601f
Lightwave, 21–22
   degree of polarization, 23
   phase front, 22
   speed of propagation, 22
   state of polarization, 22–23
Lightwave broadcasting system, 613
Line, 568
Line-amp, 150–151, 156
Linearly polarized (LP) mode, 39

Linear polarization, 22–23
Linear units, 15
Liquid crystals, 278–279
   birefringence, 280, 280f
   electric voltage, 284
   homogenous and homeotropic, 279, 279f
   polymer-dispersed, 282, 282f
   total internal reflection, 282–283, 283f
   twisted nematic, 281, 281f
Lithium niobate ($LiNbO_3$), 300, 300f
   EO coefficients, 304–305
   EO tensor matrix, 301–302
   index ellipsoid equation, 302
   longitudinal modulation, 212f, 301–302
   modulation efficiency, 306
   phase modulation, 305–306, 306f
   refractive indices, 305, 306f
   transverse modulation, 213f, 303
   waveguides, 215f, 304
Local-area network (LAN), 573f, 574
Logical link control (LLC), 559
Longitudinal EO modulator, 212f, 301–302
Long-span (LS) fiber, 65t

M

Mach-Zehnder interferometers (MZIs), 175–176, 176f, 204, 210, 292
   all-fiber MZIs, 219, 219f
   application, 221
   and electro-optic (EO) modulation, 307, 307f
   free-space configuration, 219, 219f
   and optical switching, 276, 276f
   power transfer function, 220, 221f
   transmission efficiencies, 220
Mach-Zehnder modulators (MZMs), 321, 321f
Material dispersion, 51–52
Maxwell's equations, 34, 36–37
Mean-field approximation, 166
Mechanical switch, 284–285
Medium access control (MAC) layer, 559, 581
Mega datacenters, 588–589
MEMS. *See* Micro-electro-mechanical systems (MEMS)
MEMS optical switch, 284–285, 285f
   application, 289, 290f
   crossbar fabrication, 288, 289f
   Digital Micromirror Device, 286, 286f
   3D architecture, 285, 285f
Meridional modes, 39
Meridional rays, 31, 32f
Metropolitan area network (MAN), 573f, 574

Michelson interferometer
  configurations, 222, 222*f*
  extinction ratio, 223
  power transfer function, 222–223
  transfer matrix, 222–223
Micro-bending, 46
Micro-electro-mechanical systems (MEMS), 11–12, 274, 284–285, 591
Micro Integrable Tunable Laser Assembly (μ-ITLA), 120
Micro-ring modulator, 238, 319, 321, 323–324
Mie scattering, 282
Modal dispersion, 20–21, 53, 70
Mode coupling, 55
Mode division multiplexing, 55–56
Mode-field diameter (MFD), 43
Mode-field distribution, of SMF, 44, 44*f*
Mode partition noise, 104
Modified Bessel functions, 35, 36*f*, 37
Modified Scholow-Towns formula, 103
Modulation chirp, 98, 329
Modulation efficiency, 95, 306
Modulation formats
  analog optical systems
    analog subcarrier multiplexing and optical single-sideband modulation, 473–478
    carrier-to-signal ratio, inter-modulation distortion and clipping, 478–481
    radio-over-fiber technology, 483–486
    relative intensity noise, 481–483
  bandwidth efficiency, 440
  binary NRZ and RZ modulation formats
    chromatic dispersion, 443
    fifth-order Bessel filter, 442–444, 442–443*f*
    Fourier transform, 441–442
    longdistance fiber-optic systems, 440
    normalized *Q*-values of, 444
    signal-dependent noise dominated system, 443–444
  BPSK optical systems
    coherent detection, 461, 461*f*
    LO-ASE beat noise, 462
    optical signal to noise ratio (OSNR), 462
    trans-impedance amplifier (TIA), 461
    waveform distortion, 462
  carrier-suppressed return-to-zero (CSRZ), 456–460
  DPSK optical systems
    coherent detection and direct detection, 463
    coherent homodyne detection, 463
    differential delay line-based self-homodyne detection, 465–466
    direct detection receiver, 463–464
    Mach-Zehnder interferometer, 463–465, 464*f*
  duo-binary optical modulation
    delay-and-add operation, 454, 455*f*

  eye closure penalty level, 456
  Mach-Zehnder modulator (MZM), 454, 455*f*
  time-domain optical field and optical power, 456, 456*f*
high-level PSK and QAM modulation
  additive white Gaussian noise (AWGN), 466–467
  direct detection receiver, 469
  error-vector-magnitude, 470, 472
  optical phase modulation, 466
  quadrature phase shift keying (QPSK), 469
  spectral efficiency, 466–467
M-ary and polybinary coding
  block diagram of, 450, 452*f*
  chromatic dispersion-limited fiber-optic system, 450
  controlled inter-symbol-interference (ISI), 453
  feature of, 452
  group-velocitydispersion (GVD) tolerance, 450, 451*f*
  multilevel coding scheme, 451–452
  original binary data sequence, 453
  signal-independent noise, 451
  signal spectral bandwidth, 450, 451*f*
optical system link budgeting
  OSNR budgeting, 487–489
  power budgeting, 486–487
PRBS patterns and clock recovery
  digital transmission system, 446
  frequency difference, 449
  mark-density-pattern, 447
  narrowband filter, 448
  nonlinear circuit, 449
  parameters, 445
  pattern length, 445, 445*f*
  received bit sequence, 447
  shift registers with feedback, 446, 446*f*
  time-domain waveform, 448–449, 448*f*
  truth table, 446–447, 446*f*
  voltage-controlled oscillator (VCO), 449, 449*f*
  zero-substitution-pattern, 447
transmission distance, 440
transmission medium, 440
Moore's law, 498, 587–588
MPLS. *See* Multiprotocol label switching (MPLS)
Multimode fibers (MMFs), 20–21, 38
  ITU-T G.651, 63–64
  modal dispersion, 53, 55*b*, 70
  propagation modes, 55–56
Multiple longitudinal modes, 89, 94
Multiplexers (MUXs), 10, 210, 210*f*, 239, 566
Multiplex inline optical amplifiers, 573
Multiprotocol label switching (MPLS), 572
Multiwavelength optical system, 193
MUXs. *See* Multiplexers (MUXs)
MZIs. *See* Mach-Zehnder interferometers (MZIs)

N

Near-field (NF) distribution, 44
Noise-equivalent power (NEP), 140–141, 150–151
Noise figure, 201
  of EDFA, 181, 205
  in optical amplifiers, 163–165
  of Raman amplifiers, 198, 198*f*
Nondispersion-shifted fiber (NDSF)
  chromatic dispersions, 65*f*
  specifications, 65*t*
Nonlinear Schrödinger (NLS) equation, 58–63
Non-radiative recombination, 80, 181
Non-zero dispersion-shifted fiber (NZ-DSF), 9, 64
  chromatic dispersions, 64–65, 65*f*
Normal incidence, 26
Normalized FWM efficiency, 178
Normalized propagation constant
  optical fiber, 39
  rectangle optical waveguides, 265*f*, 266
n-type semiconductor, 126–127, 128*f*
Numerical aperture, 39–43, 47–48*b*
Nyquist-frequency division multiplexing (Nyquist-FDM), 608
Nyquist pulse modulation
  digital-analog hybrid approach, 527, 527*f*
  Nyquist-FDM optical system, 526–527, 526*f*
  Nyquist filter amplitude transfer functions, 523, 524*f*
  sharp-edged spectrum, 524
  time-domain waveforms and frequency spectra of OFDM, 523, 523*f*

O

O-band, 14
O/E convertor, 125–126
ONU. *See* Optical network units (ONU)
Open-circuit voltage, 146
Optical add/drop multiplexing (OADM), 596–599, 597*f*
Optical amplifiers, 155, 201. *See also* Semiconductor optical
    amplifiers (SOAs)
  ASE noise
    in electrical domain, 161–163
    power spectral density, 160–161
  bandwidth, 158
  functions, 156*f*
  gain medium, 157, 157*f*
  noise figure, 163–165
  saturation, 159
Optical carrier 1 (OC1), 567
Optical circuit switching, 596
Optical circulators, 290–291
  applications, 253
  configuration, 253, 253*f*

FBG reflection, 253, 253*f*
  functions, 253, 253*f*
  operating principle, 254, 254*f*
Optical communication, 586–587
Optical cross connect (OXC) switching, 599–602, 599*f*
Optical directional coupler, 292
  absorption and scattering loss, 214
  directionality and reflection, 214
  insertion loss, 214
  vs. optical switches, 274, 274*f*
  power-splitting ratio, 214
  power transfer coefficient, 213
  splitting ratio, 215, 215–216*f*
  transfer matrix, 217–218
Optical distribution network (ODN), 579
Optical-electrical-optical (O-E-O) transformations, 568–569
Optical fibers, 3, 4*f*, 5, 557
  advantages, 5–6
  attenuation, 5, 5*f*, 44–48
  field distribution profile, 43–44
  guided mode, 39
  index profiles, 31, 31*f*
  Kerr effect nonlinearity, 58–63, 70–71
  LP mode, 39
  meridional modes, 39
  nonlinear Schrödinger equation, 58–63
  normalized propagation constant, 39
  numerical aperture, 39–43
  properties, 20–21
  skew modes, 39
  stimulated Brillouin scattering, 56–57
  stimulated Raman scattering, 57–58
  structure, 20, 20*f*
  wave propagation modes
    electromagnetic field theory, 34–38
    geometric optics analysis, 31–34
Optical filter, 210, 210*f*
Optical interconnection, 586–588
Optical interface, 24, 24*f*
Optical isolators, 210, 210*f*, 250, 290–291
  configuration, 250, 250*f*
  performance, 252
  polarization-independent
    configuration, 251, 251*f*
    operating principle, 251, 252*f*
Optical layer, 558, 568
Optical line terminal (OLT), 579, 581–582, 582*f*
Optical networks
  architectural design, 559
  categories, 573–574, 573*f*
  datacenter, 588–595
  elastic, 602–603

Optical networks *(Continued)*
  electronic circuit switching, 606–612
  folded bus, 560, 560*f*
  interconnection, 586–588
  layers, 557–559, 557–558*t*, 558*f*
  optical interconnection, 586–588
  optical switching and cross connection, 596–606
  passive, 579–585
  physical layer configuration, 558, 558*f*
  protection and survivability, 574–579
  synchronous optical network, 567, 567*t*
  topologies, 559–566, 560*f*
  unidirectional ring configuration, 562, 562*f*
Optical network interface (ONI), 593–594
Optical network units (ONU), 579–580, 582, 582*f*
Optical packet, 604–605, 604–605*f*
Optical packet routers, 604
Optical packet switching, 596, 604–606, 606*f*
Optical phase shift, 29, 69–70
Optical phonon, 57, 195
Optical power, 15
Optical signal-to-noise ratio (OSNR), 160, 201
  long-distance optical transmission system, 356
  N optically amplified fiber spans, 356, 356*f*
  Q-value, 357–358, 357*f*
  thermal noise and dark current noise, 357
Optical single-sideband (OSSB) modulation, 312–313, 313*f*
Optical switches, 291
  acousto-optic (AO) switch, 278, 278*f*
  advantage, 277–278
  applications, 274
  evanescent-mode coupling, 275–276, 275*f*
  liquid crystals, 278–279
    birefringence, 280, 280*f*
    electric voltage, 284
    homogenous and homeotropic, 279, 279*f*
    polymer-dispersed, 282, 282*f*
    total internal reflection, 282–283, 283*f*
    twisted nematic, 281, 281*f*
  by Mach-Zehnder interferometer configuration, 276–277, 276*f*
  vs. optical directional coupler, 274, 274*f*
  semiconductor-based, 275, 275*f*
Optical-to-electrical (OE) conversion, 172, 347, 608
Optical transmission system, 555–556
  bit error rate (BER)
    definiton, 338
    vs. Q-values, 339–347
  characteristics, 338
  electrical to optical (E/O) conversion, 338
  optical signal-to-noise ratio (OSNR)
    long-distance optical transmission system, 356

N optically amplified fiber spans, 356, 356*f*
  Q-value, 357–358, 357*f*
  thermal noise and dark current noise, 357
performance degradation
  ASE noise accumulation, 368–370
  chromatic dispersion, 361–365
  fiber nonlinearities, 372–376
  multipath interference, 370–372
  polarization mode dispersion, 365–368
physical layer, 338
physical mechanisms, 338
properties, 338
receiver sensitivity
  definition, 352–353
  direct detection receivers, 353, 353*f*
  10-Gb/s system, 354–355, 355*f*
  Q-value, 353
  waveform distortion, 355–356
wavelength division multiplexing, 358–361
XPM-induced intensity modulation in IMDD optical systems
  attenuation effects, 377
  chromatic dispersion, 377
  crosstalk waveforms, 385*f*, 386
  experimental setup, for frequency domain, 381, 381*f*
  externalcavity tunable semiconductor lasers, 381
  frequency response, 381–383, 382–383*f*
  linear propagation group delay, 377
  linear superposition, 379–380, 380*f*
  normalized crosstalk *vs.* fiber dispersion, 388, 388*f*
  normalized power crosstalk levels *vs.* dispersion compensation, 389, 389*f*
  normalized power crosstalk levels *vs.* receiver bandwidth, 386–387, 387*f*
  power transfer functions, 385–386, 386*f*
  pump-probe interaction, 376–377, 377*f*
  short fiber section, 378, 378*f*
  theoretical analysis, 376–377
  time-domain waveforms, 385, 385*f*
  transfer function, 383
XPM-induced phase modulation, 389–391
Optical wave, 20. *See also* Lightwave
Optimum coupling coefficient, 561–562, 563*f*
Orthogonal frequency division multiplexing (OFDM), 499, 608
  band-pass filter (BPF), 519
  BER, receiver bandwidth functions, 521, 521*f*
  bit error rate (BER), 519–520
  chromatic dispersion, 516–517
  cyclic prefix, 516–517, 516*f*
  data sequence mapping, 520–521, 520*f*
  double-sideband (DSB) spectrum, 514–515
  electric power spectral density, 520–521

Fourier transform, 518
frequency down-conversion, DFT for, 515, 515*f*
guard band in low frequency, 519, 519*f*
IDFT for frequency up-conversion, 514, 514*f*
intensity-modulated optical signal, 518, 518*f*
intensity modulation and direct detection, 516–517*f*, 517
with *N* subcarrier channels, 512–513, 513*f*
Nyquist sampling theorem, 517–518
optical intensity waveform, 518–519
serial-to-parallel conversion process, 519–520
spectral guard band, 512, 512*f*
spectral overlap, subcarrier channels, 512, 512*f*
spectra of the signal channel, 521–523, 522*f*
time-domain data matrix, 515, 515*f*
time-domain parallel sequences, 514, 514*f*
time-domain waveform, 521–523, 522*f*
transmission impairments, 512–513

P

Packaged fiber amplifiers, 156, 157*f*
Packets, 556–557
Packet-switched network, 556–557
Panda fiber, 66–67, 67*f*
Passive optical components, 210
    free-space optical assembly, 210
    optical directional coupler
        absorption and scattering loss, 214
        directionality and reflection, 214
        insertion loss, 214
        power-splitting ratio, 214
        power transfer coefficient, 213
        splitting ratio, 215, 215–216*f*
        transfer matrix, 217–218
    photonic integration, 211
Passive optical networks (PONs), 14, 579–585
    definition, 579
    Ethernet, 583
    gigabit, 583
    optical configurations of, 579–580, 580*f*
    time division multiplexing, 580–583, 581*f*, 583*f*, 616
    wavelength-division multiplexing, 584–585, 584–585*f*
Passive quenching, 144, 145*f*
Path, 568
PCFs. *See* Photonic crystal fibers (PCFs)
Peltier, 119
Phase front, 22
Phase noise, 102, 102*f*
Phase velocity, 48–49
Photocurrent, 129, 146
Photodetectors, 125–126
Photodiodes, 125–126, 613–614. *See also* Avalanche
        photodiode (APD)

current-voltage relationship, 135–136, 135*f*
detection speed, 133, 150
    carrier drift, 134
    junction capacitance, 134
    time domain response, 134, 135*f*
electrical circuit, of optical receivers, 136, 136*f*
noise-equivalent power (NEP), 140–141
noise sources
    dark current noise, 138
    input-referred rms noise, 138–139
    shot noise, 138
    thermal noise, 137–138
PN and PIN, 126–131
quantum efficiency, 126–131
responsivity, 126–131, 150
signal-to-noise ratio (SNR), 139–140
transimpedance gain, 136
Photon energy, 15
Photonic crystal fibers (PCFs)
    applications, 69
    cross-section, 68, 68*f*
    highly nonlinear, 69
    hollow core, 68
    large core area, 69
Photonic integration, 211
Photovoltaic (PV), 145, 151, 154
    electric power and photocurrent, 146, 147*f*
    equivalent electric circuit, 147, 148*f*
    optimum load resistance, 147
    photocurrent, 145*f*, 146
    properties, 146
P∼I curve
    laser diodes (LDs), 116–117, 116*f*
    light-emitting diodes (LEDs), 85–86
Piezo-electric transducer (PZT), 222*f*, 246
PIN diode, 125–126, 152
    advantages, 130
    properties, 131, 131*f*
    structure, 130, 130*f*
PIN-TIA module, 137
Pixel shifter circuit, 148–149, 149*f*
Planar lightwave circuits (PLCs), 10, 20, 20*f*, 30, 211,
        243, 255
    polymer based, 256
    silica-based, 255
Plane wave, 22
    polarized, 22
    reflection and refraction, 24, 24*f*
    state of polarization, 23*f*
Plastic optical fiber (POF), 69
Plesiochronous digital hierarchy (PDH), 567
p–n junction, 79–80

PN photodiode
  energy band diagram, 129*f*
  photon absorption, 129, 129*f*
Point-to-point optical transmission system, 155–156, 156*f*
Point-to-point protocol (PPP), 559
Polarization-independent optical circulators, 253, 253*f*
Polarization-independent optical isolators
  configuration, 251, 251*f*
  operating principle, 251, 252*f*
Polarization-maintaining (PM) fiber, 66, 74
  complications, 68
  output field vector, 67, 67*f*
  Panda and Bowtie, 66–67
Polarization-mode dispersion (PMD), 9–11, 53–55, 70, 608
  constant modulus algorithm, 506–507
  fiber birefringence overtime, 505
  filter transfer functions, 506–507
  recursive adaption process, 506–507
Polarization multiplexing (PM), 12–13, 12*f*, 55–56
Polarized light, 22–23
Polydimethylsiloxane (PDMS), 587–588
Polymer, 256
Polymer-dispersed liquid crystal (PDLC), 282, 282*f*
Polymer waveguides, 587–588
Polymethyl methacrylate (PMMA), 69, 587–588
PONs. *See* Passive optical networks (PONs)
Positive birefringence, 279
Post-amplifier, 155–156
Power budgeting, 486–487
Power-conversion efficiency, 114–115
Power spectral density (PSD)
  and ASE-ASE beat noise generation, 162*f*, 163
  ASE noise, 160–161
  of phase noise, 103
  and signal-ASE beat noise generation, 162, 162*f*
  single- and double-sideband modulation, 314, 315*f*
Power transfer function
  electro-absorption modulator, 326, 327–328*f*
  electro-optic modulator, 310, 311*f*
  Fabry-Perot interferometers, 224, 226, 226*f*
  intensity modulator, 308
  Mach-Zehnder interferometers, 220, 221*f*
  Michelson interferometer, 222–223
  micro-ring-resonators, 319–320, 319*f*
  semiconductor optical amplifiers, 167–169, 168–169*f*
  single-sideband modulator, 314
Preamplifier, 156
Printed circuit boards (PCB), 586–587, 587*f*
Protection channels, 575, 575*f*
p-type semiconductor, 126–127, 128*f*
Pulse broadening, 55*b*
Pump absorption efficiency, 181

Q
QPSK modulation, 316–317, 316*f*
Quadrature amplitude modulation (QAM), 12
Quadrature phase shift keying (QPSK), 10–12
Quality of service (QoS), 557, 572
Quantum efficiency, 131–133, 149
Quantum limit, 126
Quantum noise. *See* Shot noise
Quenching, 144, 145*f*
Q-value
  for binary modulated systems
    IMDD optical systems, 339–342
    receiver BER and, 342–347
  eye diagram parameterization, 348–351
  noise and waveform distortion
    signal-dependent noise, 348
    signal-independent noise, 348
  receiver sensitivity, 353
  ring-resonators, 318–319

R
Radiative recombination, 80, 181
Radio access network (RAN), 609
Radio-over-fiber technology, 483–486
Raman amplifiers, 156, 202–203
  ASE noise PSD, 197
  coupled wave equations, 195–196
  noise figure, 198
    vs. fiber length, 200, 200*f*
    forward and backward pumped, 198–199, 198*f*
    localized and distributed, 199, 199*f*
  normalized signal optical power, 196*f*, 197
  performance, 201
  relative intensity noise, 200
  scattering mechanism, 195
Raman gain coefficient, 195–196, 196*f*, 198
Rate equations, 91
  erbium-doped fiber amplifiers, 183–186
  four-wave mixing, 176–177
  steady state solutions, 92–94
Rayleigh scattering, 45, 47–48*b*
Receiver optical-to-electrical conversion technology, 347
Receiver sensitivity
  definition, 352–353
  direct detection receivers, 353, 353*f*
  10-Gb/s system, 354–355, 355*f*
  Q-value, 353
  waveform distortion, 355–356
Receiver sensitivity, 341
Reconfigurable optical add/drop multiplexer (ROADM), 596, 597–599*f*, 617

Rectangle optical waveguides
cross section, 263–264, 264*f*
field distributions
for TE modes, 267, 267*f*
for TM modes, 266–267, 266*f*
normalized propagation constant, 265*f*, 266
Reflection and refraction, 24
Brewster angle, 29–30
critical angle, 27
Fresnel reflection coefficients, 25–26
normal incidence, 26
at optical interface, 24, 24*f*
optical phase shift, 29
Refractive index
carrier-dependent, 98
LiNbO₃, 305, 306*f*
polymer, 256
silicon, 255–256, 271–272
step- and graded-index fibers, 147
Regenerator, 568–569
Relative intensity noise (RIN), 99–100
laser diodes, 99, 100*f*
and multipath interference, 372
Raman amplifier, 200
Relaxation oscillation, 98, 101*f*
Reliability, 574
Required optical signal to noise ratio (R-OSNR), 10–11
Return loss, 47–48*b*, 252
Rib waveguide, 271
Ring resonators
application, 233, 233–234*b*
configuration, 230, 230*f*, 318, 318*f*
finesse, 231–232, 232*f*
free-spectral range, 231, 320
light source modulation, 321, 321*f*
minimum power transmission, 231, 231*f*
normalized carrier density, 323, 323*f*
optical signal transmission loss, 322–323, 322*f*
photon lifetime limited modulation bandwidth, 321–322
power transfer function, 319–320, 319*f*
Q-value, 232–233, 318–319
transfer function, 322
wavelength, 319–320
ROADM. *See* Reconfigurable optical add/drop multiplexer (ROADM)
Routers, 570
Routing decision, 570
Row-shifter circuit, 148–149, 149*f*

S
S-band, 14
Scattering loss, 45

Scholow-Towns formula, 103
Second generation fiber-optic systems, 7
Section layer, 568
Self-healing, 576, 577*f*
Self-phase modulation (SPM), 60–61, 60*f*, 70–71
Sellmeier equation, 52, 74, 305
Semiconductor
carrier confinement, 83–84
direct and indirect, 81
energy band, 82, 82*f*
p–n junction and energy diagram, 79–80
refractive index, 98
spontaneous emission, 82, 83*f*
stimulated emission, 82–83, 83*f*
Semiconductor-based optical switches, 275, 275*f*
Semiconductor optical amplifiers (SOAs), 10, 156, 179, 202
all-optical signal processing
conversion process, 172, 172*f*
cross-phase modulation, 174
gain recovery, 173–174
gain saturation, 174, 175*f*
gain suppression ratio *vs.* input pulse energy, 173, 173*f*
response speed, 173
switching, 175–176, 176*f*
in butterfly package, 157*f*
vs. erbium-doped fiber amplifiers, 179
fast-gain dynamics, 169–171, 171*f*, 202
four-wave mixing
frequency conversion, 178, 179*f*
rate equations, 176–177
signal power *vs.* frequency detune, 177, 178*f*
and laser diode, 165
steady-state analysis
large-signal material gain, 166
optical gain, 167
power transfer function, 167–169, 168–169*f*
rate equation, 165–166
saturation optical power, 166
small-signal gain, 165–166
structure, 166
Shared protection scheme, 575–576, 576*f*, 616
Short-circuit current, 146
Shot noise, 138, 150–151
Side-mode suppression ratio (SMSR), 94–95
Signal-ASE beat noise, 161–162, 162*f*, 201
Signal-to-noise ratio (SNR), 126, 137
photodiodes, 139–140
Raman amplifier, 200
Silica fibers, 45, 70
attenuation, 45*f*
Raman efficiency, 57, 57*f*
Silicon dioxide (SiO₂), 20

Silicon photodiodes, 6–7
Silicon photonic ring modulator, 318, 318*f*
Silicon photonics, 211
 advantage, 270–271
 definition, 255–256
 directional coupler, 267–270, 269*f*
 edge-coupling structure, 273, 273*f*
 refractive index, 272, 272*f*
 rib waveguides, 272
 SOI wafer layer structure, 271, 271*f*
Silicon photonic structures, 211, 212*f*
Silicon ring modulator, 319–320, 329–330
Single-frequency laser, 123
 definition, 104–105
 DFB LDs, 105–108
 distributed Bragg reflector (DBR), 107, 108*f*
Single-mode fiber (SMF), 20–21, 38, 43, 73
 chromatic dispersion, 55*b*, 70
 field distribution profile, 43–44
 light source, 73
 mode-field distribution, 44, 44*f*
 polarization mode dispersion, 53
 propagation constant, 73
Single-photon detector
 biasing circuits, 144, 145*f*
 principle, 144, 144*f*
Single-sideband modulation, 313, 313*f*
 vs. double-sideband modulation, 315, 315*f*
 Jacobi-Anger identity, 314
 output optical field, 313
 power spectral density, 314, 315*f*
 power transfer function, 314
Skew modes, 39
Skew rays, 31, 32*f*
Slab optical waveguides
 magnetic fields, 257–258
 mode-cutoff condition, 259–260, 259*f*
 structure, 256, 257*f*
 symmetric and antisymmetric modes, 260–263*b*
 wave vectors, 263, 263*f*
Small-signal electrical signal model, 136, 136*f*
SMF. *See* Single-mode fiber (SMF)
Snell's law, 24
SONET. *See* Synchronous optical network (SONET)
Space-charged region. *See* Depletion region
Space optical switch, 284, 284*f*
Span switch, 578
Spatial light modulator (SLM), 193, 194*f*
Spatial optical interference, 210
Spectral hole burning, 82–83
Split-step Fourier method, 10–11
Spontaneous emission, 82, 83*f*

Spontaneous-spontaneous beat noise
 generation process, 163, 163*f*
 RF spectral density, 163
Spurious-free dynamic range (SFDR), 485–486
Square-law detection rule, 151
Square-law detector, 126
Standard single-mode fiber (SSMF), 41, 46, 64, 71
 chromatic dispersion, 64–65
 dispersion parameter, 52–53
 polarization-mode dispersion, 74
Standing wave pattern, 33, 37
Star coupler, 563, 563*f*
Star network, 563–564, 563–564*f*, 615
State of polarization (SOP), 21–22
 circular, 23
 elliptical, 23
 linear, 22–23
 of plane wave, 23*f*
Step-index fiber, 35, 72
 cross section, 39, 40*f*
 index profiles, 31, 31*f*
Step index profile, 20
Stimulated Brillouin scattering (SBS), 20–21, 56–57
Stimulated emission, 82–83, 83*f*
Stimulated Raman scattering (SRS), 20–21, 57–58, 156, 195
Stokes photons, 56–57
Stokes wave, 195–197
Stress-applying parts (SAPs), 66–67, 67*f*
Stress birefringence, 54
STS-1. *See* Synchronous transport signal level-1 (STS-1)
Surface-emitting diodes, 84, 84*f*, 86*b*
Synchronous optical network (SONET), 498, 567, 567*t*, 568*f*, 569, 612
Synchronous transport signal level-1 (STS-1), 567–569, 568–569*f*

T

TCP. *See* Transmission control protocol (TCP)
Thermal noise, 137–138
Thin-film interference filters
 applications, 240–241
 reflection and refraction, 239
 transfer function and chromatic dispersion, 240, 241*f*
 WDM demux, 242, 242*f*
Third-generation fiber-optic systems, 7
3-dB fiber directional coupler, 253, 253*f*
3-dB saturation input power, 159
Threshold, 92
Threshold condition, 89

Threshold current density, 93
Through state, 288
Time division multiplexing (TDM), 579–580
Top-of-Rack (ToR) switch, 590
Transceivers, 560, 575
Transient chirp, 98
Transimpedance amplifier (TIA), 96f, 136, 137f, 138–139
Transmission control protocol (TCP), 559, 569–571
Transmission loss, 561, 561f, 563f
Transport layer, 559
Transverse electric-field mode (TE mode):, 37
Transverse EO modulator, 213f, 303
Transverse magnetic-field mode (TM mode), 37
Transverse wave. *See* Plane wave
Traveling-wave amplifier (TWA), 165, 201–202
Truewave (TW) fiber, 65t
Truewave reduced-slope (TW-RS) fiber, 65t
Turn-on delay, 96, 97f
Twisted nematic (TN) liquid crystal, 281, 281f
Twisted wires, 4, 4f

U

U-band, 14t
Unidirectional path-switched rings (UPSRs), 577, 577–578f, 616
User datagram protocol (UDP), 571

V

Vectoral propagation constant, 21–22
Vertical-cavity surface-emitting laser (VCSEL), 78–79, 120, 586–588
  application
    array technology, 113–114, 114f
    light sources for illumination, 114–115
  vs. edge-emitting diode, 112
  output optical power, 112–113
  performance, 113, 113f
  power-conversion efficiency, 114–115
  scanning electron microscopy, 115f
  structure, 111, 111f
  threshold current and the mirror reflectivity, 112
Vertical optical coupling (VOC), 275–276
V-number, 37–38, 38f

W

Wall-plug efficiency, 124
Water absorption peaks, 45

Wave front. *See* Phase front
Waveguide dispersion, 51, 53
Wavelength, 14
Wavelength conversion, 600
Wavelength-dependent fiber coupler, 216f, 217
Wavelength division multiplexing (WDM), 8–9, 8f, 58, 417–418, 557, 564–566
  C-band, 359
  de-multiplexers, 359
  high-quality WDM multiplexers, 359
  laser diode, 358–359
  micro-electro-mechanical systems, 360, 360f
  modulation instability and impacts
    frequency components, 400
    mean-field approximation, 396
    normalized optical spectra *vs.* fiber length, 399, 400f
    normalized power spectral density, 401
    normalized RIN spectra *vs.* fiber length, 401
    performance degradation, 396
    propagation constants, 398
    signal-ASE beat noise, 405
    standard deviation of noise, 405
    system performance degradation, 405–406
    three-span fiber system, 407, 408f
    transfer matrix analysis, 397f
    transfer matrix formulation, 406–407
    Wiener-Khintchine theorem, 398
  N wavelength channels, 360
Wavelength-selective switch (WSS), 284, 284f, 591
Wavelength-space hybrid optical switching, 600, 600f
Wavelength-tunable transmitter, 576, 615
WDM. *See* Wavelength division multiplexing (WDM)
WDM-PON, 584–585, 584–585f
White noise, 137–138
Wide-area network (WAN), 573, 573f, 612
Wiener-Khintchine theorem, 103

X

XGPON, 583
XPM. *See* Cross-phase modulation (XPM)

Y

Yttrium iron garnet (YIG), 210, 250, 253

Z

Zero-chirp modulators, 310, 329

Printed in the United States
by Baker & Taylor Publisher Services